Engine Modeling and Control

Rolf Isermann

Engine Modeling and Control

Modeling and Electronic Management of Internal Combustion Engines

Springer

Rolf Isermann
TU Darmstadt
Darmstadt
Germany

ISBN 978-3-642-39933-6 ISBN 978-3-642-39934-3 (eBook)
DOI 10.1007/978-3-642-39934-3

Library of Congress Control Number: 2014943641

Springer Heidelberg New York Dordrecht London

Printed on acid-free paper

Springer is part of Springer Science+Business Media (www.springer.com)

Preface

The increasing requirements for automotive drives with internal combustion engines on reduced fuel consumption, low emissions and good driveability need continuous improvement of combustion and exhaust treatment processes and their control. This can be reached by a higher variability with an increase of actuators and sensors in addition to thermodynamic, mechanical and structural improvements. Modern engines have therefore an increasing number of manipulation variables and sensors and a complex electronic management. The design of the many control function requires good physical understanding and model-based methods taking into account mechatronic engineering principles.

The book treats as well physical-based as experimental gained engine models for gasoline (spark ignition) and diesel (compression ignition) engines and uses them for the design of the different control systems. The procedure and the workflow from theoretical and experimental modeling over simulations to calibration with test benches is systematically described and demonstrated by many examples. Not only the stationary but also the dynamic nonlinear behavior of engines is taken into account. The combustion engine models include the intake system, fuel supply and injection, combustion cycles, mechanical system, turbochargers, exhaust and cooling system and are mainly generated for real-time computation. Engine control structures and engine control development with different digital feedforward and feedback control methods, calibration, optimization and simulation tools are considered in detail. Various control systems are developed for gasoline and diesel engines with both, conventional and alternative combustion processes, based on nonlinear static and dynamic multivariable engine models and demonstrated by experiments on test benches.

The book is an introduction into the electronic engine management with many examples for engine control and it is oriented to advanced students working in control, electrical, mechanical and mechatronic engineering and will also be useful for practicing engineers in the field of engine and automotive engineering.

The author is grateful to his research associates, who have performed many theoretical and experimental research projects on the subject of this book since 1986, among them K.U. Voigt, Chr. Schmidt, St. Leonhardt, K. Pfeiffer, O. Nelles, Chr. Ludwig, St. Sinsel, M. Schüler, M. Willimowski, M. Hafner, O. Jost, J. Schaffnit,

M. Schmidt, N. Müller, S. Töpfer, J.A. Kessel, F. Kimmich, A. Schwarte, M. Vogt, A. Engber, E. Hartmanshenn, M. Weber, R. Zimmerschied, M. Kohlhase, K. v. Pfeil, M. Leykauf, S. Clever, S. Zahn, A. Schreiber, M. Mrosek, D. Böhme, H. Sequenz, Chr. Eck, A. Sidorow, M. Kowalczyk, F. Kunkel and S. Zydek.

Without their continuous work on new methods and building up and maintaining the combustion-engine test bench, measurement and computer equipment the results of this book would not have been possible. Great appreciation goes also to our precision mechanics workshop guided by A. Stark.

We also would like to thank the research organizations Forschungsgemeinschaft Verbrennungskraftmaschinen (FVV), Arbeitsgemeinschaft industrieller Forschungsvereinigungen (AiF), Deutsche Forschungsgemeinschaft (DFG), Faudi-Stiftung who supported many projects. Several results were obtained in cooperation projects with industrial companies. Among them are AVL List GmbH, Robert Bosch GmbH, Daimler AG, Adam Opel AG, Dr.-Ing. h.c. F. Porsche AG, GM Europe, and Volkswagen AG. We appreciate these cooperations strongly as they contributed positively to our own research.

I am also grateful for proofreading of some chapters by S. Clever, H. Konrad, M. Kowalczyk, F. Kunkel, H. Sequenz, H. Stuhler, S. Zahn and S. Zydek.

Finally, I would like to thank Kerstin Keller, Moritz Neeb, Lisa Hesse and especially Brigitte Hoppe for the laborious and precise text setting, Sandra Schütz for drawing many figures and Springer Verlag for the excellent cooperation.

Darmstadt, May 2014 Rolf Isermann

Contents

List of Symbols

Only frequently used symbols and abbreviations are given:

1. General letter symbols

a	parameters of differential or difference equations
b	parameters of differential or difference equations
c	spring constant, constant, concentration, stiffness
d	damping coefficient
e	equation error, control deviation $e = w - y$, number $e = 2.71828\dots$
f	fault, frequency ($f = 1/T_\mathrm{p}$, T_p period), function $f(\dots)$
g	gravitational acceleration, function $g(\dots)$, impulse response
i	integer, gear ratio, index, $\sqrt{-1}$ (imaginary unit)
j	integer, index
k	discrete number, discrete time $k = t/T_0 = 0, 1, 2, \dots$ (T_0: sampling time)
l	index, length
m	mass, order number
n	rotational speed, order number, disturbance signal
p	pressure, index, controller parameter, probability density function, process parameter
q	controller parameter
r	index, radius, reference variable, residual
s	Laplace variable $s = \delta + i\omega$, symptom, actuator position
t	continuous time
u	input signal change ΔU
v	specific volume, disturbance signal
w	speed, reference value, setpoint
x	space coordinate, state variable, concentration
y	output signal change ΔY, space coordinate, control variable change ΔY, signal

z	space coordinate, disturbance variable change ΔZ, z-transform variable $z = \exp T_0 s$
$\hat{x}$	estimated or observed variable
$\tilde{x}$	estimation error
$\bar{x}$	average, steady-state value
x_0	amplitude
x_{00}	value in steady state (identification methods)
x_d	desired value
A	area
B	magnetic flux density
C	capacitance
D	damping ratio, diameter
E	module of elasticity, energy, potential, bulk modulus
F	filter transfer function, force
G	weight, transfer function
H	magnetic field strength, height
I	electrical current, mechanical momentum, torsion, second moment of area
J	moment of inertia, loss function
K	constant, gain
L	inductance
M	torque
N	discrete number, windings number
P	power, probability
Q	generalized force, heat
R	electrical resistance, covariance or correlation function
S	spectral density, sum, performance criterion
T	absolute temperature, time constant
T_0	sampling time
U	input variable, manipulated variable (control input), voltage
V	volume
X	space coordinate
Y	output variable, space coordinate, control variable
Z	space coordinate, disturbance variable
$\mathbf{a}$	vector
$\mathbf{A}$	matrix
$\mathbf{A}^\mathrm{T}$	transposed matrix
$\mathbf{I}$	identity matrix
$\boldsymbol{\theta}$	parameter vector
$\mathbf{P}$	covariance matrix
$\boldsymbol{\psi}$	data vector
α	coefficient, angle

β	coefficient, angle
γ	specific weight, correcting factor
δ	decay factor, impulse function
ϕ	correlation function, validity function
η	efficiency
ϑ	temperature
λ	thermal conductivity, forgetting factor, failure rate
μ	friction coefficient, permeability, membership function
ν	kinematic viscosity, index
π	number $\pi = 3.14159\ldots$
ρ	density
σ	standard deviation, σ^2 variance
τ	time
φ	angle
ω	angular frequency, $\omega = 2\pi/T_\mathrm{p}$: T_p period
Δ	change, deviation
θ	parameter
Π	product
Σ	sum

2. General mathematical abbreviations

$\exp(x)$	$= e^x$
$E\{\ldots\}$	expectation of a statistical variable
dim	dimension
adj	adjoint
det	determinant
Re	real part
Im	imaginary part
$\dot{Y}$	dY/dt (first derivative)
var[]	variance
cov[]	covariance
$\mathcal{F}$	Fourier transform
$\mathcal{L}$	Laplace transform
rms (...)	root of mean squared of...

3. Letter symbols for internal combustion engines

3.1 Geometry and time (DIN 1304, ISO 31)

A	area	m^2
a	acceleration	m/s^2
b	breadth, width	m
c	absolute velocity	m/s
d	diameter	m

D	characteristic diameter	m
f	frequency	Hz
g	acceleration of free fall, gravitational acceleration	m/s^2
h	height	m
l	length	m
n	rotational speed	1/s, rpm
r	radius	m
s	actuator position	m
u	peripheral velocity	m/s
V	volume	m^3
v	specific volume	m^3/kg
w	relative velocity	m/s
ω	angular velocity	rad/s

3.2 Mechanics (DIN 1304, ISO 31)

a	specific work	J/kg
m	mass	kg
$\dot{m}$	mass flow rate	kg/s
p	pressure	Pa
C_d	orifice discharge coefficient	1
E	energy	J
F	force	N
J	moment of inertia	$kg\ m^2$
L	angular momentum	$kg\ m^2/s$
M	torque	Nm
P	power	W
W	work	J
$\nu, \nu = \eta/\rho$	kinematic viscosity	m^2/s
η	dynamic viscosity	Pa s
η	efficiency	1
ρ	mass density	kg/m^3
Π	pressure ratio	1

3.3 Thermodynamics and heat transfer (DIN 1304, ISO 31)

c_p	specific heat capacity at constant pressure	J/(kg K)
c_v	specific heat capacity at constant volume	J/(kg K)
Gr	Grashof number	1
H	enthalpy	J/kg
h	specific enthalpy	J/kg
L	characteristic length	m
n	polytropic exponent	1
Nu	Nusselt number	1
Pr	Prandtl number	1

Q	heat	J
$\dot{Q}$	heat flow rate	W
q	specific heat	J/kg
R	specific gas constant	J/(kg K)
Re	Reynolds number	1
S	entropy	J/K
s	specific entropy	J/(kg K)
T	thermodynamic temperature	K
U	internal energy	J
u	specific internal energy	J/kg
x	mass fraction	1
α	coefficient of heat transfer	W/(m^2 K)
κ	isentropic exponent	1
λ_{th}	thermal conductivity	W/(m K)
ϑ	temperature	K

3.4 Engine specific symbols

α_{ped}	accelerator pedal position	%
b_{f}	fuel consumption	kg/h
b_{sfc}	effective specific fuel consumption	g/kWh
c_{NOx}	nitrogen oxide concentration	g/m^3
c_{op}	opacity	%
c_{pa}	soot concentration	g/m^3
Δt_{pi}	timing of pilot injection	ms
$\Delta\varphi_{\mathrm{pi}}$	crank angle of pilot injection (difference angle to main injection)	°CS
ϵ	compression ratio	1
h	valve lift	m
H_{l}	lower fuel heating value	J/kg
φ	crank angle	°CS
φ_{mi}	crank angle of main injection	°CS
φ_{ign}	ignition angle	°CS
λ	air-fuel ratio (excess air factor)	1
λ_{a}	air expenditure	1
λ_{P}	connecting rod ratio	1
l_{r}	relative filling	1
L_{st}	stoichiometric air requirement	kg/kg
m_{f}	injection mass	mg/cyc
$\dot{m}_{\mathrm{f}}$	injection mass flow	kg/h
m_{mi}	main injection mass	mg/cyc
m_{pi}	pilot injection mass	mg/cyc
$\dot{m}_{\mathrm{eng,in}}$	gas flow into the engine	kg/h
$\dot{m}_{\mathrm{eng,out}}$	gas flow out of the engine	kg/h
m_{air}	air mass per cycle	mg/cyc

M_{aux}	auxiliaries torque	Nm
M_{cyl}	torque of one cylinder	Nm
M_{eng}	crankshaft mean torque at flywheel	Nm
M_{f}	friction torque	Nm
M_{g}	gas force torque	Nm
M_{i}	indicated torque	Nm
M_{l}	load torque	Nm
M_{m}	dynamic masses torque	Nm
M_{drg}	drag torque, motoring torque	Nm
M_{v}	valve train torque	Nm
$\dot{m}_{\mathrm{air}}$	air mass flow	kg/h
q_{f}	injection quantity	mm^3/cyc
q_{mi}	main injection quantity	mm^3/cyc
q_{pi}	pilot injection quantity	mm^3/cyc
r_{egr}	exhaust gas recirculation ratio	1
V_{c}	clearance volume per cylinder	m^3
V_{d}	displaced volume per cylinder	m^3
V_{D}	total displacement (all cylinders)	m^3
z	number of cylinders	1

3.5 Combustion pressure analysis

$dp_{\mathrm{cyl}}/d\varphi$	pressure gradient	Pa/°CS
$dQ_{\mathrm{f}}/d\varphi$	heat release rate	J/°CS
φ_{Q5}	location of mass fraction burned 5%	°CS
φ_{Q50}	location of mass fraction burned 50%	°CS
φ_{Q95}	location of mass fraction burned 95%	°CS
p_{mi}	mean indicated pressure	Pa
p_{m}	motored cylinder pressure (no injection)	Pa
p_{mep}	brake mean effective pressure	Pa
Q_{f}	heat release	J

3.6 Subscripts for internal combustion engines

1	state variables in front of the compressor
3	state variables in front of the turbine (in the exhaust manifold)
4	state variables after the turbine
5	state variables after the DPF
2c	state variables after the compressor
2i	state variables in the intake manifold
2ic	state variables after the intercooler
a	ambient
afi	air filter
afl	air flaps
air	air

b	burned
c	compressor
cam	camshaft
cas	crankshaft
cd	combustion duration
cg	combustion gas
cl	cooling medium, coolant
cool	cooling
cr	common rail
cyl	cylinder
dpf	diesel particulate filter
ds	delivery start
ec, EC	exhaust closes
eff	effective
eg	exhaust gas
egr	exhaust gas recirculation
egrc	EGR cooler
egrv	EGR valve
eng	engine
eo, EO	exhaust opens
es	exhaust system
eth	exhaust throttle valve
ev	exhaust valve
f	fuel, friction
geo	geometrical
H_2O	coolant water
hpegr	high pressure exhaust gas recirculation
hpp	high pressure pump
ic	intake closes
id	injection duration
igd	ignition delay
in	streaming in
int	intake
inj	injectors
io	intake opens
iv	intake valve
lpegr	low pressure exhaust gas recirculation
mc	main combustion
meas	measured
mi	main injection
mv	metering valve
oil	oil
osc	oscillating
out	streaming out
p	piston

pc	pilot combustion
pcv	pressure control valve
pi	pilot injection
r	rod
rail	rail system
red	reduced
rot	rotating
sim	simulated
soc	start of combustion
soi	start of injection
sw	swirl flap
t	turbine
tc	turbocharger
th	throttle
u	unburned
vac	vacuum system
vgt	variable geometry turbocharger
w	wall
wg	waste gate turbocharger

3.7 Abbreviations for internal combustion engines

AFR	air-to-fuel ratio
ASAM	Association for Standardization of Automation and Measuring Systems
BDC	bottom dead center
CI	compression ignition engine
CR	common rail
DOC	diesel oxidation catalyst
DPF	diesel particulate filter
EGR	exhaust gas recirculation
HFM	hot film measurement
NSC	NO_x storage catalyst
OSEK	Offene Systeme und deren Schnittstellen für die Elektronik im Kraftfahrzeug (open systems and their interfaces for the electronics in vehicles)
PF	particulate filter
PM	particulate matter
SCR	selective catalytic reduction
SI	spark ignition engine
TDC	top dead center
VGT	variable geometry turbocharger
VVT	variable valve train

4. Abbreviations for identification and signal-analysis methods

ACF	Auto Correlation Function
APRBS	Amplitude-modulated PRBS
ARMA	Auto Regressive Moving Average process
CCF	Cross Correlation Function
DFT	Discrete Fourier Transform
DSFC	Square root filtering in covariance form
DSFI	Square root filtering in information form
ELS	Extended least squares
ETA	Event Tree Analysis
FDD	Fault Detection and Diagnosis
FDI	Fault Detection and Isolation
FMEA	Failure Mode and Effects Analysis
FFT	Fast Fourier Transform
FTA	Fault Tree Analysis
HA	Hazard Analysis
HCCI	Homogeneous Charge Compression Ignition
LS	Least Squares
MIMO	Multiple Input Single Output
MISO	Multiple Input Multiple Output
MLP	Multilayer Perceptron
MTBF	Mean Time Between Failures
MTTF	Mean Time To Failure = $1/\lambda$
MTTR	Mean Time To Repair
NN	Neural Net
PCA	Principal Component Analysis
PRBS	Pseudo Random Binary Signal
RBF	Radial Basis Function
RLS	Recursive Least Squares
SISO	Single Input Single Output

1

Introduction

The increasing *electronification* and *electrification* is a dominant feature of modern automotive developments. This is demonstrated by an increasing part of electrics/electronics (E/E) of the manufacturing costs from about 20 % in 1995 to 35 % in 2012. The electrics comprise primarily the electrical energy flows to the consumers through the energy board net. Frequently, former mechanical, pneumatic or hydraulic actuated components of chassis and powertrain are replaced by electrical ones. The chassis-oriented electronics serve mainly the driving behavior, safety and comfort. Powertrain electronics are used for control functions to reach good driveability, low-fuel consumption and emissions.

These developments are possible through the increasing number of *mechatronic components* in the powertrain and the chassis. Figure 1.0.1 gives some examples for engines, drive trains, suspensions, brakes and steering systems. *Mechatronic systems* are characterized by an integration of mechanics and electronics, where the integration is between the components (hardware) and the information-driven functions (software). This development has a considerable influence on the design and operation of the *powertrain* consisting of the combustion engine and the drive train. In the case of hybrid drives this includes also the electrical motor and the battery.

1.1 Historical developments

The development of sensors, actuators and electronic control for automobiles is depicted in Fig. 1.1.1. The first mechatronic products in vehicles have been anti-lock braking (ABS, 1979), electric throttle (1986), automatic traction control (TCS, 1986) and semi-active shockabsorbers (1988), followed by the electronic stability control (ESC, 1995), electric power steering (EPS, 1996), active body control (ABC, 1999) and active front steering (AFS, 2003). These large improvements served mainly for increasing safety and comfort and required many new sensors with electrical outputs, actuators with electrical inputs and micro-controller-based electronic control units. Some further steps were *driver-assistance systems* to support the driver in performing driving maneuvers. They require sensors for the surroundings and are passive

R. Isermann, *Engine Modeling and Control*,
DOI: 10.1007/978-3-642-39934-3_1

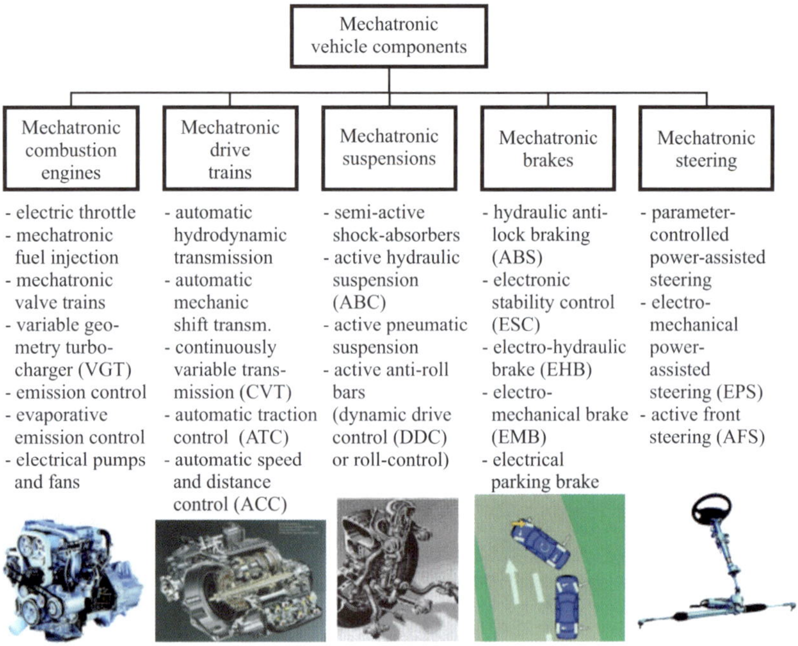

Fig. 1.0.1. Mechatronic components and systems for automobiles and engines.

by giving warnings or are active by intervening in the vehicle dynamics. Examples are parking-assistance systems which measure within the parking space and adaptive cruise control (ACC, 1999) which measures the distance and relative velocity to vehicles in front and such improve mainly the comfort and convenience of driving. Driver-assistance systems for lane-departure warning and anti-collision avoidance operate with LIDAR or RADAR and video cameras and improve primarily the safety of driving.

A common feature of these developments is the increase of electrical sensors, actuators and electronic control units, the coupling through cables and bus systems and a beginning interconnection of the decentralized control units. Some of the vehicle control systems give commands to the engine control system, as, for example, TCS, ESC and ACC.

Parallel to the increase of electronic control functions for the chassis the *engines and drive trains* have shown a similar development. This has to be seen together with the improvements of the combustion, fuel consumption and emission reductions and will be discussed in the following sections.

1.1.1 Gasoline engines (SI)

The historical development of gasoline engines during the last 50 years with view on their control is depicted in Fig. 1.1.2. Until about 1965 the engines were mechanically controlled with transistor-triggered electromechanical coil ignition. Fuel

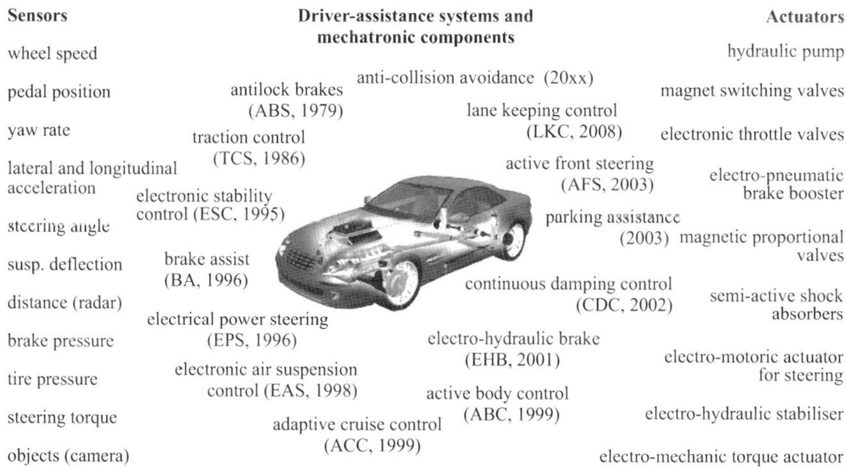

Fig. 1.1.1. Development of sensors, actuators and electronic control systems for automobiles.

injection systems for the intake manifold with electronic analog control began to replace the carburetors in 1967. Since about 1970 increasingly more functions are controlled electronically, first with transistor technology. This development required more sensors (knock sensors, air flow and air pressure sensors) with electrical outputs and actuators with electrical inputs (fuel injectors). A large influence on the developments had the state regulations and emission laws, for the United States the Clean Air Act (CARB) in California (1983) and since 1993 for US states in different tiers the laws for low emission vehicles (LEV), ultra low emission vehicles (ULEV) and super ultra low emission vehicles (SULEV). The corresponding European regulations are EURO 1 (1992), EURO 2 (1996), EURO 3 (2000), EURO 4 (2005), EURO 5 (2009), and EURO 6 (2014). These regulations were supplemented by the requirements for an on-board diagnosis in the United States OBD I (1988), OBD II (1994) etc. and EOBD (2000) for Europe.

Gasoline engines received catalytic converters with λ-control (1976) and microprocessor control in 1979. The electrical throttle was introduced in 1986, the direct injection about 1999 and since 2000 gasoline engines are supplied with variable valve trains for valve timing and lift control. Present gasoline engines are characterized by electromagnetic or piezoelectric injectors, high-pressure injection pumps (120 bar), homogeneous and stratified combustion, mechanical or turbo charging and increased specific power (downsizing). Figure 1.1.3 depicts the development of sensors and actuators for gasoline engines. Today's SI engines have about 15–25 sensors and 6–8 main manipulated variables and are controlled with a powerful microcomputer control unit (ECU) with 80–120 look-up tables and many control algorithms.

1.1.2 Diesel engines (CI)

The historical development of diesel engines with regard to their control is shown in Fig. 1.1.4. Around 1960 diesel engines had a swirl chamber, mechanically controlled

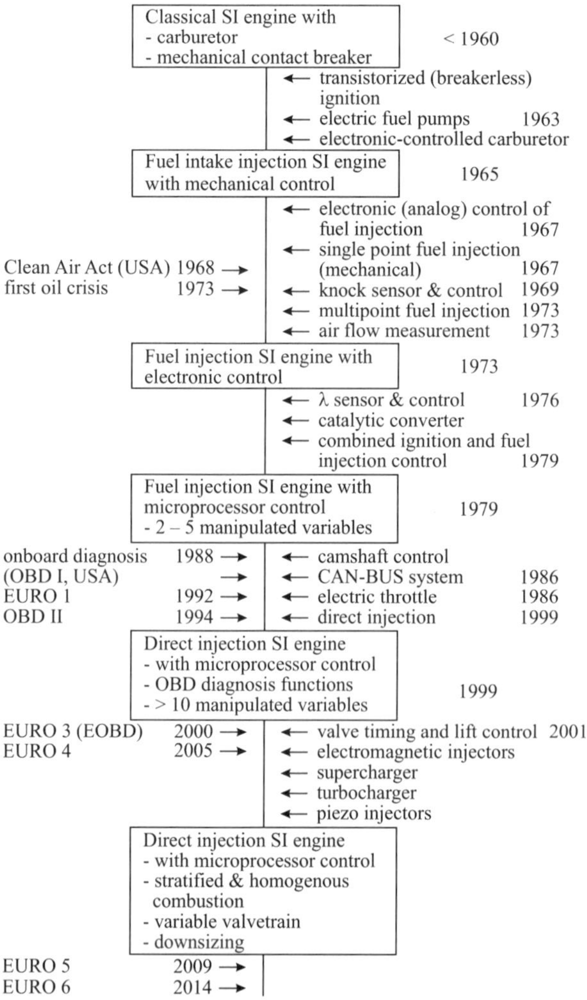

Fig. 1.1.2. Historical development of gasoline engines.

piston injection pumps and fly-weight overspeed control. Microprocessor control with direct injection and distributor pump (900 bar) and wastegate turbochargers appeared about 1989. Further steps were exhaust gas recirculation (EGR), oxidation catalyst and turbochargers with variable geometry (1992). First common-rail injection systems with direct injection (1500 bar) with VGT turbochargers reduced further fuel consumption and emissions and resulted in good dynamic torque generation. Today's diesel engines are characterized by high pressure (2000 bar), multiple common-rail injection, piezo-injectors, twin turbochargers or VGT chargers, high EGR rates, $DeNO_x$-catalyst, particulate filters with regeneration, and selective catalytic reduction (SCR).

Sensors and actuators for gasoline engines

source: Volkswagen AG

Sensors

engine speed (1967), camshaft phase

motor temperature (1967)

manifold pressure, manifold temperature (1967)

knock sensor (1969)

air-mass flap (1973)

oxygen (lambda) (1976)

airmass hot-film (1981)

ambient pressure

throttle and pedal position (1986, E-Gas)

tank pressure

ion current

combustion pressure

Actuators

electronic fuel injection (1967, D-Jetronic)

microelectronic-controlled ignition and injection (1979, Motronic)

electronic throttle (1986)

exhaust gas valve

secondary air valve (1994)

variable camshaft timing

direct injectors (1999)

variable geometry manifold

variable valve lift (2011)

turbocharging (2006)

→ 15–25 measurements
→ 6–8 manipulation variables
→ 80–120 look-up tables

Fig. 1.1.3. Sensors and actuators for gasoline engines (SI).

The development of sensors and actuators for diesel engines is summarized in Fig. 1.1.5. Present diesel engines need about 15–20 sensors, 5–9 main manipulated variables and an ECU with more than 100 look-up tables and many control algorithms.

1.2 Current engine developments

1.2.1 Gasoline engines

Current developments for the further improvement of gasoline engines are, for example, variable valve trains, downsizing and modified combustion processes.

Variable valve trains (VVT: variable valve timing) permit the improvement of the gas exchange. The conventional phase shifting of the inlet valves primarily increases the torque through early or late opening in dependence on torque and speed. In order to reduce the gas flow losses through the throttle the valves require variable timing as well as variable lift. Then the fresh air mass can be controlled by the inlet valves. In addition the residual gases can be influenced by changing the overlapping of inlet and outlet valves to improve the emissions through internal exhaust gas recirculation. Manipulation of the valve lift in two steps or continuously gives more freedom for controlling the load without throttling, see, e.g. van Basshuysen and Schäfer (2004), Braess and Seiffert (2005), Köhler and Flierl (2012). A comparison of different designs of VVT, Schulz and Kulzer (2006) shows that the fuel consumption

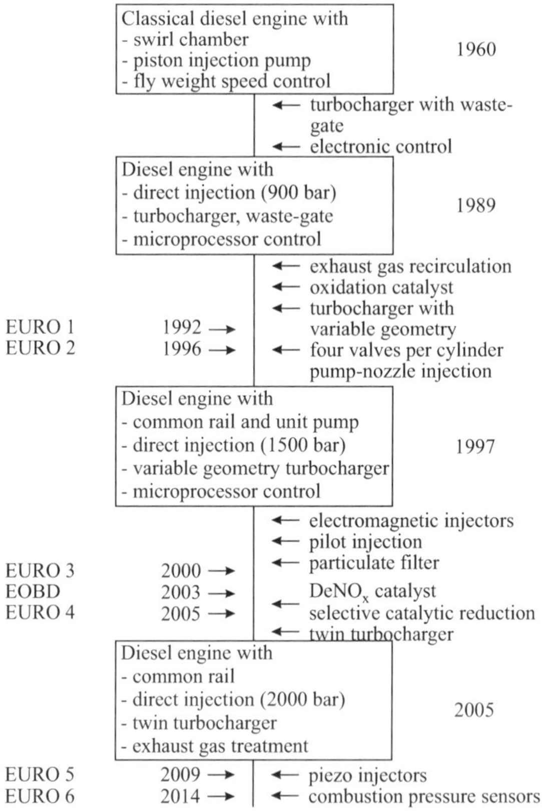

Fig. 1.1.4. Historical development of diesel engines.

can be improved with phase actuation by 3–4 %, lift switching by 8–10 %, continuously variable lift by 8–10 %, and with full variable hydraulic or electrical VVT by 14–16 %. However, the complexity is relatively high for the full variable VVT. Therefore, the first mentioned three mechanical VVT's are a good compromise.

The reduction of the displacement, i.e. *downsizing* for a given well powered vehicle leads to a smaller specific fuel consumption (less throttling) in part load, as the consumption in the torque-speed diagram shows. However, in order to increase the torque for small speeds and to reach a certain power at higher loads and speeds exhaust turbocharging or supercharging with a mechanical compressor is required. This means, for example, to reduce the displacement from 2 l to 1.3 l and an increase of the mean effective pressure from 6 to 9 bar. A comparison of different gasoline engines shows that the downsizing factor should be at least 1.3 and should be combined with a change of the operation point to higher torques by increasing the transmission ratio in the drive train (*downspeeding*) to result in a fuel reduction of about 11 %, Königstein et al (2008).

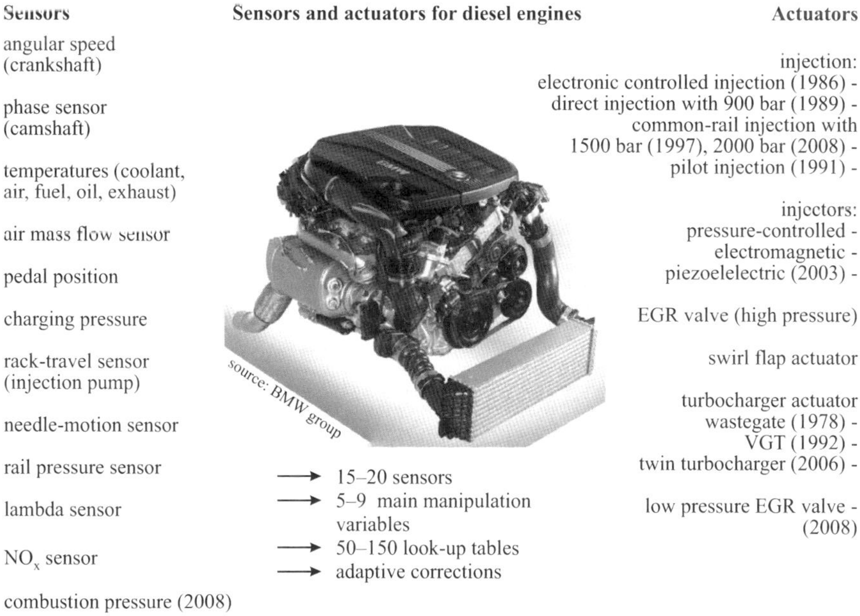

Fig. 1.1.5. Sensors and actuators for diesel engines (CI)

The optimization of the *combustion process* has of course a large influence. Compared to the conventional intake manifold injection and stoichiometric combustion with $\lambda = 1$ and three-way catalyst the direct injection into the cylinders allows considerable saving of fuel consumption for SI engines. Together with a VVT a reduction of 10 % is possible. High-pressure injectors (120 bar) with piezoelectric actuation gives a better spraying and makes a stratified, lean combustion with $\lambda > 1$ in part load possible, resulting in about 15 % fuel saving, Weingärtner et al (2007), Berns (2007). A homogeneous charge compression ignition (HCCI) with an increase of the gas temperature by increased residual gases can be obtained, for example, through early closing of the outlet valve and early injection. The combination of early closing the outlet valve and late opening of the inlet valve enables a recompression and a first injection, which can be applied for part load up to 40 %, Alt et al (2008), Backhaus (2008). However, this requires a combustion pressure measurement and control and full variable VVT. A reduction of fuel consumption of about 13–19 % is expected and a NO_x-catalyst becomes unnecessary.

1.2.2 Diesel engines

Of current interest for the further development of diesel engines are a reduction of fuel consumption, NO_x and particulates. This can be reached by further improvements of the common-rail direct injection, combustion processes, charging and exhaust-gas treatment.

Some steps for the *common-rail direct injection* are higher pressures (2200 bar) and multiple injections in order to improve the combustion, emissions and noise. Solenoid and fast piezoelectric injectors allow different combinations of pre-, main- and post-injection pulses. An increase of the exhaust-gas recirculation rate with strong cooling results in low NO_x emissions. However, a too strong increase of the EGR reduces the turbocharger power. Therefore a low-pressure EGR after the particulate filter through a cooler to the compressor inlet can be added. Then high EGR rates with a good mixture of fresh air and exhaust-gas and low temperature through an intercooler may lead to a good cylinder filling, Berns et al (2004), Hadler et al (2008). This requires several catalysts and regeneration phases and more sensors in the exhaust path, Bauer et al (2007).

A *modification of the combustion process* is the homogeneous compression ignition (HCCI), which can, e.g. be realized by an early injection with a high EGR rate in the part-load area. This leads to a strong reduction of NO_x and particulates. However, it needs a combustion feedback control with combustion pressure measurement because of the narrow possible operation limits and concentration differences in the individual cylinders, see e.g. Alt et al (2008), Backhaus (2008).

The use of *two turbochargers* with a small and large diameter enables an operation with better efficiencies, a high medium charging pressure over a larger speed range and results in improved acceleration at low speeds. The turbochargers are switched with pneumatic flaps, Steinparzer et al (2007). Also diesel engines allow a certain downsizing by increasing the specific power.

Especially large efforts go into the *exhaust treatment*, for example, through oxidation catalyst converters and particulate filters for minimization of CO, HC, NO_x and particulates. An alternative is the selective catalytic reduction (SCR) with the injection of dissolved urea, especially for heavy duty vehicles. The combination of oxidation catalyst, particulate filter, NO_x-storage catalyst and H_2S-catalyst results in a reduction of NO_x by 90 % without additives, however requires model based control and several additional sensors, and three different regeneration cycles, Hadler et al (2008).

Summarizing, *gasoline and diesel engines* show several development lines, to improve the torque generation and to decrease fuel consumption, emissions and noise. Their present development can be characterized by:

- reduction of fuel consumption and CO_2 emissions
- reduction of specific emissions (HC, CO, NO_x, particulates, dust)
- powerful exhaust gas after-treatment systems
- good driving behavior
- increased specific power (downsizing, charging)
- reduction of friction
- auxiliaries: minimization of energy consumption
- reduction of oscillations and noise.

With regard to the increasing variabilities and control functions the engines are supplied with *mechatronic components*. Figure 1.2.1 depicts some of these components.

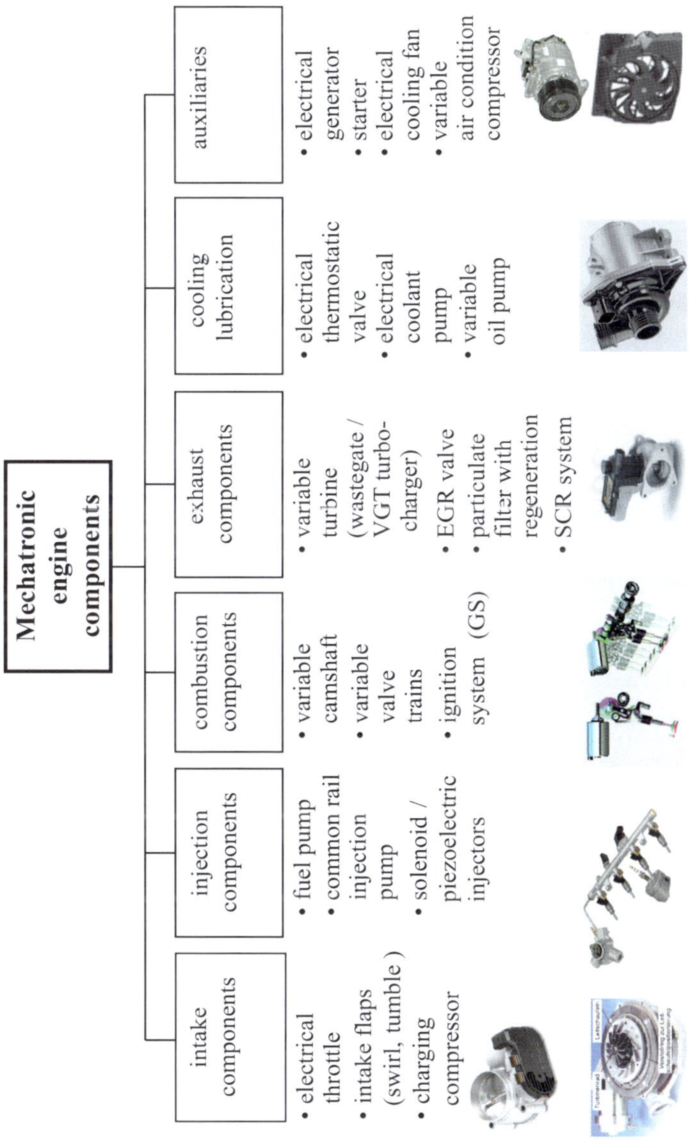

Fig. 1.2.1. Mechatronic components of internal combustion engines.

Examples are the electrical throttle in the intake system, the high-pressure common-rail injection system with solenoid or piezoelectric injectors, variable camshaft, variable valve trains and variable geometry turbochargers. The mechatronic components can be subdivided in *actuators* with electrical, pneumatic or hydraulic auxiliary energy, switching injection valves and solenoid valves and *electrical drives, pumps and fans*. Some of them are decentralized mechatronic components with local integration of sensors and electronics and some are centrally controlled by the engine control unit (ECU), see Chap. 5.

1.2.3 Alternative drives

A further step of the electrification of vehicles is the development of *hybrid drives*. One reason for this combination of combustion engines and electrical drives is a further saving of fuel consumption and emissions through the operation of the combustion engines in the ranges of better specific fuel consumption, regenerative braking and electrical driving. Micro hybrids have typically a starter/generator with one clutch and start-stop control. Mild and full hybrids in parallel configuration are characterized through a stronger starter/generator or a generator/motor with two clutches. Serial configurations operate with a combustion engine driving a generator and a separated electric drive. Full hybrids with power split systems have a separated generator and electrical motors coupled via a planetary gear. Depending on the hybridization degree the hybrid drives require a high-voltage traction board net, a DC/DC converter for the supply of the chassis board net, power electronics and a very capable battery as energy storage.

All these developments mean a strong increase of electronics and electrics and many control functions and imply a mechatronic overall design. A major influence on the efficiency has the further improvement of specific capacity, lifetime, reliability, safety and cost of the batteries, for example, in the form as NiMH or lithium-ion types. The increase of components also underlines the significance of fault diagnosis functions.

1.3 Engine control and diagnosis

The electronic engine control unit (ECU) has the task to control, optimize and supervise all relevant functions of an internal combustion engine (ICE). At first glance this comprises:

- torque generation according to the drivers accelerator pedal
- low fuel consumption
- low exhaust emissions and low noise (compliance with legislation)
- good driving behavior.

With regard to the general assembly of internal combustion engines the control functions can be dedicated to following engine subsystems or engine parts:

1. Intake system (air system)
2. Injection system
3. Fuel supply
4. Combustion and crankshaft drive
5. Lubrication
6. Exhaust gas system
7. Cooling system.

A part of the subsystems is basically controlled mechanically, as for example, the inlet and outlet valve timing and lift through the camshaft for four-stroke engines, the fuel pressure and oil pressure with overpressure relief valves and the coolant temperature with thermostatic expansion valves. Many other important variables are controlled electronically like the air flow, fuel flow, combustion, torque, speed and exhaust gas treatment. However, also the camshaft can be controlled electronically in the case of variable valve trains (VVT) and the fuel pressure for common rail direct injection systems, as well as the oil pressure. Therefore electrically commanded actuators are required which operate with electrical, pneumatic or hydraulic auxiliary energy, and sensors at the engine with electrical outputs.

The optimization of the thermodynamic, fluid dynamic and mechanical design and construction of internal combustion engines has led to an increasing number of actuators and sensors and to a strong increase of electronic control functions. Figures 1.1.3 and 1.1.5 depict the increased numbers of sensors and actuators. The electronic control unit (ECU) controls the start, idle, warm-up and normal operation. The increase of control functions is demonstrated by the size of programs and data of the digital microcomputer based system. Within the last 15 years the clock frequency of the microprocessors has increased from 12 to 150 MHz, the databus width from 8 bit to 32 bit, program storage from 32 Kbyte to 5 Mbyte, the computing time from about 10 to 300 millions instructions per second (MIPS), and the calibration labels from about 2500 to 30000 (approximate numbers, see Sect. 5.1. Within this time window of 15 years these characteristics have increased by following factors: memory: 1:100, computer power (MIPS): 1:30 and calibration parameters 1:10, compare Fig. 1.3.1.

This development was mainly influenced by the increase of variabilities to optimize the combustion and the exhaust-gas treatment. For gasoline engines this reflected first, for example, in the optimization of valve timing and lift in combination with the injections, depending on load, homogeneous or stratified operation, and λ-control. In the case of diesel engines this resulted in optimizing multiple injections, valve timing, turbo charging, exhaust-gas recirculation and particulate filters and their regeneration cycles.

Figure 1.3.2 shows a simplified signal flow diagram for gasoline engines. The engine control system has to be designed for 6–8 main manipulated variables and 5–8 measured output variables, leading to a complex nonlinear multiple-input multiple-output system. Because several output variables cannot be measured in mass production vehicles, like the torque and some emissions, as they are too costly or may not be robust enough, some output variables of the engine are controlled by *feed-forward structures*, also called *open-loop structures*. This requires that the driver's pedal position, several influencing variables like the engine speed, air mass flow, air temperature and pressure, oil and coolant temperature have to measured and feedforward control actions on the main manipulated variables have to be specified. In the case of gasoline engines *feedback control* or *closed-loop control* is used for λ-control and knock control. Diesel engines use charging pressure, air mass flow rate and overspeed feedback control, see Fig. 1.3.3. Both engine types have idling speed and coolant fluid temperature feedback control. The feedforward structures have the advantage that they do not have stability problems, but they need relatively precise engine models and sensors for the main influencing variables. The closed loop structures compensate also not measured disturbances, but their controllers must be precisely and robustly adapted to the engines nonlinear dynamic behavior. All the control functions depend strongly on the load and speed and on the operating state such as starting phase, warm-up, normal state, idling, overrun and shut-down.

The next sections give, as a first overview, some details of basic control functions for gasoline and diesel engines.

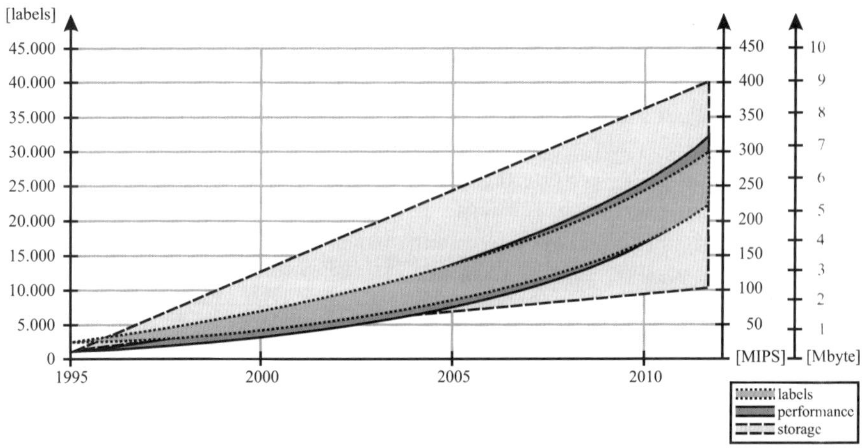

Fig. 1.3.1. Development of ECU characteristics for engine control, ETAS (2010).

1.3.1 On the control of gasoline engines

Figure 1.3.4 illustrates the main components of a direct injection gasoline engine with its actuators and sensors. To generate a certain torque at the crankshaft the gasoline engines require a very specific air-fuel mixture. For an optimal combustion they need a close to stoichiometric mixture of air and fuel in the range $0.8 < \lambda < 1.4$ and because of the mostly applied three-way catalyst the air/fuel ratio must be precisely in the range of $0.98 < \lambda < 1.02$. Therefore the air is usually throttled for part-load by a throttle valve so that the cylinders suck the air from the manifold with sub-atmospheric pressure. The air mass (air charge) taken in by the cylinders determines together with the fuel the resulting torque. It is measured by an air-mass

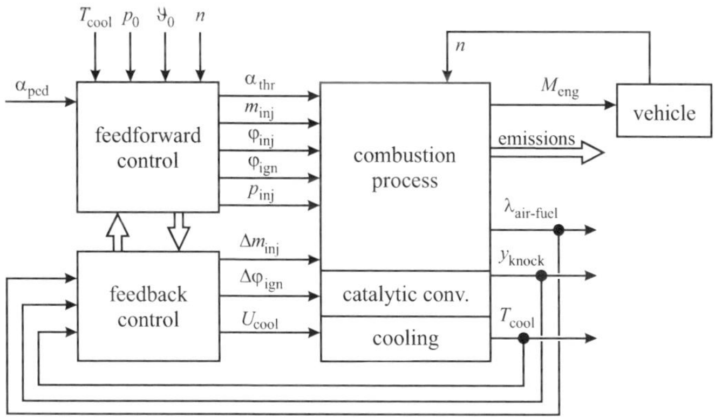

Fig. 1.3.2. Simplified control structure of a gasoline engine with intake fuel injection. 6–8 manipulated variables, 5–8 measured outputs.

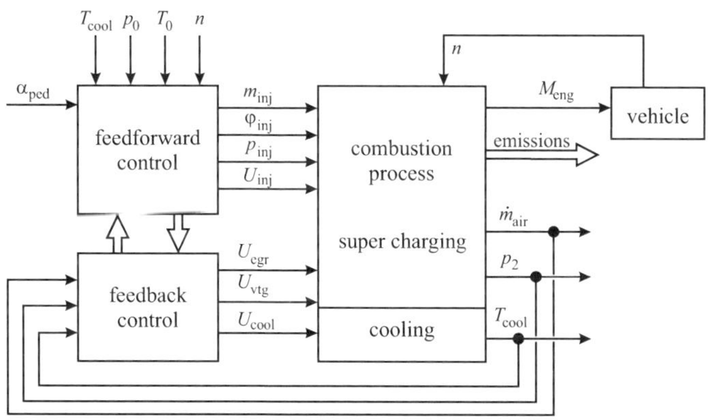

Fig. 1.3.3. Simplified control structure of a diesel engine with turbocharger. 6–8 manipulated variables, 5–8 measured outputs.

sensor or determined indirectly by a manifold pressure sensor. However, the conventional load control of the engine by an upstream throttle generates charging losses, in contrast to diesel engines. The control unit (ECU) commands the electrical actuated throttle valve depending on the drivers accelerator pedal position and adjusts the injected fuel quantity with the *fuel injection system* by feedforward control. The fuel is injected either in the intake manifold upstream the inlet valves (port injection) using a low-pressure pump (4–6 bar) or directly into the combustion chamber with controlled high-pressure fuel pumps (120–200 bar).

The *injection into the intake manifold* during the induction stroke of the cylinders may consist of several injection pulses (multi-point injection) and is applied for a conventional homogeneous combustion. The ECU converts the accelerator pedal position into a desired calculated torque value. By using various correction functions the air charge and the corresponding fuel injection mass per cylinder and an optimal

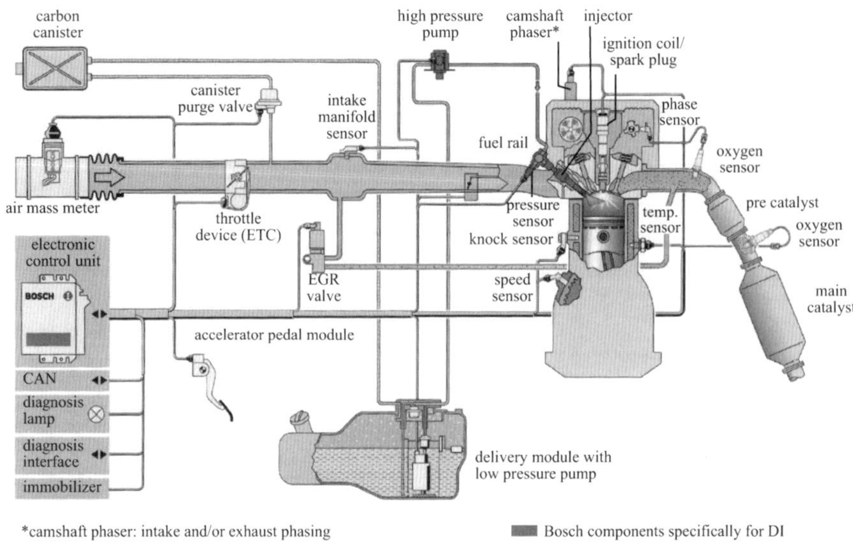

Fig. 1.3.4. Schematic of a direct injection gasoline engine with actuators, sensors and ECU for homogeneous combustion. (Bosch DI Motronic, $\lambda = 1$, by courtesy of Robert Bosch GmbH).

ignition angle relative to the top dead center of the crankshaft are calculated and the manipulated variables throttle valve angle, fuel injection angle and duration and the ignition coil activation are set by feedforward control look-up tables. The injected fuel mass is corrected by a (cascaded) lambda feedback controller with measurements of λ in the exhaust system before and after the catalyst in order to reach a stoichiometric combustion for an optimal conversion of CO, HC and NO_x in the three-way catalyst.

The *direct injection* in the cylinders has as additional manipulation the high pressure in the fuel rail and allows different types of combustion. At lower engine speed and torque a stratified cylinder charge can be reached, where in one zone a combustible air/fuel mixture cloud is generated and transferred to the spark plug. The other zone contains excess air and residual exhaust gases. The zones are controlled by a turbulence flap in the manifold. Thus, the throttle valve can be opened, avoiding throttling and charging losses, resulting in better efficiency. The fuel is injected during compression. However, the lean combustion generates larger NO_x concentrations and because of the excess air a three way catalyst cannot be used. Therefore NO_x is reduced by exhaust gas recirculation (EGR), manipulated by an EGR-valve and a NO_x accumulator-type catalytic converter is applied. At high speed and torque the engine has to run in homogeneous mode, because the generation of stratified conditions cannot be realized. Hence, this two-mode (also three-mode) combustion requires precise control functions for several variables and operates by using universal (broadband) lambda sensors with variable setpoints for the air/fuel ratio. A NO_x sensor may also be used. In addition *knock feedback control* corrects the ignition angle by using one or two knock sensors at the crankcase in order to prevent knocking combustion, which can lead to engine damages.

Based on these main control tasks Fig. 1.3.5 shows a control-oriented block structure for a gasoline engine. The overall functions of the engine control system are frequently summarized by the term *electronic engine management*. The individual feedforward and feedback control functions show a hierarchical structure which consists of higher level and lower level control functions. The higher level control functions can be grouped in control subsystems or modules for torque control, injection and air/fuel control, ignition control, knock control, air charge, and EGR control. Sublevel control functions are, for example, the position control of actuators or low fuel pressure control. Figure 1.3.5 shows seven main actuators which manipulate the air flow, the injection mass and angle, rail pressure, ignition time, camshaft phase, and recirculated exhaust gas flow. Some control functions change for different operating conditions like cold start, warming-up phase and normal operation, idling, and shut-off.

The control of *automatic transmissions* with torque converters and planetary gears or double-clutch transmissions with spur gears is mostly realized in a special electronic control unit attached to the transmission housing. The speed of the engine results together with the power consuming load, here the transmission and the vehicle.

1.3.2 On the control of diesel engines

The main components of a diesel engine with turbocharger are shown in Fig. 1.3.6. The air for diesel engines is sucked into the cylinders with slightly lower than atmospheric pressure, if no turbochargers are used and the torque is manipulated by the injected fuel mass. However, most of the modern diesel engines have turbochargers to increase the torque and the power. The air charge then depends on the charging (boost) pressure, which is controlled by the charging pressure controller, e.g. by changing the wastegate position or the position of a variable geometry turbocharger (VGT). At low load, the diesel engine operates with large excess air, i.e. with large air/fuel ratio expressed by the excess-air factor λ. Only at high loads, λ comes closer to 1. If it gets too small, the diesel engine starts smoking. The relative high compression ratio in the cylinders together with the low charging losses because of an inactive throttle valve in the intake and therefore good air supply lead to a higher efficiency compared to gasoline engines, especially at part load. However, the higher the combustion temperature and therefore the higher the thermodynamic efficiency the more NO_x gases are produced because of the oxidation of nitrogen gas contained in the air. In order to lower the combustion temperature and the NO_x concentration, the exhaust gases can be recirculated to the intake on cost of the fuel efficiency.

Modern diesel engines have either a common rail or a unit pump *injection system* with pressures until 2200 bar. The ECU commands the injected fuel mass into the cylinders according to the drivers accelerator pedal position. In order to influence the burning process with regard to optimized fuel consumption, NO_x and soot formation and the generated noise multiple injections are applied, with pre-, main-

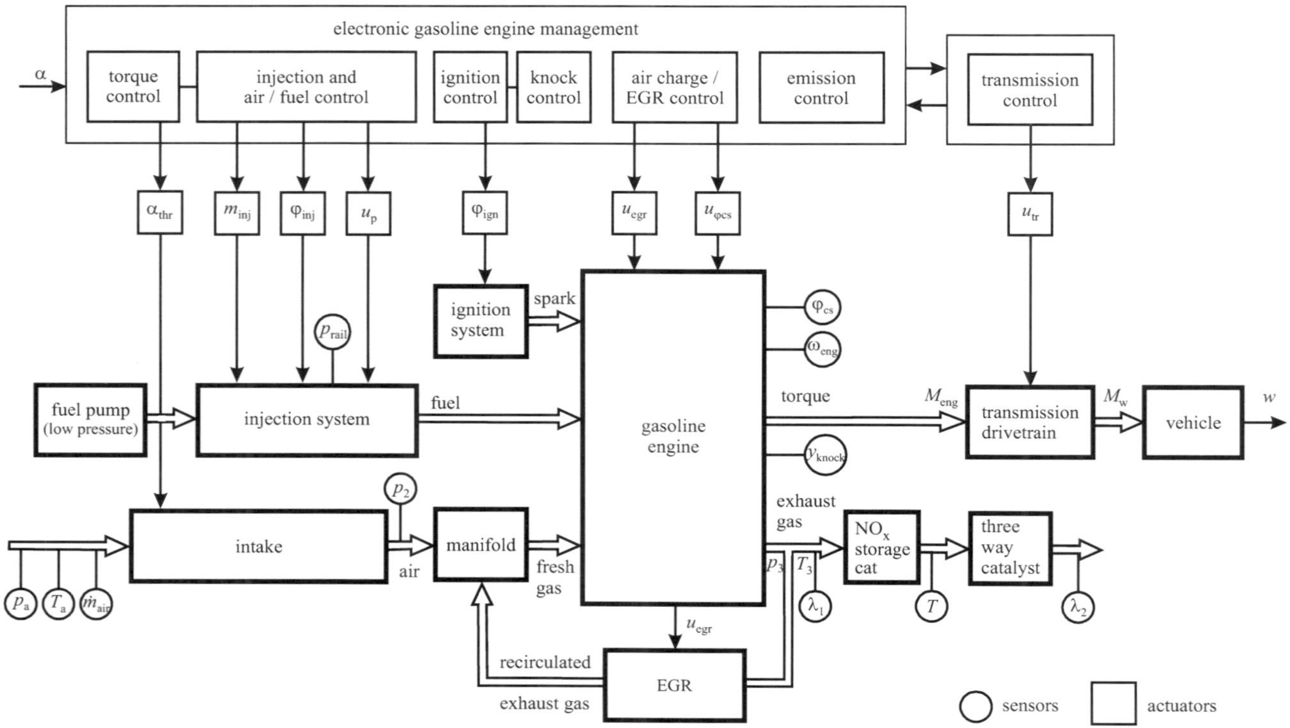

Fig. 1.3.5. Control-oriented block diagram of a gasoline engine with direct injection, actuators, control subsystems and some sensors.

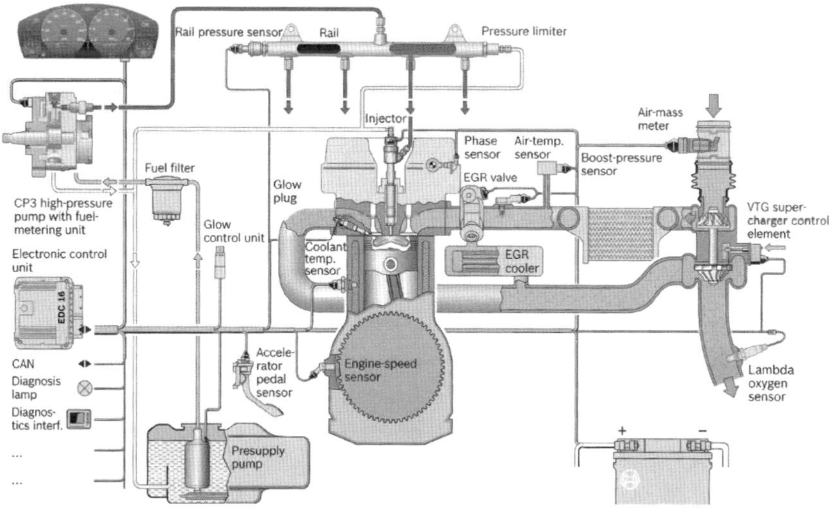

Fig. 1.3.6. Schematic of a direct injection diesel engine with VGT turbocharger, actuators, sensors and ECU. (Bosch EDC 16, by courtesy of Robert Bosch GmbH).

and after-injection pulses. The air mass flow rate in the *intake system* is indirectly controlled by the EGR valve (passenger car diesels mainly). Because the fresh gas mixture taken in by the cylinders is about constant for a certain operating point, an increasing EGR flow rate $\dot{m}_{egr}$ results in a reduction of the air flow rate $\dot{m}_{air}$. This means that the ratio $\dot{m}_{egr}/\dot{m}_{air}$ is changed. The control variable of the closed *air mass control loop* is the measured mass flow rate and its reference value is calculated from the desired reference excess air factor λ_{ref}. A further feedback control is the *charging pressure control loop*, which manipulates either a wastegate for fixed geometry turbochargers or the guide vanes of turbochargers with variable geometry (VGT). Both closed control loops are strongly coupled.

Diesel engines require a *speed controller* for the maximum speed in order to limit the speed by reducing the fuel mass. Otherwise the engine could be damaged by overspeeding. This is because a load control by intake throttling is missing. Speed control is also applied for idling, but not during the normal operating range for diesel engines in vehicles.

The *exhaust after-treatment* has the task of reducing HC, CO, NO_x and particulates. In most passenger car diesel engines HC, CO and some soot are removed by an oxidation catalyst. This may be followed by an NO_x storage catalyst (NSC), where in the loading phase NO_2 is stored within 30 to 300 s in a lean exhaust gas ($\lambda > 1$). Regeneration and removal of NO_2 take place in a rich exhaust gas ($\lambda < 1$) within 2 to 10 s. This removal or reduction of NO_2 is performed with CO and H_2 in a rich exhaust gas by retarding the injection angle and throttling the intake air. The control of the storage catalyst requires a temperature sensor and NO_x-sensor or λ-sensor. An alternative is a selective catalytic reduction (SCR), where NO_x is removed by ammonia NH_3, which stems from a liquid urea. This process operates continuously by injecting an urea/water solution ("add blue") from an extra tank, and is feedback controlled by using NO_x- and NH_3-sensors.

Emitted soot particulates can be removed by diesel particulate filters (DPF). They consist of porous ceramics or sintered metal and must be generated by soot burning with the oxygen in the exhaust gas and an increased temperature of minimum 600°C, resulting in CO_2. The regeneration is started when a criterion based on a combination of difference pressure increase and calculated soot storage from a storage model is exceeded. Then the exhaust temperature is increased by a retarded main or extra late injection and intake-air throttling, depending on the operating state of the engine. This regeneration takes about 10 to 20 minutes and the engine is controlled such, that the torque is not remarkably reduced. The control of the DPF requires a difference pressure and a temperature sensor.

Figure 1.3.7 depicts a block-oriented structure of an *electronic diesel management system*. It is, as the control system for gasoline engines, organized in a hierarchical structure which is subdivided in higher level and lower level control functions. The *higher level control functions* can be grouped into subsystems or modules for torque control, injection and fuel control, air charge and EGR control, and emission control. Lower level control functions are, for example, the position control of actuators and the control of swirl and tumble flaps. Figure 1.3.7 shows seven actuators for the air flow, injection mass and angle, rail pressure, EGR flow, camshaft phase and turbocharger. Closed-loop control is usually realized for the air flow, common rail pressure, charging pressure and in some cases for exhaust gas after-treatment. This is added by many feedforward control functions.

Beside the main control functions several *sublevel controls* are required, like position control for throttle, injection pumps, camshaft, swirl and tumble flaps, pressure control for fuel and lubrication oil. For special operating conditions or states supplemental control functions are implemented, like smoke limitation control (CI), idling-speed control, cold start-up control, and warming-up control.

Most feedforward control functions are implemented as grid-based three-dimensional look-up tables (3D-maps) or as two-dimensional characteristics. This is because of the strongly nonlinear static behavior of the IC-engines, the good interpretation and the direct programming in microprocessors with fixed point arithmetics. Some of the functions are based on physical models with correction factors, but many of the look-up tables and control algorithms are calibrated after measurements on engine test benches and with vehicles. However, as the complexity and the variants of the engines increase, *engine-model-based identification and control-design methods* are of growing importance, see Chaps. 3, 5 and 6. The control of an *automatic transmission* is, as for gasoline engines, mostly implemented in a special transmission control unit.

This concludes a first rough sketch of basic control functions for combustion engines. The individual control systems are treated in more detail throughout the book and especially in Chaps. 5, 6, 7 and 8.

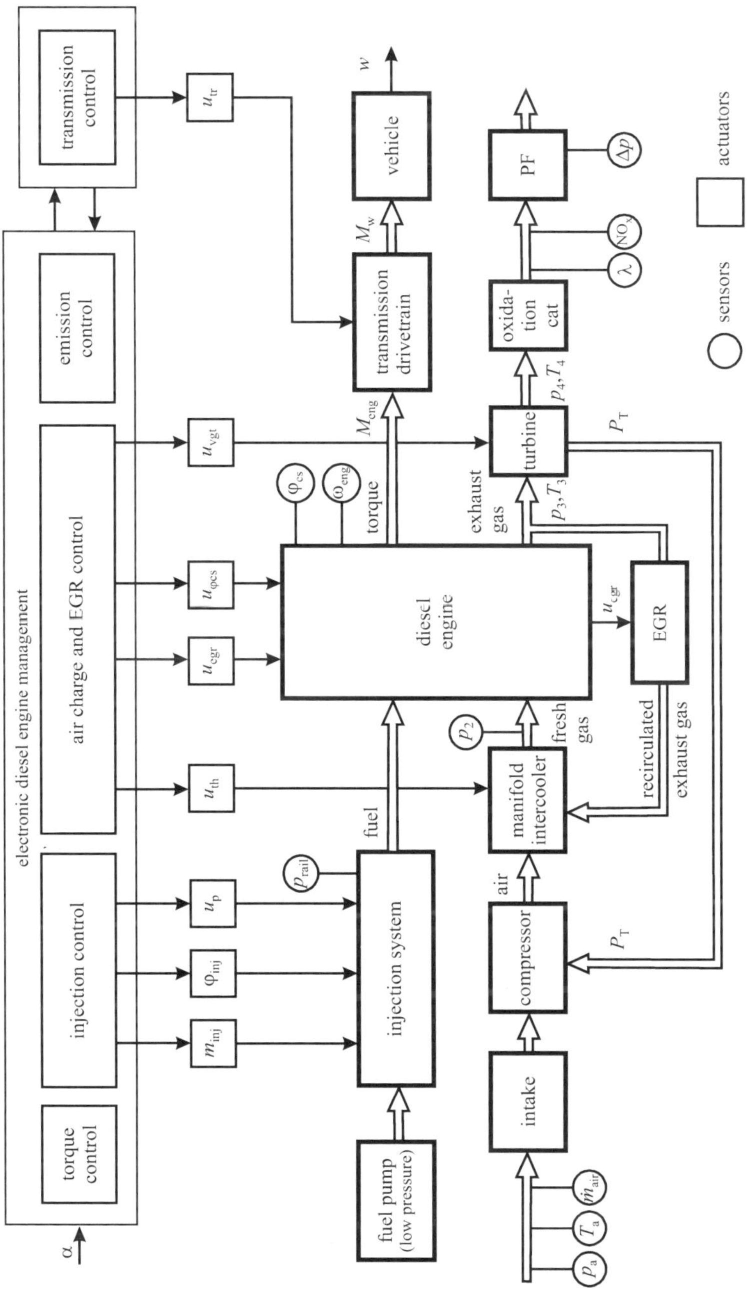

Fig. 1.3.7. Control-oriented block diagram of a diesel engine with common rail injection, actuators, control subsystems and some sensors.

1.3.3 On engine diagnosis

The strong increase of sensors, actuators, mechatronic components and electronic control units in modern vehicles have led to a large degree of complexity of both, the engine and the chassis. Therefore these developments were paralleled by implementing *on-board diagnostics* in the vehicle. Especially the requirements of the CARB (Californian Air Resources Board) for an on-board diagnosis (OBD I) of emission relevant components like sensors, actuators and ECU for the engines since 1988 and of the European Union (EOBD) since 1998 enlarged the development of diagnosis functions strongly. The used methods are, for example, limit checking of measured values, plausibility checks and testing of electric circuits. In the meantime these OBD functions take about 50 % of the ECU capacity. *Off-board diagnosis* in service stations allows to enhance fault diagnosis after cable connection with a workshop test equipment.

In order to improve the performance of fault detection and diagnosis signal-model and process-based methods are developed, Isermann (2013), Isermann (2015).

1.3.4 Development steps for engine control

Because of the increasing and multi-variable functions of internal combustion engines the development of the electronic management has become a rather involved task. Figure 1.3.8 gives a first impression of some development steps. The basic thermodynamic and mechanical design includes, e.g. the charging cycles, fuel injection, ignition, super charging, valve train, auxiliary drives, cooling, lubrication and exhaust gas after treatment. A prototype engine can then be equipped with selected actuators, sensors and a basis ECU, so that experiments on a test bench can be performed. A first basis control calibration with a basic control software then enables making systematic measurements on a test bench. Then the optimization of the many control functions can be started with the goal of minimizing fuel consumption and raw emissions. Advanced procedures use model-based methods, supported by different kinds of simulations and computer-aided design of, e.g. look-up tables for feedforward control and digital control algorithms for feedback control. An additional effort is designing and calibrating onboard diagnosis functions to satisfy legal requirements and to support service functions. A next step is the fine-tuning of control functions with vehicles and road tests. Then also the design of the shifted or automatic transmissions have to be taken into account. Automatic transmissions have their own ECU attached to or integrated into the transmission housing and also show an increasing extent of control and supervision functions, e.g. Wieczorek et al (2010). The certification with regard to emissions is usually performed on roller test benches with simulated longitudinal driving behavior of the vehicle. A more detailed workflow for the design and implementation of the control functions is given in Chap. 6.

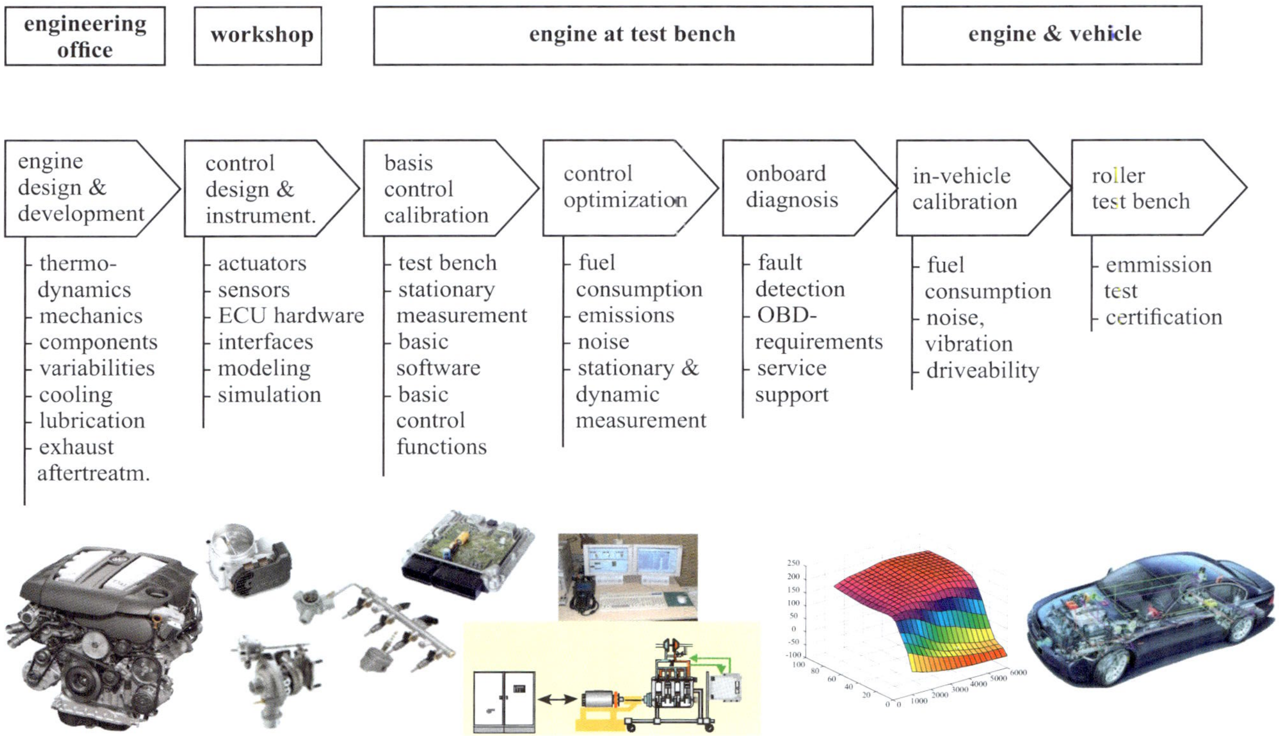

Fig. 1.3.8. Development steps from basic design to test-bench tests for the control functions of internal combustion engines. Sources of figures: VW, Bosch, Continental , IAT, BMW.

1.4 Control-oriented subdivision of combustion engines

The development of control and diagnosis functions for internal combustion engines requires a subdivision into signal-flow-oriented parts with physical input and output variables. Table 1.4.1 therefore distinguishes four *engine groups*:

A Basic engine hardware
B Electrical hardware
C Control hardware and software
D Auxiliary components.

and subdivides corresponding *engine parts* and shows the dominating physical domains. The engine parts are further detailed in Table 1.4.2 for gasoline engines and in Table 1.4.3 for diesel engines and list the main components, actuators, manipulated variables, sensor variables and auxiliary components for standard engine configurations. The components for the cooling system, electrical system and electronic control system, which are similar for both types of engines are given in Table 1.4.4.

Table 1.4.1. Engine groups and parts

Engine group	Engine parts	Physical domains
A basic engine hardware	A1 intake system	fluid dynamics, thermodynamics
	A2 injection system	fluid dynamics, mechatronics
	A3 fuel supply	fluid dynamics
	A4 combustion and mechanics	chemistry, thermodynamics, mechanics, materials
	A5 lubrication system	fluid dynamics, heat transfer
	A6 exhaust gas system, emission after-treatment	fluid dynamics, thermodynamics, chemistry
	A7 cooling system	fluid dynamics, heat transfer
B electrical hardware	B1 electrical system	low voltage electricity
	B2 ignition system	low & high voltage electricity
C control hardware and software	C1 actuators	mechatronics
	C2 sensors	physics, mechatronics
	C3 electronic control unit	computer technology, control software, algorithms
D auxiliary components	D1 filters	fluid dynamics
	D2 tank ventilation	

1.5 Contents of this book

The book is divided in three parts. After this introduction chapter, *Part I* treats basic methods for *theoretical (physical) and experimental modeling (identification)*. Especially the identification methods for obtaining mathematical models of the multivariable and nonlinear behavior of combustion engines on test benches is considered

Table 1.4.2. Main components of gasoline engines

Engine part	Components	Actuators	Manipulated variables	Sensors	Auxiliary components
A1 intake system	- intake pipe manifold - blow-by pipe	- el. throttle - intake flaps	- throttle pos. - flap pos.	- air mass flow - manifold temperature - manifold pressure	- air filter
A2 injection system	- high pressure pump (direct injection - injection valves	- metering unit - pressure control valve - injection valve coils or piezo actuat.	- valve pos. - injection duration - injection time	- fuel or rail pressure	
A3 fuel supply	- low pressure fuel pump (intake and direct inj.) - fuel filter	- canister purge valve	- valve pos.	- fuel level - fuel pressure - tank pressure	- tank ventilation - tank diagnosis pump
A4 combustion and mechanics	- combustion chamber pistons - connecting rods - crankshaft - camshaft - variable valve train - bearings - turbocharger - exhaust gas recirculation	- hydraulic actuators - swirl flaps - tumble flaps - exhaust gas recirculation valve - wastegate actuator	- camshaft position - flap pos. - exhaust gas valve pos. - inlet/outlet valve pos. - wastegate position	- engine speed - camshaft phase - knocking accelerometer - pressures - temperatures	
A5 lubrication	- oil pump - oil channels - mechanics - seals	- oil pump actuator	- oil flow	- oil pressure - oil temp. - oil level	- oil filter
A6 exhaust system	- exhaust pipes - secondary air system - catalytic conv. - muffler		- exhaust gas temperature	- exhaust gas temperature - Lambda - NO_x	
B2 ignition system	- ignition timer - ignition coils - spark plugs - ignition lock	- ignition driver stage	- ignition time/ advance angle	- knock sensor	

Table 1.4.3. Main components of diesel engines

Engine part	Components	Actuators	Manipulated variables	Sensors	Auxiliary components
A1 intake systems	- intake manifold - blow-by pipe - intercooler	- throttle plate	- throttle plate position	- air mass flow - air temp. - boost pressure	- air filter
A2 injection system	- high pressure pump - glow plugs	- metering unit - pressure control valve - injectors	- valve positions - injection duration - injection time	- fuel pressure	
A3 fuel supply	- low pressure fuel pump - fuel filter			- fuel level - fuel temperature	- tank ventilation
A4 combustion and mechanics	- combustion chamber - pistons - connecting rods - crankshaft - variable valve train - bearings - turbocharger - exhaust gas recirculation cooler	- hydraulic actuators - wastegate - turbine vanes - exhaust recirculation valve - swirl flaps	- exhaust gas flow - valve position - flap position	- engine speed - camshaft phase - camshaft position - boost pressure	
A5 lubrication	- oil pump - oil channels - mechanics - seals	- oil pump actuator	- oil flow	- oil pressure - oil temperature - oil level	- oil filter
A6 exhaust system	- exhaust pipes - catalytic oxygen converter - particulate filter - muffler		- exhaust gas temperature for regeneration	- gas temp. - lambda - NO_x - difference pressure	

Table 1.4.4. Components of cooling, electrical and electronic systems for combustion engines (gasoline and diesel engines)

Engine part	Components	Actuators	Manipulated variables	Sensors
A7 cooling system	- coolant radiator - coolant pump - coolant passages - flexible tubes	- thermostatic expansion valve - air fan - air inlet flaps	- valve position - fan speed - speed of electrical pump	- coolant temperature - coolant level
B1 electrical system	- electrical generator - electrical starter - battery - harness - fuses - CAN bus system	- power transistors - overvoltage protection	- voltages - currents - switches	- voltages - currents
C3 electronic control unit	- ECU		- electrical outputs	- electrical inputs

in detail, for the stationary as well as for the dynamic behavior. *Part II* is then devoted to describing general *mathematical models of processes and components of gasoline and diesel engines*. If possible, physically-based models are derived or semi-physical models from the intake system, through combustion, turbo charging, cooling to exhaust systems. The various methods of *engine control* are considered in *Part III*. The first chapter summarizes engine control hardware and software structures, used actuator and sensor types and mechatronic components and discusses their properties and technical data. A special chapter gives an overview of engine control development and calibration, basic methods of feedforward and feedback control including nonlinear and digital control. Then conventional and model-based calibration, numerical optimization methods and test-bench control are treated. This is followed by control-function development, also by applying rapid control prototyping and hardware-in-the loop simulation. A special section is devoted to control software development. A next chapter then considers the control of gasoline engines, from air/fuel to idle speed control. An extensive treatment follows for the control of diesel engines, from combustion modeling and control to emission control. These chapters include control functions and algorithms for gasoline engines and diesel engines, from torque to emission control, including combustion-pressure-based control.

This book takes many publications on internal combustion engine control into account and uses the results of own research and industrial cooperations at the Institute of Automatic Control of the Technische Universität Darmstadt. As an introduction into the basics of internal combustion engines the following books or lecture notes are recommended: Heywood (1988), Urlaub (1995), Cook et al (1996), Jurgen (1999), Stone (1999), Ferguson and Kirkpatrick (2001), Stotsky (2009), van Basshuysen (2009), Pischinger et al (2009), Mollenhauer and Tschöke (2010), Robert Bosch GmbH (2012), Köhler and Flierl (2012), Merker et al (2012), Pucher and Zinner (2012), Spicher (2012), Bargende (2013), Pischinger (2013), Beidl (2014), Wachtmeister (2014).

Internal combustion engine control is treated in the following books: Kiencke and Nielsen (2000), Isermann (2003), Johansson and Rantzer (2003), Guzzella and Onder (2010), Isermann (2010), Robert Bosch GmbH (2012).

References

Alt M, Grebe U, Dulzo J, Chang MF (2008) Closed loop combustion control for HCCI. In: 8. Int. Stuttgarter Symposium, Stuttgart

Backhaus R (2008) Gasoline engines with HCCI technique. Motortechnische Zeitschrift – MTZ 69(6):482–485

Bargende M (2013) Grundlagen der Verbrennungsmotoren. Lecture notes. Institut für Verbrennungsmotoren und Kraftfahrenwesen. Universität Stuttgart, Stuttgart

van Basshuysen R (2009) Gasoline engine with direct injection. Vieweg+Teubner, Wiesbaden

van Basshuysen R, Schäfer F (2004) Internal combustion engine handbook: basics, components, systems, and perspectives. SAE International, Warrendale

Bauer S, Zhang H, Pfeifer A, Wenzlawski K (2007) PKW-Dieselmotor und EU 6: Ganzheitlicher Systemansatz. In: 28. Int. Wiener Motorensymposium, Wien, Österreich

Beidl C (2014) Verbrennungskraftmaschinen I + II. Lecture notes. Technische Universität Darmstadt, Institute for Internal Combustion Engines and Powertrain Systems, Darmstadt

Berns S (2007) The gasoline challenge. In: Gasoline Systems Symposium, Bosch, Frankfurt, Germany

Berns S, Fehrmann R, Sassen K (2004) Beitrag der Diesel-Elektronik zur Leistungssteigerung und Emissionsreduzierung. In: VDI/VDE-Fachtagung AUTOREG 2004, VDI-Bericht 1828, VDI, Düsseldorf, Wiesloch, Germany

Braess H, Seiffert H (eds) (2005) Handbook of automotive engineering. SAE International, Warrendale

Cook J, Grizzle J, Sun J (1996) Automotive control systems. In: Levine W (ed) The Control Handbook, CRC Press, Boca Raton, pp 1261–1274

ETAS (2010) Measurement, electronic control calibration and diagnosis. Brochure. ETAS/COM-Sar, Stuttgart

Ferguson C, Kirkpatrick A (2001) Internal combustion engines. J. Wiley and Sons, New York

Guzzella L, Onder C (2010) Introduction to modeling and control of internal combustion engine systems, 2nd edn. Springer, Berlin

Hadler J, Rudolph F, Dorenkamp R, Stehr H, Hilzendeger J, Kranzusch S (2008) Volkswagen's new 2.0 l TDI engine for the most stringent emission standards. MTZ worldwide edition 69(5, 6):12–12, 54–59

Heywood JB (1988) Internal combustion engine fundamentals. McGraw-Hill, New York

Isermann R (ed) (2003) Modellgestützte Steuerung, Regelung und Diagnose von Verbrennungsmotoren. Springer, Berlin

Isermann R (ed) (2010) Elektronisches Management motorischer Fahrzeugantriebe. Vieweg+Teubner, Wiesbaden
Isermann R (2013) Model-based fault diagnosis of combustion engines. ASME Dynamic Systems and Control (DSC) Magazine (December):6–16
Isermann R (2015) Powertrain diagnosis. To be published. Springer, Berlin
Johansson R, Rantzer A (eds) (2003) Nonlinear and hybrid systems in automotive control. Springer, London
Jurgen R (ed) (1999) Automatic electronics handbook, 2nd edn. McGraw-Hill, New York
Kiencke U, Nielsen L (eds) (2000) Automotive control systems. For engine, driveline and vehicle. Springer, Berlin
Köhler E, Flierl R (2012) Verbrennungsmotoren, 6th edn. Vieweg, Wiesbaden
Königstein A, Grebe U, Wu K, Larsson P (2008) Differentiated analysis of downsizing concepts. MTZ worldwide edition 69(6):4–11
Merker G, Schwarz C, Teichmann R (eds) (2012) Combustion engines development. Springer, Berlin
Mollenhauer K, Tschöke H (2010) Handbook of diesel engines. Springer, Berlin
Pischinger R, Kell M, Sams T (2009) Thermodynamik der Verbrennungskraftmaschinen, 3rd edn. Springer, Wien and New York
Pischinger S (2013) Internal combustion engine fundamentals. Lecture notes. RWTH Aachen, Institute for Combustion Engines, Aachen
Pucher H, Zinner K (2012) Aufladung von Verbrennungsmotoren, 4th edn. Springer, Berlin
Robert Bosch GmbH (ed) (2012) Automotive Handbook, 12th edn. Bentley publishers, Cambridge
Schulz T, Kulzer A (2006) Überblick und Bewertung variabler Ventiltriebe. In: VDI/VDE-GMA-Fachausschuss AUTOREG
Spicher U (ed) (2012) Direkteinspritzung im Ottomotor VIII. expert Verlag, Renningen
Steinparzer F, Mattes W, Nefischer P, Steinmayr T (2007) The new BMW four-cylinder diesel engine. MTZ worldwide edition 68(11 and 12):932–943 and 24–27
Stone R (1999) Introduction to internal combustion engines, 3rd edn. SAE International and Maxmillan Press, Warrendale
Stotsky A (2009) Automotive engines. Springer, Berlin
Urlaub A (1995) Verbrennungsmotoren. Grundlagen - Verfahrenstheorie - Konstruktion, 2nd edn. Springer, Berlin
Wachtmeister G (2014) Verbrennungsmotoren. Lecture notes. Technische Universität München, Institute of Internal Combustion Engines, München
Weingärtner R, Achleitner E, Bächer E (2007) Piezo Direkteinspritzung für Benzinmotoren. In: VDI/VDE-GMA-Fachausschuss AUTOREG
Wieczorek M, Berteshofer P, Schuch B, Genzel M (2010) Growing function content of control units for automatic transmissions. ATZ worldwide 112(11):35–38

Part I

Engine Modeling and Identification Methods

2

On theoretical modeling of multiple-domain processes

The temporal behavior of multiple-domain systems, such as internal combustion engines, drivetrains and auxiliary units, can be described with the help of system theory according to uniform methods. For this, however, mathematical models have to exist for the static and dynamic behavior of the system components or the processes.

The derivation of mathematical models can take place in a theoretical (physical) or experimental way. Therefore, it is called theoretical modeling or experimental modeling or identification.

For combustion engines, mathematical models for different physical areas have to be set up and combined, e.g. for mechanics, combustion, thermodynamics, and electricity. The procedure during theoretical modeling is in principle known for the individual areas, and there also exist analogies for models between different areas. However, a basic, generally applicable methodology for theoretical modeling with an interdisciplinary view has several advantages, especially for applying computer-aided modeling. Therefore, a unified representation for modeling in different physical domains is briefly summarized in this chapter.

2.1 Theoretical and experimental modeling

The derivation of mathematical models of processes can be performed in a theoretical or experimental way. For engines and also vehicles mathematical models for different physical domains have to be set up and combined. This holds especially for mechatronic components and engines in the drivetrain. Therefore, this chapter gives an introduction into a systematic way to model multi-domain systems, also as a basis for computer-aided modeling.

The principles of theoretical modeling can follow a basic methodology, see Karnopp et al (1990), Gawthrop and Smith (1996), Isermann (2005). Fundamental equations are:

1. Balance equations for stored masses, energies, and momentum
2. Constitutive equations of special elements

R. Isermann, *Engine Modeling and Control*,
DOI: 10.1007/978-3-642-39934-3_2

3. Phenomenological equations, if irreversible processes take place
4. Entropy balance equations, if several irreversible processes are involved
5. Connection equations.

In stating these equations one has to distinguish between processes with distributed and lumped parameters. For *distributed parameters* the dependency on space and time has to be considered. This usually leads to partial differential equations. If the space dependency is negligible, the process can be considered with *lumped parameters* which leads to ordinary differential equations as a function of time. For combustion engines or drivetrains both types appear. However, one can frequently operate with lumped parameters.

By summarizing the basic equations of all process elements, one receives a *theoretical* or *physical process model* with a certain structure and certain parameters if it can be solved explicitly. Frequently, this model is extensive and complicated, so it must be simplified for further applications. The simplifications are made by linearization, reduction of the model order or approximation of systems with distributed parameters by lumped parameters when limiting on fixed locations. But also if the set of equations cannot be solved explicitly, the individual equations supply important hints for the model structure. So, e.g. balance equations are always linear and some phenomenological equations are linear in wide areas. The constitutive equations often introduce nonlinear relations.

During experimental modeling, which is also called *process identification*, one obtains the mathematical model of a process from measurements. Here, one always proceeds from *a priori* knowledge, which was gained, e.g. from the theoretical analysis or from preceding measurements. Then, input and output signals are measured and evaluated by means of identification methods in such a way that the relation between the input and output signal is expressed in a mathematical model. The result of the identification then is an *experimental model*, see Chap. 3.

Theoretical and experimental modeling mutually complete themselves. The theoretical model contains the functional description between the physical data of the process and its parameters. Therefore, one will use this model, e.g. if the process is to be favorably designed with regard to the dynamical behavior or if the process behavior has to be simulated before construction. The experimental model, on the other hand, contains parameters as numerical values whose functional relation with the physical basic data of the process remains unknown. In many cases, the real dynamic behavior can be described more exactly or it can be determined at smaller expenditure by experimentally obtained models which, e.g. is better suited for control design, the prediction of signals or for fault detection.

The following methodology for theoretical (physical) modeling is a strongly shorted version of Chap. 2 in Isermann (2005), which can be applied generally for technical systems and thus holds also for engines and vehicles. This is treated first because then the modeling of parts of the engines can be based on these general laws of modeling.

2.2 Process elements from different domains

In the following, the terms energy, matter, and information are called *quantity*. If one considers processes with *lumped parameters*, then the elements of technical processes can be classified according to the following idealized types, see Karnopp et al (1990), MacFarlane (1967), MacFarlane (1970), Isermann (2005), compare Fig. 2.2.1:

- *sources*: deliver an output quantity from a large supply, without or with losses
- *storages*: take up a quantity and deliver it in the same form
- *transformers*: take up a quantity and deliver it in the same form, without storing it
- *converters*: take up a quantity in certain form and deliver it after conversion into another form, without storing it
- *sinks*: take up an input quantity and consume it in the same or another form. Since mainly losses occur they are dissipative processes.

The first four elements are *ideal* if no losses are generated. However, *real* elements have losses. In the case of real sinks the input quantity is not always completely consumed.

The connecting lines in Fig. 2.2.1 represent the flows between the elements in the form of [quantity/time]. The arrows indicate the direction of the flows.

A further distinction of the process elements can be made with regard to their controllability with an additional auxiliary energy.

- *passive elements*: the transferred quantity is not controllable by an additional auxiliary energy. Examples are passive storages, e.g. capacitances, passive transformers as, e.g. fixed gear transmissions, or passive converters, e.g. fans with constant speed
- *active elements*: a quantity is controlled by an actuator. Thereby, an electrical or mechanical auxiliary energy usually has to supply the actuator. Examples are controllable sources as, e.g. controllable transformers as, e.g. tank with fuel pump, automatic transmissions, combustion chamber with variable valve train and EGR valves.

Active process elements are represented in Fig. 2.2.2. The storages basically show a dynamic, usually an integral, behavior. Sources, transformers, converters, and sinks can have both a mainly static transfer behavior as well as a distinct dynamic transfer behavior.

For processes with *distributed parameters*, process elements such as storages, transformers, converters, and sinks are distributed over space. By partitioning into infinitesimally small elements, one can determine process elements with lumped parameters, whose state variables are location-dependent from element to element.

Figure 2.2.3 depicts a simplified representation of a combustion engine using the introduced process elements and symbols. The inputs are a fuel mass flow and an air mass flow. A first converter generates by combustion a net heat release flow, which generates displacement power by a thermodynamic operation cycle. A second converter converts it to mechanical rotational power. A part of the combustion heat flow is delivered through the cylinder walls to the cooling system and by heat exchange to the environment. The exhaust gas components generated during the combustion pass the exhaust after treatment (catalysts, filter) and then vanish to the environment. Main storages with regard to control are the intake system, the turbocharger (not included in Fig. 2.2.3), the crank shaft rotational mass, and the cooling circuit.

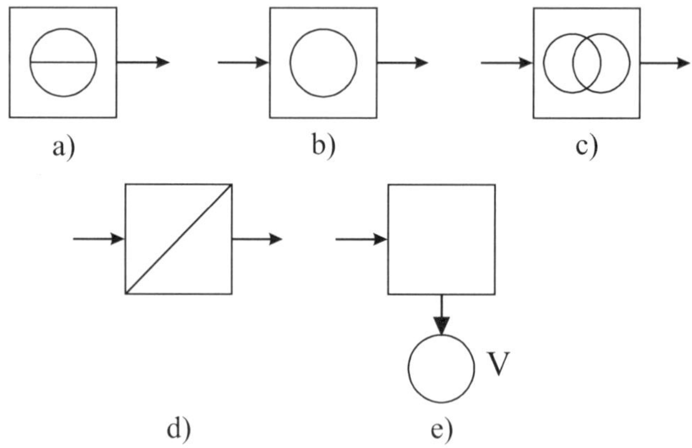

Fig. 2.2.1. Symbols of passive process elements. **a** source. **b** storage. **c** transformer. **d** converter. **e** sink (V: losses):
→ energy, matter or information flow.

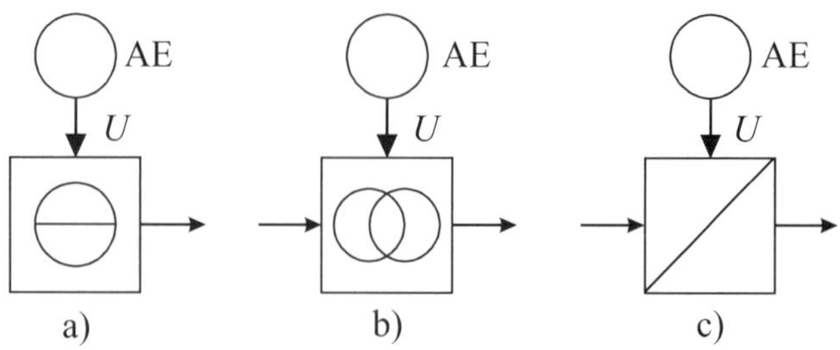

Fig. 2.2.2. Symbols of active process elements (the process element is controlled by an actuator, which is supplied by an auxiliary energy. AE: auxiliary energy; U: manipulated variable). **a** source with auxiliary energy. **b** transformer with auxiliary energy. **c** converter with auxiliary energy.

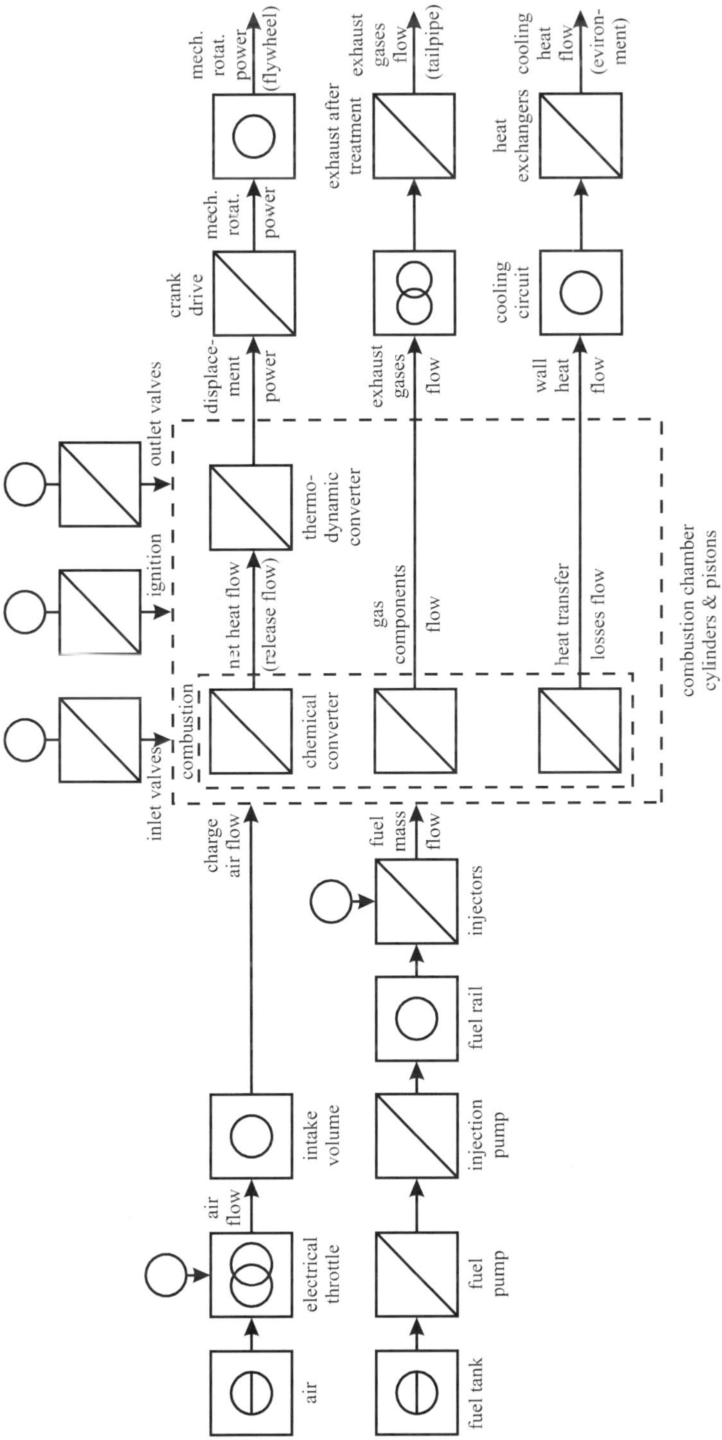

Fig. 2.2.3. Simplified schematic representation of a gasoline combustion engine with mass and energy flows and the symbols of Fig. 2.2.1 (not all losses considered)

2.3 Basic equations

After the definition of suitable cuts, the process is divided into elements such as sources, storages, transformers, converters, and sinks for energies and matters in order to set up the basic equations. These basic equations are for lumped parameter processes:

- *balance equations* (general storages, junction points)
- *constitutive equations* or *physical state equations* (sources, transformers, converters, special storages)
- *phenomenological equations* (sinks, dissipative elements).

This grouping of basic equations then applies both to processes with energy flows and matter flows.

2.3.1 Balance equations

Since the laws for conservation of mass, energy, and momentum are fundamental, they are regarded as the first type of equations. The balance equations, which are derived from these conservation laws, basically apply independently of the construction of the processes. They describe the global behavior. The mass balance applies to processes with moved matter, the energy balance to processes with all types of energy and the momentum balance to processes with moved masses.

If Q_ν describes a mass m_ν or energy E_ν, the principle of the conservation of mass or energy applied to a bounded area, where no mass or energy leaves, leads to

$$\sum_{\nu=1}^{n} Q_\nu = \text{const.} \tag{2.3.1}$$

compare Fig. 2.3.1a). If a mass or energy $\Delta Q_i(t)$ enters through the boundaries of an arbitrary control area and a mass or energy $\Delta Q_o(t)$ escapes through the boundaries in a time Δt, it holds with the conservation laws

$$\Delta Q_i(t) - \Delta Q_o(t) = \Delta Q_s(t) \tag{2.3.2}$$

where $\Delta Q_s(t)$ is the stored quantity. If (2.3.2) is divided by Δt

$$\lim_{\Delta t \to 0} \Delta Q / \Delta t = dQ/dt$$

a *generalized balance equation* for flows follows

$$\underset{\text{inflow}}{\dot{Q}_i(t)} - \underset{\text{outflow}}{\dot{Q}_o} = \underset{\text{stored flow}}{\tfrac{d}{dt} Q_s(t)} \tag{2.3.3}$$

see also Fig. 2.3.1b). Its signal flow is depicted in Fig. 2.3.2.

Balance equations for mass and energy stores thus lead to a linear integrating transfer element and cause a dynamically delayed behavior. Balance equations have to be set up for each storage. They are always linear.

The balance equations also describe the flows at the interconnection points of process elements if the storage capacity is set to zero. Balance equations are also called *continuity equations*.

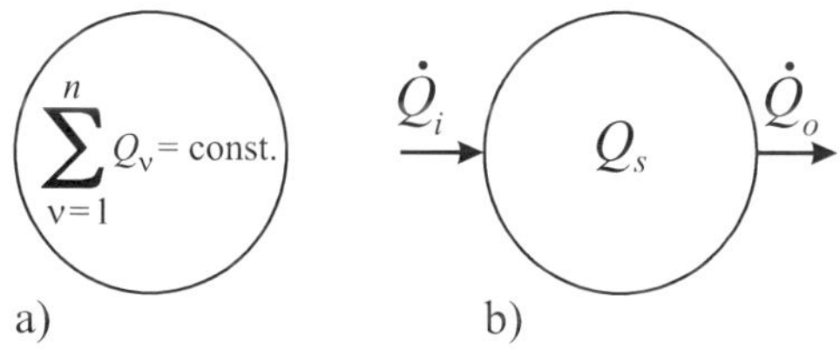

Fig. 2.3.1. General balance for a storage of energy and mass. **a** bounded area. **b** storage.

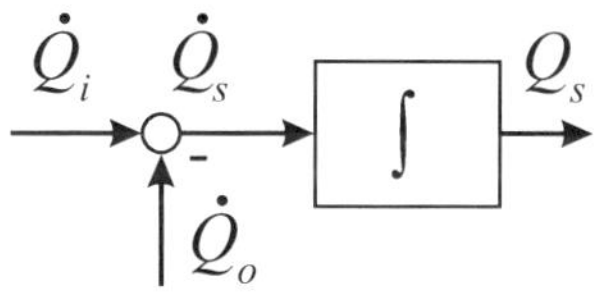

Fig. 2.3.2. Block diagram of an energy or mass storage (balance equation).

2.3.2 Constitutive equations

The coherence between input and output variables of the process elements in the form of sources, transformers, converters, sinks and also storage elements can be expressed by special physical laws in analytical form or by characteristic curves from experiments. The equations mentioned are called *constitutive equations* or *physical state equations*.

Many different physical laws that apply to the individual process elements exist. However, regarding the input/output behavior, several similarities exist.

a) Processes with energy flows

For technical processes which primarily transfer energy at their interconnections, it follows from the energy balance equation (2.3.3) without storage that the transferred energy per time interval or the power

$$P(t) = \frac{dE(t)}{dt} \tag{2.3.4}$$

is always equal at the interconnections between the process elements. Therefore, it is appropriate to determine the state variables in such a way that at the interfaces between process elements or subprocesses they describe a power. If one determines the interfaces in analogy to electrical transfer elements as a terminal pair, then the different process elements can be described as one-port systems (two-pole systems), two-port systems (four-pole systems) or generally multi-port systems (multi-pole systems), see Fig. 2.3.3. One can always distinguish two variables for a terminal pair at the transfer elements, Karnopp and Rosenberg (1975), Karnopp et al (1990), Takahashi et al (1972):

1. potential difference $e(t)$: variables such as electrical voltage, force, pressure difference appear as a difference between two terminals. They are called "effort"

2. flow $f(t)$: variables such as electrical current, velocity, volume flow enter in one of the terminals. They are called "flow".

The product of both terms is the transferred power

$$\underset{\text{power}}{P(t)} = \underset{\text{flow}}{f(t)} \quad \underset{\substack{\text{potential}\\ \text{difference}}}{e(t)}. \tag{2.3.5}$$

Here, $f(t)$ and $e(t)$ are assigned covariables, which are also called *generalized power variables*. These variables are discussed for important technical systems with energy flows in Isermann (2005). The formulation of constitutive equations with potentials and flows is especially advantageous for mechanical and hydraulic components in connection with electrical ones, i.e. for mechatronic systems.

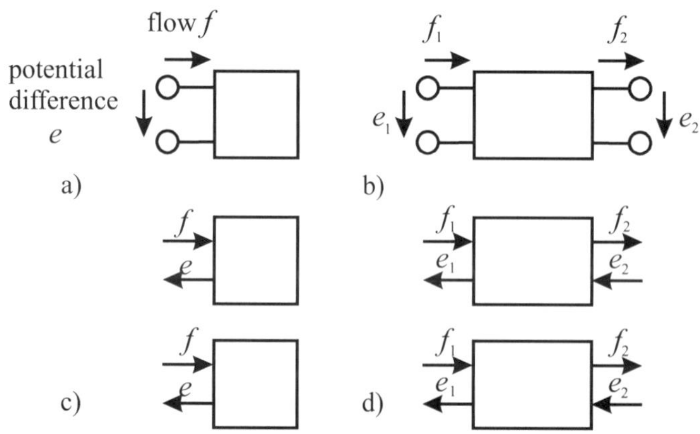

Fig. 2.3.3. Representation of one-port and two-port systems for process elements with power variables. **a** one-port system in terminal representation. **b** two-port system in terminal representation. **c** one-port system with input and output signals. **d** two-port system with input and output signals (2 of 4 possibilities).

b) Processes with different flows

The previous reflections were limited to processes that transfer exclusively energy flows. For *matter flows* (solids, liquids, steams, gases), it is practical to use the mass flow as flow $f(t)$, since the mass balance equation is a substantial basic equation. If one uses the usual variables for the potential difference $e(t)$, the product $e(t)f(t)$ is not always power $P(t)$ according to (2.3.5).

At the interconnection points between process elements, power $P(t)$ (energy flows), as well as the mass flows

$$\dot{m}(t) = \frac{dm(t)}{dt} \tag{2.3.6}$$

have to be equal between the process elements at any time. Therefore, one uses the mass flow as a further state variable at the interfaces. For incompressible matter one can also take the volume flow rate

$$\dot{V}(t) = \frac{1}{\rho}\dot{m}(t) \tag{2.3.7}$$

since the density ρ is constant.

For *heat flows* which are transferred by convection and therefore are coupled to a mass with heat capacity, the product according to Table 2.3.1 is power (in the case of convection by fluids it has to be multiplied by the specific heat). This does not apply to thermal conduction in matter if one selects the heat flow as flow f, since the heat flow itself represents power, see Table 2.3.1, last row.

For powers, it has to be determined if it is the power needed for the transportation of a mass flow, e.g. hydraulic power or the power that is transported by the mass flow, e.g. thermal energy flow.

Table 2.3.1. Variables for flow and potential difference for matter flows and heat flows

System	Flow f		Potential difference e		ef	Power
hydraulic	mass flow $\dot{m}$	$\frac{\mathrm{kg}}{\mathrm{s}}$	pressure difference p		$\dot{m}p$	$\frac{1}{\rho}\dot{m}p$
thermal						
- convection through liquids	mass flow $\dot{m}$	$\frac{kg}{s}$	temperature difference T	K	$\dot{m}T$	$\dot{m}c_pT$
- convection through gas, steam	mass flow $\dot{m}$	$\frac{\mathrm{kg}}{\mathrm{s}}$	enthalpy difference h	$\frac{\mathrm{J}}{\mathrm{kg}}$	$\dot{m}h$	$\dot{m}h$
chemical	mass flow $\dot{m}_i$	$\frac{\mathrm{kg}}{\mathrm{s}}$	concentration c_i	$\frac{\mathrm{mol}}{\mathrm{kg}}$	$\dot{m}_ic_i$	$--$
thermal						
- heat conduction	heat flow $\dot{q}$	W	temperature difference T	K	$\dot{q}T$	$\dot{q}$

These constitutive equations are required for modeling many parts of *combustion engines* because of the energy flows for chemical energy through combustion, thermodynamic energy, mechanical energy and thermal energy. Additionally, liquid and gas flows have to be modeled which transport air, fuel or oil or transport combustion and thermodynamic energy.

2.3.3 Phenomenological equations

In the case of sinks and losses in some transformers and converters, the internal process runs only in one direction and is not reversible (without additional energy supply). Examples are thermal conduction, diffusion or chemical reaction. The appropriate processes are characterized by irreversible equalization processes with increasing entropy. The reasons for the entropy increase are the dissipation of mechanical and electrical power, the exchange of mass and heat and chemical reactions, Ahrendts (1989). Thus, dissipative systems can be represented as sinks. The irreversible transients are described by phenomenological equations. One example is:

a) Fourier's law of heat conduction

$$\dot{q}_z = -\lambda \frac{\partial T}{\partial z} = -\lambda \operatorname{grad}_z T \tag{2.3.8}$$

($\dot{q}_z$ heat flow density, λ thermal conductivity, T temperature, z space coordinate).

Other examples are Fick's law of diffusion and Ohm's law of electrical current. These laws can be represented in a general form

$$\text{flow density} = -\frac{1}{\text{specific resistance}} \text{ potential gradient} \tag{2.3.9}$$

and are linear in a wide range.

Another type of phenomenological equation is the

b) Chemical reaction law

A first-order reaction $A_i \to B_i$ with c_{Ai} and c_{Bi} for the concentrations yields

$$r_z = -k c_{Ai} \left[\frac{\text{kmol}}{m^3 s}\right] \tag{2.3.10}$$

$$k = k_\infty e^{-\frac{E}{RT}} \qquad \text{Arrhenius law} \tag{2.3.11}$$

(r_z reaction rate, k reaction rate coefficient, c_{Ai} concentration of component A_i, E activation energy, k_∞ frequency factor).

c) General sinks, dissipative converters

Phenomenological equations show partially linear behavior, e.g. in the case of viscous friction or Ohm's resistances, and partially strong nonlinear behavior as in the case of throttles or friction. In the linear case, when using the one-port representation, e.g. the resistance equation applies

$$f_1 = \frac{1}{R} e_1 \tag{2.3.12}$$

where R is the resistance coefficient (The negative sign in (2.3.9) is necessary because the gradient is used with regard to the location).

Dry friction has a direction-dependent discontinuous characteristic curve

$$\begin{aligned} F &= F_{G0} \qquad \dot{z} > 0 \\ F &= -F_{G0} \qquad \dot{z} < 0. \end{aligned} \tag{2.3.13}$$

This leads to force-displacement characteristics in the form of rectangular hysteresis characteristics, whose range depends on the amplitude z_0, Isermann (2005).

In the case of combustion engines, the considered examples of phenomenological equations are required for modeling, for example, heat transfer to the cooling system, combustion, mass flows through the intake and exhaust system and all kind of frictions in the mechanical parts.

This section described the basic balance equations and constitutive equations in a general form and some phenomenological equations. The balance equations resulted from the conservation laws of physics and have, in principle, the form of (2.3.3). However, the statement of the energy balance equations for mechanical and thermodynamic processes require the consideration of some special features as treated in related textbooks and also in Karnopp et al (1990) and Isermann (2005).

2.3.4 Summary

By following the approach indicated briefly in this chapter a consistent procedure for the modeling of processes with energy and mass flows from different physical domains results. Using all particular equations, an equation system for the considered process or process part follows. From this, a signal flow diagram may be systematically composed and the state space equations as well as the differential equations for the input and output variables may be derived. If desired, a linearization around the operating point may be carried out.

The treated systematic approach and unified procedure allows not only the recognition of many similarities but is also a prerequisite for computer-aided modeling with modern software tools. It can also be applied to modeling of processes with distributed parameters with finite elements.

2.4 Time-dependent and rotation-angle-dependent models

The behavior of internal combustion engines is dominated by the reciprocating working cycle and therefore dependent on the rotation angles of the crankshaft and the camshaft. As the camshaft rotates usually with half of the speed of the crankshaft, if suffices to consider the crankshaft speed only, assuming four-stroke engines. The models considered in this book are oriented to the design of control functions in the ECU and should therefore, if possible, be applicable for real-time computations. Therefore, one has to make simplifying assumptions and only the dominant dynamic effects can be taken into account with regard to important variables of the engine, for example, the manifold pressure, torque, exhaust gas pressure and temperature, fuel consumption and emissions. Engine part processes which directly influence the single combustion like combustion pressure and temperature, air charge, fuel injection, ignition, valve phasing have usually to be described by *crank-angle-dependent models* $f(\varphi)$ (also called discrete-event models, Guzzella and Onder (2010)), Fig. 2.4.1. On the other side some engine part processes outside of the cylinders can be described by *time-dependent models* $f(t)$. The crank-angle-dependent behavior can be neglected frequently as the reciprocating behavior of the cylinder is damped because of gas or thermal storages, especially for multi-cylinder engines. Examples are the air flow, manifold pressure, turbo charger torque and speed, and emissions. If the working stroke induced fluctuations are neglected, the resulting models are *mean-value models*.

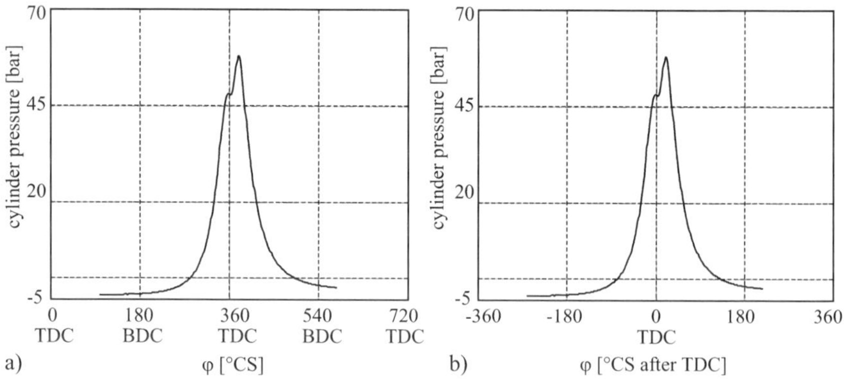

Fig. 2.4.1. Scaling of the crank angle for four-stroke engines. **a** absolute scaling: 0° CS after expansion. **b** relative scaling: 0° CS at top dead center (TDC).

The crank angle and the time are related by the engine speed

$$\varphi(t) = \omega_{\text{eng}}(t)t \tag{2.4.1}$$

and for infinitesimal small elements by

$$d\varphi = \omega_{\text{eng}}dt + td\omega_{\text{eng}}. \tag{2.4.2}$$

For constant engine speed this simplifies to

$$\begin{aligned} \varphi(t) &= \omega_{\text{eng}}t \\ d\varphi &= \omega_{\text{eng}}dt. \end{aligned} \tag{2.4.3}$$

As the mean value of the crankshaft speed during one cycle does not vary significantly, one can assume a constant engine speed for some crank-angle-dependent models.

a) Continuous models

As known from system and control theory the treatment of dynamic systems with Laplace transformation has many advantages. The Laplace transform of a continuous-time function $x(t)$ is

$$x(s) = \int_0^\infty x(t)e^{-st}dt \tag{2.4.4}$$

with the Laplace variable

$$s = \delta + i\omega \qquad [1/s]. \tag{2.4.5}$$

The functions or signals $x(t)$ have to fulfill certain conditions as $x(t) = 0$ for $t < 0$ and must lead to a convergence of $x(s)$ (integrability). A continuous-time Laplace transformation for the input signal $u(t)$ and output signal $y(t)$ leads to a transfer-function

$$G(s) = \frac{y(s)}{u(s)} \tag{2.4.6}$$

and the frequency response for $s = i\omega$

$$G(i\omega) = \lim_{s \to i\omega} G(s). \tag{2.4.7}$$

Correspondingly, a continuous angle function $x(\varphi)$ can be Laplace-transformed, Schmitt (1995)

$$x(\sigma) = \int_0^\infty x(\varphi) e^{-\sigma\varphi} d\varphi \tag{2.4.8}$$

with the Laplace variable

$$\sigma = \Delta + i\Omega \qquad [1/\text{deg}]. \tag{2.4.9}$$

Hence, an angle-dependent damping $\Delta[1/\text{deg}]$ with the angle frequency $\Omega[1/\text{deg}]$ can be defined. The function $x(\varphi)$ also has to fulfill certain convergence conditions, like $x(\varphi) = 0$ for $\varphi < 0$ and integrability.

Example 2.1 (Continuous models).
A continuous-time first-order process with dead time T_t follows the differential equation in time

$$\frac{dy(t)}{dt} + ay(t) = bu(t - T_t)$$

and the s-transfer function becomes

$$G(s) = \frac{y(s)}{u(s)} = \frac{b}{s+a} e^{-T_t s} = \frac{K}{1+Ts} e^{-T_t s}$$

with $K = b/a$, $T = 1/a$. A continuous angle first-order process with dead angle φ_t is described by a differential equation in angle

$$\frac{dy(\varphi)}{d\varphi} + a' y(\varphi) = b' u(\varphi - \varphi_t)$$

and the σ-transfer function is

$$G(\sigma) = \frac{y(\sigma)}{u(\sigma)} = \frac{b'}{\sigma + a'} e^{-\varphi_t \sigma}.$$

If the dead time T_t results from a certain dead angle φ_t, then it follows for a rotation with constant speed ω_{eng}

$$\varphi_t = \omega_{\text{eng}} T_t$$
$$T_t = \frac{1}{\omega_{\text{eng}}} \varphi_t.$$

Under the assumption of a constant dead angle φ_t the dead time changes inversely proportional to the speed and the transfer function $G(s)$ has to consider this variable dead time $T_t(\omega_{\text{eng}})$. This is not the case for $G(\sigma)$, as the dead angle is independent of the speed.

□

b) Discrete models

As the ECU samples the continuous-time sensor signal with a sampling time T_0 the theory of sampled-data systems can be used leading to *discrete-time models* see, for example Isermann (1989).

An expansion of the Laplace transform by approximation of the sampled (trapezoidal) pulses by δ-impulses results in the z-transfer function of the sampled function $x(kT_0)$ with the discrete time $k = t/T_0 = 0, 1, 2, ...$

$$x(z) = \sum_{k=0}^{\infty} x(kT_0) z^{-k} \tag{2.4.10}$$

$$z = e^{T_0 s}. \tag{2.4.11}$$

A discrete-time transfer function is then

$$G(z) = \frac{y(z)}{u(z)}. \tag{2.4.12}$$

If the crank-angle-dependent function $x(\varphi)$ is sampled with the sampling angle φ_0 resulting in the discrete angle $\kappa = \varphi/\varphi_0 = 0, 1, 2, ...$, this leads, corresponding to (2.4.10), to the ζ-transform.

$$x(\zeta) = \sum_{k=0}^{\infty} x(\kappa\varphi_0)\, \zeta^{-k} \tag{2.4.13}$$

$$\zeta = e^{\varphi_0 \sigma}. \tag{2.4.14}$$

A discrete-angle transfer function follows

$$G(\zeta) = \frac{y(\zeta)}{u(\zeta)}. \tag{2.4.15}$$

For this ζ-transformation the same rules can be applied as for the z-transformation. For example, to avoid side band effects in sampling a continuous-time function with angular frequency $\omega_{\max}$ the sampling frequency ω_0 or sampling time T_0 has to be chosen according to Shannon's sampling theorem

$$\omega_0 = \frac{2\pi}{T_0} > 2\omega_{\max} \quad \text{or} \quad T_0 < \pi/\omega_{\max}. \tag{2.4.16}$$

This corresponds to a continuous angle function with angle frequency Ω_{max} to

$$\Omega_{\max} = \frac{2\pi}{\varphi_0} > 2\omega_{\max} \quad \text{or} \quad \varphi_0 < \pi/\Omega_{\max}. \tag{2.4.17}$$

Example 2.2 (Discrete models).
The continuous-time first-order process with dead time of Example 2.1 leads with a zero-order hold to the z-transfer function

$$
\begin{aligned}
HG(z) &= \frac{y(z)}{u(z)} = \frac{b_1 z^{-1}}{1 - a_1 z^{-1}} z^{-d} \\
a_1 &= e^{-aT_0} \\
b_1 &= \frac{1}{a}(1 - a_1) \\
d &= T_t / T_0 = 0, 1, 2, \ldots
\end{aligned}
$$

and with the shifting theorem to the difference equation

$$
y(k) - a_1 y(k-1) = b_1 u(k - 1 - d).
$$

Corresponding to this procedure one obtains for the discrete-angle first-order process with dead angle and a zero-order hold

$$
\begin{aligned}
HG(\zeta) &= \frac{y(\zeta)}{u(\zeta)} = \frac{b_1' \, \zeta^{-1}}{1 - a_1' \, \zeta^{-1}} \zeta^{-d} \\
a_1' &= e^{\;a' \varphi_0} \\
b_1' &= \frac{1}{a'}(1 - a_1') \\
d' &= \varphi_t / \varphi_0 = 0, 1, 2, \ldots
\end{aligned}
$$

and the difference equation for angles follows

$$
y(\kappa) - a_1' y(\kappa - 1) = b_1' u(\kappa - 1 - d').
$$

It describes the crank-angle-dependent variable $y(\kappa)$ for a crank-angle-dependent input variable $u(\kappa)$. Also here for a constant dead angle φ_t and different speeds ω_{eng} the angle shift d' does not change, contrast to the discrete dead time d.

□

The application of this crank-angle-dependent modeling was developed and applied in Schmidt (1995) to the torque models of a 4-cylinder diesel engine. Sampling angle was selected such that the second harmonic of the ignition angle frequency could be taken into account, resulting in $\varphi_0 = 30$ deg. Then, e.g. the development of the combustion pressure and the resulting torque can be described for each sampled crank angle $\varphi(\kappa\varphi_0)$. For constant speed this corresponds to the discrete times $t_\kappa = \varphi(\kappa\varphi_0)/\omega_{\text{eng}}$.

Figure 2.4.1 depicts the scaling of the crank angle as mostly used in this book.

2.5 Semi-physical models

In general theoretical end experimental modeling complement each other. The theoretical model contains the functional description between the physical/chemical variables and includes their parameters. A considerable advantage is that the dependency of the parameters on construction data is usually explicitly known. Therefore this model is preferred for simulation before construction.

The experimental model on the other hand contains parameters as numerical values where functional relationships with physical basic process data remains unknown. However, the real stationary and dynamic behavior may be described more exactly or it can be determined at smaller expenditure by identification methods, which is, e.g. better suited for the calibration of control systems or fault detection methods.

Theoretical models are also called *"white-box models"* and experimental models *"black-box models"*. In many practical applications one has to use a suitable combination of both ways, compare Fig. 2.5.1.

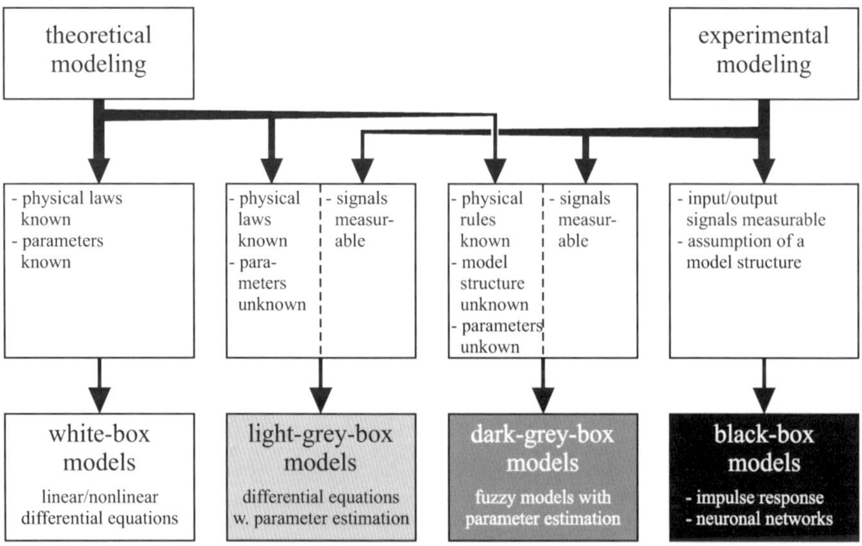

Fig. 2.5.1. Different kinds of mathematical process models.

If the physical laws are known, but the parameters not at all or not precisely enough, the parameters have to be determined experimentally, e.g. by parameter estimation methods. The resulting models can be called *"light-gray models"*. If only physical oriented if-then-rules are known, the model structure and the parameters have to be determined by experiments, leading to *"dark-gray models"*, for example by fuzzy if-then-rules and parameter adjustment. Both gray models can also be called *semi-physical models*.

These type of combined theoretical and experimental models are frequently the result in modeling combustion engines. For example the cylinder pressure is approximated by a special function, the Vibe function, and the parameters are estimated based on pressure measurements. In the case of exhaust gas emission formation only some rules are known and purely experimental approximations are mainly used.

References

Ahrendts I (1989) Technische Thermodynamik. In: Czichos H (ed) Hütte: Grundlagen der Ingenieurwissenschaften, 14th edn, Springer, Berlin

Gawthrop P, Smith L (1996) Metamodelling: bond graphs and dynamic systems. Prentice Hall, Hemel Hempstead

Guzzella L, Onder C (2010) Introduction to modeling and control of internal combustion engine systems, 2nd edn. Springer, Berlin

Isermann R (1989) Digital control systems, 2nd edn. Springer, Berlin

Isermann R (2005) Mechatronic systems – fundamentals, 2nd edn. Springer, London

Karnopp D, Rosenberg R (1975) System dynamics: a unified approach. Wiley, New York

Karnopp D, Margolis D, Rosenberg R (1990) System dynamics: a unified approach. Wiley, New York

MacFarlane A (ed) (1967) Analyse technischer Systeme. Bibl. Institut, Mannheim

MacFarlane A (ed) (1970) Dynamical system models. G.G. Harrop, London

Schmidt C (1995) Digitale kurbelwinkelsynchrone Modellbildung und Drehschwingungsdämpfung eines Dieselmotors mit Last. Dissertation Technische Hochschule Darmstadt. Fortschr.-Ber. VDI Reihe 12, 253. VDI Verlag, Düsseldorf

Schmitt M (1995) Untersuchungen zur Realisierung mehrdimensionaler lernfähiger Kennfelder in Großserien-Steuergeräten. Dissertation Technische Universität Darmstadt. Fortschr.-Ber. VDI Reihe 12, 246. VDI Verlag, Düsseldorf

Takahashi Y, Rabins M, Auslander D (1972) Control and Dynamic Systems. Addison Wesley, Menlo Park

3

Experimental modeling of engines

Several parts of engines cannot be modeled in a theoretical way only, because the mathematical formulation of parts of the processes is not precisely known or the computational expense is too large with regard to control and diagnosis applications. This holds for example for the flame propagation, pressure and temperature development of gasoline engines and for the spray development and combustion of diesel engines, both under the influence of valve-induced turbulent flow, swirl and tumble motion. Also precise and computational simple theoretical models for the emissions depending on several manipulated variables are not available.

In some cases theoretical/physical modeling is possible, however some parameters are not precisely known. Then the estimation of physically defined parameters for a given model structure can be used. Examples are the intake manifold, the internal heat transfer to the coolant or in external heat exchangers and the mechanical parts of the crankshaft drive, the turbocharger, and pumps. This means that theoretical modeling yields the basic structure of the models, but complete and precise models can, at least until now, only be obtained by experiments, mainly on test benches. The various types of these mathematical models can be described as *gray-box models* as depicted in Fig. 2.5.1. If no structure of the models is known, a priori one has to use identification methods based only on input and output measurements, leading to *black-box models*.

This chapter gives in Sects. 3.1 and 3.2 an introduction to the different methods for the experimental modeling or identification of the steady-state (stationary) and dynamic (non-stationary, transient) behavior of engines. Because these methods are based on established process identification methods of linear, nonlinear and time-varying processes known from system and control sciences, they are described first in a compact form, a shorted and adapted version of Chap. 9 in Isermann and Münchhof (2011). These basic methods are then used to describe various procedures for the identification of engines on test benches in Sects. 3.4, 3.5 and 3.7.

R. Isermann, *Engine Modeling and Control*,
DOI: 10.1007/978-3-642-39934-3_3

3.1 Identification methods for the stationary behavior of nonlinear processes

3.1.1 Grid-based look-up tables (maps)

The most direct way to determine the stationary behavior of a nonlinear process with one input u and one output y

$$y = f(u) \tag{3.1.1}$$

is to take samples in steady state for $y(k)$ and different $u(k)$ for $k = 0, 1, \ldots, N-1$ and to represent them in a graph. However, without a regression procedure the outputs $y(k)$ may be scattered because of disturbances. In the case of several inputs

$$\mathbf{u}^{\mathrm{T}} = [u_1, u_2, \ldots, u_p] \tag{3.1.2}$$

$$y = f(\mathbf{u}) \tag{3.1.3}$$

the output y can be represented in a multidimensional look-up table or map. The measurements are taken for all combinations of the inputs. Grid-based look-up tables are the most common type of nonlinear static models used in practice for one or two inputs. Especially in the field of nonlinear control, look-up tables are widely accepted as they provide a transparent and flexible representation of nonlinear relationships. Electronic control units of modern automobiles, for example, contain about 100 such grid-based look-up tables, in particular for engine and emission control. In automotive applications, due to cost reasons computational power and storage capacity are strongly restricted. Furthermore, constraints of real-time operation have to be met. Under these conditions, grid-based look-up tables represent a suitable means for the storage of nonlinear static mappings, compare the discussion in Sect. 6.5.

The models consist of a set of data points or nodes positioned on, e.g. a two-dimensional grid. Then, each node comprises two inputs u_1 and u_2. The scalar data points heights are estimates of the approximated nonlinear function at their corresponding data point position. All nodes located on grid lines, as shown in Fig. 3.1.1 are stored, e.g. in the ROM of the control unit. For model generation, usually all data point positions are fixed a priori. The most widely applied method of obtaining the data point heights is to position measurement data points directly on the grid.

If the output y has to be calculated for any input values u_1 and u_2 of a two-dimensional look-up table, an interpolation has to be made, Hoschek and Lasser (1993). In a first step the indices of the enclosing four data points have to be selected. Then, a bilinear area interpolation is performed, Schmitt (1995). For this, four areas have to be calculated. A detailed treatment of look-up tables, its identification, interpolation and extensions is considered in Sect. 6.5.

Grid-based look-up tables belong to the class of *non-parametric models*. The model structure described has the advantage that a subsequent adaptation of single data point heights due to changing environmental conditions is easy to realize. The main disadvantage of look-up tables is the exponential growth of the number of data points with an increasing number of inputs. Therefore, grid-based look-up tables are mostly restricted in practical applications to one- and two-dimensional input spaces. However, different ways do exist to extend them to multi-inputs and to use them for control as is described in Sect. 6.5.

An alternative are *parametric model representations*, like polynomial models, neural networks, or fuzzy models, which clearly require less model parameters to approximate a given input output relationship, see next sections. Therefore, the storage demand of these models is much lower. However, in contrast to area interpolation, the complexity of the computation of the output is much higher, since complex nonlinear functions for, e.g. each neuron have to be computed. On the other hand, grid-based look-up tables are not directly suitable for the identification and modeling of dynamic process behavior. Therefore, other identification methods for nonlinear dynamic processes based on parameter estimation are treated in Sect. 3.2.

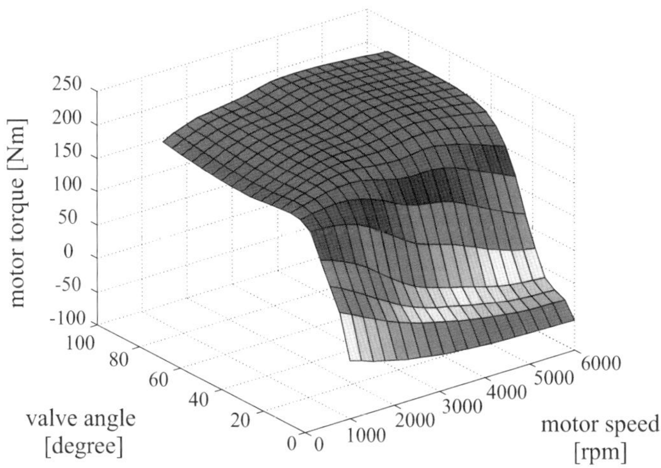

Fig. 3.1.1. Grid-based look-up table of a six-cylinder SI engine.

3.1.2 Parameter estimation for nonlinear static processes

The direct measurements of look-up tables does not include a suppression of noisy outputs. This becomes possible by assuming a parametric equation as an approximation for example in the form of a polynomial. It is required that output measurements $y(k)$ are taken for different inputs $u(k)$ spanning the range of interest, where

$$k = 0, 1, \ldots, N-1$$

is the running number of N samples. For one input and one output at sample k this yields for the usable output

$$y_u(k) = K_0 + u(k)K_1 + u^2(k)K_2 + \cdots + u^q(k)K_q \tag{3.1.4}$$

compare Fig. 3.1.2. The measured output may be contaminated by a disturbance $n(k)$

$$y(k) = y_u(k) + n(k). \tag{3.1.5}$$

Defining the vectors

$$\left.\begin{aligned} \mathbf{u}(k) &= [1\ u(k)\ u^2(k) \ldots u^q(k)] \\ \mathbf{K}^{\mathrm{T}} &= [K_0\ K_1\ K_2 \ldots K_q] \end{aligned}\right\} \tag{3.1.6}$$

leads with $k = 0, \ldots, N-1$ to a regression model

$$\mathbf{Y} = \mathbf{U}\mathbf{K} + \mathbf{n} \tag{3.1.7}$$

where $\mathbf{U}$ is a $q \times N$ matrix and $\mathbf{Y}$ a vector of dimension N. Introducing the equation error

$$\mathbf{e} = \mathbf{Y} - \mathbf{U}\mathbf{K} \tag{3.1.8}$$

where $\mathbf{U}$ and $\mathbf{Y}$ contain the measurements, resulting in an equation where the *equation error is linear in the parameters*. If the sum of squared errors is taken as a loss function

$$V = \mathbf{e}^{\mathrm{T}}\mathbf{e} = \sum_{k=0}^{N-1} e^2(k), \tag{3.1.9}$$

the minimization with

$$\frac{dV}{d\mathbf{K}} = \mathbf{0} \tag{3.1.10}$$

yields the least squares estimate, see Isermann and Münchhof (2011)

$$\hat{\mathbf{K}} = \left[\mathbf{U}^{\mathrm{T}}\mathbf{U}\right]^{-1} \mathbf{U}^{\mathrm{T}}\mathbf{Y}. \tag{3.1.11}$$

Existence of this estimate requires

$$\det = [\mathbf{U}^{\mathrm{T}}\mathbf{U}] \neq 0 \tag{3.1.12}$$

which means, the input signal $u(k)$ must change during the measurements. If $E\{n(k)\} = 0$, the parameter estimates are consistent in mean square, Isermann and Münchhof (2011). For several inputs the vectors are expanded

$$\begin{aligned} \mathbf{u}(k) = [1\ & u_1(k)\ u_1^2(k) \ldots u_1^q(k) \\ & u_2(k)\ u_2^2(k) \ldots u_2^q(k) \ldots u_p(k)\ u_p^2(k) \ldots u_p^q(k)] \end{aligned} \tag{3.1.13}$$

$$\mathbf{K}^{\mathrm{T}} = [K_{10}\ K_{11}\ K_{12} \ldots K_{1q}\ K_{21} \ldots K_{2q} \ldots K_{p1} \ldots K_{pq}]\,.$$

Applying the method of least squares for estimation of the parameters yields again (3.1.11).

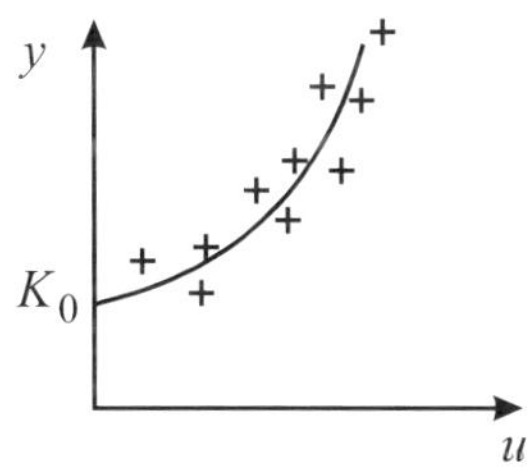

Fig. 3.1.2. Parameter estimation (regression) for a nonlinear characteristic.

3.1.3 Artificial neural networks for identification

For a general identification approach, methods of interest are those that do not require specific knowledge of the process structure and hence are widely applicable. Artificial neural networks fulfill these requirements. They are composed of mathematically formulated neurons. At first, these neurons were used to describe the behavior of biological neurons, McCulloch and Pitts (1943). The interconnection of neurons in networks allowed the description of relationships between input and output signals, Rosenblatt (1958), Widrow and Hoff (1960). In the sequel, artificial neural networks (ANNs) are considered to map input signals u to output signals y, Fig. 3.1.3. Usually, the adaptable parameters of neural networks are unknown. As a result, they have to be adapted or "trained" or "learned" by processing measured signals u and y, Hecht-Nielson (1990), Haykin (1994). This is a typical system identification problem. If inputs and outputs are gathered into groups or clusters, a classification task in connection with, e.g. pattern recognition is given, Bishop (1995). In the following, the problem of nonlinear system identification is considered (supervised learning). Thereby, the capability of ANNs to approximate nonlinear relationships to any desired degree of accuracy is utilized. Firstly, ANNs for describing the *static transfer behavior*, Hafner et al (1992), Preuß and Tresp (1994), will be investigated, which will then be extended to *dynamic behavior*, Ayoubi (1996), Nelles et al (1991), Isermann et al (1997).

Neural networks are universal approximators for static nonlinearities and are consequently an alternative to polynomial approaches. Their advantages are the need for only little a priori knowledge about the process structure and the uniform treatment of single-input and multi-input processes. In the following, it is assumed that a nonlinear system with P inputs and M outputs has to be approximated, see Fig. 3.1.3.

Fig. 3.1.3. System with P inputs and M outputs which has to be approximated by an artificial neural network.

a) Neuron model

Figure 3.1.4 shows the block diagram of a neuron. In the input operator (synaptic function), a similarity measure between the input vector $\mathbf{u}$ and the (stored) weight vector $\mathbf{w}$ is formed, e.g. by the scalar product

$$x = \mathbf{w}^{\mathrm{T}}\,\mathbf{u} = \sum_{i=1}^{P} w_i\,u_i = \left|\mathbf{w}^{\mathrm{T}}\right|\,\left|\mathbf{u}\right| \cos\phi \tag{3.1.14}$$

or the Euclidean distance

$$x = \left|\left|\mathbf{u} - \mathbf{w}\right|\right|^2 = \sum_{i=1}^{P} (u_i - w_i)^2\,. \tag{3.1.15}$$

If $\mathbf{w}$ and $\mathbf{u}$ are similar, the resulting scalar quantity x will be large in the first case and small in the second case. The quantity x, also called the activation of the neuron, affects the activation function and consequently the output value

$$y = \gamma(x - c). \tag{3.1.16}$$

Figure 3.1.5 shows several examples of nonlinear activation functions. The threshold c is a constant causing a parallel shift in the x-direction.

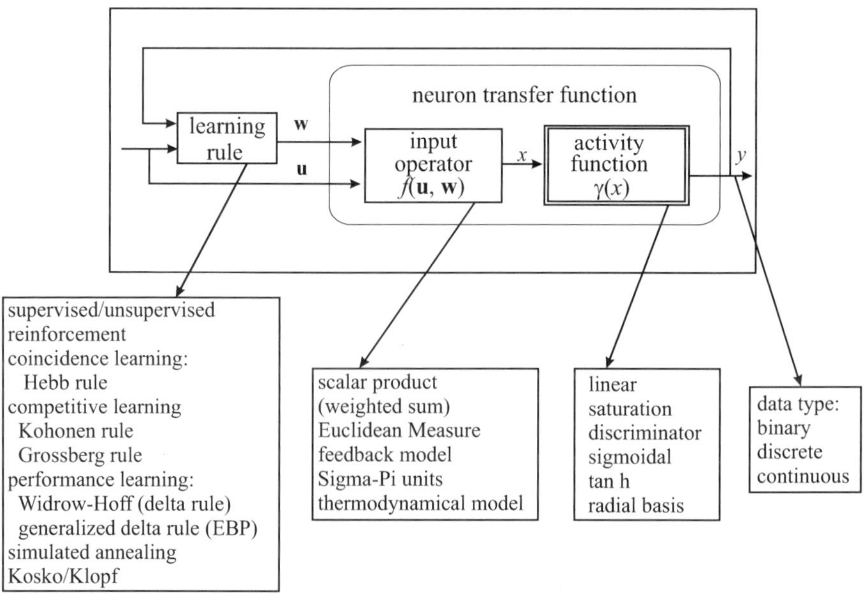

Fig. 3.1.4. General neuron model.

a) hyperbolic tangens (Tangens Hyperbolicus)

$$y = \frac{e^{(x-c)} - e^{-(x-c)}}{e^{(x-c)} + e^{-(x-c)}} = 1 - \frac{2}{1 + e^{2(x-c)}}$$

b) Sigmoidal function

$$y = \frac{1}{1 + e^{-(x-c)}}$$

c) limiter

$$y = \begin{cases} 1 & ; x - c \geq 1 \\ x - c & ; |x - c| < 1 \\ -1 & ; x - c \leq -1 \end{cases}$$

d) neutral zone

$$y = \begin{cases} 0 & ; |x - c| \leq 1 \\ x - c - 1 & ; x - c > 1 \\ x - c + 1 & ; x - c \leq -1 \end{cases}$$

e) Gauss-functions $\quad y = e^{-(x-c)^2}$

f) binary function

$$y = \begin{cases} 0 & ; x - c < 0 \\ 1 & ; x - c > 0 \end{cases}$$

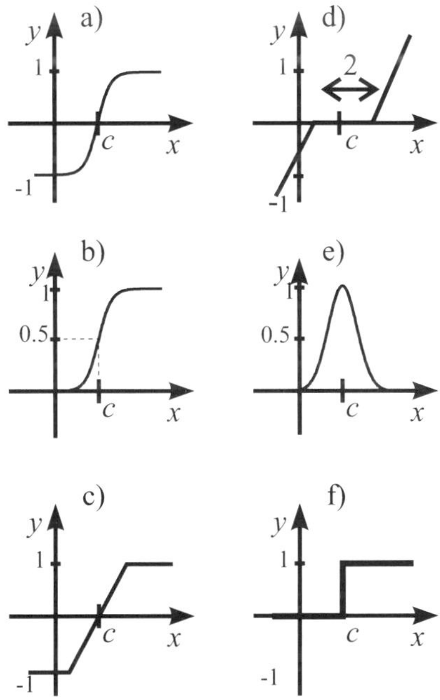

Fig. 3.1.5. Examples of activation functions.

b) Network structure

The single neurons are interconnected to a network structure, Fig. 3.1.6. Hence, one has to distinguish between different layers with parallel arranged neurons: the input layer, the first, second, ... hidden layer and the output layer. Generally, the input layer is used to scale the input signals and is not often counted as a separate layer. Then, the real network structure begins with the first hidden layer. Figure 3.1.6 shows the most important types of internal links between neurons: feedforward, backward, lateral and recurrent. With respect to their range of values, the input signals can be either binary, discrete or continuous. Binary and discrete signals are used especially for classification, while continuous signals are used for identification tasks.

c) Multi-layer perceptron (MLP) network

The neurons of a MLP network are called perceptrons, Fig. 3.1.7, and follow directly from the general neuron model, shown in Fig. 3.1.4. Typically, the input operator is realized as a scalar product, while the activation functions are realized by sigmoidal or hyperbolic tangent functions. The latter ones are a multiple of differentiable functions yielding a neuron output with $y = 0$ in a wide range. Therefore, they have a global effect with extrapolation capability. The weights w_i are assigned to the input operator and lie in the signal flow before the activation function.

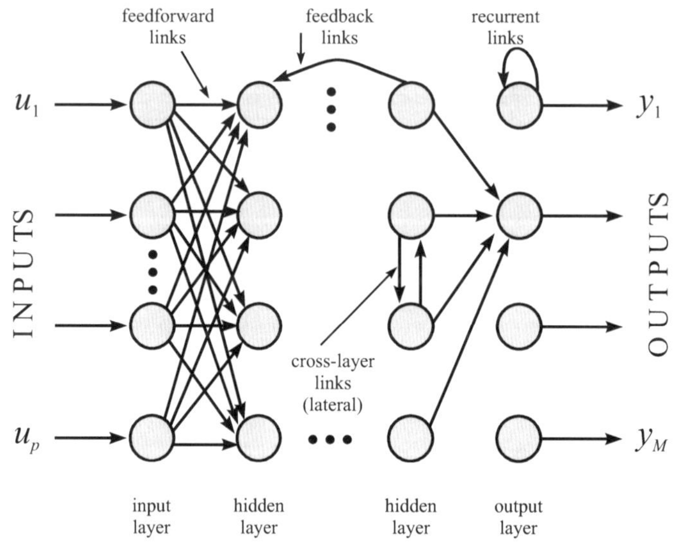

Fig. 3.1.6. Network structure: layers and links in a neural network.

The perceptrons are connected in parallel and are arranged in consecutive layers to a feedforward MLP network, Fig. 3.1.8. Each of the P inputs affects each perceptron in such a way that in a hidden layer with K perceptrons there exist $(K \cdot P)$ weights w_{kp}. The output neuron is most often a perceptron with a linear activation function, Fig. 3.1.9.

The adaptation of the weights w_i based on measured input and output signals is usually realized by the minimization of the quadratic loss function.

$$J(\mathbf{w}) = \frac{1}{2} \sum_{n=1}^{N-1} e^2(n) \tag{3.1.17}$$

$$e(n) = y(n) - \hat{y}(n)$$

where e is the model error, y is the measured output signal and $\hat{y}$ is the network output.

As in the case of parameter estimation with the least squares method,

$$\frac{dJ(\mathbf{w})}{d\mathbf{w}} = \mathbf{0} \tag{3.1.18}$$

is generated. Due to the nonlinear dependency, a direct solution is not possible. Therefore, e.g. gradient methods for numerical optimization are applied. Because of the necessary back-propagation of errors through all hidden layers, the method is called "error back-propagation" or also "delta-rule". The so-called learning rate η has to be chosen (tested) suitably. In principle, gradient methods allow only slow convergence in the case of a large number of unknown parameters.

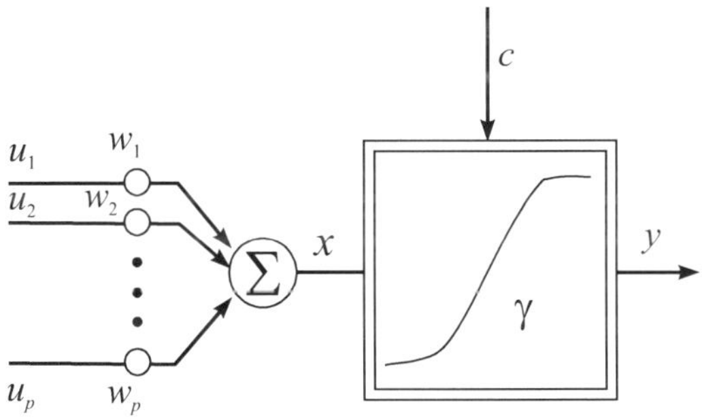

Fig. 3.1.7. Perceptron neuron with weights w_i, summation of input signals (scalar product) and nonlinear activation function.

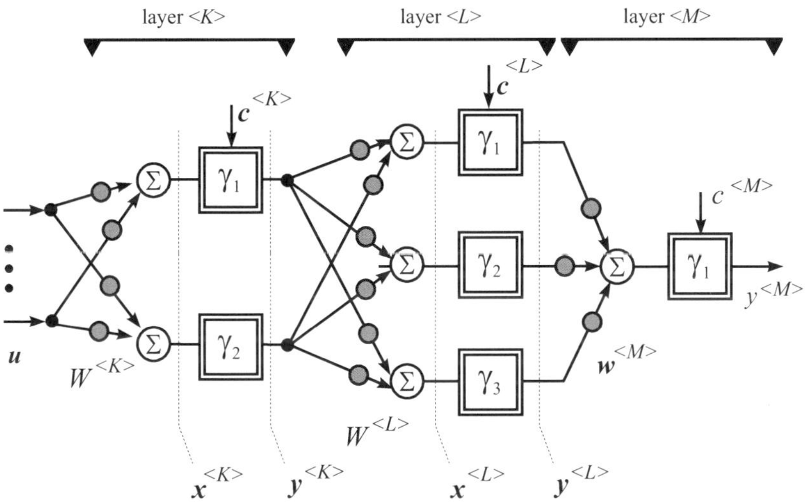

Fig. 3.1.8. Feedforward multi-layer perceptron network (MLP network). Three layers with $(2 \cdot 3 \cdot 1)$ perceptrons. $< K >$ is the first hidden layer.

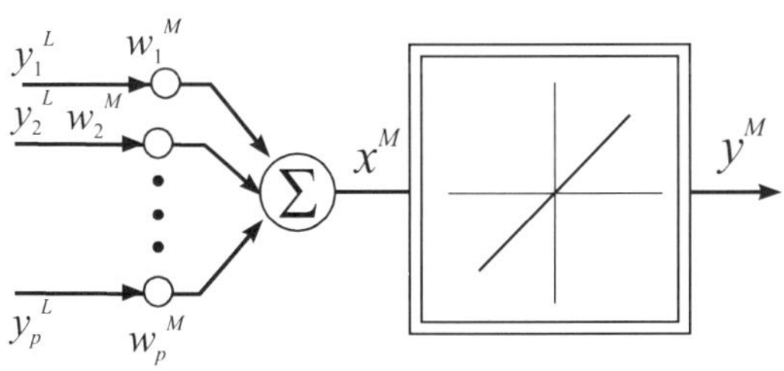

Fig. 3.1.9. Output neuron as perceptron with linear activation function.

d) Radial basis function (RBF) network

The neurons of RBF networks, Fig. 3.1.10, compute the Euclidean distance in the input operator

$$x = ||\mathbf{u} - \mathbf{c}||^2 \tag{3.1.19}$$

and feed it to the activation function

$$G_m = \gamma_m \left(||\mathbf{u} - \mathbf{c}\;||^2 \right). \tag{3.1.20}$$

The activation function is given by radial basis functions usually in the form of Gaussian functions with

$$\gamma_m = \exp\left[\frac{-1}{2} \left(\frac{(u_1 - c_{m1})^2}{\sigma_{m1}^2} + \frac{(u_2 - c_{m2})^2}{\sigma_{m2}^2} + \ldots + \frac{(u_P - u_{mP})^2}{\sigma_{mP}^2} \right) \right]. \tag{3.1.21}$$

The centers c_j and the standard deviations σ_j are determined a priori so that the Gaussian functions are spread, e.g. uniformly in the input space. The activation function determines the distances of each input signal to the center of the corresponding basis function. However, radial basis functions contribute to the model output only locally, namely in the vicinity of their centers. They possess less extrapolation capability, since their output values go to zero with a growing distance to their centers.

Usually, radial basis function networks consist of two layers, Fig. 3.1.11. The outputs γ_i are weighted and added up in a neuron of the perceptron type, Fig. 3.1.9, so that

$$y = \sum_{m=1}^{M} w_m \, \gamma_m \left(||\,\mathbf{u} - \mathbf{c}||^2 \right). \tag{3.1.22}$$

Since the output layer weights are located behind the nonlinear activation functions in the signal flow, the error signal is linear in these parameters and, consequently, the least squares method in its explicit form can be applied. In comparison to MLP networks with gradient methods, a significantly faster convergence can be obtained. However, if the centers and standard deviations have to be optimized too, nonlinear numerical optimization methods are also required.

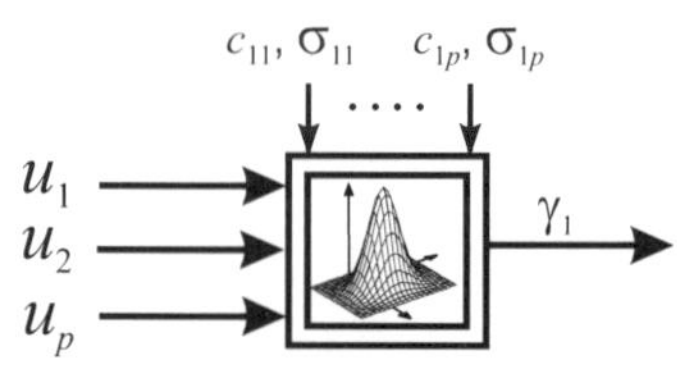

Fig. 3.1.10. Neuron with radial basis function.

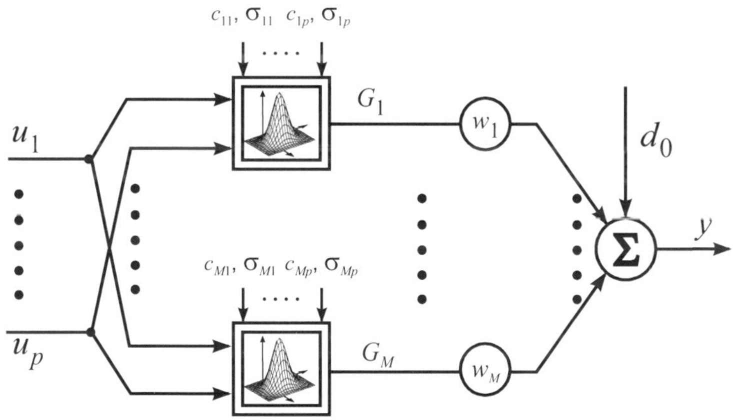

Fig. 3.1.11. Feedforward radial basis function (RBF) network.

e) Local linear model networks

The local linear model tree (LOLIMOT) is an extended radial basis function network, Nelles and Isermann (1995), Babuska and Verbruggen (1996), Isermann et al (1997), Murray-Smith and Johansen (1997), Nelles (1997), Nelles (2001), Fink (2006). It is extended by replacing the output layer weights with a linear function of the network inputs (3.1.25). Furthermore, the RBF network is normalized, so that the sum of all basis functions is one. Thus, each neuron represents a local linear model with its corresponding validity function, see Fig. 3.1.12. The validity functions determine the regions of the input space where each neuron is active. The general architecture of local model networks is extensively discussed in Murray-Smith and Johansen (1997).

Different local models result from the way of partitioning of the input space $\mathbf{u}$, e.g. grid structure, axis-orthogonal cuts, axis-oblique cuts, etc., the local model structure and the transition between submodels. Due to their transparent structure, local models offer the possibility of adjusting the model structure easier to the process structure in terms of physical-law-based relationships leading to *semi-physical models*. Such an incorporation of physical insight improves the training and the generalization behavior considerably and reduces the required model complexity in many cases.

The kind of local model network discussed here utilizes normalized Gaussian validity functions (3.1.21) and an axis-orthogonal partitioning of the input space. Therefore, the validity functions can be composed of one-dimensional membership functions and the network can be interpreted as a Takagi-Sugeno fuzzy model.

The output of the local linear model is calculated by

$$\hat{y} = \sum_{i=1}^{M} \Phi_i(\mathbf{u})\, \hat{y}_i(\mathbf{u}) \tag{3.1.23}$$

where the local linear (stationary) models are

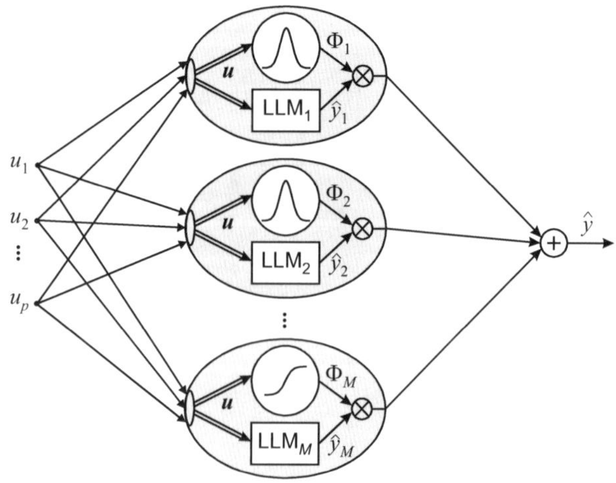

Fig. 3.1.12. Local linear model network (LOLIMOT) for the stationary behavior.

$$\hat{y}_i(\mathbf{u}) = w_{i,0} + w_{i,1}\,u_1 + \ldots + w_{i,P}\,u_P \tag{3.1.24}$$

such that

$$\hat{y} = \sum_{i=1}^{M} \Phi_i(\mathbf{u})\;(w_{i,0} + w_{i,1}\,u_1 + \ldots + w_{i,P}\,u_P) \tag{3.1.25}$$

with the normalized Gaussian validity functions

$$\Phi_i(\mathbf{u}) = \frac{\mu_i(\mathbf{u})}{\sum_{j=1}^{M}\mu_j(\mathbf{u})} \tag{3.1.26}$$

and

$$\mu_i(\mathbf{u}) = \prod_{j=1}^{p} \exp\left(-\frac{1}{2}\left(\frac{(u_j - c_{i,j})^2}{\sigma_{ij}^2}\right)\right). \tag{3.1.27}$$

The centers c and standard deviations σ act nonlinearly on the output, while the local model parameters w_i act linearly. The local linear model tree (LOLIMOT) algorithm is applied for the training. It consists of an outer loop, in which the input space is decomposed by determining the parameters of the validity functions, and a nested inner loop in which the parameters of the local linear models are optimized by local-weighted least squares estimation.

The input space is decomposed in an axis-orthogonal manner, yielding hyper-rectangles in whose centers the Gaussian validity functions $\mu_i(u)$ are placed. The standard deviations of these Gaussians are chosen proportionally to the extension of hyper-rectangles to account for the varying granularity. Thus, the nonlinear parameters $c_{i,j}$ and $\sigma_{i,j}$ are determined by a heuristic-avoiding explicit nonlinear optimization. LOLIMOT starts with a single linear model that is valid for the whole input space. In each iteration, it splits one local linear model into two new sub-models.

Only the (locally) worst performing local model is considered for further refinement. Splits along all input axes are compared and the best performing alternative is carried out, see Fig. 3.1.13.

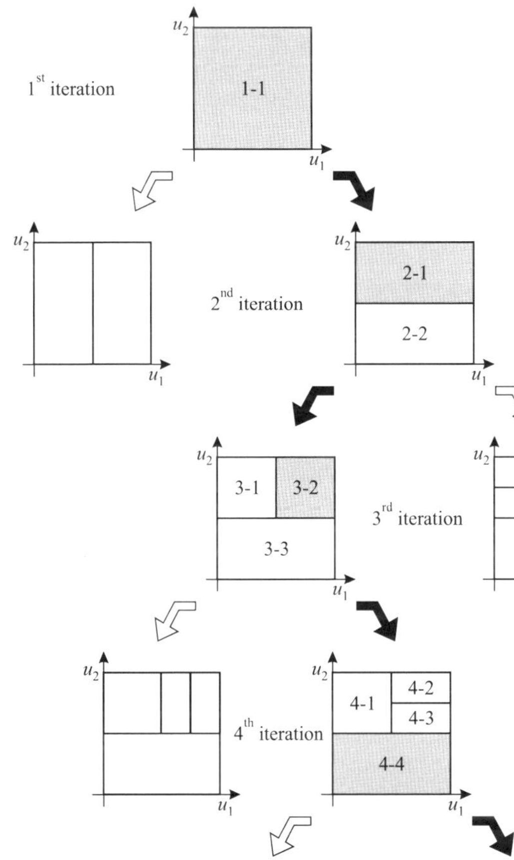

Fig. 3.1.13. Tree construction of the LOLIMOT method.

The main advantages of this local model approach are the inherent structure identification and the very fast and robust training algorithm. The model structure is adapted to the complexity of the process and an explicit application of time-consuming nonlinear optimization algorithms as for general NN is avoided.

Another local linear model architecture, the so-called *hinging hyperplane tree* (HHT), is presented in Ernst (1998), Töpfer (2002). These models can be interpreted as an extension of the LOLIMOT networks with respect to the partitioning scheme. While the LOLIMOT algorithm is restricted to axis-orthogonal splits, the hinging hyperplane trees allow an axis-oblique decomposition of the input space. These more complex partitioning strategies lead to an increased effort in model construction.

However, this feature can be advantageous in the case of strong nonlinear behavior and higher-dimensional input spaces, where axis orthogonal splits lead to many local models.

Summarizing, the three representatives of artificial neural networks are very well suited to the approximation of measured input/output data of static processes, compare also Hafner et al (1992), Preuß and Tresp (1994). For this, the training data has to be chosen in such a way that the considered input space is covered as evenly as possible with data. After the training procedure, a parametric mathematical model of the static process behavior is available. Consequently, direct computation of the output values $\hat{y}$ for arbitrary input combinations u is possible.

An advantage of the automatic training procedure is the possibility of using arbitrarily distributed data in the training data set. There is no necessity to know data at exactly defined positions, as in the case of grid-based look-up table models according to Sect. 3.1.1. This clearly decreases the required effort for measurements in practical applications.

Example 3.1 (Artificial neural network for the static behavior of a combustion engine).
As an example, the engine characteristics of a six-cylinder SI (spark-ignition) engine is used. Here, the engine torque has to be identified and is dependent on the throttle angle and the engine speed. Figure 3.1.14 shows the 433 available data points that were measured on an engine test stand.

For the approximation, an MLP network is applied. After the training, an approximation for the measurement data is shown in Fig. 3.1.15. For that purpose, 31 parameters are required. Obviously, the neural network possesses good interpolation and extrapolation capabilities. This also means that in areas with only few training data, the process behavior can be approximated quite well, Holzmann et al (1997).

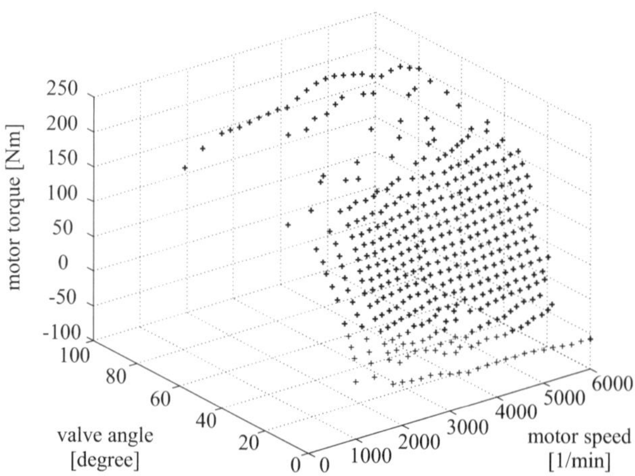

Fig. 3.1.14. Measured SI engine data (2.5 l, V6 cyl.): unevenly distributed, 433 measurement data points.

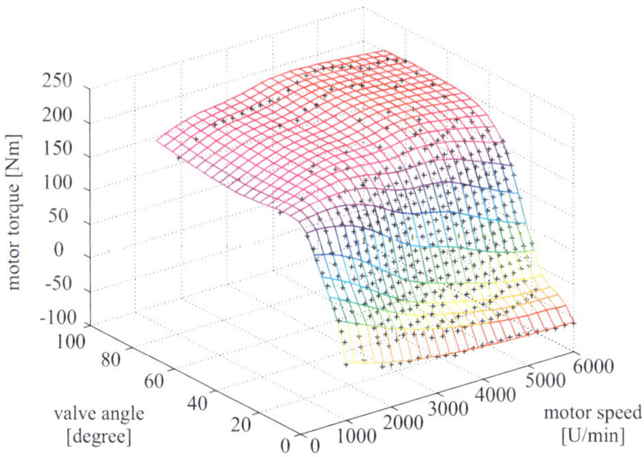

Fig. 3.1.15. Approximation of measured engine data (+) with a MLP network $(2 \cdot 6 \cdot 1)$: 31 parameters.

□

f) Local nonlinear model network

An enhancement to the presented LOLIMOT structure is a local polynomial model tree (LOPOMOT), Sequenz et al (2009), Sequenz (2013). It is composed of local polynomial models. Like for the LOLIMOT structure an axis orthogonal splitting algorithm is employed, Fig. 3.1.16a). In contrast to LOLIMOT the regressors of the local polynomial models are selected by a local formulation of Mallows' $\mathrm{C_p}$-statistic, Mallows (1973). This selection enables an adaption of the local models to the nonlinearities of the process and reduces the model error. A similar approach to LOPOMOT is presented in Bänfer and Nelles (2009). There the determination of the local polynomial degree is incorporated into the splitting algorithm. The improvement of increasing the local polynomial degree is compared to a model split and the superior is performed where the employed selection criteria are based on heuristic choices of tuning parameters.

Similar to (3.1.25) the local model structure for LOPOMOT can be written as, Sequenz (2013),

$$\hat{y}_{\mathrm{LOPOMOT}} = \sum_{j=1}^{M} \Phi_j(\mathbf{u})\, \hat{y}_j(\mathbf{u}), \tag{3.1.28}$$

see Fig. 3.1.16b).

The weighting function $\Phi_j(\mathbf{u})$ can be a Gaussian as for the LOLIMOT model, (3.1.26), or any other function that decays to 0. For example, using a pyramidal weighting function shows a compact support and is equivalent to a linear interpolation for the applied grid pattern structure. This eases the model simulation especially for an implementation on an ECU, see Sequenz (2013).

The approximation $\hat{y}_j(\mathbf{u})$ can be motivated by the Taylor series of the true function $y(\mathbf{u})$. The Taylor series in the neighborhood of $\mathbf{u}_0$ is given by

$$y(\mathbf{u}_0 + \Delta\mathbf{u}) = y(\mathbf{u}_0) + \sum_{i=1}^{o} \frac{1}{i!} \left(\frac{\partial}{\partial u_1} \Delta u_1 + \cdots + \frac{\partial}{\partial u_p} \Delta u_p \right)^i y(\mathbf{u})|_{u_0} + R_o \tag{3.1.29}$$

where $\Delta\mathbf{u} = \mathbf{u} - \mathbf{u}_0$, o is the order of the approximation and R_o is the remainder. The higher the order o the smaller the remainder R_0 and hence the more precise the approximation. However, the number of terms increases exponentially with an increase of the order. From (3.1.29) the general local polynomial approximation results

$$\hat{y}_j(\mathbf{u}) = w_{0,j} + w_{1,j}u_1 + w_{2,j}u_2 + \ldots + w_{p,j}u_p + w_{p+1,j}u_1^2 + w_{p+2,j}u_1u_2 + \ldots . \tag{3.1.30}$$

The evaluated derivatives at $\mathbf{u}_0$, $\frac{\partial_i y}{\partial u_\cdot^i}|_{\mathbf{u}_0}$ in (3.1.29, form the unknown parameters $w_{\cdot,j}$ and $\Delta u_\cdot^{\ i}$ in (3.1.29) and give the regressors $u_\cdot^{\ i}$. Given a dataset with N data points the quadratic error can be defined as

$$J = \sum_{i=1}^{N} (y(\mathbf{u}_i) - \hat{y}(\mathbf{u}_i))^2 . \tag{3.1.31}$$

The objective of a model identification is the minimization of (3.1.31). Since the true value $y(\mathbf{u}_i)$ is in general unknown, a measured value y_i at $\mathbf{u}_i$ is employed. As measurements do suffer from disturbances a variance error for the identified model results. This variance error increases with the number of parameters, in contrast to the bias error introduced by the limited flexibility of the approximation which decreases with the number of parameters, Nelles (2001). This opposite behavior is known as *bias-variance dilemma*.

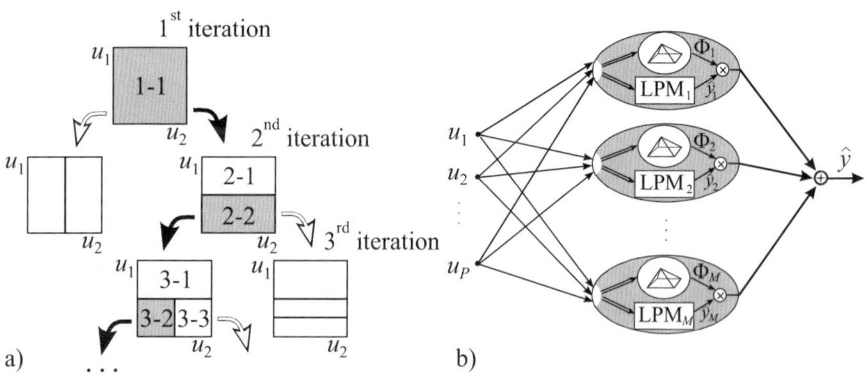

Fig. 3.1.16. Local nonlinear model network (LOPOMOT). **a** axis orthogonal splitting algorithm. **b** model structure (with pyramidical weighting functions).

To find the best trade-off between bias and variance error a selection strategy is applied. The selected regressors must be able to approximate the unknown function, while avoiding redundant regressors. (3.1.30) can therefore be written as

$$\hat{y}_j(\mathbf{u}) = w_{1,j}x_1 + w_{2,j}x_2 + \ldots + w_{n,j}x_n \tag{3.1.32}$$

$$\text{with } x_i \in \mathcal{A} = \left\{1,\ u_1,\ u_2,\ u_1u_2,\ u_1^2,\ u_1^2u_2,\ u_2^2\ldots\right\}. \tag{3.1.33}$$

$\mathcal{A}$ is the set of potential regressors and has the size $|\mathcal{A}| = k$. Since the parameters $w_{\cdot,j}$ are linear in (3.1.32), a weighted least squares algorithm is used to determine the parameters.

The applied criterion of fit to select the n significant regressors $x_i \in \mathcal{A}$ is Mallows' $\mathrm{C_p}$-statistic. It is an estimate of the scaled sum of squared errors J/σ^2 and a general formulation is given by Mallows (1973)

$$\mathrm{C_p} = \frac{\sum_{i=1}^{N}\left(y_i - \hat{y}_i\right)^2}{\hat{\sigma}^2} - N + 2(n+1). \tag{3.1.34}$$

where y_i is the measured output at $\mathbf{u}_i$, $\hat{y}_i$ the modeled output at $\mathbf{u}_i$, N the size of the dataset and $\hat{\sigma}^2$ the estimated variance. An estimation of the variance is given by the sum of squared errors with the modeled output $y_{k,i}$ using all k regressors divided by the degrees of freedom

$$\hat{\sigma}^2 = \frac{\sum_{i=1}^{N}\left(y_i - \hat{y}_{k,i}\right)^2}{N-k}. \tag{3.1.35}$$

If the order o is chosen sufficiently high, the bias error of $\hat{y}_{k,i}$ can be neglected. Then, the sum of squared errors composes only the variance error wherefore under the assumption of Gaussian error distribution it gives an unbiased estimation of the variance. Since a degree of freedom is required for this variance estimation, a 1 is added in (3.1.34) to the number of model parameters n.

The general formula presented in (3.1.34) holds for global models. However, a multi-model with several local polynomial models is applied here. Therefore, a local formulation of (3.1.34) is derived and a selection algorithm for local regressor selection is presented in Sequenz (2013). Based on the local formulation of Mallows' $\mathrm{C_p}$-statistic a criterion for the selection of the best model partition can then be presented. The local regression selection, the global model partitioning and the extension to dynamic processes is described, Sequenz (2013).

A comparison of LOLIMOT, LOPOMOT and look-up tables with regard to accuracy and computational effort is presented in Sect. 6.5.3

3.2 Process identification methods for dynamic processes

This section gives a brief introduction of some important identification methods for dynamic linear and nonlinear processes with single-input single-output signals which are relevant for engines and vehicles. Table 3.2.1 shows a survey of the most important identification methods.

Table 3.2.1. Survey of important identification methods for dynamic processes TVS: time-variant systems; MIMO: multi-input multi-output systems, NLS: nonlinear systems.

input signal / model / output signal	identi-fication method	used device	allowable distur-bances	digital computer coupling: offline	digital computer coupling: online	data processing: one shot	data processing: real-time	reachable accuracy	extendibility: TVS	extendibility: MIMO	extendibility: NLS	application example
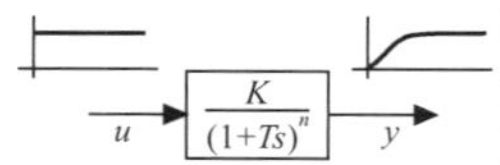 param.	determination of characteris-tic valves	recorder	very small	-	-	-	-	small	-	-	-	rough model, controller tuning
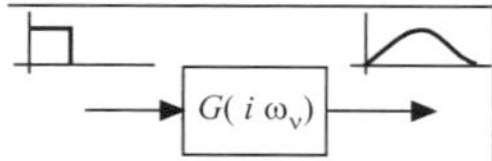 non-param.	Fourier analysis	recorder	small	X	-	X	-	medium	-	X	-	verification of theoretical models
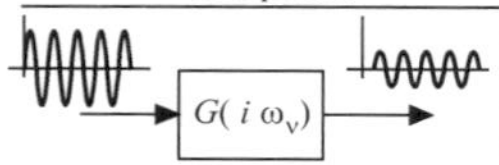 non-param.	frequency response measurement	recorder, F. resp. meas. device	medium	X	X	-	-	very large	-	X	-	verification of theoretical models
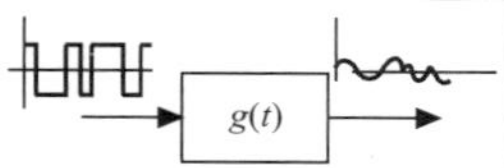 nonparam.	correlation	correlator process computer	large	- X	- X	- X	X X	large	X	X	X	detection of signal relations, time delay identification
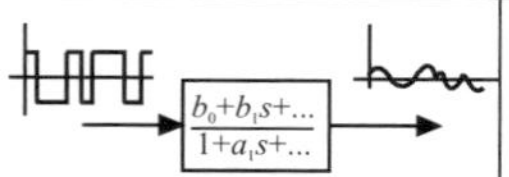 param.	model adjustment parameter estimation	analog computer process computer	small large	- X	- X	- X	X X	medium large	X X	- X	- X	analog-adaptive control design of advanced controllers, adaptive control, fault detection
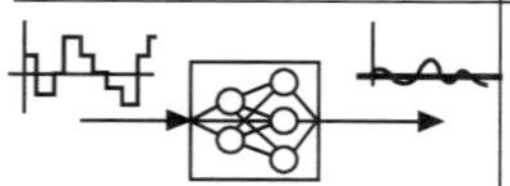 param.	neural net	process computer	small/ medium	X	X	X	-	very large	-	X	X	design of non-linear controllers, learning controllers, fault detection

In Fig. 3.2.1 those methods are extracted which can be applied for a wide range of dynamic processes with excitation signals at the input. Especially for dynamic processes the input signals have to change periodically, stochastically or in the form of special test signals. These signals may be the normal operating signals (like in servo systems, actuators or driving vehicles) or may be artificially introduced for testing, e.g. on test benches. A considerable advantage of identification methods is that with only one input and one output signal several parameters (up to about six) can be estimated, which give a detailed picture on internal process quantities.

The identification methods are described for discrete-time linear processes, both in open loop and in closed loop and then for nonlinear processes.

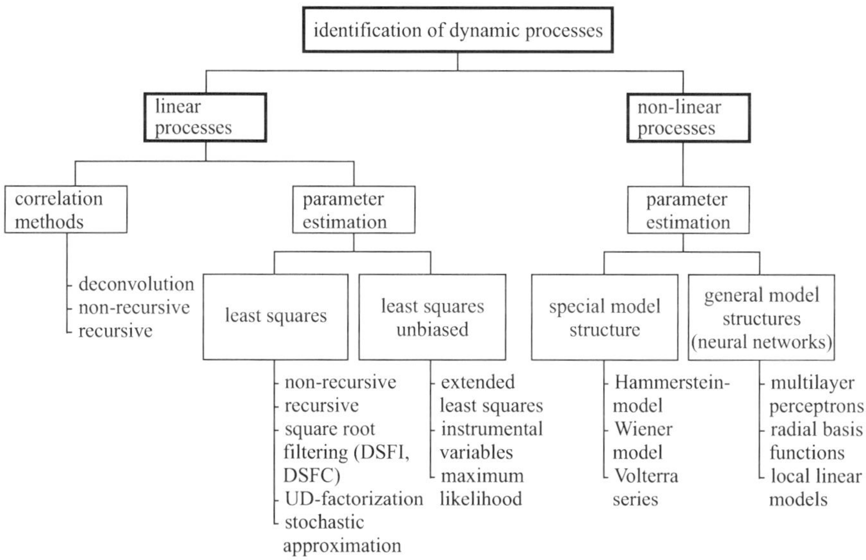

Fig. 3.2.1. Survey of some basic identification methods.

3.2.1 Correlation functions

If stationary stochastic or periodic signals act on a linear process in open loop, then the impulse response $g(\nu)$ can be determined if the autocorrelation function of the input signal and the cross-correlation function of the input and output signal are known. These correlation techniques are basic methods for the identification of linear dynamic processes and are elements of other identification methods, such as parameter estimations.

The autocorrelation function (ACF) is defined by

$$\Phi_{uu}(\tau) = E\{u(k)u(k+\tau)\} = \lim_{N\to\infty} \frac{1}{N} \sum_{k=0}^{N-1} u(k)u(k+\tau) \tag{3.2.1}$$

where $k = t/T_0$ is the discrete time with T_0, the sampling time. For finite samples of measured signals an estimate is given by

$$\hat{\Phi}_{uu}(\tau) = \frac{1}{N}\sum_{k=0}^{N-1} u(k)u(k+\tau). \tag{3.2.2}$$

For the cross-correlation function (CCF)

$$\Phi_{uy}(\tau) = E\{u(k)y(k+\tau)\} = \lim_{N\to\infty}\frac{1}{N}\sum_{k=0}^{N-1} u(k)y(k+\tau) \tag{3.2.3}$$

the estimate is

$$\hat{\Phi}_{uy}(\tau) = \frac{1}{N}\sum_{k=0}^{N-1} u(k)y(k+\tau). \tag{3.2.4}$$

Writing the CCF up to time points N and $N-1$ and subtracting both equations, one obtains the recursive form

$$\begin{array}{ccccccccc} \hat{\Phi}_{uy}(\tau,N) & = & \hat{\Phi}_{uy}(\tau,N-1) & + & \frac{1}{N+1} & [u(k-\tau)y(k) & - & \hat{\Phi}_{uy}(\tau,N-1)]. \\ \text{new} & = & \text{old} & + & \text{correction} & \text{new} & - & \text{old} \\ \text{estimate} & & \text{estimate} & & \text{factor} & \text{product} & & \text{estimate} \end{array} \tag{3.2.5}$$

For finite N the correlation function estimates contain a bias. However, this bias vanishes as $N \to \infty$. Hence, the estimates yield consistent estimates. Because also the variance converges to zero, the estimates are consistent in mean square, see for example, Isermann (1992), Isermann and Münchhof (2011).

If $E\{u(k)\} = 0$ and $E\{y(k)\} = 0$, the correlation functions for the input and output signals of a linear process are related by the convolution sum

$$\Phi_{uy}(\tau) = \sum_{\nu=0}^{\infty} g(\nu)\Phi_{uu}(\tau-\nu) \tag{3.2.6}$$

If $l+1$ values of $g(\nu)$ have to be determined and the convolution sum is truncated for $\nu > l$, then

$$\Phi_{uy}(\tau) \approx \mathbf{\Phi}_{uu}^{\mathrm{T}}\mathbf{g} \tag{3.2.7}$$

where

$$\begin{aligned} \mathbf{\Phi}_{uu}^{\mathrm{T}} &= [\Phi_{uu}(\tau)\Phi_{uu}(\tau-1)\ldots\Phi_{uu}(\tau-l)] \\ \mathbf{g}^{\mathrm{T}} &= [g(0)\; g(1)\ldots g(l)]\,. \end{aligned}$$

Now $l+1$ equations are required to get a unique solution. Therefore τ is varied within

$$-P < \tau < -P + 2l$$

and the equation system becomes

$$\mathbf{\Phi}_{uy} \approx \hat{\mathbf{\Phi}}_{uu}\cdot\hat{\mathbf{g}}. \tag{3.2.8}$$

As $\hat{\mathbf{\Phi}}_{uu}$ is a $(l+1) \times (l+1)$ square matrix the impulse response estimates result from the deconvolution equation

$$\hat{\mathbf{g}} \approx \hat{\mathbf{\Phi}}_{uu}^{-1} \mathbf{\Phi}_{uy} \tag{3.2.9}$$

where, of course, $\hat{\mathbf{\Phi}}_{uu}$ can be inverted only if

$$\det \hat{\mathbf{\Phi}}_{uu} \neq 0 \tag{3.2.10}$$

which is an identifiability condition. In other words, the process must be persistently excited (of order $l+1$).

If the input signal $u(k)$ is white noise with ACF

$$\Phi_{uu}(\tau) = \sigma_u^2 \, \delta(\tau) = \Phi_{uu}(0)\delta(\tau)$$

if follows from (3.2.6) that

$$\hat{g}(\tau) = \frac{1}{\hat{\Phi}_{uu}(0)} \hat{\Phi}_{uu}(\tau)$$

The impulse response is then proportional to the CCF. The impulse response is a non-parametric model. A considerable advantage of the correlation method is that no model structure has to be assumed. But convergence is usually slow.

If stochastic, stationary noise $n(k)$ acts on the process output, the necessary conditions for the consistent estimation of $\hat{g}(\tau)$ in mean square are the following:

- $n(k)$ and $y(k)$ are stationary
- $E\{u(k)\} = 0$
- $u(k)$ is persistently exciting
- $n(k)$ is not correlated with $u(k)$.

For more details see Eykhoff (1974), Isermann and Münchhof (2011). Correlation functions are also implicitly included in least squares parameter estimation with parametric models, Isermann (1992).

3.2.2 Parameter estimation for linear dynamic processes (method of least squares)

It is assumed that the process can be described by the linear difference equation

$$\begin{aligned} y_u(k) + a_1 y_u(k-1) + \ldots + a_m y_u(k-m) \\ = b_1 u(k-d-1) + \ldots + b_m u(k-d-m). \end{aligned} \tag{3.2.11}$$

Here,

$$\begin{aligned} u(k) &= U(k) - U_{00} \\ y_u(k) &= Y_u(k) - Y_{00} \end{aligned} \tag{3.2.12}$$

are the deviations of the absolute signals $U(k)$ and $Y_u(k)$ from the operating point described by U_{00} and Y_{00}, k is the discrete time $k = t/T_0 = 0, 1, 2, \ldots, T_0$ is the sampling time and $d = T_t/T_0 = 0, 1, 2, \ldots$ is the discrete dead-time of the process. The corresponding transfer function in the z-domain is

$$G_p(z) = \frac{y_u(z)}{u(z)} = \frac{B(z^{-1})}{A(z^{-1})}\, z^{-d} = \frac{b_1\, z^{-1} + \ldots + b_m z^{-m}}{1 + a_1\, z^{-1} + \ldots + a_m z^{-m}} z^{-d}. \tag{3.2.13}$$

The measured signal contains a stationary, stochastic disturbance

$$y(k) = y_u\,(k) + n\,(k) \;\text{ with }\; E\,\{n(k)\} = 0. \tag{3.2.14}$$

The task is to determine the unknown parameters a_i and b_i from N measured input and output signal data points, see Fig. 3.2.2.

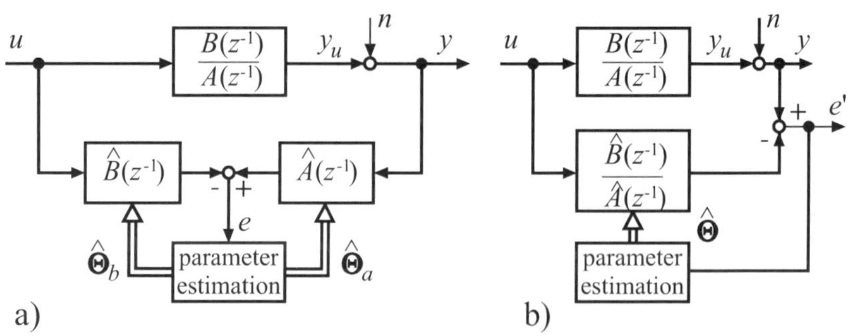

Fig. 3.2.2. Model structures for parameter estimation. **a** equation error. **b** output error.

a) Equation error methods

Let the model parameters obtained from the data up to the sample $(k-1)$ be denoted by $\hat{a}_i$ and $\hat{b}_i$. Then, (3.2.11) becomes in the presence of a disturbed output signal

$$\begin{aligned} y(k) + \hat{a}_1\, y(k-1) \ldots + \hat{a}_m\, y(k-m) \\ -\hat{b}_1\, u(k-d-1) - \ldots - \hat{b}_m u(k-d-m) = e(k) \end{aligned} \tag{3.2.15}$$

where the *equation error* (residual) $e(k)$ is introduced instead of "0" and corresponds to a generalized error, see Fig. 3.2.2. This can be seen by rewriting (3.2.15), compare Fig. 3.2.2a).

$$\hat{A}\,(z^{-1})y(z) - \hat{B}(z^{-1})\, z^{-d}\, u(z) = e\,(z). \tag{3.2.16}$$

e is linearly dependent on the parameters to be estimated (linear in the parameters).

From (3.2.15), $\hat{y}(k|k-1)$ can be interpreted as the one-step-ahead prediction, based on the measurements up to sample $(k-1)$

$$\hat{y}(k|k-1) = \boldsymbol{\psi}^{\mathrm{T}}(k)\hat{\boldsymbol{\theta}} \tag{3.2.17}$$

with the data vector

$$\begin{aligned}\boldsymbol{\psi}^{\mathrm{T}}(k) = [-y\,(k-1)\ldots - y\,(k-m)|\;u(k-d-1)\ldots\\ u(k-d-m)]\end{aligned} \tag{3.2.18}$$

and the parameter vector

$$\hat{\boldsymbol{\theta}} = [\hat{a}_1 \ldots \hat{a}_m \mid \hat{b}_1 \ldots \hat{b}_m]^{\mathrm{T}}. \tag{3.2.19}$$

Consequently, (3.2.15) can be written as

$$y(k) = \boldsymbol{\psi}^{\mathrm{T}}(k)\hat{\boldsymbol{\theta}} + e(k). \tag{3.2.20}$$

The measured signals for $k = m + d, \ldots, m + d + N$ are written in vectors, e.g.

$$\mathbf{y}^{\mathrm{T}}(m+d+N) = [y(m+d)\ldots y(m+d+N)] \tag{3.2.21}$$

Then,

$$y(m+d+n) = \boldsymbol{\Psi}(m+d+N)\hat{\boldsymbol{\theta}} + e(m+d+N) \tag{3.2.22}$$

where $\boldsymbol{\Psi}$ is a $((N+1)\times 2m)$-data matrix. Minimizing the sum of errors squared

$$V = \sum_{k=m+d}^{m+d+N} e^2\,(k) = \mathbf{e}^{\mathrm{T}}(m+d+N)\,\mathbf{e}\,(m+d+N) \tag{3.2.23}$$

yields

$$\left.\frac{dV}{d\boldsymbol{\theta}}\right|_{\boldsymbol{\theta}=\hat{\boldsymbol{\theta}}} = -2\boldsymbol{\Psi}^{\mathrm{T}}[\mathbf{y} - \boldsymbol{\Psi}\hat{\boldsymbol{\theta}}] = \mathbf{0} \tag{3.2.24}$$

for the unknown parameters. From this, the (non-recursive) estimation equation of the least squares (LS) method follows

$$\hat{\boldsymbol{\theta}} = [\boldsymbol{\Psi}^{\mathrm{T}}\boldsymbol{\Psi}]^{-1}\boldsymbol{\Psi}^{\mathrm{T}}\mathbf{y}. \tag{3.2.25}$$

The matrix

$$\mathbf{P} = [\boldsymbol{\Psi}^{\mathrm{T}}\;\boldsymbol{\Psi}]^{-1} \tag{3.2.26}$$

has the dimension (2m, 2m). The inverse exists if and only if

$$\det\,[\boldsymbol{\Psi}^{\mathrm{T}}\;\boldsymbol{\Psi}] = \det \mathbf{P}^{-1} \neq 0. \tag{3.2.27}$$

Also

$$\frac{\partial^2\,V}{\partial\,\boldsymbol{\theta}\,\partial\,\boldsymbol{\theta}^{\mathrm{T}}} = \boldsymbol{\Psi}^{\mathrm{T}}\;\boldsymbol{\Psi} \tag{3.2.28}$$

has to be positive-definite such that the loss function V has a minimum. Both requirements are satisfied if and only if

$$\det\left[\boldsymbol{\Psi}^{\mathrm{T}}\,\boldsymbol{\Psi}\right] = \det \mathbf{P}^{-1} > 0. \tag{3.2.29}$$

This condition also includes that the input signal is persistently exciting the process and that the process is stable.

From parameter-estimation methods, it is usually required that the estimate is not biased for a finite number of data samples N

$$E\left\{\boldsymbol{\theta}\,(N)\right\} = \boldsymbol{\theta}_0. \tag{3.2.30}$$

($\boldsymbol{\theta}_0$ denotes the true parameters) and is consistent in the quadratic mean

$$\lim_{N\to\infty}\; E\{\hat{\boldsymbol{\theta}}\,(N)\} = \boldsymbol{\theta}_0 \tag{3.2.31}$$

$$\lim_{N\to\infty}\; E\,[\hat{\boldsymbol{\theta}}\,(N) - \boldsymbol{\theta}_0]\;[\boldsymbol{\theta}\,(N) - \boldsymbol{\theta}_0]^{\mathrm{T}} = \mathbf{0}. \tag{3.2.32}$$

For the least squares method (3.2.31) becomes by substituting (3.2.22) into (3.2.25)

$$\begin{aligned} E\left\{\hat{\boldsymbol{\theta}}\,(N)\right\} &= \boldsymbol{\theta}_0 + E\,\left\{[\boldsymbol{\Psi}^{\mathrm{T}}\,\boldsymbol{\Psi}]^{-1}\boldsymbol{\Psi}^{\mathrm{T}}\mathbf{e}\right\} \\ &= \boldsymbol{\theta}_0 + \mathbf{b}. \end{aligned} \tag{3.2.33}$$

In order to have a vanishing bias (systematic estimation error) $\mathbf{b}$, $\boldsymbol{\Psi}^{\mathrm{T}}$ and $\mathbf{e}$ must be uncorrelated. Consequently, $e(k)$ must not be correlated and $E\left\{e(k)\right\} = 0$. The estimation is unbiased if the disturbance signal $n(k)$ is generated by the disturbance filter

$$G_v(z) = \frac{n(z)}{v(z)} = \frac{1}{A(z^{-1})} \tag{3.2.34}$$

where $v(k)$ is discrete white noise, see Fig. 3.2.3. Since this filter does not exist in practice, the least squares estimation, in general, yields biased estimates. These systematic estimation errors are the larger the greater the variance σ_n^2 of the disturbance signal is compared to the output signal σ_{yu}^2.

For the covariance matrix, the following is true if $\hat{\boldsymbol{\theta}} = \boldsymbol{\theta}_0$ (which means $e = 0$)

$$\begin{aligned} \operatorname{cov}\left[\Delta\,\boldsymbol{\theta}\right] &= E\left\{\left[\hat{\boldsymbol{\theta}} - \boldsymbol{\theta}_0\right]\,\left[\hat{\boldsymbol{\theta}} - \boldsymbol{\theta}_0\right]^{\mathrm{T}}\right\}\;\sigma_e^2\;E\;\{\mathbf{P}\} \\ &= \sigma_e^2\;E\left\{\left[\frac{1}{N+1}\boldsymbol{\Psi}^{\mathrm{T}}\;\boldsymbol{\Psi}\right]^{-1}\right\}\frac{1}{N+1} \\ &= \sigma_e^2\left\{\boldsymbol{\Phi}^{-1}(N+1)\right\}\frac{1}{N+1}. \end{aligned} \tag{3.2.35}$$

σ_e^2 is the variance of $e(k)$. $\boldsymbol{\Psi}$ is a matrix whose elements are correlation functions. For $N \to \infty$, (3.2.32) is satisfied. $E\left\{\mathbf{P}\right\}$ is proportional to the covariance matrix of the parameter estimation errors.

Because of the biased estimates for the least squares algorithm, this method can only be used for processes with no or only small disturbance signals. A big advantage of the least squares algorithm, however, is that the parameter vector $\hat{\boldsymbol{\theta}}$ can be determined in one batch calculation and no iterative methods are necessary. This is possible since the error measure employed is linear in the parameters.

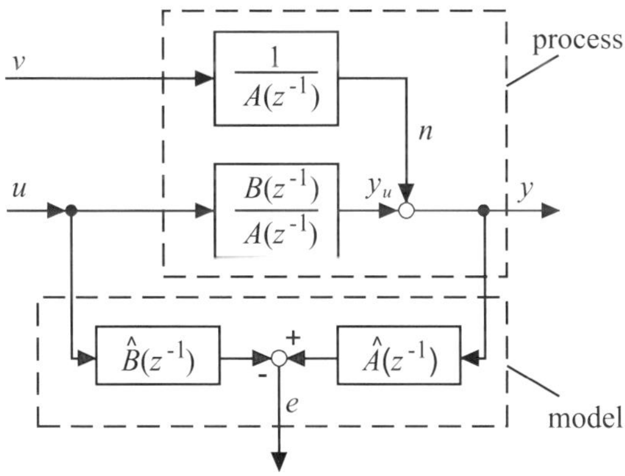

Fig. 3.2.3. Model configuration for the least-squares method with equation error (generalized error).

b) Output error methods

Instead of the equation error the output error

$$e'(k) = y(k) - y_M\left(\hat{\boldsymbol{\theta}}, k\right) \tag{3.2.36}$$

can be used, where

$$y_M\left(\hat{\boldsymbol{\theta}}, z\right) = \frac{\hat{B}(z^{-1})}{\hat{A}(z^{-1})}\, u(z) \tag{3.2.37}$$

is the model equation output, see Fig. 3.2.2b). But then no direct calculation of the parameter estimates $\hat{\boldsymbol{\theta}}$ is possible because $e'(k)$ is nonlinear in the parameters. Therefore the loss function (3.2.23) is minimized by a numerical optimization method, e.g. downhill-simplex. The computational effort is then larger, and online real-time application, in general, not possible. However, relative precise parameter estimates may be obtained, Drewelow (1990), Pfeufer (1999), Nelles (2001).

c) Recursive least squares (RLS) methods

Writing the nonrecursive estimation equations for $\hat{\boldsymbol{\theta}}(k+1)$ and $\hat{\boldsymbol{\theta}}(k)$ and subtracting one from the other, results in the recursive parameter estimation algorithm

$$\underset{\substack{\text{new}\\ \text{estimate}}}{\hat{\boldsymbol{\theta}}(k+1)} = \underset{\substack{\text{old}\\ \text{estimate}}}{\hat{\boldsymbol{\theta}}(k)} + \underset{\substack{\text{correction}\\ \text{vector}}}{\boldsymbol{\gamma}(k)} \Big[\underset{\substack{\text{new}\\ \text{measurement}}}{y(k+1)} - \underset{\substack{\text{one - step - ahead}\\ \text{prediction of the new}\\ \text{measurement}}}{\boldsymbol{\psi}^{\mathrm{T}}(k+1)\hat{\boldsymbol{\theta}}(k)}\Big]. \tag{3.2.38}$$

The correcting vector is given by

$$\begin{aligned}\gamma(k) &= \mathbf{P}(k+1)\,\boldsymbol{\psi}(k+1)\\ &= \frac{1}{\boldsymbol{\psi}^{\mathrm{T}}(k+1)\,\mathbf{P}(k)\,\boldsymbol{\psi}(k+1)+1}\,\mathbf{P}(k)\,\boldsymbol{\psi}(k+1)\end{aligned} \tag{3.2.39}$$

and

$$\mathbf{P}(k+1) = [\mathbf{I} - \gamma(k)\,\boldsymbol{\psi}^{\mathrm{T}}(k+1)]\,\mathbf{P}(k). \tag{3.2.40}$$

To start the recursive algorithm one sets

$$\begin{aligned}\hat{\boldsymbol{\theta}}(0) &= \mathbf{0}\\ \mathbf{P}(0) &= \alpha\mathbf{I}\end{aligned} \tag{3.2.41}$$

with α large ($\alpha = 100, \ldots, 1000$). The expectation of the matrix $\mathbf{P}$ is proportional to the covariance matrix of the parameter estimates

$$E\{\mathbf{P}(k+1)\} = \frac{1}{\sigma_e^2}\,\mathrm{cov}\,[\Delta\boldsymbol{\theta}(k)] \tag{3.2.42}$$

with

$$\sigma_e^2 = E\{\mathbf{e}^{\mathrm{T}}\mathbf{e}\} \tag{3.2.43}$$

and the parameter error

$$\Delta\boldsymbol{\theta}(k) = \hat{\boldsymbol{\theta}}(k) - \boldsymbol{\theta}_0. \tag{3.2.44}$$

Hence, the recursive algorithm contains the variances of the parameter estimates (diagonal elements of covariance matrix). (3.2.38) can also be written as

$$\hat{\boldsymbol{\theta}}(k+1) = \hat{\boldsymbol{\theta}}(k) + \gamma(k)\,e(k+1). \tag{3.2.45}$$

To improve the numerical properties of the basic RLS algorithms, modified versions are recommended, see Isermann and Münchhof (2011).

d) DC value estimation

As for process parameter estimation the variations of $u(k)$ and $y(k)$ of the measured signals $U(k)$ and $Y(k)$ have to be used, the DC (direct current or steady-state) values U_{00} and Y_{00} either have also to be estimated or have to be removed. Corresponding methods like differencing, averaging or estimation are described in Isermann (1992), Isermann and Münchhof (2011).

e) Extended and modified LS methods

In order to obtain unbiased parameter estimates *extended least squares methods* (ELS) exist which include parameter estimates of the disturbance model. Further the numerical properties can be improved considerably by factorization methods of the covariance matrix $\mathbf{P}$ or $\mathbf{P}^{-1}$, leading to, e.g *square root filtering* in covariance form (DSFC) or in information form (DSFI). For *time-varying processes* the RLS method can be modified by a time-depending weighting of the squared errors. The selection of a forgetting factor then determines the adaption speed to new parameters on the cost of their variance. The LS method can also be developed for *continuous-time models*. The application of the LS methods in *closed loop* requires either an external perturbation signal or special feedback controllers. For a description of these methods see, e.g. Isermann and Münchhof (2011).

3.2.3 Parameter estimation for nonlinear dynamic processes

Many processes and also engines have a nonlinear static and dynamic behavior, especially if wide range of operations are considered. One possibility is to use classical nonlinear models in combination with parameter estimation. Another method is to extend neural networks to nonlinear dynamic processes. Both ways are described shortly, assuming that the nonlinearities are continuously differentiable.

a) Parameter estimation with classical nonlinear dynamic models

Some classical nonlinear models result from expanding a linear difference equation

$$\begin{aligned} y_u(k) + a_1 y_u(k-1) + \ldots + a_m y_u(k-m) \\ = b_0 u(k) + b_1 u(k-1) + \ldots + b_m u(k-m) \end{aligned} \tag{3.2.46}$$

which can be brought to an abbreviated notation with the shift operator q^{-i}, where $q^{-i}y(k) = y(k-i)$.

$$A\left(q^{-1}\right) y(k) = B\left(q^{-1}\right) u(k). \tag{3.2.47}$$

If $A\left(q^{-1}\right)$ as well as $B\left(q^{-1}\right)$ is expanded with nonlinear terms like $u^2(k)$, $y^2(k)$, $u(k)u(k-1)$, etc., then *Kolmogorov-Gabor polynomial models* result which are, however, too complicated to apply. Volterra-series models follow without output feedback, i.e. $A\left(q^{-1}\right) = 1$. A simpler model is the *parametric Volterra model*

$$\begin{aligned} y_u(k) + a_1 y_u(k-1) + \ldots &= b_{10} u(k) + b_{12} u(k-1) + \ldots \\ &+ b_{20} u^2(k) + b_{21} u^2(k-1) + \ldots \\ &+ b_{201} u(k) u(k-1) + b_{202} u(k) u(k-2) + \ldots \end{aligned} \tag{3.2.48}$$

It has a linear feedback but ends up with many parameters and combinatorial complexity. Through specialization to a nonlinear static polynomial

$$x(k) = r_0 + r_1 u + r_2 u^2 + \ldots$$

followed by a dynamic linear model

$$y(k) = -a_1 y(k-1) + b_0' x(k) + b_1' x(k-1) + \ldots$$

a *Hammerstein model* results

$$\begin{aligned} y_u(k) + a_1 y_u(k-1) + \ldots = \\ b_{00} + b_{10} u(k) + b_{11} u(k-1) + b_{20} u^2(k) + b_{21} u^2(k-1) + \ldots \end{aligned} \tag{3.2.49}$$

which can be seen as a simplified parametric Volterra model. If a linear model is followed by a nonlinear static polynomial, a *Wiener model* is obtained which becomes, however, nonlinear in the parameters.

The Hammerstein model is suitable for nonlinear valve characteristics, the Wiener model for nonlinear sensor characteristics if the process can be assumed to be linear. For more details see Nelles (2001).

In the case of an equation error parameter estimation, the parametric Volterra model and the Hammerstein model have the advantage of being linear in the parameters. Therefore linear parameter estimation methods like LS, RLS and RELS can be applied.

b) Artificial neural networks for nonlinear dynamic processes

The memoryless static networks can be extended with dynamic elements to dynamic neural networks. One can distinguish between neural networks with external and internal dynamics, Nelles et al (1991), Isermann et al (1997), Fink (2006). ANNs with external dynamics are based on static networks, e.g. MLP or RBF networks. The discrete time input signals $u(k)$ are passed to the network through additional filters $F_i(q-1)$. In the same way, either the measured output signals $y(k)$ or the NN outputs $\hat{y}(k)$ are passed to the network through filters $G_i(q-1)$. The operator q^{-1} denotes a time shift

$$y(k) \cdot q^{-1} = y(k-1). \tag{3.2.50}$$

In the simplest case, the filters are pure time delays, Fig. 3.2.4a)

$$\hat{y}(k) = f_{NN}\ [u(k), u(k-1), \ldots, \hat{y}(k-1), \hat{y}(k-2), \ldots] \tag{3.2.51}$$

where the time-shifted sampled values are the network input signals. The structure in Fig. 3.2.4a) shows a parallel model (equivalent to the output error model for parameter estimation of linear models). In Fig. 3.2.4b), the measured output signal is passed to the network input. Then, the series-parallel model is obtained (equivalent to the equation error model for parameter estimation of linear models). One advantage of the external dynamic approach is the possibility of using the same adaptation methods as in the case of static networks. However, the drawbacks are the increased dimensionality of the input space, possible stability problems and an iterative way of computing the static model behavior, namely through simulation of the model. Then, for example, a step function is used as the input signal and one has to wait until the steady state of the model is reached.

ANNs with internal dynamics realize dynamic elements inside the model structure. According to the kind of included dynamic elements, one can distinguish between recurrent networks, partially recurrent networks and locally recurrent globally feedforward networks (LRGF), Nelles et al (1991), Nelles (2001). The LRGF networks maintain the structure of static networks except that dynamic neurons are utilized, see Fig. 3.2.5. The following can be distinguished: local synapse feedback, local activation feedback and local output feedback. The simplest case is the local activation feedback, Ayoubi (1996). Here, each neuron is extended by a linear transfer function, most often of first or second order.

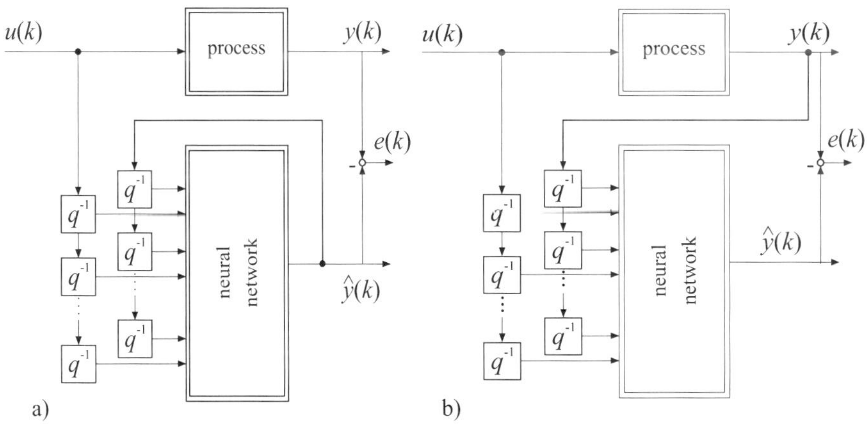

Fig. 3.2.4. Artificial neural network with external dynamics. **a** parallel model. **b** series-parallel model.

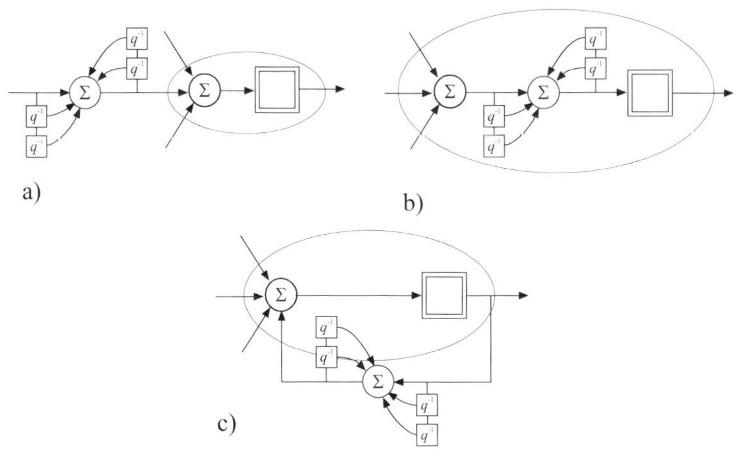

Fig. 3.2.5. Dynamic neurons for neural networks with internal dynamics. **a** local synapse feedback. **b** local activation feedback. **c** local output feedback.

Usually, MLP networks are used in LRGF structures. However, RBF networks with dynamic elements in the output layer can be applied as well, if a Hammerstein-structure of the process can be assumed, Ayoubi (1996). Usually, the adaptation of these dynamic NNs is based on extended gradient methods, Nelles et al (1991), Nelles (2001).

Based on the basic structure of ANNs, special structures with particular properties can be built. If, for example, the local linear model network (LOLIMOT) is combined with the external dynamic approach, a model structure with locally valid linear input/output models result, see Fig. 3.2.6. This method turned out to be especially suitable for combustion engines, Müller et al (2000), Hafner et al (2000b).

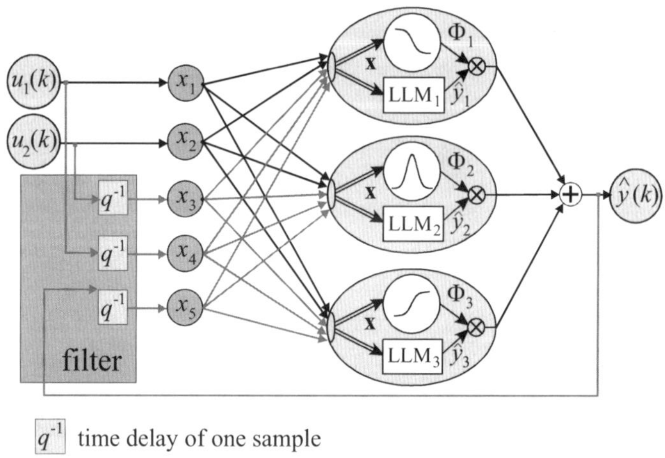

Fig. 3.2.6. Dynamic local linear model network (LOLIMOT) through expanding the structure of Fig. 3.1.12 with time shifted input and output signals.

c) Operating-point-dependent local models

Frequently, the static or dynamic behavior of processes and especially of combustion engines depends on the *operating point*, described by the variables $\mathbf{z}$. Then all the inputs have to be separated into manipulated variables $\mathbf{u}$ and operating point variables $\mathbf{z}$. By this separation local linear models can be identified with varying parameters depending on the operating point, also called *linear parameter variable models* (LPVM), Ballé (1998).

A nonlinear discrete-time dynamic model with p inputs u_i and one output y can be described by

$$y(k) = f(\mathbf{x}(k)) \tag{3.2.52}$$

with

$$\begin{aligned}\mathbf{x}^{\mathrm{T}}(k) = [&u_1(k-1), \cdots, u_1(k-n_{u1}), \cdots, u_p(k-1), \cdots, \\ &u_p(k-n_{up}), y(k-1), \cdots, y(k-n_y)]\,.\end{aligned} \tag{3.2.53}$$

For many types of nonlinearities this *nonlinear (global) overall model* can be represented as a *combination of locally active submodels*

$$\hat{y} = \sum_{i=1}^{M} \Phi_i(\mathbf{u})\, \hat{y}_i(\mathbf{u}), \tag{3.2.54}$$

see the LOLIMOT approach (3.1.23).

The validity of each submodel $\hat{y}_i$ is given by its corresponding weighting function Φ_i (also called activation or membership function). These weighting functions describe the partitioning of the input space and determine the transition between neighboring submodels as described in Sect. 3.1.3e).

According to (3.2.54) identical input spaces for the local submodels $\hat{y}_i(\mathbf{u})$ and the membership functions $\Phi_i(\mathbf{u})$ have been assumed. However, local models allow the realization of *distinct input spaces*, Fig. 3.2.7, with

$$y = \sum_{i=1}^{M} \Phi_i(\mathbf{z})\; y_i(\mathbf{x}) \tag{3.2.55}$$

if, e.g. $\mathbf{z}$ describes the operation point. The input vector $\mathbf{z}$ of the weighting functions comprises merely those inputs of the vector $\mathbf{u}$ having significant nonlinear dependencies which cannot be explained by the local submodels, e.g. for engines the torque M_{eng} and speed ω_{eng}. Especially those inputs require a subdivision into different parts. The decisive advantage of this procedure is the considerable reduction of the number of inputs in $\mathbf{z}$. Thus, the difficult task of structure identification can be simplified.

The use of separate input spaces for the local models (vector $\mathbf{x}$) and the membership functions (vector $\mathbf{z}$) for the operation point becomes more precise by considering another representation of the structure in (3.2.55). As normally local submodel approaches are assumed to being linear with reference to their parameters according to

$$y_i(\mathbf{x}) = w_{i,0} + w_{i,1}\, x_1 + \cdots + w_{in_x} x_{nx} \tag{3.2.56}$$

compare (3.1.25).(3.2.55) can be arranged to

$$\begin{aligned} y_i(\mathbf{x}) &= w_{i,0}(\mathbf{z}) + w_{i,1}(\mathbf{z})x_1 + \cdots + w_{i,n_x}(\mathbf{z})\; x_{nx} \\ \text{with } w_j(\mathbf{z}) &= \sum_{i=1}^{M} w_{ij}\; \Phi_i(\mathbf{z}). \end{aligned} \tag{3.2.57}$$

resulting in

$$y = w_0 + \sum_{j=1}^{n_x} w_j(\mathbf{z})x_j . \tag{3.2.58}$$

Thus, the specified local model approaches can be interpreted as linear-in-the-parameter relationships with *operating-point-dependent parameters* $w_j(\mathbf{z})$, whereupon these parameters depend on the input values in vector $\mathbf{z}$. Consequently, the process coefficients $w_j(\mathbf{z})$ have a physical meaning. Therefore, these models are a sort of *semi-physical models*, Töpfer et al (2002). This model structure as shown in Fig. 3.2.7 can favorably be applied for the identification of nonlinear processes with local linear models as LOLIMOT, where the nonlinear behavior is mainly determined by the operation point $\mathbf{z}$, which is for example the case in combustion engines

$$\mathbf{z}^{\mathrm{T}} = \left[M_{\text{eng}}\; \omega_{\text{eng}}\; T_{\text{oil}}\right] . \tag{3.2.59}$$

The choice of *approximate submodel structures* always requires a compromise between submodel complexity and the number of submodels. The most often applied linear submodels have the advantage of being a direct extension of the well known linear models. However, under certain conditions *more complex submodels* may be reasonable. If the main nonlinear influence of input variables can be described qualitatively by a nonlinear transformation of the input variables (e.g. $f_1(x) = (x_1^2,\ x_1 x_2, \ldots)$) based on physical knowledge, then the incorporation of that knowledge into the submodels leads to a considerable reduction of the required number of submodels, see the LOPOMOT method in Sect. 3.1.3. Generally, this approach can be realized by a pre-processing of the input variables $\mathbf{x}$ to the nonlinearly transformed variables, Fig. 3.2.8

$$\mathbf{x}^* = F(\mathbf{x}) = \left[f_1(\mathbf{x})\ f_2(\mathbf{x}) \cdots f_p(\mathbf{x})\right]^{\mathrm{T}} . \tag{3.2.60}$$

Besides those heuristically determined model structures, local model approaches also enable the incorporation of fully physically determined models. Furthermore, local models allow the employment of inhomogeneous models. Consequently, different local submodel structures may be valid within the different operating regimes.

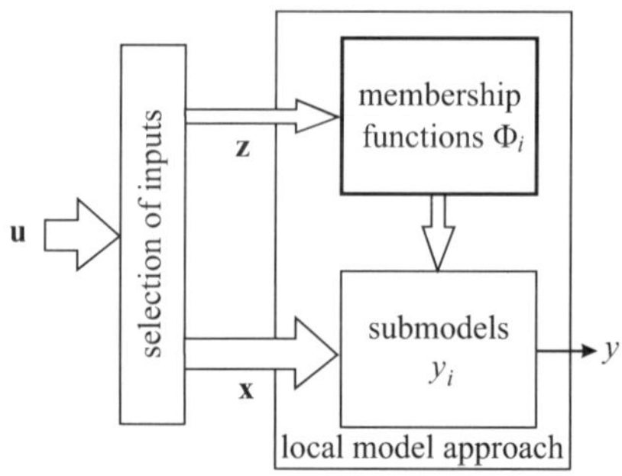

Fig. 3.2.7. Structure of local submodels with distinct inputs spaces $\mathbf{x}$ for local submodels and $\mathbf{z}$ for corresponding membership functions describing significant nonlinear effects. $\mathbf{z}$ may be the operation point.

3.3 Local and global engine models

The stationary engine behavior for a constant operation point

$$\mathbf{z}^T = \left[m_{\mathrm{f,g}}, n_{\mathrm{eng,h}}\right] \tag{3.3.1}$$

or

$$\mathbf{z}^T = \left[M_{\mathrm{eng,g}}, n_{\mathrm{eng,h}}\right] \tag{3.3.2}$$

is according to (3.1.3) and Fig. 3.1.3 for one output y_r and several inputs $\mathbf{u}^{\mathrm{T}} = [u_1, \ldots, u_p]$ represented by

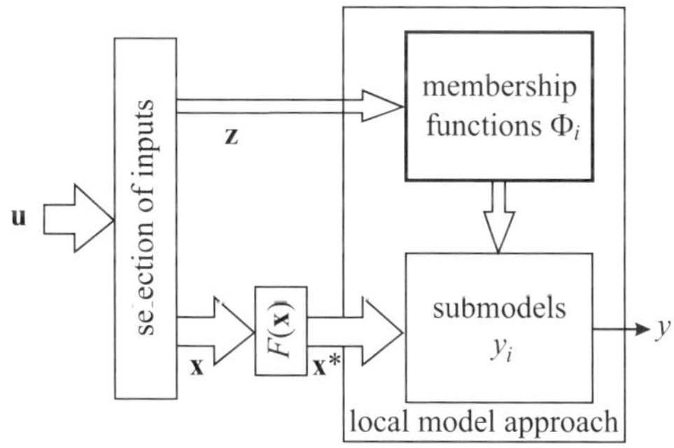

Fig. 3.2.8. Pre-processing of input variables **x** for incorporation of prior knowledge for nonlinear submodels.

$$y_r = f_{\text{loc}}(\mathbf{u}) \quad r = 1, \ldots, R. \tag{3.3.3}$$

The manipulated variables **u** then span all allowable values within the possible ranges of the actuators and the multi-input single-output model yields the corresponding output variables y_r. With regard to combustion engines this model can be linear or nonlinear and, e.g. be represented by a look-up table. Because this model is valid for a specific operating point it is called a *local stationary model*.

If for a fixed operation point **z** the input signals **u** are excited dynamically by test signals according to Fig. 3.5.1, *local dynamic models*

$$y_r(t) = f_{\text{loc}}(\mathbf{u}(t)) \tag{3.3.4}$$

are obtained which can be linear as (3.2.11) or nonlinear as expressed by (3.2.46) to (3.2.49).

The inclusion of different operation points in the model requires **z** as input. Then an expansion of (3.3.3) results in a *global stationary model*

$$y_r = f_{\text{glob}}(\mathbf{u}, \mathbf{z}). \tag{3.3.5}$$

Instead of several local stationary models one comprehensive overall model is obtained. The implementation of such a global engine model can be based on a composition of local models. Leaning on the representation of grid-based models, Sequenz (2013) proposes a global-local model structure for the stationary behavior

$$y_r = f_{\text{glob-loc}}(\mathbf{u}, \mathbf{z}) = \sum_{j=1}^{M} \phi_j(m_f, n_{\text{eng}}) f_{\text{loc}}(\mathbf{u}). \tag{3.3.6}$$

As shown in Fig. 3.3.1 the local models are composed by means of a bilinear interpolation to a global model output. If the model is evaluated at a grid point of the operating space, only one local model is active. If the output has to be evaluated between the grid points, the four surrounding models become part of the weighted interpolation sum, compare Sect. 6.5.1 and Fig. 3.3.1. This global stationary model can for example be applied for slowly changing operation points, describing then a quasi-stationary behavior.

Another case arises if the operating points $\mathbf{z}$ change fast with time, as required for real dynamic vehicle driving, where the injection mass m_{f} and the engine speed n_{eng} are functions of time. The variables describing the operating point then also become dynamic inputs, leading to a global dynamic model

$$y_r = f_{\mathrm{glob}}(\mathbf{u}(t), \mathbf{z}(t)). \tag{3.3.7}$$

Then, usually the inputs $\mathbf{u}$ also change with time.

Hence, local models are valid for selected operating points and global models include several interesting operating points. Both model types can be stationary or dynamic.

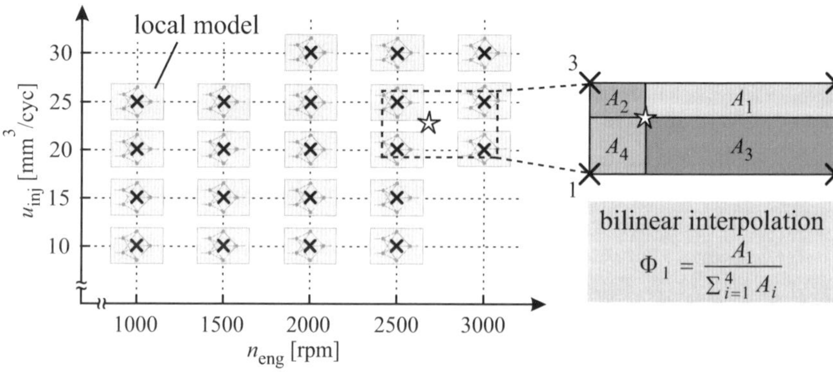

Fig. 3.3.1. Global-local stationary engine model structure. The local models are valid for fixed operating points. The global model output is determined by a bilinear area interpolation of the local models, Sequenz (2013).

3.4 Stationary identification procedures on test benches

3.4.1 Measurement procedures

A basic experimental modeling of engines requires an appropriate test bench and a computer-based development environment. Figure 3.4.1 presents a signal-flow scheme for an engine test bench with torque or speed control, via an electrical asynchronous motor (see Sect. 6.10), an electronic control unit (ECU) for the engine, a process computer for data treatment and storage and for providing reference inputs for the ECU and a bypass development system to replace or add new control functions. The used ECU may be programmed-based on a priori knowledge with similar engines.

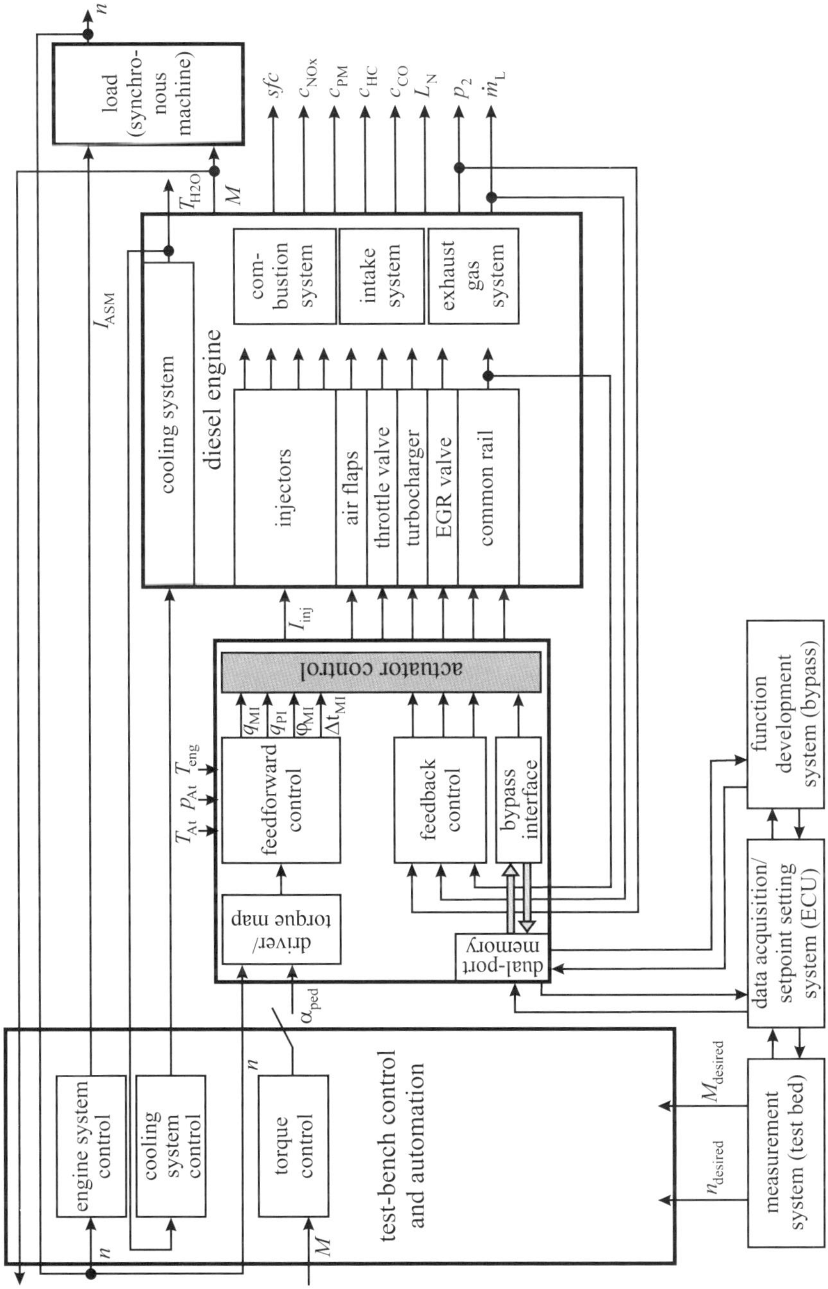

Fig. 3.4.1. Signal-flow scheme for an engine test bench with diesel engine, test-bench control, ECU input and bypass development control.

A general procedure for the experimental modeling (identification) of engines on test benches is given in Fig. 3.4.2. It begins with the *planning of the measurements*. Depending on the possibilities of the test bench, the measurement devices and the a priori knowledge of the engine, the operating points of the engine, the kind of control of the test bench (e.g. speed or torque control of the dynamometer) and stationary or dynamic measurements have to be specified. Then the determination of the *operation boundaries* has to be performed, in order to avoid damages and to include the relevant operation points. The *design* of the *test signals* for the stationary or dynamic inputs of the engine is a next step. The *measurements* for identification can then be started, taking into account the allowable operation space, many test bench specific requirements, the control of air and cooling water conditioning, and the applied method of identification. Many input and output signals are then obtained and stored. The *data* should then be *analyzed* before they are used for experimental modeling. For example, graphical tools are used to check for consistency of the data and to remove outliers or drift. The selected *identification* method can then be applied, resulting in static and dynamic engine models. This is followed by a *model analysis* with the goal of evaluating the resulting models. By applying different performance criteria the model structure can be selected and by using different plots and intersection presentations a first impression of the quality of the measured signals and the obtained models can be obtained. The influence of sensor dynamics on the models or their compensation may also be of interest. *Model validation* checks if the obtained models describe the real behavior of the engine, using e.g. performance criteria for the already measured or additional validation data. In the latter case, the input signals which were not used for the identification may be applied to a simulation with the identified models (generalization). If no extra generalization measurements are made, special resampling of data may be used. Finally, the determined engine models are used for applications, such as for the design of the engine, the design of control functions or diagnosis procedures.

Figure 3.4.3 presents a survey on measurement and identification methods for combustion engines on test benches. They can be divided in stationary, quasi-stationary and dynamic (transient) measurement. Different manipulation principles, model types and identification methods can then be selected resulting in the required mathematical engine models.

In the following, the steps for experimental modeling are described for the *stationary behavior* according to Fig. 3.4.2, taking into account the identification methods of Sect. 3.1. Sect. 3.5 then treats the experimental modeling for the *dynamic behavior*.

3.4.2 Measurement of the operation boundaries

Before systematic measurements are performed, the boundaries of the operation space (design space) for the manipulated variables have to be determined to avoid damage to the engine. The boundaries of several engine variables like peak pressure, emission concentrations, minimal and maximal speed etc. are usually known in

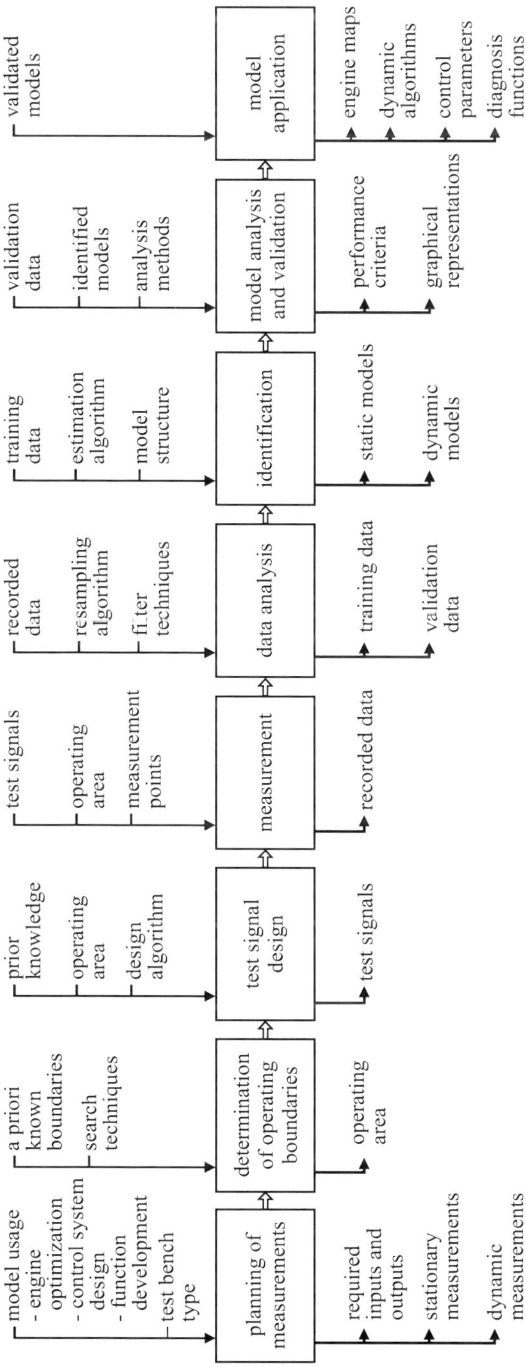

Fig. 3.4.2. Steps for the measurement and experimental modeling of engines.

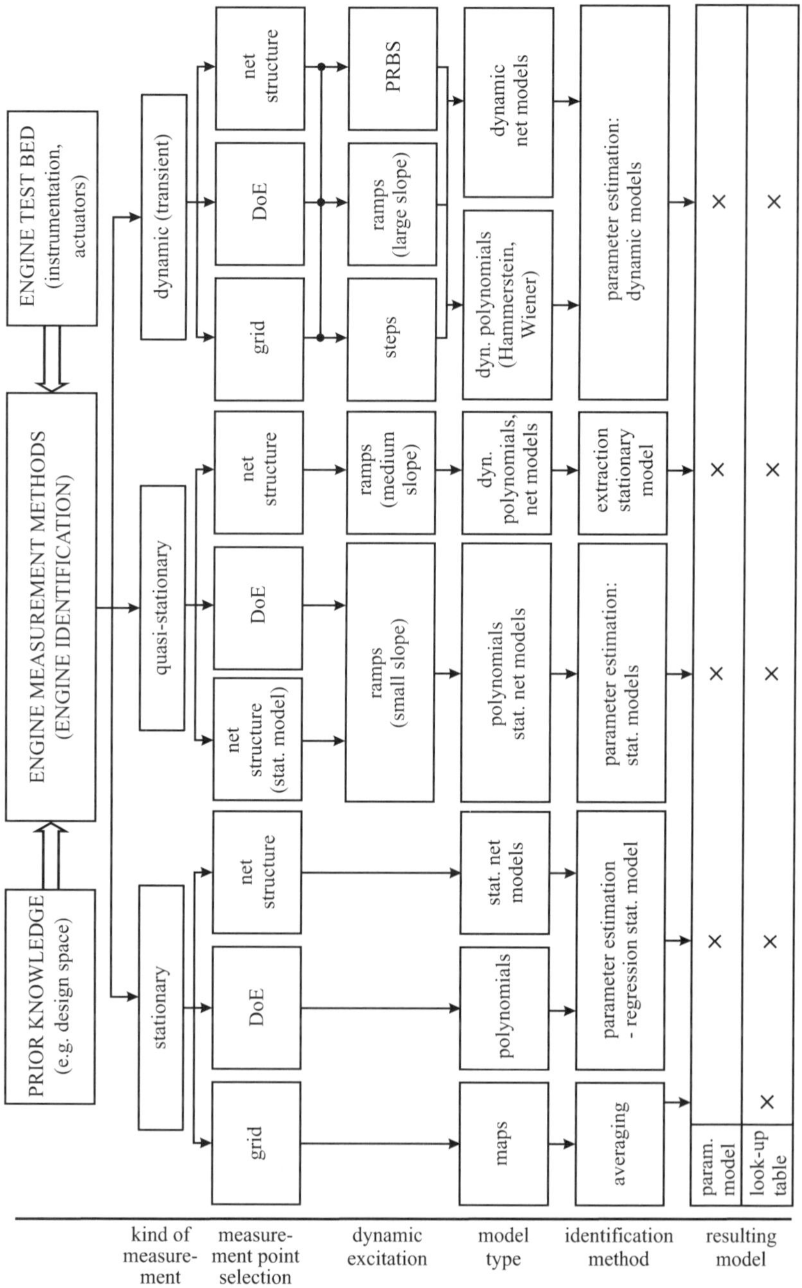

Fig. 3.4.3. Survey of different measurement and identification methods of engines on test benches.

advance, see Fig. 3.4.4. For the measurements, however, the boundaries of the corresponding manipulated variables like injection mass, ignition angle, EGR rate, etc. as indicated in Fig. 3.4.4, have to be determined by experimental search. One possibility is to start in the center of the operation space and to determine the various boundaries by changing the inputs stepwise in a star-like pattern, see Fig. 3.4.5a), Gschweitl et al (2001). However, with increasing manipulated variables this takes a long time. Another possibility is to use adaptive step sizes by predicting the boundaries with a polynomial of second order, Renninger (2004). After measurements in some few directions, the edges of the convex hull are connected and for the iterative measurements with adaptive step size they are made orthogonal to these connecting lines, see Figs. 3.4.5b) and 3.4.6. Thus, measurement time is saved.

After the determination of the operating boundaries the admissible input values and their combinations have to be determined for experimental modeling. The resulting *candidate set* should include as many input combinations as possible. A first possibility is a grid-wise selection within the boundaries, see Fig. 3.4.7a). However, depending on the form and dimension of the variation space only a few points may lie within the boundaries. This can be improved by a finer screening. But this leads to a strong increase of measurement points. An alternative is a spherical screening with n-dimensional polar coordinates, Amann and Escher (2008). Using varying radii an adaptive spherical screening can be realized, Schreiber and Isermann (2007), Schreiber and Isermann (2009). With the aid of discrete radii, candidate sets according to Fig. 3.4.7b) result, leading to a better coverage of the operation space. The adaptation to the boundaries may be based on the calculated volume, which decreases if points lie outside of the boundary.

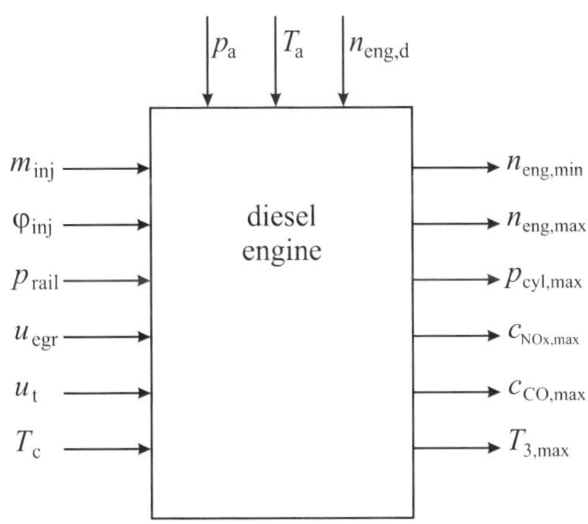

Fig. 3.4.4. Determination of the operation boundaries: limited output variables and corresponding manipulated variables (inputs) to be determined by search procedures.

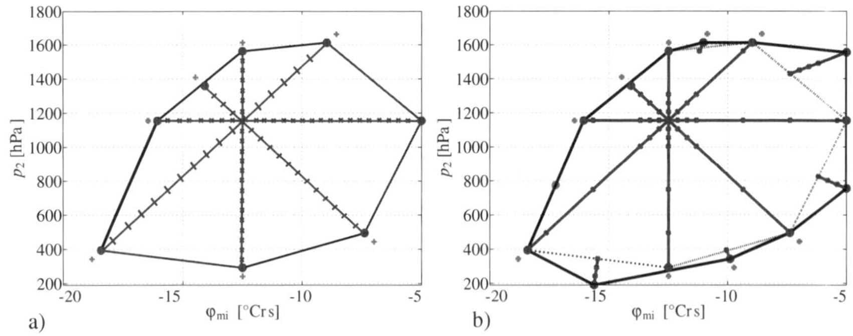

Fig. 3.4.5. Determination of operation space for the manipulated variables. **a** star-like procedure. **b** iterative-predictive procedure.

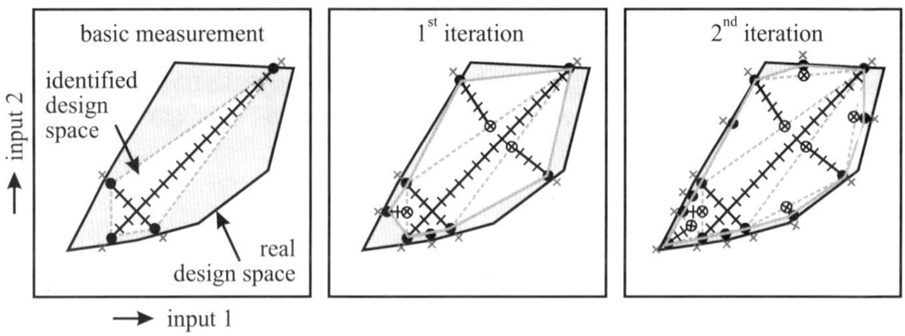

Fig. 3.4.6. Iterative-predictive determination of the operation space, Renninger (2004).

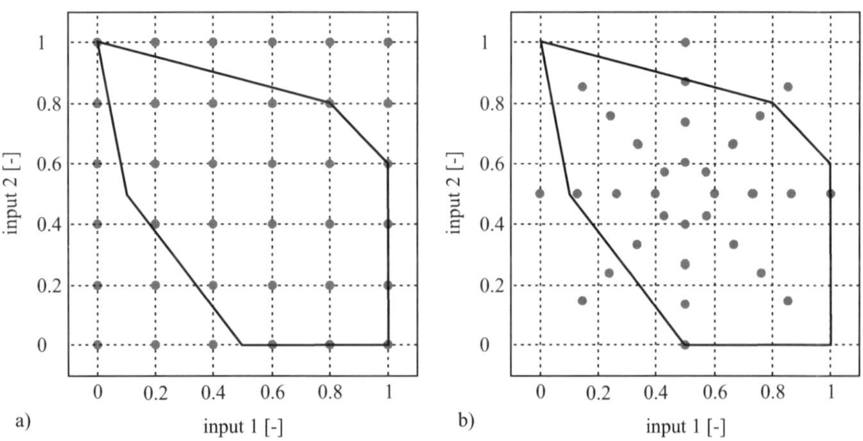

Fig. 3.4.7. Selection of measurement points within the operation boundaries. **a** grid-wise screening. **b** spherical screening.

3.4.3 Design of experiments with polynomial models

The classical way to determine the stationary behavior of engines is to apply grid measurement procedures. This means that one manipulated variable is changed stepwise with discrete values whereas the other variables are kept constant. Figure 3.4.8a) shows this for two variables to obtain a look-up table for one output. If n is the number of measurement points per variable and j the number of manipulated variables, then the resulting number of measurements becomes

$$N = n^j . \tag{3.4.1}$$

This means with $n = 4$ and $j = 3$ variables $N = 64$ measurements and for $j = 8$ variables $N = 65536$ measurements have to be made. Assuming a settling time of 2 min for passenger car engines this requires a measurement time of about 2 hours, which is acceptable, and 2184 hours or 91 days, which is no longer acceptable, see also Fig. 3.4.8b).

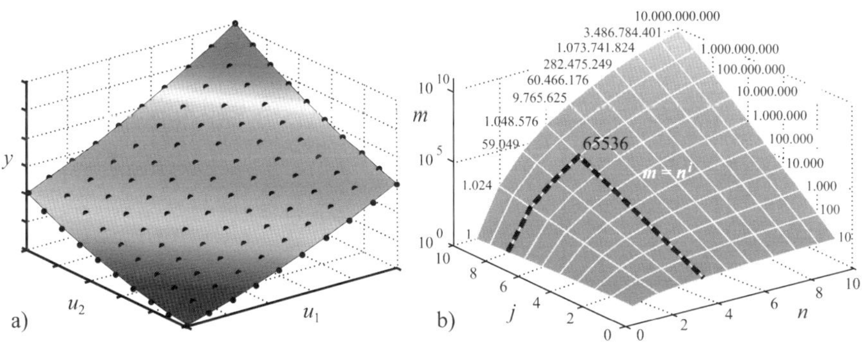

Fig. 3.4.8. Grid measurement for stationary behavior. **a** look-up table (map) $y = f(u_1, u_2)$. **b** number of measurements in dependence on discrete measurement points for each variable n and number of variables j.

To shorten the measurement time for stationary procedures the number of measurements should be as small as possible without losing too much information compared to grid measurement. This can be reached by selecting those measurements out of a candidate set which contribute most to the precision of an overall model, taking into account measurement disturbances. Statistical approaches of the so-called "Design of Experiments" (DoE) try to solve this task. In general, polynomials of type

$$y = c_0 + c_1 u_1 + c_2 u_2 + c_{11} u_1^2 + c_{22} u_2^2 + c_{12} u_1 u_2 + \ldots \tag{3.4.2}$$

are assumed. The goal is to determine the parameters $c_0, c_1, c_2, \ldots$ with as few measurement points as possible, Kleppmann (2003), Montgomery (2005). *Classical designs* place the measurement points in the input space according to fixed schemes, where the input space is often a hyper-cuboid. Figure 3.4.9 depicts some examples.

Full factorial designs place the measurements at the edges of a cube, central composite design add star points and a central point to the full factorial design.

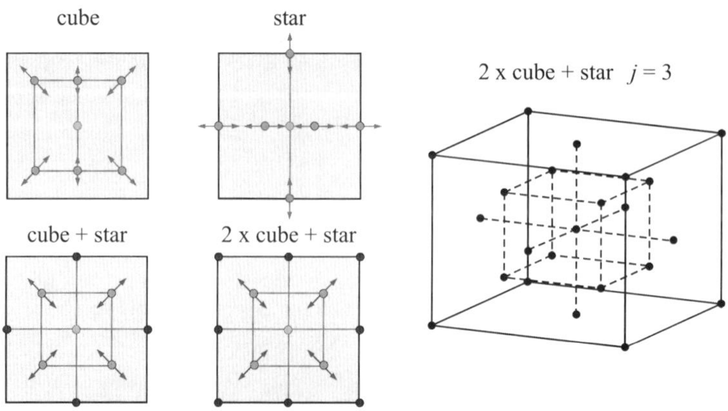

Fig. 3.4.9. Central composite design.

Optimal designs use optimization methods to determine the measurement points. Here the design space as well as the number of points is free. The least squares parameter estimation (LS) is now applied to (3.4.2), compare (3.1.11). Using a parameter vector

$$\boldsymbol{\theta}^{\mathrm{T}} = \left[c_0\ c_1\ c_2\ c_{11}\ c_{22}\ c_{12}\right] \tag{3.4.3}$$

the input data matrix

$$\mathbf{U} = \begin{bmatrix} 1 & u_1(1) & u_2(1) & u_1^2(1) & u_2^2(1) & u_1(1)u_2(1) \\ 1 & u_1(2) & u_2(2) & u_1^2(2) & u_2^2(2) & u_1(2)u_2(2) \\ \vdots & \vdots & & & & \vdots \\ 1 & u_1(N) & u_2(N) & u_1^2(N) & u_2^2(N) & u_1(N)u_2(N) \end{bmatrix} \tag{3.4.4}$$

and the output vector

$$\mathbf{y}^{\mathrm{T}} = \left[y(1)\ y(2)\ \ldots\ y(N)\right] \tag{3.4.5}$$

for N measurements of the inputs and outputs lead to the model

$$\mathbf{y} = \mathbf{U}\boldsymbol{\theta} \tag{3.4.6}$$

and the LS-solution

$$\hat{\boldsymbol{\theta}} = \left[\mathbf{U}^{\mathrm{T}}\mathbf{U}\right]^{-1}\mathbf{U}^{\mathrm{T}}\mathbf{y}. \tag{3.4.7}$$

The covariance matrix of the parameter estimates is

$$\mathrm{cov}(\boldsymbol{\theta}) = \sigma^2\left[\mathbf{U}^{\mathrm{T}}\mathbf{U}\right]^{-1} \tag{3.4.8}$$

where σ^2 is the variance of measurement noise of the output y. The covariances therefore depend on the noise which usually can be determined and the selection of the regressors in **U**. A D-optimal design now minimizes the determinant of the covariance matrix

$$\det\left[\mathbf{U}^{\mathrm{T}}\mathbf{U}\right]^{-1} \rightarrow \min$$

or maximizes

$$\det\left[\mathbf{U}^{\mathrm{T}}\mathbf{U}\right] \rightarrow \max \tag{3.4.9}$$

by selection of the measurement points, Mitchell (2000), Montgomery (2005). The maximization of the determinant is performed by a trial and error procedure, see e.g. Poland (2002) and Kötter (2007).

Figure 3.4.10 presents the results of optimal designs determined with the DETMAX algorithm from Mitchell (2000). It shows that the selection of the points depends on the assumed model and that the points are located in the outer regions of the design space. The selection of the regressors can be made by successively adding new regressors from a basic small set, and using a performance criterion, see e.g. Myers (1990), Kötter (2007).

The DoE method is a well-proven method for engine modeling and gives good results as long as the polynomials are able to approximate the behavior well. However, problems arise if the corresponding static maps are uneven and for large input spaces. The extrapolation behavior is also not good.

3.4.4 Reduced grid measurement

To obtain more measurements in the inner parts of the input space, methods of *space-filling designs* can be used. They try to cover the measurement space more equally. One method is the *latin hyper-cube sampling*, where the distances for a given number of measurement points are maximized, Santner et al (2003). However, because of the resulting random distribution this method is not recommended.

It is much simpler to generate reduced grid points, for example, five equally distributed points in a direction of stronger nonlinear behavior and four points in the direction of less nonlinear behavior. However, depending on the boundaries for grids with large grid intervals not considered areas may result. Therefore, a spherical screening, as shown in Fig. 3.4.7b), is a better alternative, Schreiber et al (2007).

A selection with regard to the operation point determined by torque and speed is frequently made based on the considered driving cycle as, e.g. FTP75 and NEDC for cars and light commercial vehicles. For trucks the measurement points for load and speed are already prescribed, see Fig. 3.4.11. Then the free manipulated variables have to be designed by other corresponding designs.

For the static measurement, the stationary behavior has to be always reached, e.g. by waiting for a steady-state which takes about two to three minutes. Usually a sequence follows in the direction of one grid line. However, to reach faster steady-states it is recommended to hold the load point (M_i, n_j) as long as possible and to change all manipulated variables first which may have a smaller thermal settling time as load changes.

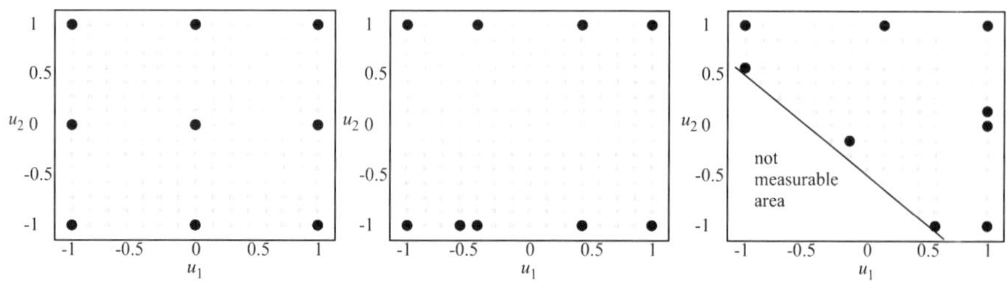

Fig. 3.4.10. D-optimal designs for different polynomial models. The gray points characterize the candidates. The black points are the selected optimal measurement points for the applied polynomials, Töpfer (2003).

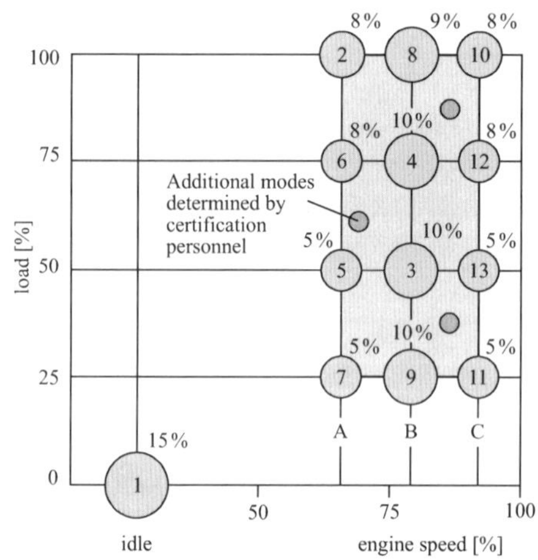

Fig. 3.4.11. Reduced screening based on driving cycles. (13-step test ESC for trucks. Numbers in % describe the distribution of steady-state operating times).

3.4.5 Neural net models for the static behavior

The neural networks (NN) described in Sect. 3.1.3 can be applied for modeling the stationary engine behavior, as for example multilayer perceptrons networks or radial basis function networks, see e.g. Kruse et al (2010), Kruse et al (2012). They also need an appropriate selection of the measurement points.

As local linear networks (LOLIMOT) are more transparent and easier to apply, e.g. for designing local linear control algorithms, this type of net models is considered in more detail. The distribution of the measurement points on the input space always needs some a priori information, which can be based on previous measurements or on a similar engine. Based on these data the structure of the net model is generated. In the case of LOLIMOT one then obtains the local linear areas automatically, as shown in Fig. 3.4.12. The regions with strong nonlinear behavior then are approximated with a larger number of local models. In order to obtain enough measurement points for each local area, a central composite design can then be used in each local model and the resulting measurement points are combined to one global design, see Fig. 3.4.12, Weber et al (2005), Zimmerschied et al (2005). Some of the points may be beyond the operation boundaries and have to be moved inside the allowable operation area. Compared to classical grid-based measurement, a considerable saving of measurement points can be reached, e.g. a reduction of the measurement time to 8 %, Martini et al (2003).

The design of the measurement points is made for each single output, if multi-input single-output (MISO) models are considered. For an overall experiment design then several measurements are put together and points with narrow distance can then be integrated to unique points, Hafner et al (2000a).

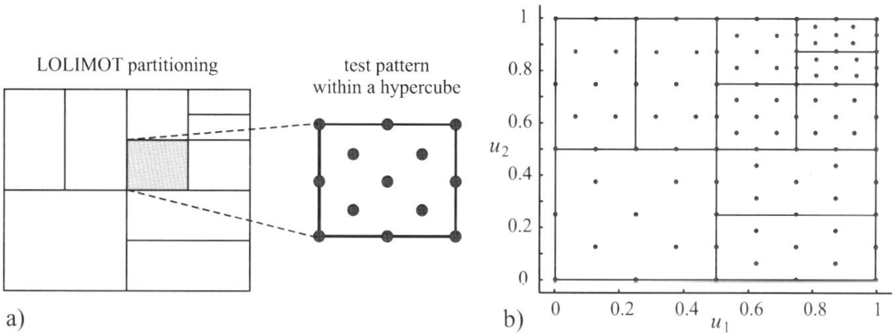

Fig. 3.4.12. Partitioning of the input space with LOLIMOT and central composite design. **a** LOLIMOT partitioning. **b** composite-designed measurement points.

3.4.6 Quasi-stationary ramp measurement (sweep mapping)

The classical way to measure the stationary or steady-state behavior of engines is to change the input variables stepwise one after another and to take some measurements of the outputs for averaging if a new steady-state is reached as described in Sect. 3.4.1. This needs about 2-3 min for passenger car engines and 4-6 minutes for truck engines for each stationary state. During the transient behavior, usually no measurements are taken, see Fig. 3.4.13a). The *lagged* behavior stems from the mechanical, fluidic and thermal dynamics of the engine and from the dynamics of the measurement equipment, compare Fig. 3.4.13b).

In order to save measurement time, the input variable can be changed slowly and continuously according to a ramp function so that the outputs follow with a little delay, Fehl (1992), Ward et al (2002), Schwarte et al (2004). For a first order dynamic system with differential equation

$$T_1 \dot{y}(t) + y(t) = K u(t) \tag{3.4.10}$$

and a ramp input with time behavior

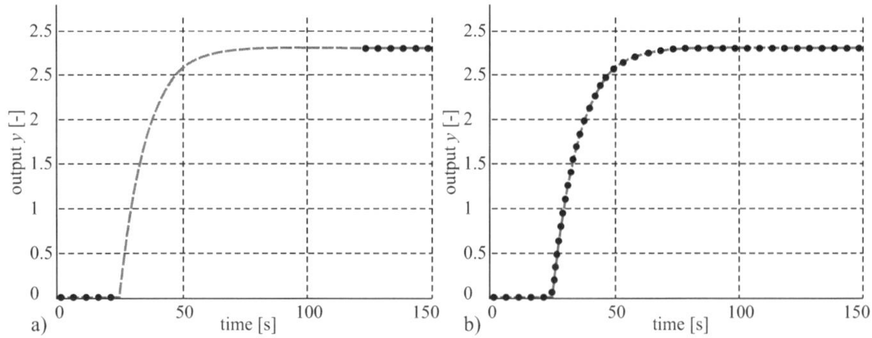

Fig. 3.4.13. Measurements for step inputs. **a** stationary behavior (steady state). **b** dynamic behavior

$$u(t) = \beta t \tag{3.4.11}$$

the output becomes

$$y(t) = K\beta t - K\beta T_1 \left(1 - e^{-\frac{t}{T_1}}\right) \tag{3.4.12}$$

compare Fig. 3.4.14, and the delay of the output caused by the time constant T_1 yields, compared to a proportional acting process with $T_1 = 0$

$$\Delta y(t) = y(t) - K\beta t = -K\beta T_1 \left(1 - e^{-\frac{t}{T_1}}\right)$$

and

$$\Delta y(\infty) = \lim_{t \to \infty} \Delta y(t) = -K\beta T_1 . \tag{3.4.13}$$

Hence, the resulting offset of the output is for $t \to \infty$ proportional to the ramp speed $\beta = du/dt$, the gain K (slope of characteristic) and the time constant T_1. If the ramp speed β is sufficiently small, the offset can be neglected. However, measurement time is saved if β is larger. If the offset then cannot be neglected, it can be compensated by a reverse measurement or by dynamic correction.

a) Offset compensation by reverse measurement

The offset due to the dynamic delay can be compensated approximately with an identical but reversed (decreasing) ramp which is symmetric to the first, increasing ramp, see Fig. 3.4.15a), Schwarte et al (2004). The nonlinear stationary behavior is then obtained by taking the mean value of the ascending and the mirrored descending slope, Fig. 3.4.15b). With two input variables, one input may be changed rampwise and the other remains constant, and then vice versa. As depicted in Fig. 3.4.16a) the inputs result in meandering sequential changes through the operation space. After reaching the starting point again, the meandering pattern is followed in the reverse direction compare Fig. 3.4.16b) and mean values of the forward and backward measurements are taken and then processed by e.g. a neural net, Schwarte et al (2004).

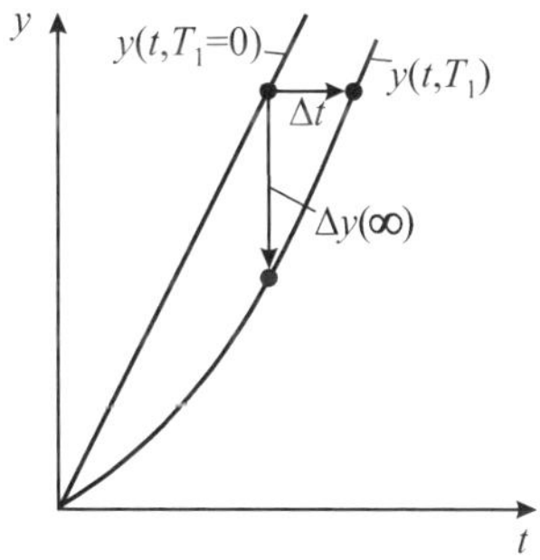

Fig. 3.4.14. Resulting offset of the output for measurement with an input ramp function and first order dynamics with time constant T_1.

This quasi-stationary measurement was applied for modeling the fuel consumption, NO_x and opacity for the inputs begin of injection and air mass. Measurements for 20 operating points (M_i, n_j) showed that 27 min were required for each operating point. A grid measurement would have allowed only 12 measurement points (e.g. 4×3) in the same time, with certainly less information content.

An alternative procedure is to simultaneously change several inputs rampwise, as shown in Fig. 3.4.17. Different ramp slopes are applied to scan the operation space in form of a zig-zag. If, for example, the measurement consists of a ramp time for u_1 with $10\,\mathrm{s}$, for u_2 with $120\,\mathrm{s}$ and for u_3 with $360\,\mathrm{s}$, the complete three-dimensional operation space is at least roughly covered. The back measurement then runs from $360\,\mathrm{s}$ to $720\,\mathrm{s}$ in the reverse direction.

However, for more than three inputs one ramp function becomes very slow, such that compared to grid measurements no measurement time is saved. This means that this quasi-stationary measurement-procedure with simultaneous ramp changes of inputs can be applied for one, two or three inputs.

b) Offset compensation by dynamic correction

If the time constant T for an assumed first order process can be identified, the output lag may be compensated by filtering the output with an approximation of the inverse of (3.4.10)

$$G_F(s) = \frac{y_{\mathrm{corr}}(s)}{y(s)} = \frac{1 + T_1 s}{1 + T_2 s} \tag{3.4.14}$$

where T_2 is a small time constant to cope with the realizability. An alternative is to shift the output

$$y_{\mathrm{corr}}(t) = y(t + \Delta t) \tag{3.4.15}$$

with

$$\Delta t = T_1 ,$$

see Fig. 3.4.14. However, frequently $y\,(u_1, u_2, \ldots)$ are nonlinear functions with operation-point-dependent K and T. Then, the offset compensation by reverse measurement may be simpler to apply.

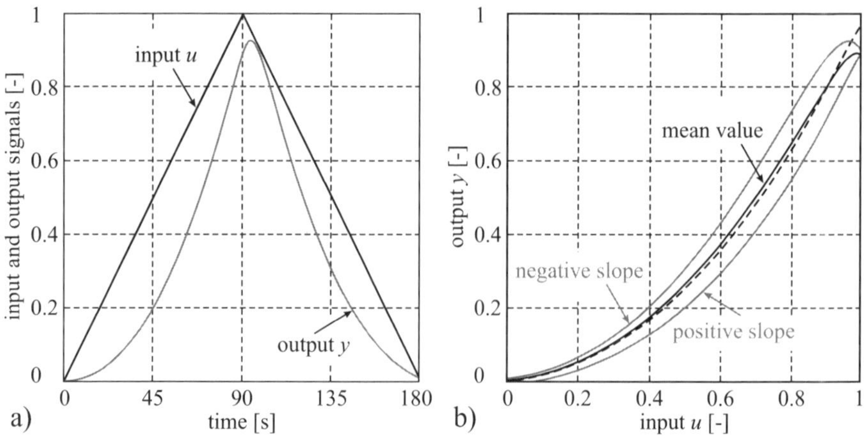

Fig. 3.4.15. Principle of quasi-stationary ramp measurement for $y(s) = u^2(s)K/(1 + Ts)$. **a** input and output time history. **b** input/output characteristics for forward and backward measurement.

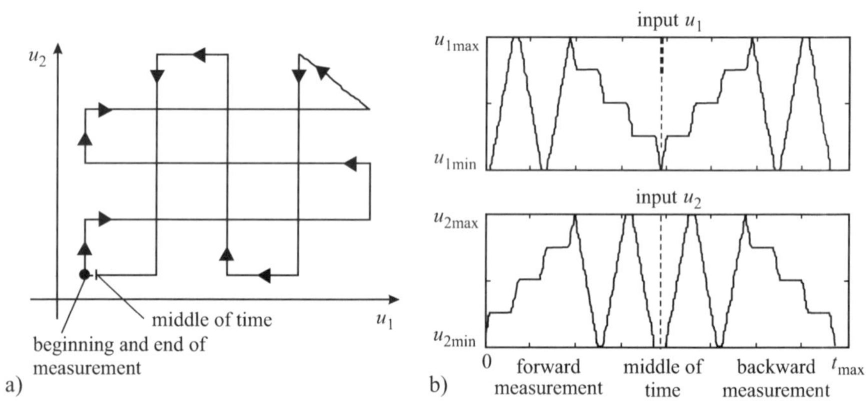

Fig. 3.4.16. Sequential quasi-stationary ramp measurement for two inputs (corresponding to a 5x5 grid). **a** sequential meandering changes of the input for the forward measurement (back measurement in reverse direction). **b** time-dependent course of the two inputs for back and forth measurement.

c) Slow dynamic ramps with neglect of an offset

If the gradient of the ramp is relatively small compared to the process dynamics, an offset of the output variable becomes small and can be neglected, especially if the offset is in the range of process and measurement disturbances. This was applied by Böhme (2013), Böhme et al (2009), to two 6 cyl, 3.8 l, gasoline engines for the *cylinder air charge determination* by changing the throttle with a ramp time of 98 s, compare Fig. 3.4.18. Measurements for the full range of operating points took 3.5 h instead of 18 h with conventional grid measurement, thus saving about 80 % of test bench occupation.

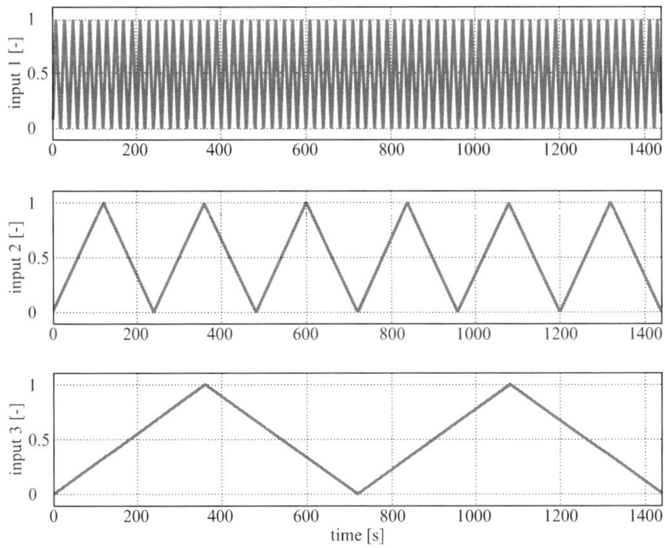

Fig. 3.4.17. Simultaneous quasi-stationary ramp measurement for three inputs.

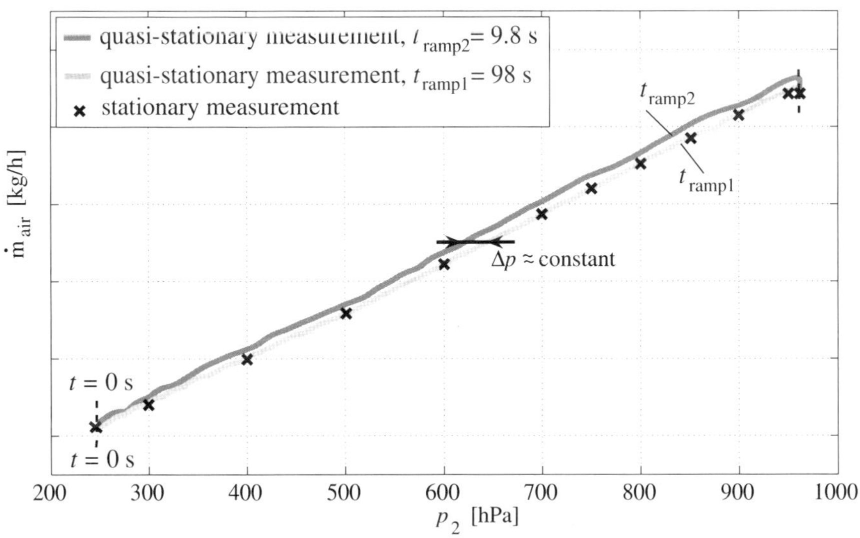

Fig. 3.4.18. Stationary and two quasi-stationary air mass flow curves with different ramp times t_{ramp}. The resulting offset for $t_{ramp} = 9.8\,\mathrm{s}$ is obvious, but can be neglected for $t_{ramp} = 98\,\mathrm{s}$, Böhme (2013).

The slow dynamic ramp method was then applied for *torque optimization* by ignition angle variations (so-called sweeps). The determination of the torque model is of special importance for controlling the torque with the ECU, e.g. by fast changing the ignition angle, see Sect. 7.1. Therefore the so-called ignition angle efficiency curve has to be calibrated, which can be approximated by a 6th order polynomial $\eta_{ign} = f(\varphi_{ign,opt} - \varphi_{ign})$, Böhme (2013).

The limitation for the speed of the ignition angle changes was in this case given by the speed control for the test bench dynamometer. For example, an ignition angle speed of $4°/\mathrm{s}$ the brake torque lags the engine torque, leading to offsets. However, with $1°/\mathrm{s}$ good results could be obtained.

An automated measurement procedure including monitoring of knocking to detect limits for early ignition, of exhaust gas temperature and HC emissions for late ignition was realized to apply the quasi-stationary measurement towards late ignition. Without loss in the quality of the results the test bench time for the full range of required operating ranges took $48\,\mathrm{h}$ instead of $240\,\mathrm{h}$ for the conventional stationary grid measurement. Here too, a saving of 80 % of test bed occupation was reached.

Leithgöb et al (2005) applied the determination of the air charge with slow ramp functions to a gasoline engine with variable valve train, saving about 50 % of test bench time. Büchel and Thomas (2009) report of about 85 % time saving for the determination of the torque model.

3.5 Dynamic identification procedures on test benches

In addition to the classical way of measuring and modeling the stationary behavior of engines, modeling of the dynamic behavior is of increasing interest. The reasons are, for example, to optimize the driveability for load changes and to reduce emissions with regard to driving cycles. About 50 % of the emissions in the NEDC or FTP75 driving cycles result from acceleration states. The influences from the non-stationary operation states on fuel consumption and emissions are mainly caused by mass, energy and momentum storage in the engine which lead to certain dynamic effects. The delays of the air system, the turbocharger, the heat transfer and emission aftertreatment are well known. These delays may lead to undershoots or overshoots in connection with dynamic control compared to the stationary variables.

Dynamic measurement procedures can be divided into local and global methods. In the case of *local measurements* the operating point of the engine, defined by the torque M_i and speed n_j, is kept constant and other control inputs like ignition time, injection time, common rail pressure, turbocharger control position, camshaft position, etc. are varied with special exciting test signals. This means that different dynamic models are generated for each operating point. When applying *global measurements* the operating point variables torque M_i and speed n_j are also manipulated with test signals by changing the injected fuel mass and the setpoint of the dynamometer speed controller, see Sect. 3.3 and Sect. 8.9 for an application. These global models are, for example, required for the optimization of control functions with regard to dynamic drive cycles and the real driving behavior of vehicles.

In the following, the experimental modeling is described taking into account the basic identification methods summarized in Sect. 3.2. It is based on Schreiber and Isermann (2009), Schreiber (2010) and Kowalczyk (2013). Some requirements on the resulting dynamic identification procedure are that they are compatible with stationary measurements, that the resulting models are suitable for the design of feedforward and feedback control, that they rely on proven algorithms and tools and are suitable for usually applied workflows.

3.5.1 Test signals and nonlinear dynamic models

a) Test signals for test benches

Some frequently used test signals for the dynamic excitation of processes are shown in Table 3.2.1. Non-periodic signals are for example step functions and pulses and periodic signals are sinusoidal functions with different frequencies to determine frequency responses. Stochastic binary signals are better replaced by pseudo random binary signals (PRBS), which are then deterministic and periodic, because of their precisely known signal parameters (e.g. ACF). Nonlinear processes require test signals with varying amplitudes. This leads to multiple step signals or amplitude modulated (APRBS). A detailed treatment of test signals is given in Isermann and Münchhof (2011).

For combustion engines on test benches mainly *step* or *ramp functions*, *PRBS* or *APRBS* are suited, see Fig. 3.5.1. However, as a Fourier analysis shows, step and ramp functions excite mainly the lower frequencies of process dynamics. However, PRBS may excite a wider frequency spectrum. PRBS are originally generated by shift registers with two feedbacks and are therefore periodic, see Fig. 3.5.2. If the (programmed) shift register has four stages, the period is $T_N = 15\lambda$ with λ the clocktime or smallest pulse width, see Isermann (1992) and Isermann and Münchhof (2011). Several PRBS can be applied simultaneously to all inputs and parameter estimation methods and allow then the identification of all paths of a MISO process.

Sinusoidal signals may also be used. However, together with the evaluation of frequency responses they basically require the assumption of linear processes and allow the estimation of only two parameters per frequency. However, using discrete frequencies one after another requires too long a measurement time. Therefore, the frequency can be changed continuously, leading to a sweep sine test signal or chirp signal, see e.g. Isermann and Münchhof (2011).

In order to excite the stationary as well as the dynamic engine behavior in a short time Fig. 3.5.1d) depicts a special *deterministic short time binary signal (DSBS)* which consists of some few rectangular pulses. It begins at a stationary point $k = 0$ and ends at a next stationary point $k = 9$. The length of the first rectangular pulse is $\lambda_{\rm min} = \mu T_0$ (with $\mu = 1, 2, 3$) according to the rules of selecting the clock-time λ of a PRBS. The second pulse length of $3\lambda_{\rm min}$, followed by a third pulse length $\lambda_{\rm max}$. If for example,

$$\lambda_{\rm max} = 4\lambda_{\rm min} = \frac{4}{3} T_{95} \tag{3.5.1}$$

is chosen and T_{95} is the 95 % settling time of a step response, a next stationary point is reached approximately. The total length of a DSBS then becomes

$$T_{\mathrm{DSBS}} = (1 + 3 + 4)\lambda_{\mathrm{min}} = 8\lambda_{\mathrm{min}}. \tag{3.5.2}$$

Therefore, it can also be called a 1-3-4 test signal. By sequencing this test signal from one stationary point to another, the selected stationary operating points are measured and the medium and lower frequencies are excited in between, see Kowalczyk (2014).

The measurement of the combustion engine dynamics includes the sensor dynamics. Therefore fast sensors are required which usually holds in the case of flow, pressure, λ and NO_x sensors. However, for temperatures or some emissions as outputs the sensor dynamics have to be extracted from the identified models, if the real engine outputs are of interest, see Sects. 3.5.3 and 8.8.1. Gas analytic emission measurement devices with large dead times and large time constants are usually not suitable for dynamic identification.

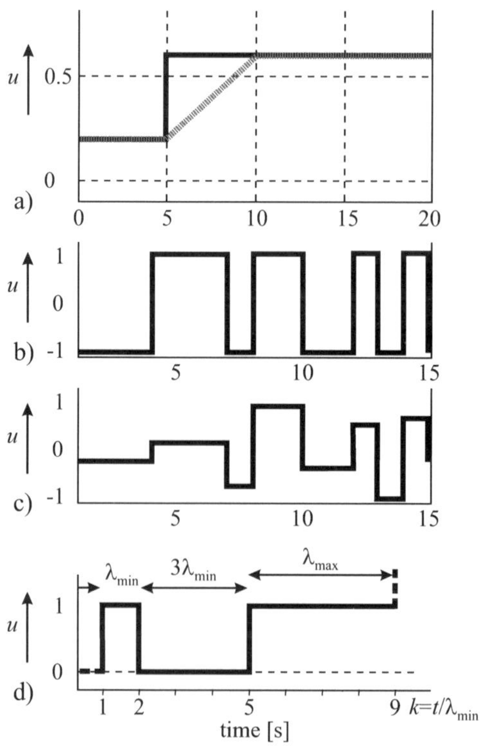

Fig. 3.5.1. Different test signals. **a** step and ramp function. **b** PRBS. **c** APRBS. **d** DSBS

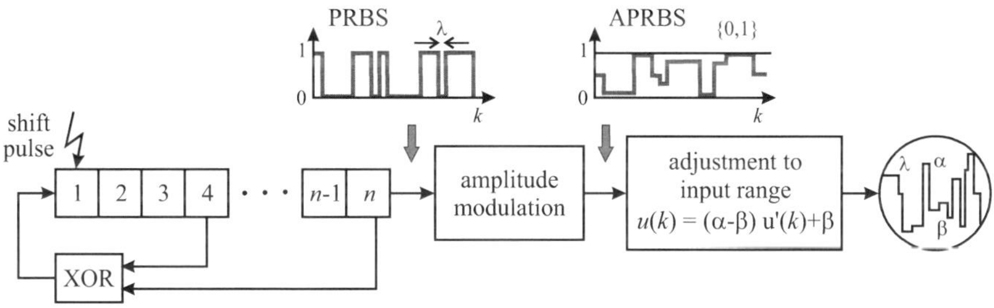

Fig. 3.5.2. Generation of PRBS and APRBS signals.

b) Type of models for engine identification

In the case of local models, a first choice for the identification of the dynamic behavior of combustion engines is to use linear models, represented by linear differential equations or with sampled data as *linear difference equations*

$$\begin{aligned} y(k) + a_1 y(k-1) + \ldots + a_m y(k-m) \\ = b_0 u(k-d) + b_1 u(k-1-d) + \ldots + b_m u(k-m-d) \end{aligned} \tag{3.5.3}$$

where $k = t/T_0$ is the discrete time, T_0 the sampling time and $d = T_t/T_0$ the discrete dead time. This type of parametric dynamic model is especially suited for parameter estimation methods. However, in many cases the engine behavior is nonlinear. As shown in Sect. 3.2.3 a first class of special *nonlinear models* follows from a direct extension of linear difference equations to *Volterra series*. With finite order they can be considered as parametric models, see (3.2.48). Through specialization simpler types follow, as the *Hammerstein model* (3.2.49), a series connection of a static polynomial nonlinearity and a linear difference equation or the *Wiener model* where the linear difference equation is followed by a static nonlinearity. One of the problems of the Volterra models is the selection of the order, the number of time shifts, resulting in combinational complexity and lacking physical interpretation. Hammerstein models may be applicable for special cases, see Sect. 3.5.4.

The second class of nonlinear models for identification are general model structures in form of *neural networks*. They are described in Sects. 3.1.3 and 3.2.3. Especially the multilayer perceptions (MLP) and the radial basis functions (RBS) allow a good approximation of nonlinear behavior. However, their structures can get complex with many adjustable parameters. In general, they do not allow a transparent interpretation of the parameters and are not close to the structure of theoretical or physical-based models.

An alternative is the identification with *local linear models*, which vary their parameters with the operating point or with the amplitudes of input signals, see Sect. 3.1.3 and 3.2.3. One representative is LOLIMOT, see Sect. 3.1.3e). For the resulting local linear models well-proven parameter estimation methods can be applied. They also allow a direct physical interpretation and the design of local linear controllers. Hence, this model type is based on a normal control engineering know-how and thus eases handling. Therefore, it is especially suitable for the identification of combustion engines. An extension with local nonlinear polynomials is LOPOMOT, see Sect. 3.1.3f).

3.5.2 Dynamic test plans

The design of dynamic test plans is much more involved than for static test plans, described in Sect. 3.4. It requires:

- selection of candidate sets
- selection of amplitudes
- sequence of measurements
- generation of dynamic test signals
- combined dynamic measurement plans.

The *selection of the candidate sets* within the given measurement boundaries can also be made for dynamic measurements as described in Sect. 3.4.4. Here too, a spherical screening allows the selection of more candidates if a reduced number of measurement points is envisaged.

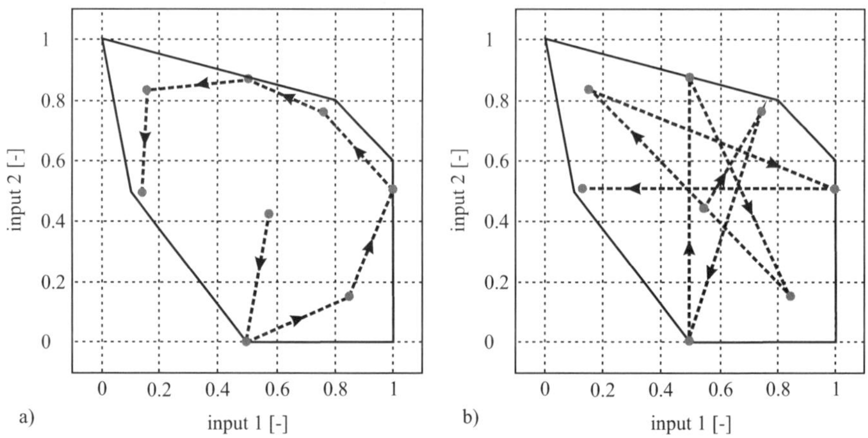

Fig. 3.5.3. Determination of the sequence of measurement points (point-to-point jumps) within the determined D-optimal measurements. **a** average point-to-point distance sorting. **b** small temperature change sorting.

Based on the selected candidate points the *dynamic test signal amplitudes* have to be determined. If step functions or PRBS are used as test signals, the amplitudes result by jumping from one measurement point to another. In order to reduce the number of jumps, a D-optimal design selection within the boundaries can be made as described in Sect. 3.4.3 for stationary measurement. This means that by maximization of the determinant of the covariance matrix for the parameter estimation of a stationary polynomial model a certain number of measurement points are obtained.

Then the *sequence for the manipulation* of the engine input variables has to be determined. This implies that a sorting order has to be found. Criteria for this sorting are e.g.:

- minimal or maximal point-to-point distance
- average point-to-point distance
- small temperature changes first.

Experiments have shown that for local modeling the average point-to-point distance should be preferred, Fig. 3.5.3a), based on the distance calculation of two candidate points I and II

$$\Delta u_{\mathrm{I,II}} = \sqrt{(u_{1,\mathrm{II}} - u_{1,\mathrm{I}})^2 + (u_{2,\mathrm{II}} - u_{2,\mathrm{I}})^2}. \tag{3.5.4}$$

For global modeling a sorting with regard to small temperature changes first are recommended, because of the smaller settling times and therefore time saving, Schreiber and Isermann (2009), see Fig. 3.5.3b). Having determined the jumping order, sequences of step functions or PRBS can be generated, thus forming the exciting test signals as shown in Fig. 3.5.1. For nonlinear processes the amplitudes of a PRBS must have not only two but several values. This results in amplitude modulated APRBS. By using step functions as inputs one obtains transient functions for each output. If enough settling time is provided, the stationary behavior follows from the steady state and the dynamic behavior from the transient course. However, step functions excite only the low frequencies. PRBS have a broader frequency spectrum and also excite medium and higher frequencies more strongly.

A final dynamic test plan (or static test plan) now follows from combining the selected operating points, the identification model and static or dynamic test signals. Table 3.5.1 shows symbols for describing the kind of combinations and Table 3.5.2 shows some suitable combinations.

Table 3.5.1. Abbreviation symbols for combining engine identification methods (examples)

Measurement point selection	Identification model	Test signal for dynamic excitation
G Reduced Grid	P Polynomial	S Step
D Design of experiments	V Volterra series	R Ramp
E ECU controller outputs	H Hammerstein model	P PRBS
L LOLIMOT structure	N Net model (general)	A APRBS
S Space filling	L LOLIMOT	
	P LOPOMOT	

Examples: GPS: Grid measurement, polynomial models and stepwise input variations
LLA: LOLIMOT structure, LOLIMOT model and APRBS

If several input signals are changed simultaneously to obtain a MISO model $y = f(u_1, u_2, \dots)$, the goal is to obtain models for each signal path $y = f(u_1)$,

Table 3.5.2. Selection of suitable combinations for dynamic engine identification (examples)

Testsignal		Design of experiments D	ECU controller outputs E	LOLIMOT structure L
Steps:	S	DNS / DHS	—	LLS
Ramps:	R	DNR / DHR	—	LLR
APRBS:	A	DLA	ELA	LLA

$y = f(u_2), \ldots$. However, each input signal is a disturbing signal for the identification of the neighbored channels. This means for example that u_2 disturbs the identification of $y = f(u_1)$ and u_1 the identification of $y = f(u_2)$. These disturbing effects can be eliminated if input signals are used which are uncorrelated. Therefore, special test signals on the basis of a PRBS were designed for the identification of multivariable processes as described in Isermann (1992), Isermann and Münchhof (2011) and Blessing (1979). The test signals result by using a Hadamard matrix and then become multi-valued signals, except the basic PRBS.

Now *some examples* of test-signal generation for dynamic identification are considered.

D-optimal amplitudes and step functions (DS)
For the case of a D-optimal selection of measurement points within the measurement boundaries and two kind of sorting orders, Fig. 3.5.3 shows how test signals with step functions are generated.

LOLIMOT structure and step functions (LLS)
Figure 3.5.4 shows the stationary characteristic of a turbocharger and the resulting local linear areas of a LOLIMOT structure. The measurement points within the local linear areas correspond to a value of 90 % of the validity functions, (3.1.26). Within each local area four step function jumps are generated from one point to another, resulting in a step function sequence as depicted in Fig. 3.5.5. Here, only one variable after another is changed so that corresponding transient functions can be observed and evaluated, Weber et al (2005).

LOLIMOT structure and APRBS (LLA)
The LOLIMOT structure is used to define maximal amplitudes within each local linear area and is used to generate a PRBS. This is then amplitude-modulated to an APRBS, see Fig. 3.5.6. The two APRBS have one PRBS sequence as a basis, according to Fig. 3.5.2.

APRBS superposition on ECU controller outputs (ELA)
If a stationary calibrated ECU already exists, a dynamic test signal can be superimposed on the manipulated variables to obtain dynamic models. Figure 3.5.7 shows an example for the superposition of an APRBS for a diesel engine, Hafner (2002).

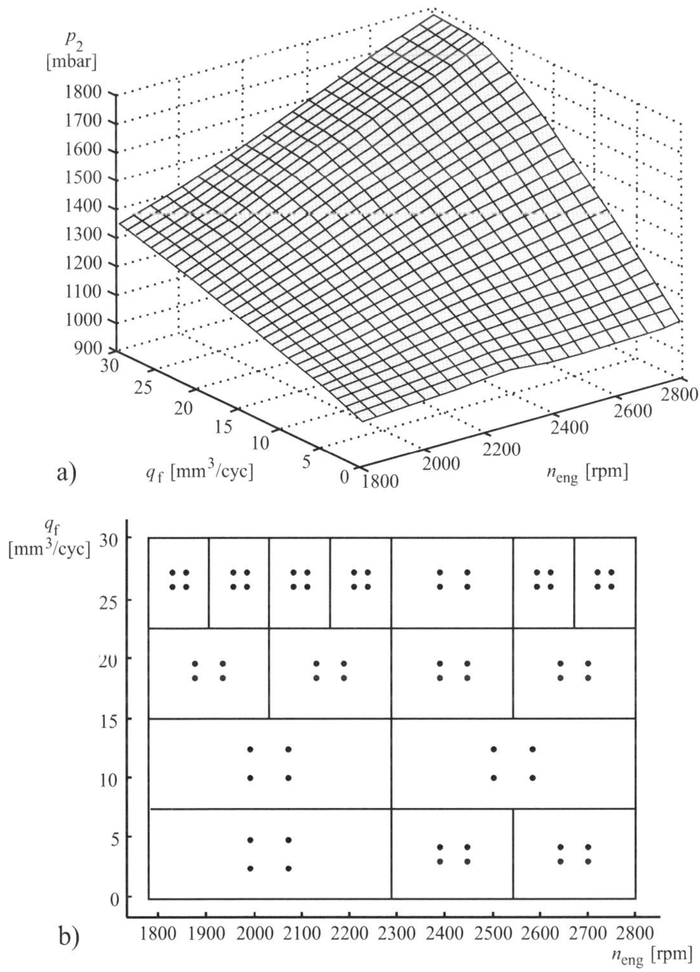

Fig. 3.5.4. a Stationary characteristic of the charging pressure of a turbocharger. **b** LOLIMOT structure with local linear areas.

3.5.3 Influence of sensor dynamics

The identified models of combustion engines on test benches include the dynamic behavior of the used sensors or measurement equipment. Their dynamic behavior is included in the identified model. If the sensor dynamics are fast, for example, as for λ-, NO_x, airflow- and pressure sensors with time constants in the range of some few to about $100\,\mathrm{ms}$, the resulting delay can mostly be neglected or has to be included for control applications. However, temperature sensors may dominate the overall dynamic behavior. The time constant of a rigid temperature sensor is, see Sect. 4.6.5,

$$T_{TS} = \frac{d_S \rho_S c_S}{4\alpha_{fw}} \tag{3.5.5}$$

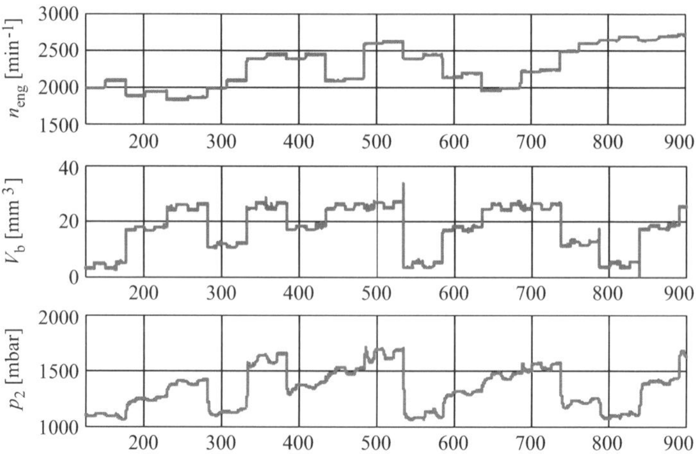

Fig. 3.5.5. Generated step function sequences for speed and injection quantity based on a LOLIMOT structure with charging pressure p_2 as output (method LLS).

where d_S is the diameter, ρ_S the density and c_S the specific heat. It is inversely proportional to the heat flow coefficient α_{fw}, which depends on the fluid speed as

$$\alpha_{fw} \sim w^m \qquad m = 0.6 \ldots 0.8. \tag{3.5.6}$$

Therefore, the time constant becomes larger for low fluid velocities. It ranges within $0.5 < T_{\mathrm{TS}} < 10\,\mathrm{s}$. The influence of temperature sensor dynamics can be approximately compensated by filtering the signals with a lead-lag element like $G(s) = (1 + T_{\mathrm{TS}}\mathrm{s})/(1 + T_1\mathrm{s})$ or with two sensors and applying a Kalman filter, Zimmerschied and Isermann (2009).

The largest influences on the the dynamic models have the gas-analytic measurement devices for the emissions HC, CO, NO_x. They usually have delay times of more than $500\,\mathrm{ms}$ and the flexible tube connections lead to dead times up to about $10\,\mathrm{s}$, see Sect. 8.9.1d). As the engine dynamics then cannot be determined anymore, the identification of dynamic models for emissions need sensors with fast dynamics which are mounted close to the engine, as is the case for λ, NO_x and opacity measurement.

3.5.4 Extract of stationary from dynamic models

In order to save the settling time of $2 \ldots 5\,\mathrm{min}$ after input variable changes, the engine can be measured with dynamic test signals and the stationary behavior can be extracted as a special case, e.g. by simulation of the dynamic models for $k \to \infty$. Another possibility is the measurement with fast-following step functions without waiting for larger settling periods. An example is shown in Fig. 3.5.8. The step functions were applied in intervals of $5 - -10\,\mathrm{s}$ for each one input. The stationary behavior followed from LOLIMOT models of first and second order for large time $k \to \infty$. A comparison with direct stationary grid measurement gave deviations of 1 to 5 %.

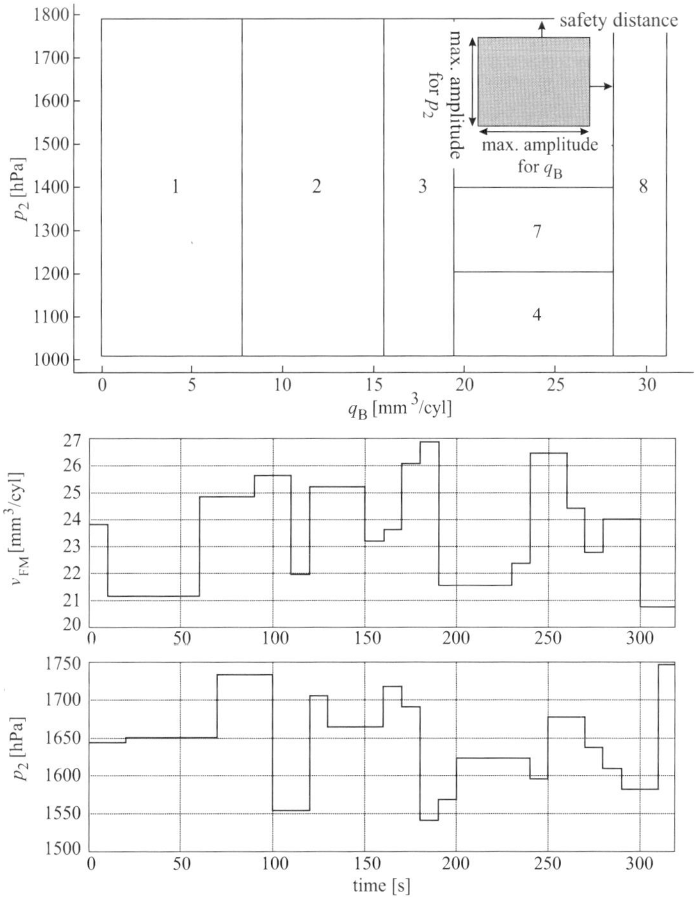

Fig. 3.5.6. Generation of APRBS test signals based on a LOLIMOT structure.

A further possibility is the use of Hammerstein models, consisting of the nonlinear polynomial

$$x = r_0 + r_1 u + r_1 u^2 \tag{3.5.7}$$

and a dynamic model

$$y(k) = -a_1 y(k-1) + b_0' + b_1' x(k-1) \tag{3.5.8}$$

resulting in a nonlinear dynamic model

$$y(k) = -a_1 y(k-1) + b_0 + b_{11} u(k-1) + b_{12} u^2(k-1). \tag{3.5.9}$$

After parameter estimation, the parameters r_i of the static model can be determined directly. The applied method was DHS with a holding time of 6 s for the step functions. Slow emission sensors were used and 85 stationary DoE measurement points

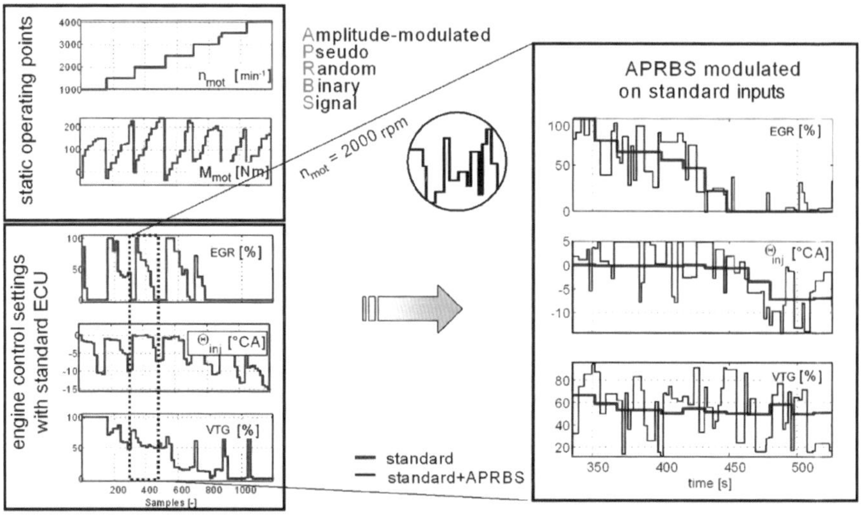

Fig. 3.5.7. Superimposed APRBS on the control outputs ("standard") of the engine ECU.

for one speed were taken, see Fig. 3.5.3. Dynamic as well as static models have shown good validation results. In comparison to stationary DoE measurements, the required measurement time was reduced from 260 min to 67 min, i.e. to 26 %.

3.6 Combined online identification procedures on test benches

The steps for conventional engine measurements and modeling are shown in Fig. 3.4.2. The procedure is performed in individual steps and with an alternation between preparations and analysis at the office and measurement at the test bench. This usually results in repetitions and includes returns, e.g. the return to the design of experiments after model analysis if too few design points in the relevant range have been measured or the repetition of the measurement after the raw data analysis if the settling time until steady-state was chosen too short. All these repetitions require additional test bench time.

A significant improvement regarding the resulting measurement time, analysis effort and quality of the models can be achieved with an online measurement and modeling procedure, Kowalczyk (2013). Thereby the engine and the modeling computer are coupled online. The online procedure operates such that the results of the preceding step are used for the design of the next step. For example, model attributes, such as the coefficient of determination R^2 or parameter variances, are used to generate new designs with excitation signal amplitudes in relevant areas. This can be achieved if all necessary inputs (e.g. a priori knowledge, identification method, analysis method) are defined before the actual measurement starts. The online procedure is characterized by the use of an automatic iterative approach, reduced user interventions during the measurement and modeling process, a fast and automatic variation

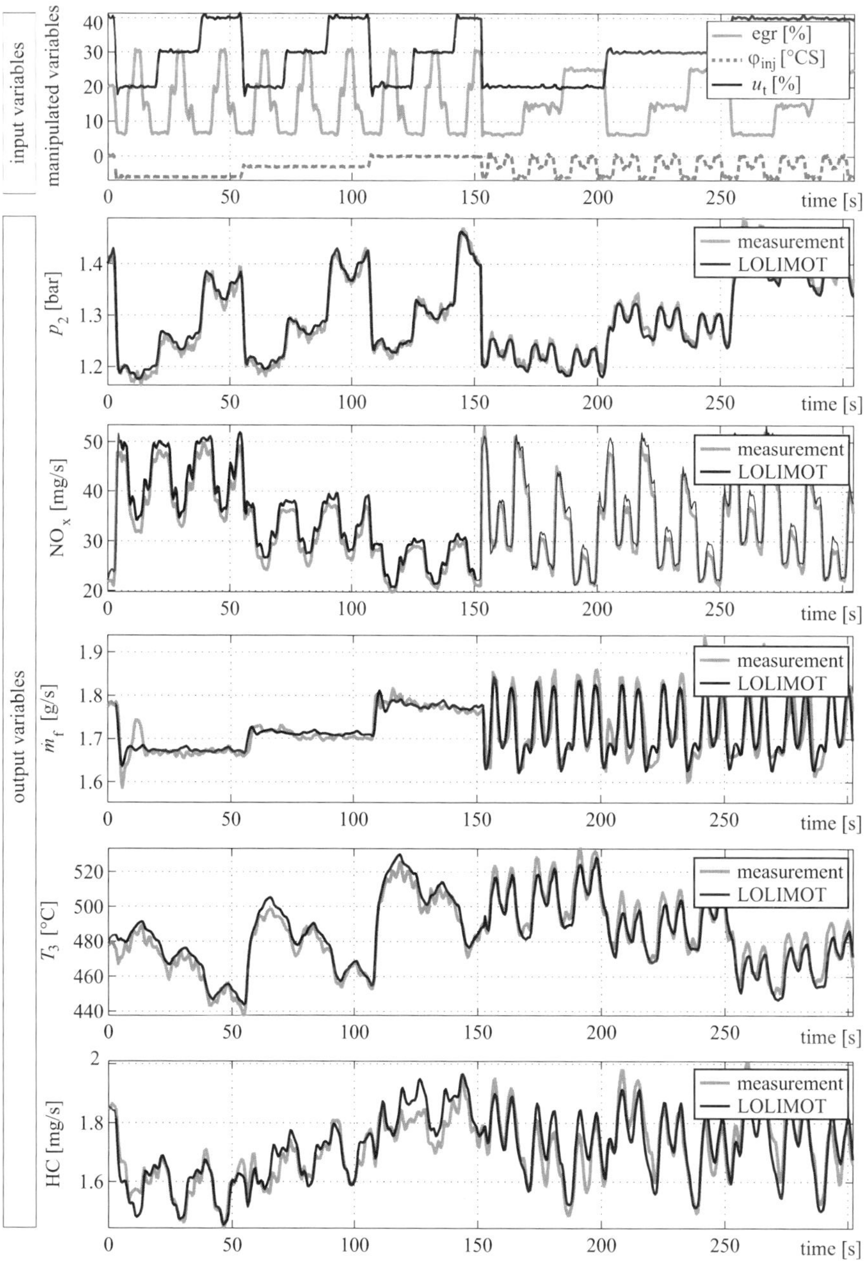

Fig. 3.5.8. Sequential step-function variations and measurement of transient functions to obtain the steady-state behavior (Identification method: GLS). Diesel engine (81 kW, EGR, VGT, VP37), $T_0 = 100$ ms, torque controlled operation, $M_{eng} = 90$ Nm, $n_{eng} = 3000$ rpm, Schüler (2001).

space determination as well as direct data processing, model identification and validation at the test bench.

3.6.1 Online iterative measurement and modeling procedure

The high degree of test-bench automation requires a safe communication of all systems involved and a monitoring of the operational limits. An example for a test bench setup which is suitable for online real-time methods is shown in Fig. 3.6.1.

The online approach requires a real-time computer for online data processing and test signal generation, i.e. an online closed loop coupled process computer. Design of Experiments tools are used for the online processing (generation of designs, data evaluation) and test bed automation tools handle the measurement process.

The combination of DoE and test bed automation tools into one methodology enables an automatic actuation, data sampling and analysis during the process operation without user interventions, see Fig. 3.6.2. During the evaluation and processing of one measurement interval, a further measurement interval is already performed. Furthermore the measurement with real-time hardware enables the system to interrupt (e.g. to stop if a limit violation occurs), and adapt (e.g. a controlled limit approach, Kowalczyk (2013)) the ongoing measurement, without stopping the whole measurement procedure and without user inputs. Through the combination of the two systems a parallelization of the measurement process and the modeling can be achieved as shown in Fig. 3.6.2 beginning with the 2nd iteration.

With the help of the parallelization, the conversion and analysis of the data of the previous measurement interval as well as model identification and analysis can be performed during the actual running measurement. This makes an evaluation afterwards in the office (as in the classical approach) unnecessary, saving the time needed for it. A further advantage of the resulting batch processing is the ability to use not only one, but several of the already developed offline analysis and identification methods.

3.6.2 Target-oriented design adaptation based on model quality

The number of required measurement points depends mainly on the kind of modeling (local or global) and the degree of nonlinearity, for the identification method (e.g. neural net or polynomial), and measurements repetitions. First, the online identification of stationary models is considered. The method for target-oriented design adaptation aims at reducing the number of needed measurement points for the overall modeling process, which means fewer measurements. It is based on the structure adaptation algorithm of local-linear model trees, see Sect. 3.1.3. After splitting the space of measured data in all dimensions, local range models are identified with the corresponding measured data and a model quality measure (e.g. coefficient of determination R^2) in each resulting orthogonal subspace. An example is Fig. 3.6.3 for the two input variables $u_1 = s_{\mathrm{egr}}$ and $u_2 = s_{\mathrm{vgt}}$. Within the actuator manipulation range for the local model of lowest quality $J_{\mathrm{adapt,min}}$, new measurement points are added (see Fig. 3.6.3 right). This may be repeated iteratively.

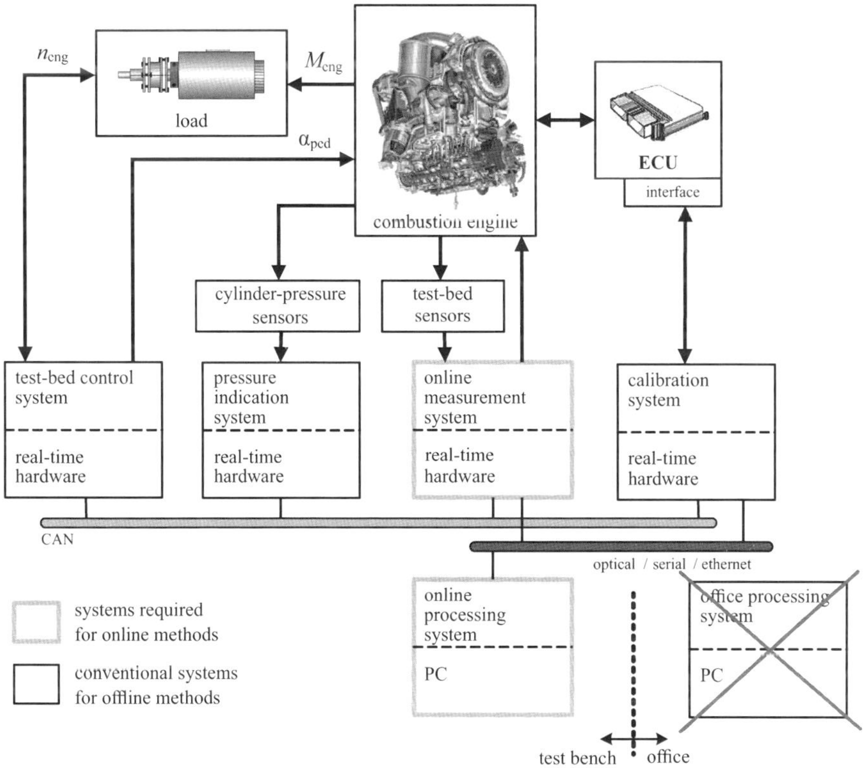

Fig. 3.6.1. Test-bench setup and signal flow as required for online real-time methods.

This procedure results in a systematic improvement of the model quality in the areas where most can be gained. The advantage in comparison to classic design generation methods is that ongoing measurement results are used online for the design point placement. Therefore, a target-oriented design adaptation is possible and new measurement points are applied only as necessary for achieving a desired model quality, reducing the amount of measured data and measurement time. The online method can be applied for the identification of stationary or dynamic models.

3.7 Model analysis and validation

After the determination or calculation of a mathematical model based on measured data with an identification method it has to be checked whether the model complies with the recorded input and output data of the real process. The first step consists of a *model analysis* which is frequently part of the identification and estimation method. For the parameter estimation methods, performance measures such as the sum of least squares of equation errors or output errors are directly used as part of the identification process and they are also used to select the model order and dead time. A

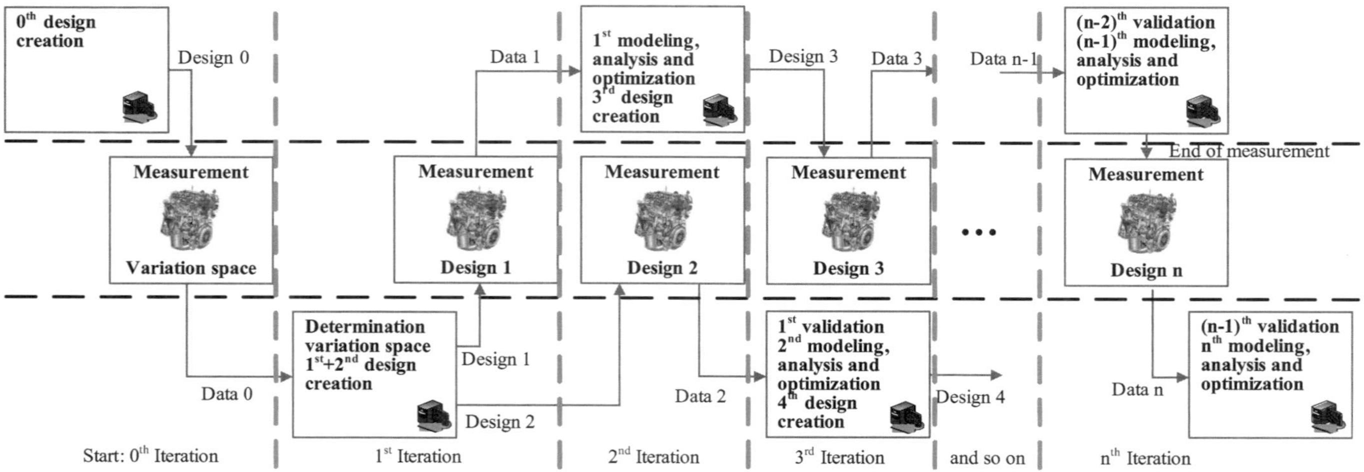

Fig. 3.6.2. Sequence of the iterative online measurements and modeling for combustion engines.

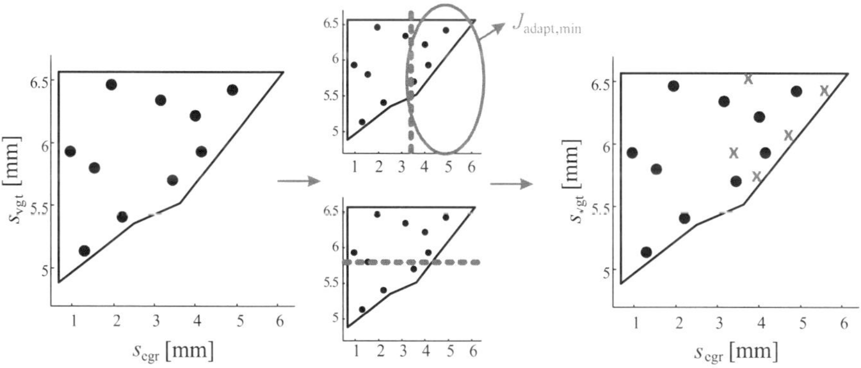

Fig. 3.6.3. Design adaptation based on the model quality. New measurement points (crosses) are placed in the range of the local model with the smallest quality criterion R^2.

next step is then the *model validation*, which can be understood as an overall quality assurance procedure of establishing evidence that the obtained model describes the measured behavior of the real process, see also Ljung (1999). This usually involves the fulfillment of a *fitness measure* or *performance criterion* for curve fitting and may include tests which show the model quality with regard to accuracy, repeatability and linearity. The validation procedure can be based on the available recorded data (single sample analysis) or additional recorded data (extra sample analysis), also called *generalization data*.

A similar procedure is called *verification*. This is understood as a quality control procedure to evaluate if the result complies with specifications imposed at the beginning of a development phase. Hence, it is more oriented to products. For general definitions and literature see, e.g. Verification and validation (2012).

3.7.1 Analysis of steady-state measurement and models

The analysis of steady-state data and models consists of four steps: (a) analysis of raw data using graphical tools, (b) application of analytical performance criteria, also for model selection, (c) graphical inspection, (d) generalization, see Kötter (2007), Kötter (2008).

a) Raw data analysis

Raw data analysis is applied before the model identification to check the signals obtained graphically by inspection for conformity. Thus, e.g. outliers can be detected and removed and drift phenomena can be recognized.

b) Analytical performance criteria

To determine the model's quality a first criterion is the sum of squared errors between measured and predicted values, Montgomery (2005)

$$SSE = \sum_{i=1}^{N} (y_i - \hat{y}_i)^2. \tag{3.7.1}$$

Two more criteria are obtained if the deviation from the mean value of the measured values $\bar{y}$ is considered.

$$SSR = \sum_{i=1}^{N} (\hat{y}_i - \bar{y})^2 \tag{3.7.2}$$

$$SST = \sum_{i=1}^{N} (y_i - \bar{y})^2 = SSR + SSE. \tag{3.7.3}$$

SSR is the sum of squares of the regression, i.e. the *squared difference* between the predicted values and the mean value of the measured values, which describes the variability explained by the model. SST is the total variability or the *sum of squares total*, i.e. the squared sum of the difference between the measured values and the mean value.

c) Diagrams for model analysis

After the identification an interactive review of the resulting models is carried out. The models are examined and possibly enhanced by iteratively changing the parameters of the identification method or the model and executing the identification again. Besides performance criteria plots should be used for analyzing the models.

The *measured vs. predicted plot* (MPP), see Fig. 3.7.1, shows for each of the observations one point, with the observed value on the abscissa and the value predicted by the model on the ordinate.

If a model explains the observed values completely and no measurement disturbances occurred during measuring, the plotted points should be located on the bisecting line. This line is usually plotted into the diagram. The distances of the plotted points from the bisecting line are an indicator for the model's quality. Points located far away from the diagonal might be outliers. Especially a group of points located away from the bisecting line indicates that the chosen model structure might not be correct. The effect of disturbances can be examined by studying the deviation from the repetition points in the measured vs. predicted plot.

d) Intersection plots for detailed analysis

The intersection plot (Fig. 3.7.2) is a possibility for the visualization of the dependence between the input variables and one of the outputs. Therefore a certain point (intersection point) in the input space has to be chosen.

Through this point and parallel to the axis of each input variable an intersection plane is drawn. These intersection planes are displayed next to each other. Additional information – like prediction intervals – are often included in this diagram. In Fig. 3.7.2 a two-dimensional example is shown. As only a part of the complete model

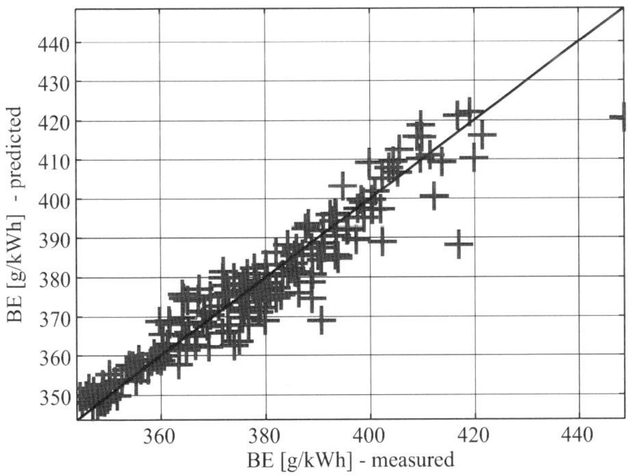

Fig. 3.7.1. Measured vs. predicted plot (MPP) for steady-state model

can be visualized at a time, an interactive possibility to select a new intersection point should be provided.

Crosshairs (coordinate axis) could be used to change the value of one variable for instance. Afterwards the plot is refreshed for the selected intersection point with the changed value. If there are only a few input variables (about 5-8) a transparent visualization is still feasible. However, for higher dimensional manipulated variables of combustion engines a complete visualization of the input space is not practicable, as the amount of required intersection points grows exponentially with the number of input variables (curse of dimensionality). Then analytical methods have to be applied.

e) Criteria for overall model quality

Besides diagrams different analytical methods for the determination of the model's quality should be considered. It is useful to consult normalized performance criteria. The *coefficient of determination* R^2 is very suitable and widely used, Montgomery (2005), Myers (1990).

$$
\begin{aligned}
R^2 &= \sum_{i=1}^{N} (\hat{y}_i - \bar{y})^2 / \sum_{i=1}^{N} (y_i - \bar{y})^2 = SSR/SST \\
&= 1 - \sum_{i=1}^{N} (\hat{y}_i - y_i)^2 / \sum_{i=1}^{N} (y_i - \bar{y})^2
\end{aligned}
\tag{3.7.4}
$$

y are the measured output values, $\hat{y}$ the predicted values and $\bar{y}$ is the average (mean) of the measured outputs. A value of $R^2 = 1$ indicates a very good agreement, i.e. a good model. A value of zero concludes that there is no description of the real system by the model at all. Consequently, the application of this criterion is straightforward.

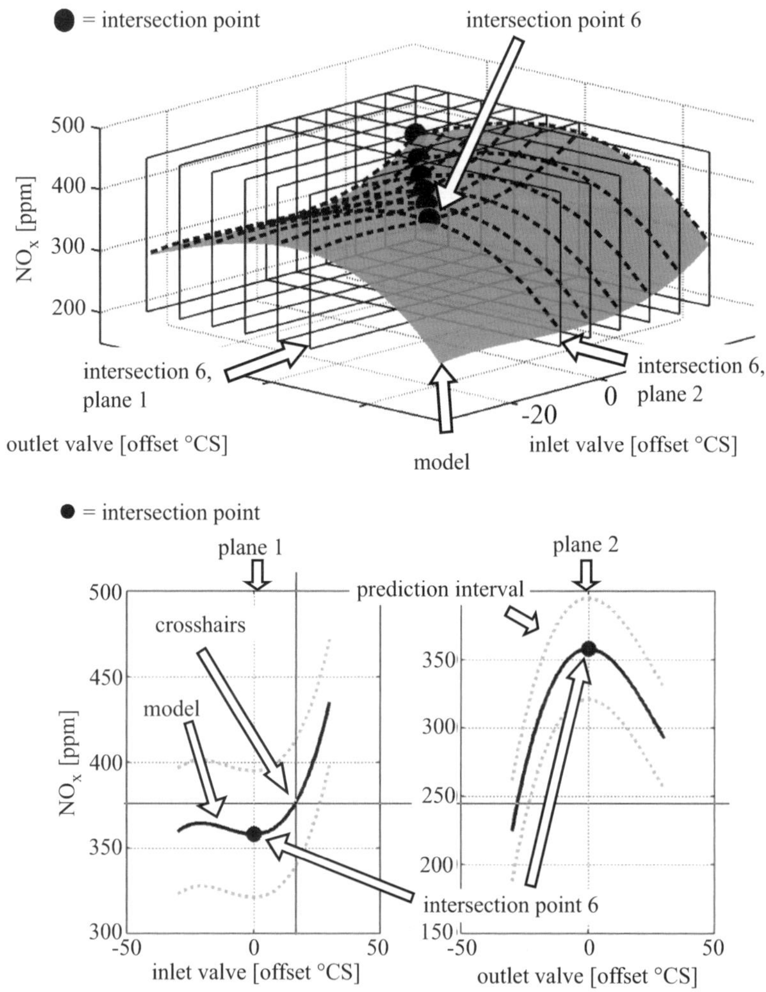

Fig. 3.7.2. Intersection plot for steady-state models

3.7.2 Analysis of dynamic measurements and models

The analysis for the identification of the dynamic behavior follows: *raw data analysis*, *model identification with model selection*, and *model analysis*. However, in contrast to steady-state models much more data has to be processed, the dynamic model order and dead time has to be selected and the dynamics of sensors and measurement equipment has to be considered. Additionally, also the steady-state models extracted from the dynamic measurements have to be validated. The following discussion assumes as example linear dynamic models in form of difference equations like

$$y(k) = b_1 u_1(k-1) + b_2 u_1(k-2) + \ldots - a_1 y(k-1) - a_2 y(k-2) - \ldots \quad (3.7.5)$$

where the model output $y(k)$ at time k depends on the input and output signals at time $k-1$ and time $k-2$ for instance.

The corresponding steady-state model is

$$y = K_1 u_1 \tag{3.7.6}$$

with

$$K_1 = \frac{\sum_{i=1}^{m} b_i}{1 + \sum_{i=1}^{m} a_i}. \tag{3.7.7}$$

The structure of the dynamic models assumed is the so-called ARX structure (autoregressive with exogenous inputs).

Besides the identification of dynamic models the analysis of the models with the continuous recordings of measurement data require new methods, as large data sets need to be processed. This begins with the raw data analysis. Only some remarks are given here. For details see e.g. Kötter (2008).

For *raw data analysis* the XT plot is usually applied, where the measured data are graphically plotted over time. Then drift can be observed and outliers can be detected.

The *analysis* of the identified model can be based on the output error

$$e(k) = y(k) - \hat{y}(k) - \hat{y}\left(\hat{\mathbf{\Theta}}, k\right) \tag{3.7.8}$$

where the measured outputs $y(k)$, the model outputs $\hat{y}\left(\hat{\mathbf{\Theta}}, k\right)$ and its error $e(k)$ are shown in an XT plot.

Again, the coefficients of determination

$$R^2 = \frac{\sum_{k=1}^{N} \left(\hat{y}(k) - \overline{y}\right)^2}{\sum_{k=1}^{N} \left(y(k) - \overline{y}\right)^2} \tag{3.7.9}$$

is a suitable criterion for an overall model quality.

If the dynamic measurements are presented in measured versus predicted plot (MPP), the plotted points carry a time stamp. For perfect measurement and model matching the points are on the bisection line. Model deviations according to (3.7.8) show up as horizontal errors, and the effect of disturbances in vertical differences. The points for steady-state measurements should be especially signed, e.g. by a* to indicate the precision of the static behavior.

For more details and examples, see Kötter (2008).

3.8 Examples for engine identification

The application of some of the described methods are now demonstrated with measurements on a dynamic test bench. The engine used is a common rail diesel engine Opel Z19 DTH, 1.9 l, 4-cyl., 4 valves, 110 kW, 315 Nm, $\epsilon = 17.5$.

a) LOLIMOT structure and identification, step functions (LLS)

The a priori knowledge for the experiment shown in Fig. 3.8.1a) was obtained through a static reduced grid measurement. Then, a static LOLIMOT was identified and step functions for the five manipulated variables were applied over a measurement time of 2.5 hrs.

The generalization in Fig. 3.8.1b) for NO_x as example shows a relatively good agreement with maximal differences of about 10 %.

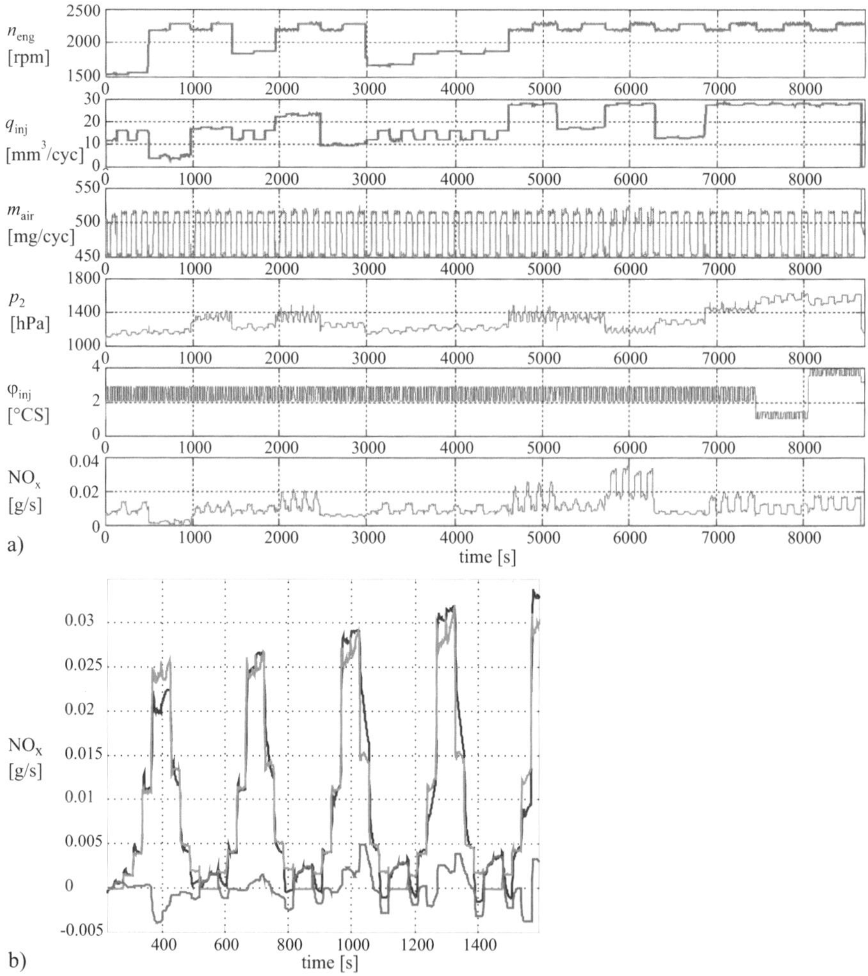

Fig. 3.8.1. Identification of a stationary NO_x-model with five inputs. Structure: LOLIMOT, Identification: LOLIMOT, Test signals: steps (LLS). **a** measured signals. **b** generalization result. Black: measured data, gray: simulated data, Schreiber (2010)

b) LOLIMOT structure and identification, APRBS (LLA)

Measurements for the identification of a NO_x model with simultaneous APRBS test signals for five inputs are depicted in Fig. 3.8.2. The selection of the amplitude was based on a basis PRBS.

The generalization results for the identification over about 45 min measurement time shows maximal differences of about 10 % which is satisfactory for dynamic exhaust gas components.

c) Local stationary torque model: Online (iterative) D-optimal operation points, polynomial model, step signal excitation (online DPD)

A torque model is identified iteratively according to Sect. 3.6, with 16 D-optimal design points per iteration. The manipulated variables are the position of the EGR valve ($0.6\,\mathrm{mm} < s_{\mathrm{egr}} < 6.3\,\mathrm{mm}$), the position of the VTG actuator ($3.7\,mm < s_{\mathrm{vtg}} < 4.8\,mm$) and the angle of main injection ($-2\,°\mathrm{CS} < \varphi_{\mathrm{mi}} < 8\,°\mathrm{CS}$), all within their allowed variation space.

The torque model is a stationary polynomial of third order, identified with the model inputs air mass flow rate $\dot{m}_{\mathrm{air}}$, boost pressure p_2, inlet temperature T_2, angle of main injection φ_{mi} and rail pressure p_{rail}. Figure 3.8.3 depicts the stationary input variables and the output variable, as well as the resulting model and the absolute error of the torque model. The total measurement time was 24 min for nine iterations with a total of 136 measured data points. The polynomial torque model obtained with an automatic regressor selection is

$$\hat{M}_{\mathrm{eng}} = w_7 p_2^3 + w_9 T_2^2 + w_{12}\varphi_{\mathrm{mi}}^2 + w_{13}\varphi_{\mathrm{mi}}^3 + w_{14}p_{\mathrm{rail}} + w_{18}\dot{m}_{\mathrm{air}}T_2 + w_{24}\varphi_{\mathrm{mi}}T_2 \tag{3.8.1}$$

with the model parameters

$$\begin{aligned}
w_7 &= 0.7013\frac{\mathrm{Nm}}{\mathrm{bar}^3}, w_9 = -6.6648\frac{\mathrm{Nm}}{\mathrm{K}^2}, w_{12} = -0.8017\frac{\mathrm{Nm}}{°\mathrm{CS}^2},\\
w_{13} &= -0.349\frac{\mathrm{Nm}}{°\mathrm{CS}^3}, w_{14} = 4.1721\frac{\mathrm{Nm}}{\mathrm{bar}}, w_{18} = 4.7938\frac{\mathrm{Nms}}{\mathrm{kgK}}\\
w_{24} &= 0.3486\frac{\mathrm{Nm}}{\mathrm{K}°\mathrm{CS}}.
\end{aligned} \tag{3.8.2}$$

The resulting model quality of the stationary torque model after the iterative online measurement procedure is $R^2 = 0.9699$, Kowalczyk (2014).

d) Global dynamic boost pressure model: Online (iterative) D-optimal operation points, LOLIMOT identification, APRBS excitation (online DLA)

A global charging pressure model is identified in two iterations with D-optimal design points. The manipulated variables are the position of the EGR valve ($0.6\,\mathrm{mm} < s_{\mathrm{egr}} < 6.3\,\mathrm{mm}$), the position of the VGT actuator ($3.7\,\mathrm{mm} < s_{\mathrm{vtg}} < 4.8\,\mathrm{mm}$) and

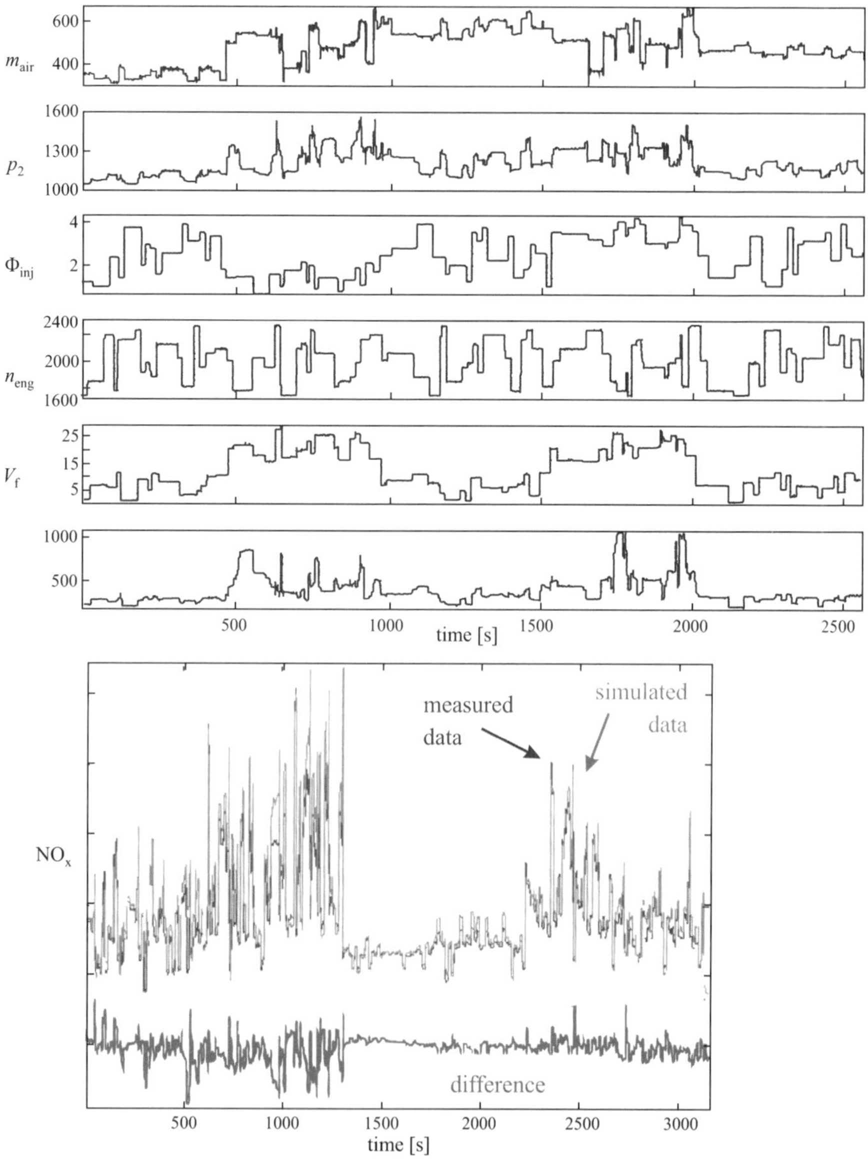

Fig. 3.8.2. Identification of a dynamic NO$_x$-model with five inputs. Structure: LOLIMOT, Identification: LOLIMOT, Test signals: APRBS. **a** measured signals. Identification quality: $J_{\mathrm{NRMSE}} = 0.085$. **b** generalization result, $J_{\mathrm{NRMSE}} = 0.339$. Black: measured date. Grey: simulated data.

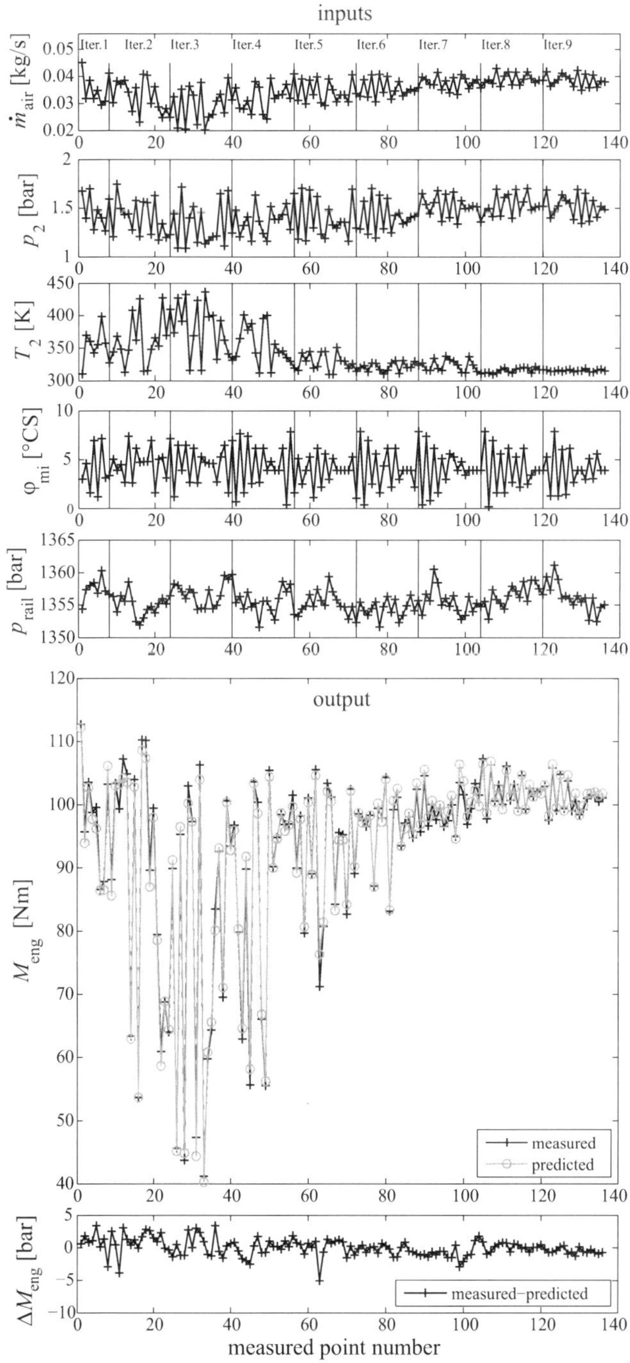

Fig. 3.8.3. Input and output variables for modeling and quality comparison for the stationary torque polynomial model of third order with an iterative identification procedure. Operation point $n_{eng} = 3000\,\text{rpm}$; $q_{mi} = 30\,\frac{\text{mm}^3}{\text{stroke}}$, Opel Z19DTH, Kowalczyk and Isermann (2014).

the operation point variables main injection quantity ($10\,\frac{\mathrm{mm}^3}{\mathrm{cyc}} < q_{\mathrm{mi}} < 30\,\frac{\mathrm{mm}^3}{\mathrm{cyc}}$) and engine speed ($1500\,\mathrm{rpm} < n_{\mathrm{eng}} < 3000\,\mathrm{rpm}$). The iterative excitation is performed with an APRBS with 2 s cycle time and eight stages, resulting in impulse lengths from 2 s to 16 s and a total signal length of 510 s. The dynamic input variables and the output variable are shown in Fig.3.8.4, as well as the resulting model and the absolute error. The total measurement time was 17 min with two iterations.

The dynamic LOLIMOT model of the boost pressure $\hat{p}_2$ consists of 17 local linear models, where the output of each dynamic local linear model is calculated by

$$\begin{aligned}\hat{p}_{2,i} =& w_{i,0}+\\ & w_{i,1}s_{\mathrm{egr}}(k) + ... + w_{i,4}s_{\mathrm{egr}}(k-3)+\\ & w_{i,5}s_{\mathrm{vtg}}(k) + ... + w_{i,8}s_{\mathrm{vtg}}(k-3)+\\ & w_{i,9}q_{\mathrm{mi}}(k) + ... + w_{i,12}q_{\mathrm{mi}}(k-3)+\\ & w_{i,13}n_{\mathrm{eng}}(k) + ... + w_{i,16}n_{\mathrm{eng}}(k-3).\end{aligned} \tag{3.8.3}$$

The resulting model quality of the boost pressure model $\hat{p}_2$ after the iterative online measurement is $R^2 = 0.9861$, Kowalczyk (2014).

e) Local dynamic boost pressure model: Online (iterative) LHS operation points, polynomial model, DSBS excitation (online LPD)

The next example shows the result for an iterative online identification for modeling the local dynamic behavior of the charging pressure. The two input variables are selected by a space-filling method and are the pulse-width of the EGR valve voltage ($0\,\% < u_{\mathrm{egr}} < 100\,\%$) and the pulse-width of the VGT actuator voltage ($0\,\% < u_{\mathrm{egr}} < 100\,\%$). They are excited with a deterministic short time binary signal DSBS. All other actuators are kept at constant values during the measurement at the operation point with $20\,\frac{\mathrm{mm}^3}{\mathrm{cyc}}$ main injection and $2000\,\mathrm{rpm}$ engine speed. The last 250 samples of the dynamic input variables and the output variable are shown in Fig.3.8.5. The absolute error of the boost pressure output is depicted for all measured samples. The total measurement time was a total of 395 s with ten iterations. The polynomial boost pressure model obtained with an automatic regressor selection is

$$\begin{aligned}\hat{p}_2(k) =& w_1 + w_4 u_{\mathrm{egr}}^2(k) + w_6 u_{\mathrm{egr}}(k) u_{\mathrm{vgt}}(k)+\\ & w_9 u_{\mathrm{egr}}^2(k-1) + w_{19} u_{\mathrm{egr}}^2(k-3) + w_{20} u_{\mathrm{vgt}}^2(k-3)+\\ & w_{22} p_2(k-1) + w_{26} p_2(k-3) + w_{27} p_2^2(k-3)\end{aligned} \tag{3.8.4}$$

with the finally identified model parameters

$$\begin{aligned}& w_1 = 0.0132\mathrm{bar}, w_4 = 0.0364\frac{\mathrm{bar}}{\mathrm{mm}^2}, w_6 = -0.0344\frac{\mathrm{bar}}{\mathrm{mm}^2},\\ & w_9 = 0.0071\frac{\mathrm{bar}}{\mathrm{mm}^2}, w_{19} = -0.0184\frac{\mathrm{bar}}{\mathrm{mm}^2}, w_{20} = 0.0059\frac{\mathrm{bar}}{\mathrm{mm}^2},\\ & w_{22} = 0.9673, w_{26} = -0.0502 \text{ and } w_{27} = 0.0468\frac{1}{\mathrm{bar}}.\end{aligned} \tag{3.8.5}$$

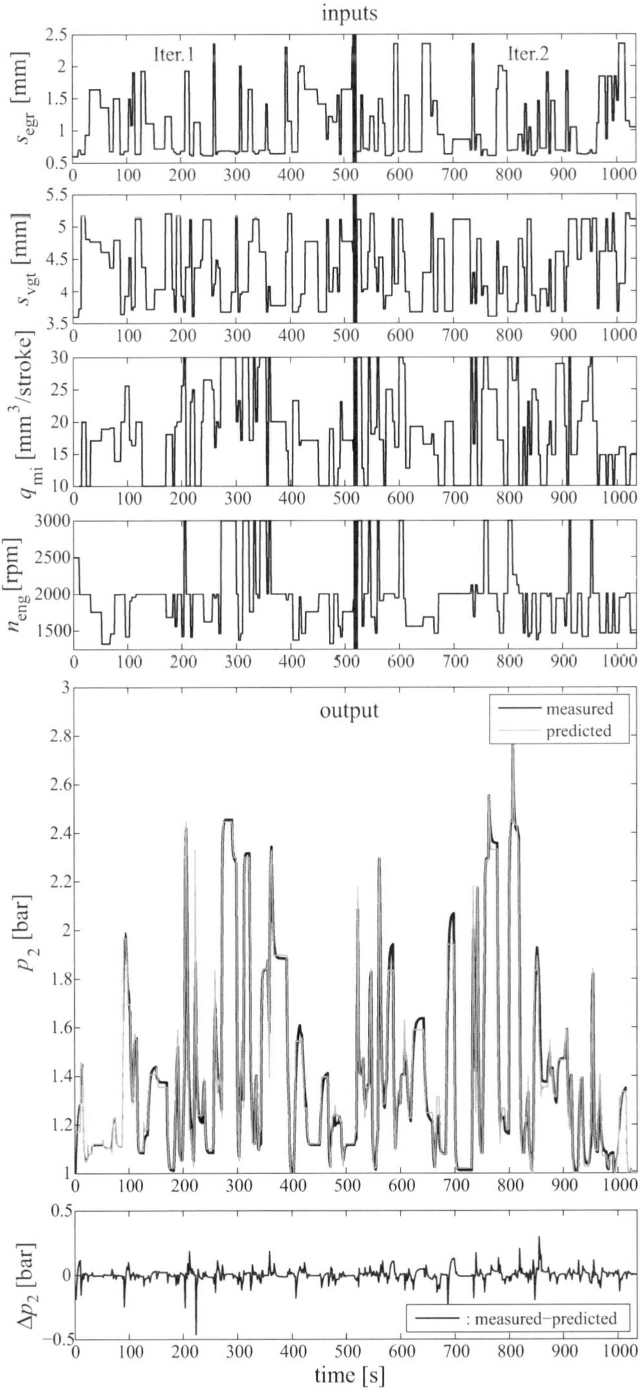

Fig. 3.8.4. Input variables for modeling and quality comparison for the global dynamic boost pressure model with an iterative identification procedure, Kowalczyk and Isermann (2014).

The resulting model quality of the dynamic boost pressure model after the iterative online measurement is $R^2 = 0.967$ for only about 6.6 min measurement time. Figure 3.8.6 shows a step response of this dynamic nonlinear model, compared to direct measurements. The stationary behavior as well as the dynamic behavior show a good agreement.

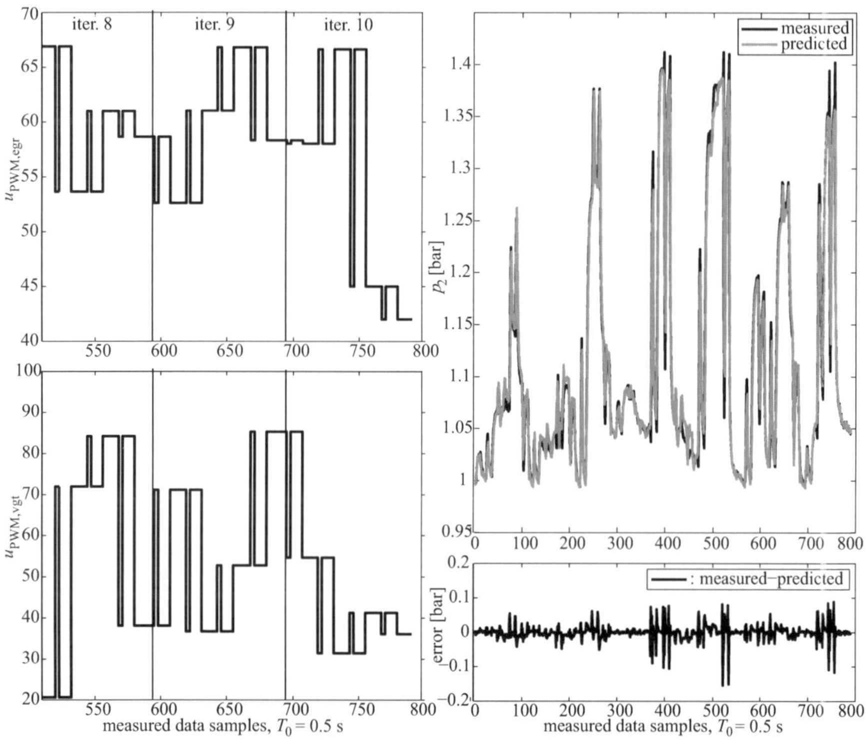

Fig. 3.8.5. Input and output variables for modeling and quality comparison for a local dynamic polynomial boost pressure model applying an iterative online identification procedure and a sequence of deterministic short time binary test signals (DSBS), Kowalczyk et al (2014). Operating point: $q_f = 20\,\text{mm}^3/\text{cyc}$, $n_{eng} = 2000$ rpm.

References

Amann H, Escher J (2008) Analysis III, 2nd edn. Birkhäuser, Basel

Ayoubi M (1996) Nonlinear system identification based on neural networks with locally distributed dynamics and application to technical processes. Fortschr.-Ber. VDI Reihe 8, 591. VDI Verlag, Düsseldorf

Babuska R, Verbruggen H (1996) An overview of fuzzy modeling for control. Control Engineering Practice – CEP 4(11):1593–1606

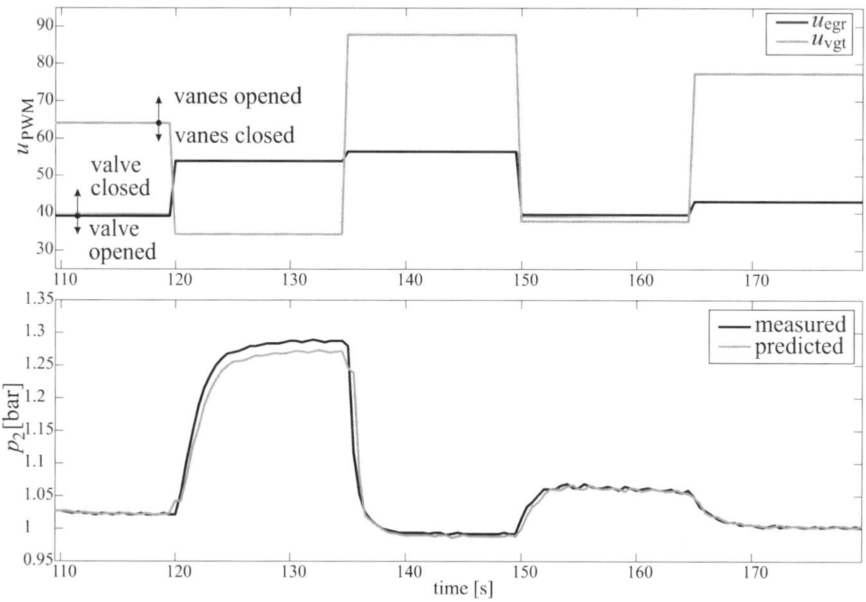

Fig. 3.8.6. Step responses of the charging pressure for step changes of two input variables at the operation point $n_{\mathrm{eng}} = 2000\,\mathrm{rpm}$ and $q_{\mathrm{mi}} = 20\,\frac{\mathrm{mm}^3}{\mathrm{stroke}}$, Kowalczyk (2014). Direct measured output compared to the simulated model output.

Ballé P (1998) Fuzzy-model-based parity equations for fault isolation. Control Engineering Practice – CEP 7:261–270

Bänfer O, Nelles O (2009) Polynomial model tree (polymot) – a new training algorithm for local model networks with higher degree polynomials. In: IEEE International Conference of Control and Automation, Christchurch, New Zealand, pp 1571–1576

Bishop C (1995) Neural networks for pattern recognition. Oxford University Press, Oxford

Blessing P (1979) Identification of input/output and noise dynamics of linear multivariable systems. In: 5th IFAC Symposium on Identification, Darmstadt, Germany

Böhme D (2013) Methoden zur effizienten Basisapplikation für Luftpfad- und Abgastemperaturmodelle von Ottomotoren. Fortschr.-Ber. VDI Reihe 12, 767. VDI Verlag, Düsseldorf

Böhme D, Isermann R, Martiny M, Kern J, Kerkau M (2009) Quasi-stationary measurement method for the ecu-functions of the cylinder charge determination and the torque model. In: 3rd International Symposium on Development Methodology, Wiesbaden, Germany

Büchel M, Thomas M (2009) Roll-out of a fast calibration approach for engine base calibration. In: 3rd Internationales Symposium on Development Methodology, AVL, Wiesbaden

Drewelow W (1990) Parameterschätzung nach der Ausgangsfehlermethode. Messen, Steuern, Regeln – MSR 33(1):15–23
Ernst S (1998) Hinging hyperplane trees for approximation and identification. In: 37th IEEE Conference on Decision and Control, Tampa, FL, USA
Eykhoff P (1974) System identification. John Wiley, London
Fehl G (1992) Entwicklung einer Meßstrategie zur kontinuierlichen Erfassung von Kennlinien auf Motorenprüfständen. PhD thesis, Technische Hochschule Darmstadt, Darmstadt
Fink A (2006) Nonlinear control based on local linear neuro-fuzzy models. Dissertation Technische Universität Darmstadt. Fortschr.-Ber. VDI Reihe 8, 1096. VDI Verlag, Düsseldorf
Gschweitl K, Pfluegl H, Fortuna T, Leithgoeb R (2001) Steigerung der Effizienz der modellbasierten Motorenapplikation durch die neue CAMEO Online DoE-Toolbox. ATZ - Automobiltechnische Zeitschrift 103(7):636–643
Hafner M (2002) Modellbasierte stationäre und dynamische Optimierung von Verbrennungsmotoren am Motorenprüfstand unter Verwendung neuronaler Netze. Dissertation Technische Universität Darmstadt. Fortschr.-Ber. VDI Reihe 12, 482. VDI Verlag, Düsseldorf
Hafner M, Müller N, Semmler I S and Skranjk (2000a) Incorporation of neuro-fuzzy knowledge for fast measurement of combustion engine maps. In: The 2000 ASME International Mechanical Engineering Congress and Exposition, Orlando, Florida, vol 1, pp 585–590
Hafner M, Schüler M, Nelles O, Isermann R (2000b) Fast neural networks for diesel engine control design. Control Engineering Practice – CEP 8:1211–1221
Hafner S, Geiger H, Kreßel U (1992) Anwendung künstlicher neuronaler Netze in der Automatisierungstechnik. Teil 1: Eine Einführung. Automatisierungstechnische Praxis – atp 34(10):592–645
Haykin S (1994) Neural networks. Macmillan College Publishing Company, Inc., New York
Hecht-Nielson R (1990) Neurocomputing. Addison-Wesley, Reading, MA
Holzmann H, Halfmann C, Germann S, Würtenberger M, Isermann R (1997) Longitudinal and lateral control and supervision of autonomous intelligent vehicles. Control Engineering Practice – CEP 5(11):1599–1605
Hoschek J, Lasser D (1993) Fundamentals of computer-aided geometric design. AK Peters, Wellesley, MA
Isermann R (1992) Identifikation dynamischer Systeme. Springer, Berlin
Isermann R, Münchhof M (2011) Identification of dynamic systems. Springer, Berlin
Isermann R, Töpfer S, Nelles O (1997) Identification with dynamic neural networks - architecture, comparisons, applications -. In: Proc. IFAC Symposium on System Identification, Fukuoka, Japan
Kleppmann W (2003) Versuchsplanung, 3rd edn. Hanser Verlag, München
Kötter H (2007) Kennfeldoptimierung. Rechnergestützte Kennfeldoptimierung. FVV Heft 831. FVV, Frankfurt
Kötter H (2008) Innovative Motorvermessung. Schnelle Motorvermessung mit optimierten Testsignalen. FVV Heft 853. FVV, Frankfurt

Kowalczyk M (2013) Online-Methoden – Optimierte schnelle Motorvermessung mit Online-Methoden zur Bestimmung von statischen und dynamischen Modellen. FVV Heft 987. FVV, Frankfurt

Kowalczyk M (2014) Online-Methoden II – Erweiterung der Online-Vermessung um aktiv lernende Optimierungsmethoden zur Emissions- und Verbrauchsreduktion. Heft 1043-2014. FVV, Frankfurt

Kowalczyk M, Isermann R (2014) Online and active learning measurement of engines. FVV conference. Report R566-2014. Madgeburg

Kowalczyk M, Isermann R, Lindner S (2014) Online-Methoden II – Erweiterung der Online-Vermessung um aktiv lernende Optimierungsmethoden zur Emissions- und Verbrauchsreduktion (Extended online-identification with active learning methods). FVV Report 1043-2014. FVV, Frankfurt

Kruse T, Kurz S, Lang T (2010) Modern statistical modeling and evolutionary optimization methods for the use in ECU calibration. In: IFAC Symposium Advances in Automotive Control, Munich, Germany

Kruse T, Kreuzinger T, Ulmer H, Lang T (2012) Use of new data-based methods for the ecu calibration on a HiL system. In: 5th Conference on Simulation and Testing for Automotive Electronics, Berlin, Germany

Leithgöb R, Bollig BM M, Henzinger F (2005) Methodology for efficient calibration of model-based ECU structures. In: 1st Internationales Symposium on Development Methodology, AVL, Wiesbaden

Ljung L (1999) System identification – theory for the user, 2nd edn. Prentice Hall, Upper Saddle River

Mallows C (1973) Some comments on c_p. Technometrics 15(4):661–675

Martini E, Voß H, Töpfer S, Isermann R (2003) Effiziente Motorapplikation mit lokal linearen neuronalen Netzen. Motortechnische Zeitschrift - MTZ 64:406–413

McCulloch W, Pitts W (1943) A logical calculus of the ideas immanent in nervous activity. Bull Math Biophys 5:115–133

Mitchell J (2000) An algorithm for the construction of D-optimal experimental designs. Technometrics 42(2):48–54

Montgomery D (2005) Design and analysis of experiments, 6th edn. John Wiley and Sons, New York

Müller N, Hafner M, Isermann R (2000) A neuro-fuzzy based method for the design of combustion engine dynanometer experiments. SAE 2000 World Congress, Detroit, Michigan, USA, March 6-9, 2000 SP-1511, pp 331-337 (12)

Murray-Smith R, Johansen T (1997) Multiple model approaches to modelling and control. Taylor and Francis, London

Myers R (1990) Classical and modern regression with applications. PWS-Kent Publishing Company, Boston

Nelles O (1997) LOLIMOT – Lokale, lineare Modelle zur Identifikation nichtlinearer, dynamischer Systeme. Automatisierungstechnik – at 45(4):163–174

Nelles O (2001) Nonlinear system identification. Springer, Heidelberg

Nelles O, Isermann R (1995) Identification of nonlinear dynamic systems – classical methods versus radial basis function networks. In: Proc. American Control Conference (ACC), Seattle, USA

Nelles O, Hecker O, Isermann R (1991) Automatic model selection in local linear model trees (LOLIMOT) for nonlinear system identification of a transport delay process. In: Proc. 11th IFAC Symposium on System Identification (SYSID), Kitakyushu, Fukuoka, Japan

Pfeufer T (1999) Modellgestützte Fehlererkennung und Diagnose am Beispiel eines Fahrzeugaktors. Fortschr.-Ber. VDI Reihe 8, 749. VDI Verlag, Düsseldorf

Poland J (2002) Modellgestützte und Evolutionäre Optinierungsverfahren für die Motorentwiclkung. Logos, Berlin

Preuß HP, Tresp V (1994) Neuro-Fuzzy. Automatisierungstechnische Praxis - at 36(5):10–24

Renninger P (2004) Schnelle Variationsraumbestimmung – Vorraussetzung für eine effektive Versuchsplanung. In: Motorenentwicklung auf dynamischen Prüfständen, Haus der Technik, Wiesbaden, Germany

Rosenblatt F (1958) The perceptron: a probabilistic model for information storage and organization in the brain. Psychological Review 65:386–408

Santner J, Williams B, Notz W (2003) The design and analysis of computer experiments. Springer, New York

Schmitt M (1995) Untersuchungen zur Realisierung mehrdimensionaler lernfähiger Kennfelder in Großserien-Steuergeräten. Dissertation Technische Universität Darmstadt. Fortschr.-Ber. VDI Reihe 12, 246. VDI Verlag, Düsseldorf

Schreiber A (2010) Dynamische Motorvermessung mit verschiedenen Methoden und Modellen. In: Isermann R (ed) Elektronisches Management motorischer Fahrzeugantriebe, Vieweg+Teubner, Wiesbaden

Schreiber A, Isermann R (2007) Identification methods for experimental nonlinear modeling of combustion engines. In: 5th IFAC Symposium on Advances in Automotive Control, Aptos, California, USA

Schreiber A, Isermann R (2009) Integrated methodology for stationary and dynamic measurement and modeling of combustion engines. In: 3rd Internationales Symposium on Development Methodology, AVL, Wiesbaden

Schreiber A, Isermann R, Vogels MS (2007) Dynamische Vermessung und Modellbildung hochdimensionaler Zusammenhänge in Verbrennungsmotoren mit dynamischen Motorenprüfständen. In: 2. Internationales Symposium für Entwicklungsmethodik, Wiesbaden, Germany

Schüler M (2001) Stationäre Optimierung der Motorsteuerung von PKW-Dieselmotoren mit Abgasturbolader durch Einsatz schneller neuronaler Netze. Fortschr.-Ber. VDI Reihe 12, 461. VDI Verlag, Düsseldorf

Schwarte A, Hack L, Isermann R, Nitzke HG, Jeschke J, Piewek J (2004) Automatisierte Applikation von Motorsteuergeräten mit kontinuierlicher Motorvermessung. In: AUTOREG 2004, VDI, Wiesloch, Germany, vol VDI-Berichte 1828

Sequenz H (2013) Emission Modelling and Model-Based Optimisation of the Engine Control. Dissertation Technische Universität Darmstadt. Fortschr.-Ber. VDI Reihe 12, 1222. VDI Verlag, Düsseldorf

Sequenz H, Schreiber A, Isermann R (2009) Identification of nonlinear static processes with local polynomial regression and subset selection. In: 15th Symposium on System Identification, SYSID 2009, Saint-Malo, France

Töpfer S (2002) Hierarchische neuronale Modelle für die Identifikation nichtlinearer Systeme. Fortschr.-Ber. VDI Reihe 10, 705. VDI Verlag, Düsseldorf

Töpfer S (2003) Statistische Versuchsplanung. In: Isermann R (ed) Modellgestützte Steuerung, Regelung und Diagnose von Verbrennungsmotoren, Springer, Berlin

Töpfer S, Wolfram A, Isermann R (2002) Semi-physical modelling of nonlinear processes by means of local model approaches. In: Proc. 15th IFAC World Congress, Barcelona, Spain

Verification and validation (2012) en.wikipedia.org/wiki; visited on May 21st 2012

Ward M, Brace CJ, Vaughan ND, Ceen R, Hale T, Kennedy G (2002) Investigation of sweep mapping approach on engine testbed. In: SAE paper number 2002-01-0615, SAE, Detroit, Michigan

Weber M, Kötter H, Schreiber A, Isermann R (2005) Model-based design of experiments for static and dynamic measurements of combustion engines. In: 3. Tagung Design of Experiments (DoE) in der Motorenentwicklung, Berlin, Germany

Widrow B, Hoff M (1960) Adaptive switching circuits. IRE WESCON Conv Rec pp 96–104

Zimmerschied R, Isermann R (2009) Nonlinear system identification of block-oriented systems using local affine models. In: 15th Symposium on System Identification, SYSID 2009, Saint-Melo, France

Zimmerschied R, Weber M, Isermann R (2005) Statische und dynamische Motorvermessung zur Auslegung von Steuerkennfeldern - eine kurze Übersicht. Automatisierungstechnik – at 53(2):87–94

Part II

Engine Models

4

General combustion engine models

As a systematic development of sophisticated control and diagnosis functions for internal combustion engines requires precise stationary and dynamic mathematical models, this chapter is devoted to treating physical-oriented models of engine components. According to Table 1.4.1 the models for the most important engine parts are considered, taking their different physical domains into account. The modeling approaches begin with the intake systems and continue with the fuel supply and injection system, and a description of the combustion cycles with combustion pressure and heat release analysis. Then the mean value and dynamic torque generation with the piston/crankshaft system, turbochargers with wastegate and variable geometry manipulation, the exhaust system with catalysts and filters and the cooling system with different kinds of heat exchangers are considered. These mathematical models lay the background for the design of the control structures in Chap. 5, the design of control functions of gasoline engines, Chap. 7 and of diesel engines, Chap. 8. For several components the basic mathematical models for gasoline and diesel engines are the same or at least similar. Where necessary, the component models were separately derived for both engine types in Chaps. 7 and 8 as, e.g. for the injection system, operating cycles, and the exhaust system.

In general, the physical/chemical-law-based models are simplified as much as possible in order to enable fast simulations (e.g. hardware-in-the-loop (HiL)) or real-time control and diagnosis functions. As most models contain parameters which are difficult to determine from construction data only, they have to be identified experimentally. The basic physical-oriented structure of the mostly nonlinear models then allow a relatively precise modeling by combining theoretical and experimental modeling approaches, resulting in so called semi-physical or gray-box models, as discussed in Chaps. 2 and 3.

R. Isermann, *Engine Modeling and Control*,
DOI: 10.1007/978-3-642-39934-3_4

4.1 Intake systems

4.1.1 Intake system for gasoline engines

For gasoline engines (or spark-ignition engines, SI) with homogeneous combustion the combustion and the torque is mainly determined by the air charge in the cylinders. This holds especially for external and internal mixture formation because of the required near stoichiometric air/fuel (A/F) ratio for homogeneous mixture. By contrast, for engines with internal mixture formation and excess air the torque is determined by varying the injected fuel as for stratified combustion.

The SI engine requires in all cases a specific air/fuel ratio to operate smoothly and efficiently. Ideally a theoretical complete combustion is possible for a mass ratio of $14.7\,\text{kg}$ of air for $1\,\text{kg}$ of gasoline. The excess-air factor λ is a measure to express deviations from the theoretical ratio $14.7 : 1$.

$$\lambda = \frac{\text{supplied air mass}}{\text{stoichiometric air mass}} = \frac{m_{\text{air}}}{m_{\text{air,sto}}}. \tag{4.1.1}$$

If $\lambda = 1$, the supplied air mass corresponds to the theoretically necessary stoichiometric air mass. For $\lambda < 1$ results an air deficiency or a rich A/F ratio, leading to a maximum power for $\lambda = 0.85 \ldots 0.95$. Excess air or a lean A/F ratio is obtained for $\lambda > 1$. This means a reduced power, but minimum specific fuel consumption around $\lambda = 1.2 \ldots 1.5$. If there is too much excess air and homogeneous combustion, the A/F mixture does not ignite, resulting in misfires. Only direct injection and charge stratification SI engines can operate with high excess-air factors up to $\lambda \approx 4$ in the part-load range.

The catalytic exhaust gas treatment with a three-way catalytic converter, however, requires exactly $\lambda = 1$. Therefore, the air mass has to be precisely determined and the fuel exactly metered before the combustion. A λ-sensor measures the oxygen-content in the exhaust gas and corrects the A/F ratio via the λ-controller.

The next sections describe the calculation of the charge air mass in the cylinders based on measurements of the air-mass flow, the manifold pressure or the throttle position. Because of the retarding effect of the volume of the intake manifold the dynamics of the air flow and air pressure play an important role.

a) Dynamics of the intake manifold

An arrangement of the components of the intake manifold with sensors and actuators is depicted in Fig. 4.1.1. Considering the engine as a pump the sucked air mass for one cylinder is theoretically, without exhaust gas recirculation (EGR),

$$\dot{m}_{\text{air,cyl,th}} = \rho_2 \dot{V}_{\text{air,cyl,th}} = \frac{p_2}{RT_2} \dot{V}_{\text{air,cyl,th}} \tag{4.1.2}$$

with the (continuously assumed) volume flow in one cylinder of a 4-stroke engine

$$\dot{V}_{\text{air,cyl,th}} = \frac{1}{2}\, V_{\text{d}}\, n \tag{4.1.3}$$

where V_d is the displacement of one cylinder and $n = n_\mathrm{eng}$ the speed. The real air mass in the cylinder is smaller because of several effects like throttling at the valves, acceleration of the air mass, valve timing, resonances etc. This is taken in account by defining a *volumetric efficiency*, Heywood (1988), Pischinger et al (2009),

$$\eta_\mathrm{v} = \frac{\dot{m}_\mathrm{air,cyl}}{\dot{m}_\mathrm{air,cyl,th}} = \frac{\dot{m}_\mathrm{air,cyl}}{\frac{1}{2}\, n\, V_\mathrm{d}\, \frac{p_2}{RT_2}} \tag{4.1.4}$$

which is determined experimentally as a look-up table in dependence on n and throttle position α_th or n and ρ_2, see Fig. 4.1.2.

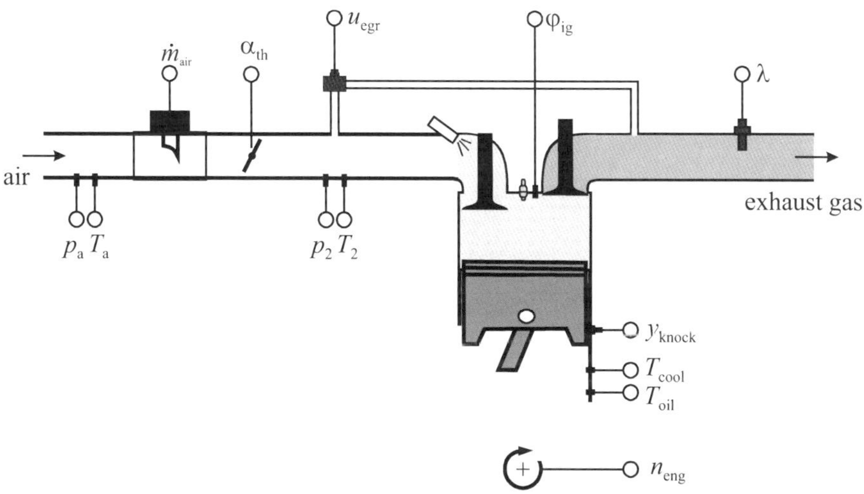

Fig. 4.1.1. Intake manifold and exhaust recirculation with actuators and sensors.

Hence, the sucked air-mass flow in the cylinder becomes without EGR (exhaust gas recirculation)

$$\dot{m}_\mathrm{air,cyl} = \eta_\mathrm{v} \frac{p_2\, V_\mathrm{d}\, n}{2\, R\, T_2} = \dot{m}_\mathrm{gas,cyl}. \tag{4.1.5}$$

For engines with EGR, fuel injection, tank ventilation and crankcase ventilation blow-by gas into the manifold influences the pressure p_2 according to the partial pressures of the gas components (Dalton's law), Müller (2003),

$$p_2 = p_\mathrm{2,air} + p_\mathrm{2,egr} + p_\mathrm{2,fuel} + p_\mathrm{2,tank} = p_\mathrm{2,air} + p_\mathrm{2,add}. \tag{4.1.6}$$

The partial pressure of the EGR for example follows from the EGR rate,

$$r_\mathrm{egr}(u_\mathrm{egr}) = \frac{p_\mathrm{2,egr}(u_\mathrm{egr})}{p_\mathrm{2,air}}\, 100\, [\%] \tag{4.1.7}$$

which has to be determined experimentally in dependence on the position u_egr of the EGR valve, pressure difference between exhaust pressure and manifold pressure and engine speed. Neglecting $p_\mathrm{2,fuel}$ and $p_\mathrm{2,tank}$ the gas mass flow with EGR yields then

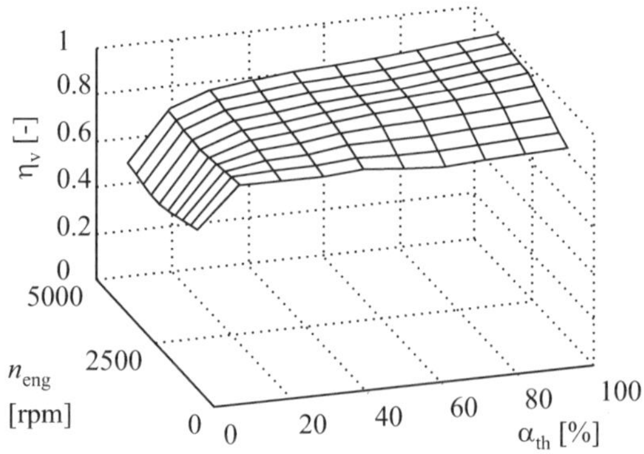

Fig. 4.1.2. Volumetric efficiency in dependence on engine speed and throttle position for a 1 l SI engine, 40 kW, Müller (2003).

$$\dot{m}_{\text{gas,cyl}} = \eta_{\text{v}}(n, \rho_2)\frac{p_2\,(r_{\text{egr}})\,V_{\text{d}}\,n}{2\,R\,T_2}. \tag{4.1.8}$$

The gas constant $R = R_{\text{air}} = 287.4$ J/kg K can be assumed the same for air and exhaust gas, Pischinger et al (2009).

Another characteristic quantity for the air mass in the cylinder is the "*relative air charge*" which follows from reference to normal condition

$$l_r = \frac{m_{\text{air,cyl}}}{m_{\text{air,norm}}} \tag{4.1.9}$$

with $m_{\text{air,norm}} = p_0\,V_{\text{d}}/R_L\,T_0,\ p_0 = 1013\,\text{bar},\ T_0 = 273\,K$ and

$$m_{\text{air,cyl}} = \dot{m}_{\text{air,cyl}}\frac{2}{n}. \tag{4.1.10}$$

l_{r} is independent on the displacement and serves for the comparison of different engines. The basis for determining the *dynamics* of the intake manifold is the mass balance

$$\frac{dm_{\text{gas,int}}}{dt} = \dot{m}_{\text{gas,in}}(t) - \dot{m}_{\text{gas,out}}(t) = \dot{m}_{\text{gas,in}}(t) - \dot{m}_{2,\text{gas}}(t) \tag{4.1.11}$$

with

$$\dot{m}_{\text{gas,in}} = \dot{m}_{\text{air,in}} + \dot{m}_{\text{add}} = \dot{m}_1 + \dot{m}_{\text{add}} \tag{4.1.12}$$

$$\dot{m}_{\text{add}} = \dot{m}_{\text{egr}} + \dot{m}_{\text{fuel}} + \dot{m}_{\text{crankcase}} + \dot{m}_{\text{tank}}$$

$$\dot{m}_{\text{gas,out}} = \dot{m}_{2,\text{gas}} = i_{\text{c}}\dot{m}_{\text{gas,cyl}} \tag{4.1.13}$$

where i_{c} is the number of cylinders. $\dot{m}_{\text{gas,in}}$ is the sum of all entering gas flows between the throttle and the intake valves and consists of the air-mass flow through the throttle $\dot{m}_{\text{air,in}}$ and all additional recirculated gas flow effects $\dot{m}_{\text{add}}$. The state equation of the gas in the intake manifold is

$$m_{\mathrm{gas,int}}\, R T_2 = p_2\, V_{\mathrm{int}} \tag{4.1.14}$$

and its first derivative becomes

$$\frac{dm_{\mathrm{gas,int}}}{dt} = \frac{V_{\mathrm{int}}}{R T_2}\,\frac{dp_2(t)}{dt}. \tag{4.1.15}$$

Introduction into (4.1.11) yields with (4.1.13) and (4.1.8)

$$\frac{2\, V_{\mathrm{int}}}{i_{\mathrm{c}} \eta_{\mathrm{v}} V_{\mathrm{d}} n}\,\frac{dp_2(t)}{dt} + p_2(t) = \frac{2\, R T_2}{i_{\mathrm{c}} \eta_{\mathrm{v}} V_{\mathrm{d}} n}\left(\dot{m}_{\mathrm{air,in}}(t) + \dot{m}_{\mathrm{add}}(t)\right) \tag{4.1.16}$$

which is a nonlinear differential equation of first order because η_{v}, $\dot{m}_{\mathrm{air,in}}(t)$, $\dot{m}_{\mathrm{add}}(t)$ and T_2 depend on the manifold pressure p_2. Assuming constant parameters for a fixed operating point the intake time constant becomes

$$T_{\mathrm{int}} = \frac{2\, V_{\mathrm{int}}}{\eta_{\mathrm{v}}\, V_{\mathrm{D}}\, n} \tag{4.1.17}$$

with $V_{\mathrm{D}} = i_{\mathrm{c}} \cdot V_{\mathrm{d}}$ the engine displacement. Hence, the time constant is the smaller, the smaller the ratio $V_{\mathrm{int}}/V_{\mathrm{D}}$ and the engine speed n. The gas mass flow $\dot{m}_{\mathrm{gas,in}}$ depends according to (4.1.12) on several flows. The air flow rate $\dot{m}_{\mathrm{air}}$ is determined by the throttle position and its difference pressure.

Figure 4.1.3 shows a scheme of the throttle plate. The plate has an elliptical form and closes with angle α_0. The cross-sectional open area $A(\alpha_{\mathrm{th}})$ for the air flow can according to Moskwa (1988) be calculated by

$$\begin{aligned} A\left(\alpha_{\mathrm{th}}\right) = &-\frac{dD}{2}\sqrt{1-\left(\frac{d}{D}\right)^2} + \frac{dD}{2}\sqrt{1-\left(\frac{d}{D}\frac{\cos\left(\alpha_0\right)}{\cos\left(\alpha_0+\alpha_{\mathrm{th}}\right)}\right)^2} \\ &+\frac{D^2}{2}\sin^{-1}\left(\sqrt{1-\left(\frac{d}{D}\right)^2}\right) - \frac{D^2}{2}\frac{\cos\left(\alpha_0+\alpha_{\mathrm{th}}\right)}{\alpha_0} \\ &\cdot \sin^{-1}\left(\sqrt{1-\left(\frac{d}{D}\frac{\cos\left(\alpha_0\right)}{\cos\left(\alpha_0+\alpha_{\mathrm{th}}\right)}\right)^2}\right). \end{aligned} \tag{4.1.18}$$

Assuming a one-dimensional and adiabatic flow it holds for the air flow, Heywood (1988), Isermann (2005),

$$\dot{m}_{\mathrm{air,th}} = \dot{m}_{1,\mathrm{th}} = c_{\mathrm{th}} A\left(\alpha_{\mathrm{th}}\right) p_{\mathrm{a}} \sqrt{\frac{2}{R T_{\mathrm{a}}}}\,\Psi\left(\frac{p_2}{p_{\mathrm{a}}}\right) \tag{4.1.19}$$

where p_{a} and T_a are the ambient pressure and temperature and Ψ is an outflow function through a contraction

$$\Psi = \left(\frac{p_2}{p_{\mathrm{a}}}\right)^{\frac{1}{\kappa}} \sqrt{\frac{\kappa}{\kappa-1}\left(1-\left(\frac{p_2}{p_{\mathrm{a}}}\right)^{\frac{\kappa-1}{\kappa}}\right)} \tag{4.1.20}$$

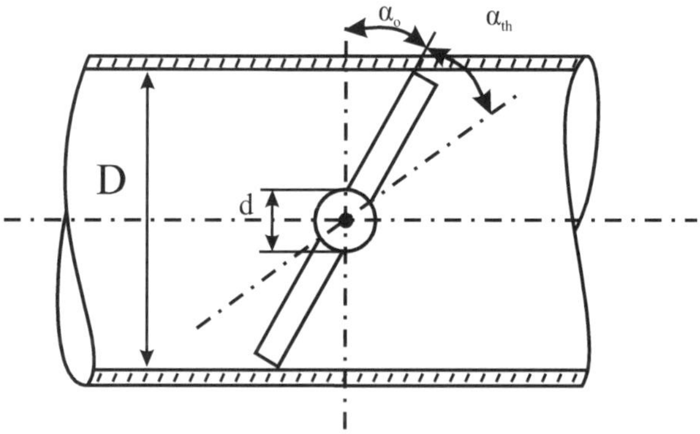

Fig. 4.1.3. Geometry of the throttle plate.

which becomes for dry air with $\kappa = 1.4$

$$\Psi = \left(\frac{p_2}{p_a}\right)^{0.714} \sqrt{3.5\left(1 - \left(\frac{p_2}{p_a}\right)^{0.285}\right)}. \tag{4.1.21}$$

This outflow function has a maximum for a critical pressure ratio

$$\left(\frac{p_2}{p_a}\right)_{\text{crit}} = \left(\frac{2}{\kappa + 1}\right)^{\frac{\kappa}{\kappa-1}} = 0.528 \tag{4.1.22}$$

for air. This leads to $\Psi_{\max} = 0.484$. Then, the sound velocity is reached. Hence, (4.1.20) is valid for subcritical pressure ratio $p_2/p_a \geq 0.528$ and for overcritical pressure ratio $p_2/p_a < 0.528$ the outflow function in (4.1.19) can be set to $\Psi = 0.484$ and the flow rate depends only on p_a, T_a, c_{th} and A which holds about for $p_2 < 0.5\,\text{bar}$. The factor c_{th} in (4.1.19) is a loss factor and models the difference to a throttle without losses, due to e.g. contraction of the flow. It has to be determined experimentally, for example, depending on the throttle angle and the engine speed, as shown in Fig. 4.1.4.

The throttling in the the throttle valve leads to a pressure drop to the intake manifold pressure p_2 and therefore also to a temperature decrease to T_2. Assuming an adiabatic change of the gas state it holds

$$p_2 v_2^{\kappa} = p_a v_a^{\kappa} \tag{4.1.23}$$

and after introducing the gas state equation $pv = RT$ it follows

$$T_2' = T_a \left(\frac{p_2}{p_a}\right)^{\frac{\kappa-1}{\kappa}} \approx T_a \left(\frac{p_2}{p_a}\right)^{0.285}. \tag{4.1.24}$$

Hence, the manifold temperature can be calculated under the conditions that no heat is transferred from the environment to the air in the manifold and additional gas flows are zero, $\dot{m}_{\rm add} = 0$. However, as these assumptions do not hold in general, T_2 is usually measured. Based on the equations for the dynamic behavior of the manifold pressure p_2 and the air flow through the throttle a signal-flow diagram can be constructed, as depicted in Fig. 4.1.5.

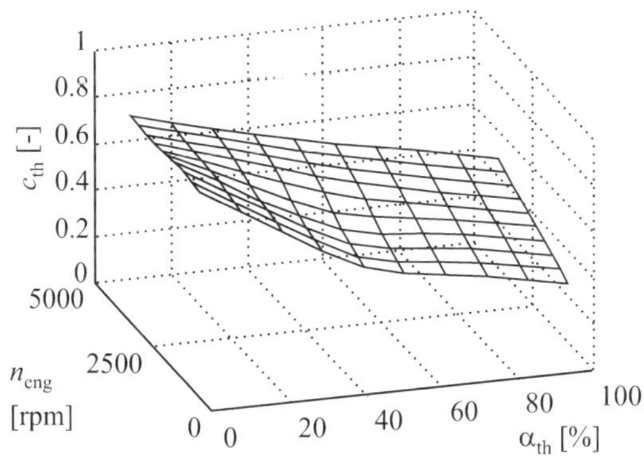

Fig. 4.1.4. Example for the loss factor $c_{\rm th}$ depending on throttle angle and engine speed, Müller (2003).

It shows that the air flow m through the throttle is a nonlinear function and depends on several operating point dependent factors. The air flow rate depends further on the manifold pressure for $(p_2/p_{\rm a}) \leq 0.528$, i.e. in the higher load range. Hence, the manifold pressure p_2 determines the gas flow rate into the cylinders due to (4.1.8). The sucked gas mass $\dot{m}_{2,\rm gas}$ into the cylinders is then delayed because of the first order low pass behavior of the intake manifold pressure with a time constant $T_{\rm int}$, see (4.1.17), which depends on the engine speed n and the volumetric efficiency $\eta_{\rm v}$, where the latter is a measure for the suction losses of the cylinders. An additional influence on the air charge is especially the mass flow rate $\dot{m}_{\rm egr}$ of the exhaust gas recirculation, which depends on the EGR-valve position, the exhaust gas pressure and the manifold pressure, see Sect. 4.1.2. Summarizing, the signal-flow diagram in Fig. 4.1.5 represents the physical based relations and shows the many nonlinear and operating point dependent parameters. Most of the parameters depend on several other variables, which can only be determined experimentally. Hence, look-up tables have to be calibrated, e.g. for $c_{\rm th}(\alpha_{\rm th}, n)$, $\eta_{\rm v}\ (\alpha_{\rm th}, n$ or $p_2)$.

b) Air charge determination

As the air charge into the cylinders $m_{\rm air}$ cannot be measured directly it has to be reconstructed based on the measurements of other variables, like air flow rate $\dot{m}_1$ before the throttle, manifold pressure p_2 or throttle position $\alpha_{\rm th}$, see also Müller (2003), Guzzella and Onder (2010).

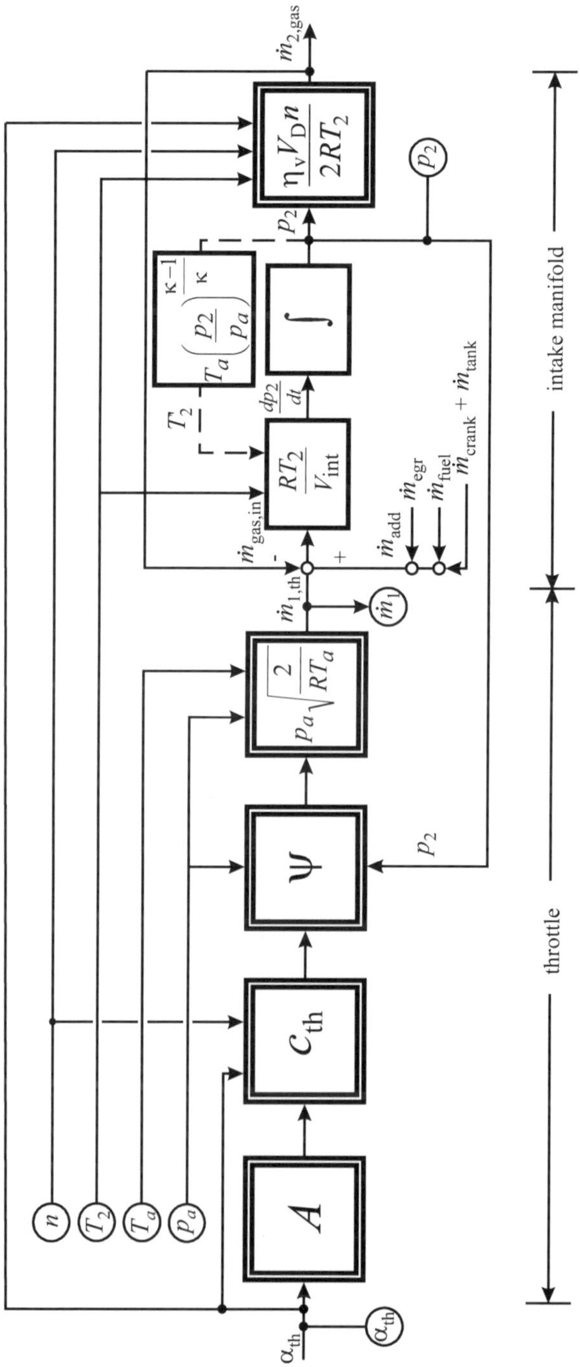

Fig. 4.1.5. Signal-flow diagram for the throttle and intake manifold. ○: measurable variables.

Air flow measurement

The most direct method is to measure the air flow rate $\dot{m}_1$ before the throttle and after the air filter, e.g. by a hot-film sensor HFM. In contrast to other flow meters, the HFM sensor directly measures the mass flow rate, possesses a wide measurement range of 1:50 and has a small time constant. In stationary condition it holds for the air flow rate into the cylinders

$$\dot{m}_{2,\mathrm{air}}(t) = \dot{m}_{\mathrm{HFM}}(t) = \dot{m}_1(t) \tag{4.1.25}$$

if no gases enter the manifold, i.e. for $\dot{m}_{\mathrm{add}} = 0$. In dynamic operation, however, $\dot{m}_{2,\mathrm{air}}$ into the cylinders is delayed by the intake manifold pressure dynamics. This can be taken into account by using (4.1.11), (4.1.12) and inserting (4.1.15) (with $\dot{m}_{\mathrm{add}} = 0$)

$$^1\tilde{\dot{m}}_{2,\mathrm{air}}(t) = \dot{m}_1(t) - \frac{V_{\mathrm{int}}}{RT_2}\frac{dp_2(t)}{dt}. \tag{4.1.26}$$

Correcting the measured air flow rate by the first derivative of the manipulated pressure allows the determination of the air charge in dynamic operation. Because of the included noise and periodicity in $p_2(t)$ this signal has to be lowpass filtered before the derivation, see Schwarte et al (2002). Further required measurements are p_2, T_2. If additional gases $\dot{m}_{\mathrm{add}}$ enter the manifold intake, (4.1.12) is valid and the manifold pressure p_2 does not only depend on $\dot{m}_{\mathrm{air,in}}$ but also on $\dot{m}_{\mathrm{add}}$, see (4.1.16). Then $p_2(t)$ and dp_2/dt change. However, the dynamic correction of (4.1.26) is approximately applicable if e.g. $\dot{m}_{\mathrm{egr}}$ is not very large compared to $\dot{m}_1(t)$.

Manifold pressure measurement

In order to save the air flow sensor, the charge-air can be determined by the manifold pressure p_2. Basically, (4.1.5) is used to determine the gas flow rate $\dot{m}_{\mathrm{gas,cyl}}$ into the cylinders. Taking the additional gases with flow rate $\dot{m}_{\mathrm{add}}$ into account it follows

$$^2\tilde{\dot{m}}_{2,\mathrm{air}}(t) = \frac{\eta_{\mathrm{v}}(n, p_2)\, V_{\mathrm{d}}\, n}{2\, R\, T_2} p_2(t) - \dot{m}_{\mathrm{add}}(t). \tag{4.1.27}$$

As the sucked gas directly depends on the dynamically measured $p_2(t)$ the dynamics of the intake manifold are already included. This air charge determination needs the measurement of p_2 and T_2, the calibrated volumetric efficiency $\eta_{\mathrm{v}}(n, p_2)$ and those for determining $\dot{m}_{\mathrm{add}}$.

Measurement of throttle position

Another possibility to circumvent an air flow sensor is to base the air charge determination on the known throttle angle α_{th}. Then, (4.1.19) can be used and with (4.1.26) follows

$$^3\tilde{\dot{m}}_{2,\mathrm{air}}(t) = c_{\mathrm{th}}(\alpha_{\mathrm{th}}, n)\, A(\alpha_{\mathrm{th}})\, p_{\mathrm{a}}\sqrt{\frac{2}{RT_{\mathrm{a}}}}\,\Psi\left(\frac{p_2(t)}{p_{\mathrm{a}}}\right) - \frac{V_{\mathrm{int}}}{RT_2}\frac{dp_2(t)}{dt}. \tag{4.1.28}$$

This method requires measurement of $\alpha_{\mathrm{th}}, p_{\mathrm{a}}, T_{\mathrm{a}}, p_2, T_2$ and calibration of $c_{\mathrm{th}}(\alpha_{\mathrm{th}}, n)$. The final air charge mass for one cylinder follows according to (4.1.10)

$$m_{\mathrm{air,cyl}} = \tilde{\dot{m}}_{2,\mathrm{air}} \frac{2}{i_{\mathrm{c}} n}. \tag{4.1.29}$$

Table 4.1.1 shows a comparison for the required sensors of the three methods. In all cases p_2, T_2 are required. Using the manifold pressure p_2 is dynamically fast but requires precise calibration of the volumetric efficiency and determination of additional gas flows $\dot{m}_{\mathrm{add}}$. The application of the throttle angle needs also ambient pressure p_{a} and temperature T_{a} and precise calibration of the throttle loss factor c_{th}. The required measurements for stationary operating conditions are summarized in Table 4.1.2. Because the dynamic compensation is not needed, p_2 or T_2 must not be measured for some cases.

Table 4.1.1. Methods for the determination of the air flow-rate into the cylinders with different sensors and dynamic compensation of intake manifold delay

Basic sensor		Additional sensors						Calibration of
		p_{a}	T_{a}	p_2	T_2	n	$\dot{m}_{\mathrm{add}}$	
air-mass flow	$\dot{m}_1$	-	-	✓	✓	-	✓	V_{int}
manifold pressure	p_2	-	-	-	✓	✓	✓	η_{v}
throttle angle $p_2/p_{\mathrm{a}} \geq 0.528$	α_{th}	✓	✓	✓	✓	✓	-	$c_{\mathrm{th}}, V_{\mathrm{int}}$
throttle angle $p_2/p_{\mathrm{a}} < 0.528$	α_{th}	✓	✓	-	✓	✓	-	$c_{\mathrm{th}}, V_{\mathrm{int}}$

Table 4.1.2. Methods for determination of the air flow-rate into the cylinders for stationary operation

Basic sensor		Additional sensors						Calibration of
		p_{a}	T_{a}	p_2	T_2	n	$\dot{m}_{\mathrm{add}}$	
air-mass flow	$\dot{m}_1$	-	-	-	-	-	✓	-
manifold pressure	p_2	-	-	-	✓	✓	✓	η_{v}
throttle angle $p_2/p_{\mathrm{a}} \geq 0.528$	α_{th}	✓	✓	✓	-	✓	-	c_{th}
throttle angle $p_2/p_{\mathrm{a}} < 0.528$	α_{th}	✓	✓	-	-	✓	-	c_{th}

A comparison of the three ways of air flow determination was made by Müller (2003). Most suited is the direct air flow measurement with dynamic compensation. The accuracy of the other methods depends much on a precise calibration of the correction factors. Average deviations of about 2.5 % could be reached.

4.1.2 Intake system for diesel engines

A typical configuration of the intake and exhaust system of a passenger car diesel engine is given in Fig. 4.1.6.

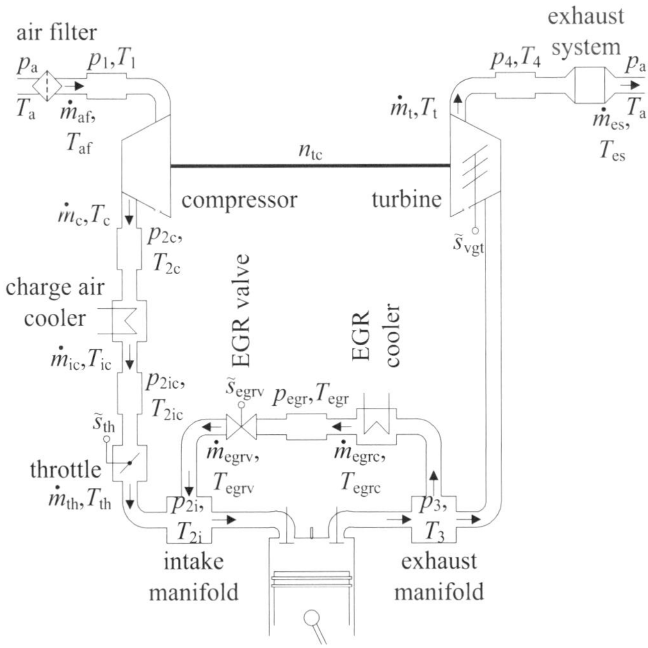

Fig. 4.1.6. Scheme of the air and exhaust path of a diesel engine with turbocharger and exhaust gas recirculation, Zahn and Isermann (2008).

The engine is equipped with an exhaust turbocharger, a charge-air cooler and an exhaust gas recirculation system. The subscript of the variables depicted in Fig. 4.1.6 indicates their point of location or measurement. The subscripts "1" and "2" specify state variables in front of and after the compressor. The subscripts "3" and "4" refer to state variables in front of and after the turbine. If a second subscript exists, it is used to detail the location of the variable.

Several approaches for the modeling of the gas path are presented in the literature. They mainly differ in their level of detail and computing time. A brief comparison is given in Table 4.1.3, Heywood (1988), Merker et al (2006), Rakopoulos and Giakoumis (2006). As distributed parameter models are computationally too time-consuming for control applications zero-dimensional or lumped parameter approaches like the "filling and emptying method" are most commonly utilized, Merker et al (2006), Pischinger et al (2009), Guzzella and Onder (2010). Within this approach the manifolds and air ducts are modeled as mass storages and valves, filters and coolers are treated as flow restrictions (orifices).

Within the air path and also exhaust path of a turbo-charged combustion engine two kind of *gas storages* can be distinguished:

a) *Gas storages*, which are characterized by a mixture of one or several gas flows, see Fig. 4.1.7. Examples are the intake manifold and the exhaust manifold.

Table 4.1.3. Comparison of air path model approaches

	Zero-dimensional	One-dimensional	3D
modeling method	quasi-steady flow models, filling and emptying method	method of characteristics, method of differences	CFD methods
flow	(quasi-)steady	unsteady	unsteady
model complexity	low	high	very high
computing time	short	medium	large
real-time capability	yes	no	no
applica-tions	parameter studies, control system design	engine design	engine design

b) *Gas pipes*, which are characterized by a large ratio $\zeta = l/d$ of length l to diameter d and therefore have a gas transportation dead time $T_\mathrm{d} = l/v$ (v: gas velocity), see Fig. 4.1.8. Examples are: air ducts in front of the compressor, between compressor and charge-air cooler, between charge-air cooler and throttle or intake manifold, exhaust recirculation path and exhaust pipes.

Additionally, *flow resistances* in the filters, valves and coolers have to be taken into account. In the following simplified models for intake systems of diesel engines are considered.

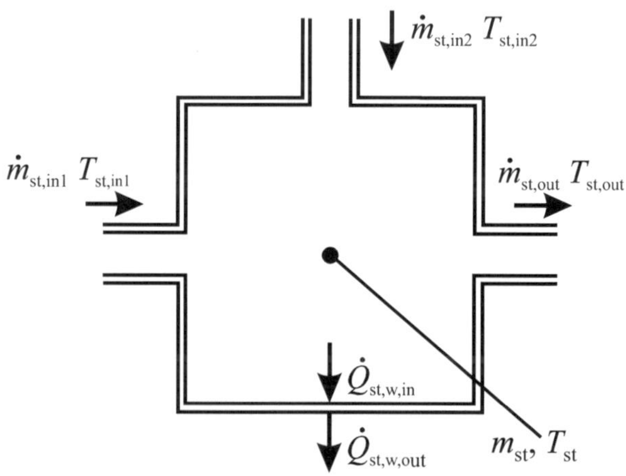

Fig. 4.1.7. Gas storage with mixture of two gas flows and one outflow.

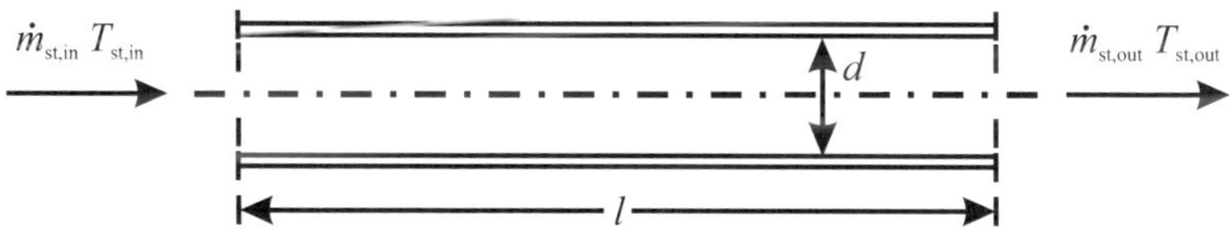

Fig. 4.1.8. Gas pipe of length l.

a) Gas storage with ideal mixture

A gas storage can be treated as a lumped parameter component, yielding a mass balance equation as (4.1.11).

$$\frac{dm_{\mathrm{gas,st}}}{dt} = \sum_{i=1}^{p} \dot{m}_{\mathrm{gas,in},i}(t) - \sum_{j=1}^{q} \dot{m}_{\mathrm{gas,out},j}(t) \tag{4.1.30}$$

or with other subscripts, compare Fig. 4.1.7

$$\frac{dm_{\mathrm{st}}}{dt} = \sum_{i=1}^{p} \dot{m}_{\mathrm{st,in},i}(t) - \sum_{j=1}^{q} \dot{m}_{\mathrm{st,out},j}(t). \tag{4.1.31}$$

Correspondingly, an energy balance can be formulated

$$\frac{dU_{\mathrm{st}}}{dt} = \sum_{i=1}^{p} h_{\mathrm{st,in},i}(t)\, \dot{m}_{\mathrm{st,in},i}(t) - \sum_{j=1}^{q} h_{\mathrm{st,out},j}(t)\, \dot{m}_{\mathrm{st,out},j}(t) - \frac{dQ_{\mathrm{st,w,in}}(t)}{dt} \tag{4.1.32}$$

with U_{st} the stored internal energy, h the specific enthalpy, and $Q_{\mathrm{st,w,in}}$ the heat transfer to the wall of the storage. For the internal energy holds

$$\frac{dU_{\mathrm{st}}}{dt} = \frac{d}{dt}\big(m_{\mathrm{st}}(t)\, u_{\mathrm{st}}(t)\big) = \frac{dm_{\mathrm{st}}(t)}{dt} u_{\mathrm{st}} + m_{\mathrm{st}} \frac{du_{\mathrm{st}}(t)}{dt} \tag{4.1.33}$$

where u_{st} is the specific internal energy.

The internal energy of a gas mixture depends on the temperature T_{st} and the air-to-fuel ratio λ_{st}

$$u_{\mathrm{st}} = u_{\mathrm{st}}(T_{\mathrm{st}}, \lambda_{\mathrm{st}}) \tag{4.1.34}$$

and can be determined according to Justi (1938) for large temperature ranges or by using the specific heat capacity and neglect of the air-to-fuel ratio

$$u_{\mathrm{st}} = c_{\mathrm{v}}(T_{\mathrm{st}})\, T_{\mathrm{st}}. \tag{4.1.35}$$

For the enthalpy in (4.1.32) holds

$$h = c_{\mathrm{p}}(T)\, T. \tag{4.1.36}$$

The state equation of the gas in the storage is

$$p_{\mathrm{st}}(t)V_{\mathrm{st}} = m_{\mathrm{st}}(t)RT_{\mathrm{st}}(t). \tag{4.1.37}$$

The heat flow to the walls of the storage is described by

$$\dot{Q}_{\mathrm{st,w,in}}(t) = A_{\mathrm{w}}\alpha_{\mathrm{gw}}\left(T_{\mathrm{st}}(t) - T_{\mathrm{w}}(t)\right) \tag{4.1.38}$$

with A_{w} the wall surface, α_{gw} the heat transfer coefficient, and T_w the wall temperature. The dynamics of the wall temperature can be approximated by

$$m_{\mathrm{w}}c_{\mathrm{w}}\frac{dT_{\mathrm{w}}}{dt} = \dot{Q}_{\mathrm{st,w,in}}(t) - \dot{Q}_{\mathrm{st,w,out}}(t) \tag{4.1.39}$$

where m_w and c_w are the mass and specific heat of the wall, assuming lumped parameters. The heat flow to the environment is

$$\dot{Q}_{\mathrm{st,w,out}} = A_{\mathrm{w}}\alpha_{\mathrm{wa}}(T_{\mathrm{w}}(t) - T_{\mathrm{a}}(t)) \tag{4.1.40}$$

with α_{wa} the heat flow coefficient to the environment and T_{a} the air temperature outside the air duct.

It is now assumed that the transported heat through the storage is much larger than the heat flow to the walls, such that $\dot{Q}_{\mathrm{st,w}} = 0$ holds. To simplify the derivations and notations only one inflow $\dot{m}_{\mathrm{st,in}} = \dot{m}_{\mathrm{in}}$ with temperature T_{in} and one outflow with $\dot{m}_{\mathrm{st,out}} = \dot{m}_{\mathrm{out}}$ and temperature T_{out} are considered. Assuming an ideal mixture in the storage, $T_{\mathrm{out}} = T_{\mathrm{st}}$ can be set.

The mass balance equation then is

$$\frac{dm_{\mathrm{st}}(t)}{dt} = \dot{m}_{\mathrm{in}}(t) - \dot{m}_{\mathrm{out}}(t) \tag{4.1.41}$$

and with

$$m_{\mathrm{st}}(t) = \frac{V_{\mathrm{st}}}{R}\frac{p_{\mathrm{st}}(t)}{T_{\mathrm{st}}(t)} \tag{4.1.42}$$

one obtains for slow temperature changes $dT_{\mathrm{st}}/dt \approx 0$

$$\frac{dp_{\mathrm{st}}(t)}{dt} = \frac{RT_{\mathrm{st}}(t)}{V_{\mathrm{st}}}\left(\dot{m}_{\mathrm{in}}(t) - \dot{m}_{\mathrm{out}}(t)\right). \tag{4.1.43}$$

The energy balance from (4.1.32) yields

$$\frac{dU_{\mathrm{st}}(t)}{dt} = h_{\mathrm{in}}(t)\,\dot{m}_{\mathrm{in}}(t) - h_{\mathrm{out}}(t)\,\dot{m}_{\mathrm{out}}(t) \tag{4.1.44}$$

or by inserting (4.1.35)

$$c_{\mathrm{v}}\frac{d}{dt}\left(m_{\mathrm{st}}(t)T_{\mathrm{st}}(t)\right) = c_{\mathrm{p}}\left(\dot{m}_{\mathrm{in}}(t)\,T_{\mathrm{in}}(t) - \dot{m}_{\mathrm{out}}(t)\,T_{\mathrm{st}}(t)\right). \tag{4.1.45}$$

The left term of (4.1.44) becomes

$$\frac{d}{dt}\big(m_{\mathrm{st}}(t)T_{\mathrm{st}}(t)\big) = m_{\mathrm{st}}(t)\frac{dT_{\mathrm{st}}(t)}{dt} + T_{\mathrm{st}}(t)\frac{dm_{\mathrm{st}}(t)}{dt} \tag{4.1.46}$$

and by replacing $m_{\text{st}}(t)$ by (4.1.42) and inserting the mass balance (4.1.41) it follows from the energy balance (4.1.44) with $\kappa = c_{\text{p}}/c_{\text{v}}$

$$\frac{dT_{\text{st}}(t)}{dt} = \frac{1}{m_{\text{st}}(t)}\left(((1-\kappa)\dot{m}_{\text{out}}(t) - \dot{m}_{\text{in}}(t))\,T_{\text{st}}(t) + \kappa \dot{m}_{\text{in}}(t)T_{\text{in}}(t)\right). \quad (4.1.47)$$

Hence, (4.1.43) and (4.1.47) describe the dynamic behavior of the pressure p_{st} and T_{st} in the storage. Both equations are mutually coupled.

To simplify (4.1.47) it is now assumed that the term $|(1-\kappa)\dot{m}_{\text{out}}| \ll |\dot{m}_{\text{in}}|$ such that

$$\frac{dT_{\text{st}}(t)}{dt} \approx \frac{\dot{m}_{\text{in}}}{m_{\text{st}}}\left(\kappa T_{\text{in}}(t) - T_{\text{st}}(t)\right). \quad (4.1.48)$$

Writing the simplified equations for two input flows and one output flow and using the symbols as in (4.1.31) and (4.1.32) results in

$$\frac{dp_{\text{st}}(t)}{dt} = \frac{R\,T_{\text{st,out}}}{V_{\text{st}}}\left[\dot{m}_{\text{st,in}}(t) - \dot{m}_{\text{st,out}}(t)\right] \quad (4.1.49)$$

$$\frac{dT_{\text{st,out}}(t)}{dt} \approx \frac{\dot{m}_{\text{st,in,1}}}{m_{\text{st}}}\kappa T_{\text{st,in1}}(t) + \frac{\dot{m}_{\text{st,in,2}}}{m_{\text{st}}}\kappa T_{\text{st,in2}} - \frac{\dot{m}_{\text{st,in}}}{m_{\text{st}}} T_{\text{st,out}}(t). \quad (4.1.50)$$

with

$$\dot{m}_{\text{st,in}} = \dot{m}_{\text{st,in,1}} + \dot{m}_{\text{st,in,2}} = \frac{\rho_{\text{st}} V_{\text{st}}}{\dot{m}_{\text{st,in}}}. \quad (4.1.51)$$

Based on (4.1.49) and (4.1.50) a signal-flow chart can be established as shown in Fig. 4.1.9. Hence, an integral behavior results for the pressure p_{st} and a first order delay for the temperature $T_{\text{st,out}}$ with regard to the mass flow inputs and temperature inputs, together with crosscouplings. The time constant for the temperature is

$$T_{\text{T1}} = \frac{m_{\text{st}}}{\dot{m}_{\text{st,in}}} = \frac{\rho_{\text{st}} V_{\text{st}}}{\dot{m}_{\text{st,in}}}. \quad (4.1.52)$$

b) Gas pipe

For a gas pipe like an air duct of length l it is assumed that the gas behaves like a plug flow leading to a transport delay for temperature changes from the input to the output in form of a dead time

$$T_{\text{d}} = \frac{l}{w} = \frac{m_{\text{st}}}{\dot{m}_{\text{st,in}}} \quad (4.1.53)$$

where w is the gas velocity. It is further assumed that the heat capacity of the pipe wall is small such that its influence can be neglected. (The precise dynamic behavior of heated or cooled pipes is treated in Sect. 4.6.2.) Hence, the energy balance of (4.1.44) becomes

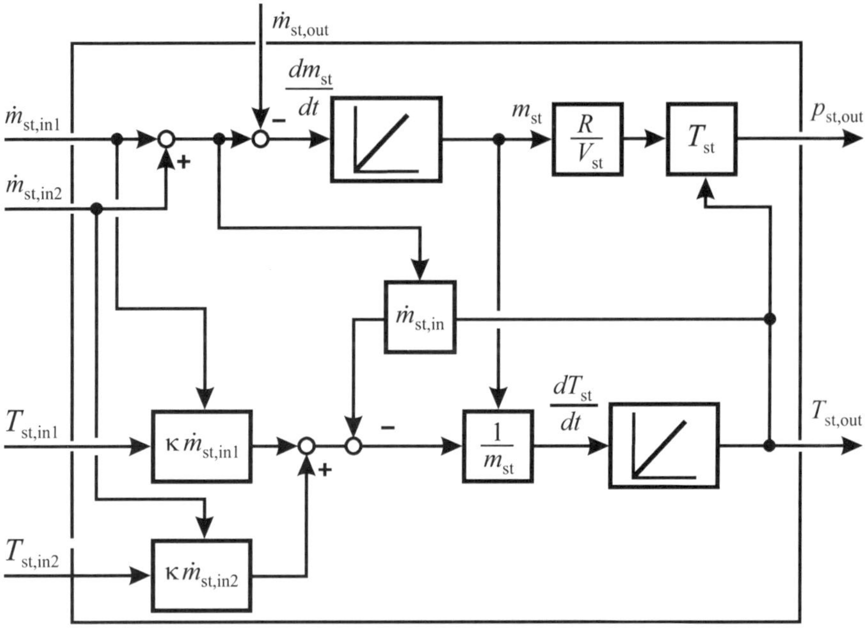

Fig. 4.1.9. Signal-flow chart for a gas storage according to Fig. 4.1.7.

$$\dot{m}_{st,out}(t)\,T_{st,out}(t) = \dot{m}_{st,in}(t)\,T_{st,in}(t-T_d) \qquad (4.1.54)$$

If mass flow changes between pipe input and output are much faster compared to temperature changes, it can be assumed that $\dot{m}_{st,out}(t) = \dot{m}_{st,in}(t)$, thus leading to a pure dead-time behavior

$$T_{st,out}(t) = T_{st,in}(t-T_d). \qquad (4.1.55)$$

The dynamic behavior of the compressible gas in the pipe follows from the simplified mass balance (4.1.43)

$$\frac{dp_{st}(t)}{dt} = \frac{R}{V_{st}} T_{st,out}(t)\left[\dot{m}_{st,in}(t) - \dot{m}_{st,out}(t)\right] \qquad (4.1.56)$$

assuming lumped parameters and $T_{st}(t) = T_{st,out}(t)$.

Finally, the signal-flow chart of Fig. 4.1.10 results. In the case of intake or exhaust manifolds partially mixtures of gases and transport dead times hold, leading to a series connection of model types of subsections a) and b). However, the corresponding characteristic time parameters are usually negligibly small.

c) Flow resistances

To model the flow resistances of the filters, valves and coolers quasi steady-state gas flows are assumed. Its inputs are the up-stream and down-stream pressure $p_{res,in}$ and

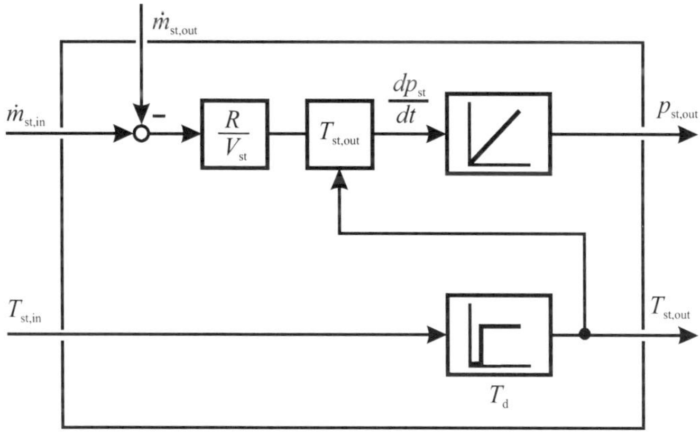

Fig. 4.1.10. Signal-flow chart for a gas pipe according to Fig. 4.1.8.

$p_{\text{res,out}}$ and the up-stream temperature T_{in}. The mass flow rate then follows for an adiabatic process according to (4.1.19)

$$\dot{m}_{\text{res,in}} = c_{\text{d}} A_{\text{ref}} \sqrt{\frac{2}{R\, T_{\text{res,in}}}}\, \psi\!\left(\frac{p_{\text{res,out}}}{p_{\text{res,in}}}\right) p_{\text{res,in}}. \tag{4.1.57}$$

The outflow function through a contraction is for *incompressible fluids*

$$\psi\!\left(\frac{p_{\text{res,out}}}{p_{\text{res,in}}}\right) = \sqrt{1 - \frac{p_{\text{res,out}}}{p_{\text{res,in}}}} \tag{4.1.58}$$

and for *compressible fluids*, see (4.1.20),

$$\psi\!\left(\frac{p_{\text{res,out}}}{p_{\text{res,in}}}\right) = \left(\frac{p_{\text{res,out}}}{p_{\text{res,in}}}\right)^{\frac{1}{\kappa}} \sqrt{\frac{\kappa}{\kappa - 1}\left(1 - \left(\frac{p_{\text{res,out}}}{p_{\text{res,in}}}\right)^{\frac{\kappa-1}{\kappa}}\right)}. \tag{4.1.59}$$

A_{ref} is a reference flow area and c_{d} a discharge coefficient accounting for friction losses and contraction effects. The determining variable for the mass flow rate is the gas velocity in the contraction. For flow resistances with a small pressure drop and low velocities like in the air filter or in coolers (4.1.58) is used, which leads to

$$\dot{m}_{\text{res,in}} = c_{\text{d}} A_{\text{ref}} \sqrt{\frac{2\, p_{\text{res,in}}}{R T_{\text{res,in}}}} \sqrt{p_{\text{res,in}} - p_{\text{res,out}}}. \tag{4.1.60}$$

Flow resistances like a throttle or the EGR valve result for nearly closed positions in large gas velocities and large pressure differences with either subcritical or overcritical pressure ratio $p_{\text{res,out}}/p_{\text{res,in}}$ as discussed in Sect. 4.1.1 for throttles.

Assuming an adiabatic change of the gas state, i.e. no heat transfer to the environment, it holds for the temperatures, see (4.1.24),

$$T_{\rm res,out} = T_{\rm res,in} \left(\frac{p_{\rm res,out}}{p_{\rm res,in}} \right)^{\frac{\kappa - 1}{\kappa}} \tag{4.1.61}$$

For small pressure differences, however, $T_{\rm res,in} = T_{\rm res,out}$ can be assumed.

The mass flow and the temperature model for a flow resistance are summarized in the signal-flow chart, Fig. 4.1.11.

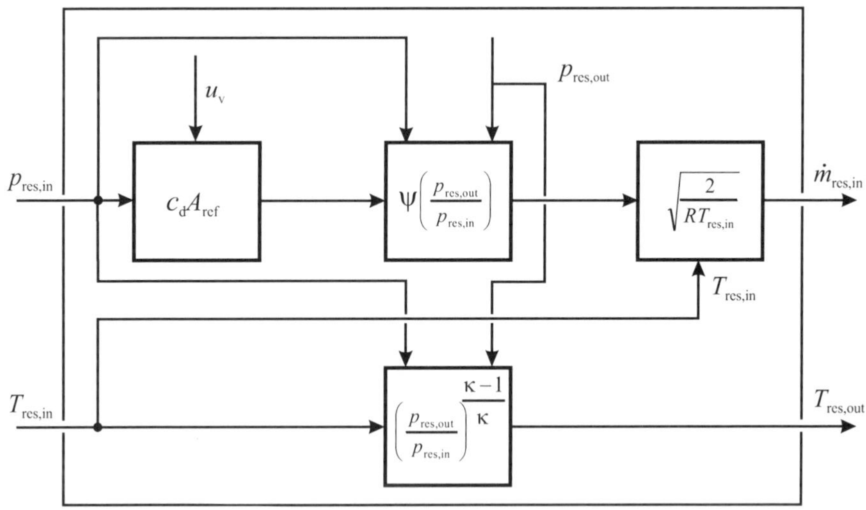

Fig. 4.1.11. Signal-flow chart for a flow resistance (compressible fluid). $\dot{m}_{\rm res}$: mass flow rate through resistance.

d) Coupled flow resistance and gas storage

Frequently, flow resistances and gas storages are coupled in series connections, where the flow resistances are the gas duct itself or a throttle. Then, a schematic as in Fig. 4.1.12 can be used as a lumped parameter approach. Two port representations for the pressures and mass flow rates and for temperatures and mass flow rates are also shown. The applied variables for the mechanical power of the flow restriction is usually the acting pressure $p_{\rm res,in}$ and the dependent variable the mass flow rate $\dot{m}_{\rm res,in}$, compare (4.1.57). For the thermal energy the applied variable is the temperature $T_{\rm res,in}$ and $\dot{m}_{\rm res,in}$ follows from the mechanical power. At the interconnection it holds

$$\begin{aligned} \dot{m}_{\rm res,in} &= \dot{m}_{\rm st,in} \\ p_{\rm res,out} &= p_{\rm st,in} = p_{\rm st} \end{aligned} \tag{4.1.62}$$

and at the exit of the gas storage

$$\begin{aligned} p_{\mathrm{st,out}} &= p_{\mathrm{st}} \\ T_{\mathrm{st,out}} &= T_{\mathrm{st}} \end{aligned} \tag{4.1.63}$$

because of the assumption of an ideal mixture in the storage.

Based on these equalities and the equations of Subsect. a) and c) the signal-flow chart of Fig. 4.1.13 can be developed. It contains two integrators with feedback for the mass and heat storage, several nonlinearities and interconnections between these two types of storages. This signal flow can be considered as a basic throttle-storage element which appears at several locations in the intake and exhaust system.

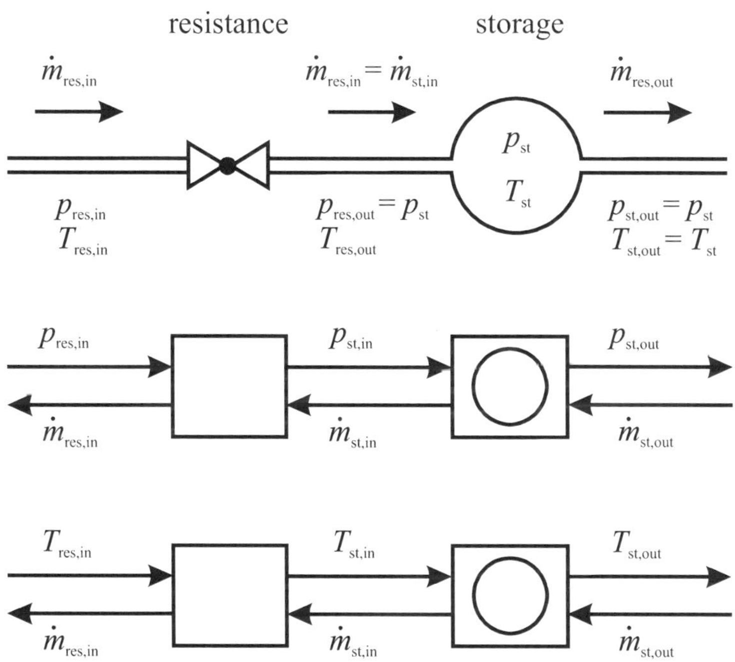

Fig. 4.1.12. a Schematic of a coupled flow resistance and gas storage. **b** Two-port representation for mechanical power. **c** Two-port representation for thermal power.

e) Intake system

The intake system of a turbo-charged diesel engine with air-charge cooler and throttle is represented (for the fluid dynamic behavior) in Fig. 4.1.14 with two-port elements, according to Sects. 2.2 and 2.3. Thus the compressor and the cylinder are converters, the air ducts and the intake manifold are storages and the air filter, the charge-air cooler are sinks. The driving forces of the intake system are generated from the compressor and the intake strokes of the cylinders. Thus the compressor pressure p_{2c} and the mass flow $\dot{m}_2$ in the combustion chambers which depends according to (4.1.5) on the engine speed n_{eng}, can be considered as applied variables.

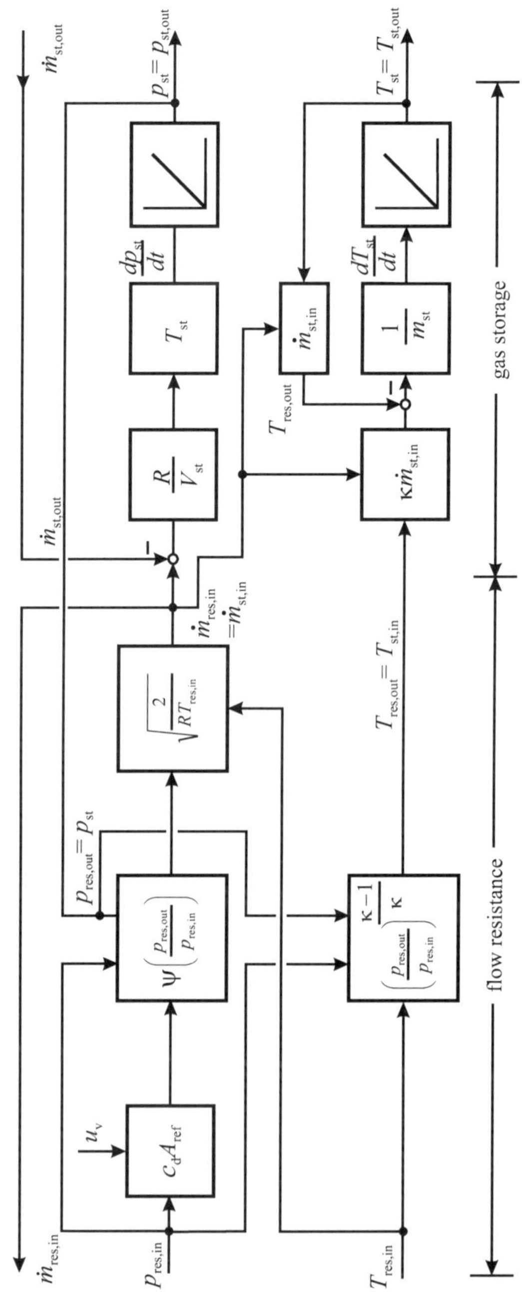

Fig. 4.1.13. Signal-flow chart for a coupled flow resistance and gas storage.

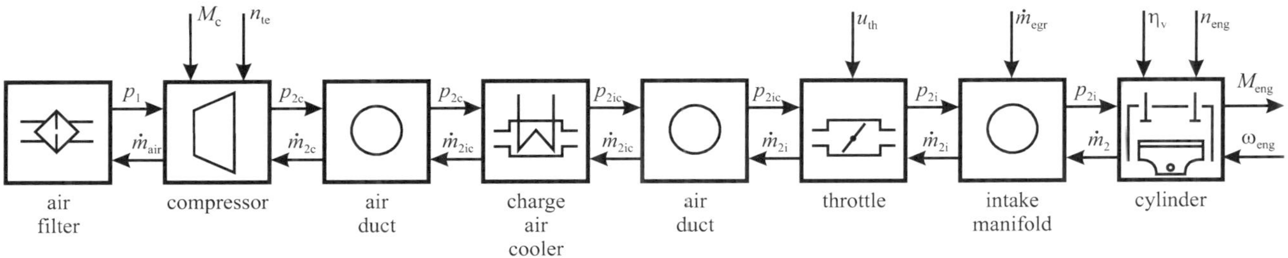

Fig. 4.1.14. Multi-port representation of the fluid dynamic behavior of the intake system for a turbo-charged diesel engine.

The sucked gas flow in the cylinder follows from (4.1.5) or (4.1.8). Thus, the manifold pressure p_{2i} together with the volumetric efficiency η_v from (4.1.4) determine the gas flow. The compressor pressure p_{2c} follows from the steady-state maps $p_{2c}(M_c, n_{tc}, \dot{m}_{2c})$.

Based on Figs. 4.1.9, 4.1.10 and (4.1.60) the signal-flow chart of Fig. 4.1.15 for the intake system follows. It consists of two gas storages and is therefore of the second dynamic order.

If the effect of the charge-air cooler resistance is neglected and only one gas storage is considered, a first order system results, Fig. 4.1.16.

The largest volumes of the intake system are usually the air ducts before and after the charge-air cooler. They determine, together with the mechanical inertia of the turbocharger rotor the dynamically delayed behavior of the charging pressure p_{2i} at the intake valves. A rough measure for a dominating time constant of the intake air path is

$$T_{it} = \frac{\dot{m}_{air}}{m_{st,air}} = \frac{\dot{V}_{air}}{V_{st,air}}. \tag{4.1.64}$$

As an example for a passenger car diesel engine the different volumes, mass flows and resulting time constants are stated in Table 4.1.4, Zahn (2012). Thus, the volume of the intake system adds up to 9.2 liters, whereas the exhaust system has about 1.4 liters. The individual time constants range between 10 and 500 ms, and the sum of time constants for the air path between compressor and intake valves, which is a measure for the overall delay, is in a range of 160 ms to 970 ms. The large time constants result for the smallest air mass flow and the small for the largest air-mass flow. These time constants have to be seen in connection with the turbocharger inertia delay, which is about 1 to 3 s, see Sect. 4.4.

A flow chart for the temperature behavior of the intake system can be developed from Figs. 4.1.9 and 4.1.10 and is depicted in Fig. 4.1.17. With the simplifications made it is determined by the dynamics of the charge-air cooler, the dead times through the air ducts and a proportional acting mixture element in the intake manifold.

Table 4.1.4. Time constants of air and exhaust path storages for an 1.9 l diesel engine, 77 kW, Zahn (2012)

Component	Pipe volume $V[\mathrm{cm}^3]$	Mass flow $\dot{m}[\mathrm{kg/h}]$	Time constant $T[\mathrm{ms}]$
air duct after air filter	2100	25 … 300	300 … 25
air duct after compressor	1700	25 … 300	250 … 35
air duct after air charge cooler	3400	25 … 300	490 … 80
intake manifold	2000	25 … 300	230 … 45
exhaust manifold	600	25 … 300	110 … 7
EGR duct	300	0 … 100	≥ 10
exhaust after turbine	500	25 … 300	70 … 4

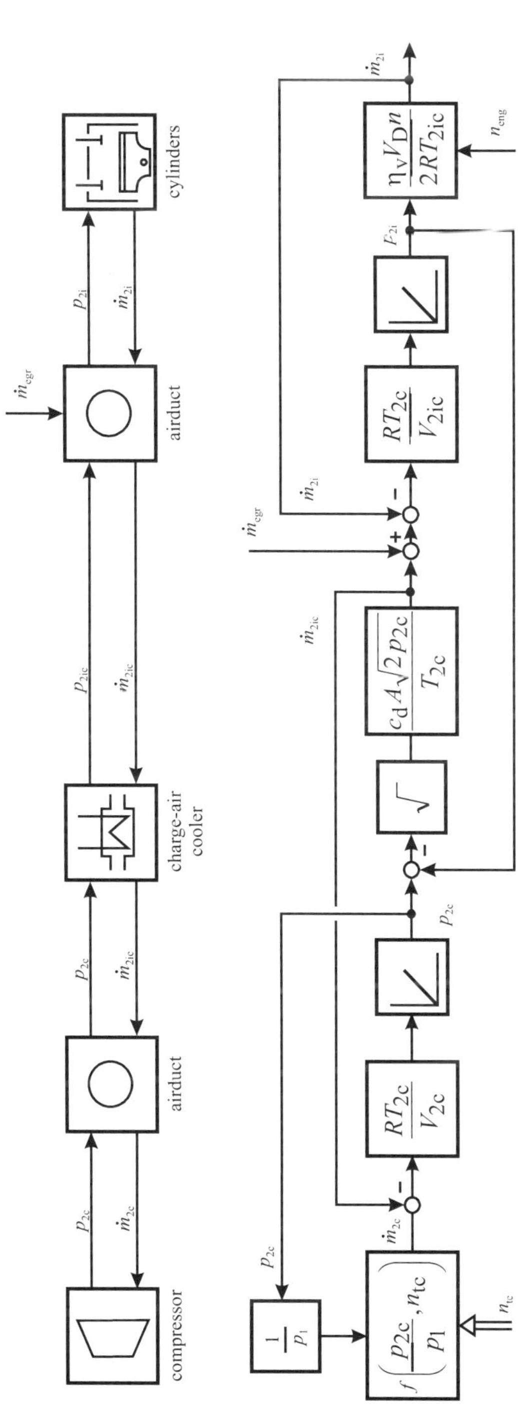

Fig. 4.1.15. Signal-flow chart for the fluid dynamic behavior of the intake system for a diesel engine with turbocharger and charge-air cooler, without throttle

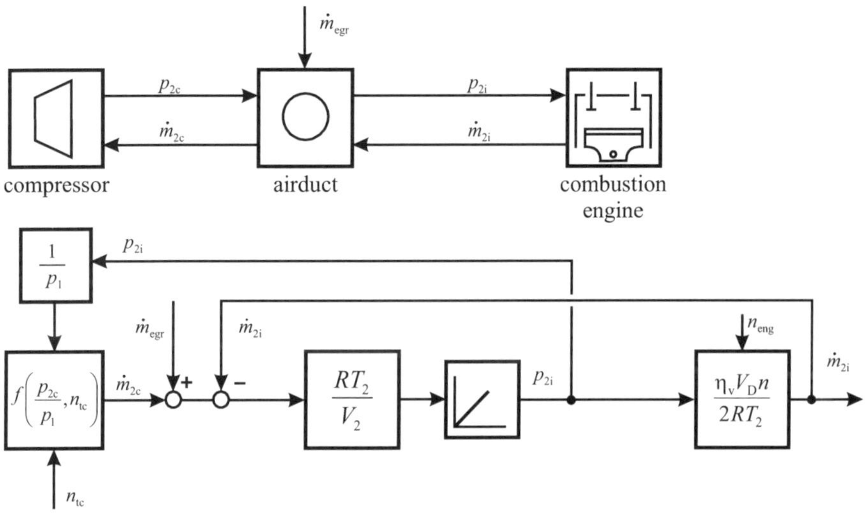

Fig. 4.1.16. Signal-flow chart for the fluid dynamic behavior of the intake system as Fig. 4.1.15 but neglect of the charge-air cooler resistance.

4.2 Combustion

The thermodynamic part of combustion engines comprise the gas exchange, fuel injection, combustion with spark ignition (gasoline) or compression ignition (diesel) and resulting work at the pistons. The following sections are an extract of the many possibilities to treat these complicated processes with regard to modeling the important parts for engine control and diagnosis of four-stroke engines.

4.2.1 Operating cycles

The four-stroke cycle consists of four phases over 720° crank angle: the intake stroke, where air (and vaporized fuel for indirect intake injection) is drawn into the cylinder, the compression stroke (and direct injected fuel), where the air/fuel mixture is compressed and ignited, either by spark plug for gasoline engines or by compression for diesel engines, and the exhaust stroke where the exhaust gases are pushed out. The gas flows of these basic strokes are controlled by the intake and exhaust valves, which are driven by the camshaft rotating with half the speed of the crankshaft. A large impact on the operating cycle has the timing of the intake and exhaust valves. Earlier or later opening or closing influences the charging process with fresh air, the internal exhaust gas recirculation and residual gas compression, to optimize fuel efficiency and exhaust emissions.

Figure 4.2.1 depicts an example for the representation of the four-stroke cycle and the pressure volume ($p - V$) diagram with measured (indicated) combustion pressure.

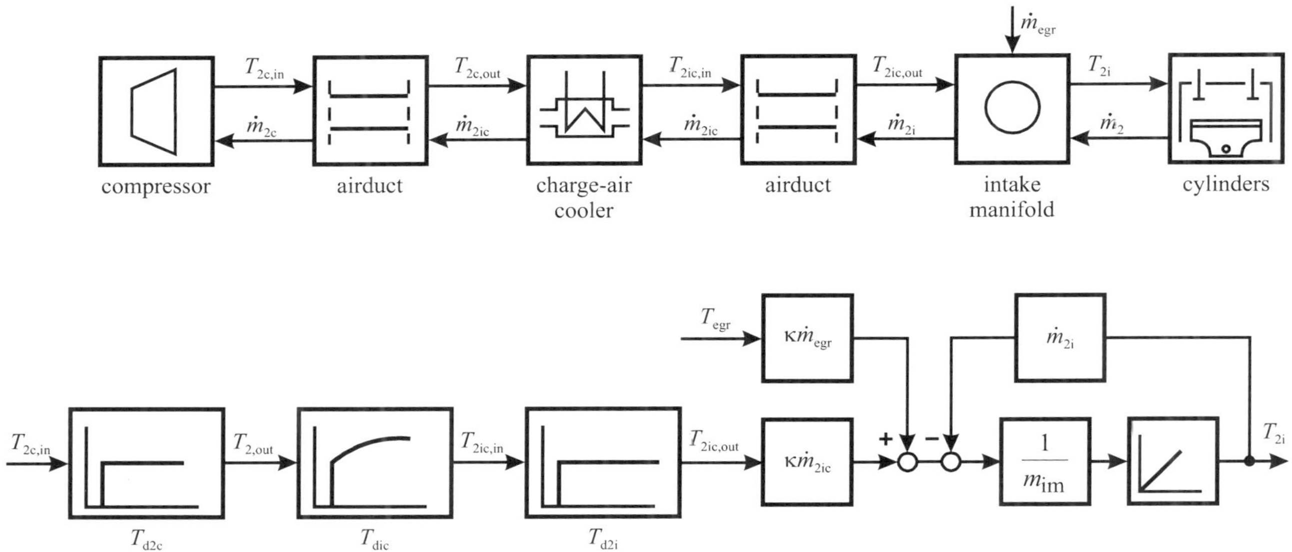

Fig. 4.1.17. Signal-flow chart for the temperature behavior of the intake system of a diesel engine with turbocharger, charge-air cooler and intake manifold.

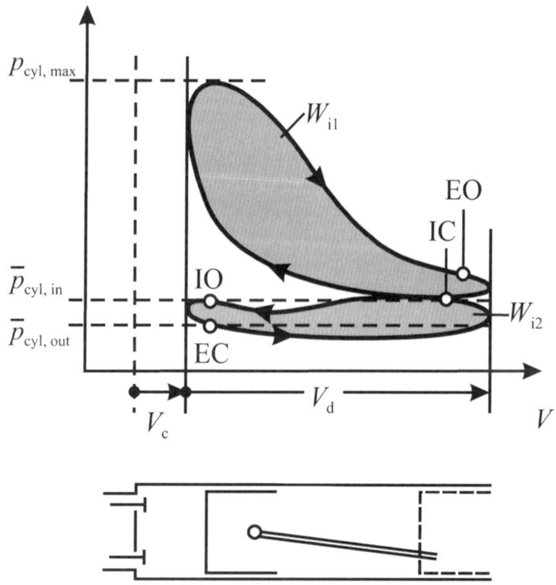

Fig. 4.2.1. $p - V$ diagram of the four-stroke cycle. IO: intake opens; IC: intake closes; EO: exhaust opens; EC: exhaust closes. $p_{\mathrm{cyl,in}}$: charging air pressure, $p_{\mathrm{cyl,out}}$: exhaust stroke pressure, $p_{\mathrm{cyl,max}}$: maximum pressure; $\overline{p}$ means average value.

a) Ideal operating cycles

For comparisons and basic thermodynamic considerations idealized cycle processes are defined. These ideal processes assume following simplifications:

- ideal gases
- gas with constant specific heat
- unthrottled air charge
- no flow losses during gas exchange
- combustion is replaced by heat flow supply
- isentropic (adiabatic) compression and expansion, no heat losses
- no friction, no gas losses.

Three idealized cycle processes can then be distinguished, see Fig. 4.2.2.

The *constant volume combustion cycle*, see Fig. 4.2.2a): isentropic adiabatic compression (1-2), isochoric heat supply with infinite speed (2-3), isentropic expansion (3-4), isochoric heat removal with infinite speed (4-1). This process is frequently used to characterize the idealized Otto process for gasoline engines.

The *constant pressure combustion cycle*, see Fig. 4.2.2b): isentropic compression (1-2), isobaric heat supply with finite combustion speed as volume increases (2-3), isentropic expansion (3-4), isochoric heat removal with infinite speed (4-1).

The *mixed constant-volume/constant pressure combustion cycle*, see Fig. 4.2.2c): isentropic compression (1-2), isochoric heat supply (2-2'), isobaric heat supply (2'-3), isentropic expansion (3-4), isochoric heat removal (4-1). This process is called the *Seiliger process* and is closer to real cycle processes as the other two. It is frequently applied for an idealized diesel process.

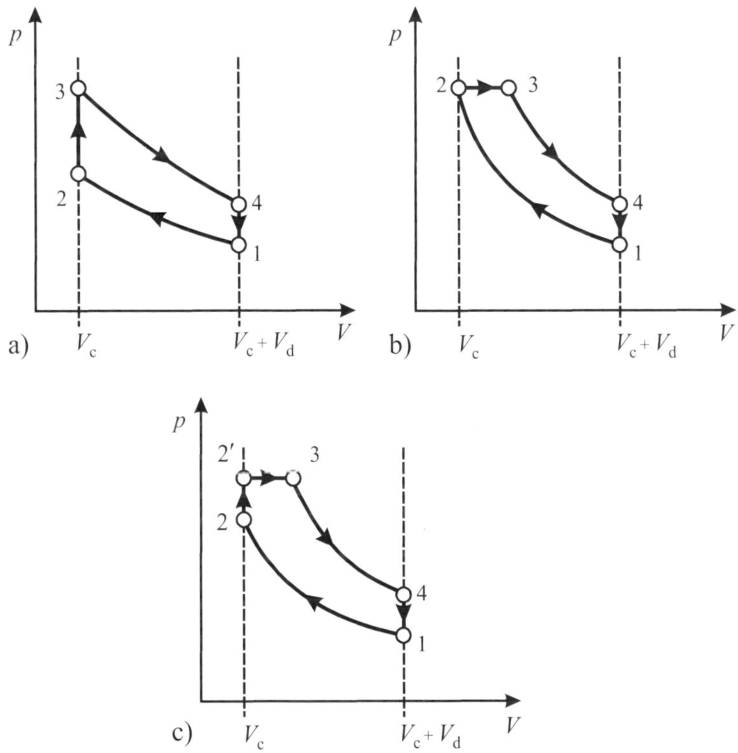

Fig. 4.2.2. Idealized $p - V$ diagrams for **a** constant volume cycle. **b** constant pressure cycle. **c** mixed constant volume, constant pressure cycle (Seiliger process).

If the supplied heat by the combustion is $q_{\rm in}$ and the removed heat $q_{\rm ex}$ during gas exchange, then the thermal efficiency can be described by

$$\eta_{\rm th} = \frac{q_{\rm in} - q_{\rm ex}}{q_{\rm in}} = 1 - \frac{q_{\rm ex}}{q_{\rm in}}. \tag{4.2.1}$$

The heat values of the *constant-volume combustion* are:

$$q_{\rm in} = c_{\rm v}(T_3 - T_2)\ ;\ q_{\rm ex} = c_{\rm v}(T_4 - T_1). \tag{4.2.2}$$

For an isentropic change of state holds

$$T^{\kappa} p^{1-\kappa} = \text{const.}$$

and after introducing the gas equation follows

$$pV = RT \cdot m$$
$$TV^{\kappa-1} = \text{const.}$$

and therefore

$$\frac{T_1}{T_2} = \left(\frac{V_2}{V_1}\right)^{\kappa-1} \quad \text{and} \quad \frac{T_4}{T_3} = \left(\frac{V_3}{V_4}\right)^{\kappa-1}. \tag{4.2.3}$$

As $V_2 = V_3$ and $V_3 = V_4$ this yields

$$\frac{T_1}{T_2} = \frac{T_4}{T_3} \tag{4.2.4}$$

and (4.2.1) becomes

$$\eta_{\text{th}} = 1 - \frac{c_\text{v}(T_4 - T_1)}{c_\text{v}(T_3 - T_2)} = 1 - \frac{T_1}{T_2} = 1 - \left(\frac{V_2}{V_1}\right)^{\kappa-1}. \tag{4.2.5}$$

Using the compression ratio

$$\varepsilon = \frac{V_\text{d} + V_\text{c}}{V_\text{c}} = \frac{V_1}{V_2} \tag{4.2.6}$$

the thermal efficiency becomes

$$\eta_{\text{thv}} = 1 - \frac{1}{\varepsilon^{\kappa-1}}. \tag{4.2.7}$$

Hence, this ideal constant-volume combustion process requires for high efficiency: large compression ratio and combustion around top dead center.

Similarly, one obtains for the *constant-pressure combustion process* with the volume ratio for the combustion and isobaric change of state

$$\varphi = \frac{V_3}{V_2} = \frac{T_3}{T_2}. \tag{4.2.8}$$

Applying (4.2.3), (4.2.6) and

$$q_{\text{in}} = c_\text{p}(T_3 - T_2) \tag{4.2.9}$$

leads with $\kappa = \frac{c_\text{v}}{c_\text{p}}$ to

$$\eta_{\text{thp}} = 1 - \frac{1}{\varepsilon^{\kappa-1}} \frac{(\varphi^\kappa - 1)}{\kappa(\varphi - 1)}. \tag{4.2.10}$$

The thermal efficiency of the constant-pressure combustion process is better for large compression ratio and gets smaller with increasing volume ratio φ.

For the third ideal cycle process, the *mixed Seiliger process* holds

$$q_{\mathrm{in}} = c_{\mathrm{v}}(T_2' - T_2) + c_{\mathrm{p}}(T_3 - T_2') \tag{4.2.11}$$

and with pressure ratio

$$\psi = \frac{p_2'}{p_2} \tag{4.2.12}$$

follows for the thermal efficiency, Grothe and Feldhusen (2007)

$$\eta_{\mathrm{ths}} = 1 - \frac{1}{\varepsilon^{k-1}} \frac{\varphi^k \psi - 1}{\psi - 1 + \kappa\psi(\varphi - 1)}. \tag{4.2.13}$$

Hence,

$$\eta_{\mathrm{thv}} > \eta_{\mathrm{ths}} > \eta_{\mathrm{thp}}. \tag{4.2.14}$$

Limitations in the realization of the cycle processes are given, e.g. by the maximal allowable pressures after compression. These are for natural aspirating engines about 50-60 bar, and with charging 70-80 bar and for diesel engines 70-80 bar or 110-200 bar with $\varphi = 1.4...1.6$. A further limit is knocking for gasoline engines if the compression ratio is too large.

b) Real operating cycles

The ideal operating cycles are theoretical reference cycles with many simplifying assumptions. As can be seen from Fig. 4.2.1 the real $p - V$ diagram shows several differences, because of the real combustion, the real gas exchange, real gases, and several losses.

All losses of a combustion engine can be taken into account by defining different efficiencies. A first efficiency is due to the real fuel conversion, because complete combustion cannot be realized. If W_{f} is the equivalent work of the supplied fuel and W_{fe} the work of the effectively burnt fuel, then the *fuel conversion efficiency* becomes

$$\eta_{\mathrm{f}} = \frac{W_{\mathrm{fe}}}{W_{\mathrm{f}}} = \frac{W_{\mathrm{fe}}}{m_{\mathrm{f}} H_{\mathrm{l}}} \tag{4.2.15}$$

where m_{f} is the injected fuel mass per stroke end H_{l} the lower heat value.

The internal (indicated) work of the real cycle can be determined with the measured (indicated) pressure p_{cyl} for one cylinder

$$W_{\mathrm{i}} = \int_{\varphi=0}^{720^\circ} p_{\mathrm{cyl}}(\varphi)\, dV(\varphi) = W_{\mathrm{i1}} + W_{\mathrm{i2}}. \tag{4.2.16}$$

It consists of the area part W_{i1} of the compression and expansion stroke and the area part W_{i2} for the gas exchange as depicted in Fig. 4.2.1. W_{i2} is negative for natural aspirating engines and positive for charged engines. If p_{cyl} is measured in dependence on time, A_{p} and s_{p} are the area and stroke (travel way) of the piston it follows with the angular speed $\omega = \frac{d\varphi}{dt}$.

$$W_{\mathrm{i}} = A_{\mathrm{p}}\omega \int_0^{\frac{4\pi}{\omega}} p_{\mathrm{cyl}}(t) \frac{ds_{\mathrm{p}}(t)}{d\varphi}\, dt. \tag{4.2.17}$$

This leads to the *cycle efficiency*

$$\eta_{\mathrm{g}} = \frac{W_{\mathrm{i}}}{W_{\mathrm{th}}} \tag{4.2.18}$$

where W_{th} is the work generated with the ideal combustion cycle. A further efficiency is due to mechanical losses because of friction of the pistons, crank drive, valve train and drive losses in the auxiliary drives as oil, water and fuel pump and alternator. These losses are expressed by the *mechanical efficiency*

$$\eta_{\mathrm{m}} = \frac{W_{\mathrm{eng}}}{W_{\mathrm{i}}} \tag{4.2.19}$$

with W_{eng} as the effective work at the flywheel.

Finally, the *effective efficiency* results comprising all efficiencies from the injected fuel to the work at the flywheel

$$\eta_{\mathrm{eff}} = \frac{W_{\mathrm{eng}}}{W_{\mathrm{f}}} = \frac{W_{\mathrm{eng}}}{m_{\mathrm{f}} H_{\mathrm{l}}} = \eta_{\mathrm{f}}\, \eta_{\mathrm{th}}\, \eta_{\mathrm{g}}\, \eta_{\mathrm{m}}. \tag{4.2.20}$$

For passenger car engines these efficiencies are in following ranges: $\eta_{\mathrm{g}} = 0.8 - 0.9$; $\eta_{\mathrm{m}} = 0.8 - 0.9$ and $\eta_{\mathrm{eff}} = 0.26 - 0.37$ for gasoline engines and $\eta_{\mathrm{eff}} = 0.32 - 0.43$ for diesel engines (direct injection).

Another frequently used quantity is the *mean indicated pressure*

$$p_{\mathrm{mi}} = \frac{W_{\mathrm{i}}}{V_{\mathrm{d}}} = \frac{1}{V_{\mathrm{d}}} \int_{\varphi=0}^{720^\circ} p_{\mathrm{cyl}}(\varphi)\, dV(\varphi) \tag{4.2.21}$$

where the volume $V(\varphi)$ follows (4.2.20). The mean indicated torque is then for a four stroke engine with the work $W_{\mathrm{i}} = p_{\mathrm{mi}} V_{\mathrm{d}} = 4\pi M_{\mathrm{mi}}$

$$M_{\mathrm{mi}} = \frac{V_{\mathrm{d}}}{4\pi} p_{\mathrm{mi}}. \tag{4.2.22}$$

4.2.2 Combustion cycle models

A further detailed analysis of the combustion in the cylinder and the generated work at the pistons is required as well for the design as for controlling the engine, as also the Seilinger cycle is only a rough approximation for the real behavior. In the following first the high pressure cycle and then the low pressure cycle is considered. It is assumed that the cylinder can be considered as a lumped parameter process, such that the gas states are not space dependent, and that the laws for ideal gases can be applied. This is also called a zero-dimensional or one-zone-modeling.

a) High pressure combustion cycle

The energy balance for a closed control volume is according to the first law of thermodynamics

$$dU = dQ - pdV \tag{4.2.23}$$

with dU the stored internal energy, dQ the supplied heat and $dL = pdV$ the performed mechanical work. In the following some notations are simplified: $V_{\text{cyl}} = V$, $p_{\text{cyl}} = p$. To obtain models in dependence on the crank angle φ the energy balance equation for a cylinder becomes for incremental changes $d\varphi$ and closed valves

$$\frac{dU(\varphi)}{d\varphi} = \frac{dQ_{\text{f}}(\varphi)}{d\varphi} - p(\varphi)\frac{dV(\varphi)}{d\varphi} \quad \frac{dQ_{\text{w}}(\varphi)}{d\varphi} + \frac{dm_{\text{f}}(\varphi)}{d\varphi}h_{\text{f}} \tag{4.2.24}$$

with

Q_{f}:	combustion energy of fuel
m_{f}:	injected fuel mass per stroke
h_{f}:	fuel enthalpy (direct injection)
H_{l}:	lower heating value of the fluid
Q_{w}:	heat transfer losses to cylinder walls.

Now, two kinds of thermal energies can be distinguished, Heywood (1988), Pischinger et al (2009),

- gross heat release: $\frac{dQ_{\text{f}}}{d\varphi}$
- net heat release: $\frac{dQ_h}{d\varphi} = \frac{dQ_{\text{f}}}{d\varphi} - \frac{dQ_{\text{w}}}{d\varphi}$.

The gross heat release describes the heat released by the combustion of the fuel and the net heat release equals the work performed at the piston because it takes the heat transfer to the cylinder walls into account. This means that from measurement of the cylinder pressure only the net heat release can be determined. For the internal energy of the gas holds

$$\frac{dU(\varphi)}{d\varphi} = m_{\text{g}}\, c_{\text{v}} \frac{dT_{\text{g}}(\varphi)}{d\varphi} \tag{4.2.25}$$

with the gas mass

$$m_{\text{g}} = m_{\text{air}} + m_{\text{res}} + m_{\text{f}} \tag{4.2.26}$$

where m_{res} is the mass of residual gases. The specific heat c_{v} (as well as c_{p}) increases with the temperature.

As a constitutive equation for the gas states the general gas equation can be used

$$pV = m_{\text{g}}\, RT_{\text{g}}. \tag{4.2.27}$$

Assuming that the gas constant R does not change and neglecting m_{f} leads to

$$m_{\text{g}}\, R\frac{dT_{\text{g}}(\varphi)}{dt} = \frac{dp(\varphi)}{d\varphi}V(\varphi) + p(\varphi)\frac{dV(\varphi)}{d\varphi}. \tag{4.2.28}$$

Inserting in (4.2.25) yields

$$\frac{dU(\varphi)}{d\varphi} = \frac{c_{\text{v}}}{R}\left(\frac{dp(\varphi)}{d\varphi}V(\varphi) + p(\varphi)\frac{dV(\varphi)}{d\varphi}\right) \tag{4.2.29}$$

and the *net heat release* becomes with (4.2.24) and (4.2.29)

$$\frac{dQ_h(\varphi)}{d\varphi} = \left(1 + \frac{c_v}{R}\right) p(\varphi)\frac{dV(\varphi)}{d\varphi} + \frac{c_v}{R}V(\varphi)\frac{dp(\varphi)}{d\varphi} - \frac{dm_f(\varphi)}{d\varphi}h_f. \quad (4.2.30)$$

The small contribution of the fuel enthalpy will be neglected. This allows to calculate the pressure in the cylinder

$$\frac{dp(\varphi)}{d\varphi} = \frac{R}{c_v V(\varphi)}\left(\frac{dQ_f(\varphi)}{d\varphi} - \frac{dQ_w(\varphi)}{d\varphi} - \left(1 + \frac{c_v}{R}\right) p(\varphi)\frac{dV(\varphi)}{d\varphi}\right). \quad (4.2.31)$$

For the average wall temperature holds according to Müller and Bertling (1971)

$$\overline{T}_w = 360 + 9\lambda_{air}^{0.4}\sqrt{n_{eng}D_{pist}} \qquad [\text{K}] \quad (4.2.32)$$

with λ_{air} air excess factor, D_{pist} [m] piston diameter, and n_{eng} [rpm].

The gas temperature follows from the gas equation

$$T_{cyl}(\varphi) = p_{cyl}(\varphi)V_{cyl}(\varphi)\frac{T_{cyl,0}}{p_{cyl,0}V_{cyl,0}} \quad (4.2.33)$$

with $T_{cyl,0}$, $p_{cyl,0}$ and $V_{cyl,0}$ at the beginning of the compression.

For the heat transfer to the walls exist several phenomenological equations, Mollenhauer and Tschöke (2010). For stationary temperature and flow the basic law is

$$\dot{Q}_w = \frac{dQ_w}{dt} = \alpha A(\overline{T}_g - \overline{T}_w) \quad (4.2.34)$$

and therefore

$$\frac{dQ_w}{d\varphi} = \frac{1}{\omega_{eng}}\alpha A(\overline{T}_g - \overline{T}_w) \quad (4.2.35)$$

with α the heat transfer coefficient and A the related area. Because the gas velocities as well as the temperature in the cylinder change very non-stationary, the heat transfer is extremely time variant and different for the phases of a combustion cycle.

According to Hohenberg (1980) the heat transfer coefficient can be approximated by

$$\alpha = 130V_d^{-0.06}p^{-0.8}\,\overline{T}_g^{-0.4}\,(\overline{v}_p + 1.4)^{0.8} \quad (4.2.36)$$

where $\overline{v}_p$ is the average piston speed, and according to Bargende (2001)

$$\alpha = 235.5V_d^{-0.073}\,p^{0.78}\,T_{cw}^{-0.477}\,\omega_{eng}^{0.78}\,\xi \quad (4.2.37)$$

with $T_{cw} = \frac{T_g+T_w}{2}$, and ξ a combustion term. A recent summary of these equations for the heat transfer to the walls is given in Lejsek et al (2010) and it is shown how the equation for the heat coefficients can be expanded for cycle individual dynamic situations, e.g. during start of the engine. The specific heat for constant volume c_v follows by using $R = c_p - c_v$ and $\kappa = c_p/c_v$ to

$$c_{\mathrm{v}} = \frac{R}{(\kappa - 1)} \tag{4.2.38}$$

For real exhaust gases, however, c_{v} depends on the temperature T_{g} and λ, Urlaub (1995). It can be approximated by

$$c_{\mathrm{v}}(T_{\mathrm{g}}) = 0.7 + 0.255 \cdot 10^{-3} \frac{T_{\mathrm{g}}}{[K]} \left[\frac{kJ}{kg\ K}\right] \tag{4.2.39}$$

for gasoline engines, Manz (1982). This dependence can be considered in (4.2.30) where T_{g} follows from (4.2.33).

The *gross heat release* can be approximated by an empirical function, the *Vibe function*, Vibe (1970)

$$Q_{\mathrm{f}}(\varphi) = Q_{\mathrm{f,tot}} \left[1 - e^{-a\varphi_{\mathrm{rel}}^{(m+1)}}\right] \tag{4.2.40}$$

for $\varphi > \varphi_{\mathrm{SOC}}$

$$\varphi_{\mathrm{rel}} = \frac{\varphi - \varphi_{\mathrm{SOC}}}{\Delta\varphi_{\mathrm{cd}}} \tag{4.2.41}$$

and

$Q_{\mathrm{f,tot}}$:	total released fuel heat
$\Delta\varphi_{\mathrm{cd}} = \varphi_{\mathrm{EOC}} - \varphi_{\mathrm{SOC}}$	combustion angle
$\varphi_{\mathrm{SOC}}, \varphi_{\mathrm{EOC}}$:	start and end of combustion
m:	formfactor
a:	Vibe factor ($a = 6.908$).

Differentiation yields

$$\frac{dQ_{\mathrm{f}}(\varphi)}{d\varphi} = a(m+1)\frac{\Delta Q_{\mathrm{f,tot}}}{\Delta\varphi_{\mathrm{cd}}} \varphi_{\mathrm{rel}}^{m}\, e^{-a\varphi_{\mathrm{rel}}^{m+1}}. \tag{4.2.42}$$

The use of the Vibe function requires that the parameters φ_{SOC}, $\Delta\varphi_{\mathrm{cd}}$ and m have to be adapted for the engine and its operation points, Sinsel (2000), Schaffnit (2002). For direct injection diesel engines a good approximation could be obtained by the superposition of two Vibe functions, Oberg (1976), Sinsel (2000) has shown how the Vibe parameters can be estimated with parameter estimation methods based on measured cylinder pressures.

After discretization of (4.2.30), (4.2.31), (4.2.35) and (4.2.42) with discrete angles $\Delta\varphi$ (e.g. 1 to 6°CS) the cylinder pressure increments $\Delta p(\varphi)$ can be calculated incrementally and summation leads to the cylinder pressure of the high pressure combustion process

$$p_{\mathrm{hp}}(\varphi) = \sum_{\varphi=\varphi_{\mathrm{IC}}}^{\varphi_{\mathrm{EO}}} \Delta p(\varphi) \tag{4.2.43}$$

where IC stands for intake closes and E0 for exhaust opens. Correspondingly, the other variables are calculated. Figure 4.2.3 shows a comparison of simulated and measured variables for a truck diesel engine.

Based on the calculated cylinder pressure the corresponding torque generated during the high pressure cycle becomes due to (4.2.22)

$$M_{\mathrm{g,hp}}(\varphi) = \frac{V_{\mathrm{d}}}{4\pi} p_{\mathrm{hp}}(\varphi). \tag{4.2.44}$$

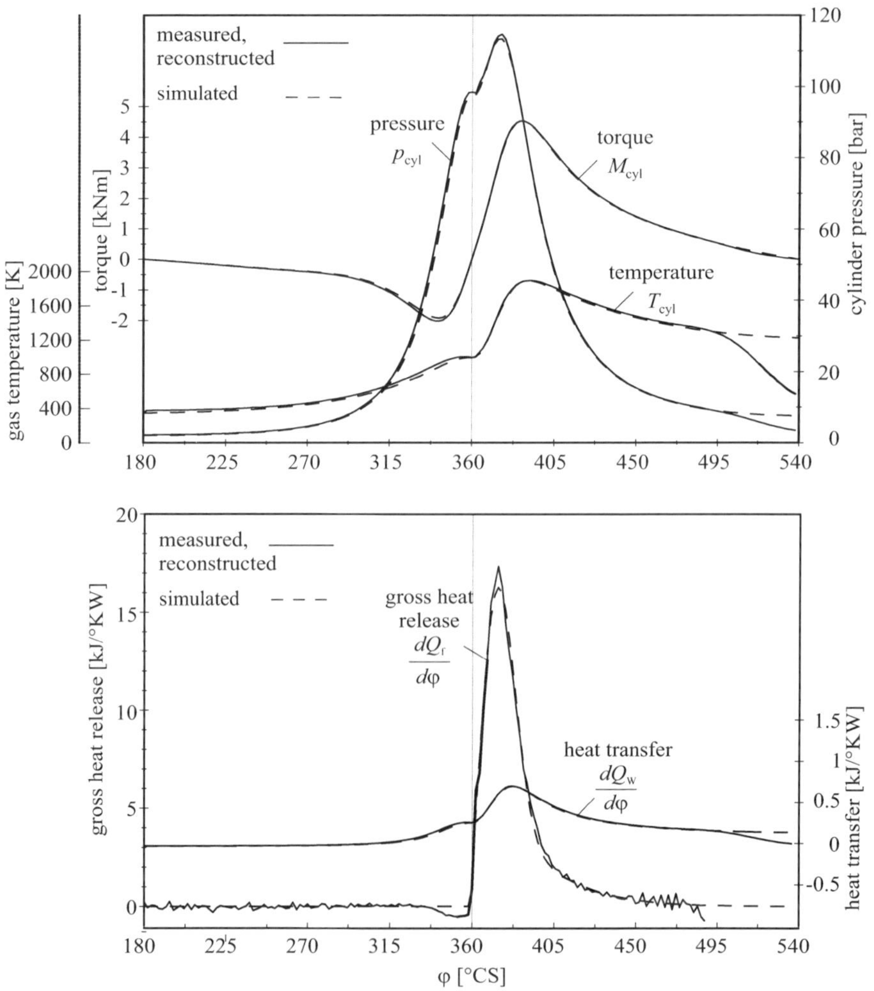

Fig. 4.2.3. Comparison of calculated and measured variables for the HP-cycle of a truck diesel engine OM 442LA, $n_{\mathrm{eng}} = 1260\,\mathrm{rpm}$, $m_{\mathrm{f}} = 176\,\mathrm{mg}$, $\varphi_{\mathrm{soi}} = 356, 5°\mathrm{CS}$, applying parameter estimation for parameters of the heat release and wall heat transfer, Sinsel (2000).

b) Low pressure combustion cycle

The low pressure combustion cycle comprises the gas exchange, that means pushing out the exhaust gases with pressure $p_{\mathrm{cyl,out}}$ and mass flow $\dot{m}_3$ and drawing in the fresh gases with pressure $p_{\mathrm{cyl,in}}$ and mass flow $\dot{m}_2$ and temperature $T_{\mathrm{cyl,in}}$. A simplified modeling of the gas exchange process will be based on semi-physical models according to Urlaub (1995), Sinsel (2000) and Schaffnit (2002).

The theoretical gas mass flow into the cylinder follows by assuming a piston pump

$$\dot{m}_{\mathrm{g,th}} = \frac{1}{2} n_{\mathrm{eng}} V_{\mathrm{d}} \rho_2 = \frac{1}{2} n_{\mathrm{eng}} V_{\mathrm{d}} \frac{p_2}{R T_2}. \tag{4.2.45}$$

Because of various flow phenomena like valve throttling and charge heating the real mass flow is smaller, which is expressed by the *volumetric efficiency coefficient*, Heywood (1988)

$$\eta_{\mathrm{v}} = \frac{\dot{m}_{\mathrm{g}}}{\dot{m}_{\mathrm{g,th}}} = \frac{\dot{m}_{\mathrm{g}}}{0.5\, n_{\mathrm{eng}} V_{\mathrm{d}} \frac{p_2}{R T_2}} \tag{4.2.46}$$

or

$$\eta_{\mathrm{v}} = \frac{m_{\mathrm{g}}}{m_{\mathrm{g,th}}} = \frac{m_{\mathrm{g}} R T_2}{p_2 V_{\mathrm{d}}}. \tag{4.2.47}$$

It depends on the engine speed, load (air-mass flow, injected fuel mass) and valve timing and can be determined experimentally on test benches or by empirical equations, e.g. Urlaub (1995). The fresh gas is then heated up in the cylinder and can be described according to Zapf (1970) by

$$T_{\mathrm{cyl,0}} = \frac{5}{6} T_2 + 85 \;\; [\mathrm{K}]. \tag{4.2.48}$$

Neglecting residual gases the cylinder pressure at the begin of the compression is

$$p_{\mathrm{cyl,0}} = \frac{m_{\mathrm{cyl,0}}\, R\, T_{\mathrm{cyl,0}}}{V_{\mathrm{c}} + V_{\mathrm{d}}} \tag{4.2.49}$$

with fresh gas mass

$$m_{\mathrm{g,0}} = \eta_{\mathrm{v}} m_{\mathrm{g,th}} = \eta_{\mathrm{v}} \frac{p_2 V_{\mathrm{d}}}{R T_2}. \tag{4.2.50}$$

The gas mass flow into one cylinder is

$$\dot{m}_2 = m_{\mathrm{cyl,0}} \frac{n_{\mathrm{eng}}}{2 \cdot 60} \tag{4.2.51}$$

and the gas mass flow out of the cylinder becomes

$$\dot{m}_3 = \dot{m}_2 + \dot{m}_{\mathrm{f}} = \frac{1}{2} (m_{\mathrm{cyl,0}} + m_{\mathrm{f}}) \frac{n_{\mathrm{eng}}}{60}. \tag{4.2.52}$$

The calculation of the work and torque for the low pressure combustion cycle requires average values for the pressures $p_{\mathrm{cyl,in}}$ and $p_{\mathrm{cyl,out}}$ during the induction and expansion stroke. Based on the average gas velocities through the valves it holds

$$p_{\text{lp}} = \begin{cases} p_{\text{cyl,in}} = p_2 \left(1 - k_{\text{in}} \overline{v}_{\text{in}}^2\right) & \text{(induction)} \\ p_{\text{cyl,out}} = p_3 \left(1 + k_{\text{out}} \overline{v}_{\text{out}}^2\right) & \text{(expansion)} \end{cases} \tag{4.2.53}$$

with averaged gas velocities

$$\overline{v}_{\text{in,out}} = \frac{A_{\text{p}}}{A_{\text{in,out}}} \overline{v}_{\text{p}} = \left(\frac{A_{\text{p}}}{A_{\text{in,out}}}\right) 4 r_{\text{CS}} n_{\text{eng}}, \tag{4.2.54}$$

see Urlaub (1995), and

A_{p}: piston area
$\overline{v}_{\text{p}}$: average piston speed
$A_{\text{in,out}}$: valve cross-section for opened valves
r_{CS}: radius of crankshaft
$k_{\text{in}}, k_{\text{out}}$: average pressure drop coefficients at the valves.

The constants have to be determined experimentally, Schaffnit (2002). Finally, the required work for the low pressure combustion cycle of one cylinder becomes with averaged values for the pressures

$$W_{\text{i2}} = (p_{\text{cyl,in}} - p_{\text{cyl,out}}) V_{\text{d}} \tag{4.2.55}$$

and the average gas exchange torque gets

$$M_{\text{g,lp}} = \frac{P_{\text{lp}}}{\omega_{\text{eng}}} = \frac{(p_{\text{cyl,in}} - p_{\text{cyl,out}}) V_{\text{d}} \frac{1}{2} \frac{\omega_{\text{eng}}}{2\pi}}{\omega_{\text{eng}}} = \frac{(p_{\text{cyl,in}} - p_{\text{cyl,out}}) V_{\text{d}}}{4\pi}. \tag{4.2.56}$$

4.2.3 Combustion model analysis

For designing cylinder individual combustion control and diagnosis systems characteristic quantities or features have to extracted from dynamic cylinder pressure measurements.

The measurement of the pressure in the cylinders yields direct information of the strokes. Therefore it is usually applied for the development and controller calibration at engine test stands. Pressure sensors are beginning to be used for series production engines, see Sect. 7.5.1. In-cylinder pressure measurement may be required by new burning processes like the homogeneous charge compression ignition (HCCI) or for lean combustion. Therefore, the generation of special features will be considered, which can be used for control as well as for diagnosis.

a) Difference pressure features

Figure 4.2.4 depicts the measured pressure curve with fuel injection and firing and the pressure curve without fuel injection, called "motored" pressure or "towed pressure". The information on the combustion lies in the difference pressure

$$\Delta p_{\mathrm{cyl}}(\varphi) = p_{\mathrm{cyl}}(\varphi) - p_{\mathrm{cyl,m}}(\varphi). \tag{4.2.57}$$

As in normal operation $p_{\mathrm{cyl,m}}$ is only measurable before fuel injection it is assumed that it is symmetrical to the top dead center (TDC) and is reconstructed by

$$p_{\mathrm{cyl,m}}(\varphi) = p_{\mathrm{cyl,m}}(-\varphi) \tag{4.2.58}$$

for $-180°\mathrm{CS} < \varphi < \varphi_{\mathrm{soi}}$ where φ_{soi} is the angle for start of the (main) injection. The missing curve segment for $|\varphi| < |\varphi_{\mathrm{soi}}|$ towards the maximum is approximated with

$$p_{\mathrm{cyl,m}}(\varphi) = a\,|\varphi|^{1.5} + b.$$

The parameters a and b are estimated on-line by the method of least squares. This leads to a good reconstruction with maximal errors of $5\,\%$, Leonhardt et al (1997) and removes sensor offsets.

For the purpose of control and diagnosis special numeric quantities of the difference pressure are required. Some examples are, compare Fig. 4.2.5

- *Center of gravity*

$$\varphi_{\mathrm{c}} = \frac{\int_{-180}^{+180} \varphi\, \Delta p_{\mathrm{cyl}}(\varphi)\, d\varphi}{\int_{-180}^{+180} \Delta p_{\mathrm{cyl}}(\varphi)\, d\varphi};\ \pi_{\mathrm{c}} = \frac{1}{2}\frac{\int_{-180}^{+180} \Delta p_{\mathrm{cyl}}^2(\varphi)\, d\varphi}{\int_{-180}^{+180} \Delta p_{\mathrm{cyl}}(\varphi)\, d\varphi} \tag{4.2.59}$$

- *Peak difference pressure*, Δp_{peak}, π_{peak}
- *Secant length*

$$\varphi_i = \varphi_{\mathrm{max},i} - \varphi_{\mathrm{min},i} \tag{4.2.60}$$

for e.g. $i = 4{,}10{,}16$ bar.

These features depend on the injected fuel mass m_{f}, begin of injection φ_{soi} and speed n_{eng}, compare Fig. 4.2.6 for the center of gravity coordinates. As also the other features show unique nonlinear functions, the corresponding maps can be approximated, e.g. by polynomials or a neural net. They serve then as features for the normal behavior in form of look-up tables and allow also the reconstruction of the injected fuel mass and injection angle even for dynamic operation. Comparison with the normal behavior then results in symptoms for fault detection in the corresponding cylinder, Leonhardt (1996), Leonhardt et al (1997), Leonhardt et al (1999), Isermann (2015).

b) Heat release analysis

In Sect. 4.2.2 it was shown that from the pressure measurement in the cylinder only the *net heat release rate* $\frac{dQ_{\mathrm{h}}}{d\varphi}$ can be determined, see (4.2.30)

$$\frac{dQ_{\mathrm{h}}(\varphi)}{d\varphi} = \frac{c_{\mathrm{v}}(T_{\mathrm{cyl}})V_{\mathrm{cyl}}(\varphi)}{R}\frac{dp_{\mathrm{cyl}}(\varphi)}{d\varphi} + \left(1 + \frac{c_{\mathrm{v}}(T_{\mathrm{cyl}})}{R}\right) p_{\mathrm{cyl}}(\varphi)\frac{dV_{\mathrm{cyl}}(\varphi)}{d\varphi}. \tag{4.2.61}$$

The small effect of the fuel $h_{\mathrm{f}}\frac{dm_{\mathrm{f}}}{d\varphi}$ is neglected. Discretizing this equation for angle increments $\Delta\varphi = \varphi_i - \varphi_{i-1}$ gives the *incremental net heat release*

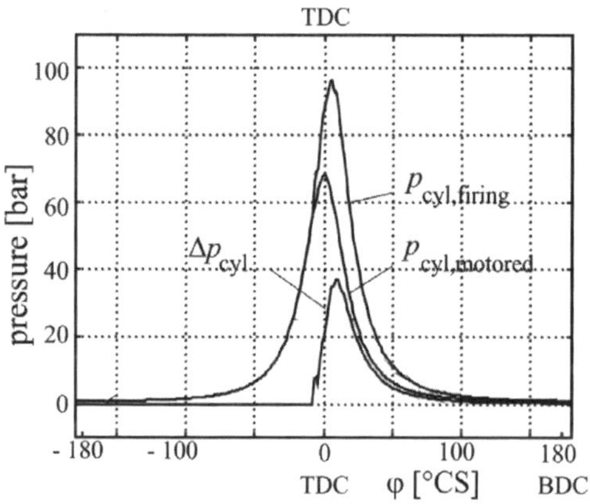

Fig. 4.2.4. Cylinder pressure curves for a diesel engine with swirl chamber, $m_{\mathrm{f}} = 10\,\mathrm{mg}$; $\varphi_{\mathrm{soi}} = 16°\,\mathrm{CS}$; $n_{\mathrm{eng}} = 2400\,\mathrm{rpm}$, 1.2 l, 40 kW.

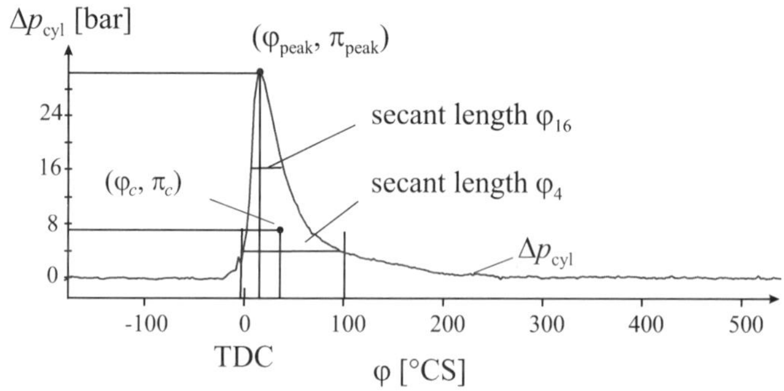

Fig. 4.2.5. Difference pressure $\Delta p_{\mathrm{cyl}}(\varphi)$ and extracted features.

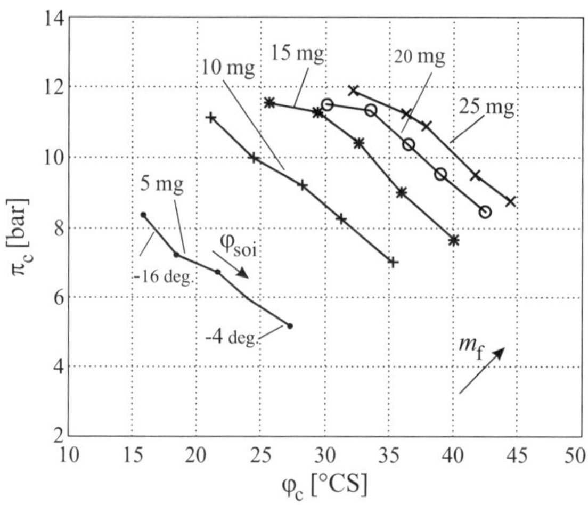

Fig. 4.2.6. Center of gravity map π_{c}, φ_{c} according to (4.2.59), $n_{\mathrm{eng}} = 2100\,\mathrm{rpm}$.

$$\Delta Q_{\mathrm{h}}(\varphi_i) = \frac{c_{\mathrm{v}}(T_{\mathrm{cyl}}(\varphi_i))}{R} V_{\mathrm{cyl}}(\varphi_i) \Delta p_{\mathrm{cyl}}(\varphi_i) + \left(1 + \frac{c_{\mathrm{v}}(T_{\mathrm{cyl}}(\varphi_i))}{R}\right) p_{\mathrm{cyl}}(\varphi_i) \Delta V_{\mathrm{cyl}}(\varphi_i) \quad (4.2.62)$$

with the differences

$$\Delta p_{\mathrm{cyl}}(\varphi_i) = p_{\mathrm{cyl}}(\varphi_i) - p_{\mathrm{cyl}}(\varphi_i - 1)$$
$$\Delta V_{\mathrm{cyl}}(\varphi_i) = V_{\mathrm{cyl}}(\varphi_i) - V_{\mathrm{cyl}}(\varphi_i - 1).$$

The *integrated net heat release* (sum heat release) is obtained by

$$Q_{\mathrm{h}\Sigma} = \sum_{\varphi=\varphi_0}^{\varphi_N} \Delta Q_{\mathrm{h}}(\varphi_i) \Delta\varphi \quad (4.2.63)$$

with $\varphi_0 = \varphi_{\mathrm{ign}}$ and $\varphi_N = \varphi_{\mathrm{E0}}$ for gasoline engines and $\varphi_0 = \varphi_{\mathrm{soi}}$ and $\varphi_N = \varphi_{\mathrm{E0}}$ for diesel engines (φ_{E0} : exhaust valve opens).

Now certain thermodynamic features can be calculated, for example, the crank angles

$$\varphi_{\mathrm{Q5}}, \quad \varphi_{\mathrm{Q50}}, \quad \varphi_{\mathrm{Q95}}$$

where the heat release values

$$Q_{\mathrm{h5}}, \quad Q_{\mathrm{h50}}, \quad Q_{\mathrm{h95}}$$

reach $5\,\%$, $50\,\%$ and $95\,\%$ of the total released heat. The specific heat release values are calculated by

$$Q_h(\varphi_{\mathrm{Qx}}) = Q_{\mathrm{x}} = \frac{x}{100}(\max Q_{\mathrm{h}} - \min Q_{\mathrm{h}}) + \min Q_{\mathrm{h}}. \quad (4.2.64)$$

Figure 4.2.7a) and b) show examples for these values. φ_{Q5} characterizes the begin of burning, φ_{Q95} the end of burning and $\Delta\varphi_{\mathrm{bd}} = \varphi_{\mathrm{Q95}} - \varphi_{\mathrm{Q5}}$ is the burning duration, see also Figs. 7.4.1, 7.5.3, 8.2.5 and 8.7.3.

The *burn rate* or *mass fraction burned* (MFB) is

$$x_{\mathrm{B}}(\varphi) = \frac{Q_{\mathrm{f}}(\varphi)}{Q_{\mathrm{f}\Sigma}} \qquad [0...1] \quad (4.2.65)$$

where $Q_{\mathrm{f}\Sigma}$ is the total gross heat release. This feature can approximately be determined from the net heat release if not absolute values but relative values are of interest. Hence

$$x_{\mathrm{B}}(\varphi) \approx \frac{Q_{\mathrm{h}}(\varphi)}{Q_{\mathrm{h}\Sigma}} \qquad [0...1] \quad (4.2.66)$$

is a good estimate for the mass fraction burned, Bargende (1995), Müller (2003).

c) Simplified equations for heat release analysis

To ease computations simplified calculations for the heat release were proposed. Hohenberg (1982) assumes between an adiabatic and an isochoric change of state two successive pressure measurements p and $p+\Delta p$ and obtains for a net heat release increment

$$\Delta Q_{\mathrm{h}}(\varphi) = \frac{c_{\mathrm{v}}}{R} V_{\mathrm{cyl}}(\varphi) \left[p_{\mathrm{cyl}}(\varphi) - (p_{\mathrm{cyl}}(\varphi) + \Delta p_{\mathrm{cyl}}(\varphi)) \left(1 + \frac{\Delta V_{\mathrm{cyl}}(\varphi)}{V_{\mathrm{cyl}}(\varphi)} \right) \right] \tag{4.2.67}$$

with κ=1.33.

Another approximation can be derived for assuming a constant volume combustion cycle, an approximation for an idealized Otto combustion process, see Fig. 4.2.2a).

With the assumption of a polytropic (adiabatic) behavior it holds for the compression stroke ($1 \to 2$)

$$p(\varphi)V^{\kappa}(\varphi) = C_1 \tag{4.2.68}$$

and for the expansion stroke ($3 \to 4$)

$$p(\varphi)V^{\kappa}(\varphi) = C_2 \tag{4.2.69}$$

where the adiabatic coefficient is $\kappa = 1 + \frac{R}{c_{\mathrm{v}}(\lambda, T)}$. Therefore the amount of energy Q_{f} released between ignition (ign) and end of combustion (eoc) ($2 \to 3$) is proportional to $(C_2 - C_1)$

$$Q_{\mathrm{f}} \sim C_2 - C_1 = p_{\mathrm{eoc}} V_{\mathrm{eoc}}^{\kappa} - p_{\mathrm{ign}} V_{\mathrm{ign}}^{\kappa}. \tag{4.2.70}$$

During the combustion the released energy is proportional to $p(\varphi)V^{\kappa}(\varphi)$. Hence, for the mass fraction burned and network release follows, McCuiston et al (1977) , Heywood (1988).

$$x_{\mathrm{B}}(\varphi) = \frac{p(\varphi)V^{\kappa}(\varphi) - p_{\mathrm{ign}} V_{\mathrm{ign}}^{\kappa}}{p_{\mathrm{eoc}} V_{\mathrm{eoc}}^{\kappa} - p_{\mathrm{ign}} V_{\mathrm{ign}}^{\kappa}} \tag{4.2.71}$$

$$Q_{\mathrm{h}}(\varphi) \approx x_{\mathrm{B}}(\varphi) Q_{\mathrm{h}\Sigma}.$$

This gives a relatively good approximation compared to detailed thermodynamic calculations, Vogt et al (2004). An optimal value for 50 % MFB is $\varphi \approx 8°$CS, Bargende (1995).

The mass fraction burned can be determined for several $\varphi_i = \varphi_1, \varphi_2, \varphi_3 \ldots$ based on measured cylinder pressures $p(\varphi_i) = p_{\mathrm{cyl}}(\varphi_i)$ for the calculation of (4.2.71). Deviations

$$\Delta x_{\mathrm{B}}(\varphi) = x_{\mathrm{B}}(\varphi_i) - x_{\mathrm{B,ref}}(\varphi_i) \tag{4.2.72}$$

indicate differences to normal reference values. This can be used for optimizing the ignition angle and for fault detection like to late or advanced burning, delayed ignition and injection faults, see Sects. 7.5 and 8.6.

Throughout this chapter the state of gas in the form

$$pv = RT \tag{4.2.73}$$

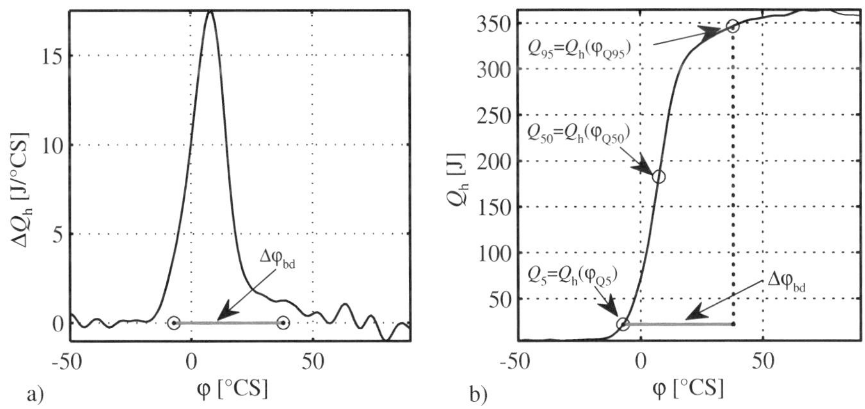

Fig. 4.2.7. a Difference heat release determined from cylinder pressure of a gasoline engine. **b** sum heat release Q_h and heat release angle φ_QX for $Q_\mathrm{h} = 5, 50$ and $95\,\%$, Kohlhase (2011).

was applied. This equation holds for pressures $p < 5\,\mathrm{bar}$, VDI heat atlas (2010), Da11. However, based on the work of van der Waal in 1873 various versions with correlation terms for higher pressure and also higher temperature were found, among them from Soave-Redlich-Kwong,

$$p = \frac{RT}{v-p}\frac{a(T)}{v^2+bv}. \tag{4.2.74}$$

Simulations show that the difference to the cylinder pressure around 60–80 bar is about maximal 2–4 %.

4.3 Mechanical system

4.3.1 Mean value torque models

The crankshaft drive transfers the combustion pressure induced forces on the pistons into a torque. For the work of the piston and the crankshaft drive follows for a four-stroke engine

$$4\pi M_\mathrm{eng} = p_\mathrm{bmep} V_\mathrm{d} \tag{4.3.1}$$

where V_d is the displacement of one cylinder, and therefore for the *brake mean effective pressure*

$$p_\mathrm{bmep} = \frac{4\pi M_\mathrm{eng}}{V_\mathrm{d}}. \tag{4.3.2}$$

This is a fictive average pressure acting on the piston to generate the torque M_eng at the clutch over one expansion stroke with volume V_d and is about 7–11 bar for gasoline engines and 8–22 bar for diesel engines. The work in the cylinder is generated by the mass of fuel m_f burnt per stroke with lower heat value H_l. This allows the calculation of the effective efficiency of the engine

$$\eta_{\text{eff}} = \frac{4\pi M_{\text{eng}}}{m_{\text{f}} H_{\text{l}}} = \frac{p_{\text{bmep}} V_{\text{d}}}{m_{\text{f}} H_{\text{l}}}. \tag{4.3.3}$$

This efficiency includes all other efficiencies like fuel conversion factor η_{f}, thermal efficiency η_{th}, efficiency of the working cycle η_{g} and mechanical efficiency η_{m}

$$\eta_{\text{eff}} = \eta_{\text{f}}\, \eta_{\text{th}}\, \eta_{\text{g}}\, \eta_{\text{m}} \tag{4.3.4}$$

compare (4.2.20). For given brake mean effective pressure p_{bmep} or given efficiency η_{eff} the mean torque at the clutch can therefore be calculated by

$$M_{\text{eng}} = \frac{p_{\text{bmep}} V_{\text{d}}}{4\pi} = \frac{m_{\text{f}} H_{\text{l}} \eta_{\text{eff}}}{4\pi}. \tag{4.3.5}$$

p_{bmep} or η_{eff} can be obtained from diagrams in dependence on speed with parameters like fuel injection based on measurements, see Fig. 4.3.1. (4.3.5) leads to a relation between brake mean effective pressure and fuel mass

$$p_{\text{bmep}} = \frac{H_{\text{l}} \eta_{\text{eff}}}{V_{\text{d}}}\, m_{\text{f}}$$

or

$$m_{\text{f}} = \frac{V_{\text{d}}}{H_{\text{l}} \eta_{\text{eff}}}\, p_{\text{bmep}} \tag{4.3.6}$$

p_{bmep} is proportional to m_{f}. However η_{eff} changes much with the operating point $(p_{\text{bmep}}, n_{\text{eng}})$. For many internal combustion engines a linear relation between the fuel consumption per time unit $b_{\text{f}}[\text{kg/h}]$ and the effective Power $P_{\text{eff}}[\text{kW}]$ can be observed at least for part load, compare Fig. 4.3.2,

$$b_{\text{f}} = c_0 P_{\text{eff}} + P_{\text{loss}} \tag{4.3.7}$$

where P_{loss} expresses the sum of friction and gas exchange losses. Introducing $m_{\text{f}} = 2\dot{m}_{\text{f}}/n_{\text{eng}} = 2b_{\text{f}}/n_{\text{eng}}$ yields

$$\begin{aligned} m_{\text{f}} &= 4\pi \left(c_0 M_{\text{eng,eff}} + M_{\text{eng,loss}}\right) \\ &= 4\pi c_0 M_{\text{eng,eff}} + m_{\text{f,loss}} \\ &= V_{\text{d}} \left(c_0 p_{\text{bmep}} + p_{\text{m,loss}}\right). \end{aligned} \tag{4.3.8}$$

These linear relationships are called *Willans-characteristics*, Greene et al (1969). The two terms distinguish in a relatively simple way between the usable effective power or torque and the various losses. $p_{\text{m,loss}}$ can be interpreted as mean effective pressure of friction and gas exchange losses.

4.3.2 Dynamic torque models

In order to model the dynamic torque generation at the clutch or flywheel the dynamic forces on the pistons and their transformation to a crankshaft torque has to be considered, see Fig. 4.3.3.

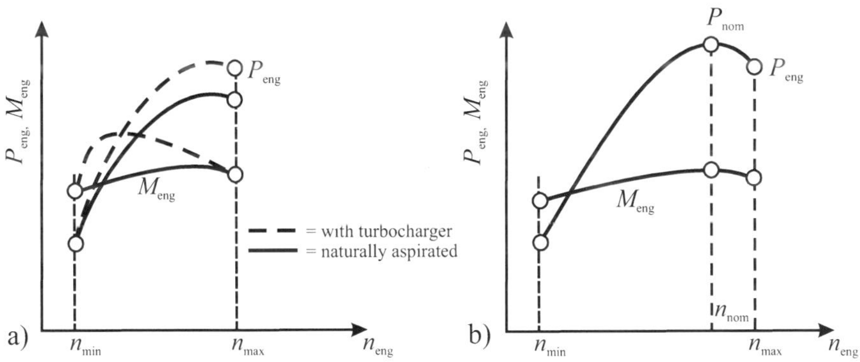

Fig. 4.3.1. Torque and power in dependence on engine speed. The brake mean effective pressure is proportional to the torque. **a** gasoline engine. **b** diesel engine.

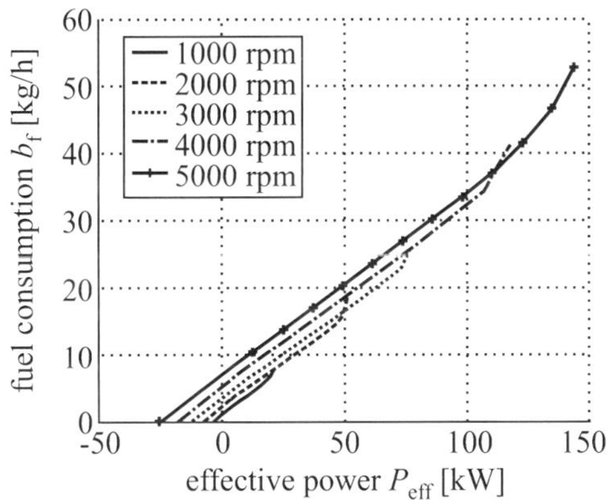

Fig. 4.3.2. Willans-characteristic for a gasoline engine, Schmidt (2003)

Assuming that the crankshaft is stiff and the moving pistons, connecting rods, crankshaft, flywheel and the primary part of the clutch can be lumped in one moment of inertia J_{eng} the torque balance of the crankshaft yields

$$J_{\text{eng}} \frac{d\omega_{\text{eng}}(\varphi)}{dt} = M_{\text{g}}(\varphi) - M_{\text{m}}(\varphi) - M_{\text{v}}(\varphi) - M_{\text{f}}(\varphi) - M_{\text{aux1}}(\varphi) - M_{\text{l}}(\varphi) \quad (4.3.9)$$

with

M_{g}	gas force torque
M_{m}	dynamic masses torque
M_{v}	valve train torque
M_{f}	friction torque
M_{aux1}	auxiliary drive torque (engine integrated)
M_{l}	load torque.

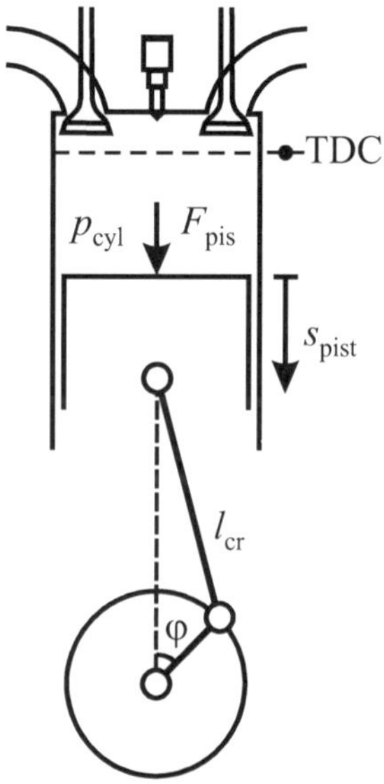

Fig. 4.3.3. Cylinder, piston, connecting rod (conrod) and crankshaft.

The required torque M_{aux1} for the auxiliary drives include the engine integrated auxiliaries (accessories) like oil pump and water pump. The torque for the auxiliary drives which can vary depending on the vehicle, like alternator, servo steering pump, air conditioning compressor, etc. are described by M_{aux2} and treated as external load. Figure 4.3.4 shows the corresponding signal-flow diagram. Hence, the engine torque at the clutch is described by

$$M_{eng}(\varphi) = M_g(\varphi) - M_m(\varphi) - M_v(\varphi) - M_f(\varphi) - M_{aux1}(\varphi) - Jeng\frac{d\omega_{eng}(\varphi)}{dt}. \tag{4.3.10}$$

Some torques can now be split into a constant and a periodic part

$$M(\varphi) = \overline{M} + \widetilde{M}(\varphi) \tag{4.3.11}$$

where the periodic part has zero mean. From (4.3.9) follows for averaged values over one operation cycle

$$J_{eng}\,\bar{\dot{\omega}}_{eng}(\varphi) = \overline{M}_g - \overline{M}_v - \overline{M}_f - \overline{M}_{aux1} - \overline{M}_l = 0 \tag{4.3.12}$$

and for the periodic acting torques

$$J_{eng}\,\tilde{\dot{\omega}}_{eng}(\varphi) = \widetilde{M}_g(\varphi) - \widetilde{M}_m(\varphi) - \widetilde{M}_v(\varphi) - \widetilde{M}_l(\varphi). \tag{4.3.13}$$

It is assumed that the friction and the auxiliary drives have no periodic parts. For stationary operating it holds $\bar{\dot{\omega}}_{eng} = 0$ in (4.3.12).

By averaging specific torques in (4.3.12) over 720°CS it follows for four-stroke engines

$$M_{\text{eng,ind}} = \overline{M}_{\text{g,720}} = \frac{1}{4\pi} \int_{\varphi_0}^{\varphi_0+4\pi} M_{\text{g}}(\varphi)\, d\varphi \qquad \text{mean indicated torque} \quad (4.3.14)$$

$$\begin{aligned} M_{\text{eng,eff}} = \overline{M}_{1,720} &= \frac{1}{4\pi} \int_{\varphi_0}^{\varphi_0+4\pi} M_l(\varphi)\, d\varphi \\ &= \frac{1}{4\pi} \int_{\varphi_0}^{\varphi_0+4\pi} M_{\text{eng}}(\varphi)\, d\varphi \qquad \text{mean effective torque} \end{aligned} \quad (4.3.15)$$

Hence, for the mean torques hold

$$M_{\text{eng,eff}} = M_{\text{eng,ind}} - \overline{M}_{\text{f}} - \overline{M}_{\text{v}} - \overline{M}_{\text{aux1}}. \quad (4.3.16)$$

Now the single torques are considered.

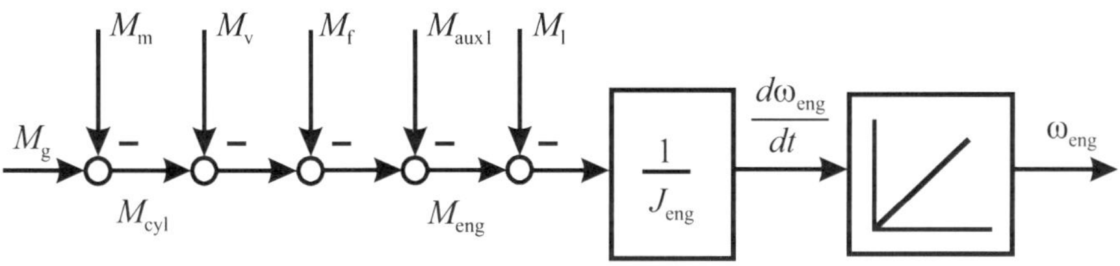

Fig. 4.3.4. Acting torques on the crankshaft.

a) Kinematics of the crankshaft drive

The travel way of a piston in dependence on the crank angle φ can be described by Fig. 4.3.3,

$$s_{\text{pist}}(\varphi) = r_{\text{cs}} \left[1 + \frac{1}{\lambda_{\text{cr}}} - \cos\varphi - \frac{1}{\lambda_{\text{cr}}} \sqrt{1 - \lambda_{\text{cr}}^2 \sin^2\varphi} \right] \quad (4.3.17)$$

with the stroke/connecting rod ratio

$$\lambda_{\text{cr}} = r_{\text{cs}}/l_{\text{cr}} \quad (4.3.18)$$

compare Heywood (1988), Robert Bosch GmbH (2011). For small λ_{cr} this can be simplified after a series expansion of the the square root to

$$s_{\text{pist}}(\varphi) = r_{\text{cs}} \left[1 - \cos\varphi + \frac{\lambda_{\text{cr}}}{4} (1 - \cos 2\varphi) \right]. \quad (4.3.19)$$

(For $\lambda_{\text{cr}} = 0.25$ the amplitude of the last term is 0.063, i.e. relatively small.)

For the cylinder volume follows

$$V_{\text{cyl}}(\varphi) = V_{\text{c}} + A_{\text{pist}}\, s_{\text{pist}}(\varphi) \quad (4.3.20)$$

where V_c is the clearance volume at TDC and A_pist the piston area with the displacement $V_\mathrm{d} = 2A_\mathrm{pist} r_\mathrm{cs}$.

The calculation of the combustion pressure p_cyl follows the pressure and heat-release analysis (4.2.31) with $V_\mathrm{cyl}(\varphi)$ according to (4.3.20). $dV_\mathrm{cyl}(\varphi)/\, d\varphi$ follows from (4.3.20) to (4.3.17) to

$$\frac{dV_\mathrm{cyl}}{d\varphi} = \frac{V_\mathrm{d}}{2}\left(\sin(\varphi) + \frac{\lambda_\mathrm{cr}\sin(2\varphi)}{2\sqrt{1-(\lambda_\mathrm{cr}\sin(\varphi))^2}}\right). \tag{4.3.21}$$

The pressure of the high pressure cycle p_cyl results from (4.2.43) and of the low pressure cycle from (4.2.53). The piston force F_pist is the result of the gas force F_g and the mass force F_m of the accelerated piston and conrod, see Fig. 4.3.5

$$\begin{aligned} F_\mathrm{pist}(\varphi) &= F_\mathrm{g}(\varphi) - F_\mathrm{m}(\varphi) \\ &= \left(p_\mathrm{cyl}(\varphi) - p_\mathrm{a}\right) A_\mathrm{pist} - \left(m_\mathrm{pist} + m_\mathrm{cr}\right) \ddot{s}_\mathrm{pist}(\varphi). \end{aligned} \tag{4.3.22}$$

m_pist is the piston mass, m_cr the oscillating part of the conrod and p_a the ambient pressure. The piston acceleration follows from (4.3.19)

$$\ddot{s}_\mathrm{pist}(\varphi) = r_\mathrm{cs}\,\omega_\mathrm{eng}^2 \left(\cos\varphi + \lambda_\mathrm{cr}\cos 2\varphi\right). \tag{4.3.23}$$

For constant speed holds $\varphi = \omega_\mathrm{eng} t$. Hence, the piston acceleration and therefore the vertical mass forces are proportional to ω_eng^2 with angular frequencies ω_eng and $2\omega_\mathrm{eng}$, Merker et al (2006). It is assumed that the engine speed is constant during one stroke and therefore no mass forces result from speed changes $\dot{\omega}$. The force from the piston to the crankshaft is, Urlaub (1995),

$$F_\mathrm{t}(\varphi) = F_\mathrm{pist}(\varphi)\sin\varphi\left(1 + \frac{\lambda_\mathrm{cr}\cos\varphi}{\sqrt{1-\lambda_\mathrm{cr}^2\sin^2\varphi}}\right). \tag{4.3.24}$$

Finally, the torque at the crankshaft through the gas forces and piston acceleration forces for one cylinder i is

$$M_{\mathrm{cyl},i}(\varphi) = F_\mathrm{t}(\varphi)\, r_\mathrm{cs} = \left(F_{\mathrm{g},i}(\varphi) - F_{\mathrm{m},i}(\varphi)\right) r_\mathrm{cs} = M_{\mathrm{g},i}(\varphi) - M_{\mathrm{m},i}(\varphi) \tag{4.3.25}$$

see Fig. 4.3.5 and can be divided in a torque $M_{\mathrm{g},i}$ from the gas forces and $M_{\mathrm{m},i}$ from the mass forces, which have zero mean.

For engines with i_c cylinders holds for each cylinder

$$M_{\mathrm{cyl},i}(\varphi) = F_\mathrm{t}\left(\varphi + (i_\mathrm{c} - 1)\,\varphi_0\right) r_\mathrm{cs} \tag{4.3.26}$$

where φ_0 is the firing distance angle to cylinder 1. For a four-cylinder engine it is, e.g. $\varphi_0 = \pi$. Their sum yields the cylinder torque

$$M_\mathrm{cyl}(\varphi) = \sum_{i=1}^{i_\mathrm{c}} M_{\mathrm{cyl},i}(\varphi) = \sum_{i=1}^{i_\mathrm{c}} \left(M_{\mathrm{g},i}(\varphi) - M_{\mathrm{m},i}(\varphi)\right). \tag{4.3.27}$$

The effective torque to the crankshaft then follows from (4.3.10)

$$M_\mathrm{eng}(\varphi) = M_\mathrm{cyl}(\varphi) - M_\mathrm{v}(\varphi) - M_\mathrm{f} - M_\mathrm{aux1}(\varphi). \tag{4.3.28}$$

In the following, the gas, mass, and friction torques are further analyzed.

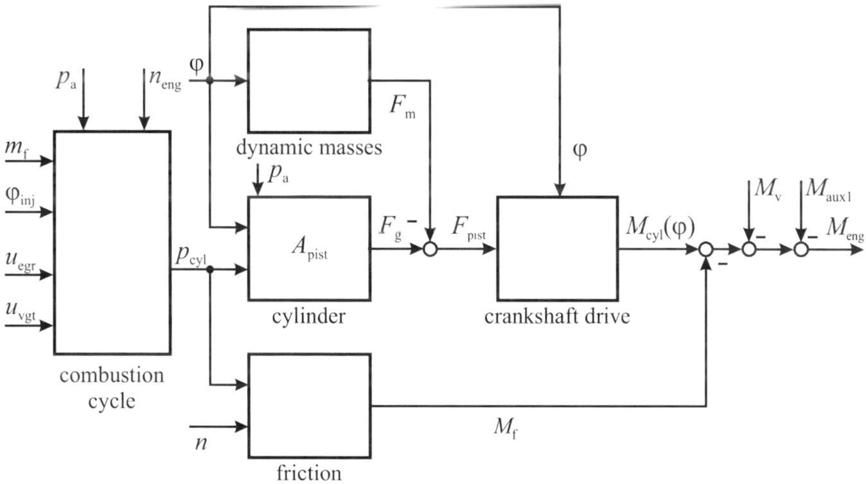

Fig. 4.3.5. Signal-flow chart for the dynamic torque of one cylinder.

b) Gas torque

The gas torque is the result of the high-pressure and low-pressure cycle in the cylinders and is the only one contributing positive work. According to (4.3.22), (4.3.24), (4.3.25) and (4.3.26) it results

$$M_{\mathrm{g}}(\varphi) = F_{\mathrm{g}}(\varphi)\, r_{\mathrm{cs}} = A_{\mathrm{pist}}\, r_{\mathrm{cs}} \sum_{i=1}^{i_{\mathrm{c}}} \left(p_{\mathrm{cyl},i}(\varphi) - p_{\mathrm{a}}\right)$$
$$\cdot \sin\left(\varphi + (i-1)\,\varphi_0\right) \left(1 + \frac{\lambda_{\mathrm{cr}} \cos\left(\varphi + (i-1)\,\varphi_0\right)}{\sqrt{1 - \lambda_{\mathrm{cr}}^2 \sin^2\left(\varphi + (i-1)\,\varphi_0\right)}}\right). \qquad (4.3.29)$$

Hence, the gas moment contains periodic components with angular frequencies ω_{eng} and $2\omega_{\mathrm{eng}}$ $(\sin^2\varphi = (1-\cos 2\varphi)/2)$. Figure 4.3.6 shows the calculated gas torque for firing and motoring operation. The cylinder pressure $p_{\mathrm{cyl}}(\varphi)$ follows from equations for the combustion cycle. A first possibility is to apply (4.2.43) for $p_{\mathrm{hp}}(\varphi)$ for the high-pressure cycle and p_{lp} from (4.2.53) for the low-pressure cycle. Another possibility is to replace the gross heat release $dQ_{\mathrm{f}}(\varphi)/d\varphi$ by a Vibe function according to (4.2.42).

A more detailed model to determine $p_{\mathrm{cyl}}(\varphi)$ by taking into account forward and backward flow through the valves is presented by Zahn (2012).

c) Mass torque

The mass torque from the oscillating piston and conrod masses can be calculated from (4.3.22), (4.3.23), (4.3.24), (4.3.26), (4.3.27)

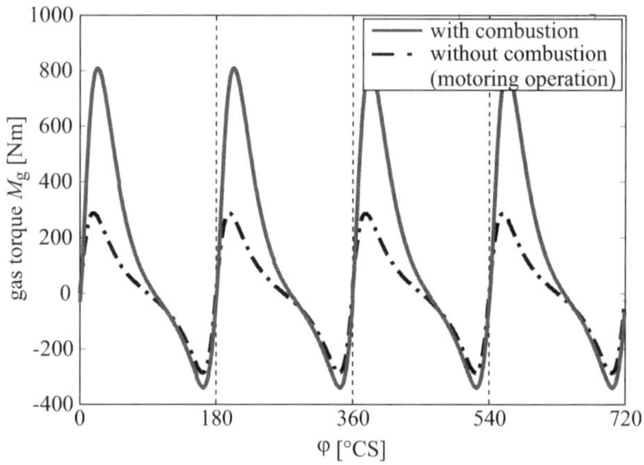

Fig. 4.3.6. Calculated gas torque from measured cylinder pressure $p_{\text{cyl},i}(\varphi)$ for a four-cylinder diesel engine 1.9l, 74 kW, Kimmich (2004).

$$\begin{aligned} M_{\text{m},i}(\varphi) = {} & (m_{\text{pist}} + m_{\text{cr}}) r_{\text{cs}}^2 \omega_{\text{eng}}^2 \left[\cos\left(\varphi + (i-1)\,\varphi_0\right) \right. \\ & \left. + \; \lambda_{\text{cr}} \cos\left(2\varphi + (i-1)\,\varphi_0\right)\right] \sin\left(\varphi + (i-1)\,\varphi_0\right) \\ & \cdot \left[1 + \frac{\lambda_{\text{cr}} \cos\left(\varphi + (i-1)\,\varphi_0\right)}{\sqrt{1 - \lambda_{\text{cr}}^2 \sin^2\left(\varphi + (i-1)\,\varphi_0\right)}}\right]. \end{aligned} \tag{4.3.30}$$

This can be simplified for a four-cylinder engine, as the mass torques of cylinders 1 and 3 as well as 2 and 4 are identical. With $\varphi_0 = \pi$ one obtains, Kimmich (2004)

$$\begin{aligned} M_{\text{m}}(\varphi) = \sum_{i=1}^{4} M_{\text{m},i}(\varphi) = 2\,(m_{\text{pist}} + m_{\text{cr}})\; r_{\text{cs}}^2\, \omega_{\text{eng}}^2 \cdot \\ \cdot \left[\sin(2\varphi) + \frac{\lambda_{\text{cr}}^2}{4} \frac{\sin(4\varphi)}{\sqrt{1 - \lambda_{\text{cr}}^2 \sin^2 \varphi}}\right]. \end{aligned} \tag{4.3.31}$$

Figure 4.3.7 depicts the mass torque for different speeds. The oscillating mass torque has zero mean and its amplitudes increase strongly with the speed because of the quadratic dependence. The resulting torque harmonics have angular frequencies $2\omega_{\text{eng}}$ (dominant magnitude) and $4\omega_{\text{eng}}$ (smaller magnitude). The influence of the mass torque compared to the gas torque can be seen from Fig. 4.3.8. The influence is negligible for $n = 1000\,\text{rpm}$. With increasing speed it increases and shows a damping of the torque amplitudes for this four cylinder engine.

d) Friction torque

The friction torque is usually determined on test benches during motoring with different speed, Fischer (2000). According to Schwarzmeier (1992) holds for an approximation of the *mean friction pressure*

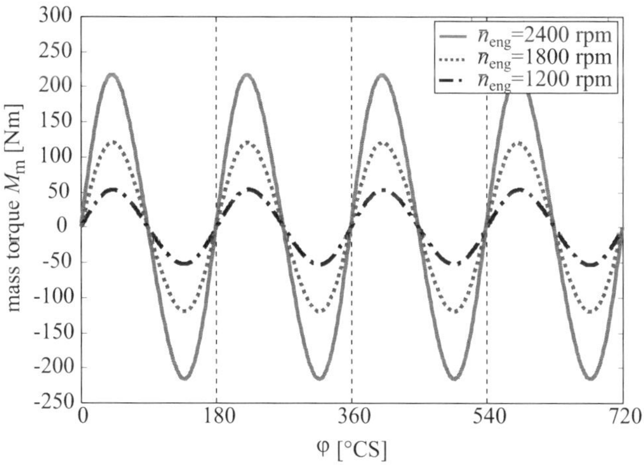

Fig. 4.3.7. Calculated mass torque for different speeds of a four-cylinder diesel engine.

$$p_{fmep} = p^*_{fmep} + k_1 \left(p^{1.35}_{bmep} - (p^*_{bmep})^{1.35} \right) + k_2 \left[n^2_{eng} - (n^*_{eng})^2 \right] \tag{4.3.32}$$

where p^*_{fmep}, p^*_{bmep} and n^*_{eng} are measured for a reference operating point, and k_1 and k_2 are determined from additional measurements, see Schaffnit (2002). The friction torque results from

$$M_{fric}(m_f, n_{eng}) = \frac{p_{fmep}(m_f, n_{eng}) V_d}{4\pi}. \tag{4.3.33}$$

The investigations from Fischer (2000) have shown that friction torque in dependence on the speed is proportional for the crankshaft bearings and quadratic for the pistons. The oil- and water-temperature have a significant influence. The friction losses in the valve train and auxiliary drives are approximately proportional to the speed. The superposition of all frictions does not show a dominating oscillating behavior.

e) Simulation of dynamic torque with split models for stationary and dynamic torque

The calculations for the combustion cycle with $p_{cyl}(\varphi)$ as output are relatively extensive. With regard to real-time computations (e.g. for HiL simulation), Sinsel (2000) developed a dynamic torque model where the stationary and dynamic contributions are separated. As shown in Fig. 4.3.9 the indicated cylinder torque $M_{eng,ind}$ and the torque at the clutch M_{eng} are determined by

$$\begin{aligned} M_{eng,ind}(\varphi) &= M_{basic}(\varphi) + M_{g,comb}(\varphi) \\ M_{eng}(\varphi) &= M_{eng,ind}(\varphi) - M_f - M_v - M_{aux1} \end{aligned} \tag{4.3.34}$$

in dependence on the operating point (m_f, n_{eng}) and relevant manipulated variables, e.g. in form of look-up-tables or a neural net.

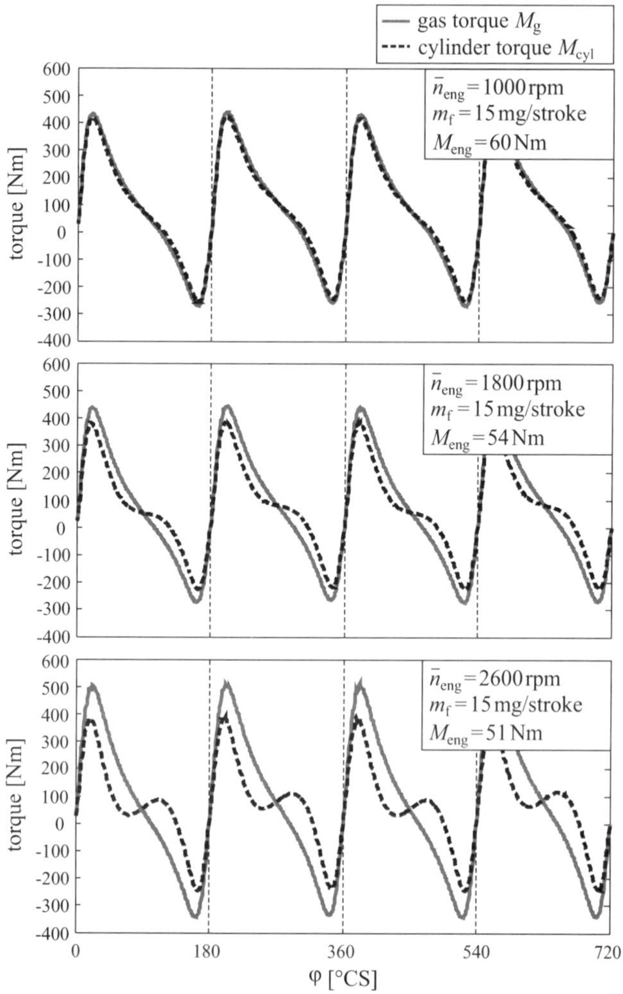

Fig. 4.3.8. Calculated gas torque $M_g(\varphi)$ with (4.3.29) based on measured cylinder torque and calculated cylinder torque $M_{cyl}(\varphi)$ including mass torque with (4.3.27) and (4.3.31) for a four-cylinder diesel engine 1.9 l, 74 kW.

The basic torque M_{basic} describes the mass and gas forces without combustion. The gas forces are determined with a simplified combustion chambers model, assuming a polytropic change of state

$$p_{cyl,basic} = p_{cyl,0} \qquad \varphi \in [0°, 180°], [540°, 720°]$$
$$p_{cyl,basic}(\varphi) = \left(\frac{V_{cyl,0}}{V_{cyl}(\varphi)}\right)^n p_{cyl,0} \qquad \varphi \in [180°, 540°] \tag{4.3.35}$$

where $p_{\text{cyl},0}$ and $V_{\text{cyl},0}$ hold at the beginning of compression and n is the polytropic coefficient, see Fig. 4.3.10. The basic torque M_{basic} follows from

$$M_{\text{basic}}(\varphi) = M_{\text{g},0}(\varphi) - M_{\text{m}}(\varphi) \tag{4.3.36}$$

according to (4.3.25). $M_{\text{g},0}(\varphi)$ is the torque of the gas forces without combustion due to (4.3.29) and (4.3.35) and $M_{\text{m}}(\varphi)$ is the dynamic mass torque obtained from (4.3.30) to (4.3.31).

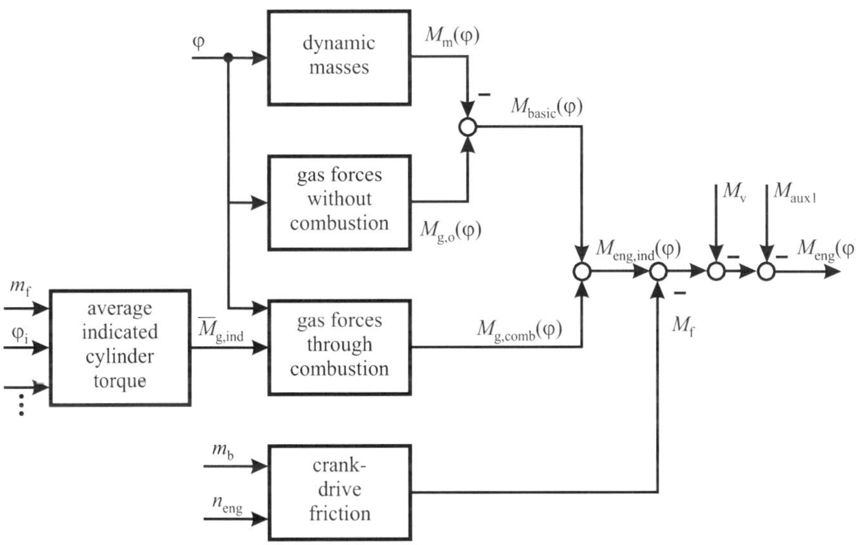

Fig. 4.3.9. Block diagram for the dynamic torque based on stationary models for the indicated torque and separated dynamic models for the combustion, gas and mass torque, Sinsel (2000), Schaffnit (2002).

Based on these torque components it follows for the motoring torque or drag torque, i.e. the torque without combustion

$$M_{\text{drag}}(\varphi) = M_{\text{g},0}(\varphi) - M_{\text{m}}(\varphi) + M_{\text{v}}(\varphi) + M_{\text{f}}(\varphi) + M_{\text{aux1}}. \tag{4.3.37}$$

The dynamic gas torque through the combustion is

$$M_{\text{g,comb}}(\varphi) = M_{\text{eng,ind}}(\varphi) - M_{\text{basic}}(\varphi) \tag{4.3.38}$$

and results from the combustion mainly during the expansion. $M_{\text{eng,ind}}(\varphi)$ is calculated based on a measured indicated cylinder pressure $p_{\text{cyl,ind}}(\varphi)$, resulting in the force on the piston

$$F_{\text{pist,ind}} = A_{\text{pist}}\, p_{\text{cyl,ind}}(\varphi) \tag{4.3.39}$$

and follows by using (4.3.24) and (4.3.25). Figures 4.3.10 and 4.3.11 show an example.

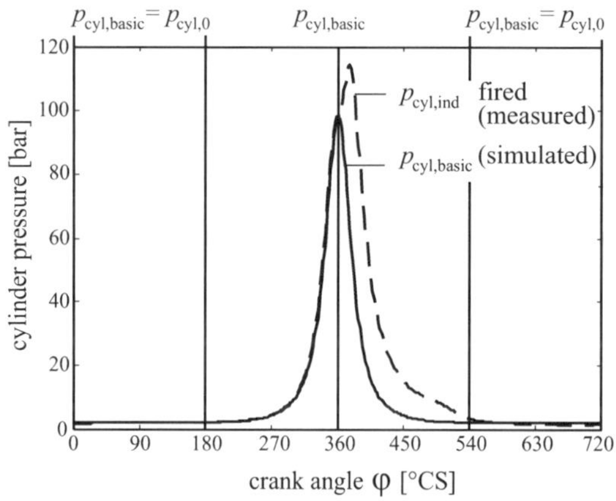

Fig. 4.3.10. Basic and measured cylinder pressure, OM 442.

To reduce the calculations the dynamic torque through combustion is, corresponding to a Vibe function (4.2.40), approximated by

$$\hat{M}_{g,comb}(\varphi) = a\,\varphi'^2\,e^{(-b\varphi')} \qquad \varphi \in [360°, 540°]$$
$$\varphi' = \varphi - 360°. \tag{4.3.40}$$

The parameters a and b have to satisfy the following two conditions:

$$\frac{1}{720°}\int_0^{720°} \hat{M}_{g,comb}(\varphi)\,d\varphi = \overline{M}_{g,ind} \tag{4.3.41}$$

as $\overline{M}_{basic} = 0$, and

$$\frac{d}{d\varphi'}\hat{M}_{g,comb}(\varphi)\bigg|_{\varphi'=\varphi_{max}} = 0 \tag{4.3.42}$$

as the position of the maximum of $\hat{M}_{g,comb}(\varphi)$ at crank angle $\varphi' = \varphi'_{max}$ is assumed to be constant and does not change with load and speed, Sinsel (2000). This leads to

$$a = \frac{4 \cdot 720° \cdot \overline{M}_{g,ind}}{(\varphi'_{max})^3}; \qquad b = \frac{2}{\varphi'_{max}}. \tag{4.3.43}$$

Figure 4.3.12 depicts the good approximation quality of (4.3.40), and Fig. 4.3.13 a comparison of the indicated torque based on measurements and calculated with (4.3.38), now including the basic torque. Figure 4.3.14 shows the resulting dynamic engine torque $M_{eng}(\varphi)$ for different speed and torque.

These models were used for the hardware-in-the-loop simulation of truck diesel engines, Sinsel (2000). Similar results were obtained by Schaffnit (2002) where the stationary torque through the high pressure combustion cycle was approximated with a LOLIMOT neural network and the inputs m_{air}, m_f, φ_{soi} and n_{eng}, also with emission models.

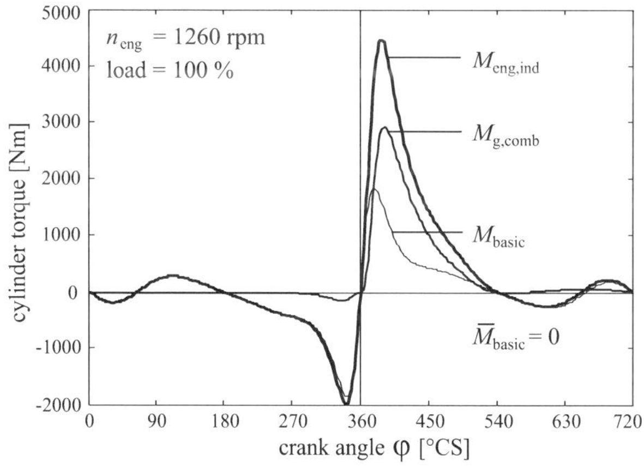

Fig. 4.3.11. Basic, indicated and combustion torque.

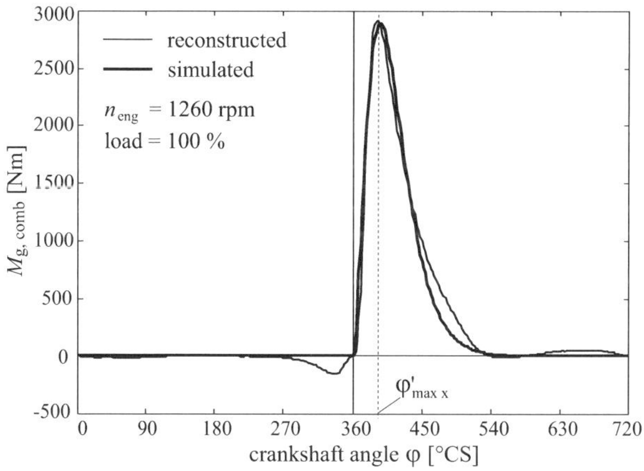

Fig. 4.3.12. Torque through combustion, simulated and derived from measurements with $\varphi' = 10°$CS.

4.3.3 Reconstruction of torque oscillations through speed measurements

The gas forces through the combustion result in an acceleration of the crankshaft between TDC and BDC. For the applied energy to the crankshaft holds for one cylinder

$$dE(\varphi) = M_g(\varphi)\, d\varphi \tag{4.3.44}$$

and for the torque through acceleration

$$M_g(\varphi) = J_{eng}\dot{\omega}_{eng}(\varphi). \tag{4.3.45}$$

(4.3.29) shows that the gas forces are mainly proportional to $\sin\varphi$. Therefore, it can be assumed that $\dot{\omega} = M_g(\varphi)/J_{eng}$ is also proportional to $\sin\varphi$. The applied energy is

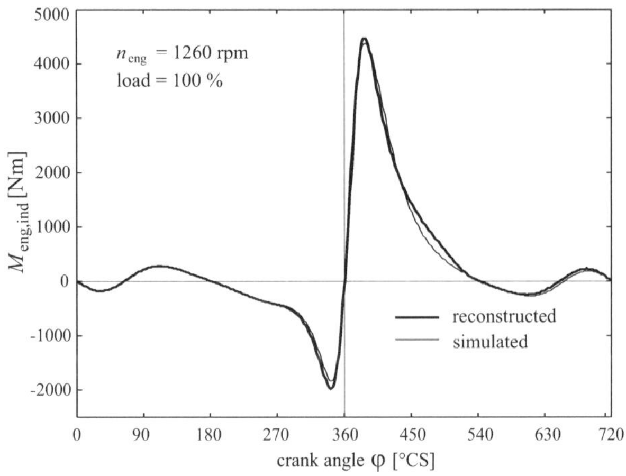

Fig. 4.3.13. Indicated cylinder torque, simulated and derived from measurements.

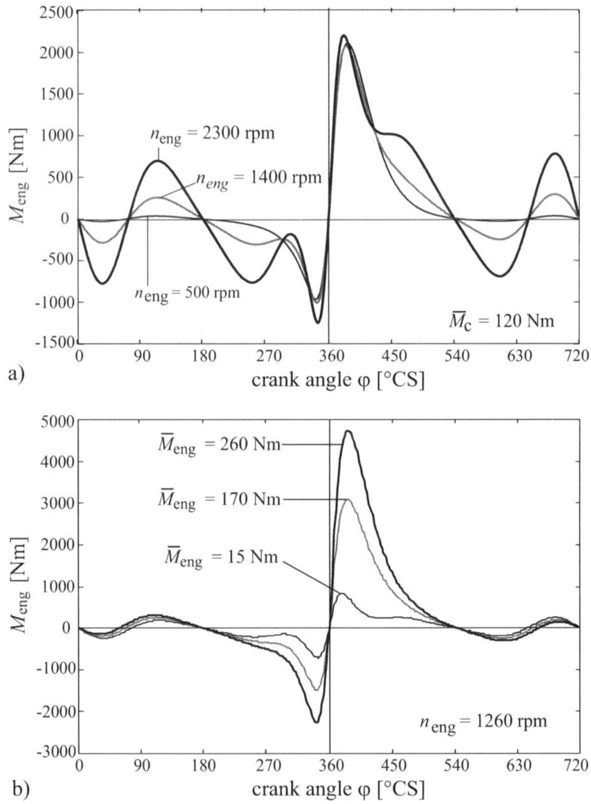

Fig. 4.3.14. Simulated dynamic crankshaft torque $M_{eng}(\varphi)$ for a eight-cylinder diesel engine (without friction torque M_f). **a** different speeds. **b** different torques.

$$\Delta E = J_{\text{eng}} \int_{\varphi(\omega_{\min})}^{\varphi(\omega_{\max})} \dot{\omega}_{\text{eng}}(\varphi)\, d\varphi \tag{4.3.46}$$

with

$\varphi(\omega_{\text{min}})$	crank angle with lowest ω_{eng}
$\varphi(\omega_{\text{max}})$	crank angle with largest ω_{eng}

compare Fig. 4.3.15. Using

$$\dot{\omega}_{\text{eng}} = \frac{d\omega_{\text{eng}}}{dt} = \frac{d\omega_{\text{eng}}}{d\varphi}\frac{d\varphi}{dt} = \frac{d\omega_{\text{eng}}}{d\varphi}\,\omega_{\text{eng}} \tag{4.3.47}$$

leads to

$$\begin{aligned} \Delta E &= J_{\text{eng}} \int_{\omega_{\min}}^{\omega_{\max}} \dot{\omega}_{\text{eng}}(\varphi)\, d\omega_{\text{eng}} = \frac{1}{2} J_{\text{eng}} \omega_{\text{eng}}^2 \Big|_{\omega_{\min}}^{\omega_{\max}} \\ &= \frac{1}{2} J_{\text{eng}} \left(\omega_{\text{eng,max}}^2 - \omega_{\text{eng,min}}^2\right). \end{aligned} \tag{4.3.48}$$

The torque increase through the applied energy is with (4.3.44)

$$\begin{aligned} \Delta M_{\text{eng}} &= \Delta E / \Delta\varphi \\ \Delta\varphi &= \varphi(\omega_{\text{eng,max}}) - \varphi(\omega_{\text{eng,min}}) \end{aligned} \tag{4.3.49}$$

If the average angular speed is

$$\overline{\omega}_{\text{eng}} = \frac{1}{2}\left(\omega_{\text{eng,max}} + \omega_{\text{eng,min}}\right) \tag{4.3.50}$$

and the amplitude of the resulting speed oscillation is noted by

$$\Delta\omega_{\text{eng}} = \frac{1}{2}\left(\omega_{\text{eng,max}} - \omega_{\text{eng,min}}\right),$$

then one obtains for the torque increase with (4.3.48), (4.3.49)

$$\begin{aligned} \omega_{\text{eng,max}} &= \overline{\omega}_{\text{eng}} + \Delta\,\omega_{\text{eng}} \\ \omega_{\text{eng,min}} &= \overline{\omega}_{\text{eng}} - \Delta\,\omega_{\text{eng}} \end{aligned}$$

for one cylinder

$$\Delta M_{\text{eng}} = \frac{2}{\Delta\varphi} J_{\text{eng}}\, \overline{\omega}_{\text{eng}} \Delta\,\omega_{\text{eng}}. \tag{4.3.51}$$

This means that the torque oscillation amplitudes generated through combustion gas forces are proportional to the average angular speed and the amplitude of the speed oscillation, Führer et al (1993). Therefore, ΔM_{eng} can be called "speed oscillation torque". If ΔM_{eng} is determined in dependence on the mean effective engine torque $\overline{M}_{\text{eng}}$ by measurements with a corresponding drive train, e.g. a model in form of a look-up table $\Delta M_{\text{eng}} = \varphi(\overline{M}_{\text{eng}}, \Delta\omega_{\text{eng}})$ can be established. Then $\overline{M}_{\text{eng}}$ can be determined by measurement of $\Delta\omega_{\text{eng}}$, Kimmich (2004). This allows to reconstruct the engine torque at the clutch if the number of cylinders is not too large ($i_{\text{c}} \leq 6$). One application is misfire detection, see concluding remarks for diesel engines.

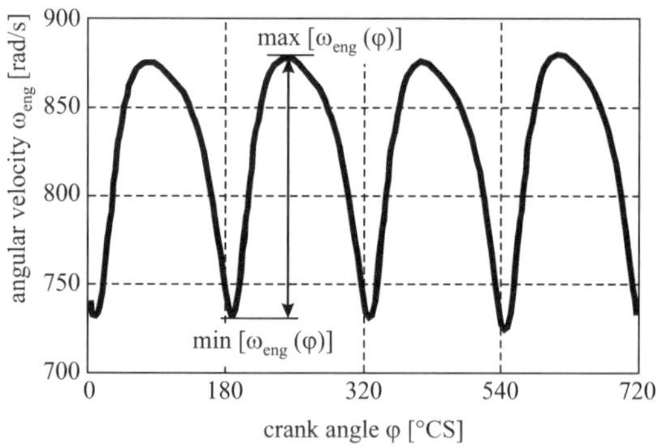

Fig. 4.3.15. Measured speed oscillations of a four-cylinder diesel engine.

4.4 Turbochargers

4.4.1 Introduction

The charged air mass in the cylinders can be increased by compressing the air flow of the intake. Because of the increased oxygen content more fuel can be injected, resulting in an increase of torque for a given displacement and engine speed. As the air heats up during compression, usually a charge-air cooler (intercooler) follows down-stream the compressor, in order to increase the air density and therefore air mass in the cylinder. One can distinguish these types of compressors:

- Mechanical driven superchargers
- Exhaust gas turbochargers
- Pressure-wave superchargers.

Mechanical superchargers are driven by the crankshaft. There exist mechanical, centrifugal and positive-displacement superchargers. As the centrifugal types require very high rotational speed, positive-displacement superchargers are used more often, e.g. for small- and medium-displacement gasoline engines which need a wide spread volumetric flow range of about 1:75. Piston-, screw-type and rotary-piston have an internal compression. The roots-type just transports the air, without internal compression. All of them have a strong dependence of the pressure on the volume flow rate and the volumetric flow is about proportional to the speed. Therefore, high pressure ratios can be generated at low flow rates, see e.g. Robert Bosch GmbH (2011).

Pressure-wave turbochargers (comprex-chargers) use pressure waves of the exhaust gas to convey energy to the intake air via a cell rotor which is mechanically driven from the crankshaft. This way the exhaust gas pulses are directly used to compress the air entrapped in the cells, without requiring mechanical power from the engine, except low bearing friction losses. This charging concept allows to be effective also for low engine speeds. However, the synchronization of the turbocharger drive from the crankshaft and the intake strokes needs a precise control.

Exhaust gas turbochargers are traditionally applied for diesel engines, in earlier time for trucks and since about 1975 also for passenger cars. They are also used for small and medium sized gasoline engines. A turbine uses the energy of the exhaust gas and drives the compressor. In contrast to mechanical superchargers the speed of the turbocharger does not directly depend on the engine speed. It is the result of a balance between turbine and compressor power.

Figure 4.4.1 depicts a typical design. The turbine is usually designed as a centripetal inward-flow or radial-flow wheel, see also Fig. 4.4.3. The exhaust gases are diverted from the outside of the spiral turbine housing to the center. A connecting shaft drives the radial compressor, where the intake air is drawn in at the center and is driven outwards by the blades of the impeller into the spiral compressor housing. Because the turbine converts the thermal energy of the exhaust gas into air compression energy, the increase of air pressure is higher than the rise of exhaust gas pressure up-stream the turbine, such delivering a positive scavenging drop over the cylinder. The compressor and its housing are made of cast aluminum alloy, the turbine wheel of steel-nickel alloys, and the turbine housing, e.g. of spheroidal cast iron.

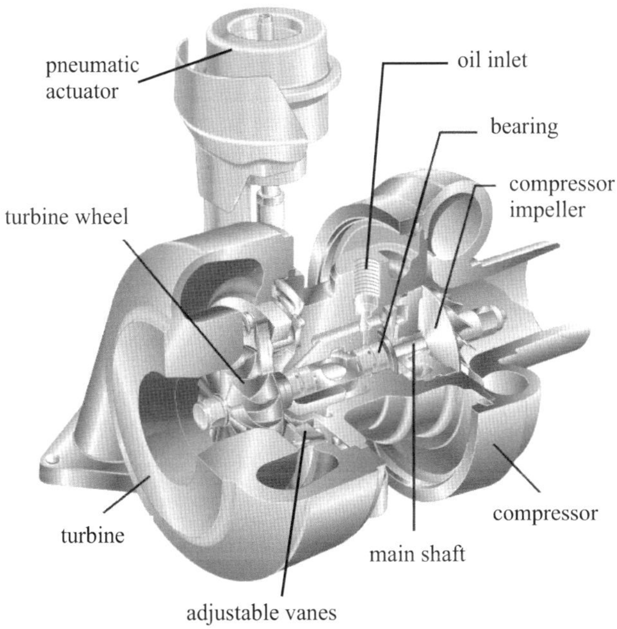

Fig. 4.4.1. Scheme of an exhaust turbocharger with variable geometry, Shaaban (2004).

The bearing housing contains the bearings and sealing elements. Radial as well as axial (thrust) bearings are designed as plain or oil bearings. They are connected to the engine's oil circuit, providing the bearings with lubrication and cooling. Piston rings at the casing openings seal the oil chamber to the exterior. For exhaust gas temperatures of up to 800°C no additional cooling is necessary. However, if the exhaust gases reach temperatures of 1000°C as is the case for gasoline engines, water-cooled bearing houses are required.

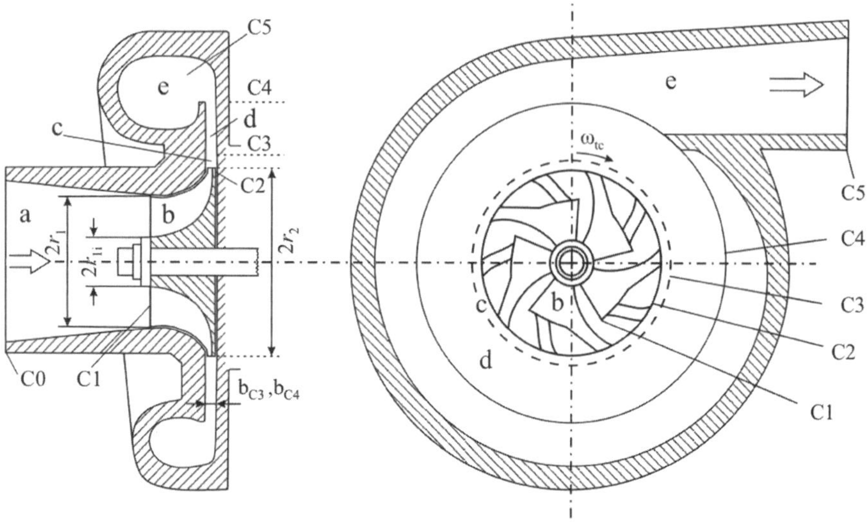

Fig. 4.4.2. Scheme of a radial flow turbo compressor, Zahn (2012). a inlet channel. b rotor. c ring slot. d diffusor. e spiral casing C0...C5 flow sections.

Because of the large speed spread of passenger-car engines, the boost pressure has to be controlled, usually by manipulation of the the turbine power. In the case of *wastegate* or *bypass control*, part of the exhaust gas flow is diverted past the turbine, thus reducing the gas flow through the turbine. This also lowers the exhaust gas back pressure and the speed of the turbocharger at high loads. At low engine load and speed, the wastegate is closed. The wastegate is manipulated by a pneumatic or electrical actuator.

A *variable turbine geometry* directs the entire exhaust gas flow through the turbine by changing the cross-section and the flow velocity to the turbine and thus the exhaust gas back pressure. This is reached either by adjustable vanes (VTG), as shown in Fig. 4.4.3, or by an adjustable sleeve (VST). Most common are the VTG turbochargers because they have a wide control range with high efficiencies. The turnable vanes are attached via control levers to an adjustable ring, which is connected to a pneumatic or electrical actuator, Fig. 4.4.3. The adjustable vanes change the gap size and flow direction to the turbine. At low engine speeds or loads a small gap forces the exhaust gas pressure and the flow velocity to increase, thus increasing the pressure difference over the turbine and the turbine speed. Therefore the turbine power increases and forces the compressor to generate a higher boost pressure. For higher engine speeds or load the gap is widened, resulting in a reduced exhaust gas back pressure and therefore smaller pressure difference over the turbine. In neutral position, the VTG turbocharger is fully opened by a spring. This can be considered a fail-safe position. A VTG turbocharger allows, compared to a fixed geometry turbocharger, a much wider operation area, resulting especially in higher turbine torques and speeds at lower loads. This way, a considerable turbo lag at low loads and speeds can be avoided.

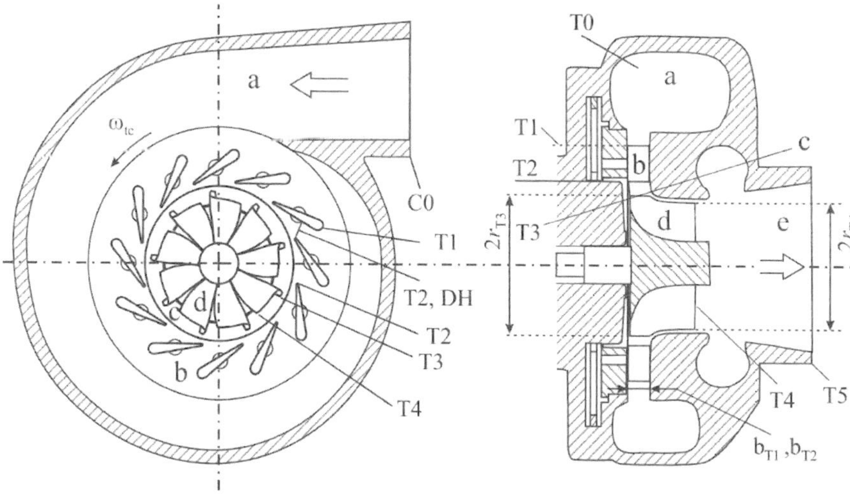

Fig. 4.4.3. Scheme of a variable-turbine geometry turbocharger (VTG), Zahn (2012). a spiral casing. b adjustable guide vanes. c bladeless ring space. d rotor with blades. e exit channel. T0...T5 flow sections.

A multi-port scheme for the energy flows of the turbocharger system is represented in Fig. 4.4.4. It consists of the compressor, the turbine and the cylinder as converters, the air-charge cooler as transformer, a throttle in the intake and an EGR valve as sinks, and air and exhaust ducts as storages. According to multi-port representations for energy flows the state variables at the interfaces between the process elements are determined in such a way that they describe a power, consisting of a potential as effort and a flow, Isermann (2005), Karnopp et al (2000). Therefore, the enthalpies are taken as effort and the mass flows as flow variables. The compressor and the intake stroke of the cylinder then generate enthalpies $h(p,T)$ which cause the corresponding mass flows of the intake system. Analogously the exhaust stroke and the turbine determine with the enthalpy of the exhaust gases the exhaust mass flow. A signal flow of all important variables is shown later.

The design of turbochargers is conventionally accompanied by the use of characteristic maps for the turbine and the compressor, which are measured on special test benches, mostly separated from each other.

To make the characteristics independent of the changing environmental conditions, the variables are given as referenced or corrected quantities by using fluid mechanical laws. The reference conditions are e.g. for the compressor at the inlet $T_{1\mathrm{ref}} = 293\,\mathrm{K}$, $p_{1\mathrm{ref}} = 0.981\,\mathrm{bar}$ and for the turbine $T_{3\mathrm{ref}} = 873\,\mathrm{K}$, also at the inlet.

As a first relative quantity the peripheral velocity of the rotor

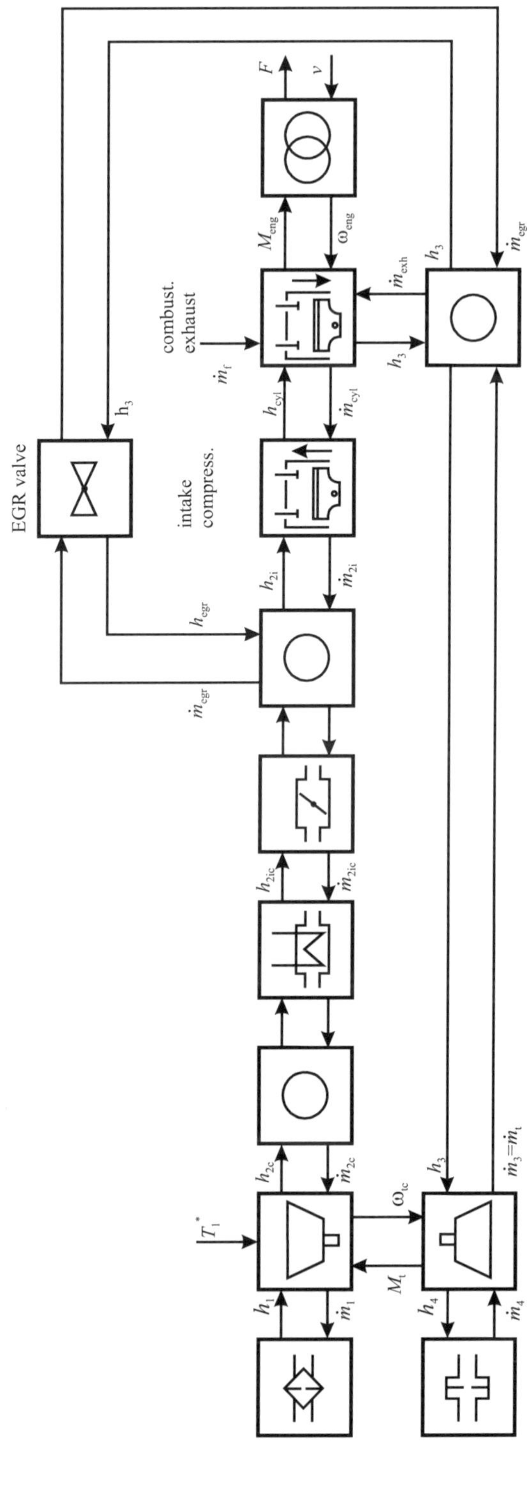

Fig. 4.4.4. Multi-port representation of the turbocharger, intake system and exhaust gas recirculation.

$$u = \pi\, d\, n_{\mathrm{tc}} = \omega_{\mathrm{tc}} r \tag{4.4.1}$$

is related to the speed of sound

$$a = \sqrt{\kappa R T} \tag{4.4.2}$$
$$(\kappa = 1.4, R = 287\,\mathrm{J/kgK})$$

which results in the Mach number

$$\mathrm{Ma} = \frac{u}{a} = \frac{\omega_{\mathrm{tc}} r}{\sqrt{\kappa R T}} = \frac{r}{\sqrt{\kappa R}} \frac{\omega_{\mathrm{tc}}}{\sqrt{T}}. \tag{4.4.3}$$

This means that the Mach number is proportional to $\omega_{\mathrm{tc}}/\sqrt{T}$ and leads to a *corrected rotational speed*

$$\omega_{\mathrm{tc,cor}} = \omega_{\mathrm{tc}} \frac{\sqrt{T_{\mathrm{ref}}}}{\sqrt{T}}. \tag{4.4.4}$$

The measured speed ω_{tc} with inlet temperature T_{in} is therefore transformed to a corrected speed $\omega_{\mathrm{tc,cor}}$ for the reference temperature $T_{\mathrm{in,ref}}$.

For the mass flow through a contraction (nozzle) holds

$$\dot{m} = A\, p_{\mathrm{in}} \sqrt{\frac{2}{R T_{\mathrm{in}}}}\, \psi \tag{4.4.5}$$

with the flow function

$$\psi = \sqrt{\frac{\kappa}{\kappa - 1}\left[\left(\frac{p_{\mathrm{out}}}{p_{\mathrm{in}}}\right)^{\frac{2}{\kappa}} - \left(\frac{p_{\mathrm{out}}}{p_{\mathrm{in}}}\right)^{\frac{\kappa+1}{\kappa}}\right]} \tag{4.4.6}$$

where p_{in} is the pressure at the inlet and the pressure at the outlet, compare (4.1.19), (4.1.20) and (4.1.57), (4.1.59). The mass flow is therefore proportional to $p_{\mathrm{in}}/\sqrt{T_{\mathrm{in}}}$ and $\psi\left(\frac{p_{\mathrm{out}}}{p_{\mathrm{in}}}\right)$. From (4.4.5) it follows

$$\dot{m} \frac{\sqrt{T_{\mathrm{in}}}}{p_{\mathrm{in}}} = A \sqrt{\frac{2}{R}}\, \psi. \tag{4.4.7}$$

This term only depends on the flow function ψ which is a function of the pressure ratio $p_{\mathrm{out}}/p_{\mathrm{in}}$. To obtain a *corrected mass flow* $\dot{m}_{\mathrm{cor}}$ for the standard conditions $p_{\mathrm{in,ref}}$ and $T_{\mathrm{in,ref}}$ it follows

$$\dot{m} \frac{\sqrt{T_{\mathrm{in}}}}{p_{\mathrm{in}}} = \dot{m}_{\mathrm{cor}} \frac{\sqrt{T_{\mathrm{in,ref}}}}{p_{\mathrm{in,ref}}}$$
$$\dot{m}_{\mathrm{cor}} = \dot{m} \frac{\sqrt{T_{\mathrm{in}}/T_{\mathrm{in,ref}}}}{p_{\mathrm{in}}/p_{\mathrm{in,ref}}} = A\, \frac{p_{\mathrm{in,ref}}}{\sqrt{T_{\mathrm{in,ref}}}} \sqrt{\frac{2}{R}}\, \psi. \tag{4.4.8}$$

This corrected mass flow can also be applied for turbo engines.

In the following sections, stationary and dynamic models of turbochargers are treated. They can be derived by thermodynamic changes of state or by fluid dynamic approaches using Euler's equation for turbomachinery.

4.4.2 Thermodynamic models of turbochargers

The overall behavior of turbomachines is usually first described by basic thermodynamic changes of state in form of idealized adiabatic compression and expansion processes, i.e. changes of state without heat transfer to and from the environment, such that an isentropic change of state $\Delta s = 0$ results.

a) Compressor

Figure 4.4.5a) shows an $h - s$ diagram for a compressor. The idealized isentropic enthalpy change is

$$\Delta h_{\mathrm{c,is}} = c_{\mathrm{pc}}(T_{2,\mathrm{is}} - T_1^*) \tag{4.4.9}$$

where the temperature $T_{2,\mathrm{is}}$ for given pressures follows for an isentropic change of state

$$\left(\frac{T_{2,\mathrm{is}}}{T_1^*}\right) = \left(\frac{p_2}{p_1}\right)^{\frac{\kappa_{\mathrm{c}}-1}{\kappa_{\mathrm{c}}}}. \tag{4.4.10}$$

The last two equations yield

$$\Delta h_{\mathrm{c,is}} = c_{\mathrm{pc}} T_1^* \left(\left(\frac{p_2}{p_1}\right)^{\frac{\kappa_{\mathrm{c}}-1}{\kappa_{\mathrm{c}}}} - 1\right). \tag{4.4.11}$$

As the compression is not isentropic in reality, because of flow friction losses which produce heat, a polytropic change of state arises with $T_2^* > T_{2,\mathrm{is}}$, see Fig. 4.4.5a), such that the required compressor enthalpy difference is

$$\Delta h_{\mathrm{c,adi}} = c_{\mathrm{pc}}(T_2^* - T_1^*). \tag{4.4.12}$$

This leads to the isentropic compressor efficiency

$$\eta_{\mathrm{c,is}} = \frac{\Delta h_{\mathrm{c,is}}}{\Delta h_{\mathrm{c,adi}}} = \frac{(T_{2,\mathrm{is}} - T_1^*)}{T_2^* - T_1^*} = \frac{\left(\frac{p_2}{p_1}\right)^{\frac{\kappa_{\mathrm{c}}-1}{\kappa_{\mathrm{c}}}} - 1}{\frac{T_2^*}{T_1^*} - 1}. \tag{4.4.13}$$

Rearranging (4.4.13) allows to determine the temperature ratio at the compressor

$$\frac{T_2^*}{T_1^*} = 1 + \frac{1}{\eta_{\mathrm{c,is}}}\left(\left(\frac{p_2}{p_1}\right)^{\frac{\kappa_{\mathrm{c}}-1}{\kappa_{\mathrm{c}}}} - 1\right) \tag{4.4.14}$$

and the required compressor power becomes

$$P_{\mathrm{c}} = \dot{m}_{\mathrm{c}} \Delta h_{\mathrm{c,adi}} = \dot{m}_{\mathrm{c}}\, c_{\mathrm{pc}}(T_2^* - T_1^*) = \dot{m}_{\mathrm{c}} \frac{c_{\mathrm{pc}}\, T_1^*}{\eta_{\mathrm{c,is}}}\left(\left(\frac{p_2}{p_1}\right)^{\frac{\kappa_{\mathrm{c}}-1}{\kappa_{\mathrm{c}}}} - 1\right). \tag{4.4.15}$$

The stationary behavior of a turbocompressor is usually represented in a compressor map

$$\left(\frac{p_2}{p_1}\right) = f\left(\dot{m}_{\text{c,cor}}, n_{\text{tc,cor}}\right) \tag{4.4.16}$$

as depicted in Fig. 4.4.6a) with corrected mass flow rate according to (4.4.8). The real mass flow then follows from

$$\dot{m}_{\text{c}} = \dot{m}_{\text{c,cor}} \frac{p_1/p_{1,\text{ref}}}{\sqrt{T_1/T_{1,\text{ref}}}}. \tag{4.4.17}$$

The compressor map shows for small mass flow rates a surge line as an operation limit. This surge line divides the map in a stable operation for larger flow rates and an unstable region for too small flow rates. Surging is characterized by a pulsating pressure and flow, resulting from local flow reversal which excite turbocharger vibrations and may lead to damage. For large mass flow rates the operating regime is limited by chocking because the flow becomes sonic. The pressure ratio shows characteristic curves for constant speed which decrease approximately quadratic with the flow rate. The curves for constant efficiencies run about parallel to the surge line. The resulting operating point of the compressor follows in connection with the corresponding operating map of the turbine. The compression process is illustrated in Fig. 4.4.5a) using an $h - s$ diagram with lines of constant pressure p_1 and p_2 and measured temperatures T_1 and T_2. For power calculation in (4.4.15) an isentropic efficiency map is used which depends on the corrected mass flow rate and speed

$$\eta_{\text{c,is}} = f(n_{\text{tc,cor}}, \dot{m}_{\text{c,cor}}) \tag{4.4.18}$$

and can be parameterized from temperature measurements, as described in detail in Sect. 4.4.5 and is shown in Fig. 4.4.7, for an example.

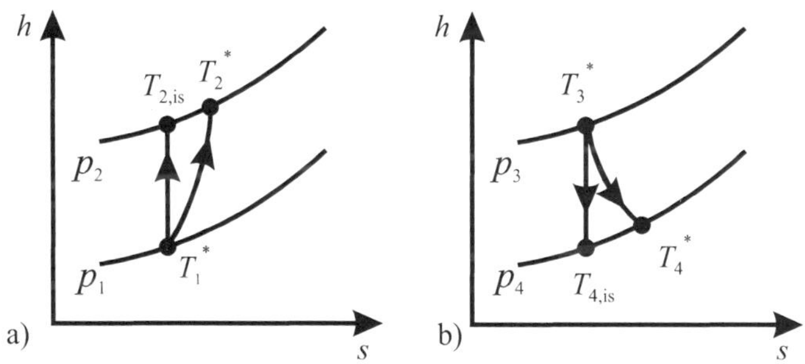

Fig. 4.4.5. Adiabatic and polytropic change of gas state. **a** compressor. **b** turbine.

Another possibility to model the compressor pressure ratio is to approximate it with quadratic polynomials, Zahn (2012),

$$\frac{p_2}{p_1}\left(\dot{m}_{\text{c,cor}}, \omega_{\text{tc,cor}}\right) = \left(\frac{p_2}{p_1}\right)_{\text{vert}} - K_{\text{c}}(\omega_{\text{tc,cor}})\left[\dot{m}_{\text{c,cor}} - \dot{m}_{\text{c,cor,vert}}(\omega_{\text{tc,cor}})\right]^2 \tag{4.4.19}$$

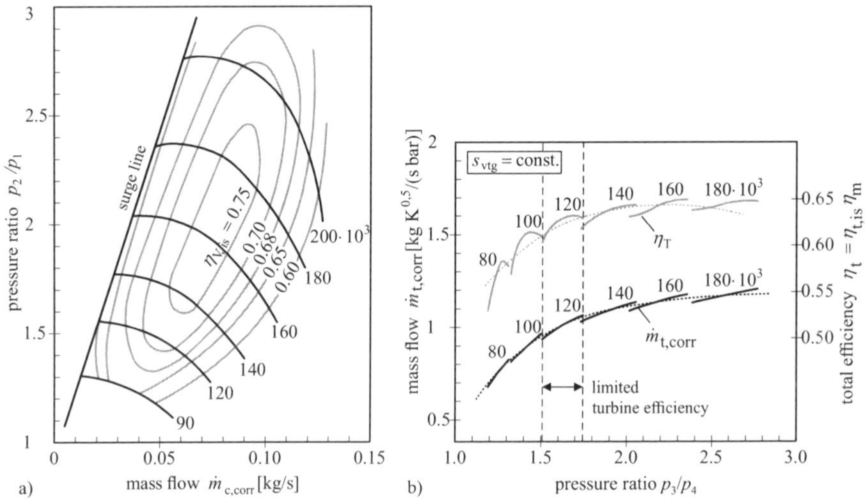

Fig. 4.4.6. Measured turbocharger characteristic maps, Mayer and Krämer (2011), Zahn (2012). **a** compressor. **b** turbine.

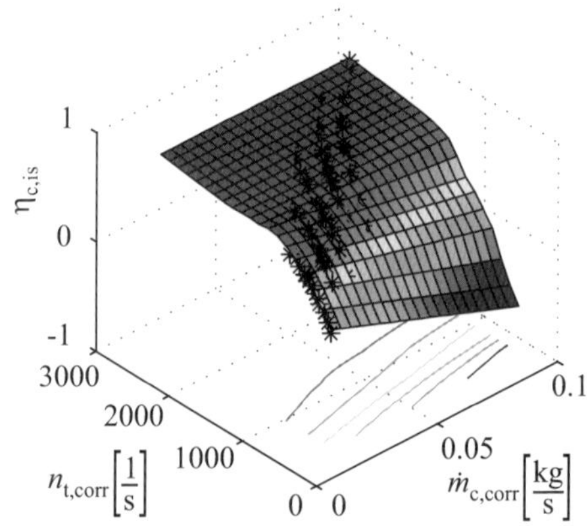

Fig. 4.4.7. Isentropic efficiency map of a compressor, approximated with a LOLIMOT net model, Sidorow et al (2011).

where $(p_2/p_1)_{\text{vert}}$ is the vertex. The vertex point represents approximately the surge line of the compressor. Hence, (4.4.19) is only valid for $\dot{m}_{\text{c,cor}} \geq \dot{m}_{\text{c,cor,vert}}$. In addition following approximations hold

$$\begin{aligned}
\dot{m}_{\text{c,cor,vert}}(\omega_{\text{tc,cor}}) &= k_1\omega_{\text{tc,cor}} + k_2\omega_{\text{tc,cor}}^2 \\
\left(\frac{p_2}{p_1}\right)_{\text{vert}}(\dot{m}_{\text{c,cor,vert}}) &= 1 + k_3\dot{m}_{\text{c,cor,vert}} + k_4\dot{m}_{\text{c,cor,vert}}^2 + k_5\dot{m}_{\text{c,cor,vert}}^3 \\
K_{\text{c}}\left(\omega_{\text{tc,cor}}\right) &= k_6 + k_7\omega_{\text{tc,cor}} + k_8\omega_{\text{tc,cor}}^2 .
\end{aligned} \tag{4.4.20}$$

Fig. 4.4.8 shows a comparison of measurements with these approximations.

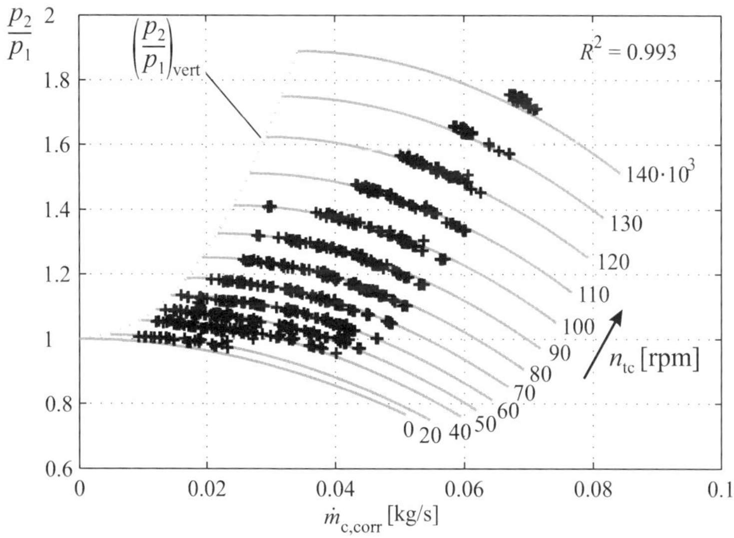

Fig. 4.4.8. Pressure ratio p_2/p_1 of a compressor. Measured values and approximation by polynomials (Zahn (2012)).

b) Turbine

Considering the $h - s$ diagram in Fig. 4.4.5b) for a turbine, the idealized enthalpy difference is

$$\Delta h_{\mathrm{t,is}} = c_{\mathrm{pt}} \left(T_3^* - T_{4,\mathrm{is}} \right) \tag{4.4.21}$$

with isentropic change of state

$$\frac{T_{4,\mathrm{is}}}{T_3^*} = \left(\frac{p_4}{p_3} \right)^{\frac{\kappa_t - 1}{\kappa_t}} . \tag{4.4.22}$$

Analogously to the compressor, it follows for an isentropic change of state

$$\Delta h_{\mathrm{t,is}} = c_{\mathrm{pt}} T_3^* \left(1 - \left(\frac{p_4}{p_3} \right)^{\frac{\kappa_t - 1}{\kappa_t}} \right) . \tag{4.4.23}$$

If the real enthalpy change is described by a polytropic change of state, with $T_4^* > T_{4,\mathrm{is}}$ it holds:

$$\Delta h_{\mathrm{t,adi}} = c_{\mathrm{pt}} \left(T_3^* - T_4^* \right) \tag{4.4.24}$$

and the isentropic turbine efficiency becomes

$$\eta_{\mathrm{t,is}} = \frac{\Delta h_{\mathrm{t,adi}}}{\Delta h_{\mathrm{t,is}}} = \frac{(T_3^* - T_4^*)}{(T_3^* - T_{4,\mathrm{is}})} = \frac{\frac{T_4^*}{T_3^*} - 1}{\left(\frac{p_4}{p_3}\right)^{\frac{\kappa_t - 1}{\kappa_t}} - 1}. \tag{4.4.25}$$

Rearranging (4.4.25) one obtains

$$\frac{T_4^*}{T_3^*} = 1 - \eta_{\mathrm{t,is}} \left(1 - \left(\frac{p_4}{p_3}\right)^{\frac{\kappa_{\mathrm{t}} - 1}{\kappa_{\mathrm{t}}}}\right) \tag{4.4.26}$$

and the generated turbine power yields

$$\begin{aligned} P_{\mathrm{t,adi}} &= \dot{m}_{\mathrm{t}}\, \Delta h_{\mathrm{t,adi}} = \dot{m}_{\mathrm{t}}\, c_{\mathrm{pt}}\, (T_3^* - T_4^*) \\ &= \dot{m}_{\mathrm{t}}\, c_{\mathrm{pt}}\, T_3^*\, \eta_{\mathrm{t,is}} \left(1 - \left(\frac{p_4}{p_3}\right)^{\frac{\kappa_{\mathrm{t}} - 1}{\kappa_{\mathrm{t}}}}\right). \end{aligned} \tag{4.4.27}$$

$\eta_{\mathrm{t,is}}$ is measured, and approximated as shown in Fig. 4.4.18. In the ideal case of an isentropic change of state the turbine power is

$$P_{\mathrm{t,is}} = \dot{m}_{\mathrm{t}}\, c_{\mathrm{pt}}\, T_3^* \left(1 - \left(\frac{p_4}{p_3}\right)^{\frac{\kappa_{\mathrm{t}} - 1}{\kappa_{\mathrm{t}}}}\right). \tag{4.4.28}$$

This corresponds to a kinetic energy of the exhaust gas which would be achieved in an isentropic expansion through a contraction

$$\dot{E}_{\mathrm{t}} = \dot{m}_{\mathrm{t}} \frac{c_{\mathrm{u,max}}^2}{2} \tag{4.4.29}$$

where $c_{\mathrm{u,max}}$ is called isentropic velocity, Whitfield and Baines (1990), Hiereth and Prenninger (2003). Hence, the ideal peripheral velocity of the exhaust gas becomes

$$c_{\mathrm{u,max}} = \sqrt{2\, c_{\mathrm{pt}}\, T_3^* \left(1 - \left(\frac{p_4}{p_3}\right)^{\frac{\kappa_{\mathrm{t}} - 1}{\kappa_{\mathrm{t}}}}\right)}. \tag{4.4.30}$$

The real turbine blade speed is now related to this maximum velocity

$$u_{\mathrm{ref}} = \frac{u}{c_{\mathrm{u,max}}} = \frac{2\,\pi r_3 \omega_{\mathrm{tc}}}{\sqrt{2\, c_{\mathrm{pt}}\, T_3^* \left(1 - \left(\frac{p_4}{p_3}\right)^{\frac{\kappa_{\mathrm{t}} - 1}{\kappa_{\mathrm{t}}}}\right)}} \tag{4.4.31}$$

and is called *isentropic velocity ratio*. As the isentropic efficiency is mainly influenced by flow losses it depends on the angle of attack of the rotor blade and can therefore be approximated by a quadratic function for constant position of guide vanes s_t, see Guzzella and Onder (2010). A parametric isentropic efficiency

$$\eta_{\mathrm{t,is,par}} = \eta_{\mathrm{t,is,max}} \left(2\frac{u_{\mathrm{ref}}}{u_{\mathrm{opt}}} - \left(\frac{u_{\mathrm{ref}}}{u_{\mathrm{opt}}} \right)^2 \right) \tag{4.4.32}$$

is introduced where the maximal efficiency is obtained for u_{opt}. The optimal values are obtained from, see Sidorow et al (2011),

$$\eta_{\mathrm{t,is,par}} = a_1 u + a_2 u^2 \tag{4.4.33}$$

with $u = \frac{u_{\mathrm{ref}}}{u_{\mathrm{opt}}}$ and $d\eta_{\mathrm{t,is}}/du = 0$. This yields

$$\eta_{\mathrm{t,is,max}} = 4a_1^2/a_2\,; \quad u_{\mathrm{opt}} = 2a_1/a_2. \tag{4.4.34}$$

Typical values are $\eta_{\mathrm{t,is,max}} \approx 0.65 \ldots 0.75$ and $u_{\mathrm{opt}} \approx 0.55 \ldots 0.65$, Guzzella and Onder (2010). For different positions of s_t, $\eta_{\mathrm{t,is,max}}$ and u_{opt} are calculated by polynomial functions

$$\eta_{\mathrm{t,is,max}}(s_t) = c_{\eta,\mathrm{t,is,max},0} + c_{\eta,\mathrm{t,is,max},1} s_{\mathrm{t}} + c_{\eta,\mathrm{t,is,max},2} s_{\mathrm{t}}^2 \tag{4.4.35}$$

$$u_{\mathrm{opt}}(s_t) = c_{\mathrm{u,opt},0} + c_{\mathrm{u,opt},1} s_{\mathrm{t}} + c_{\mathrm{u,opt},2} s_{\mathrm{t}}^2. \tag{4.4.36}$$

An example for the turbine isentropic efficiency $\eta_{\mathrm{t,is}} = f(u_{\mathrm{ref}}, s_{\mathrm{t}})$ of a turbocharger with variable geometry is shown in Fig. 4.4.9.

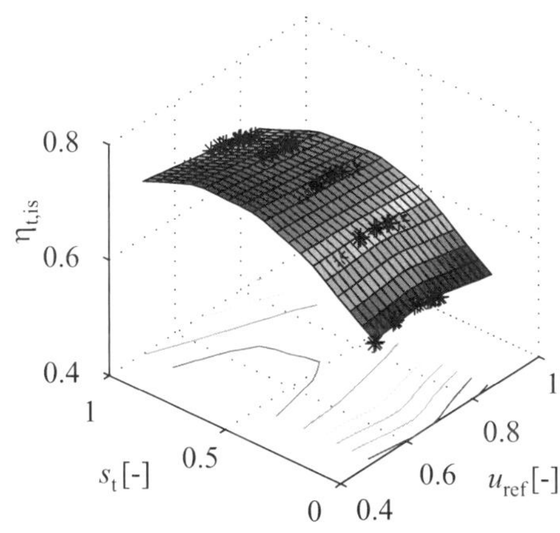

Fig. 4.4.9. Isentropic efficiency map of a VTG turbine approximated with a LOLIMOT net model, Sidorow et al (2011).

Using the parametric isentropic efficiency (4.4.32) the turbine power is calculated as follows

$$P_{\mathrm{t,par}} = \dot{m}_{\mathrm{t}}\, c_{\mathrm{pt}}\, T_3^*\, \eta_{\mathrm{t,is,par}} \left(1 - \left(\frac{p_4}{p_3} \right)^{\frac{\kappa_{\mathrm{t}}-1}{\kappa_{\mathrm{t}}}} \right). \tag{4.4.37}$$

Figure 4.4.6b) depicts the characteristic curves for a radial turbine with radially inward flow. It shows relatively limited operating regimes for constant speed. For variable speed the average characteristic corrected mass flow rate $\dot{m}_{\mathrm{t,cor}}$ asymptotically approaches a limit indicating chocking in stator nozzle blades or the rotor. The real mass flow rate follows based on measured turbine maps, Fig. 4.4.6b), according to (4.4.8) to

$$\dot{m}_{\mathrm{t}} = m_{\mathrm{t,cor}} \frac{p_3/p_{3,\mathrm{ref}}}{\sqrt{T_3/T_{3,\mathrm{ref}}}}. \tag{4.4.38}$$

Alternatively, the exhaust mass flow rate through the turbine follows that of a contraction, see (4.4.5) and (4.4.6)

$$\dot{m}_{\mathrm{t}} = A_{\mathrm{t,eff}}\, p_3 \sqrt{\frac{2}{RT_3^*}}\, \psi \tag{4.4.39}$$

$$\psi = \sqrt{\frac{\kappa_{\mathrm{t}}}{\kappa_{\mathrm{t}}-1}\left[\left(\frac{p_4}{p_3}\right)^{\frac{2}{\kappa_{\mathrm{t}}}} - \left(\frac{p_4}{p_3}\right)^{\frac{\kappa_{\mathrm{t}}+1}{\kappa_{\mathrm{t}}}}\right]} = \left(\frac{p_4}{p_3}\right)^{\frac{1}{\kappa_{\mathrm{t}}}} \sqrt{\frac{c_{\mathrm{pt}}}{R}\left[1 - \left(\frac{p_4}{p_3}\right)^{\frac{\kappa_{\mathrm{t}}-1}{\kappa_{\mathrm{t}}}}\right]} \tag{4.4.40}$$

with $A_{\mathrm{t,eff}}$ an effective cross-section of the turbine where $A_{\mathrm{t,eff}}\left(\alpha_{3,\mathrm{th}}(u_{\mathrm{vgt}})\right)$, see (4.4.84)), (4.4.85).

The common shaft bearings of the turbine and the compressor generate mechanical friction losses which can be taken into account by reducing the turbine power by a mechanical efficiency $\eta_{\mathrm{m,tc}}$

$$P_{\mathrm{t,m}} = \eta_{\mathrm{m,tc}}\, P_{\mathrm{t,par}}. \tag{4.4.41}$$

c) Turbocharger with thermodynamic models

In steady state the turbine power equals the compressor power $P_{\mathrm{t,m}} = P_{\mathrm{c}}$ and with (4.4.15), (4.4.37) and (4.4.41) it holds for the compressor pressure ratio

$$\pi_{\mathrm{c}} = \frac{p_2}{p_1} = \left[1 + \frac{\dot{m}_{\mathrm{t}}}{\dot{m}_{\mathrm{c}}} \frac{c_{\mathrm{pt}}}{c_{\mathrm{pc}}} \frac{T_3^*}{T_1^*}\, \eta_{\mathrm{tc}} \left(1 - \left(\frac{p_4}{p_3}\right)^{\frac{\kappa_{\mathrm{t}}-1}{\kappa_{\mathrm{t}}}}\right)\right]^{\frac{\kappa_{\mathrm{c}}}{\kappa_{\mathrm{c}}-1}} \tag{4.4.42}$$

with the turbocharger overall efficiency

$$\eta_{\mathrm{tc}} = \eta_{\mathrm{t,is}}\, \eta_{\mathrm{c,is}}\, \eta_{\mathrm{m,tc}} = \eta_{\mathrm{is}}\, \eta_{\mathrm{m,tc}}. \tag{4.4.43}$$

If one assumes $\dot{m}_{\mathrm{t}}/\dot{m}_{\mathrm{c}} = 1.05$, the charging pressure p_2 is increased by increasing T_3^* and p_3 in front of the turbine.

The dynamic behavior of the turbocharger results from the *balance equation* for the *angular momentum*

$$J_{\mathrm{tc}} \frac{d\omega_{\mathrm{tc}}}{dt} = M_{\mathrm{t}} - M_{\mathrm{c}} - M_{\mathrm{f}} = \eta_{\mathrm{m,tc}} M_{\mathrm{t}} - M_{\mathrm{c}} \tag{4.4.44}$$

with the torques for the turbine, compressor and friction. If the friction is dominated by viscous friction, it holds

$$M_{\rm f} = c_{\rm f}\,\omega_{\rm tc}. \tag{4.4.45}$$

It can also be taken into account by the mechanical efficiency term (4.4.38) in reducing the turbine torque. Using the relation for power $P = M\omega$ yields

$$J_{\rm tc}\,\omega_{\rm tc}(t)\frac{d\omega_{\rm tc}(t)}{dt} = P_{\rm t,par}(t) - P_{\rm c}(t) - P_{\rm f}(t) \tag{4.4.46}$$

$$P_{\rm f}(t) = c_{\rm f}\,\omega_{\rm tc}^2(t). \tag{4.4.47}$$

As the *heat flow* from the turbine to the compressor has a considerable effect on the thermodynamic variables of the turbocharger, its influence is now taken into account. The enthalpy difference Δq, see Fig. 4.4.10 caused by the heat transfer $\dot{Q}$ with

$$\Delta q = \frac{\dot{Q}}{\dot{m}} = c_{\rm p}\cdot\Delta T \tag{4.4.48}$$

is described in following for the compressor and turbine. Figure 4.4.10a) depicts for the compressor that the transferred heat $\Delta q_{\rm c,b}$ before the compression increases the air temperature from T_1 to T_1^* and the transferred heat $\Delta q_{\rm c,a}$ after the compression increases the temperature from T_2^* to T_2. The resulting enthalpy for the compressor with regard to the air temperatures including the heat transfer then is

$$\begin{aligned}\Delta h_{\rm c,dia} &= c_{\rm pc}(T_2 - T_1) = \Delta q_{\rm c,b} + \Delta h_{\rm c,adi} + \Delta q_{\rm c,a}\\ &= c_{\rm pc}\left[(T_1^* - T_1) + (T_2^* - T_1^*) + (T_2 - T_2^*)\right].\end{aligned} \tag{4.4.49}$$

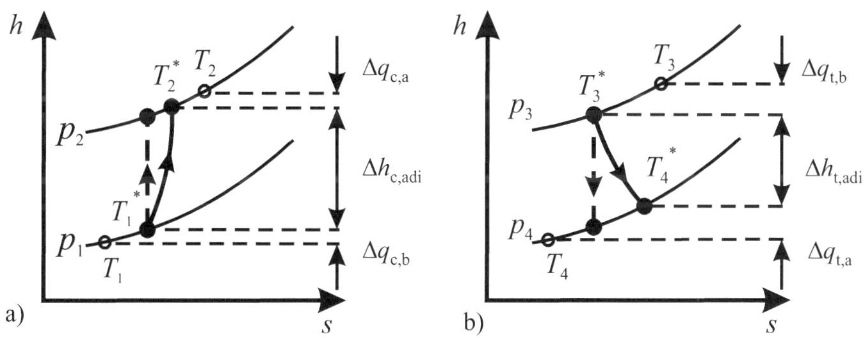

Fig. 4.4.10. Polytropic gas changes of state with heat transfer from the turbine to the compressor. **a** compressor. **b** turbine.

The corresponding influence of the heat losses on the states of the turbine are shown in Fig. 4.4.10b). Before expansion the loss of heat is $\Delta q_{\rm t,b}$ with a reduction of temperature from T_3 to T_3^* and after expansion the heat $\Delta q_{\rm t,a}$ is lost with temperature change from T_4^* to T_4. Hence the effective turbine enthalpy with regard to the exhaust gas temperatures is

$$\begin{aligned} \Delta h_{\mathrm{t,dia}} &= c_{\mathrm{pt}}(T_3 - T_4) = \Delta q_{\mathrm{t,b}} + \Delta h_{\mathrm{t,adi}} + \Delta q_{\mathrm{t,a}} \\ &= c_{\mathrm{pt}}\left[(T_3 - T_3^*) + (T_3^* - T_4^*) + (T_4^* - T_4)\right]. \end{aligned} \tag{4.4.50}$$

The heat transfer from the turbine to the charger housing and environment is relatively complicated. One has to distinguish convective heat transfer and radiation, heat conductance through the housing and shaft to the lubrication oil and the compressor and heat transfer to the environment, see e.g. Shaaban (2004), Mrosek et al (2009) and Zahn (2012). Therefore, only a part of the heat flow reaches the compressed air at the entrance and exit

$$\dot{Q}_{\mathrm{c,b}} = k_{\mathrm{c,b}}\, A_{\mathrm{c,b}}\,(T_3 - T_1) = \dot{m}_{\mathrm{c}} \Delta q_{\mathrm{c,b}} = \dot{m}_{\mathrm{c}}\, c_{\mathrm{pc}}\,(T_1^* - T_1) \tag{4.4.51}$$

$$\dot{Q}_{\mathrm{c,a}} = k_{\mathrm{c,a}}\, A_{\mathrm{c,a}}(T_3 - T_2) = \dot{m}_{\mathrm{c}} \Delta q_{\mathrm{c,a}} = \dot{m}_{\mathrm{c}}\, c_{\mathrm{pc}}\,(T_2 - T_2^*) \tag{4.4.52}$$

with $k_{\mathrm{c,b}}$ and $k_{\mathrm{c,a}}$ the overall heat transfer coefficients and $A_{\mathrm{c,b}}$ and $A_{\mathrm{c,a}}$ corresponding areas. For the turbine holds correspondingly

$$\dot{Q}_{\mathrm{t,b}} = k_{\mathrm{t,b}}\, A_{\mathrm{t,b}}\,(T_3 - T_1) = \dot{m}_{\mathrm{t}} \Delta q_{\mathrm{t,b}} = \dot{m}_{\mathrm{t}} c_{\mathrm{pt}}(T_3 - T_3^*) \tag{4.4.53}$$

$$\dot{Q}_{\mathrm{t,a}} = k_{\mathrm{t,a}}\, A_{\mathrm{t,a}}(T_4 - T_1) = \dot{m}_{\mathrm{t}} \Delta q_{\mathrm{t,a}} = \dot{m}_{\mathrm{t}} c_{\mathrm{pt}}(T_4^* - T_4). \tag{4.4.54}$$

The coefficients k_{j} and A_{j} have to be estimated experimentally. Usually, the influence of the heat transfer has to be taken into account for small mass flows but can be neglected for large mass flows, Zahn (2012).

The parameterization of isentropic efficiencies $\eta_{\mathrm{c,is}}$ and $\eta_{\mathrm{t,is,par}}$, see (4.4.18) and (4.4.32) using measured temperatures is applied together with the heat transfer models and is described elaborately in Sect. 4.4.5. Thereby the non-measured temperatures T_1^*, T_2^*, T_3^* and T_4^* are estimated which can be interpreted as a shift of the line describing adiabatic compression and adiabatic expansion in Fig. 4.4.10, Sidorow (2014). Estimated heat transfer models (4.4.51) to (4.4.54) can be used to obtain effective powers

$$P_{\mathrm{c,eff}} = \dot{m}_{\mathrm{c}} c_{\mathrm{pc}}\,(T_2 - T_1) - \dot{Q}_{\mathrm{c,b}} - \dot{Q}_{\mathrm{c,a}} \tag{4.4.55}$$

$$P_{\mathrm{t,eff}} = \dot{m}_{\mathrm{t}} c_{\mathrm{pt}}\,(T_3 - T_4) - \dot{Q}_{\mathrm{t,b}} - \dot{Q}_{\mathrm{t,a}} \tag{4.4.56}$$

which depend only on measured temperatures. These powers are now used for parameterization of the isentropic efficiencies $\eta_{\mathrm{c,is}}$ and $\eta_{\mathrm{t,is,par}}$, see (4.4.18) and (4.4.32). Then, using powers P_{c} (4.4.15) and $P_{\mathrm{t,par}}$ (4.4.37) the dynamic turbocharger speed is calculated with (4.4.46), (4.4.47).

Typical values for the specific heat for constant pressure are $c_{\mathrm{pc}} = 1080\,\mathrm{J/kgK}$ (air) and $c_{\mathrm{pt}} = 1050\,\mathrm{J/kgK}$ (exhaust gas), for the adiabatic exponent $\kappa_{\mathrm{c}} = 1.399$ (air) and $\kappa_{\mathrm{t}} = 1.361$ (exhaust gas) and for the gas constants $R_{\mathrm{c}} = 287.1\,\mathrm{J/kgK}$ (air) and $R_{\mathrm{t}} = 286.6\,\mathrm{J/kgK}$ (exhaust gas). Hence, they differ only by 3 % or less and can therefore be assumed to be equal

$$c_{pc} = c_{pt} = c_p \, ; \quad \kappa_c = \kappa_t = \kappa \, ; \quad R_c = R_t = R. \tag{4.4.57}$$

Figure 4.4.11 presents a signal-flow chart for the dynamic behavior of a turbocharger by taking the thermodyamic equations of this chapter into account.

Note that for engines with charge-air cooler $p_2 = p_{2c}$ and $T_2 = T_{2c}$ has to be taken, see Chap. 8.

4.4.3 Fluid-dynamic models of turbochargers

The derivation of turbocharger models with fluid dynamic approaches is based on Euler's equation for turbo-machinery, considering the change of the angular momentum of the gas flows and energy balances.

a) Euler's equation for turbomachinery

The considered turbo-machinery is now a turbo-compressor. The work transfer to the gas occurs in the impeller. It can be described by the change in the angular momentum with regard to the velocity components at the impeller (rotor) entry and exit, as depicted in Fig. 4.4.12. The angular momentum for a mass element dm rotating with angular speed ω is defined in the scalar case

$$dL(t) = dJ(t)\,\omega(t) \tag{4.4.58}$$

with dJ the axial moment of inertia of the mass element dm rotating with radius r

$$dJ = r^2 \,\mathrm{dm}. \tag{4.4.59}$$

Therefore, the angular moments of a mass element at the entry and exit of the impeller are

$$\begin{aligned} dL_1 &= r_1^2\,\omega\,dm \\ dL_2 &= r_2^2\,\omega\,dm. \end{aligned} \tag{4.4.60}$$

The torque through the work in the impeller follows from Newton's second law of momentum for rotation

$$M(t) = \frac{d}{dt}\,L(t). \tag{4.4.61}$$

The torque contributions at the entry and exit are

$$\begin{aligned} M_1(t) &= \frac{dL_1(t)}{dt} = r_1^2\,\omega\,\frac{dm}{dt} = r_1^2\,\omega\,\dot{m} \\ M_2(t) &= \frac{dL_2(t)}{dt} = r_2^2\,\omega\,\frac{dm}{dt} = r_2^2\,\omega\,\dot{m}. \end{aligned} \tag{4.4.62}$$

Instead of the angular velocities the absolute gas velocities are introduced

$$\omega r_1 = c_{1u} \qquad \omega r_2 = c_{2u}. \tag{4.4.63}$$

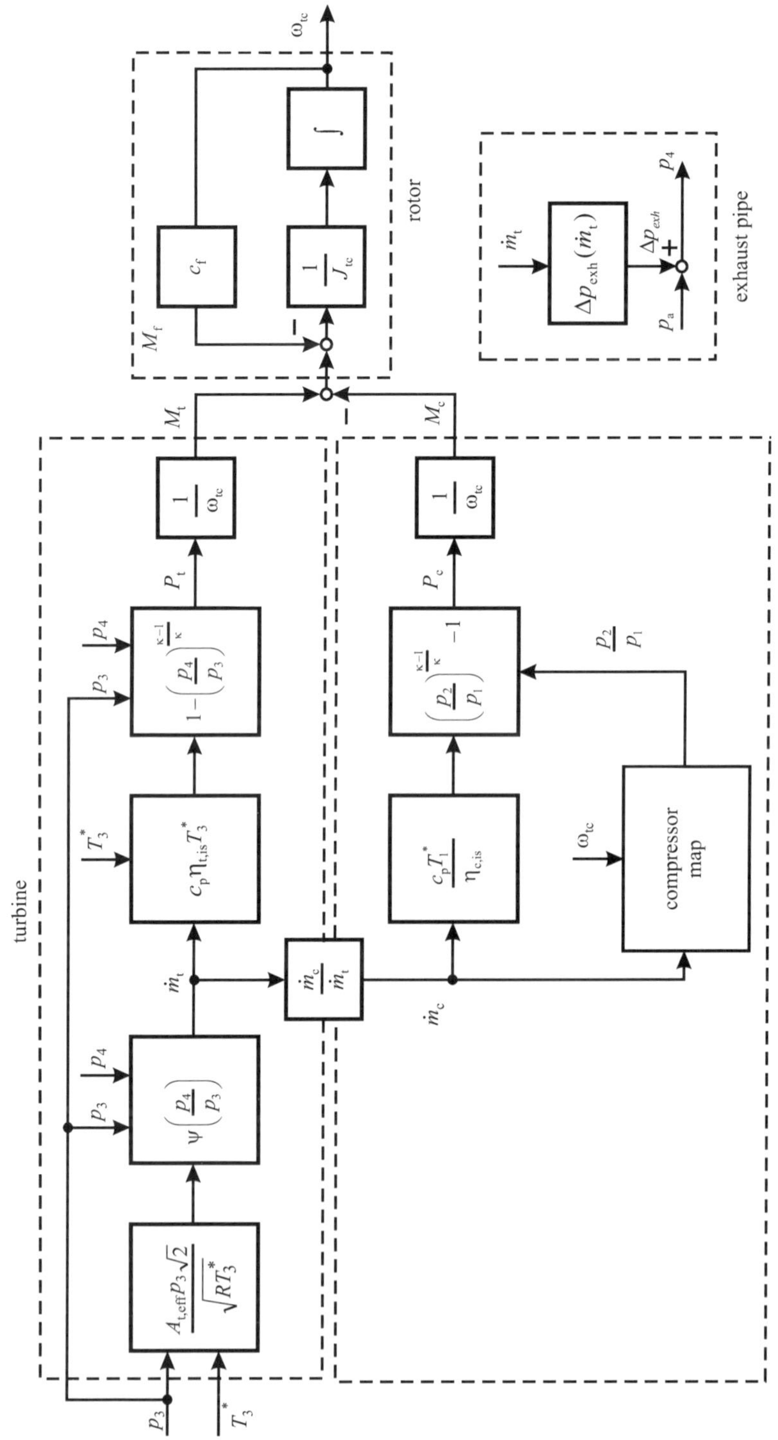

Fig. 4.4.11. Signal-flow chart for the thermodynamic model of a turbocharger (without internal heat transfer).

The required torque of the impeller then results in the stationary case

$$M_{\text{imp}} = M_2 - M_1 = \dot{m}(r_2 c_{2\text{u}} - r_1 c_{1\text{u}}) = \dot{m}\Delta(r_i c_{i\text{u}}). \tag{4.4.64}$$

Hence the torque is proportional to the change of the angular momentum from $i = 1$ to $i = 2$. For the required power follows

$$P_{\text{imp}} = M_{\text{imp}}\,\omega = \dot{m}\omega(r_2 c_{2\text{u}} - r_1 c_{1\text{u}}) \tag{4.4.65}$$

where ω is the angular speed of the impeller. Using the tangential blade velocities $u_1 = r_1\,\omega$ and $u_2 = r_2\,\omega$ leads to

$$P_{\text{imp}} = \dot{m}\,(u_2 c_{2\text{u}} - u_1 c_{1\text{u}}) \tag{4.4.66}$$

which is called *Euler equation*.

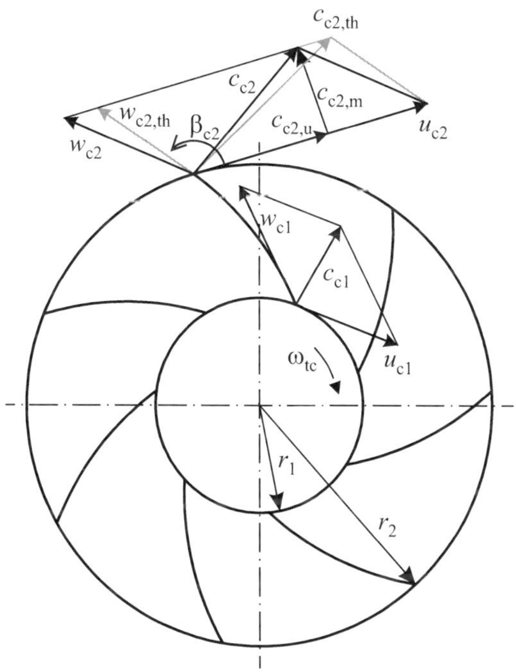

Fig. 4.4.12. Velocity diagrams of the inlet and exit for a centrifugal compressor impeller.

b) Compressor

Applying compressor specific symbols (4.4.66) yields

$$P_{\text{c}} = \dot{m}_{\text{c}}(u_{\text{c}2} c_{\text{c}2,\text{u}} - u_{\text{c}1} c_{\text{c}1,\text{u}}). \tag{4.4.67}$$

For compressors the inlet flow is frequently $c_{\rm c1u} = 0$. Hence it holds for the mechanical power

$$P_{\rm c} = \dot{m}_{\rm c}\, u_{\rm c2}\, c_{\rm c2,u}. \tag{4.4.68}$$

If it is assumed that this power is equal to the thermodynamic work and an adiabatic change of state with enthalpy $\Delta h_{\rm c,adi}$

$$P_{\rm c} = \dot{m}_{\rm c} \Delta h_{\rm c,adi}, \tag{4.4.69}$$

it holds

$$\Delta h_{\rm c,adi} = u_{\rm c2}\, c_{\rm c2,u} = r_2\, \omega_{\rm tc}\, c_{\rm c2,u}. \tag{4.4.70}$$

Theoretically, the flow leaves the impeller blades with the geometrical blade exit angle $\beta_{\rm c2}$, see Fig. 4.4.12, resulting in the theoretical peripheral component of the absolute flow velocity

$$c_{\rm c2,u,th} = u_{\rm c2} + w_{\rm c2,u,th} = u_{\rm c2} + c_{\rm c2,m} \cot \beta_{\rm c2} \tag{4.4.71}$$

where the last term becomes negative for backward-oriented blades.

The meridian component of the flow velocity follows from the mass flow rate, neglecting slit losses,

$$c_{\rm c2,m} = \frac{\dot{m}_{\rm c}}{2\pi\, \rho_{\rm 2c}^{*}\, r_2 b_{\rm c2}} \tag{4.4.72}$$

where $b_{\rm c2}$ is the blade width at exit and for the air density holds with the ideal gas state equation

$$\rho_{\rm 2c}^{*} = \frac{p_{\rm 2c}}{R T_{\rm 2c}^{*}}. \tag{4.4.73}$$

Because of the finite number of blades, the real flow does not follow the geometrical blade angle $\beta_{\rm c2}$ at the exit. Therefore, the peripheral flow velocity is smaller by a slip factor

$$c_{\rm c2,u} = \mu_{\rm c}\, c_{\rm c2,u,th} \tag{4.4.74}$$

with $\mu_{\rm c} = 0.6 \ldots 0.85$, Siglich (2008), which depends on the operation point. Hence, the real peripheral velocity is smaller by

$$c_{\rm c2,u} = c_{\rm c2,u,th} - \Delta c_{\rm c2,slip}. \tag{4.4.75}$$

The slip velocity depends linearly on the peripheral flow velocity, Stodola (1945), Zahn (2012),

$$\Delta c_{\rm c2,slip} = k_{\rm slip}\, u_{\rm c2} \tag{4.4.76}$$

and therefore

$$\mu_{\rm c} = 1 - \frac{k_{\rm slip}\, u_{\rm c2}}{u_{\rm c2} + c_{\rm c2,m} \cot \beta_{\rm c2}}. \tag{4.4.77}$$

The adiabatic enthalpy difference follows from (4.4.69) and use of (4.4.68), (4.4.71) to (4.4.74) to

$$\Delta h_{\rm c,adi} = \mu \left[(r_2\, \omega_{\rm tc})^2 + \frac{\dot{m}_{\rm c}\, \omega_{\rm tc}}{2\pi \rho_{\rm 2c}^{*}\, b_{\rm 2c}} \cot \beta_{\rm c2} \right] \tag{4.4.78}$$

and allows the calculation of the required compressor power according with (4.4.69), now combining the mechanical fluid dynamic and thermodynamic work.

c) VGT turbine

Introducing turbine specific symbols one obtains from the basic Euler equation (4.4.66) for the turbine power

$$P_{\rm t} = \dot{m}_t(u_{\rm t3}c_{\rm t3,u} - u_{\rm t4}c_{\rm t4,u}). \tag{4.4.79}$$

Assuming $c_{\rm t4u} = 0$ at the rotor exit yields

$$P_{\rm t} = \dot{m}_t\, u_{\rm t3}\, c_{\rm t3,u}. \tag{4.4.80}$$

$u_{\rm t3}$ is the circumferential blade velocity of the rotor. If this flow power is equal to the thermodynamic power

$$P_{\rm t} = \dot{m}_{\rm t}\, \Delta h_{\rm t,adi}, \tag{4.4.81}$$

it follows

$$\Delta h_{\rm t,adi} = u_{\rm t3}\, c_{\rm t3,u} = r_3\, \omega_{\rm tc}\, c_{\rm t3,u}. \tag{4.4.82}$$

The meridian component of the gas flow velocity is obtained from the exhaust gas flow rate

$$c_{\rm t3,m} = \frac{\dot{m}_{\rm t}}{2\pi\, \rho_3^*\, r_3\, b_{\rm t3}}. \tag{4.4.83}$$

In the case of a variable geometry turbine the adjustable guide vanes cause the gas flow to enter in the rotor with an angle $\alpha_{\rm t3,th}$, which is changed by the actuator position according to

$$\alpha_{\rm t3,th} = f(u_{\rm vgt}) \tag{4.4.84}$$

compare Figs. 4.4.13 and 4.4.14.

This relation depends on the kinematics of the vane guide and may be described by

$$\alpha_{\rm t3,th} = k_{\rm 1,vgt} + k_{\rm 2,vgt}\, u_{\rm vgt}. \tag{4.4.85}$$

Hence, it holds for the absolute gas velocity

$$c_{\rm t3,u,th} = c_{\rm t3,m} \cot \alpha_{\rm t3,th}. \tag{4.4.86}$$

The gas density follows from the ideal gas state equation

$$\rho_3^* = \frac{p_3}{RT_3^*} \tag{4.4.87}$$

where p_3 and T_3^* are the pressure and temperature at the entrance of the guide vanes.

Because of the finite number of guide vanes the real gas flow angle at the vane exit, Figs. 4.4.13 and 4.4.14, deviates from the geometrical angle $\alpha_{\rm t3}$ by

$$\alpha_{\rm t3} = \alpha_{\rm t3,th} - \Delta\alpha_{\rm t3}. \tag{4.4.88}$$

This leads to a change of the peripheral component of the absolute flow velocity

$$c_{\rm t3,u} = \mu_{\rm t}\, c_{\rm t3,u,th} = c_{\rm t3,u,th} - \Delta c_{\rm t3,u}. \tag{4.4.89}$$

The velocity loss can be approximated by

$$\Delta c_{\mathrm{t3u}} = (k_1 + k_2 \cot \alpha_{\mathrm{t3,th}}) + (k_3 + k_4 \cot \alpha_{\mathrm{t3,th}}) u_{\mathrm{t3}} \tag{4.4.90}$$

which is determined by experiments, Zahn (2012).

The real gas flow angle to the turbine rotor follows from

$$\cot \alpha_{\mathrm{t3}} = \frac{c_{\mathrm{t3,u}}}{c_{\mathrm{t3,m}}} = \mu_{\mathrm{t}} \frac{c_{\mathrm{t3,u,th}}}{c_{\mathrm{t3,m}}} = \mu_{\mathrm{t}} \cot \alpha_{\mathrm{t3,th}} \tag{4.4.91}$$

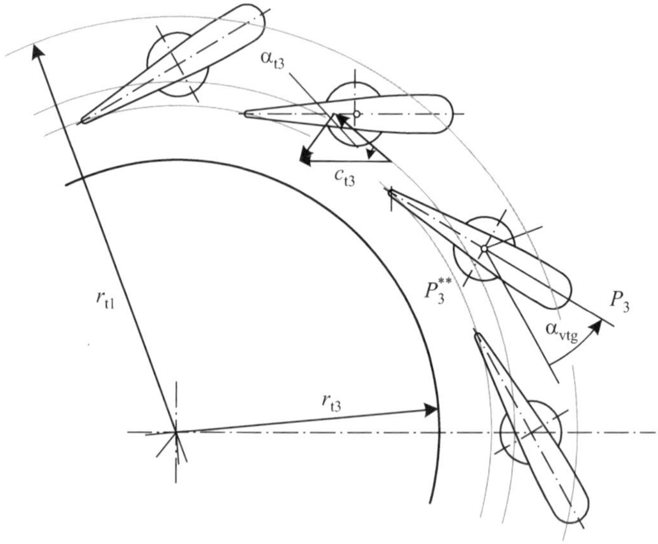

Fig. 4.4.13. Schematic of the adjustable vanes of a variable geometry radial turbine, Zahn (2012).

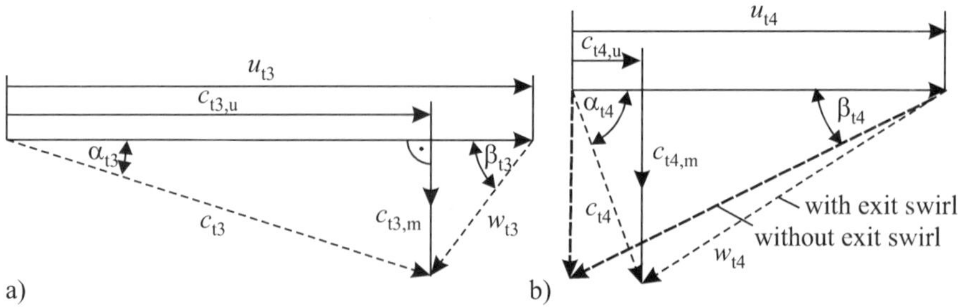

Fig. 4.4.14. Turbine velocity triangles. **a** inlet. **b** exit. u_{t3}: tangential blade velocity at rotor entrance; c_{t3}: absolute velocity at rotor entrance; α_{t3}: absolute flow angle at rotor entrance.

d) Guide vane losses

The guide vanes generate flow losses which can be approximated by an enthalpy loss

$$\Delta h_{\mathrm{vtg,loss}} = 0.5\,\xi_{\mathrm{t,bl}}\,c_{\mathrm{t3}}^2 \tag{4.4.92}$$

$$c_{\mathrm{t3}} = \sqrt{c_{\mathrm{t3,u}}^2 + c_{\mathrm{t3,m}}^2}. \tag{4.4.93}$$

For the flow loss coefficient holds

$$\xi_{\mathrm{t3}} = f(\mathrm{Re}, \alpha_{\mathrm{t3}}) \tag{4.4.94}$$

with Re the Reynolds number, see Japiske and Baines (1997), Watson and Janota (1982), Whitfield and Baines (1990), Baines (2005), Zahn (2012). These flow losses lead to a pressure drop. Assuming an isentropic change of state it holds

$$\frac{p_3^{**}}{p_3} = \left(\frac{T_3^{**}}{T_3}\right)^{\frac{\kappa_{\mathrm{t}}}{\kappa_{\mathrm{t}}-1}} \tag{4.4.95}$$

with p_3^{**} and T_3^{**} the pressure and temperature after the guide vanes at the entrance of the rotor wheel.

The exhaust gas temperature at the rotor entrance follows from

$$h_{\mathrm{t,adi}}^{**} = h_{\mathrm{t,adi}} - \Delta h_{\mathrm{vtg,loss}} = c_{\mathrm{p}} T_3^{**}. \tag{4.4.96}$$

The adiabatic enthalpy difference for the turbine rotor yields after inserting (4.4.85) to (4.4.95) in (4.4.82)

$$\Delta h_{\mathrm{t,adi}}^{**} = \frac{\dot{m}_t\,\omega_{\mathrm{tc}}\,R\,T_3^{**}}{2\pi p_3^{**}\,b_{\mathrm{t3}}}\cot\alpha_{\mathrm{t3}}. \tag{4.4.97}$$

The flow power of the turbine rotor finally follows with (4.4.73) and (4.4.86)

$$P_{\mathrm{t,rot}} = \dot{m}_{\mathrm{t}}\Delta h_{\mathrm{t,adi}}^{**} = \frac{\dot{m}_t^2\,\omega_{\mathrm{tc}}\,R\,T_3^{**}}{2\pi p_3^{**}\,b_{\mathrm{t3}}}\cot\alpha_{\mathrm{t3}}. \tag{4.4.98}$$

The turbine mass flow rate $\dot{m}_{\mathrm{t}}$ can be determined by using the contraction law (4.4.39) with variable cross-section $A_{\mathrm{t,eff}}(\alpha_{\mathrm{t3,th}})$.

Another possibility is to use an experimental approximation

$$\dot{m}_{\mathrm{t,cor}} = K_{\mathrm{a}}\left(1 - e^{-K_{\mathrm{b}}\left(\frac{p_3}{p_4}-1\right)}\right) \tag{4.4.99}$$

where $K_{\mathrm{a}} = f(\alpha_{\mathrm{t3,th}}, \omega_{\mathrm{tc}})$, $K_{\mathrm{b}} = f(\alpha_{\mathrm{t3,th}}, \omega_{\mathrm{tc}})$ are nonlinear functions. This approximation led to a good agreement with measured flow rates, Zahn (2012).

e) Turbocharger

The dynamic behavior of the turbocharger angular speed results from (4.4.46) after introducing (4.4.47), (4.4.69), (4.4.78) and (4.4.97)

$$\begin{aligned} J_{\mathrm{tc}}\,\omega_{\mathrm{tc}}\,\frac{d\omega_{\mathrm{tc}}}{dt} &= P_{\mathrm{t,rot}}(t) - P_{\mathrm{c}}(t) - P_{\mathrm{f}}(t) \\ &= \dot{m}_t\,\Delta h^{**}_{\mathrm{t,adi,rot}} - \dot{m}_c\,\Delta h_{\mathrm{c,adi}} - c_{\mathrm{f}}\,\omega^2_{\mathrm{tc}}. \end{aligned} \tag{4.4.100}$$

The exhaust gas flow rate $\dot{m}_{\mathrm{t}}$ follows from (4.4.39) or (4.4.99). The air mass flow rate $\dot{m}_{\mathrm{c}}$ is according to blow-by losses in the cylinder $\dot{m}_{\mathrm{c}} > \dot{m}_{\mathrm{t}}$ and can be approximated by $\dot{m}_{\mathrm{c}}/\dot{m}_{\mathrm{t}} = 1.05$ (assuming stationary behavior).

Finally, a signal-flow chart can be drawn, summarizing the relations for the turbine, the compressor and the rotor, see Fig. 4.4.15.

4.4.4 Wastegate turbocharger

If the turbine has a *fixed spiral housing* without adjustable vanes, the basic equation of the thermodynamic model for the turbine (4.4.27) can be used directly.

Figure 4.4.16 depicts a turbine where the turbine power is manipulated by a bypass valve, which is called wastegate. For the mass flow rate of the parallel configuration the flow law for a contraction (4.4.39) can be applied, leading to the flow rate through the turbine and the wastegate

$$\dot{m}_{\mathrm{t}} = A_{\mathrm{t,eff}}\,p_3\sqrt{\frac{2}{RT_3^*}}\,\psi\left(\tfrac{p_4}{p_3}\right) \tag{4.4.101}$$

$$\dot{m}_{\mathrm{wg}} = A_{\mathrm{wg,eff}}\,p_3\sqrt{\frac{2}{RT_3^*}}\,\psi\left(\tfrac{p_4}{p_3}\right). \tag{4.4.102}$$

With

$$\dot{m}_{\mathrm{exh}} = \dot{m}_{\mathrm{t}} + \dot{m}_{\mathrm{wg}} \tag{4.4.103}$$

and

$$\nu_{\mathrm{wg}} = \frac{A_{\mathrm{wg,eff}}}{A_{\mathrm{t,eff}}} \tag{4.4.104}$$

the turbine mass flow rate becomes

$$\dot{m}_{\mathrm{t}} = \frac{1}{1+\nu_{\mathrm{wg}}}\dot{m}_{\mathrm{exh}}. \tag{4.4.105}$$

Hence, the cross-section ratio $\nu_{\mathrm{wg}}[0, \nu_{\mathrm{wgmax}}]$ changes the turbine mass flow rate nonlinearly and should be compensated by an inverted relations within the control structure for the charging pressure.

Applying the thermodynamic model of the turbine the power $P_{\mathrm{t,adi}}$ follows from (4.4.27) with the wastegate-manipulated turbine mass flow according to (4.4.105). The stationary as well as the dynamic behavior of the turbocharger is then obtained via the balance equation for the angular momentum in (4.4.46). The inclusion of the heat transfer between turbine and compressor then requires an iterative procedure according to Sect. 4.4.5.

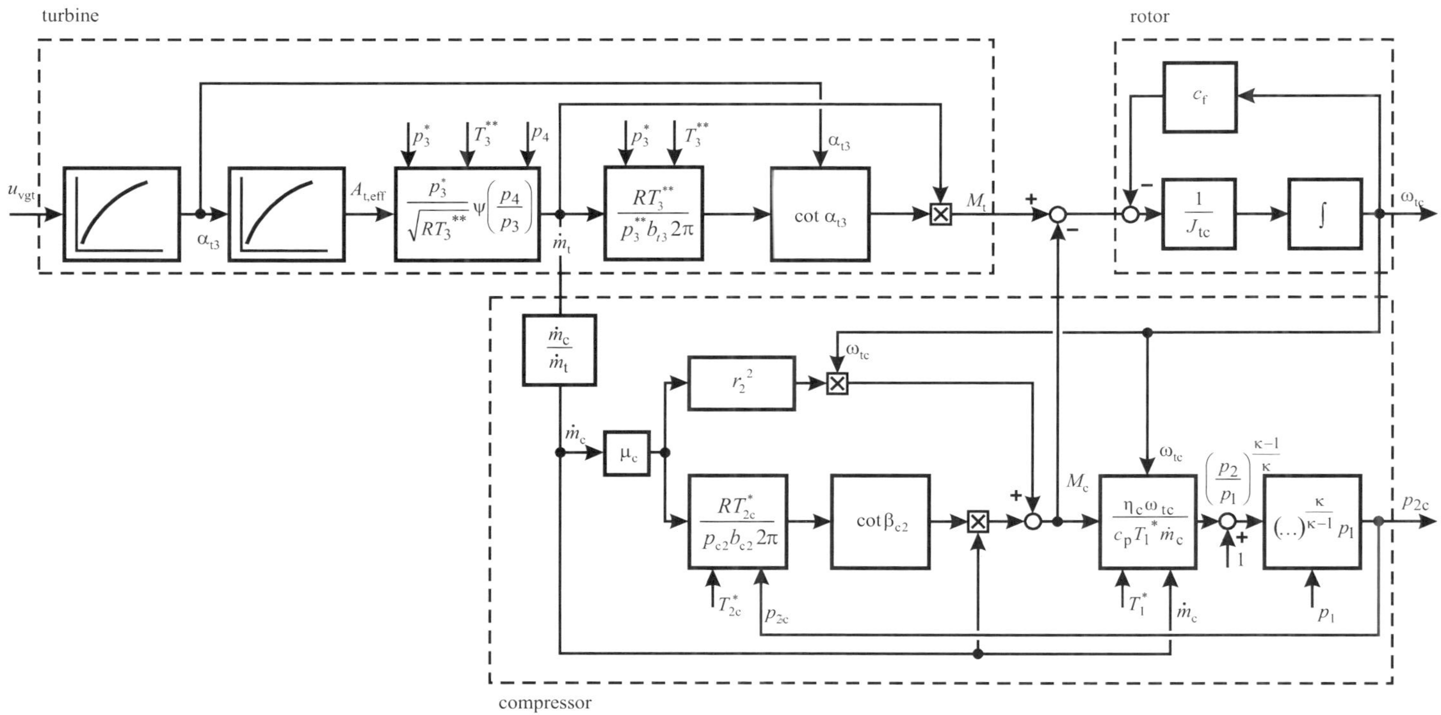

Fig. 4.4.15. Signal-flow chart for the flow dynamic and thermodynamic model of a VGT turbocharger with manipulated variable u_{vgt} as input and turbocharger speed and compressure pressure as outputs.

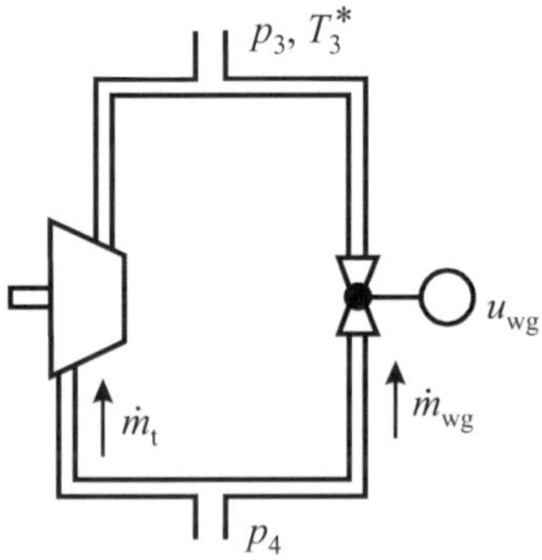

Fig. 4.4.16. Scheme of a turbine with parallel wastegate valve.

4.4.5 Identification of a nonlinear thermodynamic turbocharger model

In the following it is shown how the unknown parameters of a VTG turbocharger can be determined experimentally by parameter identification methods.

The unknown parameters of compressor and turbine models have to be estimated iteratively by minimizing the model error which is described by a mathematical norm between the calculated model outputs and stationary output measurements. Therefore, a set of stationary measurement data from the engine test bench using changes of the VTG actuator position s_t, the injection volume q_{inj} and the engine speed n_{eng} have to be obtained. The heat transfer parameter vectors $\boldsymbol{\theta}_{\mathbf{heat}}$, turbine and compressor parameter vectors $\boldsymbol{\theta}_{\boldsymbol{\eta},\mathbf{t},\mathbf{is}}$, as well as $\theta_{\eta,c,is}$ and a friction parameter θ_f to be estimated are summarized in Table 4.4.1 for the thermodynamic turbocharger model derived in Sect. 4.4.2.

Table 4.4.1. Thermodynamic turbocharger model parameters

Parameter vector	Parameters
$\boldsymbol{\theta}_{\mathbf{heat}}$	$k_{c,b}A_{c,b}$, $k_{c,a}A_{c,a}$, $k_{t,b}A_{t,b}$, $k_{t,a}A_{t,a}$
$\boldsymbol{\theta}_{\boldsymbol{\eta},\mathbf{t},\mathbf{is}}$	$c_{\eta,t,is,max,i}$, $c_{u,opt,i}$ with $i \in 0,1,2$
$\theta_{\eta,c,is}$	$\eta_{c,is}\left(n_{tc,cor}, \dot{m}_{c,cor}\right)$
θ_f	c_f

A scheme for the parameter estimation procedure is shown in Fig. 4.4.17. First the heat transfer parameters of the compressor and turbine have to be assumed with appropriate starting values. This allows to calculate the unknown temperatures T_1^*, T_{2c}^*, T_3^*, T_4^* using (4.4.51) to (4.4.54). Then the compressor parameters $\theta_{\eta,c,is}$ and the parameters of the turbine efficiency $\boldsymbol{\theta}_{\boldsymbol{\eta},\mathbf{t},\mathbf{is}}$ are estimated separately. The compressor efficiency model (4.4.18) can be directly calculated from the measurement data by using (4.4.13) and trained as static LOLIMOT model with inputs $n_{tc,cor}$ and $\dot{m}_{c,cor}$. The turbine efficiency (4.4.25) is calculated for each stationary point. Then measurement data for different constant positions of the variable turbine actuator s_t

are used. For each s_t position a pair of parameters $\eta_{\mathrm{t,is,max}}$ and u_{opt} is estimated with (4.4.33) and (4.4.34) by minimization of the least squared error. This approach is illustrated in Fig. 4.4.18a).

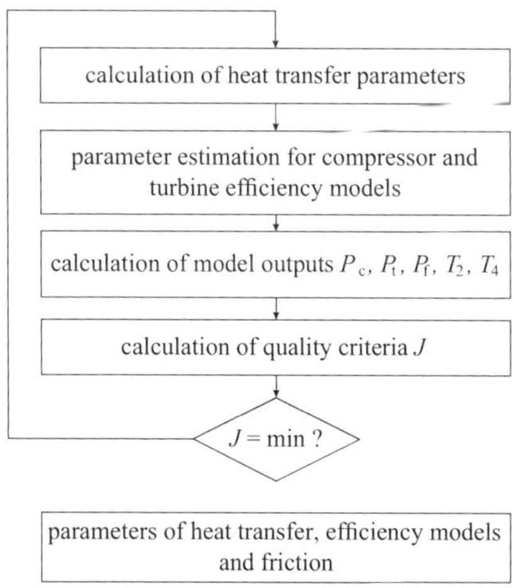

Fig. 4.4.17. Scheme for the iterative parameter estimation.

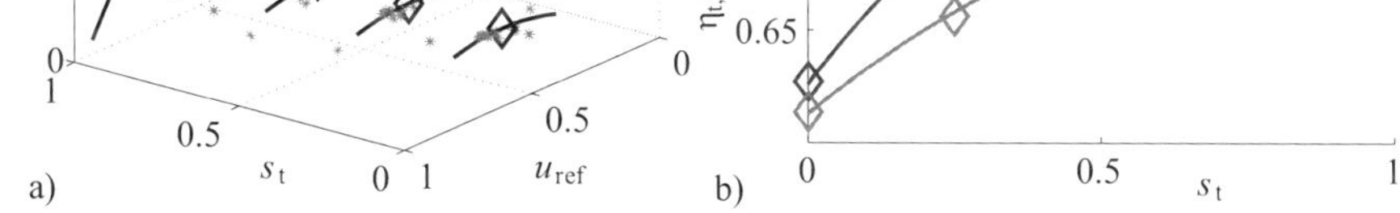

Fig. 4.4.18. Approximation of efficiency curves for constant guide vane positions. **a** approximation of parametric isentropic efficiency for constant s_t positions according to (4.4.32). **b** approximations of functions $\eta_{\mathrm{t,is,max}}(s_t)$ and $u_{\mathrm{opt}}(s_t)$ according to (4.4.35), (4.4.36).

Finally, the parameters of the polynomial functions $\eta_{\mathrm{t,is,max}}(s_t)$ and $u_{\mathrm{opt}}(s_t)$ of turbine isentropic efficiency can be estimated by use of (4.4.33), (4.4.34) and the least squares method. The results are shown in Fig. 4.4.18b). With these parameters finally the isentropic efficiency map of the turbine $\eta_{\mathrm{t,is}} = f(u_{\mathrm{ref}}, s_{\mathrm{t}})$ can be parameterized.

Knowing compressor and turbine efficiencies, the temperatures after the compressor T_2^* and after the turbine T_4^* (including heat transfer) are calculated with (4.4.14) and (4.4.26). Thus, the powers of the compressor, turbine and friction follow from (4.4.15), (4.4.37) and (4.4.47). Then an optimization criterion

$$J_{\mathrm{sum}} = w_1 \cdot J_{\mathrm{TCF}} + w_2 \cdot J_{\eta,\mathrm{c}} + w_3 \cdot J_{\eta,\mathrm{t}} + w_4 \cdot J_{\mathrm{T,2c}} + w_5 \cdot J_{\mathrm{T,4}} \quad (4.4.106)$$

is used with weights w_1 to w_5 for the performance criteria:

$$J_{\mathrm{TCF}} = \sqrt{\frac{1}{N}\sum_{k=1}^{N}\left[P_{\mathrm{t,par}}(k) + P_{\mathrm{c}}(k) + P_{\mathrm{f}}(k)\right]^2} \quad (4.4.107)$$

$$J_{\eta,\mathrm{c}} = \sqrt{\frac{1}{N}\sum_{k=1}^{N}\left[\eta_{\mathrm{c,is}}(k) \cdot f_{\mathrm{pen,c}}(\eta_{\mathrm{c,is}}(k))\right]^2} \quad (4.4.108)$$

$$J_{\eta,\mathrm{t}} = \sqrt{\frac{1}{N}\sum_{k=1}^{N}\left[\eta_{\mathrm{t,is,par}}(k) \cdot f_{\mathrm{pen,t}}(\eta_{\mathrm{t,is,par}}(k))\right]^2} \quad (4.4.109)$$

$$J_{\mathrm{T,2c}} = \sqrt{\frac{1}{N}\sum_{k=1}^{N}\left[T_2^*(k) + \frac{\dot{Q}_{\mathrm{c,a}}(k)}{c_{\mathrm{p,c}}\dot{m}_{\mathrm{c}}(k)} - T_2(k)\right]^2} \quad (4.4.110)$$

$$J_{\mathrm{T,4}} = \sqrt{\frac{1}{N}\sum_{k=1}^{N}\left[T_4^*(k) - \frac{\dot{Q}_{\mathrm{t,a}}(k)}{c_{\mathrm{p,t}}\dot{m}_{\mathrm{t}}(k)} - T_4(k)\right]^2} \quad (4.4.111)$$

(N is the number of data samples). Criterion (4.4.107) follows from the balance equation (4.4.40) with the assumption $P_t + P_c + P_f = 0$ for the steady state. Criteria (4.4.110) and (4.4.111) result from the comparison of the temperatures after the compressor T_2^* and after the turbine T_4^*, calculated with (4.4.14) and (4.4.26) and with the corresponding measured temperatures taking the heat transfer into account. The assumed penalty functions f_{pen} in criteria (4.4.108) and (4.4.109) have the form

$$f_{\mathrm{pen}}(x) = x + \gamma \cdot \left[\max\left(0, x - 0.5\right)\right]^{\nu} - \gamma \cdot \left[\min\left(0, x - 0.5\right)\right]^{\nu} \quad (4.4.112)$$

with $x \in (\eta_{\mathrm{c,is}}, \eta_{\mathrm{t,is,par}})$ and free parameters γ, ν. They are needed to penalize the efficiency values outside of the plausible range $[0, 1]$.

The parameter estimation can be formulated as a nonlinear global optimization problem with constraints in the parameters. The optimization algorithm minimizes the quality criterion (4.4.106) by variation of the parameters summarized in Table 4.4.1. The choice of weights in (4.4.106) influences the optimization result. In Sidorow (2014) the weights are chosen as $w_1 = 2$, $w_2 = 1$, $w_3 = 3$, $w_4 = 1$, $w_5 = 1$. Estimation is based on 75 stationary measurement points from a 1.9 liter CR-diesel engine with 110 kW power and GT1749MV turbocharger with variable geometry turbine. The estimated compressor or turbine power and the heat transfer flow are depicted in Fig. 4.4.19a) and b).

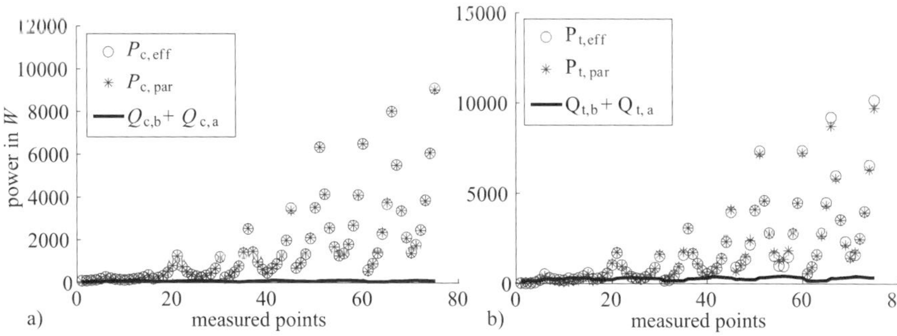

Fig. 4.4.19. Stationary compressor and turbine power and heat transfer flows obtained by nonlinear iterative parameter estimation. **a** compressor. **b** turbine.

Estimated heat transfer coefficients are summarized in Table 4.4.2. The results show that the stationary thermodynamic parameteric model for the compressor $P_{\mathrm{c,par}}$ and the turbine $P_{\mathrm{t,par}}$ according to (4.4.15), (4.4.37) matches well with the powers $P_{\mathrm{c,eff}}$ and $P_{\mathrm{t,eff}}$ determined by the measured temperatures and corrected with the heat flow losses in (4.4.55), (4.4.56). The identification results show that the heat transfer flow after the compression $\dot{Q}_{\mathrm{c,a}}$ and after the expansion in the turbine $\dot{Q}_{\mathrm{t,a}}$ can be neglected, but that the $\dot{Q}_{\mathrm{c,b}}$ before the compressor and $\dot{Q}_{\mathrm{t,b}}$ before the turbine have to be taken into account.

Table 4.4.2. Estimated turbocharger thermodynamic model parameters

Parameters	Value	Unit
$k_{\mathrm{c,b}}A_{\mathrm{c,b}}$	0.2105	W/K
$k_{\mathrm{c,a}}A_{\mathrm{c,a}}$	0	W/K
$k_{\mathrm{t,b}}A_{\mathrm{t,b}}$	0.7864	W/K
$k_{\mathrm{t,a}}A_{\mathrm{t,a}}$	0	W/K
$c_{\eta,\mathrm{t,is,max},0-2}$	[-0.29 0.34 0.62]	[-,m^{-1},m^{-2}]
$c_{\mathrm{u,opt},0-2}$	[-0.11 0.21 0.61]	[-,m^{-1},m^{-2}]

The isentropic efficiencies of compressor and turbine are shown in Fig. 4.4.20. The model for the dynamics contains the unknown parameter J_{tc} for the moment of inertia due to (4.4.46), which can be estimated from dynamic measurement data by minimization of the root-mean-square errors (RMSE) of the rotation speed. Finally, the results of the dynamic turbocharger model are shown in Fig. 4.4.21. The turbocharger speed $n_{\mathrm{tc}} = \frac{\omega_{\mathrm{tc}}}{2\pi}$ matches relatively well the measured speed with RMSE of $44.8\,1/\mathrm{s}$ or less than $1.5\,\%$ from maximum measured speed $n_{\mathrm{tc}} = 3000\,1/\mathrm{s}$.

Figure 4.4.22 shows the simulated and measured step responses of the turbocharger speed after a step in the accelerator pedal. The injection mass changes with a time constant of about 550 ms and the speed behaves like a second order system with a time delay of 350 ms and slope time of 2 s which includes the fuel injection delay.

The step response of the thermodynamic turbocharger model to an excitation with the turbine vane position is depicted in Fig. 4.4.23. The time constant of the electrical actuator is about 350 ms and the resulting time delay of the speed is 250 ms and the slope time 1 s of this higher engine speed, including the actuator delay. In both cases the agreement of modeled and the measured dynamic behavior is rather good.

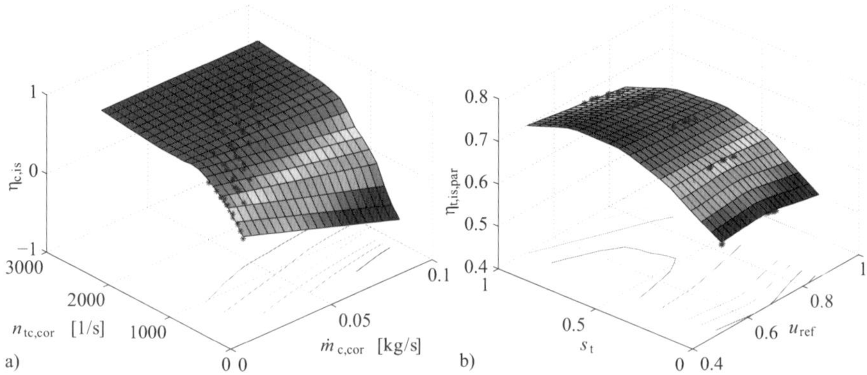

Fig. 4.4.20. Estimated efficiencies. **a** compressor. **b** turbine.

4.5 Exhaust system

The exhaust system discharges the exhaust gas at an appropriate part of the vehicle and supports the reduction of exhaust gas noise by mufflers (silencers). In addition the emissions from the internal combustion engine are reduced by catalytic converters and particulate filters. Hence, the exhaust system consists of the exhaust manifold, components for exhaust gas treatment, components for sound absorption and connecting pipes.

Mufflers are usually located in the middle and rear section of the exhaust pipe. They smooth the exhaust gas pulsations and reduce the noise as much as possible. The location is usually optimized in order to avoid resonances of the oscillating gas system. *Reflection mufflers* have chambers of different lengths and are connected by pipes. Thus resonators consisting of chambers and pipes and the diversion of exhaust gases dampen the noise especially at low frequencies.

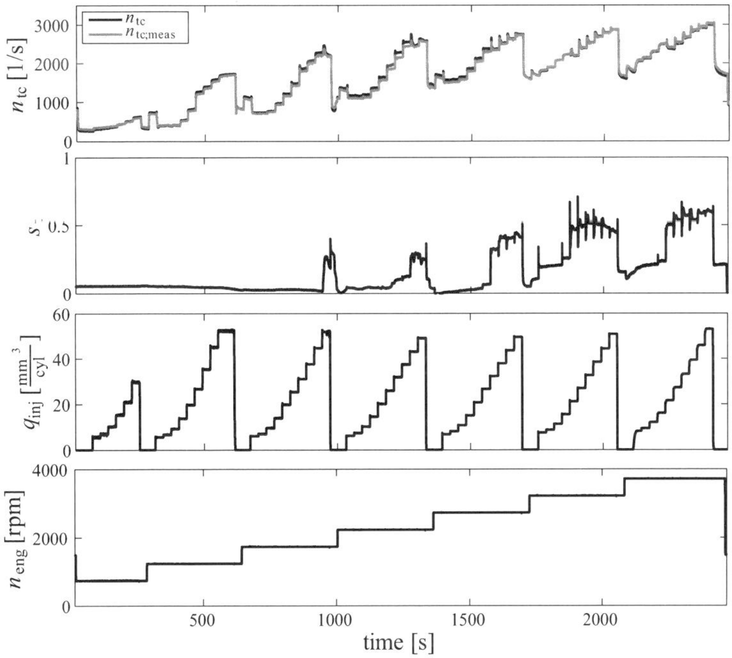

Fig. 4.4.21. Modeled and simulated turbocharger speed with the thermodynamic model (s_{vgt} changes according to the control by the series ECU).

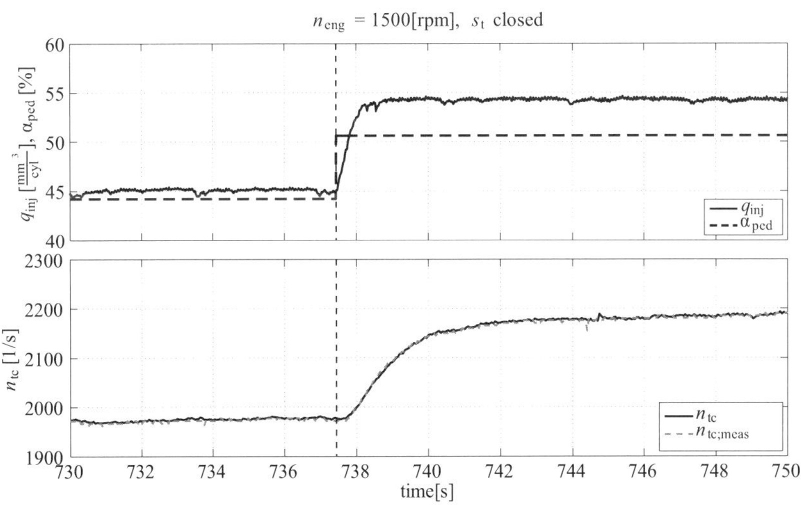

Fig. 4.4.22. Comparison of modeled and measured dynamic behavior for the thermodynamic turbocharger model: turbocharger speed response to a step of the accelerator pedal α_{ped} at $n_{eng} = 1500\,\mathrm{rpm}$ for s_t closed.

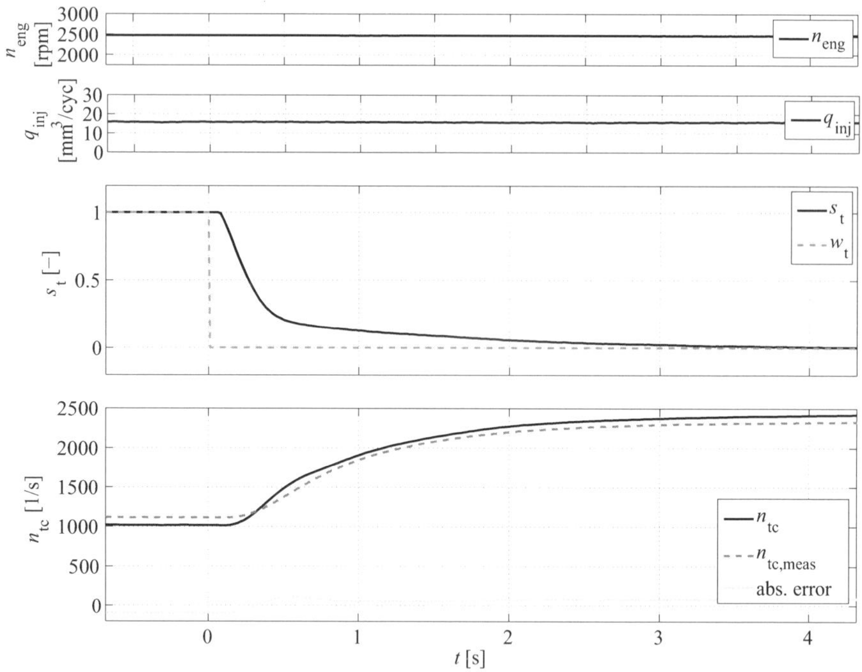

Fig. 4.4.23. Comparison of modeled and measured dynamic behavior for the thermodynamic turbocharger model: turbocharger speed response to a step in the turbine vane position s_t at $n_{eng} = 2500\,\text{rpm}$, w_t: reference value for position control loop, q_{inj}= const.

Absorption mufflers usually have one chamber and a perforated tube in the center. The sound is damped by the absorption material, e.g. long-fiber mineral wool, in a relatively broad frequency band at higher frequencies. Combinations of reflection and absorption mufflers also exist.

Catalytic converters consist of a ceramic or metallic substrate or monolith, and steel housing with an inflow and outflow funnel. The substrate is made with a large number of fine, parallel channels which are covered with a catalytic coating. Catalytic converters reduce the main pollutants CO, HC, and NO_x and convert them into CO_2 and H_2O. For high requirements a primary catalytic converter and a main catalytic converter are used, see e.g. Sect. 7.2

Particulate filters keep the solid particulates (soot) of the exhaust gases back, especially for diesel engines. Mostly ceramic filters are used with a large number of parallel channels. Because the channels are open and closed alternatively, the exhaust gases have to flow through the porous walls of a honeycomb structure. More details are given in Sect. 8.9.

Figure 4.5.1 shows examples for the exhaust system of gasoline and diesel engines and possible placement of sensors. Primary catalytic converters are installed close to the engine in order to quickly reach the operating temperature (light off) after start. Particulate filters are also mounted close to the engine, to allow high temperature regeneration.

As the emission reduction and control is treated in the chapters for the control of gasoline and diesel engines, this section considers the flow and temperature behavior of the exhaust system.

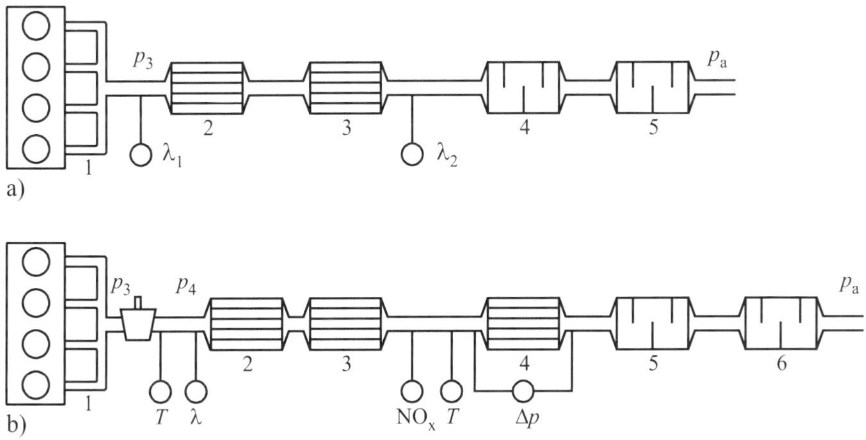

Fig. 4.5.1. Exhaust systems (examples). **a** Gasoline engine: 1 manifold, 2 primary catalytic converter, 3 main catalytic converter, 4 first muffler, 5 second muffler; **b** Diesel engine: 1 manifold, 2 oxidation catalytic converter, 3 NO_x storage catalyst, 4 particulate filter, 5 first muffler, 6 second muffler.

4.5.1 Flow behavior of exhaust pipes

The flow characteristic of the exhaust system influences the back exhaust gas pressure at the outlet valve of the engine. The exhaust gas flow is periodic according to the period of the outflow valves and consists of pulses. However, with regard to the flow resistance it can be treated as a quasi-steady flow, Jakisch (1996). The flow resistance stems mainly from the components like mufflers, catalytic converter and particulate filters, and also the connecting pipes.

The type of flow can assumed to be turbulent because the flow velocities for car engines with displacement 1 to 2 l, speed from 1000 to 6000 rpm and temperatures between 150 and 900°C are in the range of $5 \leq v_{\mathrm{exh}} \leq 180\,\mathrm{m/s}$, Jakisch (1996). This results in Reynold numbers $8000 \leq Re \leq 65000$ where

$$Re = \frac{d\,v_{\mathrm{exh}}}{\nu} \tag{4.5.1}$$

with d the tube diameter and ν the kinematic viscosity. Usually turbulent flow in tubes can be assumed for $Re > 4000$.

The pressure drop of a fluid in a tube then follows from

$$\Delta p = p_{\text{in}} - p_{\text{out}} = \frac{\lambda_{\text{f}}}{d} \frac{\rho_{\text{exh}} v_{\text{exh}}^2}{2} l \tag{4.5.2}$$

with λ_{f} the friction coefficient, ρ_{exh} the exhaust gas density and l the tube (section) length.

The exhaust gas velocity follows at a specific point i of the exhaust pipe from

$$\dot{m}_{\text{exh}} = A_{\text{exh}} \rho_{\text{exh}} v_{\text{exh}} \tag{4.5.3}$$

with

$$\dot{m}_{\text{exh}} = \dot{m}_{\text{air}} + \dot{m}_{\text{f}} \tag{4.5.4}$$

and the gas-state equation

$$\rho_{\text{exh}} = \frac{p_{\text{exh}}}{R_{\text{exh}} T_{\text{exh}}} \tag{4.5.5}$$

to

$$v_{\text{exh}} = \frac{R_{\text{exh}} T_{\text{exh}}}{A_{\text{exh}} p_{\text{exh}}} \dot{m}_{\text{exh}}. \tag{4.5.6}$$

Therefore, the velocity changes along the tube with regard to the inlet velocity

$$\frac{v_{\text{exh}}}{v_{\text{in}}} = \frac{T_{\text{exh}} p_{\text{in}}}{T_{\text{in}} p_{\text{exh}}} \frac{A_{\text{in}}}{A_{\text{exh}}}. \tag{4.5.7}$$

Usually the velocity decreases down-stream. For a longer pipe section with linear temperature decrease it holds, Kalide (1980), Willimowski (2003)

$$\Delta p_{\text{exh}} = \frac{\lambda}{d_{\text{exh}}} \frac{\rho_{\text{exh}} v_{\text{in}}^2}{2} \frac{\overline{T}}{T_{\text{in}}} \frac{p_{\text{in}}}{p_{\text{exh}}} l \tag{4.5.8}$$

where $\overline{T}$ is an average temperature.

However, if pressure and temperature changes are not too large (4.5.2) can be used, such that for a tube section i it holds

$$\Delta p_i = \frac{\lambda_i}{d_i} \frac{\rho_i v_i^2}{2} l_i \tag{4.5.9}$$

with ρ_i and v_i according to (4.5.5) and (4.5.6).

The pressure drops of additional components with unsteady flows like mufflers, tube bents or filters are modeled by

$$\Delta p_j = \zeta_j \rho_j \frac{v_j^2}{2} \tag{4.5.10}$$

where $\zeta_j = \lambda_j l_j / d_j$ is a resistance coefficient for the specific component. Typical values for the chambers of reflection and absorption mufflers are $0.2 \leq \zeta \leq 2$, Jakisch (1996).

Applying the mass flow $\dot{m}_{\text{exh}} = A_j \rho_j v_j$ with A_j the cross-section area and the gas state equation $\rho_j = p_j / R T_j$ one obtains

$$\Delta p_j = \zeta_j \frac{\dot{m}_{\mathrm{exh}}^2}{2A_j^2\,\rho_j} = \zeta_j \frac{\dot{m}_{\mathrm{exh}}^2\,RT_j}{2A_j^2\,p_j}. \tag{4.5.11}$$

If the components are connected in series, it holds for the pressure drop of the exhaust system with $p_{\mathrm{out}} = p_{\mathrm{a}}$, p_{a} the atmospheric pressure,

$$\begin{aligned}\Delta p_{\mathrm{exh}} &= p_{\mathrm{in}} - p_{\mathrm{out}} = \sum_i \frac{\lambda_i}{d_i}\frac{\rho_i\,v_i^2}{2}\,l_i + \sum_j \zeta_j\,\rho_j\,\frac{v_j^2}{2} \\ &= \dot{m}_{\mathrm{exh}}^2\left[\sum_i \frac{\lambda_i}{2d_i}\frac{RT_i}{A_i^2\,p_i}\,l_i + \sum_j \frac{\zeta_j}{2}\frac{RT_j}{A_j^2\,p_j}\right]\end{aligned} \tag{4.5.12}$$

which can be approximated by

$$\Delta p_{\mathrm{exh}} = c_{\mathrm{Rexh}}\,\dot{m}_{\mathrm{exh}}^2. \tag{4.5.13}$$

(4.5.12) or (4.5.13) allows to calculate the exhaust gas pressure p_{in} at the beginning of the exhaust pipes. It depends quadratically on the mass flow and is inverse proportional to the square of the cross-section areas. Without turbocharger it is $p_{\mathrm{in}} = p_3$ i.e. the back pressure after the outlet valves. In the case of turbochargers $p_{\mathrm{in}} = p_4$ is valid, i.e. the pressure after the turbine. Measured and calculated back pressures with the models given before for a 12 cylinder gasoline engine without turbocharger in the whole operating area $n = 650\ldots5500\,\mathrm{rpm}$ and $M_{\mathrm{eff}} = 0\ldots450\,\mathrm{Nm}$ in Willimowski (2003) show a good agreement. At low speed and torque the back pressure in this case is close to $1\,\mathrm{bar}$ and increases up to $1.65\,\mathrm{bar}$ for high speed and torque.

4.5.2 Thermal behavior of exhaust pipes

To simplify the considerations it is assumed that an exhaust pipe of length l is considered in which exhaust gas with temperature $T_{\mathrm{in}} = T_3$ flows in and leaves it with T_{out}. The exhaust pipe can be considered as a cooled pipe, such that the mathematical models for heat exchangers and especially that of the "heated tube", see Sect. 4.6, can be applied. The heat transfer from the exhaust pipe to the environment is by convection and by radiation. Depending on the local air flow around the pipe the heat transfer by convection is a mixture of parallel flow and cross flow with a specific outer heat flow, see (4.6.2),

$$\dot{q}_{\mathrm{2w,c}} = \frac{\dot{Q}_{\mathrm{2w,c}}}{A_{\mathrm{2w}}} = \alpha_2(\vartheta_{\mathrm{w2}} - \vartheta_{\mathrm{F2}}) \tag{4.5.14}$$

where the heat transfer coefficient depends on the air velocity. For the heat transfer through radiation it holds

$$\dot{q}_{\mathrm{2w,r}} = \frac{\dot{Q}_{\mathrm{2w,r}}}{A_{\mathrm{2w}}} = \sigma_2(T_{\mathrm{w2}}^4 - T_{\mathrm{F2}}^4) \tag{4.5.15}$$

with σ_2 the radiation coefficient, $T_{F2} = 273° + \vartheta_{F2}$ the temperature of the surrounding air and $T_{w2} = 273° + \vartheta_{w2}$ the wall temperature.

The radiation coefficient for black radiators is the Stefan–Boltzmann constant

$$\sigma = 5.6704 \cdot 10^{-8}\,\mathrm{W/m^2\,K^4}. \tag{4.5.16}$$

For non-black radiators an emission degree $\epsilon(T)$ provides a correction

$$\sigma_2 = \epsilon(T)\sigma. \tag{4.5.17}$$

The emission degree for the so called gray radiators is $0.012 \leq \epsilon \leq 0.98$.

According to the VDI heat atlas (2010) the heat radiation density of a black radiator is for $T_{w2} = 373, 523, 723, 923, 1173\,\mathrm{K}$ $q_{w2} = 314, 1097, 4242, 15493, 41151, 107343\,\mathrm{W/m^2}$. Hence, a significant part of heat radiation can be expected for temperatures $T_{w2} > 723 \ldots 923\,\mathrm{K}$ or $T_{w2} > 450 \ldots 650°\mathrm{C}$.

This means that the specific heat flow through radiation is larger after the exhaust manifold and then decreases proportionally to T_{w2}^4 with the wall temperature. Hence, in a first section of length l_{I} of the exhaust pipe the heat transfer is through radiation and convection. For the second section of length l_{II} it can be assumed that the heat transfer is mainly by convection, see Fig. 4.5.2. The ratio of the specific heat flow $\dot{q}_{2\mathrm{w,r}}/\dot{q}_{2\mathrm{w,c}}$ depends on the coefficients α_2 and σ_2 and the wall temperature T_{w2}.

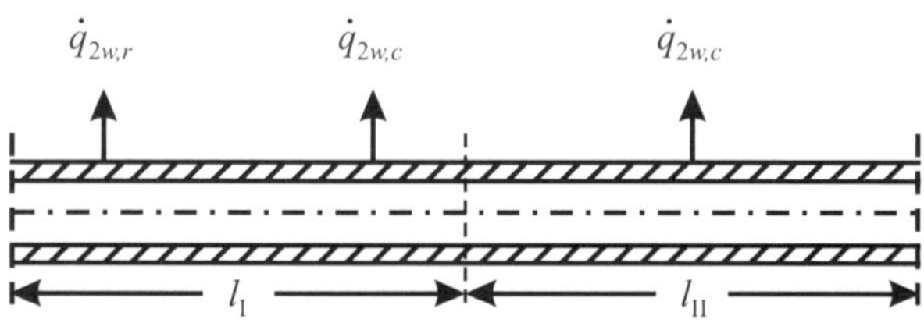

Fig. 4.5.2. Exhaust pipe, divided in two sections. Section I: heat transfer by radiation and convection. Section II: heat transfer mainly by convection.

To determine the steady-state temperature profile in longitudinal direction of the tube segments heat exchanger models for parallel flow can be used, e.g. (4.6.9) or (4.6.11). A detailed simulation of the temperature behavior of an exhaust pipe with a catalytic converter by numerical solution of partial differential equations is described in Büchner et al (2001).

As *dynamic models* for a heated tube usually assume a constant specific heat flow over the considered length, the exhaust pipe is divided in several sections i, each with averaged constant specific heat flows on the outside of the tube

$$\dot{q}_{2\mathrm{w},i} = \dot{q}_{2\mathrm{w,r},i} + \dot{q}_{2\mathrm{w,c},i} \tag{4.5.18}$$

and on the inner side, between the exhaust gas and the tube wall

$$\dot{q}_{1\mathrm{w},i} = \alpha_{1\mathrm{w},i}(\vartheta_1 - \vartheta_{\mathrm{w}}) \tag{4.5.19}$$

compare (4.6.20) and (4.6.21).

Then a section of length l_i has the characteristic fluid parameter (4.6.32)

$$\kappa_{F,i} = \frac{4\alpha_{1,w,i}\, l_i}{d_1\, \rho_1\, c_{p1}\, v_i} \tag{4.5.20}$$

using the symbols of Sect. 4.6.4. The specific heat-flow coefficient $\alpha_{1,w,i}$ on the inner side depends on the gas velocity according to Nusselt's law (4.6.22). v_i follows from (4.5.6). The dynamic behavior of the exhaust gas output temperature $\Delta T_{out,i}$ depends on changes of the input temperature $\Delta T_{in,i}$, exhaust gas velocity Δv_i and specific heat flow $\Delta \dot{q}_{2w,i}$. Figure 4.5.3 shows a corresponding signal-flow scheme. Dynamic models of such a heated or cooled pipe are treated in Sect. 4.6.4 as well as simplified models. The input/output behavior of the temperatures of one tube section i is approximated by the transfer function (4.6.36).

$$\widetilde{G}_{T,i}(s) = \frac{\Delta T_{out,i}}{\Delta T_{in,i}} = \left(a_i + \frac{b_i}{1 + T_{b,i}s}\right) e^{-T_{t,i}s} \tag{4.5.21}$$

for $\kappa_{F,i} \leq 1.5$. If $\kappa_{F,i} > 1.5$, several elements have to be connected in series, see (4.6.36). For changes of the exhaust gas velocity Δv_i and specific heat flow $\Delta \dot{q}_{2w,i}$ the second order models (4.6.37) and (4.6.38) can be used.

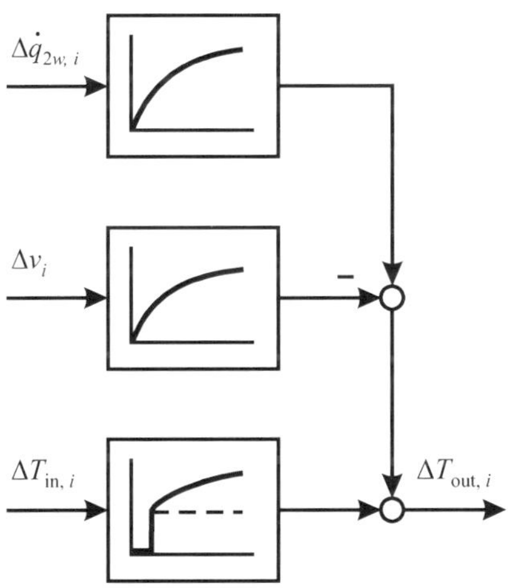

Fig. 4.5.3. Signal flow of the output temperature of an exhaust pipe section i for changes of the input temperature, exhaust gas velocity and specific heat flow.

Example 4.1. A tube section with the following data

$$l = 1\,\text{m}; \quad d_1 = 0.03\,\text{m}; \quad d_2 = 0.026\,\text{m}; \quad \rho_1 = 1.29\,\text{kg/m}^3;$$
$$c_{p1} = 1005\,\text{J/kg K}; \quad v = 10\ldots100\,\text{m/s}; \quad \alpha_{1,w} = 33\ldots200\,\text{J/s m}^2\text{K}$$

yields the fluid parameter

$$\kappa_{\mathrm{F}} = 0.34 \ldots 0.20.$$

Hence, one element is required to approximate the input/output temperature behavior according to (4.5.21). The dead time is

$$T_{\mathrm{t}} = l/v = 0.1 \ldots 0.01\,\mathrm{s} = 100 \ldots 10\,\mathrm{ms}.$$

The corresponding inner tube wall time constant according to (4.6.35) becomes with $\rho_{\mathrm{w}} = 7900\,\mathrm{J/kg\,K}$ and $c_{\mathrm{w}} = 510\,\mathrm{J/kg\,K}$

$$T_{\mathrm{w1}} = 229 \ldots 38\,\mathrm{s}$$

and for the simplified model (4.5.21) one obtains

$$T_{\mathrm{b}} = 267 \ldots 42\,\mathrm{s}.$$

Hence, the tube-section transfer function can be approximated for $v = 10\,\mathrm{m/s}$ by

$$\widetilde{G}_{\mathrm{T}}(s) = \left(0.71 + \frac{0.29}{1 + 267s}\right) e^{-0.1s}$$

and for $v = 100\,\mathrm{m/s}$ by

$$\widetilde{G}_{\mathrm{T}}(s) = \left(0.82 + \frac{0.18}{1 + 42s}\right) e^{-0.01s}.$$

Hence, the transfer behavior is mainly proportional with an additional slow first order time lag element of small gain, as shown in Fig. 4.5.3.

□

For the *experimental identification* of the dynamic behavior of exhaust gas temperatures suitable temperature sensors are required as, e.g. thermocouples or resistance thermometers with small time constants. Their time constants can be calculated with (4.6.42)

$$T_{\mathrm{s}} = \frac{\rho_{\mathrm{s}}\, d_{\mathrm{s}}\, c_{\mathrm{s}}}{4\alpha_{\mathrm{fs}}}. \tag{4.5.22}$$

They depend strongly on the heat transfer coefficient α_{fs} which is proportional to the gas velocity v^m, $m = 0.6 \ldots 0.8$, see (4.6.22). As these sensor time constants are in the range of several seconds they influence the identification of exhaust temperature models considerably.

In order to reconstruct the real exhaust gas temperature the sensor dynamics have to be compensated. However, this requires an exact dynamic temperature sensor model. Such a model can be experimentally obtained by using two sensors with different time constants and eliminating the input signal for the exhaust gas temperature by parameter estimation of the two sensor time constants. Zimmerschied and Isermann (2010) have shown how it is possible to reconstruct the real exhaust gas temperature after estimation of the sensor time constants with the local linear net model parameter estimation method LOLIMOT in dependence on air mass flow and injected fuel mass for a CR diesel engine by inverted sensor models or extended Kalman filtering. The time constants of two thin thermocouples with diameters of 0.5 and $1\,\mathrm{mm}$ are in the range of $0.5 \ldots 1\,\mathrm{s}$ and $1.6 \ldots 3.3\,\mathrm{s}$ for $n_{\mathrm{eng}} = 1000 \ldots 3000\,\mathrm{rpm}$.

4.6 Heat exchangers

Internal combustion engines contain several heat exchangers as for example radiators for cooling systems, oil coolers within the oil circuit, charge-air coolers after a supercharger or turbocharger, exhaust gas recirculation coolers and heat exchangers for internal heating. Therefore engine modeling requires mathematical models of the corresponding heat exchanging devices. After a short view on the different types of heat exchangers different models for the stationary and dynamic behavior are considered. The behavior of models of the stationary behavior for counter and parallel flow are described. This is followed by simplified dynamic heat-exchanger models with lumped parameters and dynamic models for a basic element, a heated tube with distributed parameters. The simple lumped parameter models with ordinary differential equations approximate the behavior of heat exchangers with relatively short tube lengths, whereas a treatment with distributed parameters and therefore partial differential equations are needed for heat exchangers with long tube lengths. At the end, relatively simple models result as approximative solutions. These models are suited for the simulation of the stationary and dynamic behavior and for the design of control systems.

4.6.1 Heat exchanger types

Heat exchangers transfer heat between two or more media. They are required for cooling or heating at various locations within or around the combustion engine. The fluids are usually liquids or gases and appear in following combinations:

- liquid-liquid: oil-to-coolant coolers
- liquid-gas: coolant-to-air coolers (radiators), coolant-to-air heaters
- gas-liquid: exhaust gas-to-coolant coolers, air conditioning condensors
- gas-gas: charge air-to-air coolers (intercoolers).

Heat exchangers are built in a variety of designs. Frequently applied designs for engines are *tubular or cross-flow heat exchangers* or *disk heat exchangers*, see Fig. 4.6.1. With regard to the flow direction one distinguishes counter flow, parallel flow and cross flow.

Coolant radiators are an example for cross flow and are made of aluminum. They are assembled as braced radiators or mechanically joined radiators. The tubes are round, oval or flat-oval with fins. The distribution of the coolant over the radiator-block results through radiator tanks which are made of fiberglass-reinforced polyamides or of aluminum. Intercoolers with air cooling have a similar design.

Oil-to-coolant coolers are frequently characterized by a disk-stack design. The individual aluminum disks have turbulence inserts between the disks. The disks fit together and form a casing. Passages connect the channels formed by the disks such that coolant and oil flow through alternate channels, preferably in counter flow.

Exhaust gas recirculation coolers consist typically of some short tubes which are cooled through the coolant flowing outside of the tubes in cross or counter flow.

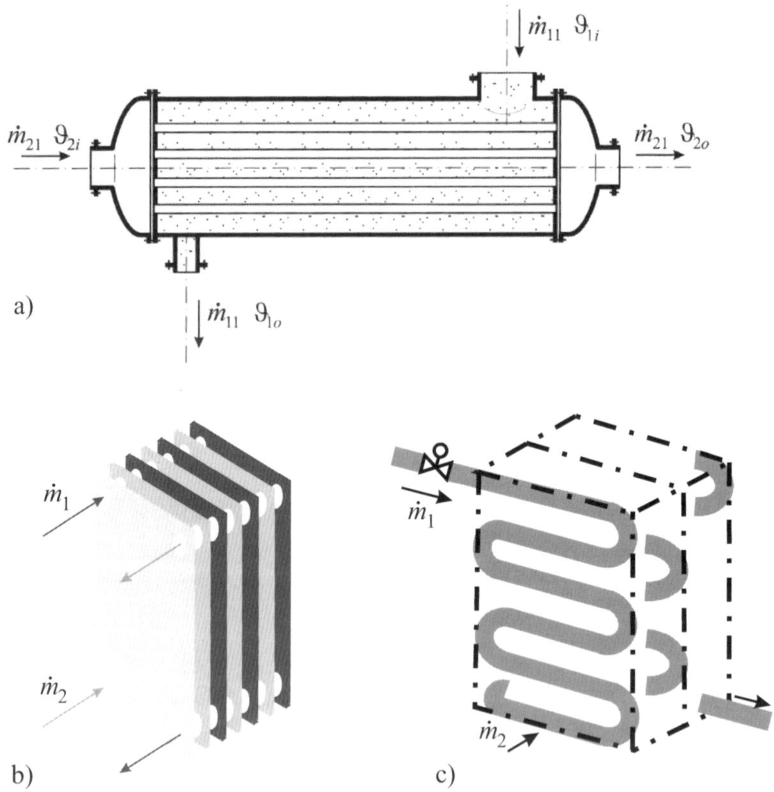

Fig. 4.6.1. Heat exchanger designs. **a** tubular heat exchanger. **b** disk heat exchanger. **c** cross flow heat exchanger.

The stationary behavior of heat exchangers is based on the heat transfer to and from the separating wall. Therefore, heat exchangers have basically distributed parameters as many location dependent heat transfer elements exist. However, as the heat exchangers for combustion engines are relatively small, the application of some simplifying assumptions allow to derive overall heat exchanger models with lumped parameters. More precise models are obtained by a distributed parameter approach, where the dynamics of heated tubes are considered. Therefore, different models of heat exchangers will be considered.

In this chapter, the symbols in Table 4.6.1 are mainly used.

4.6.2 Heat exchanger models for stationary behavior

Heat exchanging fluids are usually separated by a wall. If the heat transfer is through convection, the stationary specific heat flow is, compare Fig. 4.6.2

$$\dot{q}_{1\mathrm{w}} = \frac{\dot{Q}_{1\mathrm{w}}}{A_{1\mathrm{w}}} = \alpha_1 \left(\vartheta_{\mathrm{F}1} - \vartheta_{\mathrm{w}1}\right) \tag{4.6.1}$$

Table 4.6.1. Symbols used for heat exchangers

A area	α heat transfer coefficient
c_p specific heat capacity at constant pressure	ϑ temperature (instead of T: time constant)
d diameter	ρ density
	λ thermal conductivity
l length of tubes	
k overall heat transfer coefficient	Subscripts:
$\dot{m}$ mass flow rate	$_1$ primary side of heat exchanger
$\dot{q}$ specific heat flow $\dot{q} = \dot{Q}/A$	$_2$ secondary side of heat exchanger
$\dot{Q}$ heat flow	$_\mathrm{w}$ wall
r vaporization heat, residual	$_\mathrm{s}$ steam
s tube wall thickness, Laplace variable $s = \sigma + i\omega$	$_\mathrm{i}$ inlet
T time constant, thermodynamic temperature	$_\mathrm{o}$ outlet
v velocity	$_\mathrm{f}$ fluid
z tube length coordinate	A bar like $\overline{m}$ means steady state.

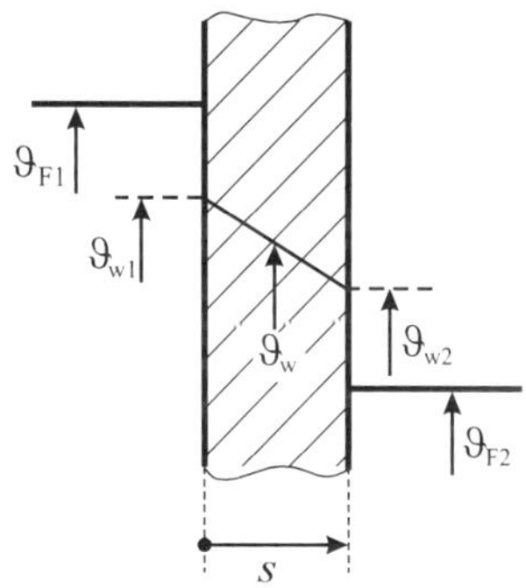

Fig. 4.6.2. Heat transfer through a wall.

with $A_{1\mathrm{w}}$ the surface area and α_1 the heat transfer coefficient. Similarly, it holds for the other side of the wall

$$\dot{q}_{2\mathrm{w}} = \frac{\dot{Q}_{2\mathrm{w}}}{A_{2\mathrm{w}}} = \alpha_2 \left(\vartheta_{\mathrm{w}2} - \vartheta_{\mathrm{F}2}\right). \tag{4.6.2}$$

The conductive heat transfer through the wall is (Fourier's law)

$$\dot{q}_\mathrm{w} = \frac{\dot{Q}_\mathrm{w}}{A_\mathrm{w}} = \frac{\lambda}{s} \left(\vartheta_{\mathrm{w}1} - \vartheta_{\mathrm{w}2}\right) \tag{4.6.3}$$

with λ the thermal conductivity coefficient and s the wall thickness. In the stationary case all heat flows are identical

$$\dot{q}_{1\mathrm{w}} = \dot{q}_{2\mathrm{w}} = \dot{q}_\mathrm{w}.$$

The overall heat transfer then follows if $A_{1\mathrm{w}} = A_{2\mathrm{w}} = A_\mathrm{w}$ as

$$\dot{q}_{12} = \frac{\dot{Q}_{12}}{A_{\mathrm{w}}} = k\left(\vartheta_{\mathrm{F1}} - \vartheta_{\mathrm{F2}}\right) \tag{4.6.4}$$

with the overall heat transfer coefficient

$$k = \frac{1}{\frac{1}{\alpha_{1\mathrm{w}}} + \frac{s}{\lambda} + \frac{1}{\alpha_{2\mathrm{w}}}}. \tag{4.6.5}$$

Hence the smallest value of $\alpha_{1\mathrm{w}}$ or $\alpha_{2\mathrm{w}}$ dominates the overall heat transfer.

The heat transfer through a *tubular heat exchanger* as shown in Fig. 4.6.3 with massflows $\dot{m}_1$ and $\dot{m}_2$, where the heat transfer is from the fluid with $\dot{m}_1$ to the fluid with $\dot{m}_2$, and specific heat capacity c_{p1} and c_{p2} is defined as

$$\dot{Q}_{12} = k\,A\,\Delta\vartheta_{\mathrm{m}} \tag{4.6.6}$$

where ϑ_m is the average temperature difference

$$\Delta\vartheta_m = \frac{\Delta\vartheta_{\mathrm{la}} - \Delta\vartheta_{\mathrm{sm}}}{\ln\left(\Delta\vartheta_{\mathrm{la}}/\Delta\vartheta_{\mathrm{sm}}\right)} \approx \frac{1}{2}\left(\Delta\vartheta_{\mathrm{la}} + \Delta\vartheta_{\mathrm{sm}}\right) \tag{4.6.7}$$

and $\Delta\vartheta_{\mathrm{la}}$ and $\Delta\vartheta_{\mathrm{sm}}$ the larger and smaller temperature difference according to Fig. 4.6.4.

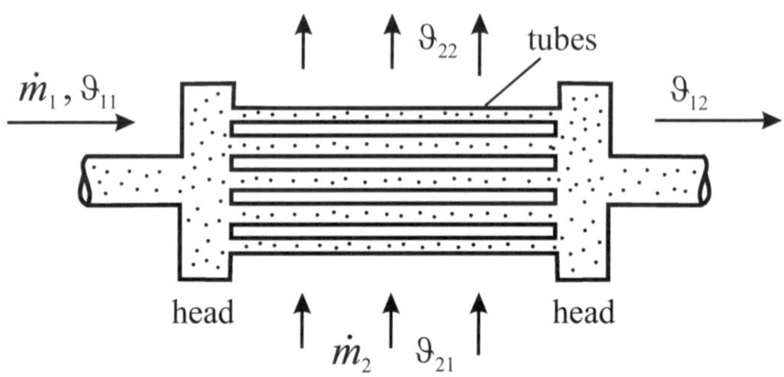

Fig. 4.6.3. Scheme of a tubular heat exchanger for liquid-gas, crossflow.

The temperature profiles in longitudinal direction show an exponential course. It holds for example under ideal conditions for the fluid 1 which increases the temperature in counter flow

$$\vartheta_{12} = \vartheta_{21} - \frac{1 - \frac{\dot{m}_1 c_{\mathrm{p1}}}{\dot{m}_2 c_{\mathrm{p2}}}}{1 - \frac{\dot{m}_1 c_{\mathrm{p1}}}{\dot{m}_2 c_{\mathrm{p2}}} \exp\left[\left(\frac{1}{\dot{m}_2 c_{\mathrm{p2}}} - \frac{1}{\dot{m}_1 c_{\mathrm{p1}}}\right) kA\right]} \left(\vartheta_{21} - \vartheta_{11}\right) \tag{4.6.8}$$

and for parallel flow

$$\vartheta_{12} = \vartheta_{21} - \frac{1 - \exp\left[\left(\frac{1}{\dot{m}_2 c_{\mathrm{p2}}} - \frac{1}{\dot{m}_1 c_{\mathrm{p1}}}\right) kA\right]}{1 - \frac{\dot{m}_1 c_{\mathrm{p1}}}{\dot{m}_2 c_{\mathrm{p2}}} \exp\left[\left(\frac{1}{\dot{m}_2 c_{\mathrm{p2}}} - \frac{1}{\dot{m}_1 c_{\mathrm{p1}}}\right) kA\right]} \left(\vartheta_{21} - \vartheta_{11}\right). \tag{4.6.9}$$

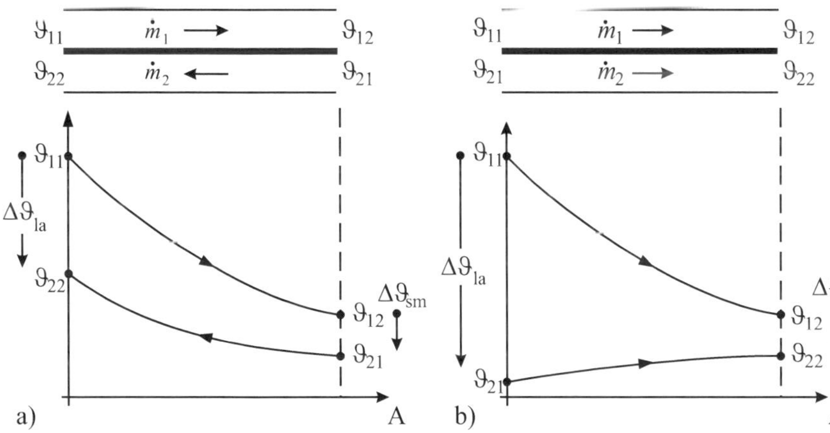

Fig. 4.6.4. Temperature profiles in tubular heat exchangers. **a** counter flow. **b** parallel flow.

For details see VDI heat atlas (2010), Grothe and Feldhusen (2007), Holman (1976).

For heat exchangers with small heat exchange area A and therefore $\exp x \approx 1+x$ these equations can be approximated for counter flow

$$\begin{aligned}\vartheta_{12} &= \vartheta_{21} - \frac{1-\frac{\dot{m}_1 c_{p1}}{\dot{m}_2 c_{p2}}}{1-\frac{\dot{m}_1 c_{p1}}{\dot{m}_2 c_{p2}}\left[1+\left(\frac{1}{\dot{m}_2 c_{p2}}-\frac{1}{\dot{m}_1 c_{p1}}\right)kA\right]}(\vartheta_{21}-\vartheta_{11}) \\ &= \vartheta_{21} - K_{2c}(\vartheta_{21}-\vartheta_{11}) \qquad (4.6.10)\end{aligned}$$

and for parallel flow

$$\begin{aligned}\vartheta_{12} &= \vartheta_{21} - \frac{1-\left[1+\left(\frac{1}{\dot{m}_2 c_{p2}}-\frac{1}{\dot{m}_1 c_{p1}}\right)kA\right]}{1-\frac{\dot{m}_1 c_{p1}}{\dot{m}_2 c_{p2}}\left[1+\left(\frac{1}{\dot{m}_2 c_{p2}}-\frac{1}{\dot{m}_1 c_{p1}}\right)kA\right]}(\vartheta_{21}-\vartheta_{11}) \\ &= \vartheta_{21} - K_{2p}(\vartheta_{21}-\vartheta_{11}). \qquad (4.6.11)\end{aligned}$$

Hence, the change of the outlet temperature $\Delta\vartheta_{12}$ after a change of the inlet temperature $\Delta\vartheta_{11}$ of the fluid with increasing temperature becomes for counter flow

$$\Delta\vartheta_{12} = K_{2c}\Delta\vartheta_{11} \qquad (4.6.12)$$

and for parallel flow

$$\Delta\vartheta_{12} = K_{2p}\Delta\vartheta_{11}. \qquad (4.6.13)$$

The gains K_{2c} and K_{2p} depend essentially on the mass flows and specific heats. For other relationships of inlet and outlet temperatures see, e.g. Grothe and Feldhusen (2007).

4.6.3 Heat exchanger models for the dynamic behavior with lumped parameters

As an example a tubular heat exchanger as in Fig. 4.6.3 is considered. In many cases the dynamic behavior of the outlet temperature ϑ_{12} of fluid 1 in dependence on its inlet temperature ϑ_{11}, the mass flow $\dot{m}_1$ and the heat flow $\dot{q}_{2\mathrm{w}}$ is of interest. This leads to the three transfer functions shown in Fig. 4.6.5.

If the whole heat exchanger is considered as a lumped heat storage with average temperature ϑ_{12}, then the heat balance yields

$$m_h c_h \frac{d\vartheta_{12}}{dt} = \dot{m}_1 c_{\mathrm{p1}}(\vartheta_{11} - \vartheta_{12}) + \underbrace{\alpha_2 A_2(\vartheta_{21} - \vartheta_1')}_{\dot{Q}_2 = \dot{q}_{2\mathrm{w}} A_2} \tag{4.6.14}$$

where $m_h c_h$ is the thermal capacity of the tubes and the fluid, such that $m_\mathrm{h} c_\mathrm{h} = m_\mathrm{t} c_\mathrm{t} + m_1 c_{\mathrm{p1}}$ and averaged temperatures of the fluid are taken for the heat transfer

$$\vartheta_1' = \frac{1}{2}(\vartheta_{11} + \vartheta_{12})$$

and for the wall temperature of the tubes $\vartheta_\mathrm{w} = \vartheta_{\mathrm{w1}} = \vartheta_{\mathrm{w2}} \approx \vartheta_1'$ is simplified.

For small temperature changes $\Delta\vartheta$ this leads to the first order differential equation

$$\begin{aligned}\frac{m_h c_h}{\left(\dot{m}_1 c_{\mathrm{p1}} + \frac{\alpha_2 A_2}{2}\right)} \frac{d\vartheta_{12}(t)}{dt} + \Delta\vartheta_{12}(t) \\ = \frac{\dot{m}_1 c_{\mathrm{p1}} - \frac{\alpha_2 A_2}{2}}{\left(\dot{m}_1 c_{\mathrm{p1}} + \frac{\alpha_2 A_2}{2}\right)} \Delta\vartheta_{11}(t) + \frac{\alpha_2 A_2}{\left(\dot{m}_1 c_{\mathrm{p1}} + \frac{\alpha_2 A_2}{2}\right)} \Delta\vartheta_{21}(t)\end{aligned} \tag{4.6.15}$$

with the time constant

$$T_{\mathrm{h2}} = \frac{m_\mathrm{h} c_\mathrm{h}}{\dot{m}_1 c_{\mathrm{p1}} + \frac{\alpha_2 A_2}{2}}. \tag{4.6.16}$$

This first order model with one time constant is a rough estimate and can be used to describe the dynamics after changes $\Delta\vartheta_{11}(t)$ or $\Delta\vartheta_{21}(t)$.

4.6.4 Heat exchanger models for the dynamic behavior with distributed parameters

Modeling the dynamic behavior of a heated tube serves as a basic element to derive and understand the temperature dynamics of heat exchangers. In many cases, it is of interest how the output temperature $\vartheta_{1\mathrm{o}}(t)$ depends dynamically on the input temperature $\vartheta_{1i}(t)$, the velocity $v_1(t)$ and the specific heat flow $\dot{q}_{2\mathrm{w}}(t)$, see Fig. 4.6.6. The following description follows Acklin and Läubli (1960), Profos (1962), Isermann (1969), Isermann (1971).

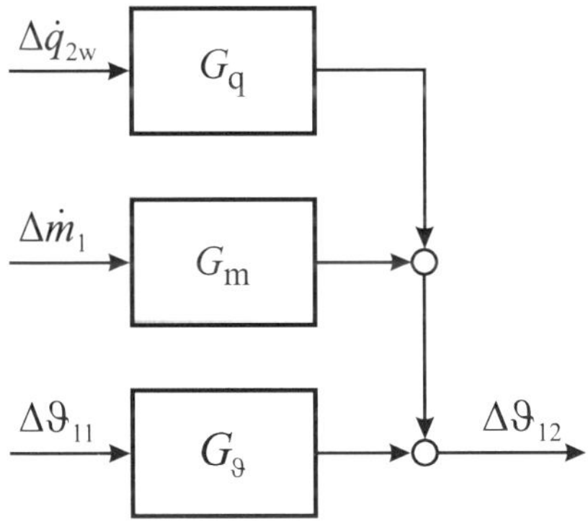

Fig. 4.6.5. Transfer functions for the outlet temperature of a tubular heat exchanger.

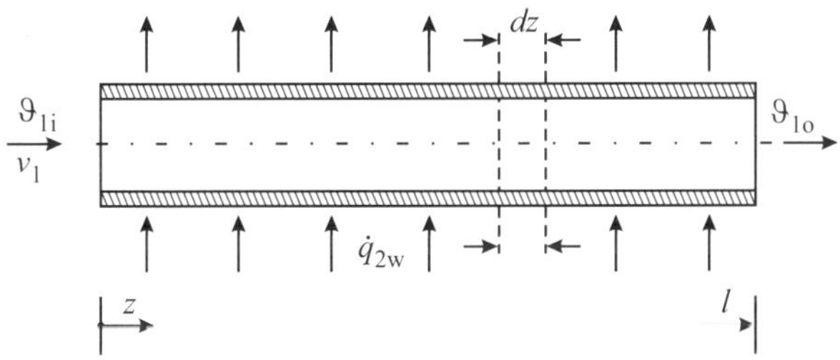

Fig. 4.6.6. Scheme of a heated tube.

a) Heated tube with distributed parameters

It is assumed that the geometric dimensions and the specific heat flow $\dot{q}_{2\mathrm{w}}(z)$ are constant along the length coordinate z and that turbulent flow holds internally, further ideal mixture perpendicular to the z-axis.

As the tube is a distributed parameter system where the fluid temperature $\vartheta_1(z,t)$ depends on the location z and the time t an infinitesimal small element is considered, see Fig. 4.6.7. It consists of two heat storages, the inner fluid and the tube wall. Therefore two balance equations are stated:

- enthalpy balance equation for the flowing fluid:

$$\frac{Dh_1(z,t)}{Dt} = \frac{1}{A_1\rho_1}\frac{\partial\dot{Q}_{1\mathrm{w}}}{\partial z}\ ;\ d\dot{Q}_{1\mathrm{w}} = \dot{q}_{1\mathrm{w}}\pi d_1 dz \tag{4.6.17}$$

$$\frac{\partial h_1(z,t)}{\partial t} + v_1(t)\frac{\partial h_1(z,t)}{\partial z} = \frac{\pi d_1}{A_1\rho_1}\dot{q}_{1\mathrm{w}}(z,t) \tag{4.6.18}$$

- heat balance of the tube wall element:

$$\dot{q}_{2\mathrm{w}}\ \pi\ d_2\ dz - \dot{q}_{1\mathrm{w}}\ \pi\ d_1\ dz = A_\mathrm{w}\ \rho_\mathrm{w}\ c_\mathrm{w}\ \frac{\partial\vartheta_\mathrm{w}}{\partial t}. \tag{4.6.19}$$

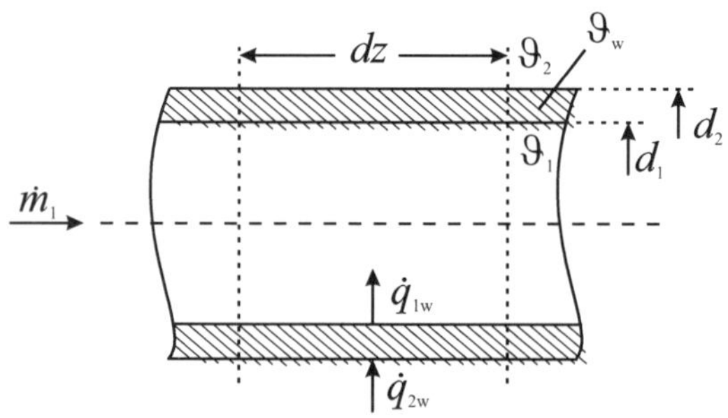

Fig. 4.6.7. Tube element.

After assuming small changes of the temperature $\Delta\vartheta(t)$ and other variables and introducing the heat transfer equations

$$\dot{q}_{1\mathrm{w}} = \alpha_{1\mathrm{w}}(\vartheta_{\mathrm{w}} - \vartheta_1) \tag{4.6.20}$$

$$\dot{q}_{2\mathrm{w}} = \alpha_{2\mathrm{w}}(\vartheta_2 - \vartheta_{\mathrm{w}}) \tag{4.6.21}$$

with Nusselt's law for the heat transfer coefficient

$$\alpha_{1\mathrm{w}} = \overline{\alpha}_{1\mathrm{w}} \left(\frac{v_1}{\overline{v}_1}\right)^m \tag{4.6.22}$$

$$\Delta\alpha_{1\mathrm{w}} = \overline{\alpha}_{1\mathrm{w}}\, m\, \frac{\Delta v_1}{\overline{v}_1}$$

($m = 0.8$ for turbulent flow, $\overline{v}$ means steady state value) and for the enthalpy

$$\Delta h_1 = c_{\mathrm{p}1}\, \Delta\vartheta_1 \tag{4.6.23}$$

it follows

$$\begin{aligned} \frac{\partial\vartheta_1(z,t)}{\partial t} + \overline{v}_1 \frac{\partial\vartheta_1(z,t)}{\partial z} &= \frac{1}{T_{\mathrm{f}}} \left(\Delta\vartheta_{\mathrm{w}}(z,t) - \Delta\vartheta_1(z,t)\right) \\ &\quad - \frac{1}{T_{\mathrm{f}}} \left(\overline{\vartheta}_{\mathrm{w}} - \overline{\vartheta}_1\right)(m-1)\frac{\Delta v_1}{\overline{v}_1} \end{aligned} \tag{4.6.24}$$

$$\begin{aligned} \frac{\partial\vartheta_{\mathrm{w}}(z,t)}{\partial t} &= \frac{1}{\alpha_{2\mathrm{w}}\, T_{\mathrm{w}2}} \Delta\dot{q}_{2\mathrm{w}}(t) - \frac{1}{T_{\mathrm{w}2}} \Delta\vartheta_{\mathrm{w}}(z,t) \\ &\quad + \frac{1}{T_{\mathrm{w}1}} \left(\Delta\vartheta_1(z,t) - \Delta\vartheta_{\mathrm{w}}(z,t)\right) \\ &\quad - \frac{1}{T_{\mathrm{w}1}} m \left(\overline{\vartheta}_{\mathrm{w}} - \overline{\vartheta}_1\right) \frac{\Delta v_1(t)}{\overline{v}_1}. \end{aligned} \tag{4.6.25}$$

Herewith three parameters are defined:

$$T_\text{f} = \frac{d_1\,\rho_1\,c_\text{p1}}{4\,\overline{\alpha}_\text{1w}} \qquad \text{(fluid time constant)} \tag{4.6.26}$$

$$T_\text{w1} = \frac{A_\text{w}\,\rho_\text{w}\,c_\text{w}}{\pi\,d_1\,\overline{\alpha}_\text{1w}} \qquad \text{(inner tube wall time constant)} \tag{4.6.27}$$

$$T_\text{w2} = \frac{A_\text{w}\,\rho_\text{w}\,c_\text{w}}{\pi\,d_2\,\overline{\alpha}_\text{2w}} \qquad \text{(external tube wall time constant)} \tag{4.6.28}$$

(4.6.24) and (4.6.25) are now first Laplace transformed with regard to the time t in the s-domain and then with regard to the location z in the ζ-domain. Laplace back-transformation from the ζ-domain into z-domain with setting of $\Delta\vartheta_1(z=0,s) = \Delta\vartheta_\text{1i}(s)$; $\Delta\vartheta_\text{1i}(z=l,s) = \Delta\vartheta_\text{1o}(s)$ and elimination of the tube wall temperature leads to following three transfer functions, Isermann (1969)

$$G_\vartheta(s) = \frac{\Delta\vartheta_\text{1o}(s)}{\Delta\vartheta_\text{1i}(s)} = e^{-T_t s}\,e^{-\kappa_\text{f}\frac{T_\text{w1}s+\eta}{T_\text{w1}s+\eta+1}} \tag{4.6.29}$$

$$\begin{aligned} G_\text{q}(s) &= \frac{\Delta\vartheta_\text{1o}(s)}{\Delta\dot{q}_\text{2w}(s)} \\ &= \frac{\left(\overline{\vartheta}_\text{w} - \overline{\vartheta}_1\right)d_2}{\bar{\dot{q}}_\text{2w}d_1}\,\frac{1/\eta}{T_\text{w2}\,T_\text{f}\,s^2 + \left(T_\text{f}\,\frac{1+\eta}{\eta} + T_\text{w2}\right)s + 1}\,\left(1 - G_\vartheta(s)\right) \end{aligned} \tag{4.6.30}$$

$$\begin{aligned} G_\text{v} &= \frac{\Delta\vartheta_\text{1o}(s)}{\Delta v_1(s)} \\ &= -\frac{\left(\overline{\vartheta}_\text{w} - \overline{\vartheta}_1\right)}{\overline{v}_1}\,\frac{\frac{1}{\eta} + (1-m)(1+T_\text{w1}s)}{T_\text{w2}\,T_\text{f}\,s^2 + \left(T_\text{f}\frac{1+\eta}{\eta} + T_\text{w2}\right)s + 1}\,\left(1 - G_\vartheta(s)\right). \end{aligned} \tag{4.6.31}$$

Three new parameters are

$$\kappa_\text{f} = \frac{4\,\alpha_\text{1w}\,l}{d_1\,\rho_1\,c_\text{p1}\,v_1} \qquad \text{(fluid parameter)} \tag{4.6.32}$$

$$\eta = \frac{\alpha_\text{2w}\,d_2}{\alpha_\text{1w}\,d_1} \qquad \text{(convective heating parameter)} \tag{4.6.33}$$

$$T_t = l/v \qquad \text{(dead time)} \tag{4.6.34}$$

and additionally from (4.6.27):

$$T_\text{w1} = \frac{\left(d_2^2 - d_1^2\right)\,\rho_\text{w}\,c_\text{w}}{4\,\alpha_\text{1w}\,d_1} \qquad \text{(inner tube wall time constant)} \tag{4.6.35}$$

The parameters κ_f and η define the form of the resulting frequency response loci or the transient functions and T_w1 and T_w2 the frequency or time scale. Figures 4.6.8 and 4.6.9 show the corresponding transient functions and frequency response loci. The temperature behavior is for small κ_1 (e.g. short tube length) approximately a step plus a first order lag and tends to a higher order lag behavior for large κ_1 (large tube length). Both, the heating/temperature and the velocity/temperature behavior show approximately a second order lag behavior, however, with opposite sign, see also Fig. 4.6.10. Table 4.6.2 shows typical ranges of the characteristic parameters for heat exchangers. If the influence of the wall temperature on the specific heat on the external tube is negligible, as it can be be assumed for heating through radiation, it can be set $\alpha_{2\mathrm{w}} = 0$ or $\eta = 0$.

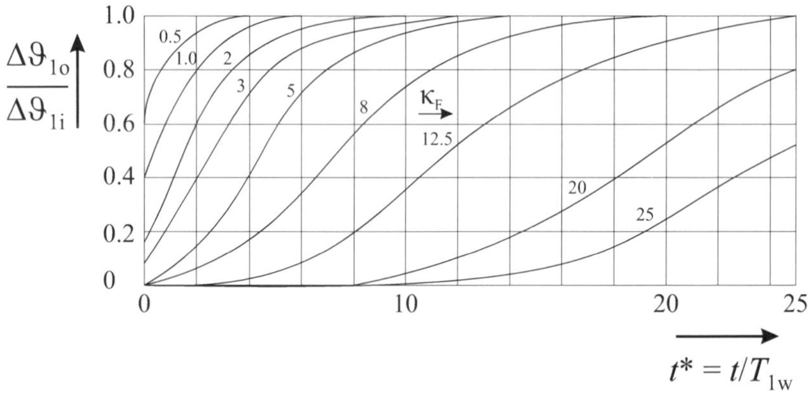

Fig. 4.6.8. Transient functions for $G_\vartheta(s)$ with κ_f as parameter (heating through radiation, $\eta = 0$) and $T_\mathrm{t} = 0$, Profos (1962).

Table 4.6.2. Characteristic parameters of heated tubes, Isermann (1969)

	Liquid	Gases, steam
Primary flow 1 (inner tube)	$0.1 < \kappa_\mathrm{f} < 20$ $0.2 < T_\mathrm{w1}/T_\mathrm{f} < 0.7$	$0.1 < \kappa_\mathrm{f} < 20$ $20 < T_\mathrm{w1}/T_\mathrm{f} < 100$
secondary flow 2 (outer tube)	gas $0 < \eta < 0.2$ (water/ air heater)	gas, steam $0 < \eta < 1$ (steam superheater)
	liquid, condensating steam $0.5 < \eta < 3$ (steam/ water heat exchanger)	liquid $0.5 < \eta < 10$

b) Simplified models of heated tubes

Approximation of the transcendental transfer functions with rational transfer function and time delay allows simpler expressions, as was shown in Acklin and Läubli (1960), Läubli (1961), Isermann (1969). For the change of the input temperature the tube is divided into n small sections of length Δl. Then the following approximation holds

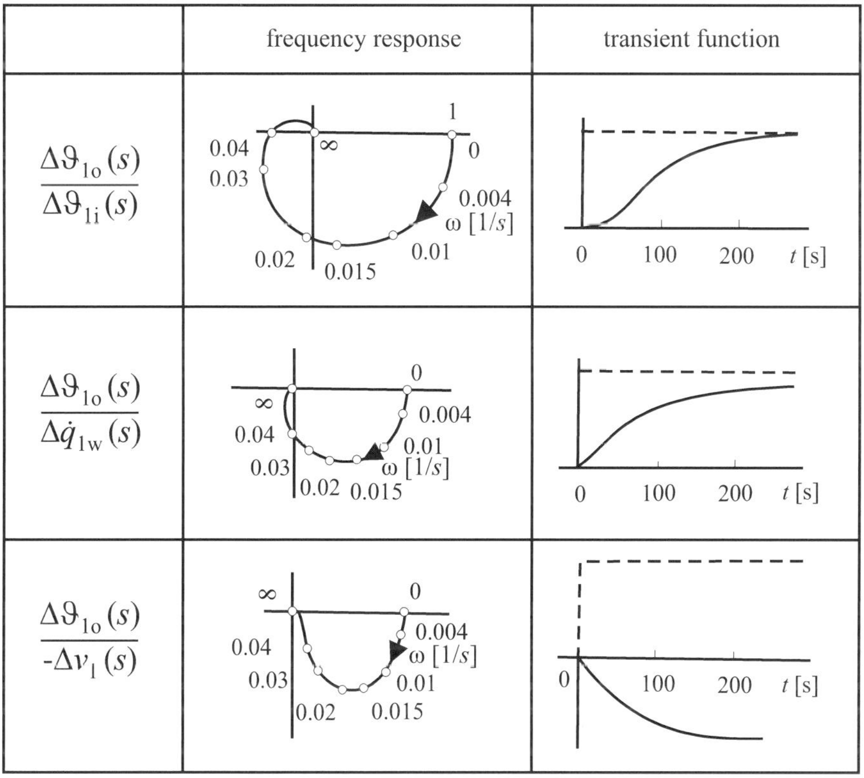

Fig. 4.6.9. Frequency responses and transient responses of a long heated tube, Profos (1962).

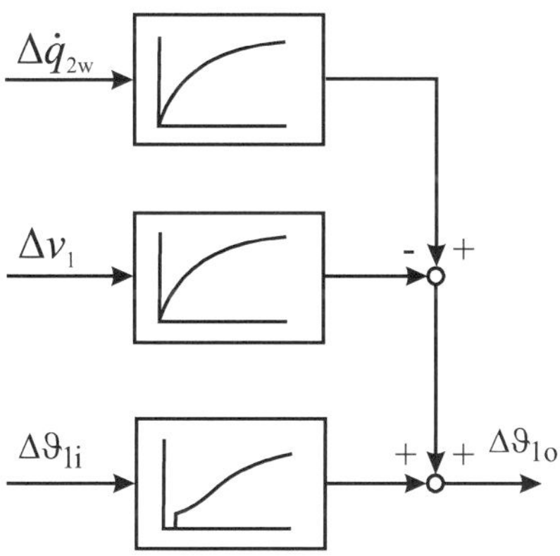

Fig. 4.6.10. Signal-flow diagram of a heated tube for small changes of variables.

$$\widetilde{G}_{\vartheta}(s)=\frac{\Delta\vartheta_{1\mathrm{o}}(s)}{\Delta\vartheta_{1\mathrm{i}}(s)}=\left(a+\frac{b}{1+T_{\mathrm{b}}\,s}\right)^{n}\;e^{-T_{t}s} \tag{4.6.36}$$

$$a=e^{-\Delta\kappa_{\mathrm{f}}}$$

$$b=e^{-\Delta\kappa_{\mathrm{f}}\frac{\eta}{1+\eta}}-a$$

$$T_{\mathrm{b}}=\frac{\Delta\kappa_{\mathrm{f}}}{(1+\eta)^{2}}\frac{a+b}{b}T_{\mathrm{w1}}$$

$$n=\frac{\kappa_{\mathrm{f}}}{\Delta\kappa_{\mathrm{f}}}=1,2,3,\ldots$$

$$\Delta\kappa_{\mathrm{f}}=\frac{4\alpha_{1\mathrm{w}}\Delta l}{d_{1}\rho_{1}c_{\mathrm{p1}}v_{1}}$$

$$\frac{\Delta\kappa_{\mathrm{f}}}{\kappa_{\mathrm{f}}}=\frac{\Delta l}{l}\quad\text{with }\Delta\kappa_{\mathrm{f}}\leq 1.5.$$

This means, that (4.6.29) was approximated for small $\Delta\kappa_{\mathrm{f}}\leq 1.5$ by a first order rational transfer function (4.6.36). For larger κ_{f} these transfer functions are connected in series. The other two transfer functions for total length l result in

$$\widetilde{G}_{\mathrm{q}}(s)=\frac{\Delta\vartheta_{1\mathrm{o}}(s)}{\Delta\dot{q}_{2\mathrm{w}}(s)}=\frac{G_{\mathrm{q}}(0)}{(1+T_{1}s)\,(1+T_{2}s)} \tag{4.6.37}$$

$$T_{1}=\frac{1}{\eta}\,(1-\psi)\,T_{\mathrm{w1}}$$

$$T_{2}=\frac{1}{\eta}\,T_{\mathrm{w1}}$$

$$\widetilde{G}_{\mathrm{v}}(s)=\frac{\Delta\vartheta_{1\mathrm{o}}(s)}{\Delta v(s)}=\frac{G_{\mathrm{v}}(0)}{(1+T_{1}s)}\left\{\frac{\frac{1}{\eta}}{1+T_{2}s}+(1-m)\right\}. \tag{4.6.38}$$

The gains are

$$G_{\vartheta}(0)=\psi=e^{-\kappa_{\mathrm{f}}\,\frac{\eta}{1+\eta}}\qquad(\text{if } c_{\mathrm{p1i}}=c_{\mathrm{p1o}})$$

$$G_{\mathrm{q}}(0)=\frac{\overline{\vartheta}_{\mathrm{w}}-\overline{\vartheta}_{1}}{\bar{\dot{q}}_{2\mathrm{w}}}\,\frac{d_{2}}{d_{1}}\,\frac{1}{\eta}(1-\psi)$$

$$G_{\mathrm{v}}(0)=-\frac{\overline{\vartheta}_{\mathrm{w}}-\overline{\vartheta}_{1}}{\overline{v}_{1}}\left(\frac{1}{\eta}+1-m\right)(1-\psi).$$

These approximations hold for liquid flows in the inner tube with $(0.2<T_{\mathrm{w1}}/T_{\mathrm{f}}<0.7)$ and liquids and condensing steam for the secondary flow $(0.5<\eta<3)$. For other parameter combinations see Isermann (1969).

The output temperature is usually measured by a *temperature sensor*, e.g. a resistance thermometer or a thermocouple with a transfer function described in Sect. 4.6.5.

c) Simplified models of heat exchangers

Based on the models of a heated tube and other elements the overall model of a heat exchanger can now be given. The signal-flow diagram of a *tubular heat exchanger* is depicted in Fig. 4.6.11 for the case that the inlet temperature is a manipulated variable. In addition to the heated tube dynamics following elements have to be considered: the head of the heat exchanger as a mixing storage, a transport deadtime $T_t = l/v$ and the temperature sensor dynamics. This means that a higher order delay system results. However, usually the heat exchanger and the sensor time constants are dominating, compare e.g. Isermann (1970), Isermann and Jantschke (1970).

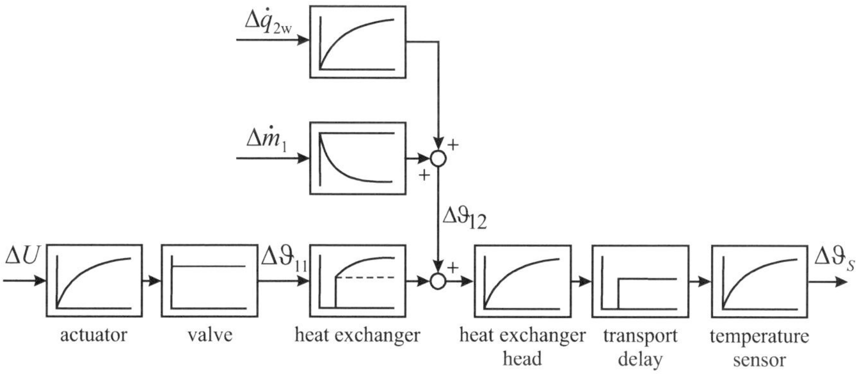

Fig. 4.6.11. Signal-flow diagram for a tubular heat exchanger with manipulation of the inlet temperature by a mixing valve.

Cross flow heat exchangers are more difficult to model. For the dynamic behavior of the inner fluid the derived equations of the heated tube can be applied. To describe the dynamics of the outlet temperature ϑ_{22} of the second fluid the integral over the wall temperature can be modeled to obtain an average temperature over the outlet heat exchanger area. It was shown in Isermann and Jantschke (1970) that a simplified model is

$$
\begin{aligned}
G_{\vartheta\vartheta}(s) &= \frac{\Delta\vartheta_{22}(s)}{\Delta\vartheta_{11}(s)} = \frac{K_{22}}{1+T_{22}s} \\
K_{22} &= \frac{\dot{m}_1 c_{\mathrm{p1}}}{\dot{m}_2 c_{\mathrm{p2}}}(1-\psi) \\
T_{22} &= \frac{(1-\psi)}{\eta}(T_{\mathrm{f}}+T_{\mathrm{w1}})
\end{aligned}
\tag{4.6.39}
$$

with ψ as for (4.6.38).

4.6.5 Dynamic models of temperature sensors

Temperatures are usually measured with resistance thermometers or thermocouples which can be considered as heat exchanging elements. Assuming that the complete sensor component consisting of the sensor element and its protective tube can be considered as a lumped heat storage the following heat balance equation holds

$$m_{\mathrm{s}} c_{\mathrm{s}} \frac{d\vartheta_{\mathrm{s}}(t)}{dt} = A_{\mathrm{s}} \alpha_{\mathrm{FS}} (\vartheta_{\mathrm{f}}(t) - \vartheta_{\mathrm{s}}(t)) \tag{4.6.40}$$

$$\frac{m_{\mathrm{s}} c_{\mathrm{s}}}{A_{\mathrm{s}} \alpha_{\mathrm{FS}}} \frac{d\vartheta_{\mathrm{s}}(t)}{dt} + \vartheta_{\mathrm{s}}(t) = \vartheta_{\mathrm{f}}(t). \tag{4.6.41}$$

ϑ_{s} and ϑ_{f} are the temperatures of the sensor and the fluid. For the sensor mass holds

$$m_{\mathrm{s}} = \pi d^2 l \rho_{\mathrm{s}} / 4$$

and the heat transfer surface is

$$A_{\mathrm{s}} = \pi d l.$$

The sensor time constant then becomes

$$T_{\mathrm{s}} = \frac{m_{\mathrm{s}} c_{\mathrm{s}}}{A_{\mathrm{s}} \alpha_{\mathrm{fs}}} = \frac{\rho_{\mathrm{s}} d_{\mathrm{s}} c_{\mathrm{s}}}{4 \alpha_{\mathrm{fs}}}. \tag{4.6.42}$$

Hence, the time constant of a temperature sensor is proportional to its diameter d and inverse proportional to the heat transfer coefficient α_{fs}. The heat transfer coefficient α_{fs} depends on the fluid velocity v. Applying Nusselt's law (4.6.22) with a known $\overline{\alpha}_{\mathrm{fs}}(\overline{v})$ for a certain fluid velocity $\overline{v}$ yields

$$\frac{\alpha_{\mathrm{fs}}}{\overline{\alpha}_{\mathrm{fs}}} = \left(\frac{v}{\overline{v}}\right)^m \tag{4.6.43}$$

with $m = 0.6 \ldots 0.8$ for turbulent flow. Thus the sensor time constant becomes

$$T_{\mathrm{s}} = \frac{\rho_{\mathrm{s}} d_{\mathrm{s}} c_{\mathrm{s}}}{4\, \overline{\alpha}_{\mathrm{fs}}} \frac{\overline{v}^m}{v^m} \tag{4.6.44}$$

and changes inverse proportional to the fluid velocity v^m. Therefore, the fluid velocity v has a dominating effect on the dynamics of the temperature sensor.

For example, the time constants of relatively thin thermocouples within the exhaust pipe of a CR diesel engine were determined in the range of 0.5… 1 s and 1.6… 3.3 s for $n_{\mathrm{eng}} = 1000 \ldots 3000\,\mathrm{rpm}$, Zimmerschied and Isermann (2010). However, series exhaust temperature sensors with larger robust design have shown time constants in the range of 4… 16 s for high and low exhaust gas mass flows.

If the temperature after a heat exchanger is measured and used e.g. for control, then the sensor dynamics have to be included in the overall dynamics, because the sensor time constant is frequently in the same magnitude or even larger than the heat exchanger time constants.

4.7 Cooling system

In order to protect the combustion gas contacted materials from overheating and to retain the lubrication effect of the engine oil the engine has to be cooled. For passenger cars and heavy-duty vehicles water cooling has become standard. The coolant consists of water, antifreeze and anticorrosion inhibitors. As about one third of the heat provided by the fuel has to be removed to the environment a precise functioning control and supervision of the cooling system are important requirements. Therefore, models for the cooling system will be considered. The following sections describe different models of the cooling system. The cooling system is considered as a lumped parameter process to obtain simplified dynamic models as in Sect. 4.6.2 for heat exchangers. The models may be required for simulations and for control design of the coolant system and as a basis for the thermomanagement.

Figure 4.7.1 depicts a scheme of a typical, basic cooling system. The coolant pump is either driven by a V-belt and therefore the speed is given through the engine speed or by an electric motor, which allows independent speed control. After cold start the thermostatic valve opens to the bypass only in order to reach the operation coolant temperature fast. The thermostatic controller then opens the way to the radiator such that forced external cooling begins. The valve is a double-acting disk, in which both sides are opened allowing a mixture of uncooled and cooled coolant flow to the engine such that according to the setpoint of, e.g. a thermostat a certain operating temperature at the entrance to the engine is reached. The setpoint of the thermostat may be changed through a heating resistor to raise the operating temperature at part-load and to reduce it at higher load. This allows to safe fuel, to reduce wear and improves the heating of the vehicle interior. In the case of electrical driven pumps the speed is also changed and optimized with regard to operating requirements.

As the temperature time behavior of the coolant system is much slower than, e.g. the torque generation, the coolant circuit is frequently in a dynamic state during driving with different load. Therefore the dynamics are of interest for normal driving cycles. First the cooling flow is considered, and then the coolant temperature.

4.7.1 Cooling flow models

The cooling circuit flow as depicted in Fig. 4.7.1 is generated by the coolant pump which is traditionally driven by the belt for the auxiliary units and therefore its speed depends on the engine speed. However, if it is driven by an electrical motor the speed can be controlled independently from the engine speed and optimized with regard to the required heat transport to the environment. The volume flow in the cooling circuit in steady-state is determined by the pressure flow characteristics of the pump and the flow resistances of the coolant valve and the coolant circuit.

The coolant pumps are usually either radial or axial centrifugal pumps. The generated difference pressure $\Delta p_{\mathrm{p}} = p_{\mathrm{p,out}} - p_{\mathrm{p,in}}$ or the delivery head $H = \Delta p_{\mathrm{p}}/\rho_{\mathrm{c}} g$ in dependence on the volume flow $\dot{V}_{\mathrm{p}}$ and speed ω_{p} can be approximated by

$$\Delta p_{\mathrm{p}} = \rho_{\mathrm{c}}\, g\, H = \rho_{\mathrm{c}}\, g \left(h_{\mathrm{nn}} \omega_{\mathrm{p}}^2 - h_{\mathrm{nv}} \omega_{\mathrm{p}} \dot{V}_{\mathrm{p}} - h_{\mathrm{vv}} \dot{V}_{\mathrm{p}}^2 \right) \tag{4.7.1}$$

see Dixon (1966), Pfleiderer and Petermann (2005), Isermann and Münchhof (2011).

The pressure/flow characteristic of the cooling circuit depends on the flow resistance of the pipes, engine block, radiator, oil cooler, air heater and position u_1 of

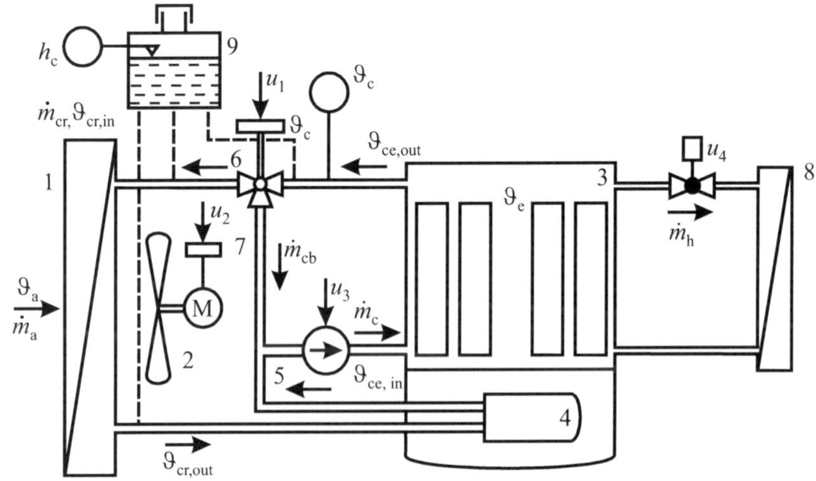

Fig. 4.7.1. Schematic of an engine cooling system. 1 coolant radiator, 2 fan, 3 combustion engine, 4 oil cooler, 5 coolant pump, 6 thermostatic valve, 7 bypass, 8 cabin heat exchanger, 9 coolant expansion tank. u_1 manipulated variable for thermostatic valve, u_2 manipulated variable for fan speed, u_3 manipulated variable for pump speed, u_4 manipulated variable for cabin heating, ϑ_c coolant temperature, h_c coolant level.

the coolant valve. These components are connected together in a serial and parallel configuration. Modeling and simulation of the coolant flow is rather complicated such that one- and three-dimensional fluid dynamic calculations are required to describe the various effects. The realization of measurements along the many passages with a mixture of laminar flow, turbulent flow and dead water circulation is hardly possible. Fluid dynamic simulations are shown in Zieher et al (2011). However, to develop simplified models of the cooling for the purpose of control it is assumed that pressure drops may be described by turbulent flow and therefore approximated by

$$\Delta p_c = \zeta_c(u_1)\rho_c \frac{v_c^2}{2} = \zeta_c(u_1)\frac{\rho_c}{2A_c^2}\dot{V}_p^2 = h_{rr}(u_1)\dot{V}_p^2 . \tag{4.7.2}$$

The resistance coefficients ζ_c depends on various positions of the thermostatic valve and valves, e.g. for cabin heating, oil temperature control, etc. Hence, the resistance characteristic may change considerably. Together with the changing speed of the coolant pump many different operating points result as intersection points of both characteristics, as shown in Fig. 4.7.2.

A comparison of the measured and with (4.7.1) approximated pump characteristic is depicted in Fig. 4.7.3a) for different pump speeds. Figure 4.7.3b) shows the corresponding flow characteristic of the cooling circuit.

To describe the dynamic behavior of the coolant mass flow it is assumed that the coolant circuit is simplified to a representative pipe with cross-section A_c and length l_c, resulting in a representative coolant mass $m_c = A_c\rho_c l_c$. The momentum balance equation then becomes

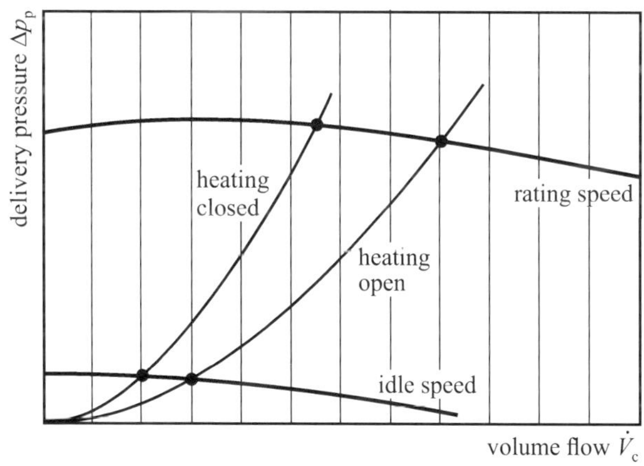

Fig. 4.7.2. Pressure/flow characteristics of a coolant pump and coolant circuit, van Basshuysen and Schäfer (2004).

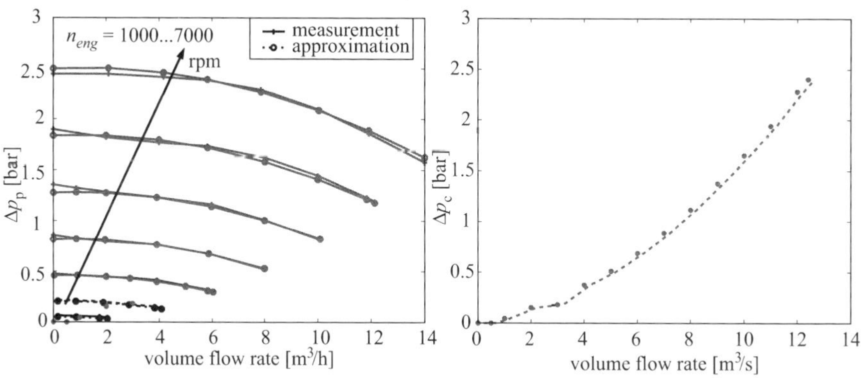

Fig. 4.7.3. **a** Measured and approximated pump characteristic, **b** Measured cooling circuit characteristic, Schmidt (2003).

$$m_c \frac{dv_c(t)}{dt} = A_c \left(\Delta p_p(t) - \Delta p_c(t) \right) \tag{4.7.3}$$

or

$$T_I \frac{d\dot{m}_c(t)}{dt} = \Delta p_p(t) - \Delta p_c(t)$$

with

$$v_c = \frac{\dot{m}_c}{A_c \rho_c}$$
$$T_I = \frac{m_c}{A_c^2 \rho_c}.$$

A corresponding signal flow is depicted in Fig. 4.7.4. It shows the integration forward path stemming from the mass flow balance equation and a negative feedback path resulting from the resistance characteristic $\Delta p_c(\dot{m}_c, u_1)$ (heating valve closed).

The steady-state pump characteristic $\Delta p_{\mathrm{p}}\,(\omega_{\mathrm{p}}, \dot{m}_{\mathrm{c}})$ acts in a forward path and determines the pump pressure in dependence on the pump speed ω_{p} and the mass flow $\dot{m}_{\mathrm{c}}$. Hence, a nonlinear proportional acting behavior results after changes of the pump speed or valve position, where the parameters depend on the operating point, Isermann (2005).

For small changes of the pump speed the dynamic behavior can be linearized around the operation point and yields after insertion of (4.7.1) and (4.7.2) in (4.7.3) with volume flow $\dot{V}_{\mathrm{c}} = \dot{m}_{\mathrm{c}}/\rho_{\mathrm{c}}$.

$$T_{\mathrm{c}}\frac{d\dot{V}_{\mathrm{p}}(t)}{dt} + \Delta\dot{V}_{\mathrm{p}}(t) = K_{\mathrm{c}}\Delta\omega_{\mathrm{p}}(t) \tag{4.7.4}$$

with gain and time constant

$$K_{\mathrm{c}} = \frac{\Delta\dot{V}_{\mathrm{p}}}{\Delta\omega_{\mathrm{p}}} = \frac{\rho_{\mathrm{c}} g(2h_{\mathrm{nn}} - h_{\mathrm{nv}}\bar{\dot{V}}_{\mathrm{p}})}{2h_{\mathrm{rr}}\bar{\dot{V}}_{\mathrm{p}}} \tag{4.7.5}$$

$$T_{\mathrm{c}} = \frac{l_{\mathrm{c}}\rho_{\mathrm{c}}}{2A_{\mathrm{c}}h_{\mathrm{rr}}\bar{\dot{V}}_{\mathrm{p}}} = \frac{1}{2A_{\mathrm{c}}^2 h_{\mathrm{rr}}}\frac{m_{\mathrm{c}}}{\bar{\dot{V}}_{\mathrm{p}}}. \tag{4.7.6}$$

Hence, the dynamic behavior of the coolant volume flow can be described by a first order differential equation. The time constant is proportional to the moved coolant mass and inverse proportional to the volume flow. If an electrical driven coolant pump is used, the dynamics of the pump lead to a further first order differential equation, see Isermann and Münchhof (2011), Sect. 6.1.2. The dynamic model of the coolant flow $\dot{m}_{\mathrm{c}}(t)$ describes the behavior mainly for changes of the pump speed. In the case of the coolant temperature control with thermostatic valve only, the coolant mass flow $\dot{m}_{\mathrm{c}}$ then appears as a dynamically changing parameter in (4.7.18) and in the signal-flow chart, Fig. 4.7.6. For belt-driven pumps this parameter changes with the engine speed and for electrically-driven pumps with the controlled pump speed.

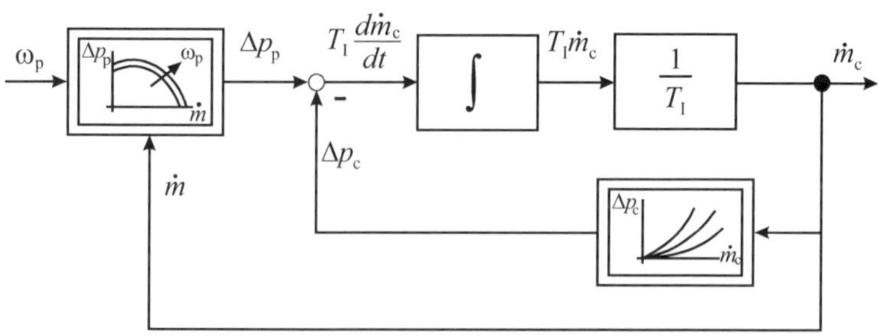

Fig. 4.7.4. Signal-flow scheme for the dynamic behavior of the coolant flow and the pump speed $U = \omega_{\mathrm{p}}$ as input.

4.7.2 Cooling temperature models

a) First order model with lumped parameters

Because of the different heat exchangers and the varying coolant flows modeling of the coolant system is quite involved. However, in order to get a first rough estimate the whole coolant circuit is considered as a lumped parameter system, where ϑ_e is a mean temperature of the engine. It is assumed that the interior heating is closed. An energy balance equation then yields

$$m_e c_e \frac{d\vartheta_e(t)}{dt} = \dot{H}_{\mathrm{e,in}}(t) - \dot{Q}_{\mathrm{ea}}(t) - \dot{Q}_{\mathrm{r}}(t) - P_{\mathrm{eng}}(t) - \dot{H}_{\mathrm{e,out}}(t) \tag{4.7.7}$$

with the following quantities

m_{e}	representative mass of all coolant relevant engine components (engine block, coolant, oil, ...)
c_{e}	representative specific heat capacity (average value)
$\dot{H}_{\mathrm{e,in}}$	enthalpy flow from the combustion
$\dot{H}_{\mathrm{e,out}}$	enthalpy flow of the exhaust gases, leaving the cylinders
$\dot{Q}_{\mathrm{ea}}$	heat flow from the engine block to the environment
$\dot{Q}_{\mathrm{r}}$	heat flow of the radiator from the coolant to the environment
P_{eng}	delivered mechanical power at the flywheel

The heat flow for an ideal combustion is

$$\dot{H}_{\mathrm{e,in}} = \dot{m}_{\mathrm{f}} H_{\mathrm{l}} \tag{4.7.8}$$

where $\dot{m}_{\mathrm{f}}$ is the fuel mass flow and H_{l} the lower heat value of the fuel. For the mechanical power at the flywheel holds

$$P_{\mathrm{eng}} = M_{\mathrm{eng}}\omega_{\mathrm{eng}} = H_{\mathrm{e,in}}\eta_{\mathrm{eff}} \tag{4.7.9}$$

with the effective efficiency of the engine

$$\eta_{\mathrm{eff}} = \frac{P_{\mathrm{eng}}}{\dot{H}_{\mathrm{e,in}}} = \frac{M_{\mathrm{eng}}\,\omega_{\mathrm{eng}}}{\dot{m}_{\mathrm{f}}\,H_{\mathrm{l}}} = \frac{4\pi M_{\mathrm{eng}}}{m_{\mathrm{f}} H_{\mathrm{l}}} \tag{4.7.10}$$

see (4.2.20), which depends on the operating state

$$\eta_{\mathrm{eff}} = f(u_{\mathrm{inj}}, n_{\mathrm{eng}}).$$

This effective efficiency comprises all efficiencies from the injected fuel to the work at the flywheel, including the cycle efficiency, which takes the removed heat during gas exchange into account, see e.g. (4.2.1). m_{f} is the completely burned fuel per combustion cycle. The enthalpy flow of the exhaust gases is

$$\dot{H}_{\mathrm{e,out}} = c_{\mathrm{eg}}\dot{m}_{\mathrm{eg}} T_3. \tag{4.7.11}$$

The heat flow from the engine block to the environment can be approximated by

$$\dot{Q}_{\rm ea} = \alpha_{\rm ea} A_{\rm ea} (\vartheta_{\rm e} - \vartheta_{\rm a}) \tag{4.7.12}$$

with $\alpha_{\rm ea}$ a heat transfer coefficient and $A_{\rm ea}$ the engine's outer surface. For the heat flow of the radiator to the air holds approximately

$$\dot{Q}_{\rm r} = k_{\rm r} A_{\rm r} (\vartheta_{\rm c} - \vartheta_{\rm a}) \tag{4.7.13}$$

where $k_{\rm r}$ is the overall heat transfer coefficient, $A_{\rm r}$ the surface of the radiator and $\vartheta_{\rm c}$ the measured coolant temperature at the thermostatic valve. The basic equations and parameters for the heat transfer are explained in Sect. 4.6.2, (4.6.1) to (4.6.5). Inserting these equations into (4.7.7) yields

$$\begin{aligned} m_{\rm e} c_{\rm e} \frac{d\vartheta_{\rm e}(t)}{dt} &= (1 - \eta_{\rm eff}) \dot{m}_{\rm f} H_{\rm l} - \alpha_{\rm ea} A_{\rm ea} (\vartheta_{\rm e}(t) - \vartheta_{\rm a}(t)) - k_{\rm r} A_{\rm r} (\vartheta_{\rm c}(t) - \vartheta_{\rm a}(t)) \\ &\quad - c_{\rm eg} \dot{m}_{\rm eg} T_3 . \end{aligned} \tag{4.7.14}$$

Two simplifying assumptions can now be made. Assuming that the heat transfer flow is $\dot{Q}_{\rm r} = 0$ because of a closed control valve during warming up, leads to a *warming up time constant* of the engine temperature

$$T_{\rm e1} = \frac{m_{\rm e} c_{\rm e}}{\alpha_{\rm ea} A_{\rm ea}} \tag{4.7.15}$$

see also Guzzella and Onder (2010). This time constant can be used to obtain a rough estimate of the engine temperature during the warming up phase. If it is assumed that $\vartheta_{\rm e} = \vartheta_{\rm c}$, (4.7.14) yields an overall time constant for the engine plus radiator temperature

$$T_{\rm e2} = \frac{m_{\rm e}^* c_{\rm e}^*}{\alpha_{\rm ea} A_{\rm ea} + k_{\rm r} A_{\rm r}} . \tag{4.7.16}$$

Then m_e^* and c_e^* include the radiator. k_r depends strongly on the air speed through the radiator, i.e. vehicle speed and/or fan speed.

b) Second order model with lumped parameters

To obtain a more detailed model of the coolant temperature dynamics the engine block and the radiator are now considered as two lumped parameter components and the coolant valve as a three-way valve, see Fig. 4.7.1.

Figure 4.7.5 depicts the corresponding energy flows of both components including the effect of changing the position u_1 of the coolant valve. The corresponding energy flows are discussed in the sequel.

Coolant valve bypass and radiator

At first the effect of u_1 on the temperatures $\vartheta_{\rm ce,in}$ and $\vartheta_{\rm cr,out}$ is considered. It is assumed that the mass flow rate through the radiator is a function of the valve position

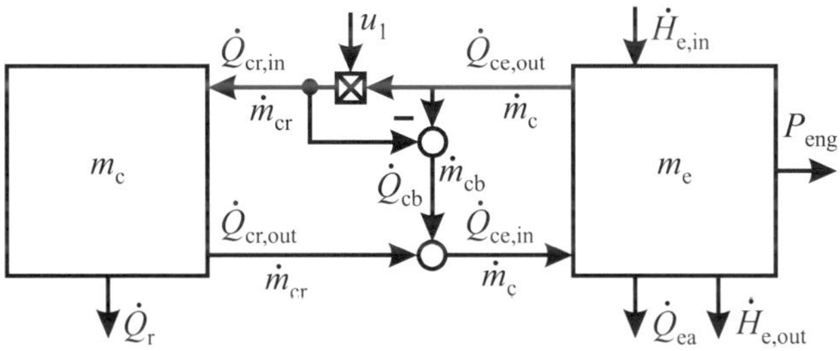

Fig. 4.7.5. Energy flows of the coolant radiator and the engine block with the manipulation variable u_1 of the coolant valve.

$$\dot{m}_{cr}(u_1) = \dot{m}_c f(u_1) \tag{4.7.17}$$

where $f(u_1)$ describes the flow characteristic of the valve and $\dot{m}_c$ is the mass flow rate of the coolant through the engine. To simplify notations a linear characteristic is assumed

$$\dot{m}_{cr}(u_1) = \dot{m}_c u_1 \tag{4.7.18}$$

with $u_1 = 1$ for an opened and $u_1 = 0$ for a closed valve to the radiator. A mass flow balance for the valve leads to

$$\dot{m}_c = \dot{m}_{cr} + \dot{m}_{cb} \tag{4.7.19}$$

with $\dot{m}_{cr} = \dot{m}_c u_1$ and $\dot{m}_{cb} = (1 - u_1)\,\dot{m}_c$ the mass flow rates through the radiator and the bypass. It is first assumed that the coolant mass flow rate $\dot{m}_c$ is constant.

The heat balance at the mixing point of the bypass flow and the radiator flow yields, neglecting the oil cooler,

$$\dot{Q}_{ce,in} = \dot{Q}_{cr,out} + \dot{Q}_{cb} = \dot{Q}_{cr,out} + (\dot{Q}_{ce,out} - \dot{Q}_{cr,in}) \tag{4.7.20}$$

or by introducing (4.7.18)

$$\dot{m}_c c_c \vartheta_{ce,in} = \dot{m}_c c_c \vartheta_{cr,out} u_1 + \dot{m}_c c_c \vartheta_c (1 - u_1) \tag{4.7.21}$$

$$\vartheta_{ce,in} = u_1 \vartheta_{cr,out} + (1 - u_1)\vartheta_c = \vartheta_c - u_1(\vartheta_c - \vartheta_{cr,out}) \tag{4.7.22}$$

where $\vartheta_{ce,in}$ is the coolant entrance temperature of the engine block, $\vartheta_{cr,out}$ the coolant temperature of the radiator output, $\vartheta_c = \vartheta_{ce,out}$ is the temperature at the coolant valve and c_c is the specific heat value of the coolant. Hence, the position u_1 of the valve changes $\vartheta_{ce,in}$ in a multiplicative, i.e. nonlinear way. It holds $\vartheta_{ce,in} = \vartheta_c$ for $u_1 = 0$ and $\vartheta_{ce,in} = \vartheta_{cr,out}$ for $u_1 = 1$.

It is now assumed that the dynamics of the coolant temperature in the radiator is small compared to the coolant temperature in the engine block in order to simplify the relations.

Thus, the steady-state heat balance of the radiator becomes

$$\dot{Q}_r = \dot{m}_{cr} c_c (\vartheta_{cr,in} - \vartheta_{cr,out}) = k_r A_r (\overline{\vartheta}_{cr} - \vartheta_a). \tag{4.7.23}$$

With $\vartheta_{cr,in} = \vartheta_{\mathrm{c}}$ and assuming a mean temperature in the radiator it holds

$$\overline{\vartheta}_{cr} = \frac{1}{2}(\vartheta_{cr,in} + \vartheta_{cr,out}) = \frac{1}{2}(\vartheta_{\mathrm{c}} + \vartheta_{cr,out}). \tag{4.7.24}$$

The outlet temperature of the radiator then follows from (4.7.23)

$$\vartheta_{\mathrm{cr,out}} = \frac{2\frac{\dot{m}_{\mathrm{cr}} c_{\mathrm{c}}}{k_{\mathrm{r}} A_{\mathrm{r}}} - 1}{2\frac{\dot{m}_{\mathrm{cr}} c_{\mathrm{c}}}{k_r A_r} + 1} \vartheta_{\mathrm{c}} + \frac{2}{\frac{2\dot{m}_{\mathrm{cr}} c_{\mathrm{c}}}{k_{\mathrm{r}} A_{\mathrm{r}}} + 1} \vartheta_{\mathrm{a}}. \tag{4.7.25}$$

Introducing

$$b_{\mathrm{r}} = \frac{\dot{m}_{\mathrm{c}}\, c_{\mathrm{c}}}{k_{\mathrm{r}} A_{\mathrm{r}}} \tag{4.7.26}$$

and (4.7.18) yields for $0 < u_1 \leq 1$

$$\vartheta_{\mathrm{cr,out}} = \frac{2b_{\mathrm{r}} u_1 - 1}{2b_{\mathrm{r}} u_1 + 1} \vartheta_{\mathrm{c}} + \frac{2}{2b_{\mathrm{r}} u_1 + 1} \vartheta_{\mathrm{a}}. \tag{4.7.27}$$

Dynamic behavior of the engine block

The engine block temperature ϑ_{e} is now considered as a lumped representative mean temperature as for a heat exchanger in Sect. 4.6.3. According to Fig. 4.7.5, the heat balance of the engine block is described by

$$m_{\mathrm{e}} c_{\mathrm{e}} \frac{d\vartheta_{\mathrm{e}}(t)}{dt} = \dot{H}_{\mathrm{e,in}}(t) - P_{\mathrm{eng}}(t) + \dot{Q}_{\mathrm{ce,in}}(t) - \dot{Q}_{\mathrm{ce,out}}(t) - \dot{Q}_{\mathrm{ea}}(t) - \dot{H}_{\mathrm{e,out}}(t) \tag{4.7.28}$$

where m_{e} and c_{e} are the mass and specific heat capacity of the engine block. The heat flows in and out of the engine block are

$$\begin{aligned} \dot{Q}_{\mathrm{ce,in}} &= \dot{m}_{\mathrm{c}} c_{\mathrm{c}} \vartheta_{\mathrm{ce,in}} \\ \dot{Q}_{\mathrm{ce,out}} &= \dot{m}_{\mathrm{c}} c_{\mathrm{c}} \vartheta_{\mathrm{ce,out}} = \dot{m}_{\mathrm{c}} c_{\mathrm{c}} \vartheta_{\mathrm{c}}. \end{aligned} \tag{4.7.29}$$

This leads with (4.7.9) and (4.6.6) to

$$m_{\mathrm{e}} c_{\mathrm{e}} \frac{d\vartheta_{\mathrm{e}}}{dt} = (1 - \eta_{\mathrm{eff}}) \dot{H}_{\mathrm{e,in}} + \dot{m}_{\mathrm{c}} c_{\mathrm{c}} (\vartheta_{\mathrm{ce,in}} - \vartheta_{\mathrm{c}}) - \alpha_{\mathrm{ea}} A_{\mathrm{ea}} (\vartheta_{\mathrm{e}} - \vartheta_{\mathrm{a}}) - \dot{H}_{\mathrm{e,out}}(t). \tag{4.7.30}$$

To obtain a relation between the coolant exit temperature of the engine block $\vartheta_{\mathrm{ce,out}} = \vartheta_{\mathrm{c}}$ and ϑ_{e} the transferred heat flow from the engine block to the coolant is simplified by assuming a stationary relation (neglecting wall dynamics because the wall masses are included in m_{e})

$$\dot{Q}_{\mathrm{ec}} = k_{\mathrm{ec}} A_{\mathrm{ec}} \left(\vartheta_{\mathrm{e}} - \frac{1}{2}(\vartheta_{\mathrm{c}} + \vartheta_{\mathrm{ce,in}}) \right) = \dot{m}_{\mathrm{c}} c_{\mathrm{c}} (\vartheta_{\mathrm{c}} - \vartheta_{\mathrm{ce,in}}) \tag{4.7.31}$$

which results in

$$\vartheta_c = b_{e1}\vartheta_e - b_{e2}\vartheta_{ce,in} \tag{4.7.32}$$

with

$$\begin{aligned} b_{e1} &= \frac{2}{1+2b_e} \\ b_{e2} &= \frac{1-2b_e}{1+2be} \\ b_e &= \frac{\dot{m}_c c_c}{k_{ec}A_{ec}}. \end{aligned} \tag{4.7.33}$$

Introducing in (4.7.30) yields for the engine block temperature

$$\begin{aligned} &m_e c_e \frac{d\vartheta_e(t)}{dt} + (\alpha_{ea}A_{ea} + \dot{m}_c c_c b_{e1})\vartheta_e(t) \\ &= (1-\eta_{eff})\dot{H}_{e,in}(t) + \dot{m}_c c_c (1+b_{e2})\vartheta_{ce,in}(t) + \alpha_{ea}A_{ea}\vartheta_a(t) - \dot{H}_{e,out}(t). \end{aligned} \tag{4.7.34}$$

Hence, a *thermal engine block time constant* becomes

$$T_{ec} = \frac{m_e c_e}{\alpha_{ea}A_{ea} + \dot{m}_c c_c b_{e1}}. \tag{4.7.35}$$

The engine inlet temperature $\vartheta_{ce,in}$ follows from (4.7.22) with $\vartheta_{cr,out}$ according to (4.7.27), both depending on the coolant valve position u_1.

Dynamic behavior of the radiator

The radiator and the coolant in the radiator are considered as a lumped parameter process with representative values for the mass m_c and specific heat capacity c_c of the tubes and the fluid. The heat flow balance then yields, compare Fig. 4.7.5 and Sect. 4.6.3

$$m_c c_c \frac{d\vartheta_{cr,out}(t)}{dt} = \dot{Q}_{cr,in}(t) - \dot{Q}_{cr,out}(t) - \dot{Q}_r(t). \tag{4.7.36}$$

Applying for the heat flows

$$\begin{aligned} \dot{Q}_{cr,in} &= \dot{m}_{cr} c_c \vartheta_c = \dot{m}_c u_1 c_c \vartheta_c \\ \dot{Q}_{cr,out} &= \dot{m}_{cr} c_c \vartheta_{cr,out} = \dot{m}_c u_1 c_c \vartheta_{cr,out} \\ \dot{Q}_r &= k_r A_r \left(\frac{1}{2}(\vartheta_c + \vartheta_{cr,out}) - \vartheta_a\right) \end{aligned} \tag{4.7.37}$$

and (4.7.24) for $\overline{\vartheta}_{cr}$ with the abbreviation (4.7.26) leads to

$$\frac{2m_c c_c}{k_r A_r}\frac{d\vartheta_{cr,out}(t)}{dt} + (2b_r u_1 + 1)\,\vartheta_{cr,out}(t) = (2b_r u_1 - 1)\vartheta_c(t) + 2\vartheta_a(t) \tag{4.7.38}$$

and a *coolant radiator time constant*

$$T_{cr} = \frac{2m_c c_c}{(2b_r u_1 + 1)\,k_r A_r}. \tag{4.7.39}$$

The signal flow in Fig. 4.7.6 shows how the two heat storages, the engine block, and the radiator are coupled to a dynamic second order system. The manipulated variable u_1 of the expansion valve has a multiplicative and therefore a nonlinear effect on several variables and parameters because it changes the coolant flow through the radiator and the bypass. In order to result in equations that are not too complicated, several simplifying assumptions were made. The signal-flow diagram also includes the case when the coolant flow $\dot{m}_c$ is used as a manipulated variable.

The described model is only an approximation of the real behavior, because several effects have been simplified, e.g. the deadtimes for temperatures in the tubes and the real dynamics of the radiator, according to the heat exchanger models in the previous sections. Also other cooling heat exchangers have to be taken into account, as e.g. the oil coolers of the engine and of the automatic transmission, the EGR cooler and the interior heater.

The equations show that the static and dynamic behavior depends also on the coolant pump speed and therefore coolant flow $\dot{m}_c$ directly and indirectly via the heat transfer coefficients k_{ec} and k_r. The time constants T_{ec} and T_{cr} are approximately inverse proportional to $\dot{m}_c$ and therefore the smaller the larger the coolant flow rate.

Another detailed thermal model is shown by Guzzella and Onder (2010) and Cortona et al (2002) where the heat transfer in the engine is more analyzed. Other contributions are Arici et al (1999) and Shibata et al (1993).

Because of the many parameters not precisely known for the representative masses, surface areas and heat transfer coefficients, it is recommended to determine the *cooling-system behavior* in addition experimentally with suitable identification methods. The theoretically, with many simplifying assumptions derived models give hints for the structure of the nonlinear identification models.

A simplification of (4.7.34) can be obtained by neglecting the dynamic behavior of the radiator and by assuming small changes Δu_1 around a steady-state $\overline{\vartheta}_e = \text{const.}$ This yields after linearization with

$$\Delta\vartheta_{ce,in} = -(\vartheta_c - \vartheta_{cr,out})\Delta u_1 = -\Delta\overline{\vartheta}_{cr}\Delta u_1 \tag{4.7.40}$$

assuming $\Delta\overline{\vartheta}_{cr} = \text{const.}$ and $\dot{H}_{e,out} = \text{const.}$, to

$$m_e c_e \frac{d\vartheta_e}{dt} + (\alpha_{ea}A_{ea} + \dot{m}_c c_c b_{e1})\,\Delta\vartheta_e(t) = -\dot{m}_c c_c (1 + b_{e2})\Delta\overline{\vartheta}_{cr}\Delta u_1(t) \tag{4.7.41}$$

and with (4.7.32)

$$\Delta\vartheta_c \approx b_{e1}\Delta\vartheta_e \tag{4.7.42}$$

$$m_e c_e \frac{d\vartheta_c}{dt} + (\alpha_{ea}A_{ea} + \dot{m}_c c_c b_{e1})\,\Delta\vartheta_c(t) = -\frac{\dot{m}_c c_c}{b_{e1}}(1 + b_{e2})\Delta\overline{\vartheta}_{cr}\Delta u_1(t) \tag{4.7.43}$$

or

$$T_{ec}\frac{d\vartheta_c(t)}{dt} + \Delta\vartheta_c(t) = -K_{ec}\Delta u_1(t) \tag{4.7.44}$$

with the gain

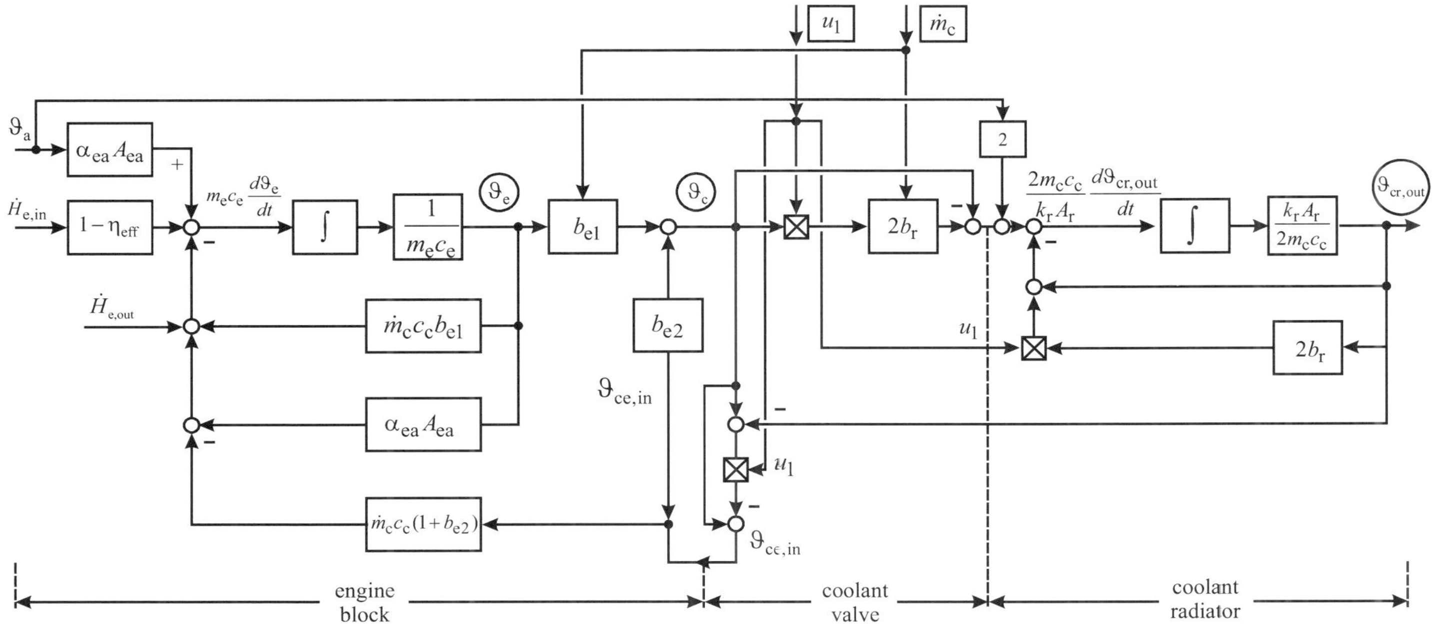

Fig. 4.7.6. Signal-flow chart for the engine block temperature ϑ_e, the coolant temperature ϑ_c at the valve and the radiator exit temperature $\vartheta_{cr,out}$ with u_1 as the manipulated variable of the valve and $\dot{m}_c$ the coolant mass flow dependent on pump speed.

$$K_{\text{ec}} = -\frac{\dot{m}_{\text{c}} c_{\text{c}} (1 + b_{\text{e2}}) \Delta \overline{\vartheta}_{\text{cr}}}{b_{\text{e1}} (\alpha_{\text{ea}} A_{\text{ea}} + \dot{m}_{\text{c}} c_{\text{c}} b_{\text{e1}})} \tag{4.7.45}$$

and the time constant T_{ec} as (4.7.35).

This first order model for the *dynamic behavior of the coolant temperature* may be sufficient for designing a coolant temperature controller with manipulation by the coolant valve. However, many variables influence the parameters, such as $\dot{m}_{\text{c}}$, u_1, m_{f}, ϑ_{a} and v_{air}, where v_{air} is the air velocity through the radiator. Additionally, the time constant of the thermostatic expansion valve has to be taken into account with a first order model according to (4.6.41).

The lumped parameter models may be improved by assuming a dead time T_{ed}, taking into account the traveling time of coolant temperature changes from the coolant entrance to the exit of the engine block. The dead time has to approximate different dead times, as the flow through the cylinder space and cylinder head has different travel ways l_{ce} and velocities v_{ce}, Laramee et al (2005).

An approach for estimating an average dead time is

$$T_{\text{ed}} = \frac{\overline{l}_{\text{ce}}}{\overline{v}_{\text{ce}}} = \frac{V_{\text{ce}}}{\dot{V}_{\text{ce}}} \tag{4.7.46}$$

where $\overline{l}_{\text{ce}}$ and $\overline{v}_{\text{ce}}$ and the coolant volume in the engine block V_{ce} and its volume flow $\dot{V}_{\text{ce}}$ are representative values.

Therefore, following simplified model for small changes of the coolant valve may be used for control design

$$G_{\text{u}\vartheta\text{c}}(s) = \frac{\Delta\vartheta_c(s)}{\Delta u_1(s)} = -\frac{K_{\text{ec}}}{(1 + T_{\text{ec}} s)} e^{-T_{\text{ed}} s}. \tag{4.7.47}$$

Changes of the pump speed do not include a significant dead time, i.e. lead to dynamically less delayed changes of the engine outlet temperature. Sect. 7.10 describes various ways to develop coolant temperature controls.

The described modeling of cooling systems tries to approximately present the basic behavior in order to obtain hints for the design of the components and the control system. In order to result in relatively precise approaches computer aided flow and heat transfer simulation tools have to be used, as e.g. shown by Müller et al (1998), Zieher et al (2011), Karl and Feldhaus (2008). However, because of the many required and not precisely known parameters the theoretical analysis has to be supplemented by experiments.

4.8 Lubrication system

The lubrication system serves to lubricate and cool the pairings of the crankshaft drive, the valve train and other moving components and to remove local contaminants, wear particles and combustion residues. The most frequently used forced-feed lubrication is based on a positive displacement oil pump (mostly gear pump or vane pump), which delivers a defined volumetric oil flow rate from the oil sump to an oil cooler and and oil filter, see Fig. 4.8.1. The oil volume flow rate is about 30 to 35 l/h per kW engine power, Eberan-Eberhorst (2010). As the oil pressure increases with the speed of the pump, a pressure relief or pressure control valve limits the oil pressure to a maximum of for example 5 bar to prevent damage of the oil cooler, oil filter and seals. The oil cooler is cooled by the engine coolant or by air. The oil filter can be circumvented by a pressure released bypass valve in the case of strong contamination if, e.g. the pressure difference exceeds 0.8...2.5 bar. The main oil filter is a full-flow filter which filters the complete oil flow of the engine and holds back particles in the range of 0.5 to 500 μm which could cause damage or wear. This filter is regularly exchanged as part of the oil service. Additionally bypass filters can be provided, which hold back superfine particles, such as abrasive and soot particles. These bypass filters are mainly used for diesel engines and are limited to an oil flow of maximal 10 %.

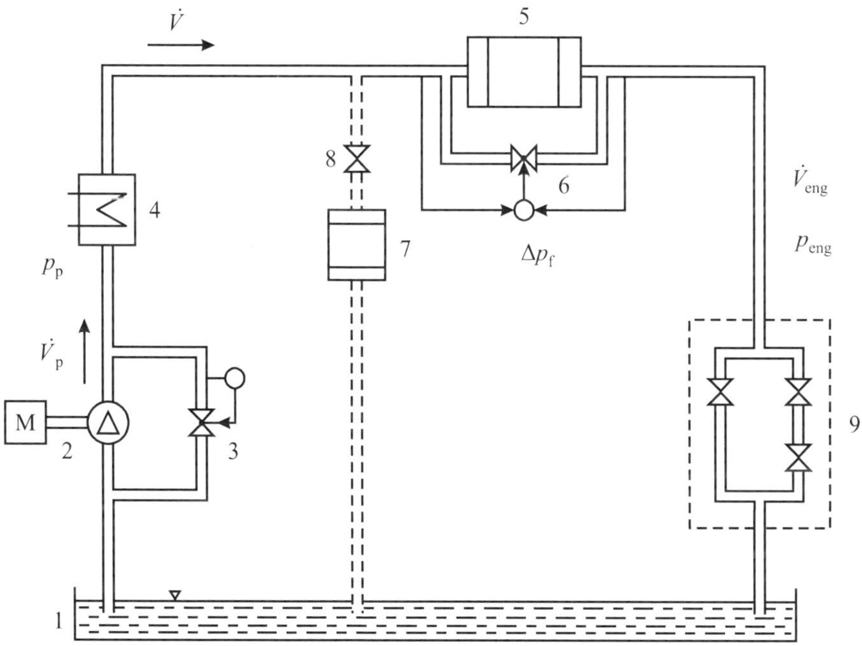

Fig. 4.8.1. Scheme of a lubrication oil circuit. 1 oil pan, 2 oil pump, 3 pressure control valve, 4 oil cooler, 5 full-flow filter, 6 bypass valve, 7 bypass filter, 8 throttle, 9 engine oil flow passages.

The oil first flows to an oil gallery and then through many oil ducts to the crankshaft, connecting rods, piston pins, camshaft, valves, turbocharger, etc. and then flows back unpressurized to the oil sump in the oil pan underneath the crank case, see e.g. Mollenhauer and Tschöke (2010), Robert Bosch GmbH (2011).

An alternative to this wet-sump lubrication is the dry-sump lubrication, when the oil is gathered in an external oil reservoir. This system gives more freedom for the design of the engine and functions also for high lateral accelerations.

For steady-state conditions, the oil volume flow rate and the oil pressures depend on the flow characteristics of the pump, the filters and the engine. The delivered flow of the pump $\dot{V}_{\mathrm{p}}$ is mainly determined by the pump speed ω_{p} and the flow absorbed by the engine $\dot{V}_{\mathrm{eng}}$ depends on the oil pressure p_{eng}, the engine speed ω_{eng} and the oil temperature T_{oil}. In the following the behavior of the pressures and flows is considered with simplifying assumptions. Because the oil pump sucks the oil from the oil pan with atmospheric pressure, it holds for the pump difference pressure $\Delta p_{\mathrm{p}} = p_{\mathrm{p}}$. The design and basic equations for gear and vane pumps are described in Ivantysyn and Ivantysynova (1993), Reinhardt (1992), Eberan-Eberhorst (2010), Berg (2009), Mollenhauer and Tschöke (2010), Affenzeller and Gläser (1996). Some aspects of modeling the oil circuit can be found, e.g. Reulein (1998).

The delivered theoretical volume flow $\dot{V}_{\mathrm{p}}$ of a positive displacement pump, as e.g. a gear pump is

$$\dot{V}_{\mathrm{p,th}} = V_{\mathrm{p,th}}\,\omega_{\mathrm{p}} \tag{4.8.1}$$

where V_{th} is the theoretical (ideal) displacement volume per turn. The required driving power becomes

$$P_{\mathrm{p,th}} = \dot{V}_{\mathrm{p,th}}\,p_{\mathrm{p}} = V_{\mathrm{p,th}}\,\omega_{\mathrm{p}}\,p_{\mathrm{p}}. \tag{4.8.2}$$

The real, effective volume flow is lowered by internal leak flows $\dot{V}_{\mathrm{p,l}}$, see Fig. 4.8.2a), which depend on the generated pressure p_{p} and speed ω_{p}

$$\dot{V}_{\mathrm{p,eff}}\left(p_{\mathrm{p}},\omega_{\mathrm{p}}\right) = V_{\mathrm{p,th}}\,\omega_{\mathrm{p}} - \dot{V}_{\mathrm{p,l}}\left(p_{\mathrm{p}},\omega_{\mathrm{p}}\right) = V_{\mathrm{p,th}}\,\omega_{\mathrm{p}}\,\eta_{\mathrm{vol}} \tag{4.8.3}$$

and can be taken intro account by a volumetric efficiency η_{vol}, see Isermann (2005). This efficiency may also comprise charging losses, which reduce $V_{\mathrm{p,th}}$, e.g. for higher pump speeds.

The corresponding power loss through the internal leakages is

$$P_{\mathrm{p}} = \dot{V}_{\mathrm{p,l}}\,p_{\mathrm{p}} \tag{4.8.4}$$

leading to a reduced delivery power

$$\begin{aligned} P_{\mathrm{p,vol}} &= P_{\mathrm{p,th}} - P_{\mathrm{p,l}} = \left(V_{\mathrm{p,th}}\,\omega_{\mathrm{p}} - \dot{V}_{\mathrm{p,l}}\right)p_{\mathrm{p}} \\ &= V_{\mathrm{p,th}}\,\omega_{\mathrm{p}}\,p_{\mathrm{p}}\,\eta_{\mathrm{vol}} = P_{\mathrm{p,th}}\,\eta_{\mathrm{vol}}. \end{aligned} \tag{4.8.5}$$

Pump internal flow losses yield a pressure drop $\Delta p_{\mathrm{p,l}}\left(\omega_{\mathrm{p}}\right)$ such that the effective pump difference pressure becomes

$$p_{\mathrm{p,eff}} = p_{\mathrm{p}}(\omega_{\mathrm{p}}) - \Delta p_{\mathrm{p,l}}(\omega_{\mathrm{p}}). \tag{4.8.6}$$

Figure 4.8.2b) shows qualitatively a characteristic of the effective pump pressure, where the pump operates against a constant throttle resistance. This characteristic may be approximated by

$$p_{\mathrm{p,eff}} = p_{\mathrm{p0}} + c_{\mathrm{p1}}\omega_{\mathrm{p}}. \tag{4.8.7}$$

The corresponding pump delivery power to the fluid is then reduced

$$P_{\mathrm{p,vol,l}} = P_{\mathrm{p,vol}} - \dot{V}_{\mathrm{p,th}} \Delta p_{\mathrm{p,l}} \tag{4.8.8}$$

and can be taken into account by a hydraulic loss efficiency η_{h}

$$P_{\mathrm{p,vol,l}} = P_{\mathrm{p,vol}}\, \eta_{\mathrm{h}} = P_{\mathrm{p,th}}\, \eta_{\mathrm{vol}}\, \eta_{\mathrm{h}}. \tag{4.8.9}$$

The bearing and gear friction losses are covered by a mechanical efficiency η_{m} such that the effective pump delivery power to the pumped fluid becomes

$$P_{\mathrm{p,eff}} = P_{\mathrm{p,vol,l}}\, \eta_{\mathrm{m}} = P_{\mathrm{p,th}}\, \eta_{\mathrm{vol}}\, \eta_{\mathrm{h}}\, \eta_{\mathrm{m}} = P_{\mathrm{p,th}}\, \eta_{\mathrm{tot}}. \tag{4.8.10}$$

The overall efficiency η_{tot} covers all losses. The required shaft driving power is therefore

$$\begin{aligned} P_{\mathrm{p,in}} &= \frac{1}{\eta_{\mathrm{tot}}} P_{\mathrm{p,eff}} = \frac{1}{\eta_{\mathrm{tot}}} \dot{V}_{\mathrm{p,eff}}(p_{\mathrm{p}}, \omega_{\mathrm{p}}) p_{\mathrm{p,eff}}(\omega_{\mathrm{p}}) \\ &= \frac{1}{\eta_{\mathrm{h}}\, \eta_{\mathrm{m}}} V_{\mathrm{p,th}} \omega_{\mathrm{p}} \left(p_{\mathrm{p0}} + c_{\mathrm{p1}}\omega_{\mathrm{p}}\right). \end{aligned} \tag{4.8.11}$$

It increases linearly as well as quadratically with the pump speed.

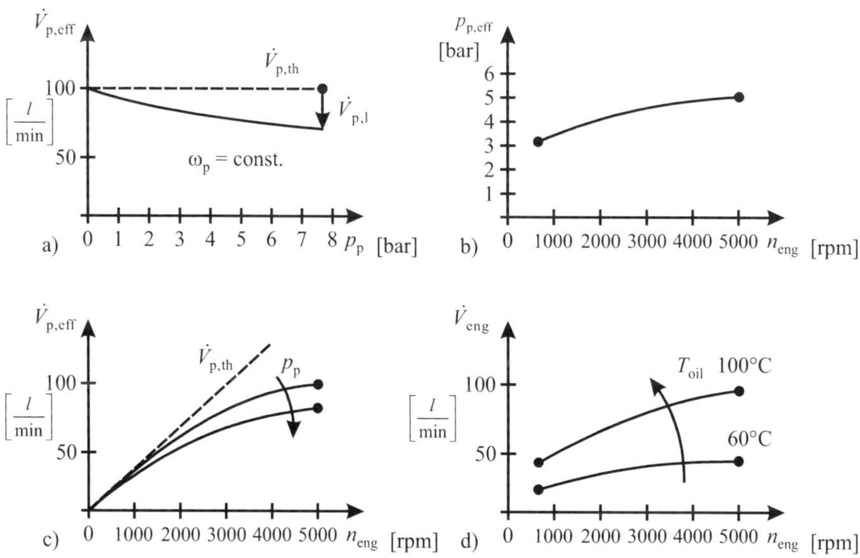

Fig. 4.8.2. Characteristics of lubrication circuit components (qualitative). **a** pump flow $\dot{V}_{\mathrm{p}}(p)$. **b** pump pressure $p_{\mathrm{p}}(n_{\mathrm{eng}})$. **c** pump flow $\dot{V}_{\mathrm{p}}(n_{\mathrm{eng}})$. **d** engine absorption flow $\dot{V}_{\mathrm{eng}}(n_{\mathrm{eng}})$.

The oil flow absorbed by the engine is assumed to be determined by a turbulent flow which approximately follows the flow law through a contraction

$$\dot{V}_{\text{eng}}(p_{\text{eng}}, \omega_{\text{eng}}) = c_{\text{eng}}(\omega_{\text{eng}}, T_{\text{oil}}) \sqrt{\frac{2}{\rho_{\text{oil}}}} \sqrt{p_{\text{eng}}}. \tag{4.8.12}$$

Since the flow also depends on the engine speed, the coefficient c_{eng} changes with ω_{eng}. As the oil temperature has a large influence on the oil viscosity, it has a considerable effect on c_{eng}, and the absorbed oil flow through the engine behaves as depicted in Fig. 4.8.2d).

p_{eng} is the acting pressure at the engine entrance (gallery) and is reduced by flow losses through the filter, oil cooler and channels

$$p_{\text{eng}} = p_{\text{p}} - \Delta p_{\text{f}}\left(\dot{V}_{\text{p,eff}}\right). \tag{4.8.13}$$

The volume flow delivered by the pump $\dot{V}_{\text{p}}$ may be reduced by the flow rate $\dot{V}_{\text{cv}}$ of an active pressure control valve

$$\dot{V}_{\text{p}}(t) = \dot{V}_{\text{p,eff}}(t) - \dot{V}_{\text{cv}}(u_1, t) \tag{4.8.14}$$

with

$$\dot{V}_{\text{cv}}(u_1) = \frac{1}{\sqrt{c_{\text{v}}(u_1)}} \sqrt{p_{\text{p,eff}}}, \tag{4.8.15}$$

see Fig. 4.8.3. The dynamic behavior of the pressures in the oil circuit can be determined by considering the mass balance equation taking a volume storage into account. The dynamics of the flows follow from the balance equation for the moved oil mass.

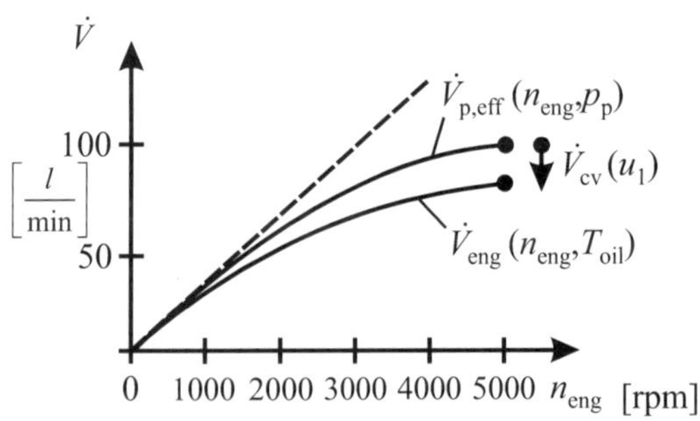

Fig. 4.8.3. Flow characteristic for the pump delivery and the engine absorption.

a) Volume storage

The dynamic behavior of the pressures in the oil circuit follows from the cooperation of the mass flows. Therefore the *mass balance equation* between the delivered mass flow of the pump $\dot{m}_{\text{p}}(t) = \rho_{\text{oil}} \dot{V}_{\text{p}}(t)$ and the absorbed mass flow of the engine $\dot{m}_{\text{eng}}(t) = \rho_{\text{oil}} \dot{V}_{\text{eng}}(t)$ is considered, taking into account a mass storage $m_{\text{pe}} =$

$\rho_{oil} V_{pc}$ between pump, oil cooler, oil filter, main gallery and distribution ducts to the lubrication points, leading to the volume balance equation

$$\frac{d}{dt} V_{pe}(t) = \dot{V}_p(t) - \dot{V}_{eng}(t). \tag{4.8.16}$$

To consider the varying storage volume of this volume balance equation, an oil reservoir V_{pe} is assumed, now taking into account a compressible fluid with bulk modulus

$$\beta = -V_{pe}(\frac{\partial p}{\partial V}) \tag{4.8.17}$$

which leads to a relation between pressure change and volume change $\Delta p = \Delta V \, \beta / V$ or

$$\frac{dp}{dt} = \frac{\beta}{V_{pe}} \Delta \dot{V}. \tag{4.8.18}$$

A flow difference between pump and engine thus becomes for closed pressure control valve with $\dot{V}_{cv} = 0$

$$\frac{dp_{eng}(t)}{dt} = \frac{\beta}{V_{pe}} \left(\dot{V}_{p,eff}(\omega_p, t) - \dot{V}_{eng}(\omega_{eng}, p_{eng}, t) \right). \tag{4.8.19}$$

Insertion of (4.8.3) and (4.8.12) in (4.8.19) results in

$$\frac{V_{pe}}{\beta} \frac{dp_{eng}(t)}{dt} = \left(V_{p,th} \, \omega_p(t) \, \eta_{vol} - c_{eng}(\omega_{eng}, T_{oil}) \sqrt{\frac{2}{\rho_{oil}}} \sqrt{p_{eng}(t)} \right). \tag{4.8.20}$$

Linearization around an operating point for pump speed changes yields, assuming c_{eng} as constant,

$$T_p \frac{dp_{eng}(t)}{dt} + \Delta p_{eng}(t) = K_p \Delta \omega_p(t) \tag{4.8.21}$$

with gain and time constant

$$\begin{aligned} K_p &= \frac{V_{p,th} \eta_{vol}}{a_1}, \\ T_p &= \frac{V_{pe}}{\beta \, a_1}, \\ a_1 &= \frac{c_{eng}(\omega_{eng}, T_{oil})}{\sqrt{2 \rho_{oil} \, \overline{p}_{eng}}}. \end{aligned} \tag{4.8.22}$$

Thus, the pressure at the engine entrance behaves, after pump speed changes and with simplifying assumptions, as a linear first order system with a small time constant T_p. However, as the parameters depend on ω_{eng} and $\omega_p = i_p \, \omega_{eng}$, with i_p the pump gear ratio, the dynamic equation is nonlinear in reality. The time constant is the larger, the larger the storage volume, the smaller the bulk modulus and the flow resistance coefficient of the engine.

If a variable pump with manipulated displacement volume $V_{\text{th}}(up)$ is applied (as for a gear pump with axial shifted gear wheels), it follows from (4.8.20) after linearization for constant speed ω_{eng}

$$T_{\text{p}}\frac{dp_{\text{eng}}(t)}{dt} + \Delta p_{\text{eng}}(t) = K_{\text{pV}}\Delta V_{\text{p,th}} \tag{4.8.23}$$

with

$$K_{\text{pV}} = \frac{\overline{\omega}_{\text{p}}\eta_{\text{vol}}}{a_1}. \tag{4.8.24}$$

b) Moved mass storage

The momentum balance of the moved oil mass yields

$$m_{\text{oil}}\frac{dw_{\text{p}}(t)}{dt} = A_{\text{p}}\left(p_{\text{p,eff}}(\omega_{\text{p}},t) - \Delta p_{\text{f}}(\dot{V}_{\text{eng}},t) - p_{\text{eng}}(\omega_{\text{eng}},t)\right) \tag{4.8.25}$$

with the oil mass of a representative sectional area A_{p} and length l_{oil}

$$m_{\text{oil}} = \rho_{\text{oil}}\,V_{\text{oil}} = \rho_{\text{oil}}\,A_{\text{p}}\,l_{\text{oil}}, \tag{4.8.26}$$

the pressure loss in the oil filter and oil channels

$$\Delta p_{\text{f}} = c_{\text{f}}\dot{V}_{\text{eng}}^2 \tag{4.8.27}$$

and the oil speed at the pump exit with sectional area A_{p}

$$w_{\text{p}} = \frac{1}{A_{\text{p}}}\left(\dot{V}_{\text{p,eff}} - \dot{V}_{\text{cv}}\right). \tag{4.8.28}$$

It is assumed that the pump pressure characteristic can be approximated by (4.8.7), compare Fig. 4.8.2 b) and the engine absorption flow and pressure by

$$\dot{V}_{\text{eng}} = \dot{V}_{\text{eng,0}} + c_{\text{e1}}\,\omega_{\text{eng}} \tag{4.8.29}$$

$$p_{\text{eng}} = p_{\text{eng,0}} + c_{\text{e2}}\,\omega_{\text{eng}}, \tag{4.8.30}$$

compare Fig. 4.8.2 d). All the coefficients, especially those for the engine absorption flow, also depend on the oil temperature T_{oil}.

The steady-state of the volume flow is determined by the volume balance equation (4.8.16). If the pressure control valve is closed, the effective pump flow equals the absorbed engine flow. When the delivered pump flow exceeds the absorbed engine flow, the pressure increases and the control valve opens to release a part $\dot{V}_{\text{cv}}$ of the volume flow, see Fig. 4.8.3.

The dynamic behavior can be discussed by inserting the various approximations for the pressure in (4.8.25). With the assumption of a closed pressure control valve, this leads to

$$\rho_{\text{oil}}\,l_{\text{oil}}\frac{d\dot{V}_{\text{p}}(t)}{dt} = A_{\text{p}}\left(p_{\text{p0}} + c_{\text{p1}}\,\omega_{\text{eng}} - c_{\text{f}}\dot{V}_{\text{p}}^2 - p_{\text{eng,0}} - c_{\text{e2}}\,\omega_{\text{eng}}\right). \tag{4.8.31}$$

Assuming small changes of ω_{eng} and $\dot{V}_{\text{p}}$ around a steady state $(\overline{\dot{V}}_{\text{p}}, \overline{\omega}_{\text{eng}})$ yields for the *volume-flow dynamics*

$$T_{\text{v}} \frac{d\dot{V}_{\text{p}}(t)}{dt} + \Delta\dot{V}_{\text{p}}(t) = K_{\text{v}}\, \Delta\omega_{\text{eng}}(t) \tag{4.8.32}$$

with gain and time constant

$$\begin{aligned} K_{\text{v}} &= \frac{(c_{\text{p1}} - c_{\text{e2}})}{2\, c_{\text{f}}\, \overline{\dot{V}}_{\text{p}}} \\ T_{\text{v}} &= \frac{\rho_{\text{oil}} V_{\text{oil}}}{2\, c_{\text{f}}\, A_{\text{p}}^2\, \overline{\dot{V}}_{\text{p}}} . \end{aligned} \tag{4.8.33}$$

The gain depends on the difference of the slopes of the pump and the engine pressure characteristics.

The time constant is the smaller the smaller the oil volume and the larger the oil flow. As experiments show, the time constant is in the range of 1 to 3 s.

The design of the oil pump and its characteristic is usually performed such that for idle speed and high oil temperatures the oil pressure is large enough to enable a good lubrication of all engine components. However, the oil pressure and the oil flow rate then become too large for high engine speed for mechanically driven pumps without variabilities. Therefore, the oil pressure is usually controlled with a by-pass valve, as depicted in Fig. 4.8.1. In order to save the required power *variable oil pumps* were developed which allow to manipulate the oil flow-rate, e.g. by electro-hydraulic valves. In the case of gear pumps the gear wheels are shifted relatively to each other and for vane pumps the eccentricity can be changed, see Sect. 7.11.

4.9 Drive-train dynamics and surge damping

The dynamic behavior of the drive train including the engine, clutch, transmission, propeller shaft, drive shaft and wheels has a considerable effect on the torque transfer, component stress and oscillations. Resulting driveline oscillations which are forced through abrupt positive and negative engine torque changes may lead to unwanted and uncomfortable bucking oscillations, also known as driveline jerking. This section considers simplified models of the drive train and anti-jerk controllers for surge-damping.

The drive train of a rear-driven vehicle is depicted in Fig. 4.9.1. It is assumed that a manual or automatic gear shift transmission with clutch is considered and that the components of the drive train can be represented as in Fig. 4.9.2. Two dominating elasticities are assumed, one at the clutch and one at the drive shaft between the differential and the rear wheels. This leads to a mass-spring-damper system with following turning and moving masses: engine and primary part of the clutch, secondary part of the clutch, transmission, propeller shaft and differential, wheels and vehicle mass.

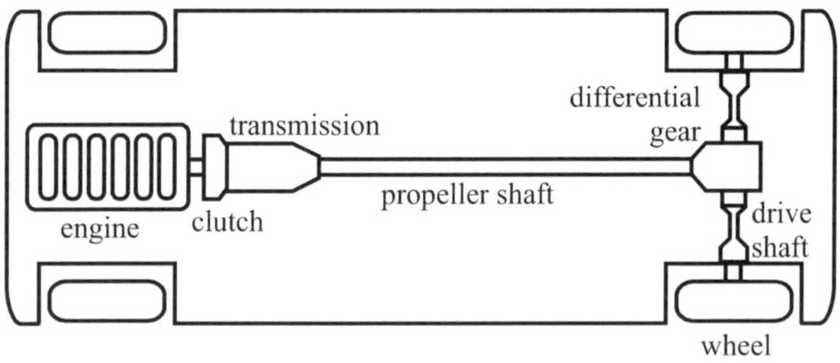

Fig. 4.9.1. Drive-train components for a rear-driven car.

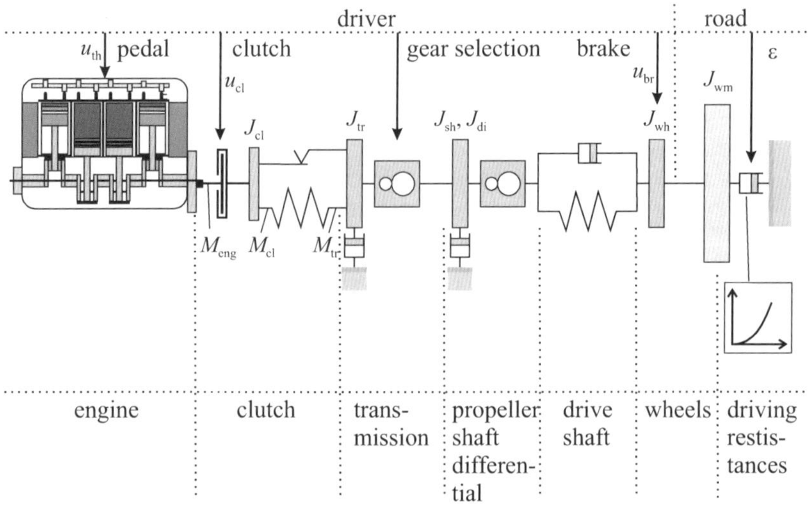

Fig. 4.9.2. Schematic of a vehicle drive train (simplified).

It is shown in Kiencke and Nielsen (2000) that a main contribution to drive train dynamics stems from the drive shaft which is subject to a large torsion, due to the high transmission rates of the shifted gear and the differential gear. The transmission, propeller shaft and differential are assumed to be stiff. Then the elements of the drive train can be modeled as follows:

Engine: The mean value torque balance at the flywheel yields according to (4.3.9) and (4.3.10)

$$J_{\text{eng}}\,\ddot{\varphi}_{\text{eng}}(t) = \overline{M}_{\text{eng}}(t, n_{\text{eng}}, m_{\text{f}}) - M_{\text{cl,f}}(t) \tag{4.9.1}$$

with

J_{eng} :	moment of inertia of the engine
$\overline{M}_{\text{eng}}$:	mean engine torque at clutch, operating point dependent
$M_{\text{cl,f}}$:	clutch torque during slipping

Clutch and transmission:
The transmitted torque of the clutch depends on its state: open, engaged or slipping. During slipping dry friction can be assumed:

$$M_{\mathrm{cl,f}}(u_{\mathrm{cl}}, \Delta\omega_{\mathrm{cl}}, t) = M_{\mathrm{cl,0}}(u_{\mathrm{cl}})\,\mathrm{sign}(\omega_{\mathrm{eng}}(t) - \omega_{\mathrm{cl}}(t)) \tag{4.9.2}$$

with

$M_{\mathrm{cl,0}}$:	dry friction coefficient of the clutch
u_{cl} :	clutch pedal position
$\Delta\omega_{\mathrm{cl}} = \omega_{\mathrm{eng}} - \omega_{\mathrm{cl}}$:	difference speed engine/clutch.

As the clutch possesses spring-damper elements the torque balance at the output of the clutch becomes

$$\begin{aligned} M_{\mathrm{cl}}(u_{\mathrm{cl}}, \Delta\omega_{\mathrm{cl}}, t) &= M_{\mathrm{cl,f}}(u_{\mathrm{cl}}, \Delta\omega_{\mathrm{cl}}, t) - M_{\mathrm{cl,c}}(\Delta\varphi_{\mathrm{cl}}, t) - M_{\mathrm{cl,d}}(\Delta\dot{\varphi}, t) \\ &= M_{\mathrm{cl,f}}(u_{\mathrm{cl}}, \Delta\omega_{\mathrm{cl}}, t) - M_{\mathrm{cl,cd}} \end{aligned} \tag{4.9.3}$$

with torque generated by the clutch springs and dampers

$$\begin{aligned} M_{\mathrm{cl,c}}(\Delta\varphi_{\mathrm{cl}}, t) &= c_{\mathrm{cl}}(\varphi_{\mathrm{cl}} - \varphi_{\mathrm{tr}}) \\ M_{\mathrm{cl,d}}(\Delta\dot{\varphi}_{\mathrm{cl}}, t) &= d_{\mathrm{cl}}(\dot{\varphi}_{\mathrm{cl}} - \dot{\varphi}_{\mathrm{tr}}) \\ M_{\mathrm{cl,cd}} &- M_{\mathrm{cl,c}} + M_{\mathrm{cl,d}} \end{aligned}$$

φ_{cl} :	clutch output angle
φ_{tr} :	transmission input angle
c_{cl} :	clutch spring constant
d_{cl} :	clutch damping coefficient.

Note that the clutch usually contains different springs in series, which leads to a nonlinear characteristic $M_{\mathrm{cl,c}}(\Delta f_{\mathrm{cl}})$, see e.g. Kiencke and Nielsen (2000).

Hence, in the slipping phase the torque balance at the clutch yields

$$J_{\mathrm{cl}}\ddot{\varphi}_{\mathrm{cl}}(t) = M_{\mathrm{cl,f}}(u_{\mathrm{cl}}, \Delta\omega_{\mathrm{cl}}, t) - M_{\mathrm{cl,cd}}(t) \tag{4.9.4}$$

with

J_{cl} :	moment of clutch inertia on transmission side

and for the transmission, shaft and differential follows

$$J_{\mathrm{tr,di}}(i_{\mathrm{tr}})\,\ddot{\varphi}_{\mathrm{tr}} = M_{\mathrm{cl,cd}}(t) - M_{\mathrm{tr,d}}(t) - \frac{1}{i_{\mathrm{tr}}\,i_{\mathrm{di}}} M_{\mathrm{ds,cd}} \tag{4.9.5}$$

with

$J_{\mathrm{tr,di}}$:	moment of inertia of transmission, shaft and differential
$M_{\mathrm{tr}} = M_{\mathrm{cl,cd}}$:	torque at transmission input
$M_{\mathrm{di}} = M_{\mathrm{ds,cd}}$:	torque at drive shaft input

and transmission ratios and torque ratios

$$i_{\mathrm{tr}} = \frac{\dot{\varphi}_{\mathrm{tr}}}{\dot{\varphi}_{\mathrm{sh}}} = \frac{\text{input angular speed transmission}}{\text{shaft angular speed}}$$
$$i_{\mathrm{di}} = \frac{\dot{\varphi}_{\mathrm{sh}}}{\dot{\varphi}_{\mathrm{di}}} = \frac{\text{shaft angular speed}}{\text{differential speed at output}}$$
$$M_{\mathrm{sh}} = i_{\mathrm{tr}}\, M_{\mathrm{tr}}\,; \quad M_{\mathrm{di}} = i_{\mathrm{di}}\, M_{\mathrm{sh}}.$$

If the shifted transmission has two axles with moment of inertia $J_{\mathrm{tr},1}$ and $J_{\mathrm{tr},2}$, the shaft has J_{sh} and the differential J_{di} then the lumped overall moment of inertia is

$$J_{\mathrm{tr,di}} = J_{\mathrm{tr},1} + \frac{1}{i_{\mathrm{tr}}^2}\left(J_{\mathrm{tr},2} + J_{\mathrm{sh}} + J_{\mathrm{di}}\right). \tag{4.9.6}$$

The friction torque of the transmission (bearings, oil) is assumed to be proportional to the angular speed

$$M_{\mathrm{tr,d}} = d_{\mathrm{tr}}\, \dot{\varphi}_{\mathrm{tr}}. \tag{4.9.7}$$

Drive shaft, wheels and vehicle
The torque at the differential output is for two drive shafts

$$\begin{aligned} M_{\mathrm{di}} &= i_{\mathrm{di}}\, M_{\mathrm{sh}} = i_{\mathrm{tr}}\, i_{\mathrm{di}}\, M_{\mathrm{tr}} \\ &= 2c_{\mathrm{ds}}\left(\varphi_{\mathrm{di}}(t) - \varphi_{\mathrm{wh}}(t)\right) + 2d_{\mathrm{ds}}\left(\dot{\varphi}_{\mathrm{di}}(t) - \dot{\varphi}_{\mathrm{wh}}(t)\right) \end{aligned} \tag{4.9.8}$$

and for the torque balance of two wheels with dynamic roll radius r_{dyn} one obtains

$$2\, J_{\mathrm{wh}}\, \ddot{\varphi}_{\mathrm{wh}}(t) = M_{\mathrm{ds,cd}}(t) - F_{\mathrm{x}}(t)\, r_{\mathrm{dyn}} \tag{4.9.9}$$

with the torque of the driveshaft

$$M_{\mathrm{ds,cd}}(t) = 2\, c_{\mathrm{ds}}\left(\varphi_{\mathrm{di}}(t) - \varphi_{\mathrm{wh}}(t)\right) + 2\, d_{\mathrm{ds}}\left(\dot{\varphi}_{\mathrm{di}}(t) - \dot{\varphi}_{\mathrm{wh}}(t)\right) \tag{4.9.10}$$

and the longitudinal vehicle drive force for a road with slope angle γ

$$F_{\mathrm{x}}(t) = m_{\mathrm{veh}}\ddot{x}_{\mathrm{veh}}(t) + m_{\mathrm{veh}}\, g\, \sin\epsilon + F_{\mathrm{ro}}(t) + F_{\mathrm{air}}(t) \tag{4.9.11}$$

where the rolling resistance force for one wheel is

$$F_{\mathrm{ro},j} = c_{\mathrm{ro}} F_{\mathrm{z},j} = \left(c_{\mathrm{ro},0} + c_{\mathrm{ro},1} v_{\mathrm{veh}} + c_{\mathrm{ro},4} v_{\mathrm{veh}}^4\right)_j F_{\mathrm{z},j} \tag{4.9.12}$$

and for four wheels

$$F_{\mathrm{ro}} = \sum_{j=1}^{4} F_{\mathrm{ro},j} \tag{4.9.13}$$

with vertical wheel forces

$$F_{\mathrm{z},j} = m_j\, g \tag{4.9.14}$$

and the air drag force

$$F_{\mathrm{air}} = c_{\mathrm{w}} A_{\mathrm{w}} \frac{\rho_{\mathrm{air}}}{2} v_{\mathrm{veh}}^2 \tag{4.9.15}$$

with

c_{w} :	air resistance coefficient
A_{w} :	effective area (maximum vehicle cross-section).

Hence, by insertion into (4.9.9) the torque balance for two wheels is, assuming no slip between tire and road and therefore $\dot{x}_{\mathrm{veh}} = r_{\mathrm{dyn}} \omega_{\mathrm{wh}} = r_{\mathrm{dyn}} \dot{\varphi}_{\mathrm{wh}}$

$$J_{\mathrm{wm}}\, \ddot{\varphi}_{\mathrm{wh}}(t) = M_{\mathrm{ds,cd}}(t) - r_{\mathrm{dyn}}\left(m_{\mathrm{veh}}\, g\, \sin \epsilon + F_{\mathrm{ro}}(t) + F_{\mathrm{air}}(t)\right) \tag{4.9.16}$$

with replacement moment of inertia for two wheels and mass of the vehicle

$$J_{\mathrm{wm}} = 2\, J_{\mathrm{wh}} + m_{\mathrm{veh}} r_{\mathrm{dyn}}^2 . \tag{4.9.17}$$

Figure 4.9.3 shows a signal-flow scheme of the considered drive train for the clutch in slipping operation, summarizing all the relations. If the clutch is fully engaged, the engine torque acts directly on the secondary side of the clutch, such leading to the signal-flow scheme of Fig. 4.9.4. Then, a *three-mass system* with two elasticities results. Figure 4.9.5 depicts a simulation result after a step change in the engine torque. With a step-change magnitude of $50\,\mathrm{Nm}$, the torsion damper in the clutch stays under traction. The torque at clutch as well as the engine angular speed show weakly damped buckling (surging) oscillations of about $5\,\mathrm{Hz}$.

If the drive system is considered as a *two-mass system* with the main elasticity at the drive shafts, an approximation of the undamped natural frequency can be calculated directly. The ratio of inertias of the engine, clutch, gear, propeller shaft and differential are lumped together, where the dominating term comes from the engine

$$J_1 \approx J_{\mathrm{eng}} . \tag{4.9.18}$$

The driven moment of inertia of the wheels and the vehicle follows from (4.9.17)

$$J_2 = J_{\mathrm{wm}} . \tag{4.9.19}$$

Both masses are connected by the overall transmission ratio $i = i_{\mathrm{tr}}\, i_{\mathrm{di}}$, the stiffness and elasticity of the drive shaft $c_2 = 2\, c_{\mathrm{ds}}$ and $d_2 = 2\, d_{\mathrm{ds}}$.

Then, following torque balances can be stated, compare Fig. 4.9.6, with $M_1 = M_{\mathrm{eng}}$ and $\varphi_1 = i \varphi_1'$

$$J_1\, \ddot{\varphi}_1 = M_1 - \frac{1}{i}\left[c_2(\varphi_1' - \varphi_2) + d_2(\dot{\varphi}_1' - \dot{\varphi}_2)\right] \tag{4.9.20}$$

$$J_2\, \ddot{\varphi}_2 = c_2(\varphi_1' - \varphi_2) + d_2(\dot{\varphi}_1' - \dot{\varphi}_2) - M_2 . \tag{4.9.21}$$

Introducing the torsion angle

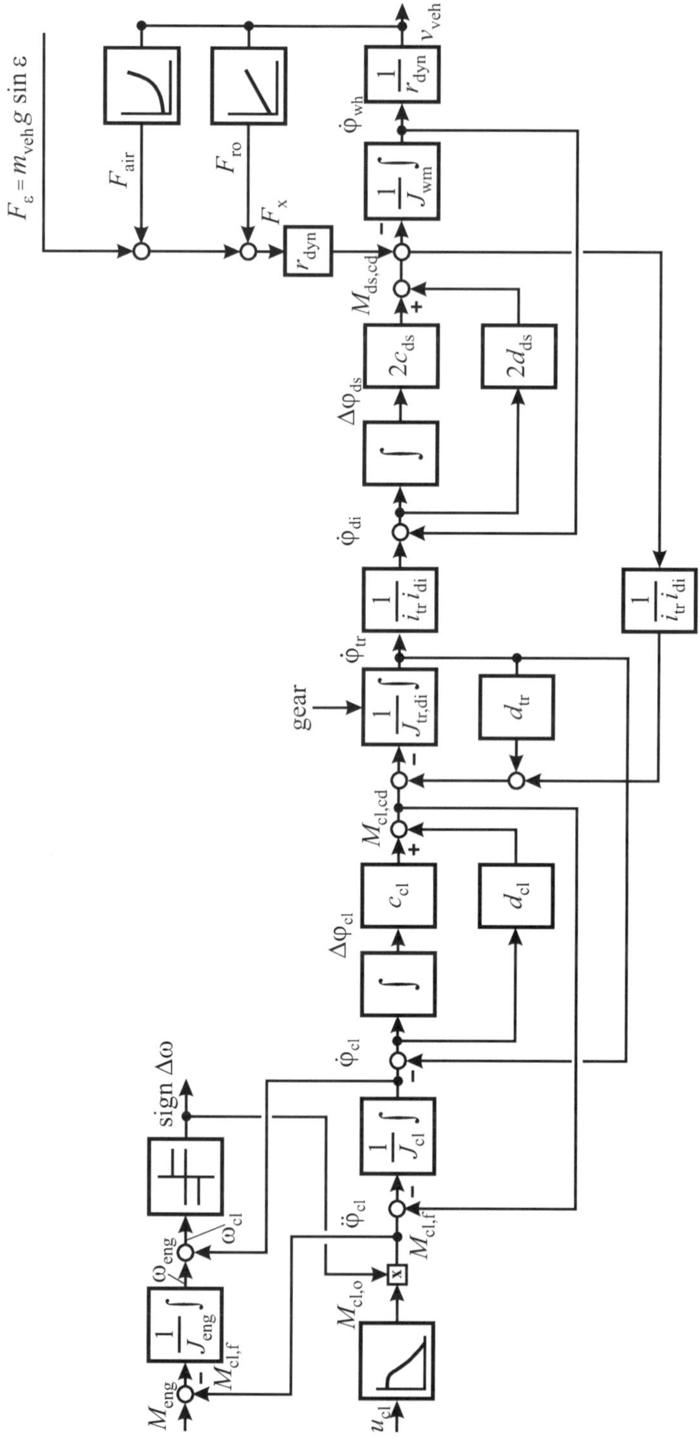

Fig. 4.9.3. Signal-flow scheme for the dynamic behavior of the drive train with shifted transmission and clutch in slipping operation.

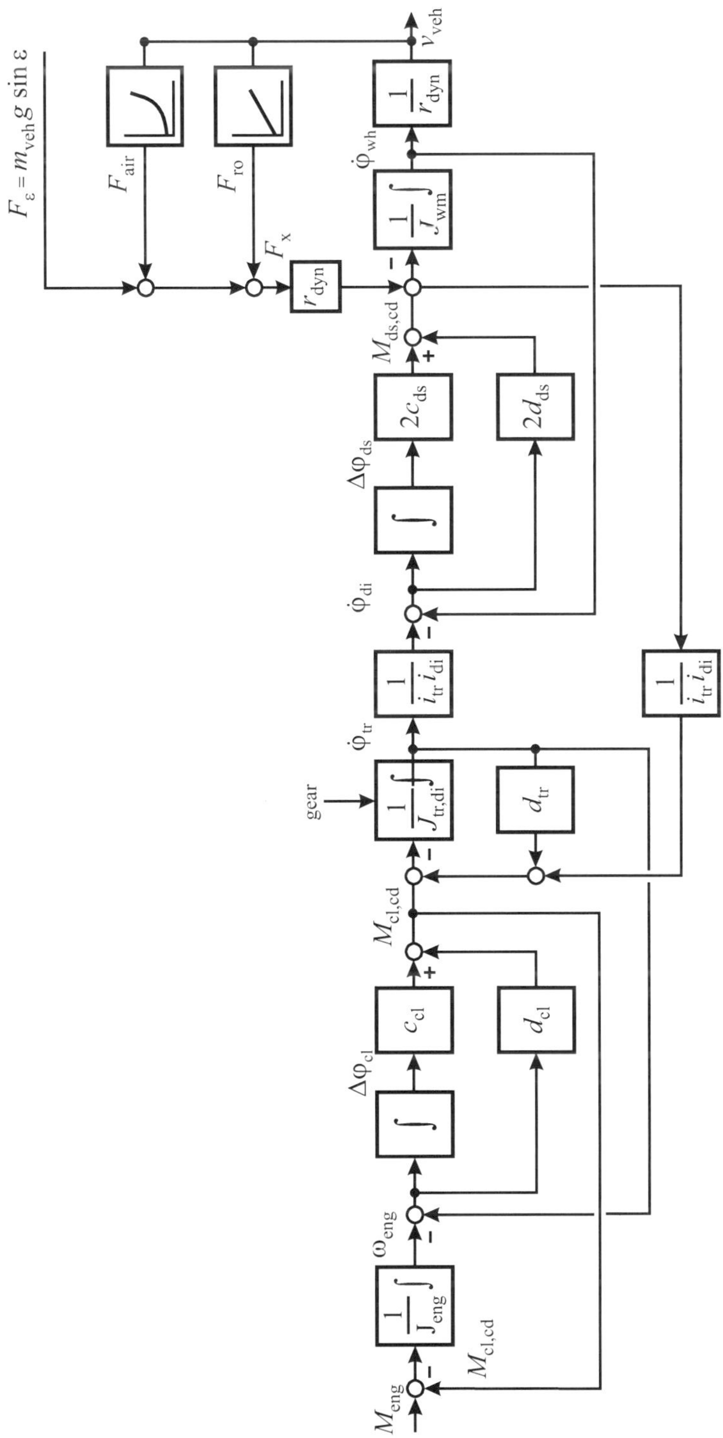

Fig. 4.9.4. Signal-flow scheme for the dynamic behavior of the drive train with shifted transmission and engaged clutch.

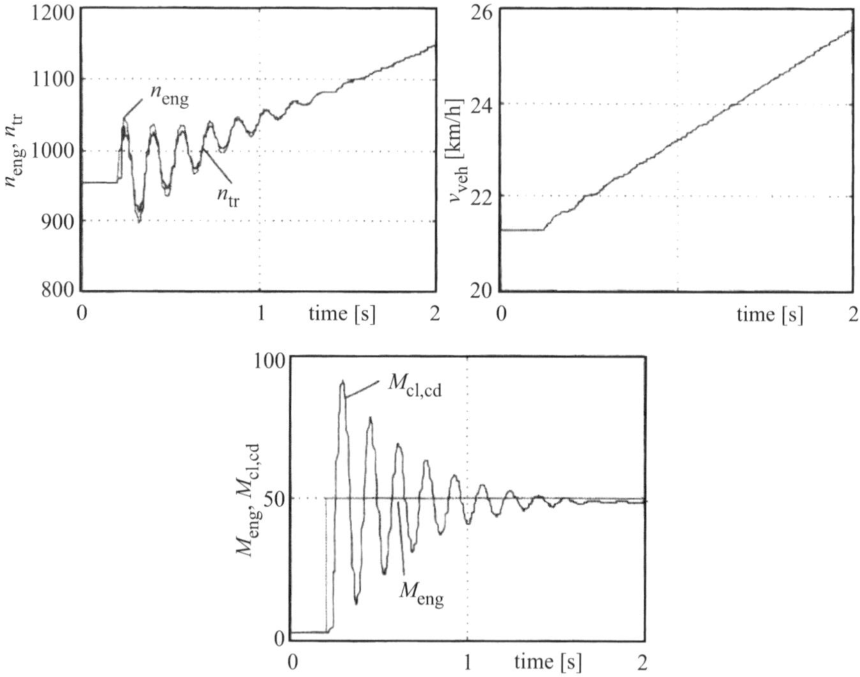

Fig. 4.9.5. Simulated buckling oscillations after a positive step change of the engine torque and third gear, Pfeiffer (1997). Passenger car $m_{veh} = 1100\,\mathrm{kg}$, $P = 51\,\mathrm{kW}$, $M_{eng,max} = 133\,\mathrm{Nm}$.

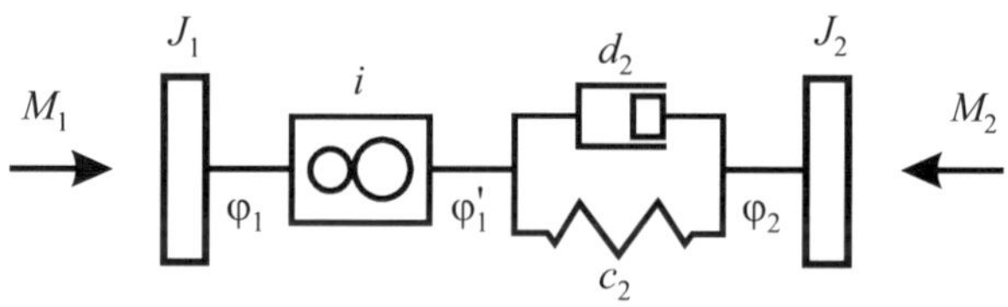

Fig. 4.9.6. Simplified representation of the drive train as two-mass system.

$$\Delta\varphi = \varphi_1' - \varphi_2 \tag{4.9.22}$$

and subtracting (4.9.20) and (4.9.21) leads to, compare Isermann (2005)

$$J_{\text{eff}}\Delta\ddot{\varphi}(t) + d_2\Delta\dot{\varphi}(t) + c_2\Delta\varphi(t) = k_1 M_1(t) + k_2 M_2(t) \tag{4.9.23}$$

with

$$J_{\text{eff}} = \frac{J_1 J_2 i^2}{J_2 + i^2 J_1}$$
$$k_1 = \frac{J_2 i}{J_2 + i^2 J_1}$$
$$k_2 = \frac{J_1}{J_2 + i^2 J_1}.$$

The undamped natural frequency of the drive train then becomes

$$\omega_0^2 = \frac{c_2}{J_{\text{eff}}} = c_2 \frac{J_2 + i^2 J_1}{J_1 J_2 i^2} = c_2 \left(\frac{1}{J_1 i^2} + \frac{1}{J_2} \right). \tag{4.9.24}$$

Applying the data of a compact car, as used for the simulation of Fig. 4.9.5, the frequencies of Table 4.9.1 result, Pfeiffer (1997).

Table 4.9.1. Undamped natural buckling oscillations for the vehicle of Fig. 4.9.5, Pfeiffer (1997)

Gear	1	2	3	4	5
f_0 [Hz]	2.9	4.6	6.3	8.0	8.9

The transmission ratio has a significant influence on the buckling frequency. The lower the gear the lower the buckling frequency. Typical frequencies are in the range of 2 Hz for the 1st gear and 9 Hz for the 5th gear, Reik et al (1990).

Backlash in the transmission, shaft or differential has a further influence on the buckling oscillations. This leads to larger amplitudes and smaller damping, Pfeiffer (1997).

The buckling oscillations can be damped by appropriate feedback controllers in the ECU with the engine speed as the preferred controlled variable.

The oscillations $\Delta\varphi$ of the drive shaft are transferred to the crankshaft angle by $\Delta\varphi_{\text{eng}} = i\Delta\varphi$ such that a second order differential equation can be stated for the flywheel angles too. This also holds for the engine speed $\Delta\omega_{\text{eng}} = \Delta(\Delta\varphi_{\text{eng}}/\Delta t) = \Delta\dot{\varphi}_{\text{eng}}$

$$a_2 \Delta\ddot{\omega}_{\text{eng}}(t) + a_1 \Delta\dot{\omega}_{\text{eng}}(t) + a_0 \Delta\omega_{\text{eng}}(t) = b_0 \Delta M_{\text{eng}}(t). \tag{4.9.25}$$

If a feedback controller of PD-type

$$\Delta M_{\text{eng}}(t) = -k_{\text{p}} \Delta\omega_{\text{eng}}(t) - k_{\text{d}} \Delta\dot{\omega}_{\text{eng}}(t) \tag{4.9.26}$$

is introduced, the characteristic equation becomes

$$a_2 \Delta\ddot{\omega}_{\text{eng}}(t) + (a_1 + b_0 k_{\text{d}}) \Delta\dot{\omega}_{\text{eng}}(t) + (a_0 + k_0 k_{\text{p}}) \Delta\omega_{\text{eng}}(t) = 0. \tag{4.9.27}$$

The undamped natural frequency and the damping ratio then are

$$\begin{aligned}\omega_0 &= \frac{\sqrt{a_0 + k_0 k_\mathrm{p}}}{\sqrt{a_2}} \\ \zeta &= \frac{a_1 + b_0 k_\mathrm{d}}{2\sqrt{a_0 + k_0 k_\mathrm{p}}} .\end{aligned} \tag{4.9.28}$$

Hence, the buckling oscillations can be damped by properly selecting k_d and k_p.

The engine torque has to be changed fast with correct phasing to compensate for the buckling oscillations. For gasoline engines this can be realized by changing the ignition angle and for diesel engines by the injection mass, which both introduce a dead-time of at least one stroke into the loop. As the amplitudes of the engine speed are relatively small compared to the speed signal's value, first a low pass filter is applied to eliminate higher frequent disturbance signals which is followed by a high pass filter to eliminate the lower frequencies. Both filters then form a band pass filter which can be approximated by a DT_2-element (derivative and second order low pass), Schmidt (1995). The results of a PD-feedback control and an internal model control, realized as discrete-time and discrete-angle algorithm in dependence on the selected gear is shown by Schmidt (1995). Extensive investigations for drive train modeling and control are described in Kiencke and Nielsen (2000). Several simulations and measurements are shown for speed control and oscillation damping of trucks, also during gear shifting.

A further study shows a state controller with a Kalman filter to estimate the difference angle $\Delta\varphi$, possible backlash, engine and wheel speed based on measurements of engine and wheel speed, and to damp the buckling oscillations by a state controller feedback designed as model predictive controller, Behrendt et al (2009), see also Engelhardt (2011). An extension to automatic transmissions with opened lock-up clutch is described in Behrendt et al (2011). The calibration of an anti-jerk controller by using a control prototyping development environment is reported in Engelhardt and Looman (2009).

At least premium vehicles have *two-mass flywheels* which consist of a smaller primary and a larger secondary rotational mass connected by a spring-damper system. Their application results in a considerable reduction of torsion oscillations from the engine to the drive line, Reik et al (1990), Walter and Kiencke (2007). One reason is that the resonance frequency through the large secondary rotational mass is shifted to values smaller than the idling speed. In that case, the starter has to be relatively strong to move the engine fast through the low resonance frequency, Reik et al (1998). Buckling drive line oscillations can still occur, because they depend on the drive line elasticities. Because of the good damping of the two-mass flywheel the excitation from the engine side is smaller.

References

Acklin L, Läubli F (1960) Die Berechnung des dynamischen Verhaltens von Wärmetauschern mit Hilfe von Analog-Rechengeräten. Technische Rundschau

Affenzeller J, Gläser H (1996) Lagerung und Schmierung von Verbrennungsmotoren. Springer, Wien
Arici O, Johnson J, Kulkarni A (1999) Cooling system simulation. part i: model development. In: SAE paper number 1999-01-240, SAE, Detroit, Michigan
Baines N (2005) Radial turbines: An integrated design approach. In: 6th European Conf. on Turbomachinery - Fluid Dynamics and Thermodynamics, Lille, France
Bargende M (1995) Most optimal location of 50 percent mass fraction burned and automatic knock detection components for automatic optimization of SI engine calibrations. MTZ worldwide 56(10)
Bargende M (2001) Ein Gleichungsansatz zur Berechnung der instationären Wandwärmeverluste im Hochdruckteil von Ottomotoren. Dissertation. Technische Hochschule Darmstadt, Darmstadt
van Basshuysen R, Schäfer F (2004) Internal combustion engine handbook: basics, components, systems, and perspectives. SAE International, Warrendale
Behrendt H, Zemke S, Grotjahn M (2009) Model predictive anti-jerk control of vehicle driveline oscillations. In: 7. Intern. Symposium Steuerungsysteme für den Antriebsstrang, Berlin, Germany
Behrendt H, Zemke S, Trabelsi A, Ortmaier T (2011) Model-based predictive anti-jerk control of powertrains with automatic transmissions. In: 8. Symp. Powertrain Control Systems for Motor Vehicles, Berlin, Germany, pp 76–81
Berg M (ed) (2009) Ölkreislauf im Verbrennungsmotor III. expert-Verlag, Renningen
Büchner S, Laries SS, degen A, Donnerstag A, Held W (2001) A modular numerical simulation tool predicting catalytic converter light-off by improved modeling of thermal management and conversion characteristics. In: SAE Technical paper Series, Warrendale, PA, 2001-01-0940
Cortona E, Onder C, Guzzella L (2002) Engine thermomanagement with electrical components for fuel consumption reduction. International Journal of Engine Research 3(3):157–170
Dixon S (1966) Fluid mechanics, thermodynamics of turbomachinery. Pergamon Press, Oxford
Eberan-Eberhorst C (ed) (2010) Schmierung von Verbrennungsmotoren, 3rd edn. expert verlag, Renningen
Engelhardt T (2011) Entwurf einer strukturumschaltenden Zustandsregelung zur Schwingungsdämpfung eines Kfz-Antriebsstranges. In: Automotive Colloquium, VDI, München, Germany
Engelhardt T, Looman C (2009) Modellbasierter Steuerungs-und Regelungsentwurf für Komfortfunktionen im Antriebsstrang, 1. In: Automobiltechnisches Kolloquium, München
Fischer G (2000) Expertenmodell zur Berechnung der Reibungsverluste von Ottomotoren. Dissertation Technische Universität Darmstadt. dissertation.de, Berlin
Führer J, Sinsel S, Isermann R (1993) Erkennung von Zündaussetzern aus Drehzahlsignalen mit Hilfe eines Frequenzbereichsverfahrens. In: Proceedings 13. Tagung Elektronik im Kraftfahrzeug, Essen, Germany
Greene A, Lucas G, Young S (1969) The testing of internal combustion engines. English Universities P., London

Grothe KH, Feldhusen J (eds) (2007) DUBBEL. Taschenbuch für den Maschinenbau, 22nd edn. Springer, Berlin

Guzzella L, Onder C (2010) Introduction to modeling and control of internal combustion engine systems, 2nd edn. Springer, Berlin

Heywood JB (1988) Internal combustion engine fundamentals. McGraw-Hill, New York

Hiereth H, Prenninger P (2003) Aufladung der Verbrennungskraftmaschine. Springer, Wien, New York

Hohenberg G (1980) Experimentelle Erfassung der Wandwärme von Kolbenmotoren. Habilitationthesis. Technische Universität Graz, Graz

Hohenberg G (1982) Der Verbrennungsverlauf. In: 4th Wiener Motorensymposium. Fortschr.-Ber. VDI Reihe 6, Nr. 103, VDI, Düsseldorf, pp 71–88

Holman J (1976) Heat transfer. McGraw-Hill, New York

Isermann R (1969) Einfache mathematische Modelle für das dynamische Verhalten beheizter Rohre. Journal Wärme 75:89–94

Isermann R (1970) Mathematical models of steam-heated heat exchangers (in German). Regelungstechnik und Prozess-Datenverarbeitung 18:17–23

Isermann R (1971) Einfache mathematische Modelle für das dynamische Verhalten beheizter Rohre. Neue Technik (4):13–20

Isermann R (2005) Mechatronic systems – fundamentals, 2nd edn. Springer, London

Isermann R (2015) Combustion-engine diagnosis. To be published. Springer, Berlin

Isermann R, Jantschke H (1970) Dynamic behavior of water- and steam-heated crossflow heat exchanger in air conditioning units (in German). Regelungstechnik und Prozess-Datenverarbeitung 18:115–122

Isermann R, Münchhof M (2011) Identification of dynamic systems. Springer, Berlin

Ivantysyn J, Ivantysynova M (1993) Hydrostatische Pumpen und Motoren, Konstruktion und Berechnung. Vogel, Würzburg

Jakisch T (1996) Widerstandsbeiwerte durchströmter Schalldämpferkomponenten. PhD thesis, Universität Kaiserslautern, Darmstadt

Japiske D, Baines N (1997) Introduction to turbomachinery. Concepts ETI Inc., Norwich VT.

Justi E (1938) Spezifische Wärme, Enthalpie, Entropie und Dissoziation technischer Gase. Springer, Berlin

Kalide W (1980) Einführung in die technische Strömungslehre, 5th edn. Carl Hanser Verlag, München

Karl C, Feldhaus U (2008) CFD-simulation of a coolant circuit of a truck diesel engine. Motortechnische Zeitschrift – MTZ 69(2):117–123

Karnopp D, Margolis D, Rosenberg R (2000) System dynamics: a unified approach, 3rd edn. Wiley, New York

Kiencke U, Nielsen L (eds) (2000) Automotive control systems. For engine, driveline and vehicle. Springer, Berlin

Kimmich F (2004) Modellbasierte Fehlererkennung und Diagnose der Einspritzung und Verbrennung von Dieselmotoren. Dissertation Technische Universität Darmstadt. Fortschr.-Ber. VDI Reihe 12, 549, VDI Verlag, Düsseldorf

Kohlhase M (2011) Brennraumdruckbasiertes Motormanagement für Otto- und Dieselmotoren zur Verbrauchs- und Emissionsreduktion. Dissertation Technische Universität Darmstadt. Fortschr.-Ber. VDI Reihe 12, 743, VDI Verlag, Düsseldorf

Laramee R, Garth C, Doleisch H, Schneider J, Hanser H, Hagen H (2005) Visual analysis and exploration of fluid flow in a cooling jacket. In: IEEE Visualization, Minneapolis, MN, USA, pp 623–630

Läubli R (1961) Zum Problem der Nachbildung des Verhaltens von Dampferzeugern auf Rechenmaschinen. Technische Rundschau 2:35–42

Lejsek D, Kulzer A, Kufferath A, Hohenberg G, Bargende M (2010) Calculation of the wall heat transfer during start-up of SI engines. MTZ worldwide 71(4):22–29

Leonhardt S (1996) Modellgestützte Fehlererkennung mit neuronalen Netzen - Überwachung von Radaufhängungen und Diesel-Einspritzanlagen. Dissertation Technische Hochschule Darmstadt. Fortschr.-Ber. VDI Reihe 12, 295. VDI Verlag, Düsseldorf

Leonhardt S, Schwarz R, Isermann R (1997) Real-time supervision of the diesel engine injection process. In: 1997 SAE International Congress and Exposition, Cobo Center, Detroit, Michigan, USA

Leonhardt S, Müller N, Isermann R (1999) Methods for engine supervision and control based on cylinder pressure information. IEEE/ASME Transactions on Mechatronics 4(3):235–245

Manz P (1982) Schnelle Indizierauswertung. In: 4th Wiener Motorensymposium. Fortschr. Ber. VDI Reihe 6, Nr. 103, VDI, Düsseldorf, pp 71–88

Mayer M, Krämer G (2011) Abgasturbolader, 6th edn. Verlag moderne industrie AG, Landsberg/Lech

McCuiston F, Lavoie G, Kauffmann C (1977) Validation of a turbulent flame progation model for a spark ignition engines. In: SAE Technical paper Series, Warrendale, vol 86, pp 200–221

Merker G, Schwarz C, Stiesch G, Otto F (2006) Simulating combustion. Springer, Berlin

Mollenhauer K, Tschöke H (2010) Handbook of diesel engines. Springer, Berlin

Moskwa J (1988) Automotive engine modeling for real time control. Phd-Thesis. Massachusetts Institute of Technology, Dept. of Mechanical Engineering, Massachusetts, MA

Mrosek M, Zahn S, Isermann R (2009) Parameter estimation for physical based air path models of turbocharged engines - an experience based guidance. In: 9th International Conference on Engines and Vehicles, SAE, Naples, Italy

Müller H, Bertling H (1971) Programmierte Auswertung von Druckverläufen in Ottomotoren. Fortschr.-Ber. VDI Reihe 6, 30. VDI, Düsseldorf

Müller N (2003) Adaptive Motorregelung beim Ottomotor unter Verwendung von Brennraumdruck-Sensoren. Dissertation Technische Universität Darmstadt. Fortschr.-Ber. VDI Reihe 12, 545. VDI Verlag, Düsseldorf

Müller U, Klingebiel F, Kahlstorf U (1998) Computer simulation of cooling circuits. Automobiltechnische Zeitschrift – ATZ 100(7–8):552–556

Oberg HJ (1976) Die Darstellung des Brennverlaufs eines schnelllaufendne Dieselmotors durch zwei überlagerte Vibe-Funktionen. Dissertation. Braunschweig

Pfeiffer K (1997) Fahrsimulation eines Kraftfahrzeuges mit einem dynamischen Motorenprüfstand. Dissertation Technische Hochschule Darmstadt. Fortschr.-Ber. VDI Reihe 12, 336, VDI Verlag, Düsseldorf

Pfleiderer C, Petermann H (2005) Strömungsmaschinen, 7th edn. Springer, Berlin

Pischinger R, Kell M, Sams T (2009) Thermodynamik der Verbrennungskraftmaschinen, 3rd edn. Springer, Wien and New York

Profos P (1962) Die Regelung von Dampfanlagen. Springer, Berlin

Rakopoulos C, Giakoumis E (2006) Review of thermodynamic diesel engine simulations under transient operating conditions. In: SAE World Congress 2006. Paper 2006-1-0884, Detroit, MI, USA

Reik W, Albers A, Schnurr M (1990) Torsionsschwingungen im Antreibsstrang von Kraftfahrzeugen. In: 4. Internationales LuK-Kolloquium, Baden-Baden, Germany

Reik W, Seebacher R, Kooy A (1998) Das Zweimassenschwungrad. In: 6. Internationales LuK-Kolloquium, pp 69–94

Reinhardt G (1992) Schmierung von Verbrennungskraftmaschinen. expert Verlag, Renningen

Reulein C (1998) Simulation des instationären Warmlaufverhaltens von Verbrennungsmotoren. Doctoral Thesis Technische Universität München. Hieronymus-Print, München

Robert Bosch GmbH (ed) (2011) Automotive Handbook, 8th edn. Bentley publishers, Cambridge

Schaffnit J (2002) Simulation und Control Prototyping zur Entwicklung von Steuergerätefunktionen für aufgeladene Nutzfahrzeug-Dieselmotoren. Dissertation Technische Universität Darmstadt. Fortschr.-Ber. VDI Reihe 12, 492. VDI Verlag, Düsseldorf

Schmidt C (1995) Digitale kurbelwinkelsynchrone Modellbildung und Drehschwingungsdämpfung eines Dieselmotors mit Last. Dissertation Technische Hochschule Darmstadt. Fortschr.-Ber. VDI Reihe 12, 253. VDI Verlag, Düsseldorf

Schmidt M (2003) Maßnahmen zur Reduktion des Energieverbrauchs von Nebenaggregaten im Kraftfahrzeug. Dissertation Technische Universität Darmstadt. Fortschr.-Ber. VDI Reihe 12, 557. VDI Verlag, Düsseldorf

Schwarte A, Kimmich F, Isermann R (2002) Model-based fault detection and diagnosis for diesel engines. MTZ worldwide 63(7-8)

Schwarzmeier M (1992) Der Einfluss des Arbeitsverlaufs auf den Reibmitteldruck von Dieselmotoren. PhD thesis, Technische Universität München, München

Shaaban S (2004) Experimental investigation and extended simulation of turbocharger non-adiabatic performance. PhD thesis, Universität Hannover, Hannover

Shibata Y, Shimonosono H, Yamai Y (1993) New design of cooling system with computer simulation and engine compartment simulation. In: SAE paper number 931075, SAE, Detroit, Michigan

Sidorow A (2014) Model-based fault diagnosis of the air path and turbocharger of a diesel engine. Internal Report. Institute of Automatic Control, Technische Universität Darmstadt

Sidorow A, Isermann R, Cianflone F, Landsmann G (2011) Comparison of a turbocharger model-based on isentropic efficiency maps with a parametric approach based on Euler's turbo-machinery equation. In: Preprints of the 18th IFAC World Congress, Milano, Italy

Siglich H (2008) Technische Fluidmechanik, 6th edn. Springer, Berlin, Heidelberg, New York

Sinsel S (2000) Echtzeitsimulation von Nutzfahrzeug-Dieselmotoren mit Turbolader zur Entwicklung von Motormanagementsystemen. Logos, Doctoral thesis. University of Technology, Darmstadt. Berlin

Stodola A (1945) Dampf- und Gasturbinen, 5th edn. Springer, Berlin

Urlaub A (1995) Verbrennungsmotoren. Grundlagen - Verfahrenstheorie - Konstruktion, 2nd edn. Springer, Berlin

VDI heat atlas (2010) VDI Verlag, Düsseldorf

Vibe I (1970) Brennverlauf und Kreisprozess von Verbrennungsmotoren. VEB-Verlag Technik, Berlin

Vogt M, Müller N, Isermann R (2004) Online adaptation of grid-based look-up tables using a fast regressor technique. Transactions of the ASME 126(Dec):732–739

Walter A, Kiencke U (2007) Realtime reconstruction of instanteneous engine and driveline torques. Motortechnische Zeitschrift – MTZ 67(6):24–27

Watson N, Janota M (1982) Turbocharging the internal combustion engine. The Macmillan Press, London, Basingstoke

Whitfield A, Baines N (1990) Design of radial turbomachines. Longman, Harlow, Essex

Willimowski M (2003) Verbrennungsdiagnose von Ottomotoren mittels Abgasdruck und Ionenstrom. Dissertation Technische Universität Darmstadt. Shaker Verlag, Aachen

Zahn S (2012) Arbeitsspielaufgelöste Modellbildung und Hardware-in-the-Loop-Simulation von Pkw-Dieselmotoren mit Abgasturbolader. Dissertation Technische Universität Darmstadt. Fortschr.-Ber. VDI Reihe 12, 760. VDI Verlag, Düsseldorf

Zahn S, Isermann R (2008) Modelling and hardware-in-the-loop simulation of diesel engines for development and test of future engine management system. In: FISITA World Automotive Congress, München, Germany

Zapf H (1970) Untersuchungen zur Vorausberechnung der Ladungsendtemperatur in Viertakt-Dieselmotoren. Motortechnische Zeitschrift – MTZ 31(12)

Zieher F, Ennemoser A, Petutschnig H (2011) Cae-simulation of cooling systems. ATZ extra, Automotive engineering partners pp 78–83

Zimmerschied R, Isermann R (2010) Nonlinear time constant estimation and dynamic compensation of temperature sensors. Control Engineering Practice 18:300–310

Part III

Engine Control

5

Engine control structure and components

The electronic engine control represents the center for commanding different combustion engine functions. It consists of the electronic control unit (ECU), the sensors and actuators, and the communication system. In the following, a view on the structures of ECU's and an overview of control components like sensors and actuators and some auxiliary units are given.

5.1 Engine control structures and function blocks

The functions of the engine control can be described by considering the hardware structure, the software structure and the hierarchically organized control functions.

5.1.1 Hardware structure

Figure 5.1.1 depicts a block-oriented scheme for the ECU-hardware structure with connected sensors and actuators.

The electronic control unit (ECU) is based on one or two microcontrollers. They consist of:

- *Microprocessor*. It contains the central processing unit (CPU) and has an own control unit for the execution of instructions from the program memory, and a logic unit (ALU) to handle arithmetic and logic operations. The CPU performs the flow control with address management and a command decoder. Processing happens in one cycle (10 ns for 100 MHz) towards the external memory.

- *Memories*. Microprocessors operate with nonvolatile and volatile memories. Random access memories (RAM) are write/read volatile memories and part of the microprocessor as main memory. Their content is lost if the voltage is switched off. Static RAM (SRAM) retain the information until it is overwritten or the power supply is cut off. Dynamic RAM (DRAM) have a lower power consumption, but must be rewritten after reading and after certain time intervals to compensate for the self-discharge.

R. Isermann, *Engine Modeling and Control*,
DOI: 10.1007/978-3-642-39934-3_5

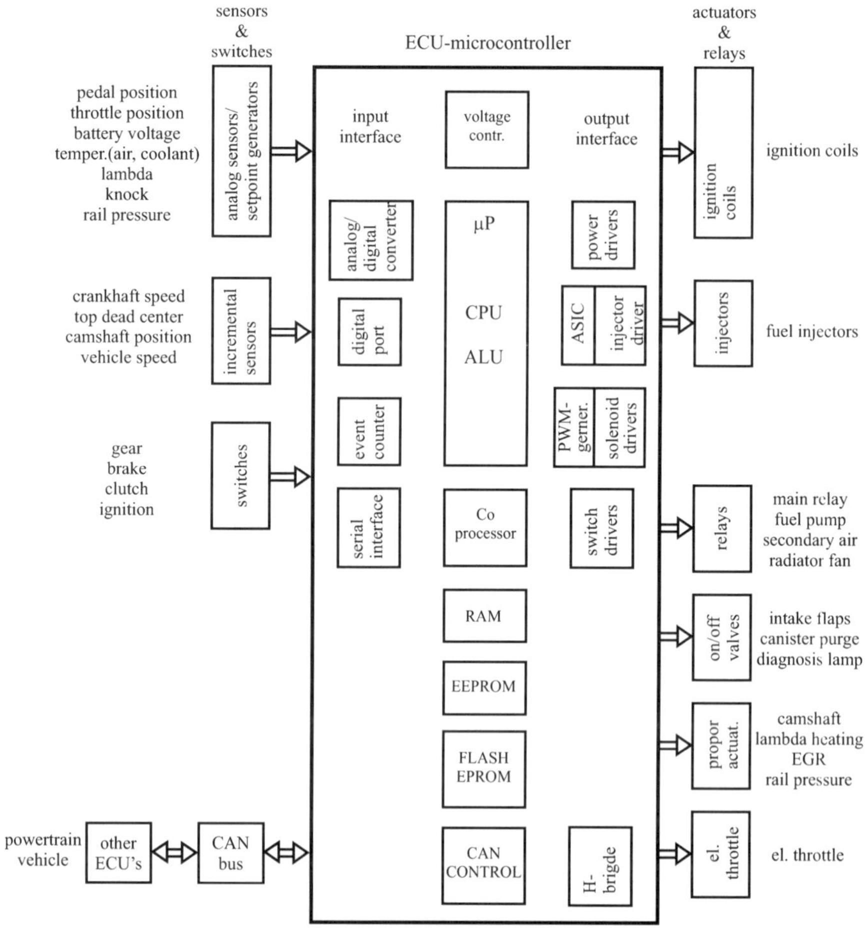

Fig. 5.1.1. Hardware structure of an electronic control unit (ECU) for gasoline engines.

Read-only memories (ROM) are nonvolatile. They store all programs and data that will not change and are retained even in the absence of a voltage. ROM's contain usually program code for control functions and constant data, programmed by the manufacturer. If programmed by the user, programmable read-only memories (PROM) are applied.

Some types of read-only memories can be erased and reprogrammed. Erasable read-only memories (EPROM) are erased with UV-radiation and newly programmed. Electrical erasable read-only memories (EEPROM) can be electrically erased and reprogrammed in the ECU with a separate device. An advantage is that each line of memory can be individually rewritten. A next development are Flash EPROM's. They allow the erasure, or flashing of complete memory storage by electrical pulses and reprogramming, by using a programming device. If the flash memory is part of the ECU, flashing can be applied on-board. There-

fore, flash memories have become standard as they allow the modification of nonvolatile storage during the engines life-cycle.
The arithmetic unit and the control unit of a microprocessor operate via an internal bus with the different types of memories. RAM is used to store variable data. Program instructions and permanent data are hold in the ROM types.
A coprocessor may be coupled through the internal bus with the main microprocessor and is at least partially used to monitor the main processor functions and for tasks of expansions. The main microprocessor is interconnected with the peripheral memories (RAM, ROM) and the input/output interfaces by the address and data bus with a parallel 8-, 16- or 32-bit structure.

- *Input interface*. The input modules provide a link between external sensors or switches and the microprocessor. They receive the analog, binary or digital input signals and transmit them to the microprocessor. The electrical analog input signals may be filtered, are converted to a uniform voltage, and pass a multiplexer, hold element and an analog-to-digital converter (ADC). Digital inputs are acquired word-wise. For incremental sensor inputs the number of signal changes is processed by counting circuits. A bus interface allows the communication with other electronic control units (e.g. by CAN-bus), as for the automatic transmission, and chassis control (ESC/TRC).

- *Output interface*. The output interface serves as a link to the actuators and relays. Based on the drivers' and sensors' inputs the microcomputer calculates the manipulated variables to the engine. The low-level signal outputs of the microprocessor are amplified by the driver stages to the level required by the various actuators. Usually the drivers may supply enough power for the actuators. In the case of large currents as for the radiator fan the driver may switch a relay.
Special power drivers are required for an inductive ignition system. Based on the calculated ignition time the ignition driver stage cuts the ignition coil to generate the high voltage for the spark plugs.
The injected fuel quantity depends on the opening duration of the fuel injector and the fuel pressure. The microprocessor delivers a digital signal for the beginning and duration of the injection. In the case of high-pressure common-rail direct injection system the injection duration is very small.
A special application specific circuit (ASIC) therefore generates a dynamic triggering signal to the DC/DC converter in the ECU to generate a booster voltage of, e.g. 65 V, which results in a boost current peak followed by a holding current to drive the fuel injectors.
Solenoid drivers for proportional acting electromagnetic actuators are controlled with a high frequent pulse-width-modulated (PWM) voltage. The PWM signal is, e.g. generated with a special ASIC. An H-bridge with diodes allows boost voltages for fast field changes in both directions.
The position of the electrical throttle is manipulated via an H-bridge to allow a controlled DC motor speed and position change in both directions.

Figure 5.1.1 summarizes the various components of the ECU microcontroller and shows the various sensors and switches at the input side and the different actuators and relays at the output side which manipulate the engine.
Table 5.1.1 gives a survey of the development of the technical data for gasoline engine ECU's over the last 20 years. Within this time the clock frequency increased from 12 to 150 MHz, the data bus width from 8 bit to 32 bit and the program storage of the EPROM's from 32 kbyte to 2 Mbyte (approximate numbers). The computing performances increased from about 10 to 300 MIPS, see Fig. 1.3.1. The increase of input variables (sensors) and manipulated output variables led to a large expansion of control functions and therefore to a strong increase of calibration parameters (labels), within the last 10 years from about 4.000 to 16.000, see Table 5.1.2. A similar impression gives the development of diesel engines, see Table 5.1.3. Figure 5.1.2 summarizes these developments for gasoline and diesel engines also in graphical form for the last decade. Two examples of the ECU hardware design are depicted in Fig. 5.1.3.

Table 5.1.1. Development of some technical data for gasoline ECUs

year		1990	1995	1998	2001
ECU type		Bosch M 1.8	Bosch M 4.3	Bosch ME 7.0	Bosch ME 7.0.1
Micro controller		80535	80 C51	C167	C167
data bus width		8	16	16	32
clock frequency	MHz	12	20-40	24	40
RAM	kB	8	>30	>50	136
EEPROM	kB	32			
FLASH EPROM	MB		0.5-0.8	0.512	1
Connector pins		60	112-154	120	120
analog inputs	8	32-82			
digital inputs	40	>46			

year		2003	2006	2008
ECU type		Bosch ME 9.0	Bosch MED 17	Continental EMS 2 Tricore
Micro controller		MPC 555	Tricore TC 1796	
data bus width		32	32	
clock frequency	MHz	56-66	40-150	40-150
RAM	kB	160	352	144
EEPROM	kB			
FLASH EPROM	MB	0.5-2	1.5-6	2-8
Connector pins		154	56-154	87-204
analog inputs	8		44	16-36
digital inputs	40			10-52

Table 5.1.2. Development of the number of engine control inputs and required calibration labels, Klar and Klages (2009)

	1997	2002	2004	2007
manipulated variables	5	8	10	14
ECU calibration labels	4000	9000	12000	16000

Table 5.1.3. Development of ECUs for diesel engines from Bosch, see also Figs. 1.3.1 and 5.1.2

		1997	2001	2003	2006	2009
ECU-type		EDC 15	EDC 16	EDC 16+	EDC 17	EDC 17+
microprocessor		C167	MPC 555	MPC 563 MPC 565	TC 1767 TC 1797 (Tricore)	
data bus width	bit	16	32	32	32	32
clock frequency	MHz	20	40	56/66	80/150	80-180
program storage (EPROM)	MB	0.256	1.5-4	1.5-4	1.5-6	2-8
computing power	MIPS	10	45	45-80	90-200	
connector pins		112-134	112-154	112-154	94-222	94-222
calibration parameters		4500	7000-10000	7000-10000	15000-20000	

5.1.2 Software structure

The software structure of an electronic control unit for engines is highly complex. It is hierarchically and modularly organized and must comply with strong real-time requirements. The *engine control* has to be seen in connection with the overall powertrain control of the vehicle. Figure 5.1.4 shows a schematic where the engine control itself is part of the powertrain management and vehicle motion control. The *powertrain management* includes also the transmission (automatic or manual shifted gear) and possibly electrical drives (e.g. starter/generator for hybrid drives). The *vehicle motion control* comprises the powertrain and the chassis in form of traction control (TCS), electronic stability control (ESC), adaptive cruise control (ACC), see Fig. 1.1.1. In this way software function blocks are defined which can be treated as closed units with defined interfaces. These function software blocks may be partially integrated in the engine ECU, like TCS and ACC, or implemented in other ECU's, like for the automatic transmission or the brake system (ESC).

A first gross subdivision of the *engine control software* can be made by a differentiation between basis software and application software, Stuhler et al (2010), see Fig. 5.1.5. The *basis software* serves in lower levels to organize the data flow from the input components to the microcomputer and to the output components. This includes the reading of sensor data or setpoints via the analog, digital and binary inputs or sending command signals to the various driver stages, as well as communication and component diagnosis tasks. Therefore the basis software is hardware-oriented, ECU specific and independent on the application programs of the ECU.

ECU performance development gasoline

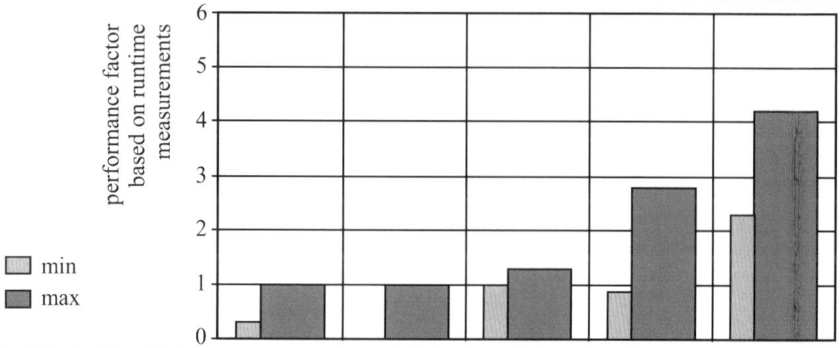

SOP	1998	2001	2003	2006	2009
ECU type	M(E)7	ME(D)9		ME(D)17	
word width [bit]	16	32		32	
flash memory [MByte]	0.5 - 0.8	0.5 - 2		1.5 - 6	2 - 8
connector pins	112 - 154	154		56 - 154	56 - 211
a) clock frequency [MHz]	20 - 40	40	56 - 66	40 - 150	80 - 180

ECU performance development diesel

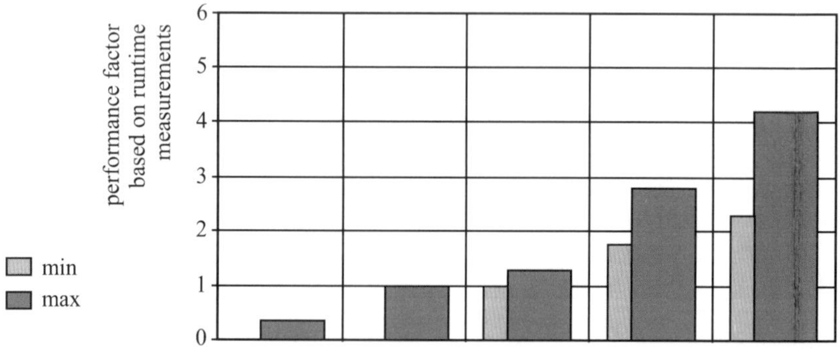

SOP	1998	2001	2003	2006	2009
ECU type	EDC15	EDC16(+)		EDC17(+)	
word width [bit]	16	32		32	
flash memory [MByte]	0.3	1.5 - 4		1.5 - 6	2 - 8
connector pins	112 - 154	112 - 154		94 - 222	
b) clock frequency [MHz]	20	40	56 - 66	80 - 150	80 - 180

Fig. 5.1.2. Development of ECU performance (1998–2009), courtesy to Robert Bosch GmbH. **a** gasoline engine. **b** diesel engine (SOP: start of production).

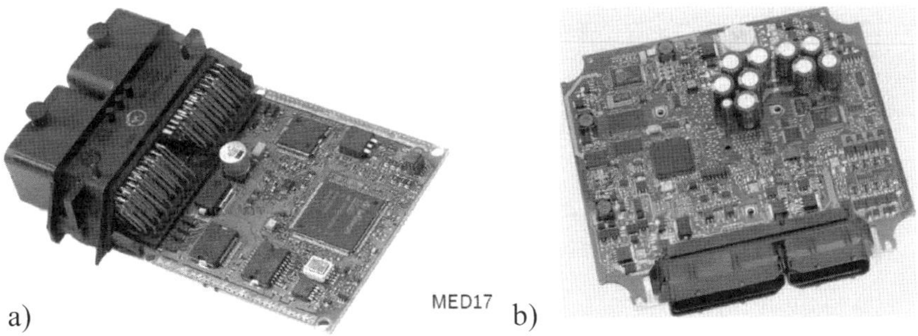

Fig. 5.1.3. Examples for ECU hardware design. **a** Motronic MED 17, Scholl (2007). **b** ECU for light commercial engines, Continental Automotive GmbH (2011) (Freescale MPC 555, 40 MHz, including drivers).

The *application software* operates in higher levels, and executes the engine control and some vehicle control functions. It receives the input values via the basis software functions of the input components and sends the control outputs to the basis software of the output components. The application software is in principle hardware independent. It is structured according to the main engine components like fuel system, air system, ignition, combustion, cooling, exhaust system and to the vehicle motion, see also Fig. 5.1.4, and it is function-oriented. These main components comprise control function blocks like injection control, ignition control and torque control for the combustion or engine charge control and EGR control for the air system. These control functions are considered in the next section and treated in more detail in Chaps. 7 and 8 for gasoline and diesel engines.

For the development of the software special tools are applied. They can be divided in function development tools, software implementation tools and calibration tools. The overall procedure follows a V-development scheme, which will be described in Chap. 6. The software implementation is performed either with manual programming or by automatic code generation.

The software function development begins for example by using MATLAB/Simulink and transfer to language C with a precompiler. The C-program is then turned into machine code with special compilers. This procedure is, e.g. realized with the industrial software systems and development environments like Target Link from dSpace or ASCET from ETAS, see Chap. 6.

5.1.3 Engine control function blocks

The overall control of internal combustion engines consists of discrete-event or crank-angle or time dependent control functions. After turning the ignition key the ECU has to be started. It passes through different *discrete ECU-states*, like booting, initializing the controllers, testing of hardware and software and the engine starting program as shown in Fig. 5.1.6, block I. The start command then lets the starter turn

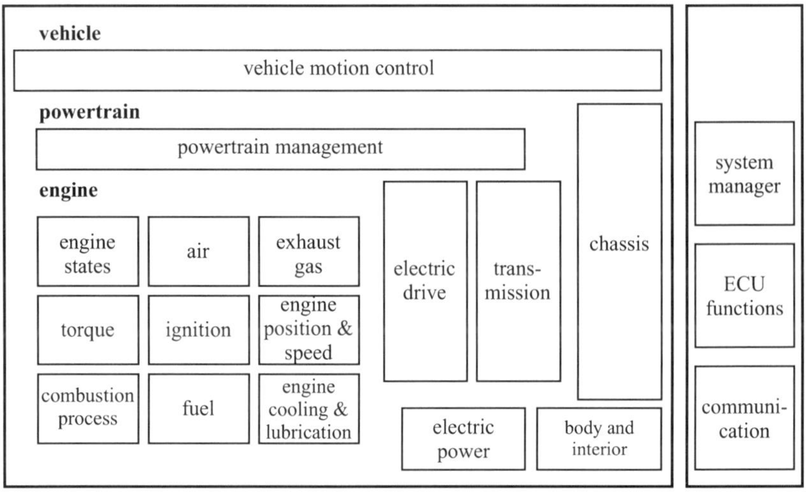

Fig. 5.1.4. Hierarchical structuring of engine, powertrain and vehicle control system into function blocks, Eisath and Brandl (2007).

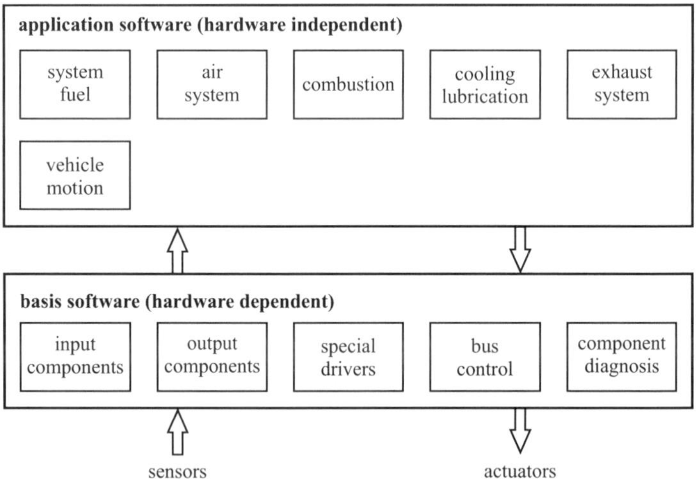

Fig. 5.1.5. Software structure of an ECU, Stuhler et al (2010).

and starts the engine according to a special starting program. Hence, the ECU follows a certain boot-up program and thereby goes through different discrete states.

The engine operation also distinguishes different *discrete engine states*, see Fig. 5.1.6, block II. For starting, a cold-start and warm-start phase has to be distinguished. This is followed by a warming-up phase. Also for idling speed control a cold or warm state has to be provided. After warming-up the engine reaches the normal operation state. The different engine states require certain set-points or correcting factors for the engine control functions. They can be represented by state diagrams or state machines with nodes and edges. The transition from one state to another depends on conditions, which are for example the ambient and the coolant or oil temperature. Other discrete-events are the shifting of gears, the regeneration of catalysts or particulate filters or the limp-home mode, and finally the shut-off of the engine.

The *main control functions* of combustion engines are either *crank-angle dependent* or *time dependent*, see Fig. 5.1.6, block III. According to the batch process characteristic of the combustion the injection and ignition control depend on the crank-angle position and are therefore based on the pulses of the crankshaft tooth wheel sensor. Other control functions depend only on time, like the air flow and EGR flow control, emission (lambda) control, fuel supply control and coolant control. Torque control is with regard to injection and ignition crank-angle dependent and with regard to air and EGR flow depends on continuous-time. These control functions together with corresponding actuators and sensors were already presented in Figs. 1.3.5 and 1.3.7.

A further group are the *diagnosis functions*, see block IV. They are based on available sensors and apply special testing routines for the sensors, the ECU and the actuators and generate outputs to the malfunction/failure memory, to the cockpit in switching warning lamps and to the limp-home control module (discrete engine state). The evaluation of sensor data can be considered as continuous in time, whereas testing procedures and diagnosis with alarming are discrete events. Especially because of the increase of emission related on-board diagnosis (OBD) functions the extent of diagnosis functions has reached up to 50 % of all ECU software functions.

The arrangement of the discrete states of the ECU and the engine and the continuous control and diagnosis functions in Fig. 5.1.6 shows a *hierarchical structure* within the functions of the ECU. After the boot-up program of the ECU and the starting command through the starting key the cold- or warm-start program comes in action. The main input is then the driver's accelerator pedal and after idling, the various engine control functions start to operate. The engine passes through the warming-up phase and the control functions obtain mainly temperature dependent correcting values and setpoints. Also gear shifting and exhaust after-treatment influence, e.g. setpoints of control functions. Therefore these discrete engine states are one level above the continuous control functions.

The diagnosis functions can be seen for signal processing as parallel to the control functions but have with regard to a safe driving function of the automobile a lower level significance, because they only indicate and store faults and do no intervene in the engine operation, except in the case of drastic protection needs, for example, if a limp-home mode has to be triggered.

A modern engine control system contains more than 80 function blocks, more than 400 control functions and the software comprises more than 3 million code and program lines, Stuhler et al (2010).

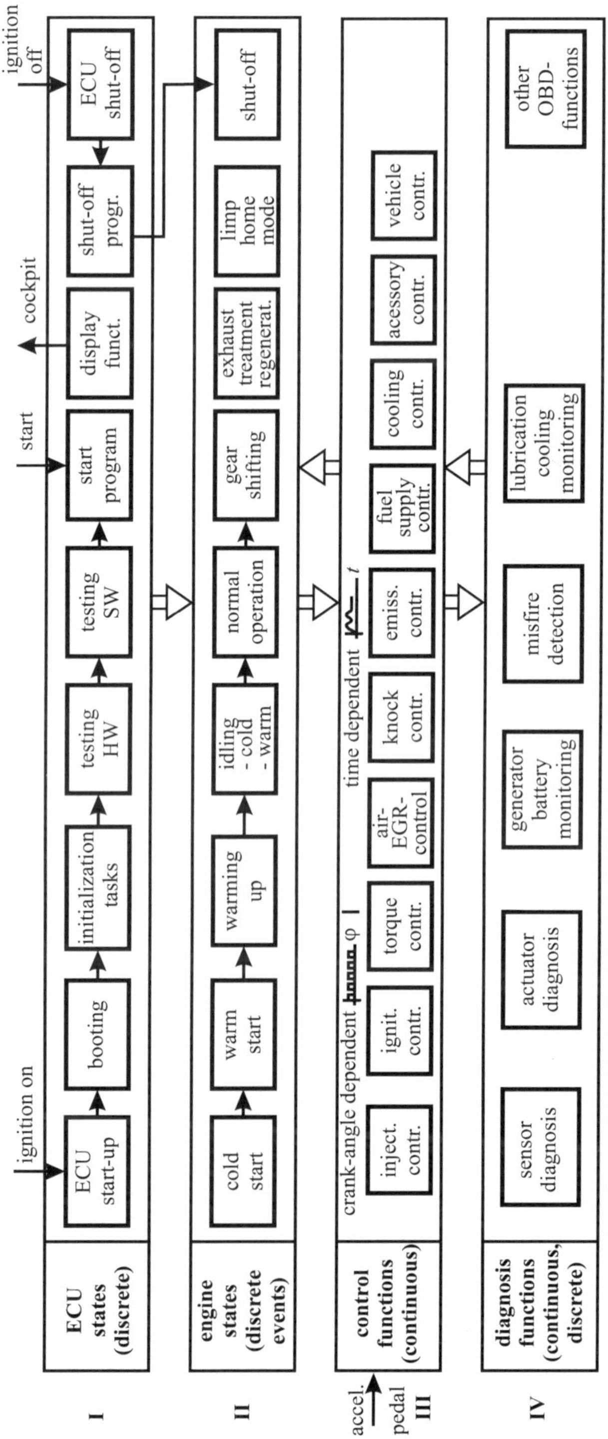

Fig. 5.1.6. Discrete event states, control and diagnosis functions for a gasoline engine.

5.2 Actuators and drives

Actuators represent the link between the signal level of the ECU and the combustion engine. The basic structure of a mechatronic actuator is shown in Fig. 5.2.1. It consists of the actuator drive, the mechanical actuator component and a sensor, with an energy flow in the forward direction and an electronic feedback controller. The actuator generally manipulates process energy or matter flow. The required actuator drive needs an auxiliary energy, which may be electrical, pneumatic or hydraulic. If no feedback of signals is applied, the actuator is controlled feedforward, as depicted in Fig. 5.2.2 a). A precise actuation requires in general a cascaded control, consisting of a minor inner controller, e.g. for the position or a current and a main outer controller, e.g. the air flow or the charging pressure see Fig. 5.2.2b). The minor controller then compensates the friction or backlash, increases the actuation speed, and weakens the effect of a nonlinear characteristic for the main controller. Pneumatic actuators for engines are usually applied as low-cost devices and have then no position feedback on cost of manipulation precision.

Table 5.2.1 gives a survey of actuators and drives for combustion engines. They are divided with regard to their action for the air system, combustion, exhaust system, cooling and lubrication and for auxiliaries. The following types can be distinguished:

A) actuators with electrical, pneumatic or hydraulic auxiliary energy
B) on/off injection valves and magnetic switching valves
C) electrical drives, pumps and fans

The properties of the actuators of type A are stated in Table 5.2.2. They have quite different properties with regard to power, power/weight ratio, translatory or rotational motion and manipulation dynamics and precision, Isermann (2000), Isermann (2005). These features and their advantages and disadvantages account for their different use for combustion engines, see Buch et al (2010). A further classification can be made with regard to the control structure of these mechatronic components as:

(1) *Decentralized mechatronic components*
- actuators and drives with local integration of sensors, electronics and amplifiers
- examples: fuel injection pumps, electrical throttle, EGR valve, electrical generator

(2) *Centrally controlled actuators and switches*
- directly controlled by the central control unit (ECU)
- examples: pneumatic actuators, electrically actuated flaps, shut-off valve, EGR cooler by-pass valve

(3) *Centrally controlled drives*
- directly controlled by the ECU
- examples: fuel pump, oil pump, secondary air pump, radiator fan, starter

Hence, for class (1) a mechatronic hardware-oriented and function-oriented integration is realized. The components of class (2) and (3) are directly controlled from the ECU. Frequently, the manipulated position or current is not measured and

Table 5.2.1. Mechatronic actuators and drives for combustion engines SI: gasoline engine, CI: diesel engine

actuator type	components	air system	injection system	combustion system	exhaust system	cooling, lubrication	auxiliaries
A	electrical actuators	• el. throttle • flaps (charge, manifold)	• low-pressure valve • high-pressure valve (rail)	• valve lift (SI)	• VTG-actuator • EGR valve	• coolant flow valve • heating flow valve	
	pneumatic actuators	• flaps (charge, motion)			• wastegate • VTG-actuator • EGR valve		
	hydraulic actuators			• camshaft positioner			
B	injection valves		• electro-magnetic • piezoelectric		• urea injection (SCR), (CI) • diesel fuel (PF regeneration)		
	magnetic switching valves		• shutoff valve (CI)		• EGR-bypass valve • SCR-dosing unit (CI)		
C	electrical drives					• el. fan	• el. generator • el. starter
	electrical pumps	• air vacuum pump	• fuel pump		• secondary air pump (SI) • urea pump (CI)	• coolant pump	• air condition compressor
	mechanical pumps	• compressor	• injection pump		• turbine (VTG)	• oil pump	• air condition compressor • hydraulic steering pump

Table 5.2.2. Some properties of actuators for combustion engines (type A)

auxiliary supply energy	design	auxiliary energy potential	power [W]	specific power [W/kg]	suitability for movement		actuation performance • dynamics • precision	advantages	disadvantages
					translational	rotatory			
electrical	electromagnet	12V, 24V		50	good	medium	• fast • medium	simple structure	• nonlinear • hysteresis • holding current
	piezoelement	130 - 160 V		500	good	–	• very fast • good	simple design	• small stroke • cost
	electromotor • DC, BLDC • torque	12V, 24V	<300	40-130	medium	good	• medium • good	• with gear rot. or linear • without gear	• gear • small travel
pneumatic	diaphragm	vacuum 0.1 - 0.8 bar	<100	5 - 25	good	medium	• medium	simple design	• vacuum pump • pressure pump • nonlinear • hysteresis
	cylinder	pressure 6-8 bar	>1000	200 - 400	good	medium	• medium		
hydraulic	cylinder rotating motor	3-5 bar		1000 - 2500	good medium	medium good	• fast • good	large forces	• systems expense • cost

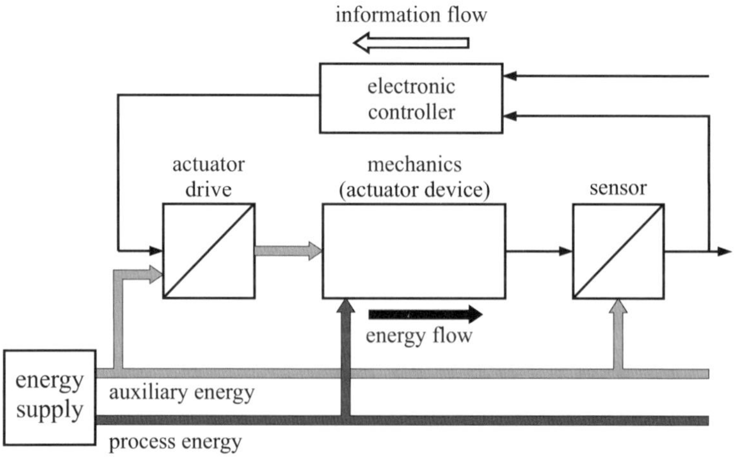

Fig. 5.2.1. Mechatronic actuator with integrated feedback control.

therefore a feedback to the ECU does not exist. These components are functionally integrated with the ECU, but not with regard to the hardware. Figure 5.2.3 shows some examples. The development of these actuators and drives shows an increase of electrical actuators for smaller actuator power (e.g. EGR valve and VTG actuator), a preservation of hydraulic actuator for large actuator power (e.g. camshaft actuator).

Pneumatic actuators are increasingly substituted by electrical actuators because of the drop of intake vacuum pressure for gasolines with no or relative opened throttles and no sufficient position precision in the case of, e.g. VTG actuators. Examples of some electrical or mechanical driven pumps are depicted in Fig. 5.2.4. They are usually centrally controlled.

5.3 Sensors

The measurement of physical and chemical quantities of combustion engines occurs with specially developed sensors, see e.g. van Basshuysen and Schäfer (2004), Robert Bosch GmbH (2007a), Robert Bosch GmbH (2011). Also here an increasing integration of the sensor element with electronics can be observed, compare Fig. 5.3.1. Different integration steps can be distinguished, Robert Bosch GmbH (2011):

(1) Sensor element with signal evaluation and analog-digital converter (ADC) within the ECU (conventional)
(2) Sensor element with analog electronics at the sensor (analog connection to the ECU)
(3) Sensor element with digital electronics at the sensor (binary signals or digital bus connection to the ECU)

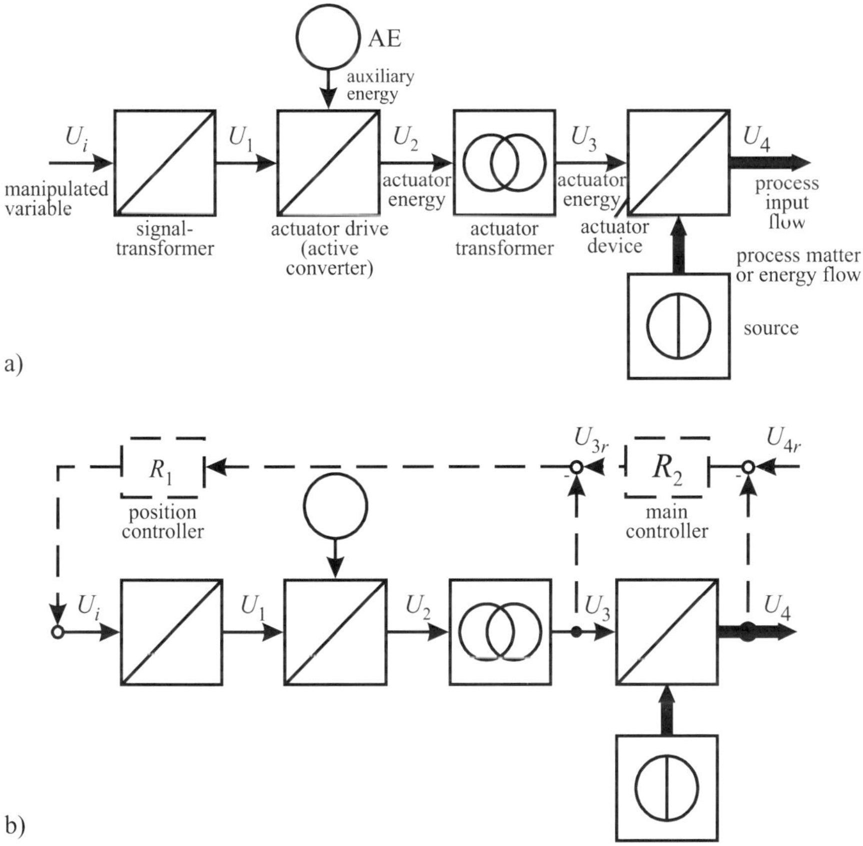

Fig. 5.2.2. Basic structures of actuators. **a** feedforward controlled. **b** cascaded feedback controlled.

(4) Sensor element with signal conditioning. ADC and microcomputer at the sensor (digital bus connection)

Table 5.3.1 summarizes the properties of different sensors. The sensors for speed and temperature have a classical realization (1). Integrated versions of step (2) are, e.g. angle and speed sensors. The integration steps (3) and partially (4) are demonstrated by pressure and flow sensors. The outputs of the sensors are either electrical voltages or currents in analog, value proportional or frequency proportional form. The evaluation electronics with auxiliary power supply from the board net generates for the integration step (2) a signal which is transferred to the ECU after filtering, correction of the characteristic and amplification. If the sensor element is added by digital electronics with an integrated circuit (IC) or application specific integrated circuit (ASIC), then integration step (3) is reached, and after ADC more sophisticated evaluation algorithms can be applied, followed by a binary or digital noise insensitive signal bus transfer to the ECU, see Fig. 5.3.1. The next integration step

a) electronic throttle valve (1) or (2)

b) general purpose actuator for intake flaps or VTG-turbocharger with return spring and integrated position control electronics (2)

c) electrical exhaust valve, max. 700°C (2)

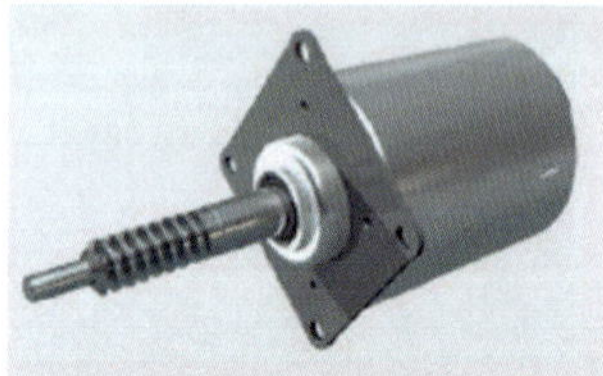

d) variable valve lift actuator (excentric valve lift shaft), 355 W (2)

e) electrical EGR valve (1)

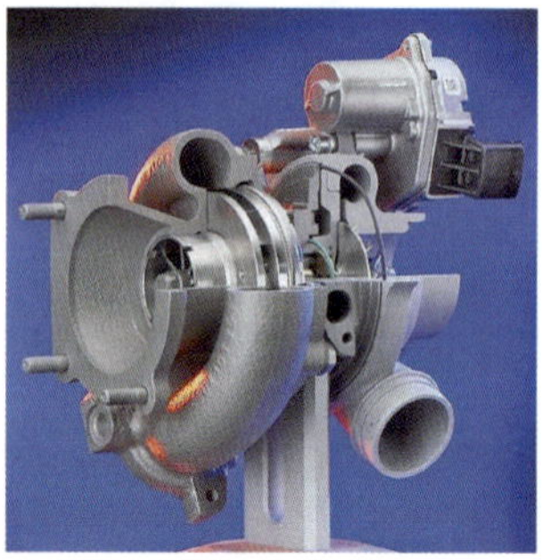

f) variable geometry turbocharger (2)

Fig. 5.2.3. Examples for engine actuators: (1) Decentralized controlled, (2) Centrally controlled. Sources for **a**, **b**, **c**, **d** Continental Automotive GmbH (2011), **e** Klotzbach et al (2003), **f** Gabriel et al (2007).

(4) is realized if the sensor is integrated with an ADC and a microprocessor with a PROM. Then, e.g. disturbance signals can be measured too and compensated, digital filters, static and dynamic corrections can be provided, series tolerances can be reduced through individual calibration, etc. These highly integrated sensors are also called "intelligent" or "smart" sensors, Robert Bosch GmbH (2007a).

The hardware integration of the electronics with the sensor element is usually performed as hybrid technology, i.e. with other semiconductor elements or with ASIC's, or monolithic in form of micro electronic mechanical systems (MEMS), which is a combination of microelectronics and micromechanics. The advantages of this higher integration are a relief of the ECU, uniform, flexible and bus oriented interfaces, higher precision through disturbance compensation, individual calibra-

a) electrical cooling pump, brushless dc motor, 200 W, 12 V, 32 A, 9000 l/h, 800 hPa

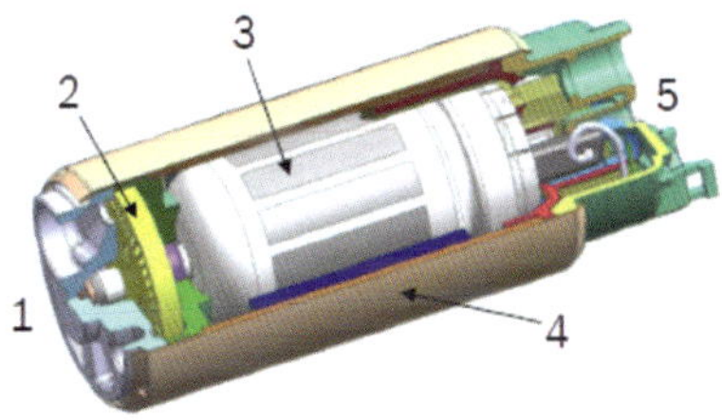

b) fuel pump: 1 suction side, 2 impeller, 3 dc motor, 4 housing, 5 pressure side

c) commonrail high pressure pump, 2 radial pistons, max. 2000 bar, max. 4000 rpm

d) electrical cooling fan

Fig. 5.2.4. Examples for electrical or mechanical driven pumps and a fan for engines, mostly centrally controlled. Sources for **a**, **c** Continental Automotive GmbH (2011), **b** Scholl (2007), **d** Robert Bosch GmbH, Electrical Drives.

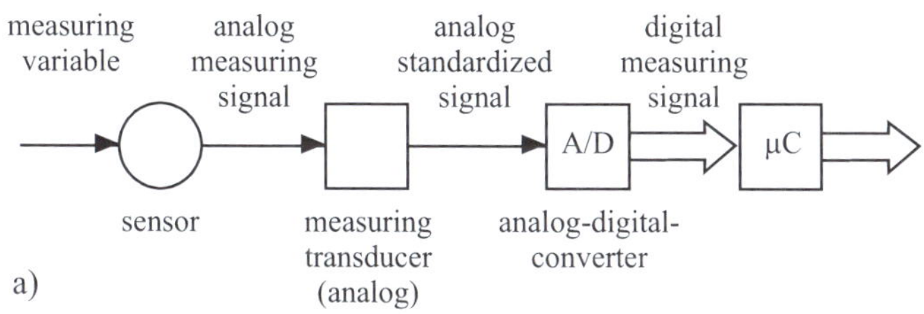

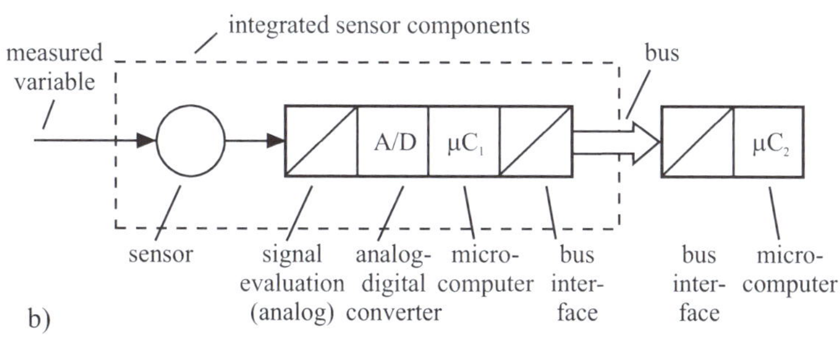

Fig. 5.3.1. Integration of sensor technology. **a** conventional measuring chain with digital processing in the ECU. **b** integrated sensor components with digital processing electronics.

Table 5.3.1. Sensor designs and integration with electronics for combustion engines (examples)

measured quantity	measurement location (examples)	measurement range (examples)	measurement principle (examples)	technology	integration step		
					none	analog electronics	digital electronics
mechanical							
speed	crankshaft	30-7000 rpm	Induct., increm.	magnetic coil	X	–	–
displacement, angle	camshaft	45°	Induct., Halleff.		–	X	–
acceleration	engine block	3 -25 kHz	piezoelectrical	seism. Mass	X	–	–
low pressure	air path	0.1 - 10 bar	piezoresisitive	Si-diaphragm	–	–	X
high pressure	rail	100 - 2200 bar	resistive	steel diaphragm	–	–	X
flow rate	air path	- 1200 kg/h	thermal	resistances	–	–	X
oil level	oil sump	10 - 100 mm	capacitive, ultrasonic		X	–	X
thermal							
temperature	air, coolant,	-40 - 150 °C	resistive	NTC, PTC	X	–	–
	exhaust gas	-40 - 1100 °C	thermocouple		X	–	–
chemical							
oxygen	exhaust path	λ: 0.8 - 1.2, 0.7-8	galv. oxygen cell	solid electrolyte	X	–	–
NO_x	exhaust path		galv. active electrodes	ZrO_2 ceramics	X	–	–
electrical							
voltage	control unit		ADC		–	–	X
current	control unit		measurement shunt		–	X	X

tion and noise free signal transfer. Figure 5.3.2 depicts some sensors which can be classified to various integration steps.

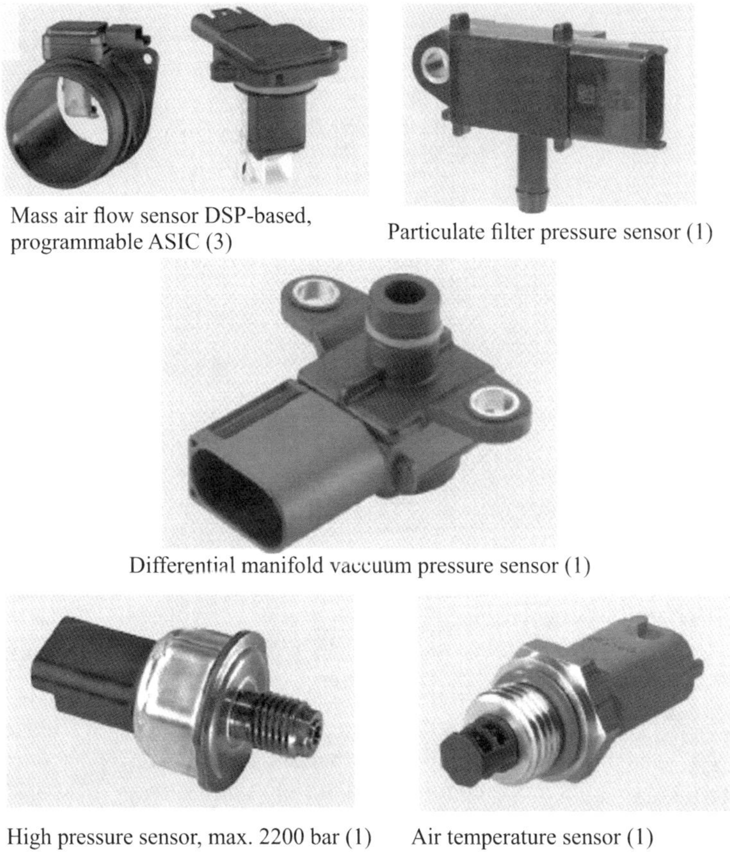

Fig. 5.3.2. Examples of sensors for engines, integration steps (1) to (4) (see text), Continental Automotive GmbH (2011).

5.4 Mechatronic components

Some components of internal combustion engines show an integration of mechanics with electronics and information processing, which is called mechatronics. This integration is between the components (hardware) and the information-driven functions (software) in order to find an optimal balance between the basic mechanical structure, sensor and actuator implementation. One can then distinguish an energy flow (feedforward) and an information flow (feedback), see Fig. 5.4.1, Isermann (2005).

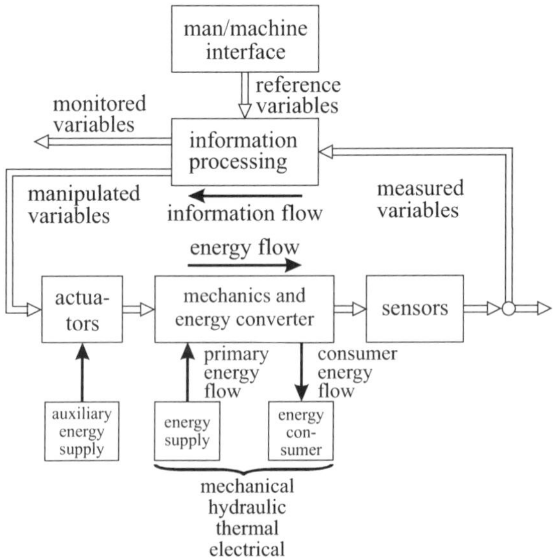

Fig. 5.4.1. Integration of mechanics with electronics and information processing to a mechatronic system.

The mechatronic components can be divided in actuators, drives and pumps. Detailed descriptions of their structure are, e.g. given in van Basshuysen and Schäfer (2004), Robert Bosch GmbH (2007a), Robert Bosch GmbH (2011). Examples of *mechatronic actuators* are, see also Fig. 5.2.3:

- *Electrical throttle*: A DC motor drives the throttle over a gear against a return spring. The position control (usually in the ECU) is based on a potentiometer. A second potentiometer is used for fault detection.

- *Electrical EGR valve*: The EGR valve or flap may be driven by a solenoid or a DC motor, also with integrated electronic position control see, e.g. van Basshuysen and Schäfer (2004).

Some examples of *mechatronic pumps and drives* are, see also Fig. 5.2.4:

- *High-pressure fuel injection.* The high-pressure piston pumps, the rail and the injection are highly integrated systems. Different types of pumps for gasoline and diesel engines with one, three and more pistons were developed over the years, Robert Bosch GmbH (2011). Well-known types for diesel engines are axial-piston distributor pump (1962), analog electrically controlled axial-piston distributor pump (1986), unit pump system (1995), see Fig. 5.4.2a), radial piston distributor pump (1996). These pump systems are characterized by the fact, that the required injection pressure is generated at the moment of each injection and operates together with pressure-controlled injection valves and can for electronically controlled systems be influenced by a solenoid valve. Common-rail systems allow to separate the pump pressure generation from injection quantity and time, see Fig. 5.4.2b). Typical pumps for common-rail are 3 or 1 cylinder radial piston pumps. The basis structure is the same for diesel and gasoline engines, but the pressure is for diesel engines with up to 2000 bar much higher than the maximum 120 bar for gasoline engines. Figure 5.4.3 shows examples for injectors of gasoline engines. The digital control of the pump and the magnetic or piezo-electric driven injection is performed in the ECU. A detailed description and modeling is treated in Sect. 5.5

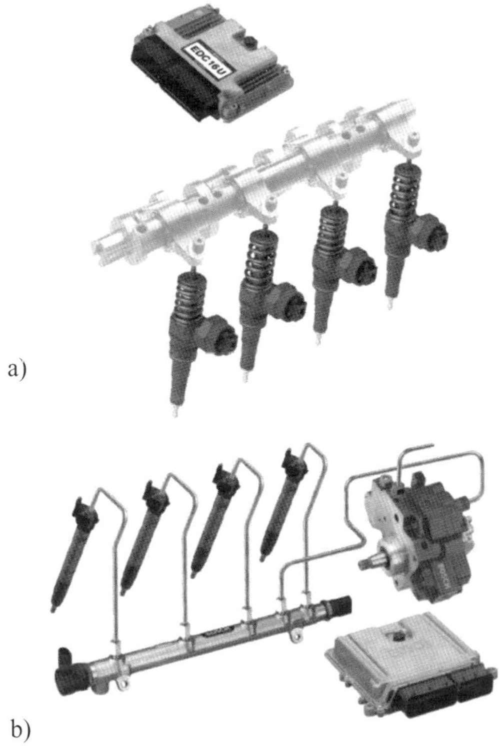

Fig. 5.4.2. Examples of injection systems for diesel engines, Robert Bosch GmbH (2011). **a** unit injector system. **b** common-rail injection system.

- *Radiator fan*. In order to control the engine cooling the radiator fans are for passenger cars driven by a DC motor or brushless DC motor, up to 850W. They are usually speed-controlled with multistage relays and series resistors or continuously with power electronics. The speed control may be integrated in the power electronics or in the ECU. This reduces the number of cables and saves space, Kaiser et al (2000), Schöner (2004).

Port injector (DEKA IV)
1 – 6 bar, min 1.2 ms,

Solenoid direct injector (SDI)
100 bar, min 300 μs

Piezodirect injection (PDI)
50 – 200 bar, min 150 μs, 190 V

Fig. 5.4.3. Examples for gasoline injectors, Continental Automotive GmbH (2011).

- *Electrical coolant pump.* Traditionally, the coolant pump is driven via the V-belt together with other auxiliaries. The generated pressure and flow is then dependent on the engine speed. In order to adjust the required coolant flow to the load, the engine and ambient air temperature a thermostatic expansion valve controls the coolant temperature by changing the valve position and the mass flow to the radiator, see Sects. 4.6 and 7.10. However, this throttling results in energy losses. Electrically driven coolant pumps with engine-independent speed control can be adjusted to the required power without throttling valves. In addition they can be controlled flexibly with regard to fast warming-up, increased cabin heating and cooling after stop of the engine. A typical design consists of an axial-radial impeller, a brushless DC motor and integrated electronics with a special ASIC for speed control. The rotor with permanent magnets is separated from the stator coils by a split tube and therefore no seals at moving parts are required. The electrical power ranges from 200 to 1100 W with a volume flow of 800 to 7.500 l/h, Genster and Stephan (2004). Another actuation principle for a radial centrifugal cooling pump is to manipulate variable vanes after the rotor, a design for engines of commercial vehicles, Wickerath et al (2011).

  As these examples show, some of them are decentralized mechatronic components with local integration of sensors and electronics and some are centrally controlled by the ECU, see also Schöner (2004).

5.5 Fuel supply and injection system

The fuel supply and injection system has developed to an important mechatronic system for both gasoline and diesel engines. Therefore, it is considered and modeled in some more detail. For intake-manifold injection a low-pressure system is sufficient, whereas for direct injection a high-pressure system is required, which is fed by the (standard) low-pressure system, Robert Bosch GmbH (2011).

5.5.1 Low-pressure fuel-supply system

a) Low-pressure fuel-supply configurations

The standard solution for the transport of the fuel from the tank to the engine is performed by an electrical fuel pump. The first electronic gasoline injection system used electric pumps outside of the tank (in-line pumps). The present designs, however, have the pump inside the tank (in-tank pumps), as part of a fuel delivery module. This module is a cylindrical chamber and houses, compare Fig. 5.5.1, e.g. the pump and its DC motor, a suction jet pump for continuous filling, an intake filter to protect the pump, a non-return valve, a pressure relief valve (e.g. 7 bar), a fuel filter at the pressure side, a pressure regulator for returnless supply and a fuel level sensor.

The electrical fuel pump has to deliver about 60 to 250 l/h with pressures of 3 to 6.5 bar. Standard types have permanent magnet-excited DC motors with copper commutator and carbon brushes. The pump designs are either of positive displacement type, like roller-cell pumps or internal-gear pumps, or flow-type pumps, with a rotating ring consisting of peripheral vanes. Both pumps may generate pressures up to 5 bar and have an efficiency of 20–25 %.

Low-pressure fuel supply systems with or without fuel return are distinguished. The *standard configuration*, as depicted in Fig. 5.5.1a), has a mechanic pressure regulator at the fuel rail supplying the injectors at the intake manifold. The diaphragm-based pressure regulator keeps the pressure difference between the fuel and the intake or ambient pressure constant by opening or closing a valve to a return line. Thus a certain excess fuel flow returns back to the tank, which has the drawback that the fuel heats up, generating fuel vapor.

A *returnless configuration* is obtained if the mechanic pressure regulator is located at the fuel tank or intake delivery module. Then a special return line from the engine to the tank is saved and the fuel heating is smaller.

A further improvement is to manipulate the speed of the electrical pump with regard to the required fuel flow. This is called a *control demand fuel supply system* and uses a pressure sensor with electrical output after the pump, as shown in Fig. 5.5.1b). The measured fuel pressure is transmitted to the ECU of the engine, where a digital controller manipulates the armature voltage of the DC motor, e.g. by a pulse-width-modulation. This electronic pressure control system has several advantages: electrical energy is saved, fuel heating is further reduced, the pressure reference value can be increased for warm-start and cold-start, the measured pressure can be taken into account for injection timing and fuel-system diagnostics are possible.

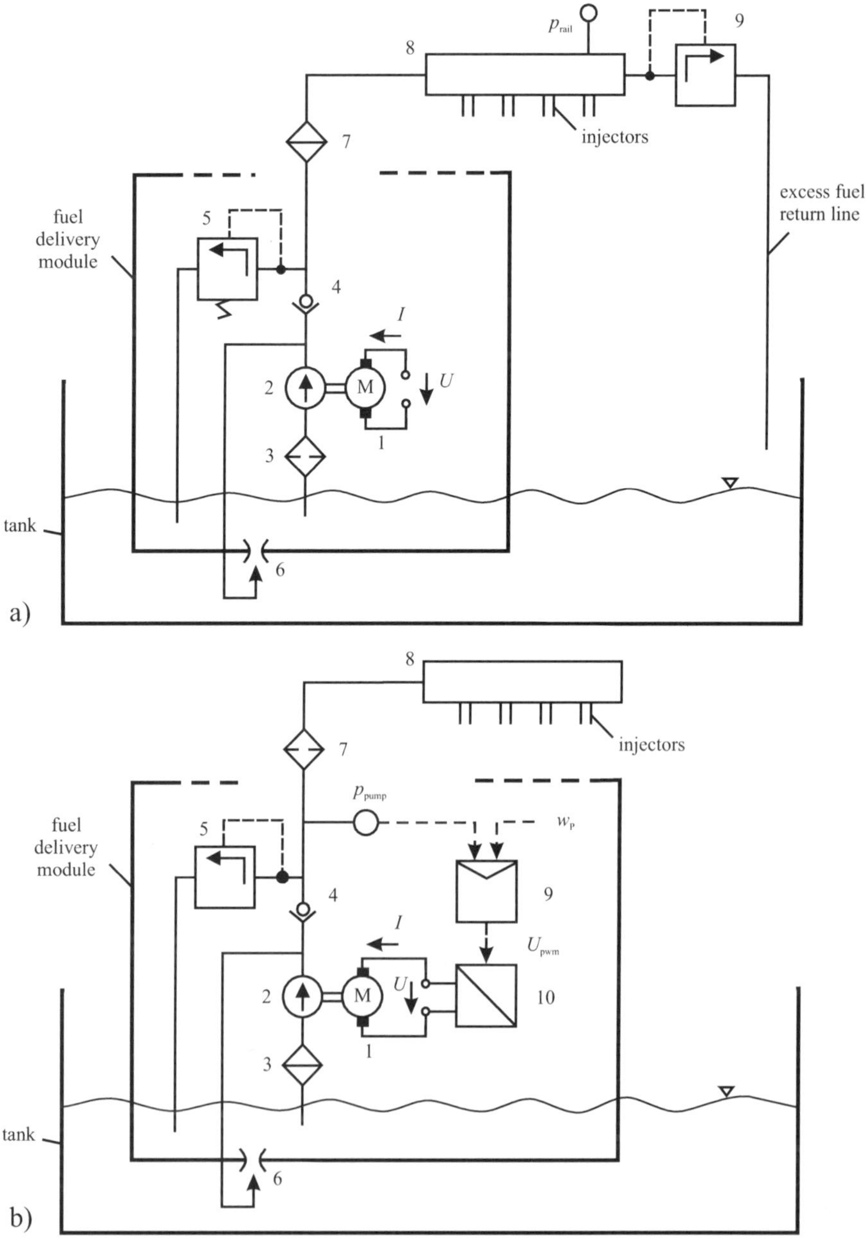

Fig. 5.5.1. **a** Fuel supply for intake manifold injection, with fuel return. 1 DC motor; 2 pump; 3 filter (sieve); 4 non-return valve; 5 pressure relief valve; 6 jet pump; 7 filter; 8 fuel rail; 9 pressure regulator. **b** fuel supply for intake manifold injection, with demand (pressure) control. 1 DC motor; 2 pump; 3 filter (sieve); 4 non-return valve; 5 pressure relief valve; 6 jet pump; 7 filter; 8 fuel rail (without return line); 9 pressure controller (ECU); 10 pulse width modulator for pump voltage.

b) Models of a low-pressure fuel supply system

To investigate the possibilities of signal or process model based methods for pump control design and for fault detection the static and dynamic models of a fuel supply system are required. Figure 5.5.2 shows a two-port signal-flow diagram for the relevant elements. Based on the models given in Isermann (2005), Sect. 6.6.2, the following equations describe the behavior.

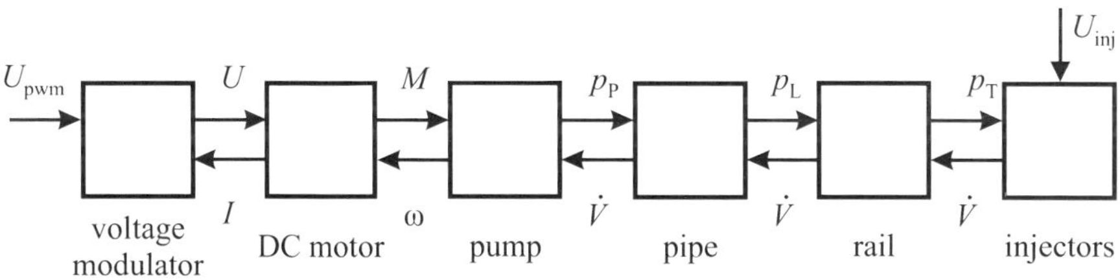

Fig. 5.5.2. Two-port signal-flow diagram of a low-pressure fuel-supply system.

The mechanical part of the DC motor follows

$$J_P \frac{d\omega(t)}{dt} = M_{el}(t) - M_F(t) - M_P(t) \tag{5.5.1}$$

where J_P is the moment of inertia of motor and pump. For the friction torque holds

$$M_F = M_{F1}\omega. \tag{5.5.2}$$

As the time constant of the armature current is negligible small compared to the mechanical time constant it holds

$$R_A I_A(t) = U_A(t) - \Psi\omega(t) \tag{5.5.3}$$

$$M_{el} = \Psi I_A(t). \tag{5.5.4}$$

If the pump is of flow-type the basic behavior of centrifugal pumps for incompressible fluids can be assumed, leading to the pressure difference (proportional to the delivery head)

$$\Delta p_P(t) = h_{nn}\omega^2(t) - h_{nv}\omega(t)\dot{V}(t) - h_{vv}\dot{V}^2(t) \tag{5.5.5}$$

where $\dot{V}(t)$ is the volume flow. This equation includes flow and flow frictional losses.

The pressure difference between the pump pressure and the intake-manifold pressure depends on the throttling losses in the non-return valve, fuel filter, tubes and injectors. Assuming that the related pressure losses are quadratic in the volume flow it holds

$$\begin{aligned} \Delta p_T(t) &= a_f \frac{d\dot{V}(t)}{dt} + h_{fT}\dot{V}^2(t) \\ a_f &= \frac{\rho l}{A}. \end{aligned} \tag{5.5.6}$$

Based on these equations a signal-flow diagram of the fuel supply system can be developed. However, a further simplification can be made if it is assumed that the pressure differences at the pump and the tube system are equal, i.e. $\Delta p_{\mathrm{P}}(t) = \Delta p_{\mathrm{T}}(t)$ and thus neglecting the influence of a low intake manifold pressure. Then, (5.5.5) and (5.5.6) with $a_{\mathrm{f}} = 0$ (small dynamics) can be set equal, resulting in

$$\begin{aligned} \dot{V}(t) &= \kappa\omega(t) \\ \kappa &= \frac{-h_{\mathrm{nv}} + \sqrt{h_{\mathrm{nv}}^2 + 4\left(h_{\mathrm{fT}} + h_{\mathrm{vv}}\right) h_{\mathrm{nn}}}}{2\left(h_{\mathrm{fT}} + h_{\mathrm{vv}}\right)} . \end{aligned} \tag{5.5.7}$$

The pump difference pressure becomes

$$\begin{aligned} \Delta p_{\mathrm{P}}(t) &= \tilde{h}_{\mathrm{nn}}\omega^2(t) \\ \tilde{h}_{\mathrm{nn}} &= h_{\mathrm{nn}} - \kappa h_{\mathrm{nv}} - \kappa^2 h_{\mathrm{vv}} . \end{aligned} \tag{5.5.8}$$

The required power of the pump is

$$P_{\mathrm{P}} = \dot{V}\Delta p_{\mathrm{P}} . \tag{5.5.9}$$

Further it holds

$$P_{\mathrm{P}} = M_{\mathrm{P}}\,\omega . \tag{5.5.10}$$

Hence, the required torque of the pump is

$$M_{\mathrm{P}} = \frac{\dot{V}}{\omega}\Delta p_{\mathrm{P}} . \tag{5.5.11}$$

Inserting (5.5.8) yields

$$M_{\mathrm{P}}(t) = \tilde{h}_{\mathrm{nn}}\dot{V}(t)\omega(t) = \tilde{h}_{\mathrm{nn}}\kappa\omega^2(t) . \tag{5.5.12}$$

Based on these equations a simplified signal-flow results, as shown in Fig. 5.5.3. Hence, the electrical part of the DC motor and the mechanical part of the DC motor and pump show (approximately) linear behavior whereas the fluid part of the pump and the tubes with valves and injectors introduce nonlinear behavior.

5.5.2 High-pressure fuel supply and injection system for gasoline engines

Direct injection gasoline systems require a high-pressure injection system to compress the fuel to 50...120 bar. Typical pump designs are triple-plunger pumps or single-plunger pumps, Robert Bosch GmbH (2011). The *triple-plunger pump* is basically a radial-piston pump with three plungers at an offset angle of 120° driven by the camshaft. It is filled by a low-pressure pump with admission pressure of 3...5 bar through an inlet valve and compresses the fuel volume until the spring-loaded outlet valve opens. The three plungers allow a delivery overlap, such reducing pressure pulses in the rail. The delivery quantity is proportional to the camshaft speed. If the setpoint of the rail pressure is reached, the rail pressure controller opens

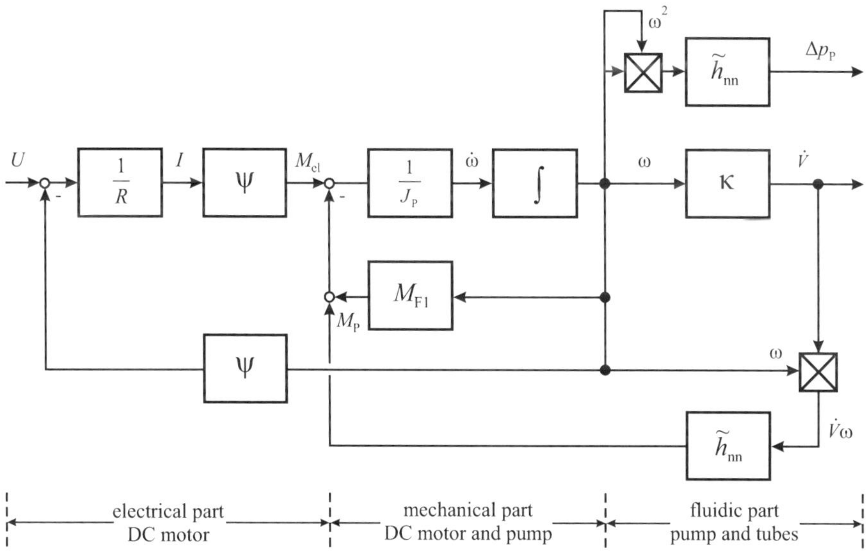

Fig. 5.5.3. Simplified signal-flow diagram of a low-pressure fuel supply system.

the rail pressure control valve and excess fuel is returned to the intake side of the high-pressure pump at reduced pressure level, see Fig. 5.5.4a).

A *single-plunger pump* possesses an electrically switchable delivery-quantity control valve, which is activated such that only a definite fuel quantity is pressurized. Then only the required fuel quantity is delivered to the rail, which is specified by the rail pressure control and a return line is not necessary, see Fig. 5.5.4b). This design also saves mechanical power and avoids an increase of the fuel temperature.

5.5.3 High-pressure common-rail fuel supply and injection system for diesel engines[1]

This section considers the function and modeling of a common-rail system for diesel engines in more detail, Clever (2011). Figure 5.5.5 shows a scheme of a CR-system as it is typical for passenger car diesel engines. It accumulates the fuel under high pressure in the common-rail in order to supply the injectors time-independently and with a large scale of pressure levels, Robert Bosch GmbH (2007b), Hammer et al (2014). The fuel is delivered from the tank and pre-compressed by an electrical fuel pump to about 0.5 MPa (5 bar). The fuel's actual temperature is measured in the fuel filter. After filtration the fuel flows to the tooth belt driven high-pressure pump. The volume flow into the pump is controlled by the metering valve. Its supply current is measured by the ECU. In the high-pressure pump the fuel is compressed to the common rail pressure level (30-180 MPa or 300-1800 bar) and then discharged to the rail. From there it is injected into the cylinders each time an injector opens. The common

[1] Extract from Clever (2011)

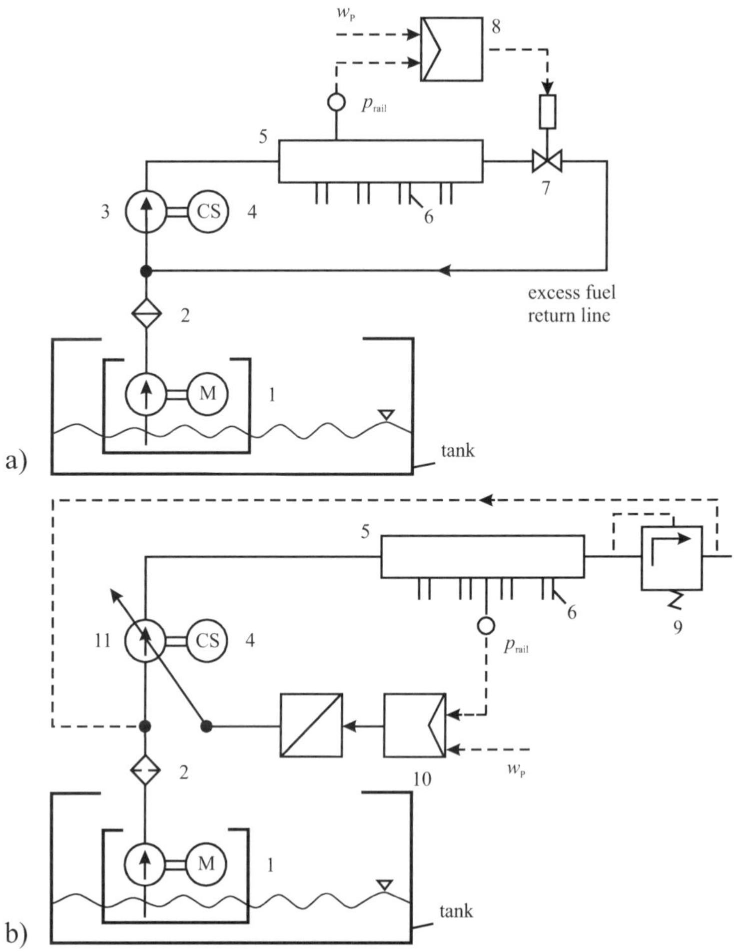

Fig. 5.5.4. **a** High-pressure fuel supply system with pressure control valve and fuel return. **b** high-pressure fuel supply system with rail pressure control by an adjustable pump (control demand). 1 fuel delivery module with low-pressure pump; 2 filter; 3 high-pressure pump; 4 camshaft; 5 high-pressure fuel rail; 6 high-pressure injectors; 7 pressure control valve and relief valve; 8 rail pressure controller; 9 rail pressure relief valve; 10 rail pressure controller; 11 adjustable high-pressure pump.

rail pressure is measured by a high-pressure sensor and controlled by the pressure control valve. Its supply current is measured, too. Thus, the essential variables are:

T_f fuel temperature measured in the fuel filter
I_{mv} supply current of the metering valve
I_{pcv} supply current of the pressure control valve
n_{eng} engine speed
u_{inj} desired injection quantity
p_{cr} common rail pressure.

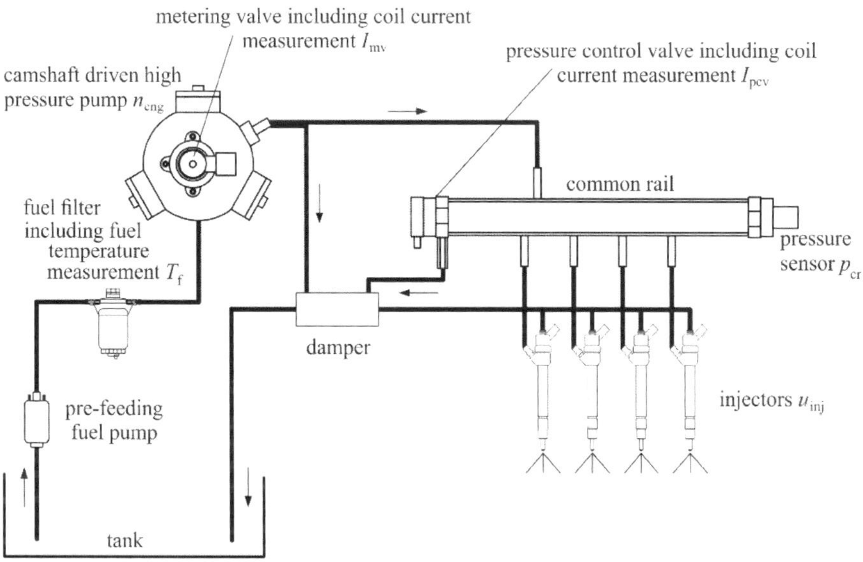

Fig. 5.5.5. Scheme of the common rail injection system with measurable variables.

a) Common rail pressure generation

To model the dynamic behavior of the common rail pressure a lumped parameter system is assumed. The mass balance of the common rail is given by

$$\dot{m}_{hpp} - \dot{m}_{pcv} - \sum_{j=1}^{4} \dot{m}_{inj,j} = \dot{m}_{cr} \tag{5.5.13}$$

with

$\dot{m}_{hpp}$ fuel mass flow from the three pistons of the high-pressure pump into the common rail
$\dot{m}_{pcv}$ fuel mass flow through the pressure control valve
$\sum_{j=1}^{4} \dot{m}_{inj,j}$ fuel mass flow to the four injectors
$\dot{m}_{cr}$ stored fuel mass flow accumulated inside the common rail.

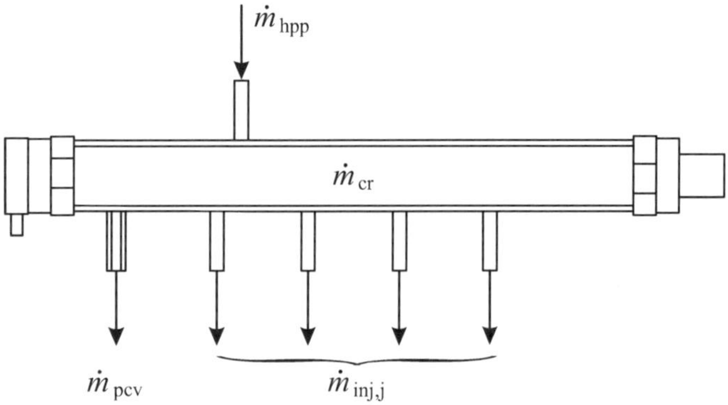

Fig. 5.5.6. Mass flow *into* and *from* the common rail.

The fuel mass flows *into* and *from* the common rail are illustrated in Fig. 5.5.6.

The change of the fuel mass accumulated inside the common rail can also be expressed by its change of density. Assuming that the volume of the common rail remains constant it holds

$$\dot{m}_{\text{cr}} = V_{\text{cr}} \cdot \frac{d\rho_{\text{cr}}}{dt} \tag{5.5.14}$$

with

V_{cr}	volume of the common rail
ρ_{cr}	fuel density.

By means of the bulk modulus E, the compressibility is expressed by

$$dp = \frac{E}{\rho} \cdot d\rho. \tag{5.5.15}$$

Applying (5.5.13) to 5.5.15 and expressing the mass flow by means of its corresponding volume flow, it follows

$$\frac{dp_{\text{cr}}}{dt} = \frac{E_{\text{cr}}}{V_{\text{cr}}} \cdot \left(\dot{V}_{\text{hpp}} - \dot{V}_{\text{pcv}} - \sum_{j=1}^{z} \dot{V}_{\text{inj},j} \right) \tag{5.5.16}$$

with

E_{cr}	pressure- and temperature-dependent bulk modulus
$\dot{V}_{\text{hpp}}$	volume flow through the high-pressure pump
$\dot{V}_{\text{pcv}}$	volume flow through the pressure control valve
$\sum_{j=1}^{z} \dot{V}_{\text{inj,j}}$	volume flow to z injectors.

The different volume flows excite the pressure inside the common rail periodically. This results in a special characteristic of the high-pressure sensor signal. In stationary operation points the signal characteristic is mainly determined by the periodic delivery of the fuel and the periodic injection processes. The special characteristic of the common rail pressure signal is addressed in Clever (2011).

The considered high-pressure pump is a 3-cylinder radial piston pump, Fig. 5.5.7, Robert Bosch GmbH (2007b). The pump consists of three pump elements located concentrically around the drive shaft. The upward motion of the pistons of the pump elements is performed by an eccentric, that is located on the drive shaft. The downward motion of the pistons is executed by means of the force of a return spring, that is attached to each of the pistons. During the downward motion, fuel is floating through an inlet valve into each of the pump elements. In the following step the fuel is compressed by the upward motion of the pistons and the inlet valve is closing. As soon as the pressure inside the pump cylinders exceeds the pressure level inside the common rail, the discharge valve opens and the fuel streams into the common rail. The drive shaft is actuated by the crank shaft driven tooth belt. The transmission ratio between the drive shaft and the crank shaft is 2:3. This means that during each injection, or each 180°CS, one of the three pump elements is delivering fuel. The relation between the rotation angle of the eccentric shaft (unit: °EA) and the crank shaft is given by

$$\chi = \frac{2}{3} \cdot \varphi \tag{5.5.17}$$

with

χ rotation angle of the eccentric shaft in °EA.

On the low-pressure side, the delivery rate of the high-pressure pump can be influenced by a throttle valve (metering valve mv), Fig. 5.5.8. The metering valve is located in front of the inlet valve of the high-pressure pump and controls the fuel mass delivered by the pump.

The volume flow through the metering valve can be described by following throttle equation

$$\dot{V}_{\mathrm{mv}} = \alpha_{\mathrm{mv}} \cdot A_{\mathrm{mv}} \cdot \sqrt{\frac{2\Delta p_{\mathrm{mv}}}{\rho}} \tag{5.5.18}$$

with

$\dot{V}_{\mathrm{mv}}$ fuel volume flow through the metering valve
α_{mv} flow coefficient
A_{mv} opening cross section
Δp_{mv} pressure difference between valve inlet and valve outlet.

The volume flow delivered by pumps that are throttled at their suction side show a characteristic behavior of fuel delivery. Figure 5.5.9 illustrates the characteristic volume flow delivered by such a high-pressure pump with a constant throttling and constant common rail pressure with respect to the pump speed, see Welschof (1992).

Two main areas of the pump's behavior can be identified:

(1) The volume flow raises linearly up to a limiting speed.

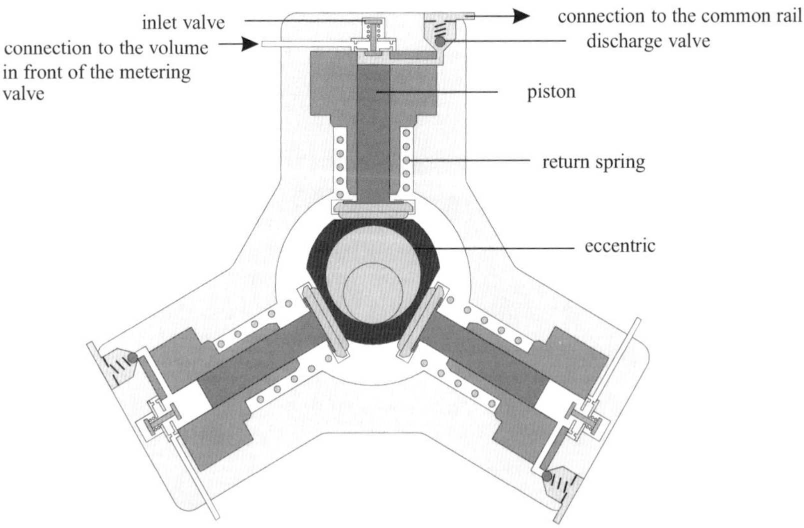

Fig. 5.5.7. Sectional view of the 3-cylinder radial piston pump CP1H by Bosch.

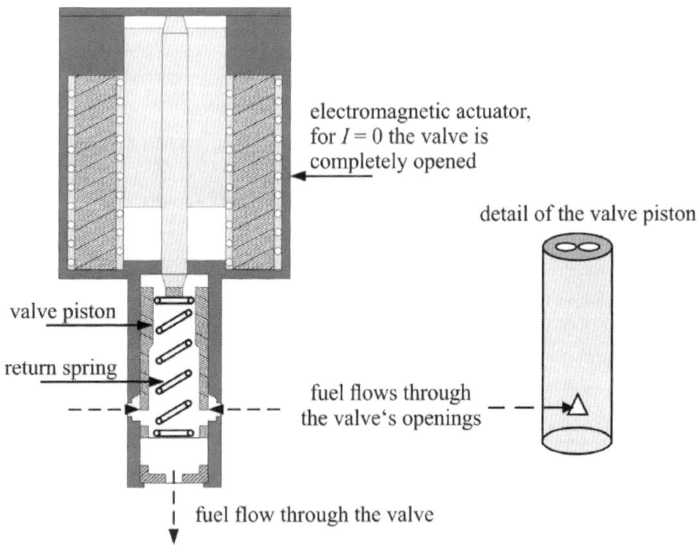

Fig. 5.5.8. Sectional view of the metering valve.

(2) The volume flow decreases slightly for higher speeds.

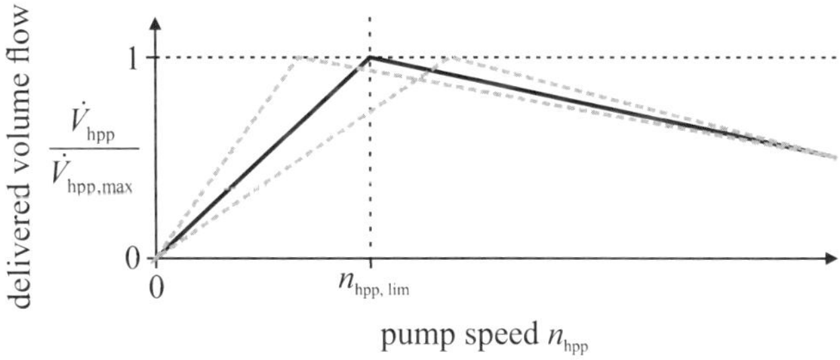

Fig. 5.5.9. Illustration of the characteristic volume flow delivered by a high-pressure pump with a constant throttling at its suction side and constant common rail pressure. The gray dashed lines indicate the influence of different restrictions in front of the high-pressure pump on the limiting speed. The smaller the restriction the smaller the limiting speed.

This behavior can be explained with a detailed view at the different phases of a single delivery cycle, Welschof (1992). For this purpose, the delivery cycle is divided into three phases:

Phase 1: Starting at the top dead center, the piston moves downwards, the outlet valve closes and the pressure in the piston chamber decreases fast. Gas releases from the remaining fuel, filling the increasing volume.

Phase 2 (pump speed lower than limiting speed): The inlet valve opens and fuel flows through the valve into the piston chamber. The inlet valve keeps open until the whole volume of the piston chamber is filled with fuel at the bottom dead center. The previously released gas has been dissolved again. This holds if the pump speed is lower than the limiting speed, Fig. 5.5.9.

Phase 2 (limiting speed is exceeded): The inlet valve opens and fuel flows through the valve into the piston chamber. If the limiting speed of the pump is exceeded, the bottom dead center is reached before the complete volume of the piston chamber is filled with fuel. Thus, previously released gas remains in the piston chamber.

Phase 3 (pump speed lower than limiting speed): The piston moves upwards and the pressure inside the piston chamber increases fast. The inlet valve closes immediately. Once the common rail pressure level is exceeded, the outlet valve opens and the compressed fuel is discharged in the common rail.

Phase 3 (limiting speed is exceeded): The piston starts moving upwards. Previously released gas is still in the piston chamber. The pressure inside the piston chamber increases slowly. The inlet valve remains open and fuel flows inside the chamber, until the complete gas is dissolved. Thus, not the maximum available volume of the piston chamber is filled with fuel. Hence, the delivered fuel volume decreases. The rotation angle of the eccentric shaft from the bottom dead center till the desired (common rail) pressure level is reached is called *non-delivery angle* χ_{nd}.

During each revolution, each pump element delivers an amount of fuel that is equal to its effective fuel delivery volume. Assuming a cosine stroke curve with volume change, the displacement volume of a particular pump element with respect to the rotation angle of the eccentric shaft is given by (see Fig. 5.5.10)

$$V_{\text{hpp},i}(\chi_i) = \frac{V_{\text{hpp,max}}}{2}\left(1-\cos(\chi_i)\right), \chi_i := \chi + (i-1)\frac{2}{3}\cdot 180°\text{EA} \quad (5.5.19)$$

with

$\frac{dV_{\text{hpp,i}}}{d\chi_i}$	displacement volume of the pump element i per $d\chi_i$
$V_{\text{hpp,max}}$	cylinder capacity of one pump element.

The change of the displacement volume is found to

$$\frac{\frac{dV_{\text{hpp},i}}{d(\chi_i)}}{d\chi_i} = \frac{V_{hpp,max}}{2}\sin(\chi_i), \chi_i := \chi + (i-1)\frac{2}{3}\cdot 180°\text{EA} \quad (5.5.20)$$

with

$\frac{\frac{dV_{\text{hpp},i}}{d(\chi_i)}}{d\chi_i}$	change of the displacement volume of the pump element i.

If it is assumed that the fuel delivery of a single cylinder starts if the non-delivery angle is reached and ends at its top dead center (180°EA), the fuel volume flow delivered by the pump with respect to the rotation angle of the eccentric shaft can be expressed as:

$$\frac{dV_{\text{hpp}}}{d\chi} = \begin{cases} 0 & \text{for } -180°\text{EA} < \chi_i < \chi_{\text{nd},i} \\ \sum_{i=1}^{3}\frac{V_{\text{hpp,max}}}{2}\sin\chi_i & \text{for } \chi_{\text{nd},i} \le \chi_i \le 180°\text{EA}. \end{cases} \quad (5.5.21)$$

The relation between the time domain and the rotation angle of the eccentric shaft can be expressed with (5.5.17) and (5.5.20) as:

$$\varphi = \frac{360°\text{CS}}{60\frac{\text{s}}{\text{min}}}\cdot n_{\text{eng}}\cdot t \quad (5.5.22)$$

$$\chi = \frac{2}{3}\cdot\frac{360°\text{CS}}{60\frac{\text{s}}{\text{min}}}\cdot n_{\text{eng}}\cdot t \quad (5.5.23)$$

$$\chi_i = \frac{2}{3}\cdot\frac{360°\text{CS}}{60\frac{\text{s}}{\text{min}}}\cdot n_{\text{eng}}\cdot t + (i-1)\frac{2}{3}\cdot 180°\text{EA} \quad (5.5.24)$$

$$\frac{d\chi}{dt} = \frac{2}{3}\cdot\frac{360°\text{CS}}{60\frac{\text{s}}{\text{min}}}\cdot n_{\text{eng}}. \quad (5.5.25)$$

The fuel volume flow delivered by the pump is finally given by

$$\frac{dV_{\mathrm{hpp}}}{dt}=\frac{2}{3}\cdot\frac{360°\mathrm{CS}}{60\frac{\mathrm{s}}{\min}}\cdot n_{\mathrm{eng}}\begin{cases}0 & \text{for } -180°\mathrm{EA}<\chi_i<\chi_{\mathrm{nd},i}\\ \sum_{i=1}^{3}\frac{V_{\mathrm{hpp,max}}}{2}\sin\chi_i & \text{for } \chi_{\mathrm{nd},i}\leq\chi_i\leq 180°\mathrm{EA}\end{cases} \tag{5.5.26}$$

with

$\frac{dV_{\mathrm{hpp}}}{dt}$ fuel volume flow delivered by the high-pressure pump.

Using the non-delivery angle the difference between the real and the theoretical volume flow is considered. An idealized illustration of the discontinuous fuel delivery described by (5.5.26) is shown in Fig. 5.5.10.

Hence the fuel pump operates with "phase-modulated control". The volume flow streaming from the high-pressure pump, therefore, primarily depends on the engine speed, the opening cross section of the metering valve, the fuel pressure in the low-pressure system and the bulk modulus of the fuel inside the piston chambers. A block diagram of the main influencing variables on the generation of the volume flow through the high-pressure pump is depicted in Fig. 5.5.11 where p_{lp} represents the low pressure of the pre-compressed fuel.

b) Volume flow through the pressure control valve

The pressure control valve is an electromagnetically impelled spherical valve, Fig. 5.5.12. The volume flow through the pressure control valve, primarily depends on the opening cross section of the valve and the pressure inside the common rail, Fig. 5.5.13. The volume flow though the valve can also be expressed by means of a throttle equation and is given by:

$$\dot{V}_{\mathrm{pcv}}=\alpha_{\mathrm{pcv}}\cdot A_{\mathrm{pcv}}\cdot\sqrt{\frac{2\Delta p_{\mathrm{pcv}}}{\rho}} \tag{5.5.27}$$

with

$\dot{V}_{\mathrm{pcv}}$ fuel volume flow through the pressure control valve
α_{pcv} flow coefficient
A_{pcv} opening cross section
Δp_{pcv} pressure difference between valve inlet and valve outlet ($\Delta p_{\mathrm{pcv}}\approx p_{\mathrm{cr}}$).

c) Volume flows to the injectors

By means of the z solenoid valve injectors, fuel is injected into the individual cylinders. The initiation of the injections, their duration as well as the quantity of injections is controlled by the ECU. The injector and in particular the space around

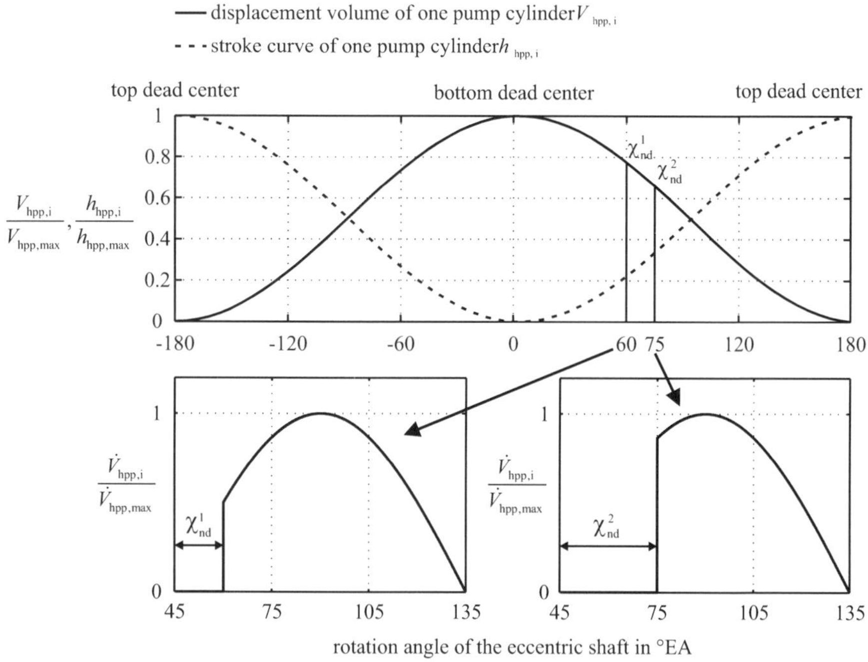

Fig. 5.5.10. Illustration of the discontinuous fuel delivery of the high-pressure pump. The figure shows the cosinusoidal stroke and corresponding volume of a single pump cylinder with respect to the rotation angle of the eccentric shaft. Additionally, two different non-delivery angles are marked (60°EA and 75°EA). For these angles, the corresponding volume flows are shown. All values are normalized.

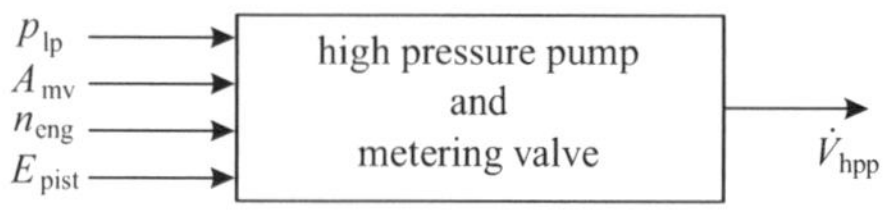

Fig. 5.5.11. Main influencing variables on the generation of the volume flow through the high-pressure pump.

the injector needle and the fuel control chamber (FCC) is supplied with fuel that is directly streaming from the common rail.

The function of the solenoid valve injector, Fig. 5.5.14, can be divided into three operation modes:

Injector closed: The solenoid valve is not actuated. The fuel pressure levels inside the control chamber as well as on the pressurized surface are equal. The compressive force acting from the control chamber (FCC) upon the needle plus the force from the return spring exceed the pressure force that acts on the pressurized surface. Thereby the needle is pressed on the jet port.

Injector opening: After the solenoid valve is actuated, the fuel streams out of the control chamber (FCC) into the return line and the pressure decreases. The

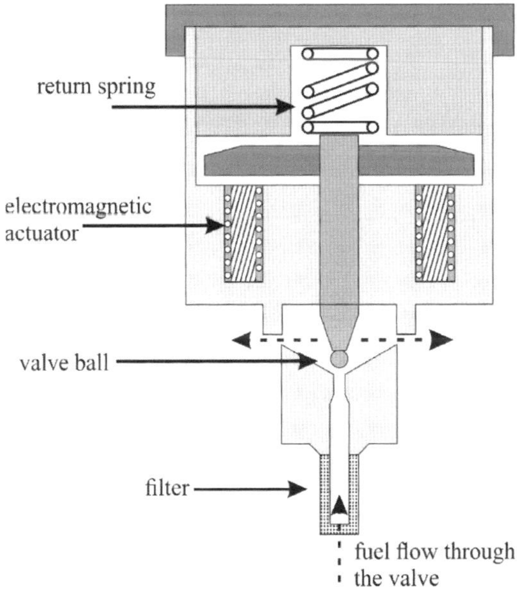

Fig. 5.5.12. Sectional view of the pressure control valve.

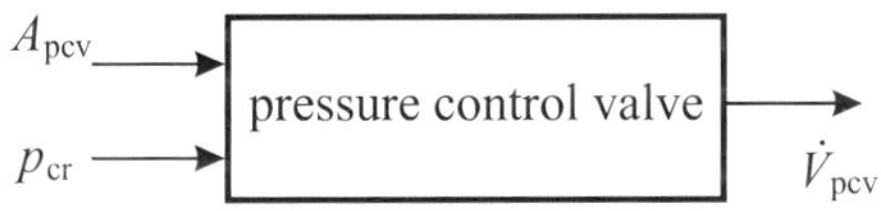

Fig. 5.5.13. Main influencing variables on the volume flow through the pressure control valve.

throttling in front of the control chamber is designed such that the decrease in pressure is not compensated by the feed from the common rail. After a short period of time the pressure, that acts upon the pressurized surface exceeds the opposing forces. The injector opens and remains in this position for the duration of the actuation of the solenoid valve.

Injector closing: As soon as the solenoid valve closes through a valve spring and therefore the volume flow through the valve is being stopped, the pressure inside the control chamber is built up again. Once the resulting compressive force, in addition to the force from the return spring, starts to exceed the opposing force, the injector closes again.

The volume flow to one of the injectors is composed of the injected and the return flow quantity, which in turn depend on the opening duration of the injectors and return flow of the solenoid valve. As a measure for the volume flow, it is feasible to use the desired injection quantity, Fig. 5.5.15. For a more detailed specification of the processes inside the injectors see Seykens et al (2005).

The influence of the periodic operations of the high-pressure pump and the injectors on the rail pressure is investigated in Clever (2011) and used for fault detection.

Based on the component models of the injection system a signal-flow chart for the pressure generation in the common rail can be established, as depicted in Fig. 5.5.16.

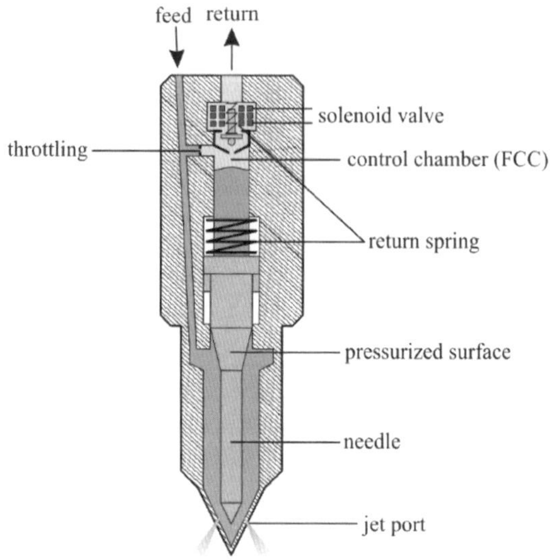

Fig. 5.5.14. Setup of a solenoid valve injector.

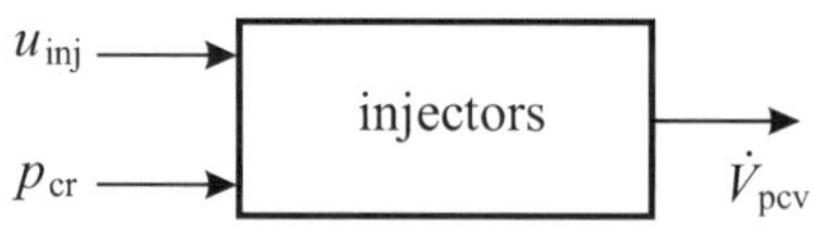

Fig. 5.5.15. Main influencing variables on the volume flow through the solenoid valve injector.

d) Thermodynamic properties of diesel fuel

The thermodynamic properties of diesel fuel like its liquid density and liquid velocity of sound are important for modeling the common rail injection system. Diesel fuel is a mixture of several hydrocarbons mainly paraffin, naphthalene and aromatics, see e.g. Mollenhauer and Tschöke (2010). Its thermodynamic properties depend on the composition of the used fuel. This composition and the fuel's thermodynamic properties respectively, depend on different factors like its base material (e.g. pure oil or renewable primary products). However, differences in the composition of the diesel fuel cannot be recognized in practical approaches. Thus, it is mandatory to approximate the different diesel fuels using equations derived for a representative

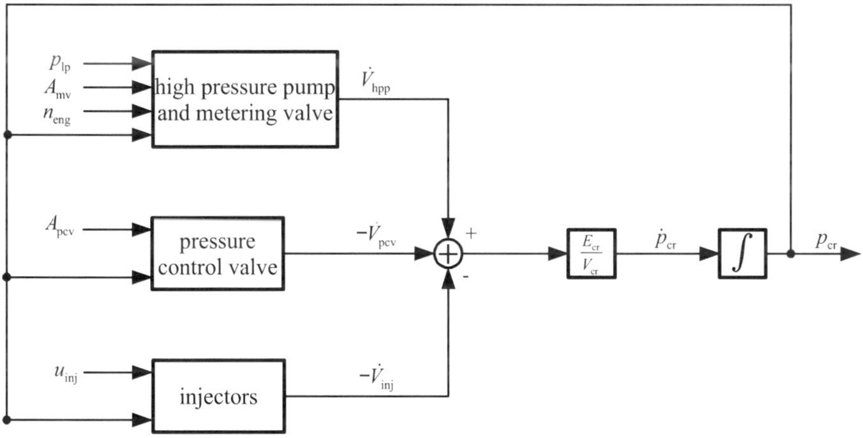

Fig. 5.5.16. Signal-flow chart for the common rail pressure.

fuel composition. There are some publications containing such equations, see e.g. Thiemann (1988) and Keppeler (1997). One publication containing a comprehensive review of the existing data and an approximation of the thermodynamic properties in the needed pressure range up to 2000 bar is Kolev (2007). The equations for the required approximations are reviewed in Clever (2011). In Fig. 5.5.17 the bulk modulus and the fuel's density with respect to the fuel's pressure for different temperatures are shown exemplarily.

The considered models for the high-pressure common-rail injection system are used for model-based fault diagnosis as shown by Clever (2011).

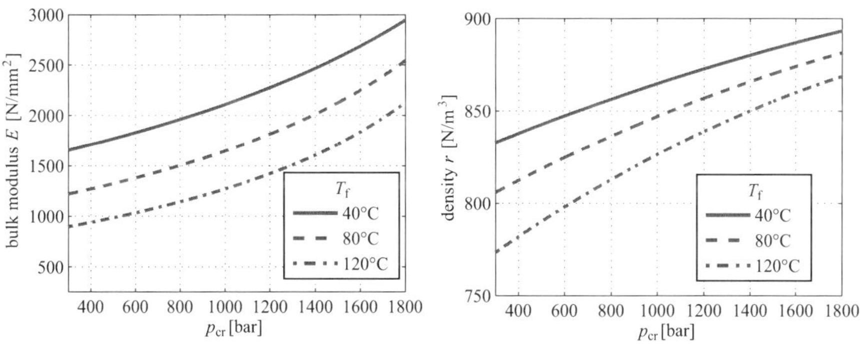

Fig. 5.5.17. Bulk modulus and density of diesel fuel with respect to the fuel's pressure and temperature, see also Kolev (2007) and Clever (2011).

5.6 Variable valve timing system

Valve timing for the inlet and outlet valves is an important variability to improve combustion, torque and emissions. Because of the relatively large torque peaks, avoidance of backlash and damping properties until now mainly hydraulic actuation is applied.

5.6.1 Camshaft phasing

The camshaft timing or positioning is performed relatively to the crankshaft with a positioning range of 40 to $60°$CS (corresponds to 20 to $30°$CAM of the camshaft). An electro-hydraulic servo valve manipulates the oil flow to and from the camshaft actuator. This actuator has, e.g. a rotary design with a sliding vane wheel or an axial piston/cylinder design with a helical teeth gear, see e.g. Heisler (1995), Hannibal et al (2002), van Basshuysen and Schäfer (2007), Kopp (2006), Meinig and Bohner (2013). Thus, the hydraulic camshaft actuation can be sketched as depicted in Fig. 5.6.1. The housing of the camshaft actuator with its chain or belt driven pulley, the stator, is connected with two oil channels to the servo control valve. The rotor is attached to the camshaft and turns via sliding vanes relatively to the pulley depending on the pressure difference in two oil chambers. The servo-control valve is connected with the engine oil circuit and actuated with PWM voltage from the ECU. Therefore, it operates like a proportional valve. The spool of the servo valve operates against a spring which holds it without electrical actuation in a basis position. Then the full oil pressure moves the vanes of the rotor to one restriction, such that in the case of the inlet camshaft a late valve timing is obtained. Without oil pressure the actuator is mechanically locked, leading to a mechanical stiff connection. The camshaft phasing control loop consists of the angle sensor, the servo valve and the camshaft actuator, as shown in Fig. 5.6.1. The camshaft phase $\Delta\varphi_{\mathrm{cam}}$ is usually determined from the difference of angular position sensor signals of the camshaft and crankshaft, Hannibal et al (2002).

5.6.2 Models and control of a hydraulic camshaft phasing system

As an example the camshaft phasing system with a sliding vane actuator of Fig. 5.6.1 is considered. The derived models lean on Münchhof (2006) and Isermann (2005). The magnetic force F_{m} of an electromagnet depending on the applied current I_{a} is generally a nonlinear function $F_{\mathrm{m}}(z)$ depending on the design of the armature and its counter part. However, it can be assumed that its characteristic can be approximated at least piecewise by a linear relation.

$$F_{\mathrm{m}} = c_{\mathrm{z}}(z) I_{\mathrm{a}}. \tag{5.6.1}$$

The spool operates against a spring with force

$$F_{\mathrm{s}} = c_{\mathrm{s}} z_{v}. \tag{5.6.2}$$

Neglecting dynamics and the magnetic hysteresis it holds $F_\mathrm{m} = F_\mathrm{s}$ and the position of the spool follows to

$$z_v = \frac{c_\mathrm{z}}{c_\mathrm{s}} I_\mathrm{a} = k_\mathrm{I} I_\mathrm{a} \tag{5.6.3}$$

for an applied current and with $U_\mathrm{a} = R I_\mathrm{a}$

$$z_v = \frac{c_\mathrm{z}}{c_\mathrm{s} R} U_\mathrm{a} = k_\mathrm{U} U_\mathrm{a} \tag{5.6.4}$$

for an applied voltage. Usually this voltage is manipulated by pulse width modulation (PWM) and changing the duty cycle.

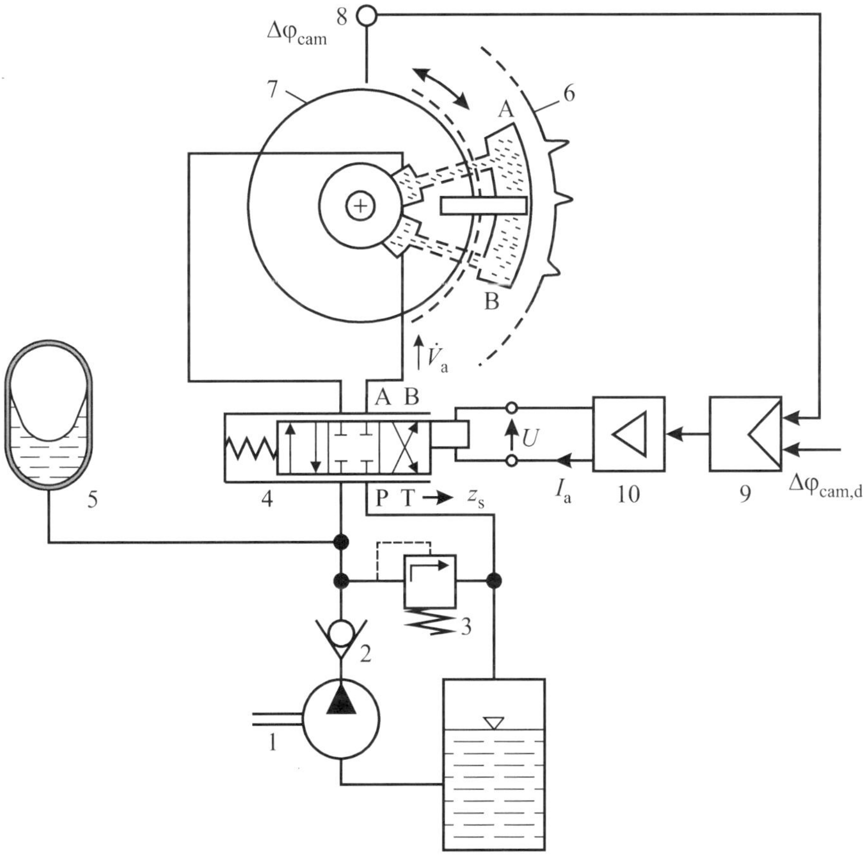

Fig. 5.6.1. Scheme of a hydraulic camshaft phasing system. 1 oil pump, 2 nonreturn valve, 3 pressure-relief valve, 4 4/3-servo control valve with connections to oil supply P, oil sump T, chamber A and B, 5 accumulator (optional), 6 pulley (stator), chain or belt driven, 7 rotor attached to camshaft, 8 phase sensor, 9 camshaft phase controller (ECU), 10 amplifier, driver (ECU).

The spool has overlapped valves. Therefore, the two ways to the actuator chambers A and B are closed in the center position $\Delta\varphi_{\text{cam}} = 0$, such that the position of the phasing actuator is frozen and the enclosed fluid has the same pressure $p_{\text{A}} = p_{\text{B}}$ on both sides. After the spool passes the valve lapping with $z_v > z_{v0}$, e.g. the orifice opens to the supply pressure with $p_{\text{s}} = 1\ldots5$ bar and the other orifice B opens to the oil sump with $p_{\text{a}} \approx 1$ bar. The volume flows through the valve orifices can with $z'_v = z_v - z_{v0}$ be described by

$$\dot{V}_{\text{j}}\left(z'_v\right) \underset{j=A,B}{=} \alpha_{\text{D}} A_v\left(z'_v\right)\sqrt{\frac{2}{\rho}}\sqrt{|\Delta p_v|}\text{sign}\left(\Delta p_v\right) \quad (5.6.5)$$

with α_{D} a contraction number, A_v the effective opening area and the pressure difference

$$\Delta p_{v\text{A}} = p_{\text{s}} - p_{\text{A}} \text{ and } \Delta p_{v\text{B}} = p_{\text{B}} - p_{\text{a}} \quad (5.6.6)$$

where p_{a} is the ambient pressure. For the volume flows holds

$$\dot{V}_{\text{A}} = -\dot{V}_{\text{B}} \quad (5.6.7)$$

if leakages over the vanes are neglected.

Figure 5.6.2 depicts the volume flow characteristic $\dot{V}(I_{\text{a}})$ for a spool valve. In the center position around $I_{\text{a}} = I_{\text{a0}}$ the volume flows $\dot{V}_{\text{A}}$ and $\dot{V}_{\text{B}}$ are zero. The oil is then trapped in the chambers and the rotor is fixed in a certain position. The characteristics also show a hysteresis behavior which originates from the electromagnet and dry friction of the spool valve.

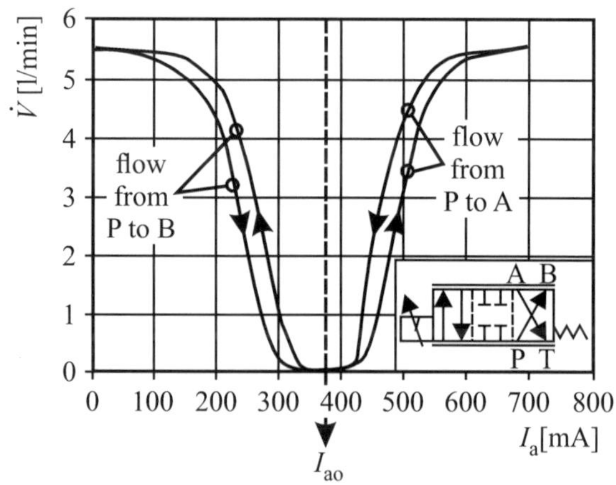

Fig. 5.6.2. Volume flow characteristic of a 4/3 proportional servo valve for oil pressure p_{s} until 5 bar, Hannibal et al (2002).

Changes of the armature current $\Delta I_{\text{a}}(t) = I_{\text{a}}(t) - I_{\text{a0}}$ lead to a change $\Delta z'_v(t) = z_v(t) - z_{v0}$ of the spool valve and corresponding flows $\Delta\dot{V}_{\text{A}}(t)$ and

$\Delta \dot{V}_{\mathrm{B}}(t)$. The resulting pressure changes $\Delta p_{\mathrm{A}}(t)$ and $\Delta p_{\mathrm{B}}(t)$ and consequently the pressure difference

$$\Delta p_{\mathrm{rot}}(t) = \Delta p_{\mathrm{A}}(t) - \Delta p_{\mathrm{B}}(t) \tag{5.6.8}$$

generate a rotor torque ΔM_{cam} which turns the rotor and the camshaft with a change of the angular speed $\Delta \dot{\varphi}_{\mathrm{cam}}(t)$.

The flow through the spool valve can after linearization be approximated by

$$\begin{aligned} \dot{V}_{\mathrm{A}} &= k_{\dot{\mathrm{V}}} z_v' - k_{\mathrm{c}} p_{\mathrm{A}} \\ \dot{V}_{\mathrm{B}} &= -\left(k_{\dot{\mathrm{V}}} z_v' - k_{\mathrm{c}} p_{\mathrm{B}}\right) \end{aligned} \tag{5.6.9}$$

with the coefficients following from valve characteristics

$$k_{\dot{\mathrm{V}}} = \frac{\partial \dot{V}_{\mathrm{j}}}{\partial z_v} \; ; \; k_{\mathrm{c}} = \frac{\partial \dot{V}_{\mathrm{j}}}{\partial p_{\mathrm{A}}} \tag{5.6.10}$$

which depend on the supply pressure p_{s}.

The dynamic behavior of the oil chamber follows from a *mass balance* of the oil

$$\frac{dm_{\mathrm{A}}(t)}{dt} = \dot{m}_{\mathrm{A,in}}(t) - \dot{m}_{\mathrm{A,out}}(t) \tag{5.6.11}$$

where the mass inflow is with (5.6.9)

$$\dot{m}_{\mathrm{A,in}} = \dot{V}_{\mathrm{A}} \rho_{\mathrm{A}} = \left(k_{\dot{\mathrm{V}}} z_v' - k_{\mathrm{c}} p_{\mathrm{A}}\right) \rho_{\mathrm{A}} \tag{5.6.12}$$

$$\dot{m}_{\mathrm{A,out}} = \dot{V}_{\mathrm{A}} \rho_{\mathrm{A}} = A_{\mathrm{a}} r_{\mathrm{a}} \Delta \dot{\varphi}_{\mathrm{cam}} \rho_{\mathrm{A}}. \tag{5.6.13}$$

Herewith A_{a} is the effective area of the vanes, v_{a} is a representative radius and the oil volume in chamber A is

$$V_{\mathrm{A}} = A_{\mathrm{a}} r_{\mathrm{a}} \Delta \varphi_{\mathrm{cam}}. \tag{5.6.14}$$

For the stored mass holds

$$\frac{dm_{\mathrm{A}}(t)}{dt} = \frac{d}{dt}\left(V_{\mathrm{A}} \rho_{\mathrm{A}}(t)\right) = V_{\mathrm{A}} \frac{d\rho_{\mathrm{A}}(t)}{dt} = V_{\mathrm{A}} \frac{\rho_{\mathrm{A}}}{\beta} \frac{dp_{\mathrm{A}}}{dt} \tag{5.6.15}$$

where β is the compressibility module (bulk modulus)

$$\beta = -V \left(\frac{dp}{dV}\right)_{\mathrm{r=const}}. \tag{5.6.16}$$

Inserting these equations in (5.6.11) yields

$$\frac{A_{\mathrm{a}} r_{\mathrm{a}} \Delta \varphi_{\mathrm{cam}}}{k_{\mathrm{c}} \beta} \frac{dp_{\mathrm{A}}(t)}{dt} + p_{\mathrm{A}}(t) = \frac{k_{\dot{\mathrm{V}}}}{k_{\mathrm{c}}} z_v'(t) - \frac{A_{\mathrm{a}} r_{\mathrm{a}}}{k_{\mathrm{c}}} \Delta \dot{\varphi}_{\mathrm{cam}}(t). \tag{5.6.17}$$

Hence, the pressure in chamber A behaves as a first order delay with *chamber time constant*

$$T_{\mathrm{p}} = \frac{A_{\mathrm{a}} r_{\mathrm{a}} \Delta\varphi_{\mathrm{cam}}}{k_{\mathrm{c}} \beta} \tag{5.6.18}$$

which is small for small $\Delta\varphi_{\mathrm{cam}}$ and large β, i.e. large oil stiffness.

Correspondingly it holds for chamber B

$$\frac{A_{\mathrm{a}} r_{\mathrm{a}} \Delta\varphi_{\mathrm{cam}}}{k_{\mathrm{c}} \beta} \frac{dp_{\mathrm{B}}(t)}{dt} + p_{\mathrm{B}}(t) = -\frac{k_{\dot{\mathrm{V}}}}{k_{\mathrm{c}}} z_v'(t) + \frac{A_{\mathrm{a}} r_{\mathrm{a}}}{k_{\mathrm{c}}} \Delta\dot{\varphi}_{\mathrm{cam}}(t). \tag{5.6.19}$$

The *angular momentum balance* of the rotor with the attached camshaft for changes $\Delta\varphi_{\mathrm{cam}}$ becomes

$$J_{\mathrm{cam}} \Delta\ddot{\varphi}_{\mathrm{cam}}(t) = A_{\mathrm{a}} r_{\mathrm{a}} \Delta p_{\mathrm{rot}}(t) - M_{\mathrm{cam}}\left(\Delta\dot{\varphi}(t)\right) \tag{5.6.20}$$

where M_{cam} is the driving torque of the camshaft. This torque oscillates strongly. For a 4 cyl. engine the peak values are $+20$ Nm and -10 Nm for $n_{\mathrm{eng}} = 1000$ rpm and $+30$ Nm and -22 Nm for $n_{\mathrm{eng}} = 6000$ rpm, see Hannibal et al (2002). An average value is $+8$ Nm. Hence it is assumed that for the relative turning of the rotor with camshaft and the stator the torque can be approximated my a Coulomb and viscous term.

$$M_{\mathrm{cam}}\left(\Delta\varphi\right) = M_{\mathrm{cam},0} \mathrm{sign} \Delta\dot{\varphi}_{\mathrm{cam}} + M_{\mathrm{cam},1} \Delta\dot{\varphi}_{\mathrm{cam}}. \tag{5.6.21}$$

If (5.6.21) is inserted in (5.6.20), small deviations $\Delta\dot{\varphi}_{\mathrm{cam}}$ are considered and the Coulomb friction is neglected, it follows

$$\frac{J_{\mathrm{cam}}}{M_{\mathrm{cam},1}} \Delta\ddot{\varphi}_{\mathrm{cam}}(t) + \Delta\dot{\varphi}_{\mathrm{cam}}(t) = \frac{A_{\mathrm{a}} r_{\mathrm{a}}}{M_{\mathrm{cam},1}} \Delta p_{\mathrm{rot}}(t). \tag{5.6.22}$$

Thus, the angular speed of the rotor is a first order delay with time constant

$$T_{\mathrm{cam}} = \frac{J_{\mathrm{cam}}}{M_{\mathrm{cam},1}}. \tag{5.6.23}$$

Combining these equations leads to a signal flow chart in Fig.5.6.3. It shows an integral behavior $\Delta\varphi_{\mathrm{cam}}\left(I_{\mathrm{a}}(t)\right)$ with two nonlinear first order delays of the actuator chambers and rotor with camshaft. Therefore, it is a dynamic system of 3rd order. Usually it can be assumed that the pressure in the chamber opening to the oilsump reaches the ambient pressure fast, such that $p_{\mathrm{B}} = p_{\mathrm{a}}$. If for the case of motion the torque generating chamber is considered and the relations are linearized, one obtains a signal flow as in Fig. 5.6.4.

Hence, the dynamic model for the camshaft positioning can be described by the transfer function

$$G_{\mathrm{U}\varphi}(s) = \frac{\Delta\varphi_{\mathrm{cam}}(s)}{\Delta U_{\mathrm{a}}(s)} = \frac{K_{\mathrm{U}\varphi}}{(1 + T_{\mathrm{p}} s)(1 + T_{\mathrm{cam}} s)\, s} \tag{5.6.24}$$

and if it can be assumed that $T_{\mathrm{p}} \ll T_{\mathrm{cam}}$ it holds

$$G_{\mathrm{U}\varphi}'(s) = \frac{K_{\mathrm{U}\varphi}'}{(1 + T_{\mathrm{cam}} s)\, s}. \tag{5.6.25}$$

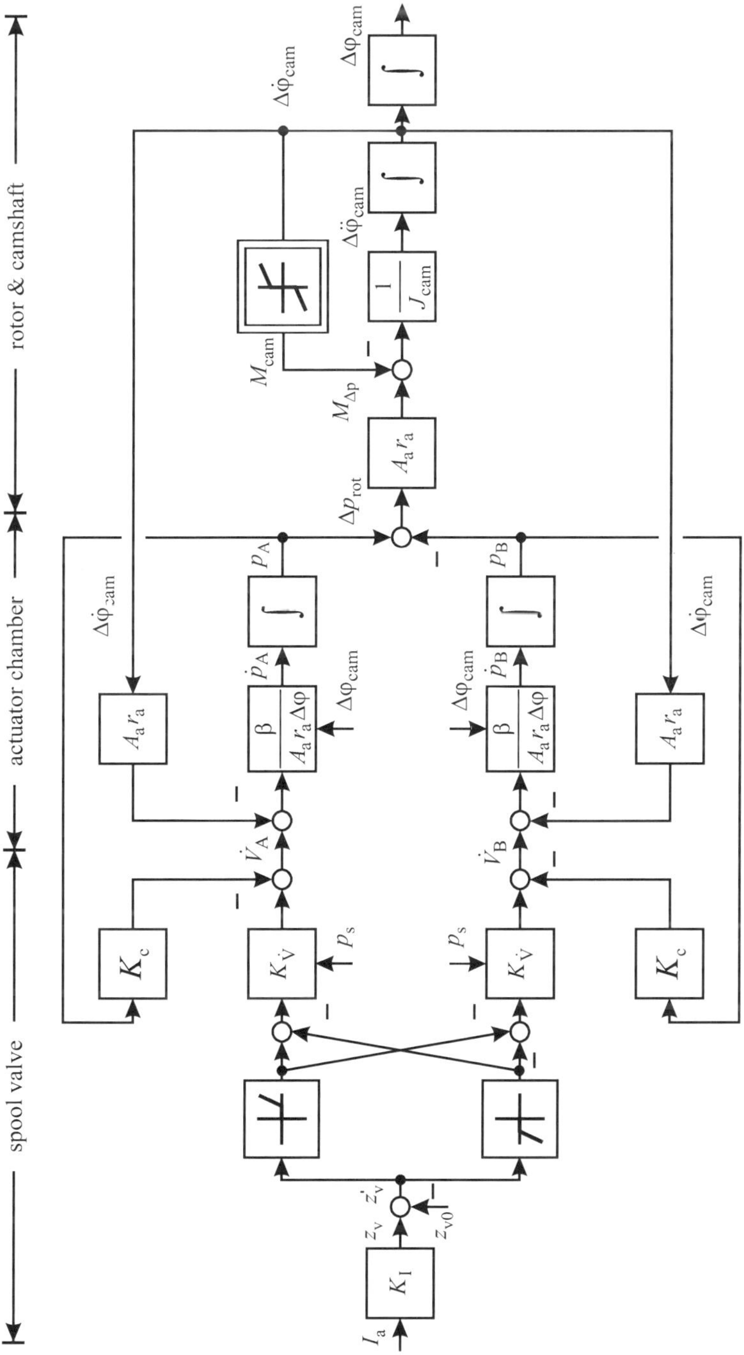

Fig. 5.6.3. Signal-flow chart of a hydraulic camshaft phasing system with sliding vane actuator.

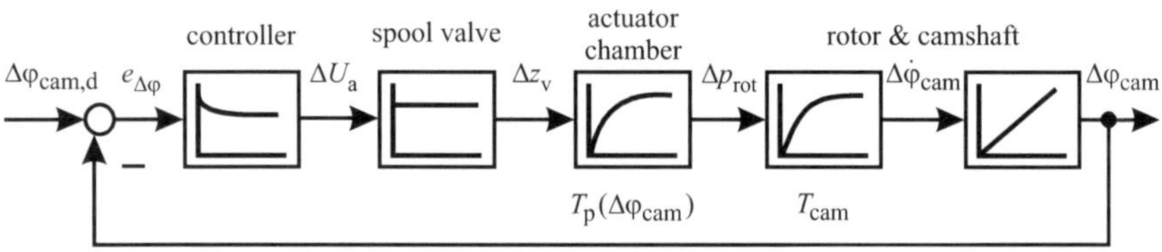

Fig. 5.6.4. Camshaft phase control loop with simplified, linear signal flow for the hydraulic camshaft system.

Therefore, the dynamic behavior can be approximately described as an integrating element with first order delay. Because of the integral behavior the phase position controller can be proportional acting, and therefore, e.g. a P or PD controller with parameters depending on oil temperature and engine speed can be applied.

References

van Basshuysen R, Schäfer F (2004) Internal combustion engine handbook: basics, components, systems, and perspectives. SAE International, Warrendale

van Basshuysen R, Schäfer F (2007) Handbuch Verbrennungsmotor, 4th edn. Vieweg, Wiesbaden

Buch A, Klöckner S, Bernarding E, Eid T (2010) Electric actuators for modern internal combustion engines. MTZ worldwide 71(12)

Clever S (2011) Model-based fault detection and diagnosis for a common rail engine. Diss. Technische Universität Darmstadt. Fortschr.-Ber. VDI Reihe 8, 1202. VDI Verlag, Düsseldorf

Continental Automotive GmbH (2011) Powertrain products. Benefits and technical information. Continental, Regensburg

Eisath C, Brandl A (2007) Mastering data management in increasingly complex systems: efficient data handling within the calibration process at Siemens VDO. In: 2nd International Symposium on Develpoment Technology, Wiesbaden, Germany

Gabriel H, Jacob S, Münkel U, Rodenhäuser H, Schmalzl HP (2007) The turbocharger with variable turbine geometry for gasoline engines. MTZ worldwide 68(2):7–10

Genster A, Stephan W (2004) Always at the correct temperature – thermal management with electric coolant pump. MTZ worldwide 65(11):6–7

Hammer J, Raff M, Naber D (2014) Advanced diesel fuel injection equipment– a never ending BOSCH story. In: 14. Internationales Stuttgarter Symposium, Springer, pp 31–45

Hannibal W, Knecht A, Stephan W (2002) Nockenwellenverstellungen für Ottomotoren. Die Bibliothek der Technik 247, Verlag Moderne Industrie, Landsberg/Lech

Heisler H (1995) Advanced engine technology. Society of Automotive Engineers, Warrendale

Isermann R (2000) Mechatronic systems: concepts and applications. Transactions of the Institute of Measurement and Control 22(29–55)
Isermann R (2005) Mechatronic systems – fundamentals, 2nd edn. Springer, London
Kaiser T, Staiger W, Orthmann G, Huebner P (2000) Thermal optimisation of mechatronically integrated power electronics for engine cooling fan using brushless technology. Purdue Engineering Student Council (PESC) '00, Galway, Ireland
Keppeler S (1997) Common-Rail-Einspritzsystem für den direkteinspritzenden Dieselmotor. Dissertation. Aachen
Klar H, Klages B (2009) Calibration methods and tools for drive train electronics. In: 3rd Int. Symposium on Development Technlogy, Wiesbaden, Germany, pp 178–189
Klotzbach P, Sari O, Dismon H (2003) Innovative electronically controlled exhaust gas recirculation valve. MTZ worldwide 64(9):10–12
Kolev N (2007) Multiphase flow dynamics, vol 3. Springer, Berlin
Kopp C (2006) Variable Ventilsteuerung für Pkw-Dieselmotoren mit Direkteinspritzung. PhD thesis, Otto-von-Guericke-Universität Magdeburg
Meinig U, Bohner J (2013) New camshaft phaser module for automobile engines. MTZ worldwide 74(7-8):48–52
Mollenhauer K, Tschöke H (2010) Handbook of diesel engines. Springer, Berlin
Münchhof M (2006) Model-based fault detection for a hydraulic servo axis. Dissertation Technische Universität Darmstadt. Fortschr.-Ber. VDI Reihe 8, 1105. VDI Verlag, Düsseldorf
Robert Bosch GmbH (ed) (2007a) Automotive electrics, automotive electronics. J. Wiley, Chichester, UK
Robert Bosch GmbH (ed) (2007b) Diesel-Speichereinspritzsystem Common Rail, 4th edn. Christiani, Konstanz
Robert Bosch GmbH (ed) (2011) Automotive Handbook, 8th edn. Bentley publishers, Cambridge
Scholl H (2007) Gasoline sytems for a cleaner future. In: Gasoline Systems Symposium, Robert Bosch GmbH, Frankfurt
Schöner H (2004) Automotive mechatronics. Control Engineering Practice 12:1343–1351
Seykens X, Somers L, Baert R (2005) Detailed modeling of common rail fuel injection process. Journal of Middle European Construction and Design of Cars 3:30–39
Stuhler H, Ricken V, Diener R (2010) Aufbau und Anpassung der Motorsteuerungs-Software für Otto-und Dieselmotoren. In: Isermann R (ed) Elektronisches Management motorischer Fahrzeugantriebe, Vieweg-Teubner, Wiesbaden, chap 2, pp 38–66
Thiemann W (1988) Zur Bestimmung des Einspritzverlaufs bei Dieseleinspritzanlagen mit Gleichraumentlastung. Motortechnische Zeitschrift - MTZ 49(5):207–217
Welschof B (1992) Analytische Untersuchungen über die Einsatzmöglichkeit einer sauggedrosselten Hydraulikpumpe zur Leistungssteuerung. Dissertation. Aachen
Wickerath B, Fournier A, Durand J, Brömmel A (2011) Fully variable mechanical coolant pump for commercial vehicles. MTZ worldwide 72(1):42–46

6

Engine-control methods and calibration

The design and implementation of engine control functions has developed into a sophisticated and labor-intensive procedure. This is for many reasons, among them the high multi-variable complexity of engine control, the high performance requirements of suppliers, manufacturers and customers, and legislative certification limits for fuel consumption and emissions, and competition.

The following sections summarize some general procedures for control-function design, control-software generation, required computers and software tools, and test benches.

6.1 Engine-oriented electronic control design

The design and implementation of electronic control and diagnosis systems is highly interrelated with the design of the mechanics, thermodynamics and fluid mechanics of the internal combustion engines. It belongs to the design of mechatronic systems and requires a systematic development across the classical boundaries. With regard to the timeline of the workflow a simultaneous or concurrent engineering in different domains has to be performed, see e.g. VDI 2206 (2003), Isermann (2005).

6.1.1 V-development model

The development of the electronic engine control system can, according to the design of the control software functions for mechatronic systems, be divided into

(1) control-system design
- control-function development
- control-software development

(2) control-system integration
- component integration
- calibration
- performance testing

R. Isermann, *Engine Modeling and Control*,
DOI: 10.1007/978-3-642-39934-3_6

This design procedure is highly interrelated and requires many iterative steps and special development tools. It can favorably be represented in a so called V-model, see Fig. 6.1.1, which covers all aspects from the analysis of the user requirements to acceptance tests, see e.g. VDI 2206 (2003), Schäuffele and Zurawka (2005), Isermann (2005), BRD (1997) with origins in Böhm (1979), STARTS Guide (1989), Bröhl (1995), Droeschel and Wiemers (1999).

A corresponding V-model can be given for the hardware development of the electronic control unit (ECU) which is assumed here to exist already.

An alternative to the V-model is, e.g. the waterfall-model, e.g. Royce (1970) which is organized sequentially in one direction with recursions. The V-model has the intention that the results, documents and tests of the right branch correspond to the development procedures of the left branch. A further developed version with more flexibility is the V-model XT, BRD (2004) as discussed in Borgeest (2008).

Some important steps of the V-model for the development of control and software functions can be described as follows:

(1) Requirements
- user requirements
- definition of general (overall) functions and data (rated values) of the final product (ECU)
- general solution outline
- development and manufacturing costs
- timely development and milestones
- result: requirements document (does not include technical implementation)

(2) Specifications
- definition of the product (ECU) that fulfills the requirements
- partitioning in manageable modules for control and diagnosis
- specification of features and data of the modules
- consideration of the sources, tools and limitations for the development and final manufacturing and maintenance
- specification of hardware data
- specification of used software, compilers and development systems
- result: specification document

(3) Control system design
- detailed partitioning into electronic, mechanic, hydraulic, pneumatic and thermal components with their auxiliary power supplies
- detailed fixing of type of sensors and actuators and their data
- detailed data of interfaces between ECU, sensors and actuators
- task distribution between sensors and actuators with integrated electronics and ECU
- specification of power-related data
- hardware design: data of microprocessors, data-storages, interfaces, bus systems, cabling and plug systems
- control engineering design:
  - definition of sensor inputs and outputs to the actuators

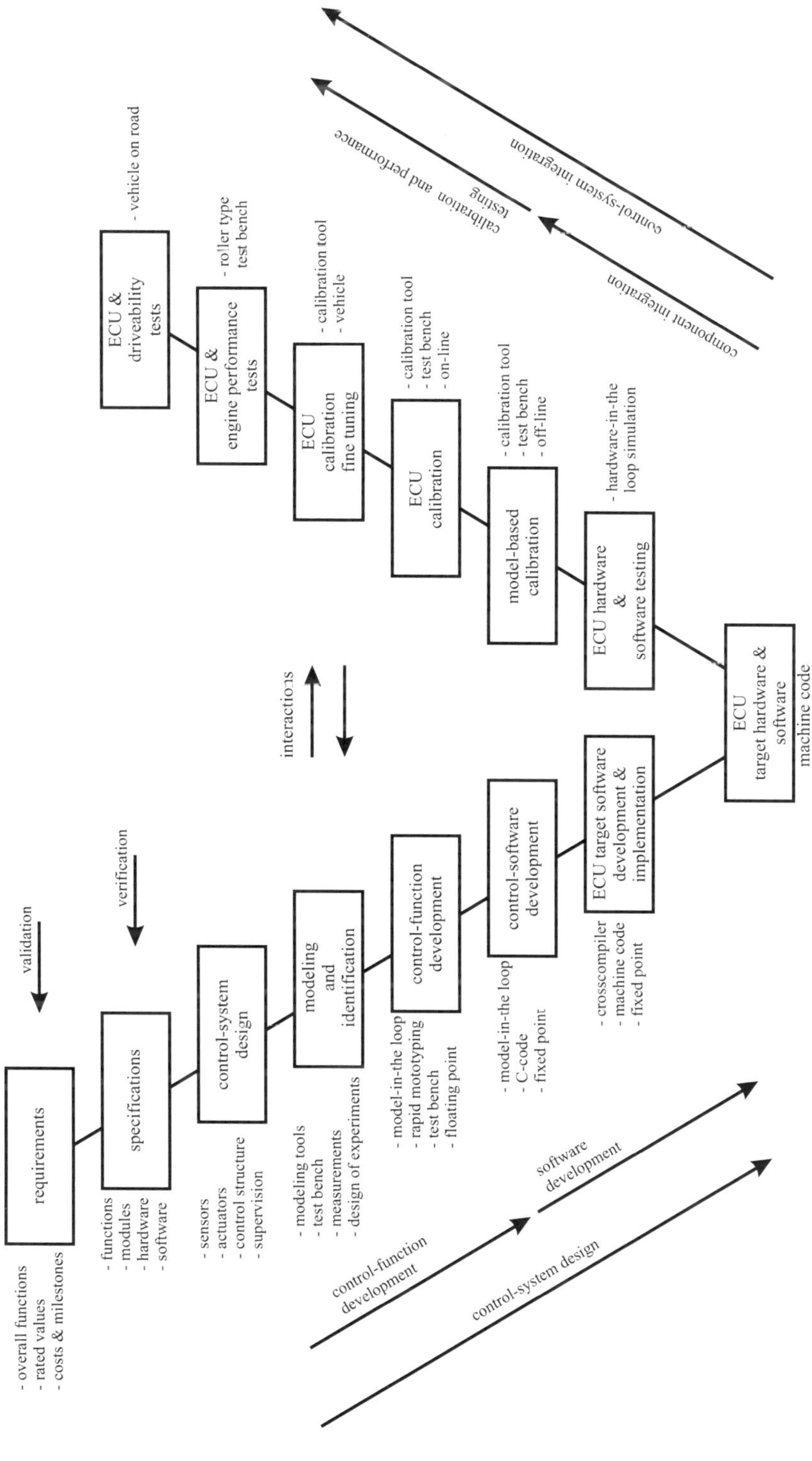

Fig. 6.1.1. V-model for the development of control functions, control software, calibration and performance tests.

  - control-system structure: feedforward control (open loop) and feedback control (closed loop)
  - required engine and component models
  - model-based design
  - calibration (parametrization) methods
  - supervision and diagnosis functions
- reliability and safety issues; FMEA (fault mode and effect analysis) studies for sensor, actuator and ECU faults and failures
- result: control-system design document

(4) Modeling and identification
- required mathematical models of engine, sensors, actuators, transmission, vehicle
- theoretical/physical modeling
- experimental modeling
- use of modeling/identification tools
- measurement procedures for test benches, design of experiments
- kind of models: stationary (look-up tables, polynomials, neural networks); dynamic (differential equations, neural networks)
- result: engine and component models

(5) Control-function development
- hierarchical control structure
- computer-supported design, manual design
- ECU states: from start-up to shut-off
- engine states (discrete): from cold start to shut-off
- control functions: crank-angle-dependent, time-dependent
- sampling times and word length
- supervision and diagnosis functions
- model-in-the-loop simulation: engine models and ECU models
- rapid control prototyping with development ECU, by-pass computer and test bench
- result: control structure and control algorithms

(6) Control-software development
- software architecture layers, modules
- software-component interfaces
- high-level language: selection, floating point (e.g. C-code, MATLAB/Simulink)
- availability of compilers for target software
- implementation of control functions and modules into software structure
- standardization and reuse of software modules
- code optimization
- testing of software modules with, e.g. model-in-the-loop simulation
- rapid control prototyping with by-pass computer and test-bench experiments
- result: control software (modules) in high-level language

(7) ECU-target software development and implementation
- transfer of high-level language control software into machine code with fix point arithmetic. Use of cross compiler

- test of control functions with simulated engine
- software-in-the-loop simulation
- hardware-in-the-loop simulation if real-time functions with components are of interest
- result: implemented control software in target ECU

(8) ECU hardware and software testing
- the ECU with target hardware and software undergoes intensive function tests
- integration tests with simulated sensor signals and actuators in real time
- hardware-in-the-loop simulation with simulated sensor output signals, simulated or real actuators and real-time simulated comprehensive engine model
- automated test-runs
- testing of reaction to extreme speeds and loads (outside of normal operation)
- reliability and safety tests
- tests for electromagnetic compatibility (EMC)
- result: verification that the ECU meets its specifications

(9) Calibration of the control functions
- Free parameters of control algorithms, characteristic curves or look-up tables (maps) are adapted to the real engine, drivetrain and vehicle
- supported by calibration tool with editors at implementation level or physically-defined level
- off-line calibration with interrupted engine operation
- on-line calibration with running engine
- basic calibration of the stationary behavior
- dynamic calibration of the dynamic behavior
- calibration by manual optimization
- calibration by optimization with engine models
- result: calibrated ECU control functions

(10) Calibration with fine-tuning
- final fine-tuning of free parameters with the transmission and vehicle
- driving experiments with the target vehicle
- use of calibration tools, offline or online
- result: adapted ECU functions to transmission and vehicle. Verification of specifications.

(11) Final ECU, driveability tests and field tests
- final engine control functions are tested with the target vehicle
- performance tests for different loads and environmental conditions (summer, winter, weather)
- driveability tests
- result: validation of requirements

6.1.2 Workflow for control development and calibration

The *workflow for the control-function development* is depicted in Fig. 6.1.2 in more detail, also showing the use of software tools, computers and test-bench experiments.

It begins with physical and/or experimental modeling. The control development then comprises engine simulation, control-function development and optimization, supported by an ECU prototype or prototyping computer for probing the control functions with the engine on the test bench. The next steps are then the software development and testing for the target ECU by using personal computers. Frequently, the ECU hardware and software testing is performed with hardware-in-the loop simulation, connecting the ECU with real actuators and sensor interfaces.

A corresponding *workflow for the model-based calibration of the control functions* is illustrated in Fig. 6.1.3. Based on the already gained physical and/or experimental engine models the model-based calibration of the control functions is carried out. First the basic stationary control functions, as e.g. reference variables dependent on the operation point, and then the calibration of the dynamic control functions, by using dynamic engine and drive train models, is executed.

These procedures are carried out either with personal computers or special calibration computers. The last step is a fine tuning with the vehicle on a roller test bench or on the road.

The following sections consider the model-based control-function development and calibration in more detail.

6.2 Basic control structures

First, a look at some basic control-system structures which are suitable for combustion engine control is given. To describe briefly the principles, the components are assumed to have linear dynamic behavior and are treated in continuous time with transfer functions $G(s)$ where s is the Laplace variable $s = \delta + i\omega$. Initially, basic feedforward and feedback control systems are viewed, then their combinations and interconnected control systems, like cascaded control and two-variable control. The extension to multi-variable systems, to nonlinear behavior and to the resulting digital-control algorithms with sampled signals is then discussed subsequently.

Because some important output variables of internal combustion engines are not measured as, e.g. the torque and emissions, feedforward control systems play an important role, see Fig. 6.2.1a). An advantage then is that stability problems are avoided. But relatively precise process models are required and parameter changes of the engine and external disturbances are not automatically compensated.

If engine variables are measured and a relatively precise control is required, then feedback control systems are applied, see Fig. 6.2.1b), as e.g. for the air mass flow, the charging pressure or the exhaust gas excess air factor λ. In order to reach a good control performance and to avoid instability, dynamic engine models are required for a systematic design.

The controlled processes of combustion engines are characterized by mostly nonlinear behavior and low dynamic order, e.g. $n = 2$ or 3, sometimes with dead time and they are strongly dependent on the operation point. As the real-time requirements with sampling times of about $10 \ldots 20\,\mathrm{ms}$, 5 to 10 manipulated variables and $5 \ldots 8$ controlled variables, are quite strong, the final control algorithms have to be

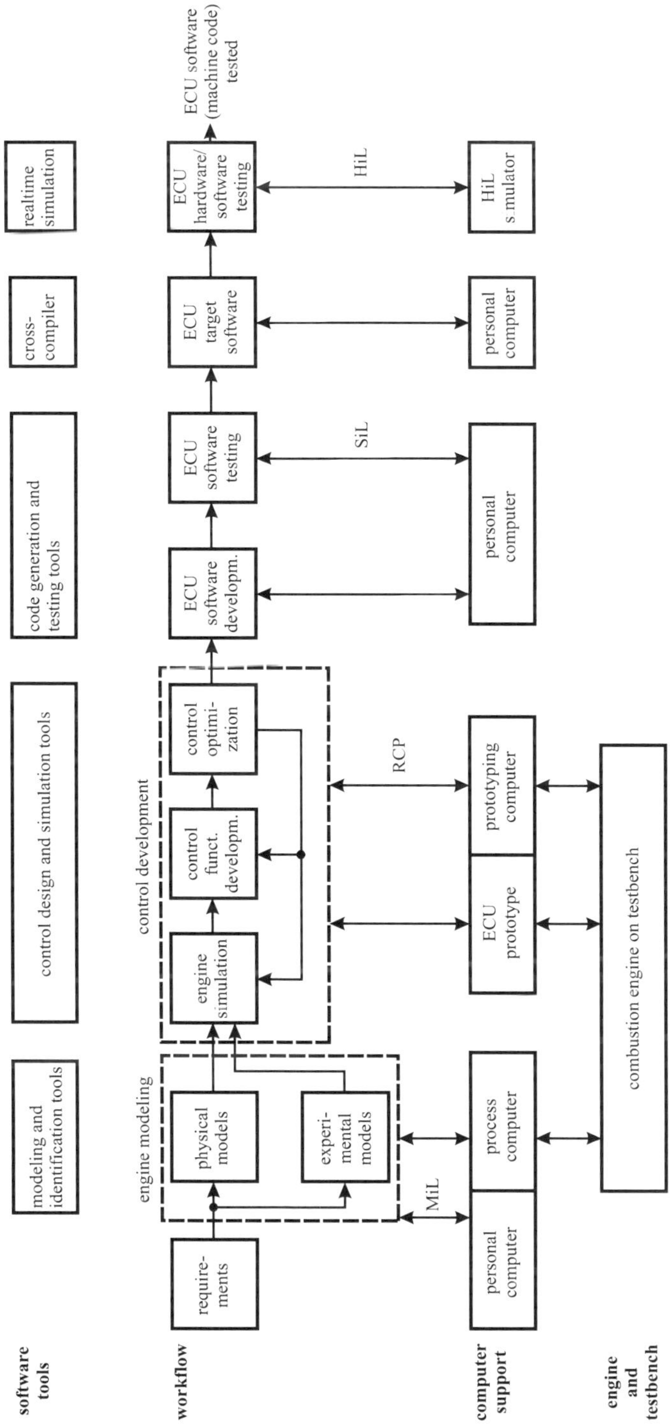

Fig. 6.1.2. Workflow for control-function development.

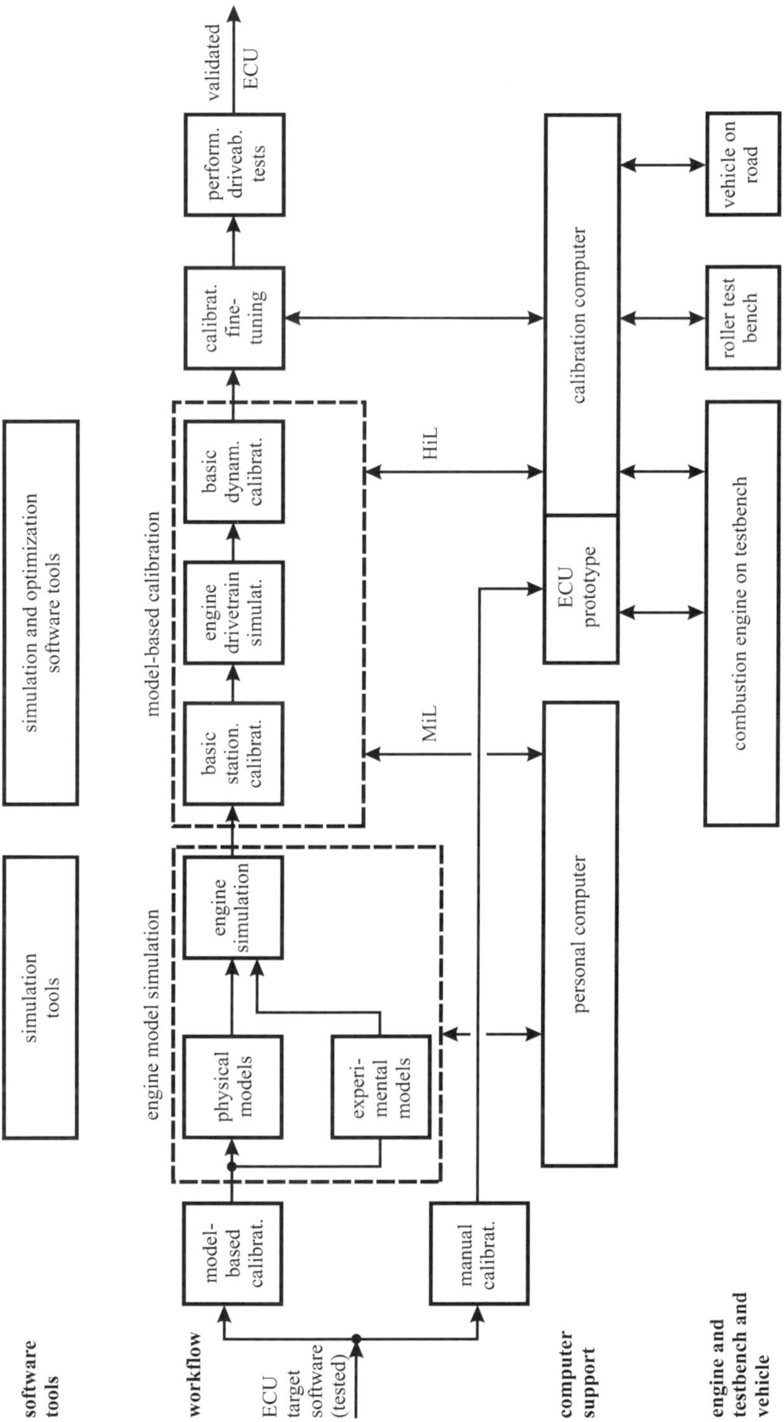

Fig. 6.1.3. Workflow for model-based control-function calibration.

of small computational expense and should be easy to understand and to calibrate. Therefore the following short summary concentrates on those control designs which satisfy practical requirements for engine control.

A detailed treatment for the design of control systems is available in many textbooks on control theory and practice, e.g. Kuo (1995), Åström and Wittenmark (1997), Ogata (2008), Dorf and Bishop (2010), Föllinger (2013). Therefore, the next sections first summarize some basic control systems as a starting point for the design of special, mostly nonlinear control systems for gasoline and diesel engines, treated in Chaps. 7 and 8. The assumed linear behavior in the next sections can only be considered as a basis for more sophisticated designs and to discuss some basic properties.

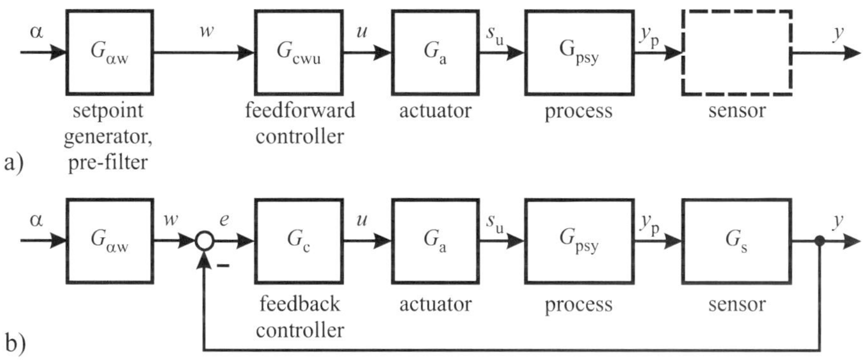

Fig. 6.2.1. Basic control structures. **a** Feedforward control (sensor not available for series production engine), **b** Feedback control. α command variable, $w = y_d$ setpoint, reference variable, or desired control variable, y controlled variable, u manipulated variable, s_u actuator position, y_p process output.

6.3 Linear feedforward control

An elementary feedforward control is depicted in Fig. 6.2.1a). A setpoint generator yields the setpoint variable or reference variable $w(t)$ based on a command variable $\alpha(t)$ (e.g. the accelerator pedal). The goal is that the process output $y(t)$ follows the setpoint variable $w(t)$ as directly as possible. If $y(t)$ is not measurable, a feedback control cannot be applied. If now a model of the actuator $G_a(s)$ and the process $G_{psy}(s)$ is known, a feedforward control can be designed. An ideal response would be $y(t) = w(t)$. However, as the actuator and process always have a certain dynamic delaying behavior, this requirement is not realizable. Assuming that the process has the following transfer function

$$G_{psy}(s) = \frac{y(s)}{s_u(s)} = K_{psy}\frac{1 + b_1 s + \ldots + b_m s^m}{1 + a_1 s + \ldots + a_n s^n} \tag{6.3.1}$$

and for the actuator holds a proportional behavior

$$G_\mathrm{a}(s) = \frac{s_\mathrm{u}(s)}{u(s)} = K_\mathrm{a} \tag{6.3.2}$$

the overall transfer function of the process becomes

$$G_\mathrm{p}(s) = \frac{y(s)}{u(s)} = G_\mathrm{a}(s) G_\mathrm{psy}(s) \tag{6.3.3}$$

with the gain

$$K_\mathrm{p} = K_\mathrm{a} K_\mathrm{psy}. \tag{6.3.4}$$

Using this overall transfer function of the process, Fig. 6.3.1 results.

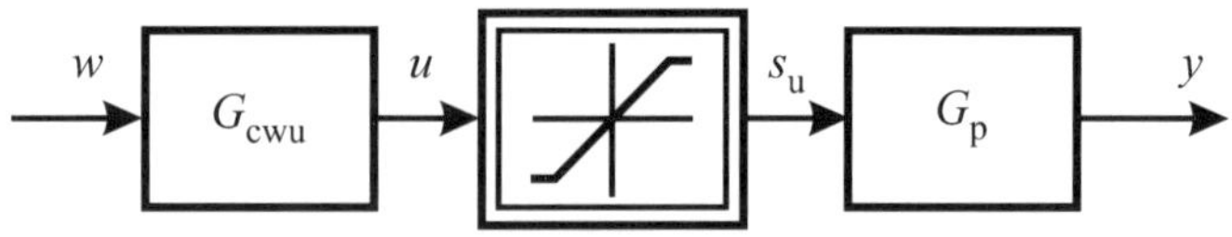

Fig. 6.3.1. Feedforward control for a process with limited manipulation range.

a) Static feedforward control

If the actuator and process dynamics are not taken into account and actuator and process gain are known, a static feedforward controller

$$G_\mathrm{cwu}(s) = K_\mathrm{cwu} = 1/K_\mathrm{p} \tag{6.3.5}$$

can be used to satisfy the stationary behavior such that the overall response becomes

$$G_\mathrm{wy}(s) = \frac{y(s)}{w(s)} = \frac{1 + b_1 s + \ldots + b_m s^m}{1 + a_1 s + \ldots + a_n s^n} \; (m < n) \tag{6.3.6}$$

i.e. a delayed dynamic response with gain 1.

b) Dynamic feedforward control

An ideal feedforward control is obtained by

$$\frac{y(s)}{w(s)} = G'_\mathrm{cwu}(s) G_\mathrm{p}(s) = 1 \tag{6.3.7}$$

which leads to a dynamic feedforward control with an inverted process model

$$G'_{\text{cwu}}(s) = \frac{u(s)}{w(s)} = \frac{1}{G_{\text{p}}(s)}. \tag{6.3.8}$$

Introducing (6.3.1) and (6.3.2) yields with the assumption of a minimum phase process

$$G'_{\text{cwu}}(s) = \frac{1}{K_{\text{p}}} \frac{1 + a_1 s + \ldots + a_n s^n}{1 + b_1 s + \ldots + b_m s^m}. \tag{6.3.9}$$

Because this transfer function would require exact derivatives of the output signal of order n it is not realizable. Therefore, a realizability filter or reference shaping filter

$$G_{\text{rf}}(s) = \frac{1}{(1 + T_{\text{r}} s)^n} \tag{6.3.10}$$

has to be applied. The realizable feedforward controller then becomes

$$G_{\text{cwu}} = G_{\text{p}}^{-1}(s) G_{\text{rf}}(s) = \frac{1}{K_{\text{p}}} \frac{1 + a_1 s + \ldots + a_n s^n}{1 + b_1 s + \ldots + b_m s^m} \frac{1}{(1 + T_{\text{r}} s)^n} \tag{6.3.11}$$

such that the overall behavior yields

$$G_{\text{wy}}(s) = \frac{y(s)}{w(s)} = G_{\text{cwu}}(s) G_{\text{p}}(s) = G_{\text{rf}}(s) = \frac{1}{(1 + T_{\text{r}} s)^n}. \tag{6.3.12}$$

The time constant T_{r} of the realizability filter is now a design parameter.

Example 6.1 (Dynamic feedforward control for a manipulated variable as input).
As an example a second order low-pass process is considered with transfer function

$$G_{\text{p}}(s) = \frac{y(s)}{u(s)} = \frac{K_{\text{p}}}{(1 + T_1 s)(1 + T_2 s)}.$$

The realizable feedforward controller then results in

$$G_{\text{cwu}}(s) = \frac{u(s)}{w(s)} = \frac{(1 + T_1 s)(1 + T_2 s)}{K_{\text{p}} (1 + T_{\text{r}} s)^2}$$

if an actuator saturation (Fig. 6.3.1) is not considered. The overall behavior follows as

$$G_{\text{wy}} = \frac{1}{(1 + T_{\text{r}} s)^2}.$$

If $T_{\text{r}} = T_2$ is selected, the feedforward controller becomes

$$G''_{\text{cwu}}(s) = \frac{1}{K_p} \frac{1 + T_1 s}{1 + T_{\text{r}} s}$$

which is a proportional-derivative (PD) element with a time lag. A simplification of $G'_{\text{cwu}}(s)$ by assuming $T_{\text{r}} \ll T_1$ is

$$G''_{\text{cwu}}(s) = K_{\text{cwu}} (1 + T_{\text{D}} s)$$

which is a proportional-derivative feedforward controller with $K_{\mathrm{cwu}} = 1/K_{\mathrm{p}}$. The derivative time T_{D} can then be parameter-optimized or just selected as $T_{\mathrm{D}} = T_1$, which yields

$$G_{\mathrm{wy}} = \frac{y(s)}{w(s)} = \frac{1 + T_{\mathrm{D}}s}{(1 + T_1 s)(1 + T_2 s)} = \frac{1}{1 + T_2 s}$$

resulting in an overall behavior of first order.

□

The dynamic feedforward controllers improve the dynamic overall response compared to the static feedforward controller (6.3.3). However, if T_{r} is selected too small or the derivative time T_{D} too large, the response of the manipulated variable $u(t)$ becomes too large and may exceed the restricted range of the manipulating variable.

A further method to design a dynamic forward controller is to assume a feedback controller for the design and to use its manipulated variable u_{m} for feedforward control as shown in Fig. 6.3.2a), Pfeil (2011). The transfer behavior of the feedforward controller then becomes (without considering an actuator saturation)

$$G_{\mathrm{cwu}}(s) = \frac{u_{\mathrm{m}}(s)}{w(s)} = \frac{u(s)}{w(s)} = \frac{G_{\mathrm{c}}(s)}{1 + G_{\mathrm{c}}(s)G_{\mathrm{p,m}}(s)} \tag{6.3.13}$$

where $G_{\mathrm{c}}(s)$ is a feedback controller. For the (simulated) feedback controller $G_{\mathrm{c}}(s)$ a P or PD controller can be used. In this case the process behavior is indirectly taken into account with its process model $G_{\mathrm{p,m}}(s)$ and the dynamics of the feedforward correspond to an assumed closed-loop control.

c) Dynamic feedforward control with actuator saturation

The actuators usually have a limited manipulation range and a saturation function has to be taken into account as depicted in Fig. 6.3.2. The considered proportional acting feedforward controllers then reach a limited process input $s_{\mathrm{u}}(t)$ for larger inputs $w(t)$ thus limiting the dynamic response of the process output. Figure 6.3.3 shows the responses with the classical feedforward controller $G_{\mathrm{cwu}}(s)$ of Example 6.1. If the time constant T_{r} of the realizability filter is reduced, the process input becomes larger. However, if the manipulated variable reaches the saturation, a further increase of T_{r} makes the response slower. This means that the design of the feedforward controller should take the saturation into account.

One way is to simulate a closed loop corresponding to (6.3.13), but with a model of the saturation function, as depicted in Fig. 6.3.2. The feedback controller $G_{\mathrm{c}}(s)$ should have proportional behavior but no integral term to avoid wind-up of u_{m}.

The dynamics can be further improved if the process model is described by a state-space representation

$$\dot{\mathbf{x}}(t) = \mathbf{A}\mathbf{x}(t) + \mathbf{b}u(t) \tag{6.3.14}$$

$$y(t) = \mathbf{c}^{\mathrm{T}}\mathbf{x}(t) \tag{6.3.15}$$

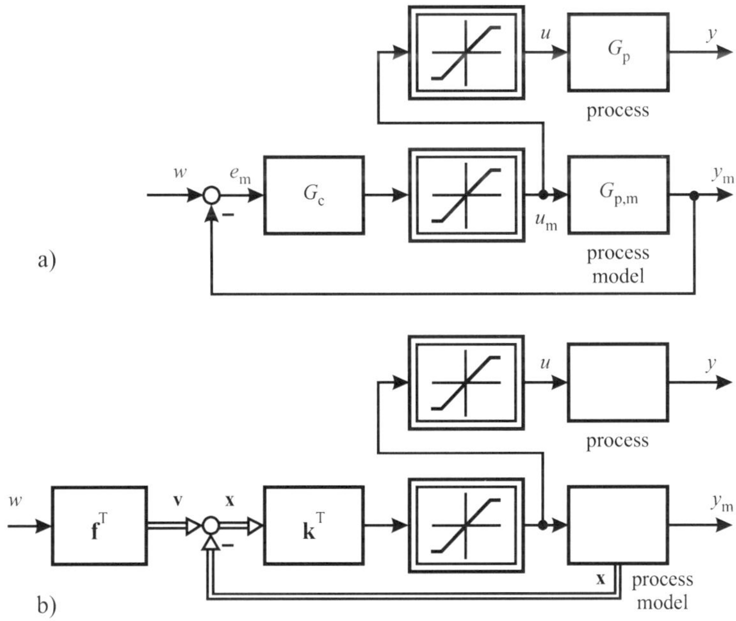

Fig. 6.3.2. Feedforward control for a process with input saturation by using a simulated feed back control system. **a** parameter-optimized controller. **b** state-space controller.

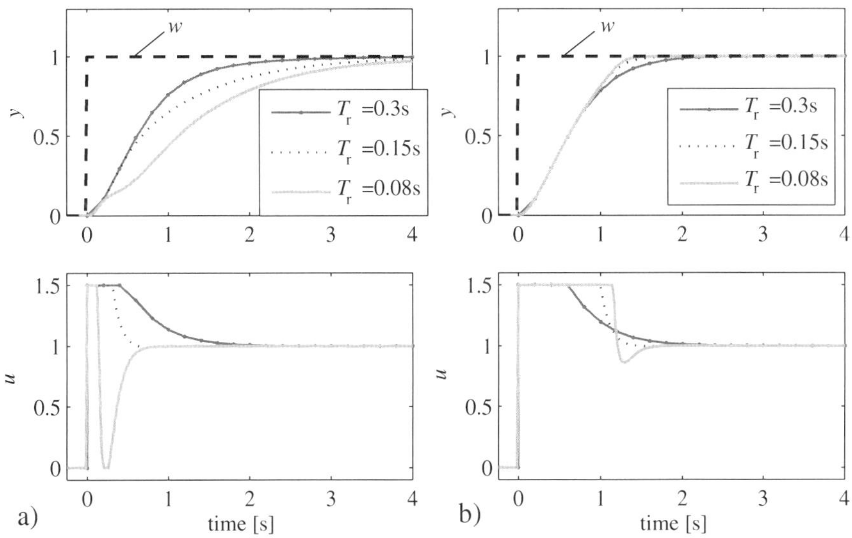

Fig. 6.3.3. Control performance of two feedforward structures for a second order process model with $T_1 = 1\,\mathrm{s}$ and $T_2 = 0.2\,\mathrm{s}$, Pfeil (2011). **a** classical feedforward control. **b** state-space-reference shaping filter.

and a state controller

$$u(t) = \mathbf{k}^{\mathrm{T}}\mathbf{x}(t) + fw(t) \tag{6.3.16}$$

is applied, Pfeil (2011). As all states of the process model are accessible, a complete state feedback can be used which results in optimal damped responses after appropriate design with, e.g. pole placement. f can be considered as a prefilter to avoid offset control deviations in steady-state. Figure 6.3.3 shows a result if the poles are placed by

$$\lambda_{\mathrm{r1}} = \lambda_{\mathrm{r2}} = -1/T_{\mathrm{r}} \tag{6.3.17}$$

for different values. The control behavior is significantly better than with classical feedforward controllers.

6.4 Linear feedback control

According to the development of control engineering practice and control theory many different feedback controllers are available. They can be classified, e.g. into linear and nonlinear, deterministic and stochastic, single-variable and multi-variable controllers, Åström and Wittenmark (1997). Linear controllers can be divided into parameter-optimized and structure-optimal controllers, Isermann (1989). In the case of *parameter-optimized controllers*, the controller structure, i.e. the form and the order of the controller equation is given and the free parameters are adjusted with regard to the controlled process, using an optimization criterion or tuning rules. For example, PID controllers belong to this category. Controllers are called *structure optimal* if both the control structure and the controller parameters are designed optimally according to the structure and the parameters of the process model. Examples are pole-zero-cancelation controllers, state-controllers, internal-model controllers or predictive controllers.

Assuming linear behavior the transfer function of a closed loop according to Fig. 6.2.1b) and the reference variable $w(t)$ as input is

$$G_{\mathrm{wy}}(s) = \frac{y(s)}{w(s)} = \frac{G_{\mathrm{c}}(s)G_{\mathrm{p}}(s)}{1 + G_{\mathrm{c}}(s)G_{\mathrm{p}}(s)} \tag{6.4.1}$$

with the process model, including the actuator and the output sensor

$$G_{\mathrm{p}}(s) = G_{\mathrm{a}}(s)\, G_{\mathrm{psy}}(s)\, G_{\mathrm{s}}(s) \tag{6.4.2}$$

and the control deviation (control error)

$$e(t) = w(t) - y(t) \tag{6.4.3}$$

and $y(t)$ the controlled variable and $u(t)$ the manipulated variable. Stability requires that the poles of

$$1 + G_{\mathrm{c}}(s)\, G_{\mathrm{p}}(s) = 0 \tag{6.4.4}$$

lie in the left half of the s-plane.

The choice of a measure for the control performance plays a central role in controller design. A relatively widely used performance criterion is the general quadratic criterion

$$I_1 = \int_0^\infty [e^2(t) + r\Delta u^2(t)]dt \tag{6.4.5}$$

with

$$\Delta u(t) = u(t) - u(\infty) \tag{6.4.6}$$

the deviation of the manipulated variable from a steady-state value $u(\infty)$, i.e. the control effort, which corresponds for state-space control to

$$I_2 = \int_0^\infty [\mathbf{x}^{\mathrm{T}}\mathbf{Q}\mathbf{x}(t) + r\Delta u^2(t)]dt. \tag{6.4.7}$$

The following sections briefly describe the design of some linear controller types which are applicable for engine control. For more details the reader is referred to, e.g. Isermann (1989), Kuo (1995), Franklin and Powell (2006), Ogata (2008), Dorf and Bishop (2010).

6.4.1 Parameter-optimized controllers (POC)

The basic equation of a linear PID controller in continuous time is

$$u(t)|_{\mathrm{PID}} = K\left[e(t) + \frac{1}{T_{\mathrm{I}}}\int_0^t e(\tau)d\tau + T_{\mathrm{D}}\frac{de(t)}{dt}\right] \tag{6.4.8}$$

with the parameters K gain, T_{I} integration time, T_{D} derivative time (lead time), see e.g. DIN 19226 (1994). By omitting terms, one obtains a PI, PD, or P controller

$$u(t)|_{\mathrm{PI}} = K\left[e(t) + \frac{1}{T_{\mathrm{I}}}\int_0^t e(\tau)d\tau\right] \tag{6.4.9}$$

$$u(t)|_{\mathrm{PD}} = K\left[e(t) + T_{\mathrm{D}}\frac{de(t)}{dt}\right] \tag{6.4.10}$$

$$u(t)|_{\mathrm{P}} = Ke(t). \tag{6.4.11}$$

In this definition the gain influences all terms. Another way to represent a PID controller is to use independent parameters of the three terms

$$u(t) = K_{\mathrm{P}}e(t) + K_{\mathrm{I}}\int_0^\infty e(\tau)d\tau + K_{\mathrm{D}}\frac{de(t)}{dt} \tag{6.4.12}$$

with $K_P = K$, $K_{\mathrm{I}} = K/T_{\mathrm{I}}$, $K_{\mathrm{D}} = KT_{\mathrm{D}}$.

The corresponding transfer functions are

$$\begin{aligned} G_{\mathrm{c,PID}}(s) &= \frac{u(s)}{e(s)} = K[1 + \frac{1}{T_{\mathrm{I}}s} + T_{\mathrm{D}}s] = K_{\mathrm{P}} + K_{\mathrm{I}}\frac{1}{s} + K_{\mathrm{D}}s \\ &= \frac{K(1 + T_{\mathrm{c1}}s)(1 + T_{\mathrm{c2}}s)}{s} \end{aligned} \tag{6.4.13}$$

with

$$T_{c1} = \frac{T_I}{2}\left[1 + \sqrt{1 - 4\frac{T_D}{T_I}}\right]$$

$$T_{c2} = \frac{T_I}{2}\left[1 - \sqrt{1 - 4\frac{T_D}{T_I}}\right].$$

Hence, the PID controller possesses two real zeros, if $T_D < T_I/4$.

The controller parameters K, T_I, T_D have to be adjusted to the controlled process, to obtain an appropriate control performance. This can be reached by the following procedures:

- *parameter optimization* by using a process model and an optimization method. Analytical solutions are only possible for low-order process models. Therefore mostly numerical optimization methods have to be applied.
- *tuning rules* to obtain approximately a best possible behavior, based on measured transient functions or oscillation tests at the stability limit.
- *trial-and-error methods*, which start with small parameter values and increase the values stepwise until the closed loop becomes weakly damped.

As actuators usually reach a saturation the output of the integral term of controllers continues to grow (wind-up) until the sign of the control error changes and integration turns back. Since this leads to large overshoots, an anti-wind-up technique has to be implemented, e.g. by an additional feedback with a dead zone around the integrator, see e.g. Franklin and Powell (2006).

6.4.2 Internal model control (IMC)

A direct way to include a process model in a control structure is the internal-model-control (IMC) principle. A process model is arranged in parallel to the process, see Fig. 6.4.1a), such that output deviations

$$\Delta y(t) = y(t) - y_m(t) \tag{6.4.14}$$

are fed back. The feedback element $G_{fb}(s)$ generates the manipulated variable according to

$$u(s) = G_{fb}(s)\,(w(s) - \Delta y(s))\,. \tag{6.4.15}$$

If the process model matches the real process behavior

$$G_{p,m}(s) = G_p(s) \tag{6.4.16}$$

and process disturbances are $n(t) = 0$, the output deviation is $\Delta y(t) = 0$ and changes of the reference variable act to the process in a feedforward manner. The selection

$$G_{fb}(s) = G_p^{-1}(s) \tag{6.4.17}$$

then leads to a compensation of the process transfer function, such that $y(s) = w(s)$, compare (6.3.7). If the process disturbances are not zero, $n(t) \neq 0$, then the disturbance is reconstructed as $\Delta y(t) = n(t)$ and the feedback with (6.4.17) results in a perfect compensation of the disturbance at the process output.

However, as this inverted process model is usually not realizable a realizability filter (low-pass filter)

$$G_{\mathrm{rf}}(s) = \frac{1}{\prod_{r=1}^{q}(1 + T_{\mathrm{r}}s)} \tag{6.4.18}$$

is introduced, such that the feedback element becomes

$$G_{\mathrm{fb}}(s) = G_{\mathrm{p}}^{-1}(s)\, G_{\mathrm{rf}}(s). \tag{6.4.19}$$

The order q is usually chosen such that it is at least equal to the order of the process model denominator $q \geq m$.

In order to describe the internal model controller in the classical way, the process model G_{pm} is considered as a part of the controller, as shown in Fig. 6.4.1b), and yields

$$G_{\mathrm{c,IMC}} = \frac{u(s)}{e(s)} = \frac{G_{\mathrm{fb}}(s)}{1 - G_{\mathrm{fb}}(s)\, G_{\mathrm{p,m}}(s)} = \frac{1}{G_{\mathrm{p}}(s)} \frac{G_{\mathrm{rf}}(s)}{1 - G_{\mathrm{rf}}(s)}. \tag{6.4.20}$$

The closed-loop transfer function for setpoint changes then becomes with $G_{\mathrm{p}}(s) = G_{\mathrm{p,m}}(s)$

$$G_{\mathrm{wy}}(s) = \frac{y(s)}{w(s)} = G_{\mathrm{rf}}(s). \tag{6.4.21}$$

Hence, the closed-loop behavior is only determined by the realizability filter and can therefore be designed in a direct way by choosing the filter. A corresponding transfer function for the disturbance behavior is

$$\frac{y(s)}{n(s)} = 1 - G_{\mathrm{rf}}(s) \tag{6.4.22}$$

and also depends only on the realizability filter.

In reality it cannot be assumed that the real process and the process model are identical. Then according to

$$G_{\mathrm{p}}(s) = G_{\mathrm{p,m}}(s) + \Delta G_{\mathrm{p,m}}(s) \tag{6.4.23}$$

a difference transfer function $\Delta G_{\mathrm{p,m}}(s)$ appears. An analysis of the closed-loop behavior shows that the resulting closed loop may become weakly damped or even unstable if the process poles are close to the imaginary axis. Therefore the application has to be restricted to well-damped process behavior.

Further the controller itself should be stable, such that non-minimum phase systems must be excluded. Therefore following conditions have to be met for the design of the IMC controller

- The process $G_{\mathrm{p}}(s)$ may not contain poles and zeros with positive real part, i.e. the process must be stable and minimum phase.

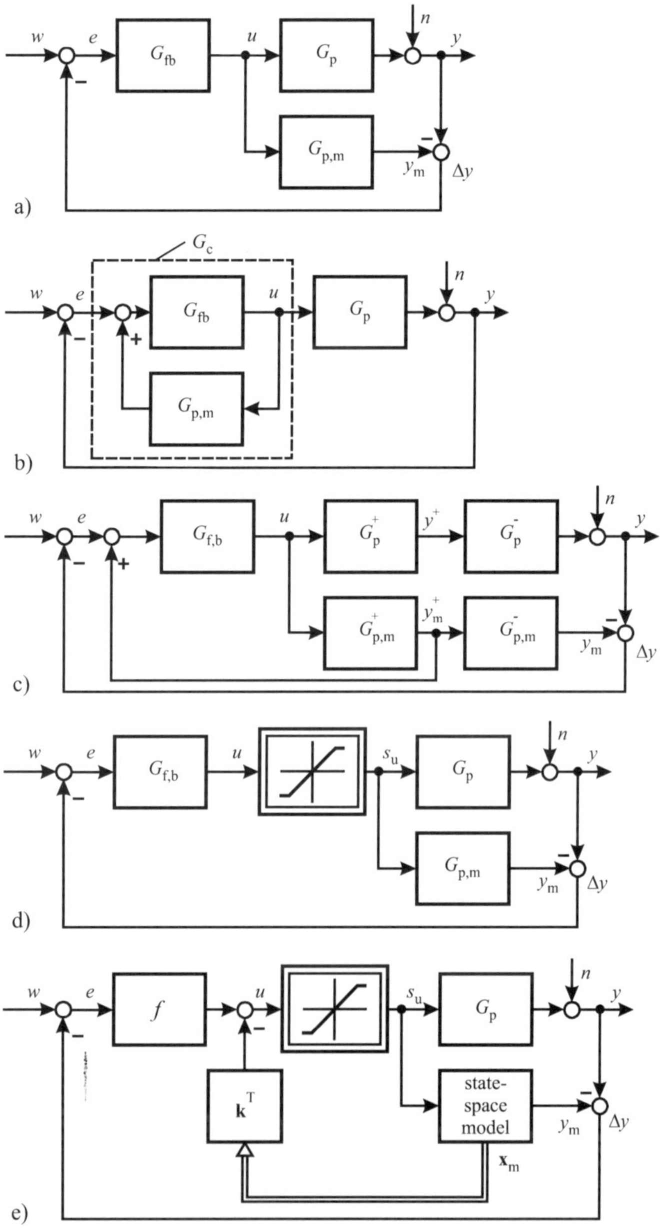

Fig. 6.4.1. Internal model control structures. **a** structure with parallel model. **b** resulting controller structure. **c** with non-minimum phase process part. **d** with actuator saturation. **e** with actuator saturation and state model.

- Realizability of the IMC controller implies that the term $G_{\mathrm{rf}}(s)/(1 - G_{\mathrm{rf}}(s))$ adds additional poles, such that its pole excess $(m - n)$ (m number of poles, n number of zeros) has to be greater or equal the pole excess of the process, see e.g. Isermann (1989).

To extend the design of the IMC controller, the process model is split in a term $G^{+}_{\mathrm{p,m}}(s)$ which is invertible (stable and well-damped) and a term $G^{-}_{\mathrm{p,m}}(s)$ which is not invertible (non-minimum phase, dead time)

$$G_{\mathrm{p,m}}(s) = G^{+}_{\mathrm{p,m}}(s)\, G^{-}_{\mathrm{p,m}}(s) \tag{6.4.24}$$

and the inversion is restricted to $G^{+}_{\mathrm{p,m}}(s)$, such that

$$G_{\mathrm{fb}}(s) = (G^{+}_{\mathrm{p,m}}(s))^{-1} G_{\mathrm{rf}}(s) \tag{6.4.25}$$

see Fig. 6.4.1c).

In the case of a process with dead time it holds for the model

$$G_{\mathrm{p,m}}(s) = G^{+}_{\mathrm{p,m}}(s)\, e^{-T_{\mathrm{d}} s} \tag{6.4.26}$$

and as shown only the output of the invertible term $G^{+}_{\mathrm{p,m}}(s)$ is used to form the IMC controller. However, Δy is determined with the model output including the dead time and compared with the reference variable.

Actuator restriction can be taken into account by adding this nonlinearity in the parallel model, but not in the feedback element (6.4.19) because its inversion becomes infinite, see Fig. 6.4.1d). If the process can be described by a Hammerstein model, e.g. by a second order nonlinearity at the input and this nonlinearity can be inverted, it can be included in the controller design.

Figure 6.4.1e) shows an IMC structure where the parallel model is represented in state-space form and the calculated states are fed back. This is called *model-state feedback control*, Mhatre and Brosilow (2000), Wright and Kravaris (2006). The design of the feedback vector $\mathbf{k}^{\mathrm{T}}$ and the prefilter f follows from (6.3.14) to (6.3.16), see Pfeil (2011).

Example 6.2 (Internal model control).
Applying the design equations (6.4.19) and (6.4.20) yields for the second order process from Example 6.1

$$G_{\mathrm{fb}}(s) = \frac{1}{K_{\mathrm{p}}} \frac{(1+T_1 s)(1+T_2 s)}{(1+T_{\mathrm{r}})^2}$$
$$G_{\mathrm{c,IMC}}(s) = \frac{1}{2K_{\mathrm{p}}T_{\mathrm{r}}} \frac{(1+T_1 s)(1+T_2 s)}{s} \frac{1}{1+\frac{T_{\mathrm{r}}}{2}s}.$$

Hence, the resulting controller shows PID behavior with a first order lag.

□

6.4.3 Cancelation controller (CAC)

Another way to include the process model directly is to design a cancelation controller. Then the closed-loop transfer function for the reference variable $G_{\mathrm{wy}}(s)$ is specified, see (6.4.1), and then solved for the controller

$$G_{\mathrm{c,canc}}(s) = \frac{u(s)}{e(s)} = \frac{1}{G_{\mathrm{p}}(s)} \frac{G_{\mathrm{wy}}(s)}{1 - G_{\mathrm{wy}}(s)}. \tag{6.4.27}$$

Comparison with (6.4.20) shows that the specified transfer function is identical to the realizability filter, as also expressed by (6.4.21). Therefore the same restrictive conditions hold for the design.

The selection of $G_{\mathrm{wy}}(s)$ can be a low-pass process as (6.4.18) or a more weakly damped oscillating process with conjugate complex poles if some overshoot is desired. For a first order process and first order reference transfer function, a PI controller follows directly. For a second order process a PID controller follows, see Example 6.2. (A detailed treatment of discrete-time cancelation controllers is given in Isermann (1989)).

6.4.4 State-space control

State-space control is based on the state-space model of the process which is for a single-input single-output (SISO) process

$$\dot{\mathbf{x}}(t) = \mathbf{A}\mathbf{x}(t) + \mathbf{b}u(t) \tag{6.4.28}$$

$$y(t) = \mathbf{c}^{\mathrm{T}}\mathbf{x}(t) + du(t). \tag{6.4.29}$$

The vector $\mathbf{x}(t)$ represents the internal states of the process and its order is equal to the order n of the process transfer function. The state representation can be selected in different forms, such as the controllable or observable canonical form. A state controller feeds back the states of the process according to

$$u(t) = -\mathbf{k}^{\mathrm{T}}\mathbf{x}(t), \tag{6.4.30}$$

see Fig. 6.4.2. The determination of the gains $\mathbf{k}^{\mathrm{T}}$ can be based on pole placement or on quadratic performance criteria, solving a matrix-Riccati equation.

Since the states are usually not measurable, a state observer is necessary to reconstruct the state variables $\mathbf{x}(t)$, based on measurements of $u(t)$ and $y(t)$. The state observer uses the process model parallel to the process and the difference between the measured output $y(t)$ and the observer output $\hat{y}(t)$ is fed back to correct the observer states

$$\Delta\dot{\mathbf{x}}(t) = -\mathbf{h}^{\mathrm{T}}\left(y(t) - \hat{y}(t)\right). \tag{6.4.31}$$

The observer equation then becomes

$$\dot{\hat{\mathbf{x}}} = \mathbf{A}\hat{\mathbf{x}}(t) + \mathbf{b}u(t) + \mathbf{h}^{\mathrm{T}}\left(y(t) - \mathbf{c}^{\mathrm{T}}\hat{\mathbf{x}}(t)\right). \tag{6.4.32}$$

The feedback gains $\mathbf{h}^{\mathrm{T}}$ can be designed similar to the state feedback, e.g. by pole placement.

In the control law (6.4.30) the state $\mathbf{x}(t)$ is now replaced by the observed state

$$u(k) = -\mathbf{k}^{\mathrm{T}}\hat{\mathbf{x}}(t) \tag{6.4.33}$$

see Fig. 6.4.4. Hence, the basic state feedback changes the dynamics of the process, e.g. makes it faster and well stabilized and damped. Different additions do exist to introduce a comparison of the controlled output $y(t)$ with the reference variable $w(t)$. Figure 6.4.4 shows a simple solution where an integrator is used to generate a correction of the manipulated variable

$$\Delta u(t) = k_{\mathrm{I}} \int (w(t) - y(t))\, dt \tag{6.4.34}$$

to avoid steady-state control errors, corresponding to the integral part of a PID controller.

Based on the multi-variable process model (MIMO)

$$\dot{\mathbf{x}}(t) = \mathbf{A}\mathbf{x}(t) + \mathbf{B}\mathbf{u}(t) \tag{6.4.35}$$

$$y(t) = \mathbf{C}\mathbf{x}(t) + \mathbf{D}\mathbf{u}(t). \tag{6.4.36}$$

the state controller becomes

$$\mathbf{u}(t) = -\mathbf{k}^{\mathrm{T}}\mathbf{x}(t) \tag{6.4.37}$$

where $\mathbf{k}$ is a gain matrix. For more details refer to the rich literature, e.g. Ogata (2008), Dorf and Bishop (2010). Discrete-time versions of state control are treated, e.g. Ackermann (1983), Isermann (1989), Isermann (1991), Isermann et al (1992), Kuo (1995), Åström and Wittenmark (1997), Franklin and Powell (2006).

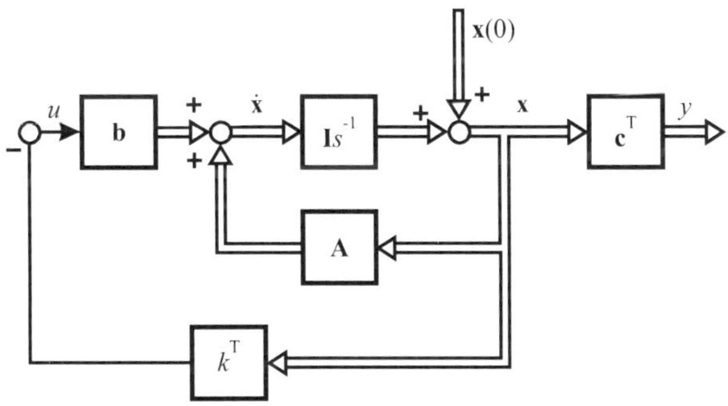

Fig. 6.4.2. Process model in state-space form and state-feedback controller.

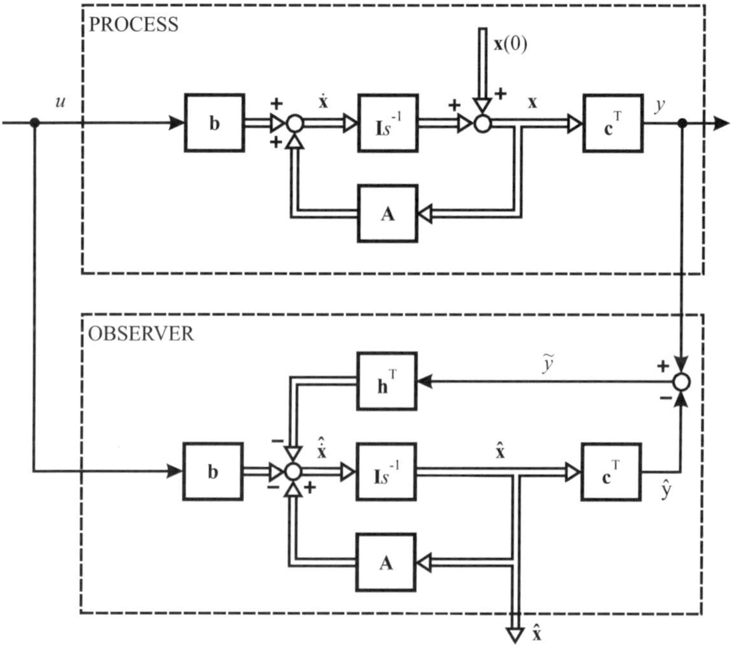

Fig. 6.4.3. Process model with state observer.

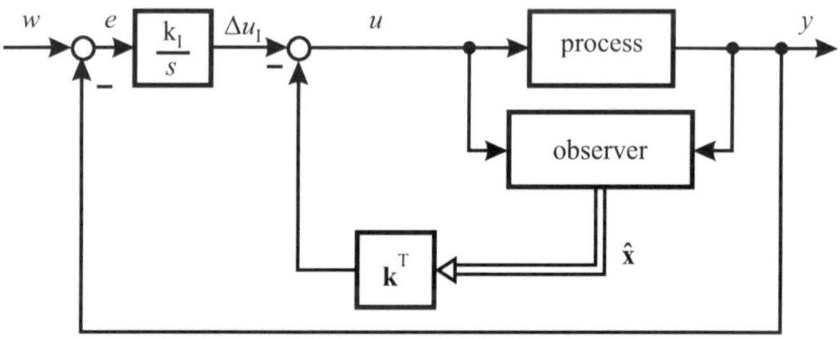

Fig. 6.4.4. State controller with observer.

6.4.5 Interconnected control systems

The behavior of single closed loops can be significantly improved by adding one ore more inner loops, by combination with feedforward control and by optimal coupling of neighbored multi-variable closed loops. The resulting control systems are called *interconnected control systems*.

a) Cascade control systems

If one variable y_2 between a process part 1 and a process part 2 is measurable, it can be used to form an inner closed loop by adding a minor (auxiliary) controller in cascade to the standard outer closed loop with the main controller, as depicted in Fig. 6.4.5. The advantages are

(1) Disturbances n_2 acting on process part 1 are already controlled by the minor controller and do not influence the main control variable y_1.
(2) Parameter changes of process part 1 are attenuated and do not have to be taken into account in the parametrization of the main controller.
(3) The behavior of the main control variable y_1 becomes faster if the minor loop leads to faster eigenvalues than those of the process part 1 without the minor controller.

More details and examples are given in Isermann (1991). This cascade control system is especially advantageous if process part 1 is an actuator and the actuator position y_2 can be measured. In addition to the mentioned advantages also nonlinear effects, like position-dependent actuator gain or hysteresis are compensated by the minor controller. Examples are the position control of VTG turbochargers or the EGR valve.

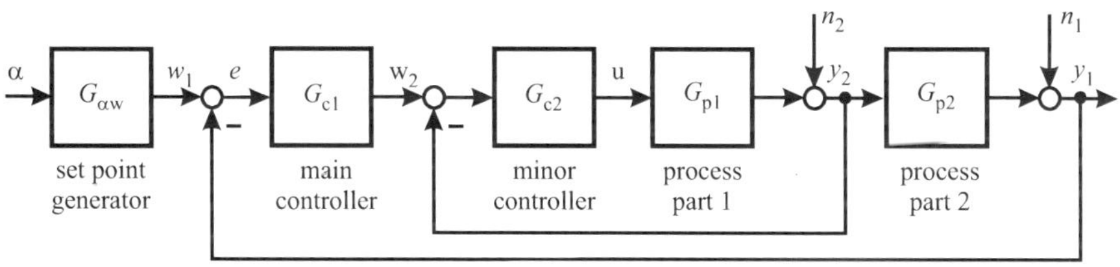

Fig. 6.4.5. Cascaded control system with inner loop and outer loop.

b) Combinations of feedback and feedforward control

Figure 6.4.6a) depicts a process with a transfer function

$$G_{\mathrm{pv}}(s) = \frac{n(s)}{v(s)} \tag{6.4.38}$$

for an *external disturbance variable* $v(t)$ *as input* and an usually different transfer function

$$G_{\mathrm{p}}(s) = \frac{y(s)}{u(s)} = G_{\mathrm{a}}(s)\, G_{\mathrm{psy}}(s) \tag{6.4.39}$$

for the manipulated variable $u(t)$. If $v(t)$ can be measured, then an ideal feedforward control is obtained by

$$G'_{\mathrm{cvu}}(s) = \frac{u_{\mathrm{ff}}(s)}{v(s)} = -G_{\mathrm{p}}^{-1}(s)\, G_{\mathrm{pv}}(s). \tag{6.4.40}$$

Analogously to the dynamic feedforward control described in Sect. 6.3b) a realizability filter $G_{\mathrm{rf}}(s)$ according to (6.3.10) has to be added to cope with the inversion of $G_{\mathrm{p}}(s)$, such that the realizable feedforward control becomes

$$G_{\mathrm{cvu}}(s) = -G_{\mathrm{p}}^{-1}(s)\, G_{\mathrm{rf}}(s)\, G_{\mathrm{pv}}(s). \tag{6.4.41}$$

This holds for a minimum phase process. For non-minimum processes only the invertible process part can be compensated, see (6.4.24) and (6.4.25).

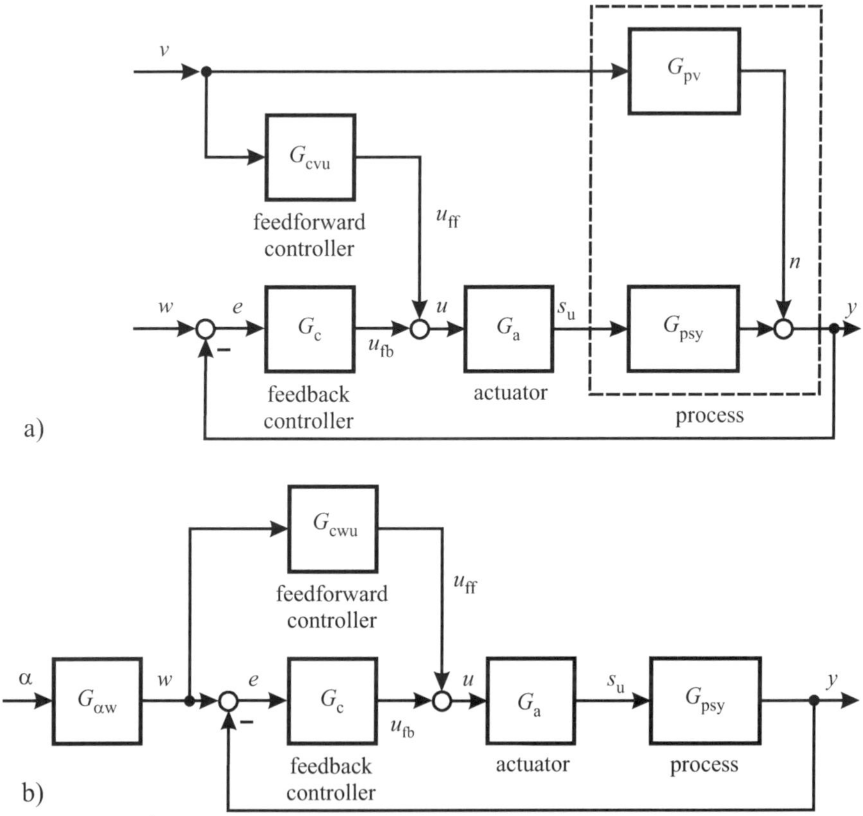

Fig. 6.4.6. Combinations of feedback and feedforward control for. **a** external process disturbances. **b** Setpoint changes w.

Example 6.3 (Dynamic feedforward control for a disturbance variable as input).
As in Example 6.1 a second order process $G_{\mathrm{p}}(s)$ is assumed and a disturbance behavior

$$G_{\mathrm{pv}}(s) = \frac{K_{\mathrm{pv}}}{1 + T_3 s}.$$

Then, a realizable feedforward controller becomes

$$G_{\mathrm{cvu}}(s) = -\frac{(1+T_1 s)(1+T_2 s)}{K_{\mathrm{p}}(1+T_{\mathrm{r}1} s)(1+T_{\mathrm{r}2} s)} \frac{K_{\mathrm{pv}}}{(1+T_3 s)}.$$

If $T_{\mathrm{r}1} = T_2$ is selected, it follows

$$G_{\mathrm{cvu}}(s) = -\frac{K_{\mathrm{pv}}}{K_{\mathrm{p}}} \frac{(1+T_1 s)}{(1+T_{\mathrm{r}2} s)(1+T_3 s)}$$

which is a proportional-derivative (PD) element with two lag time constants. A simplification is obtained by assuming $T_{\mathrm{r}2} \ll T_1$

$$G''_{\mathrm{cvu}}(s) = -\frac{K_{\mathrm{pv}}}{K_{\mathrm{p}}} \frac{(1+T_1 s)}{(1+T_3 s)}$$

resulting in a PD element with a time lag.

□

If the disturbance behavior $G_{\mathrm{pv}}(s)$ is dynamically faster than the manipulated behavior $G_{\mathrm{p}}(s)$, the feedforward controllers approximately result in a proportional-derivative (PD) behavior and if both have the same order of the denominator approximately P behavior is sufficient, assuming low-pass processes.

In many cases, the overall control performance can be improved significantly by by combining feedback and feedforward control, as illustrated in Fig. 6.4.6a). In the case of a disturbance $v(t)$ the closed loop just has to compensate remaining control errors resulting from the non-ideal feedforward control and other disturbances. However, the feedforward control should be designed by taking the closed-loop behavior into account, as shown in Isermann (1991), Chap. 17, e.g. by a sequential procedure with mutual optimization of the feedforward and feedback controller.

The feedforward control can also be applied for *reference variables or setpoints as inputs* of a closed loop, as shown in Fig. 6.4.6b). This case belongs to *servo control systems*, where a drive system has to follow precisely setpoint changes. In this context the control of combustion engines and vehicles can be considered as a servo-control problem. An ideal feedforward control, by by-passing the controller G_{c} is then obtained from

$$\frac{y(s)}{w(s)} = G_{\mathrm{cwu}}(s)\, G_{\mathrm{p}}(s) = 1 \tag{6.4.42}$$

and thus

$$G'_{\mathrm{cwu}}(s) = \frac{u_{\mathrm{ff}}(s)}{w(s)} = G_{\mathrm{p}}^{-1}(s) \tag{6.4.43}$$

or with a realizability filter

$$G_{\mathrm{cwu}}(s) = G_{\mathrm{p}}^{-1}(s)\, G_{\mathrm{rf}}(s). \tag{6.4.44}$$

In case of an almost perfect feedforward control the control error $e(t)$ is very small for reference variable changes and just simple feedback controllers are good enough to reach a good control performance. (This will be used in Sect. 8.5 for charging pressure and air mass control).

Example 6.4 (Dynamic feedforward control for a reference variable as input). Using the second order process of Example 6.1 a realizable feedforward controller becomes

$$G_{\mathrm{cwu}}(s)=\frac{u_{\mathrm{ff}}(s)}{w(s)}=\frac{(1+T_1s)(1+T_2s)}{K_{\mathrm{p}}(1+T_{\mathrm{r}1}s)(1+T_{\mathrm{r}2}s)}$$

which is the same as in Example 6.1. Applying $T_{\mathrm{r}1}=T_2$ leads to

$$G_{\mathrm{cwu}}(s)=\frac{1}{K_{\mathrm{p}}}\frac{(1+T_1s)}{(1+T_{\mathrm{r}2}s)}$$

i.e. a PD element with a time lag.

□

Because of the resulting differential (D) behavior the signal $dw(t)/dt$ is proportionally acting on the manipulated variable and therefore $G_{\mathrm{cwu}}(s)$ is also called a "speed adding" by-pass element. An additional effect for the case of a ramp input $dw(t)/dt=$ const. is that the otherwise arising offset $e(t)=w(t)-y(t)$ vanishes for $t\to\infty$.

c) Multi-variable control

If closed control loops are mutually coupled, as shown in Fig. 6.4.7, their interactions have to be taken into account. There exist a great variety of multi-variable process structures. However, in many cases these structures can be transformed into some basic standard canonical structures, such as P-, V-, or H-canonical structures, Mesarovic (1960), Fisher and Seborg (1976).

Figure 6.4.7 depicts a P-canonical structure for a two-input two-output process. For example, the coupled charging pressure and air mass control of diesel engines shows this structure, see Fig. 8.5.2. The process transfer functions can then be stated in matrix notation

$$\begin{aligned}\begin{bmatrix}y_1(s)\\y_2(s)\end{bmatrix}&=\begin{bmatrix}G_{\mathrm{p}11}(s)&G_{\mathrm{p}21}(s)\\G_{\mathrm{p}12}(s)&G_{\mathrm{p}22}(s)\end{bmatrix}\begin{bmatrix}u_1(s)\\u_2(s)\end{bmatrix}\\\mathbf{y}&=\qquad\mathbf{G}(s)\qquad\mathbf{u(s)}.\end{aligned}\tag{6.4.45}$$

The properties of two-variable control systems with main controllers $G_{\mathrm{C}11}(s)$, $G_{\mathrm{C}22}(s)$ and coupling controllers $G_{\mathrm{C}12}(s)$, $G_{\mathrm{C}21}(s)$ are discussed in Isermann (1991). They depend especially on the *dynamic coupling factor*

$$\kappa(s)=\frac{G_{12}\,G_{21}}{G_{11}\,G_{22}}.\tag{6.4.46}$$

For example, the resulting eigenvalues depend on this coupling factor. For a negative static coupling $\kappa_0(0)<1$ the gain of one process part, e.g. G_{11}, increases by closing the other loop with $G_{\mathrm{C}22}(s)$ and the gain of $G_{\mathrm{C}11}(s)$ must be reduced. Another property is that if the coupling elements are fast, then a fast coupled loop 2 has a stronger effect on y_1 than a slow one. Also the location of external disturbances and a different or simultaneous change of reference variable have an influence on the control behavior. Depending on these properties the main controllers reinforce or counteract each other. If they counteract (disturb) each other, the coupling controller should be designed in a decoupling way. However, because of the many possibilities, multi-variable control systems should be designed for the special case and for the main external exciting variables. A rigorous treatment of multi-variable control systems is possible in state space representation. The discussed transfer-function approach has the advantage to be more transparent.

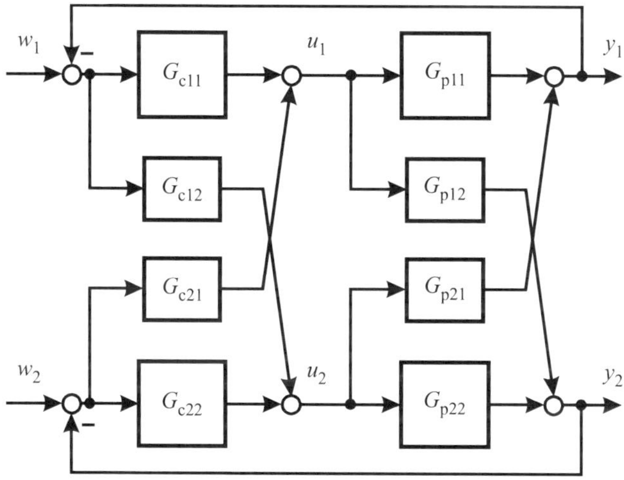

Fig. 6.4.7. Two-variable process in P-canonical structure, main controllers and interaction (coupling) controllers.

6.5 Nonlinear static engine feedforward control

The multi-variable structure of internal combustion engines requires the implementation of nonlinear multi-input multi-output (MIMO) control functions or at least several multi-input single-output (MISO) control algorithms. Therefore it is of interest in which digital-computer-oriented structure these control functions can be realized. As these control functions are usually developed after measurement of the engine's behavior (engine identification), multi-variable engine models are obtained in a first step. In a second step they are used for the design of multi-variable control functions. Hence, the control-algorithm design has to be considered together with the engine model structures. In order to obtain a modular, transparent structure, multi-input single-output (MISO) representations are preferred and considered subsequently.

The most frequently used representations for the stationary nonlinear behavior of engines are look-up tables or grid-based maps. Alternatives are polynomials or net based models, like special neuronal networks. Dynamic feedforward control is required for the dynamically delayed engine process parts, like the air path, EGR path and turbochargers. This section considers basic feedforward control functions.

6.5.1 Look-up tables

Look-up tables are the most commonly applied representations of linear and nonlinear stationary (static) relations between one or two input variables and one output variable. As already noted in Sect. 3.1.1, their advantages are the simplicity, easy readability, and the relatively low computational demand for storing and reading. The identification consists in the simplest case in storing the measured outputs at grid positions or grid nodes, representing input data. However, they are basically restricted to one or two inputs. For higher-dimensional representations usually several low-dimensional look-up tables have to be combined.

In the following one- and two-dimensional look-up tables, their interpolation and representation with basis functions are considered. A parameter estimation allows to determine look-up table presentations for the cases that the measurements are not made at the grid positions, the number of grid points has to be reduced or for noisy measurements. Then, an extension to multi-inputs and methods for the optimization of the grid structure are described.

a) One-dimensional look-up tables

A nonlinear characteristic for one output y dependent on one input u

$$y = f(u) \tag{6.5.1}$$

is represented by a one-dimensional map at specified discrete inputs u_i

$$u_1 = c_1\,,\ u_2 = c_2\,,\ \ldots\,,\ u_M = c_M \tag{6.5.2}$$

such that the function (6.5.1) is sampled and results in values $y_i = w_i$

$$w_1 = f(c_1)\,,\ w_2 = f(c_2)\,,\ \ldots\,,\ w_M = f(c_M) \tag{6.5.3}$$

as depicted in Fig. 6.5.1. These discrete outputs w_i are also called heights and are stored in the memory of the ECU for the input axis points $u_i = c_i$. They result, for example, directly from measurements or indirectly from the evaluation of an intermediate representation. If output values are required between the stored input axis points, a linear interpolation with two neighboring values is applied.

$$y(u) = \frac{w_{i-1}(c_i - u) + w_i(u - c_{i-1})}{c_i - c_{i-1}} \tag{6.5.4}$$

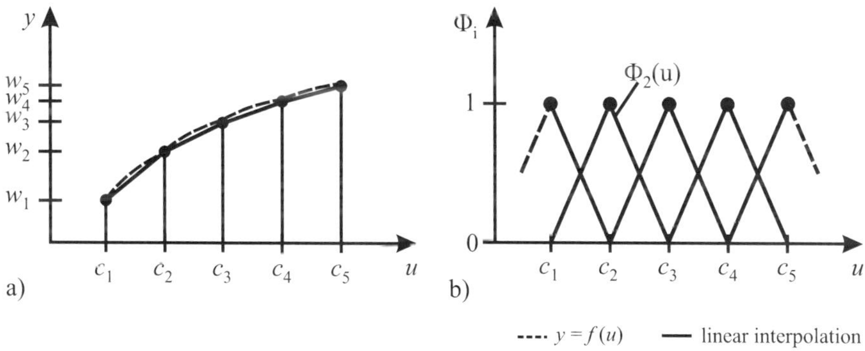

Fig. 6.5.1. One-dimensional look-up table $y = f(u)$. **a** sample points and linear interpolation, **b** basis functions.

such that $y(u_{i-1}=c_{i-1})=w_{i-1}$ and $y(u_i=c_i)=w_i$ and $c_{i-1} \leq u \leq c_i$. This means that the characteristic between the input axis points is approximated by straight lines, see Fig. 6.5.1a).

Another representation of the one-dimensional look-up table follows after introducing triangular basis (or weighting) functions, such that

$$y(u) = \sum_{i=1}^{M} w_i \phi_i(u, \mathbf{c}) \tag{6.5.5}$$

where

$$\mathbf{c}^{\mathrm{T}} = \begin{bmatrix} c_1 \; c_2 \; \dots \; c_M \end{bmatrix} \tag{6.5.6}$$

are the discrete input values and the basis functions are

$$\phi_i(u, \mathbf{c}) = \begin{cases} (u - c_{i-1})/(c_i - c_{i-1}) & \text{for } c_{i-1} \leq u \leq c_i \\ (u - c_{i+1})/(c_i - c_{i+1}) & \text{for } c_i < u \leq c_{i+1} \\ 0 & \text{otherwise.} \end{cases} \tag{6.5.7}$$

These basis functions are then triangles with the peak $\phi_i(u, c_i) = 1$ at the center c_i as shown in Fig. 6.5.1b). They determine the weighted influences of the height w_i at c_i on the output value $y(u)$.

In order to determine the output $y(u)$ for $c_{i-1} \leq u \leq c_i$ two neighboring values w_{i-1} and w_i are weighted according to

$$\begin{aligned} y(u) &= w_{i-1}\phi_{i-1}(u, c_i, c_{i-1}) + w_i\phi_i(u, c_i, c_{i-1}) \\ &= w_{i-1}\frac{c_i - u}{c_i - c_{i-1}} + w_i\frac{u - c_{i-1}}{c_i - c_{i-1}}. \end{aligned} \tag{6.5.8}$$

The output y therefore depends only on two basis functions and for the sum of the weights follows

$$\sum_{i=1}^{M} \phi_i(u, \mathbf{c}) = 1. \tag{6.5.9}$$

As (6.5.8) and (6.5.4) are identical, the triangular basis functions provide a linear interpolation between two output heights w_i and w_{i-1}.

The characteristic in Fig. 6.5.1 is represented with equidistant input axis points c_i. In order to adjust the resolution to a nonlinear course of $y = f(u)$ the axis points can also be distributed non-uniformly. The triangular basis functions then become asymmetrical, see e.g. Nelles (2001).

b) Two-dimensional look-up tables and their optimization

The one-dimensional look-up table is extended to two inputs by using a grid with two inputs u_1 and u_2, see Fig. 6.5.2. The usually nonlinear function

$$y = f(u_1, u_2) \tag{6.5.10}$$

is represented by a two-dimensional map at the discrete input grid points (nodes) $\mathbf{u} = [u_{1,i}\ u_{2,j}]$

$$\begin{aligned} u_{1,1} &= c_{1,1}\ ,\ u_{1,2} = c_{1,2}\ ,\ \ldots\ ,\ u_{1,M1} = c_{1,M1} \\ u_{2,1} &= c_{2,1}\ ,\ u_{2,2} = c_{2,2}\ ,\ \ldots\ ,\ u_{2,M2} = c_{2,M2} \end{aligned} \tag{6.5.11}$$

such that the function (6.5.10) is sampled and results in values (heights)

$$y_{i,j} = w_{i,j} \qquad i = 1 \ldots M_1,\ j = 1 \ldots M_2. \tag{6.5.12}$$

Hence, $N = M_1 M_2$ output values have to be stored for two inputs and one output. If the output values have to be determined between the grid points a bilinear interpolation is applied, see Fig. 6.5.2, Hoschek and Lasser (1993), Schmitt (1995). Using the surrounding output values the four grid points are weighted with the opposite areas according to

$$\begin{aligned} y(u_1, u_2) =& (w_{i,j}\, a_{i+1,j+1} + w_{i+1,j}\, a_{i,j+1} + w_{i,j+1}\, a_{i+1,j} + w_{i+1,j+1}\, a_{i,j}) \\ & /(a_{i+1,j+1} + a_{i,j+1} + a_{i+1,j} + a_{i,j}) \end{aligned} \tag{6.5.13}$$

with the areas 1 to 4

$$\begin{aligned} a_{i+1,j+1} &= (c_{1,i+1} - u_1)(c_{2,j+1} - u_2) \\ a_{i,j+1} &= (u_1 - c_{1,i})(c_{2,j+1} - u_2) \\ a_{i+1,j} &= (c_{1,i+1} - u_1)(u_2 - c_{2,j}) \\ a_{i,j} &= (u_1 - c_{1,i})(u_2 - c_{2,j}). \end{aligned} \tag{6.5.14}$$

A linear interpolation cannot be realized because the four heights generally do not lie on a plane. However, the axis parallel cuts through the bilinear interpolation are linear functions, Nelles (2001).

The two-dimensional look-up table can also be represented in the form of basis functions

$$y(\mathbf{u}) = \sum_{i=1}^{M1} \sum_{j=1}^{M2} w_{i,j} \phi_{i,j}(\mathbf{u}, \mathbf{c}). \tag{6.5.15}$$

Each output value obtains a basis function with the same dimension as the input grid, similar to the one-dimensional case. For $p = 2$ inputs the basis functions are two dimensional pyramids on the interpolation area with the vertices $w_{i,j}$ at $u_{i,j} = c_{i,j}$, see Fig. 6.5.3. They are only non-zero within the interpolation area $c_{1,i} < u_1 < c_{1,i+1}$, $c_{2,j} < u_2 < c_{2,j+1}$.

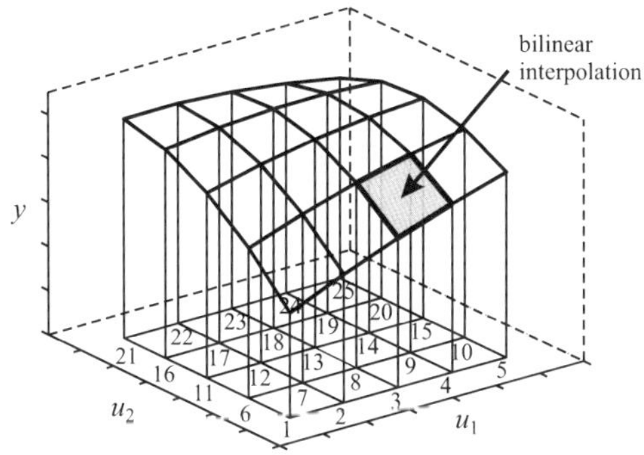

Fig. 6.5.2. Two-dimensional look-up table for two inputs $y = f(u_1, u_2)$.

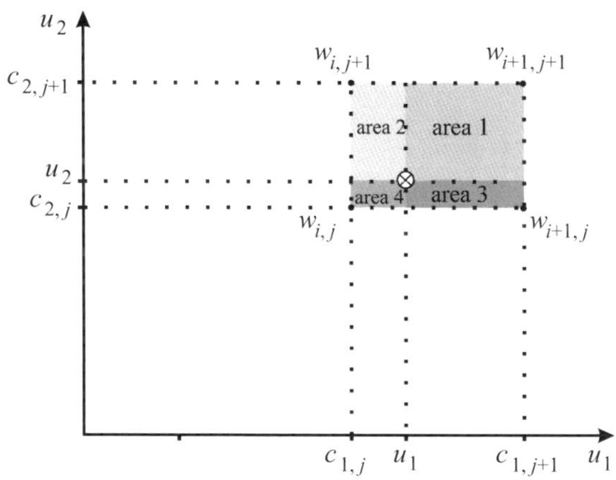

Fig. 6.5.3. Scheme for the bilinear interpolation of a two-dimensional look-up table.

The accuracy of look-up tables representations depends on the number of grid points. For equal samples $M_1 = M_2$ on each of the two axes they grow quadratically with the number of positions per input axis. In the case of smooth surfaces a small number of data points is sufficient, while for stronger nonlinear behavior a finer grid has to be chosen. The area interpolation is based on the assumption that all data point heights are available in the whole range. However, this condition is sometimes not satisfied, such that extrapolation may be required.

If the measurements of the output $y(u_1, u_2)$ are precisely at the nodes of the assumed grid $u_{1,i} = c_{1,i}$ and $u_{2,j} = c_{2,j}$, then the measured outputs $w_{i,j} = y_{i,j}$ can be stored directly and represent the look-up table. However, if the number of nodes has to be reduced to save memory space or if the measurements are noisy the heights have to be optimized by applying parameter estimation. This is also required if the measurements $y(\mathbf{u})$ are not available on the nodes of the assumed grid for any u_1 or u_2. Therefore the *optimization of the look-up table heights* is considered subsequently.

It is now assumed that L measurements of the process output are made

$$\mathbf{y}_{\mathrm{p}}^{\mathrm{T}}(\mathbf{u}) = \left[y(1), y(2), \ldots, y(L)\right] \tag{6.5.16}$$

for Q different combinations

$$\mathbf{U}^{\mathrm{T}} = \left[\mathbf{u}(1), \mathbf{u}(2), \ldots, \mathbf{u}(L)\right] \tag{6.5.17}$$

of the input

$$\mathbf{u}^{\mathrm{T}} = \left[u_{1,i}\ u_{2,j}\right] \tag{6.5.18}$$

and the look-up table heights $w_{i,j}$ have to be estimated.

The basis-function representation relates the given inputs with the measured outputs according to (6.5.15) or if the nodes are now numbered sequentially as shown in the grid of Fig. 6.5.2

$$\mathbf{c}_N^{\mathrm{T}} = \left[c_1, c_2, \ldots, c_N\right] \tag{6.5.19}$$

with $N = M_1 M_2$ by

$$y(\mathbf{u}(i)) = \sum_{\nu=1}^{N} w_\nu \phi_\nu(\mathbf{u}(i), \mathbf{c}). \tag{6.5.20}$$

Turning into vector notation yields

$$\mathbf{y} = \mathbf{X}\mathbf{w} \tag{6.5.21}$$

with the regression matrix

$$\mathbf{X} = \begin{bmatrix} \phi_1(\mathbf{u}(1), \mathbf{c}) & \ldots & \phi_N(\mathbf{u}(1), \mathbf{c}) \\ \phi_1(\mathbf{u}(2), \mathbf{c}) & \ldots & \phi_N(\mathbf{u}(2), \mathbf{c}) \\ \vdots & & \vdots \\ \phi_1(\mathbf{u}(L), \mathbf{c}) & \ldots & \phi_N(\mathbf{u}(L), \mathbf{c}) \end{bmatrix} \tag{6.5.22}$$

and the parameter vector

$$\mathbf{w}^{\mathrm{T}} = \left[w_1, w_2, \ldots, w_N\right]. \tag{6.5.23}$$

Taking into account output disturbances

$$\mathbf{y}_{\mathrm{p}} = \mathbf{y}_{\mathrm{u}} + \mathbf{n} \tag{6.5.24}$$

the process model becomes

$$\mathbf{y}_{\mathrm{p}} = \mathbf{X}\mathbf{w} + \mathbf{n} \tag{6.5.25}$$

which leads to an equation error

$$\mathbf{e} = \mathbf{y}_{\mathrm{p}} - \mathbf{X}\hat{\mathbf{w}} \tag{6.5.26}$$

where $\hat{\mathbf{w}}$ are the look-up table parameters to be estimated. As the equation error $\mathbf{e}$ is linear in the parameters $\hat{\mathbf{w}}$, minimization of the loss function

$$V = \mathbf{e}^{\mathrm{T}}\mathbf{e} \tag{6.5.27}$$

(sum of squared errors) allows to apply the least squares parameter estimation in the direct form

$$\hat{\mathbf{w}} = [\mathbf{X}^{\mathrm{T}}\mathbf{X}]^{-1}\mathbf{X}^{\mathrm{T}}\mathbf{y}_{\mathrm{p}} \tag{6.5.28}$$

see e.g. Isermann and Münchhof (2011). One assumption for a unique solution is that the number of measurements L is equal or larger than the number N of parameters: $L \geq N$. However, the regression matrix is sparse, as each row of $\mathbf{X}$ contains only $2^? = 4$ non-zero elements. According to the bilinear interpolation, see (6.5.15) and (6.5.13), only four basis functions are activated, see Fig. 6.5.3. If the measurement data does not fill the whole input space, some basis functions will not be activated and $\mathbf{X}$ will contain zero columns and becomes singular. In this case the corresponding basis functions and heights have to be removed.

c) Regularization of look-up table identification

Depending on the process and the measurement design, there might be grid points in the lattice structure which are not or only weakly excited. This can either be due to constraints of the process if several grid points are not reachable or due to a poor distribution of the measurement points. If grid points are not excited, the corresponding heights cannot be estimated. If grid points are weakly excited, the input matrix $\mathbf{X}$ is ill-conditioned and sharp differences in the grid point heights can occur, as can be seen in the identified two-dimensional base look-up table in Fig. 6.5.4a) for a torque model.

To avoid these sharp differences and to enable the estimation of not excited grid points, a regularization is introduced, Sequenz (2013). This regularization smooths the look-up table and estimates the heights of not excited grid points by penalizing the differences to neighboring grid point heights.

The differences of grid point heights to neighboring grid point heights can be penalized by appending several rows to the input matrix $\mathbf{X}$ of (6.5.22) and the output vector $\mathbf{y}$ of (6.5.16) by

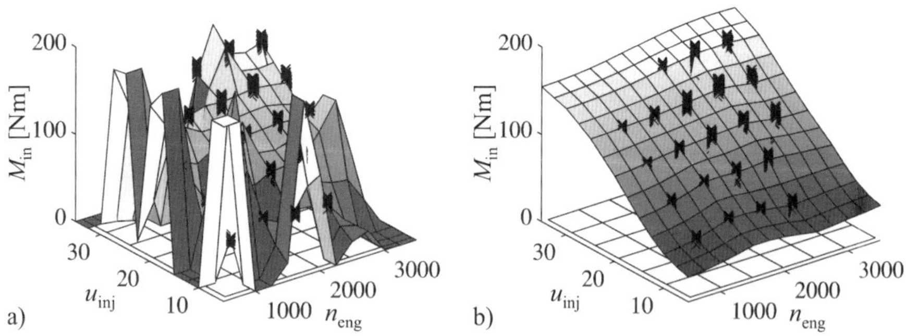

Fig. 6.5.4. **a** Identified look-up table for the torque of a 1.9 l CR-Diesel engine without regularization and **b** with regularization. Training data is plotted as crosses. Sharp differences occur in regions with insufficient excitation, due to missing measurements or poor grid point distributions. The distribution of grid points is indicated by the lattice in the n_{eng}, u_{inj}-plane. The engine speed n_{eng} and the injection quantity u_{inj} are given in rpm and mm^3/cyc respectively, Sequenz (2013).

$$\mathbf{X}_{\text{reg}} = \begin{pmatrix} \mathbf{X} \\ \lambda\mathbf{\Gamma} \end{pmatrix}, \quad \mathbf{y}_{\text{reg}} = \begin{pmatrix} \mathbf{y}_{\text{P}} \\ \mathbf{0} \end{pmatrix} \tag{6.5.29}$$

with the additional rows

$$\lambda\mathbf{\Gamma} = \begin{pmatrix} 2\lambda & -\lambda & 0 & 0 & \dots & -\lambda & 0 & 0 & \dots \\ -\lambda & 3\lambda & -\lambda & 0 & \dots & 0 & -\lambda & 0 & \dots \\ 0 & -\lambda & 3\lambda & -\lambda & \dots & 0 & 0 & -\lambda & \dots \\ \vdots & \vdots & \vdots & \vdots & \dots & \vdots & \vdots & \vdots & \dots \end{pmatrix} \tag{6.5.30}$$

and the regularization parameter

$$\lambda \in [0, \infty]. \tag{6.5.31}$$

The degree of regularization depends on the parameter λ. The larger λ is chosen the smoother the look-up table. $\lambda = 0$ corresponds to the non-regularized solution. In the two-dimensional case, each grid point has two, three or four neighbors, depending on the location of the regarded grid point. A corner grid point has two neighbors, a grid point on the edge three and a grid point in the interior of the lattice four neighbors. (6.5.30) shows the regularization for the first three grid points starting with the grid point in the lower left corner. Applying this regularization, the identification of grid point heights remains a linear problem, wherefore the solution is given analog to (6.5.28) by

$$\hat{\mathbf{w}}_{\text{reg}} = \left[\mathbf{X}_{\text{reg}}^{\text{T}} \mathbf{X}_{\text{reg}}\right]^{-1} \mathbf{X}_{\text{reg}}^{\text{T}} \mathbf{y}_{\text{reg}}. \tag{6.5.32}$$

It is further possible to use different smoothing parameters in each dimension, λ_1 and λ_2 respectively. An interior grid point of the lattice is then smoothed by

$$\mathbf{X}_{\text{reg}} = \begin{pmatrix} \dots & \dots & \dots & \dots & \dots & \dots & \dots & \dots & \dots \\ \dots & -\lambda_2 & \dots & -\lambda_1 & 2\lambda_1 + 2\lambda_2 & -\lambda_1 & \dots & -\lambda_2 & \dots \\ \dots & \dots & \dots & \dots & \dots & \dots & \dots & \dots & \dots \end{pmatrix} \text{ and } \mathbf{y}_{\text{reg}} = \begin{pmatrix} \dots \\ 0 \\ \dots \end{pmatrix}. \tag{6.5.33}$$

The performance of the regularization can be seen in Fig. 6.5.4. The sharp differences of the non-regularized look-up table are smoothed. The regions where no measurements are available are also well extrapolated. This illustrates the necessity of a regularization, if the measurements are not given on the grid points. The non-regularized solution then highly reacts on noisy measurements.

The desired degree of smoothness is different for each process. It can be adjusted by the heuristic determination of the regularization parameter λ. Since the regularization of (6.5.29) leads to a linear solution, (6.5.32), several values of λ can easily be tested. The selection of λ does however depend on the number of measurements since the influence of the additional rows in $\mathbf{X}_{\text{reg}}$ decreases with an increase of the number of measurements. It is therefore recommended, Pukelsheim (2006), to calculate λ via the determination of the measurement independent regularization parameter $\widetilde{\lambda} \in [0, \infty]$ by

$$\lambda = \widetilde{\lambda} \sqrt{\frac{\text{trace}(\mathbf{X}^{\mathrm{T}}\mathbf{X})}{\text{trace}(\mathbf{\Gamma}^{\mathrm{T}}\mathbf{\Gamma})}}. \tag{6.5.34}$$

A closer look at the additional rows in (6.5.30) further shows, that these correspond to a directional derivative of second order in each dimension. Hence, the regularization penalizes the bending of the look-up table and can therefore also be interpreted as Tikhonov regularization, see Press et al (2007), Tikhonov and Arsenin (1977).

d) Multi-dimensional look-up tables

The look-up-table representation can mathematically be extended to several input variables u_ν, $\nu = 1, \dots, p$,

$$\mathbf{y} = f(\mathbf{u}) = f(u_1, u_2, \dots, u_p). \tag{6.5.35}$$

The output then depends on all combinations of the input values such that the number N of stored values for M_i samples per input grows exponentially

$$N = \prod_{i=1}^{p} M_i. \tag{6.5.36}$$

Figure 6.5.5 indicates this exponential growth assuming the same number for all axis points, such that $N = M^p$. Hence, the look-up-table representation is just feasible for a relatively low number of inputs.

The basis function representation for $p > 2$ then follows (6.5.20), but as the grid vector (6.5.19) has the dimension N it grows exponentially according to (6.5.36). For the parameter estimation (6.5.28) with (6.5.16) and (6.5.22) can be applied, however, expanded with the higher dimension N. The number of measurements also has to be $L \geq N$.

Because multi-dimensional look-up tables are not transparent and not suitable for easy interpretable graphical presentations, in practice, the look-up tables are limited to two dimensions. An alternative that allows more than two inputs is to combine several 2D-look-up tables.

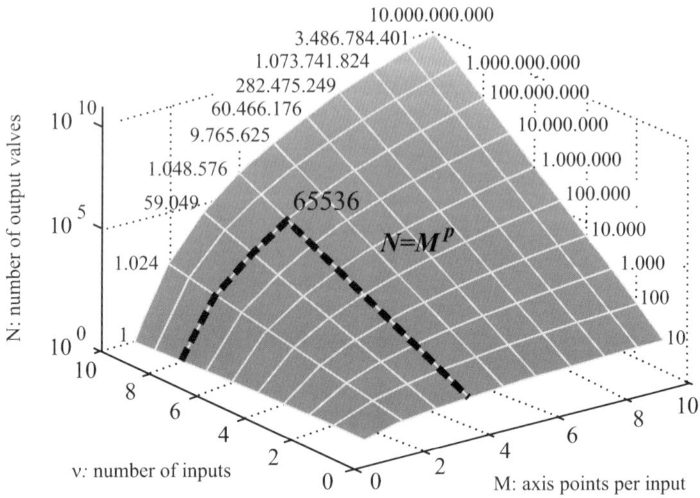

Fig. 6.5.5. Number of look-up table output values N, in dependence on the number p of inputs u_ν and the number M of discrete axis points.

e) Optimization of the grid

The equations for the look-up table output (6.5.13) and (6.5.15) show that the model output is linear in the heights $w_{i,j}$ and nonlinear in the grid positions **c**. Therefore the heights could be optimized using linear methods, as treated with LS estimation according to (6.5.28) for given and fixed grid positions **u**. However, if the grid positions **c** have to be optimized, this becomes a nonlinear optimization problem.

The optimization of the grid positions may be required if the measurement data are not evenly distributed over the input space or if the number of stored grid heights should be reduced to save memory space, Nelles and Fink (2000a). Nelles and Fink (2000b) have developed a nested optimization, where the variation of position and optimization of the heights are performed sequentially (LUTOP toolbox), see also Fink (2006).

f) Representation of multi-dimensional functions with two-dimensional look-up tables

Multi-dimensional static functions can be represented by appropriate combinations of two-dimensional look-up tables. One may distinguish additive, multiplicative and

$$y = f(u_1, u_2, u_3, u_4) \tag{6.5.37}$$

is considered as an example, Sequenz (2013). An *additive structure* results if several two-dimensional look-up tables are added with their outputs, see Fig. 6.5.6. Each look-up table has two inputs and one output such that, for example

$$y = y_{\text{base}} + \Delta y_1 + \Delta y_2 = f_{\text{base}}(u_1, u_2) + f_{\text{corr1}}(u_1, u_3) + f_{\text{corr2}}(u_1, u_5). \tag{6.5.38}$$

One input can be the same for several look-up tables, expressing the dependence on one variable of the operation point like $u_1 = \omega_{\text{eng}}$. The identification is performed first with the two inputs resulting in the best overall model performance, e.g. u_1, u_2. This look-up table is used as a basis to determine a first residual

$$r_1 = y - f_{\text{base}}(u_1, u_2). \tag{6.5.39}$$

Then the best input combination (u_1, u_i) or (u_1, u_j) is determined and r_1 is used to identify a first correction look-up table, e.g.

$$\Delta y_1 = f_{\text{corr1}}(u_1, u_3) \tag{6.5.40}$$

by a separate experiment or calculation changing the inputs u_2 and u_3 for each grid point u_1, u_2. This is repeated with other inputs, thus leading to the additive structure depicted in Fig. 6.5.6 (with a fifth input u_3).

An alternative is to identify (6.5.37) with a simultaneous excitation of all inputs and to determine $f_{\text{base}}(u_1, u_2)$ as an approximation by regression. Then, (6.5.40) is applied and the first correction look-up table is specified, etc.

For a *multiplicative structure* several two-dimensional look-up tables are multiplied at their outputs, see Fig. 6.5.7. The base look-up table

$$y_1 = f_{\text{base}}(u_1, u_2) \tag{6.5.41}$$

is then multiplied with the output of the first correction look-up table

$$\Delta y_1 = f_{\text{corr1}}(u_2, u_3). \tag{6.5.42}$$

The residual for the determination of the first correction look-up table is

$$r_1 = y / f_{\text{base}}(u_1, u_2). \tag{6.5.43}$$

This procedure is repeated with other correction look-up tables thus leading to a multi-dimensional function

$$y = y_{\text{base}} \cdot \Delta y_1 \cdot \Delta y_2 = f_{\text{base}}(u_1, u_2) \cdot f_{\text{corr1}}(u_1, u_3) \cdot f_{\text{corr2}}(u_1, u_4). \tag{6.5.44}$$

A *nested structure* is obtained if the output of one look-up table $y_1 = f_{\text{base}}(u_1, u_2)$ is used as input for a correcting look-up table $y_2 = f_{\text{corr1}}(y_1, u_3)$ together with a new input u_3, as depicted in Fig. 6.5.8. This yields

$$y_2(u_1, u_2, u_3) = f_{\text{corr1}}(f_{\text{base}}(u_1, u_2), u_3). \quad (6.5.45)$$

A next input u_4 then needs a further correction look-up table $y_3 = f_{\text{corr2}}(y_2, u_4)$ and the output then becomes

$$y(u_1, u_2, u_3, u_4) = f_{\text{corr2}}(f_{\text{corr1}}(f_{\text{base}}(u_1, u_2), u_3), u_4). \quad (6.5.46)$$

A disadvantage of this structure is that the outputs $y_1, y_2, \ldots$ have to be subdivided in new discrete points.

A further possibility is a *global-local structure* or a look-up table in look-up table structure as shown in Fig. 6.5.9. This means that at the grid points of a global base look-up table

$$y = f_{\text{base}}(u_1, u_2) \quad (6.5.47)$$

another local look-up table is arranged

$$\Delta y_1 = f_{\text{corr1}}(u_3, u_4). \quad (6.5.48)$$

The base look-up table in basis function representation according to (6.5.20) is then described by

$$y = \sum_{\nu=1}^{N} \phi_\nu(u_1, u_2) w_\nu \quad (6.5.49)$$

and the grid points w_ν are not constants but depend on another look-up table $w_\nu = \Delta y_1$ due to (6.5.48). This leads to

$$y(u_1, u_2, u_3, u_4) = \sum_{\nu=1}^{N} \phi_\nu(u_1, u_2) f_{\text{corr1},\nu}(u_3, u_4). \quad (6.5.50)$$

The outputs of the local look-up tables are thus weighted by a bilinear interpolation.

This representation is suited for combustion engines where $u_1 = u_{\text{inj}}$ and $u_2 = n_{\text{eng}}$ determine the global operating point and u_3 and u_4 are additional, local-manipulated variables such as $u_3 = u_{\text{vgt}}$ and $u_4 = u_{\text{egr}}$.

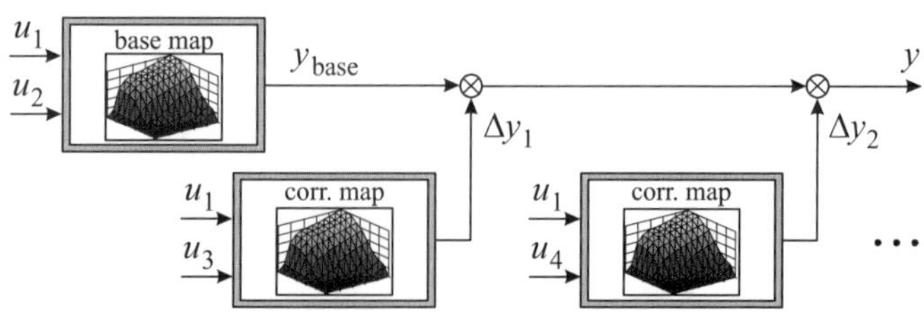

Fig. 6.5.6. Additive structure of multi-dimensional functions with 2D-look-up tables.

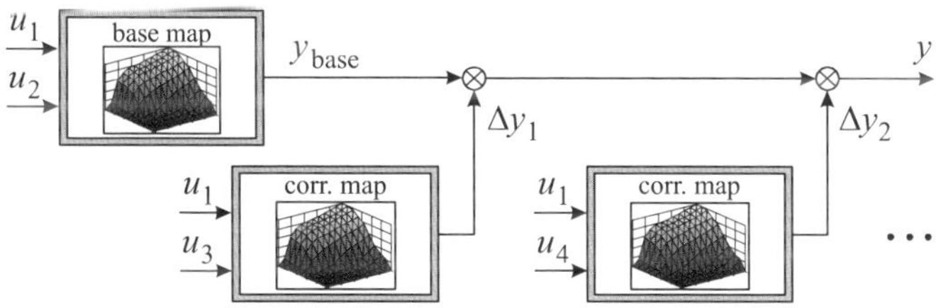

Fig. 6.5.7. Multiplicative structure of multi-dimensional functions with 2D-look-up tables.

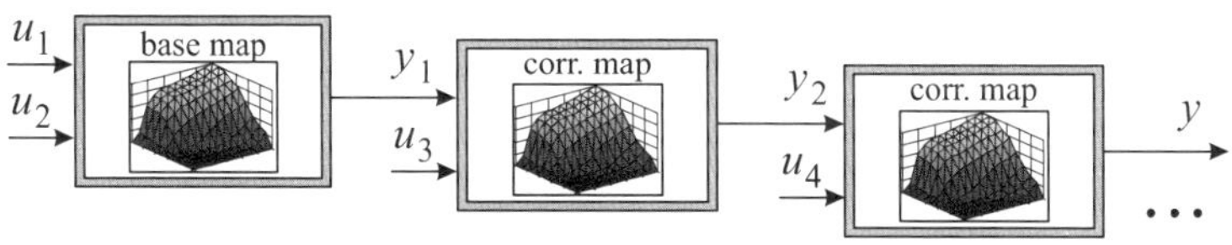

Fig. 6.5.8. Nested structure of multi-dimensional functions with 2D-look-up tables.

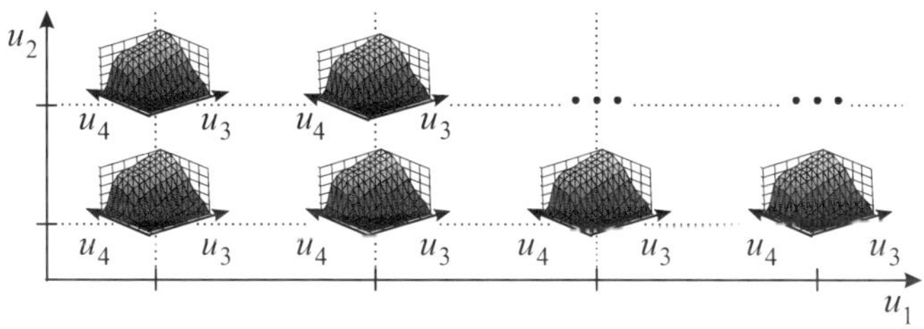

Fig. 6.5.9. Global-local structure. At each grid point (u_1, u_2) of the global look-up table local look-up tables with own grid points (u_3, u_4) are placed.

g) Inversion of look-up tables

The application of nonlinear static models $y = f(u_1, u_2)$ frequently requires an inversion of the inputs and outputs like

$$u_2 = f^{-1}(u_1, y) \tag{6.5.51}$$

for the feedforward control of, e.g. the injection mass $u_2 = u_{\text{inj}}$ in dependence on the required torque $y = M_{\text{eng}}$ and speed $u_1 = \omega_{\text{eng}}$. An unique inversion is possible under the condition that the look-up table heights w_i, see (6.5.3), for fixed u_1 are a monotonic function

$$w_{i,j} < w_{i,j+1} \quad (\text{or } w_{i,j} > w_{i,j+1}) \qquad \forall j \in 1 \dots M_2. \tag{6.5.52}$$

A first possibility to invert a two-dimensional look-up table is to represent the data of the look-up table in the form of (6.5.51) and to *identify a new inverted look-up table*

$$u_2 = g(u_1, y) \tag{6.5.53}$$

with the least squares estimation method (6.5.28).

Another way is an *inverse evaluation of characteristics*. For a given input u_1, the look-up table yields the characteristic $y = f_{u1}(u_2)$. If u_1 is not on the selected grid points, a linear interpolation according to (6.5.4) has to be applied. With the assumption of a monotonic function the values of the inverted characteristic

$$u_2 = f_{u1}^{-1}(y) \tag{6.5.54}$$

can directly be determined after discretizing y and stored with different values for u_1 as inverted look-up table (6.5.51).

The identification of a new inverted look-up table may include some small approximation errors because of the applied parameter-estimation method, depending on the number of grid points and the available heights y. The inverse evaluation of characteristics does not include estimation errors and is therefore a precise inversion.

h) Use of inverted look-up tables for feedforward control

If the output y is the controlled variable, like the torque of a *gasoline engine*

$$M_i = f(m_{\text{air}}, n) = f(\mathbf{u}) \tag{6.5.55}$$

and $u_1 = m_{\text{air}}$ (manipulated variable) and $u_2 = n$ (operating point) are two main inputs, compare Fig. 7.1.1, then the torque model $M(\mathbf{u})$ has to be inverted to determine the desired air charge

$$m_{\text{air,d}} = f^{-1}(M_i, n) \tag{6.5.56}$$

see Fig. 7.1.3. In this look-up table model the torque M and the speed n are the two inputs and the air mass $m_{\text{air,d}}$ is the output (reference variable) for the feedforward control of the torque, compare Fig. 6.5.10 and the inverted look-up table blocks in Figs. 7.1.3 and 7.1.5.

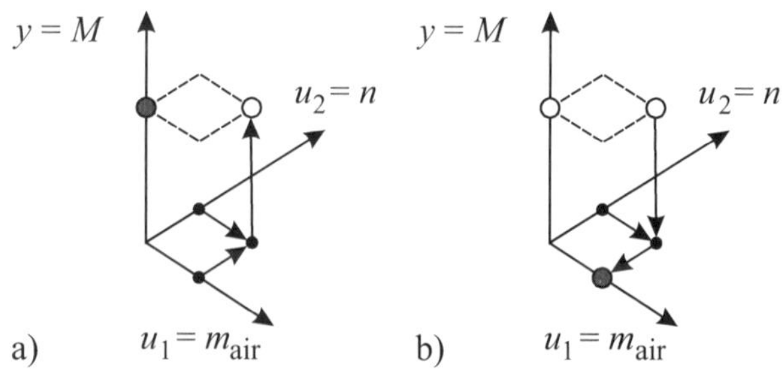

Fig. 6.5.10. Static feedforward control with look-up tables. **a** engine model $M = f(m_{\text{air}}, n)$. **b** feedforward control $m_{\text{air}} = f^{-1}(M, n)$.

The other main manipulated variables are the result of an optimization applying, e.g. multi-criteria as described in Sects. 6.8 and 6.9 and are stored in look-up tables

$$m_{\text{f}}(m_{\text{air}}, n),\ \varphi_{\text{f}}(m_{\text{air}}, n),\ \varphi_{\text{ign}}(m_{\text{air}}, n)$$

see also Sect. 7.1. The torque model (6.5.55) is valid for these optimal values. If there are deviations from these optima, correction look-up tables are used as shown in Fig. 7.1.1.

The inner torque, which is required to determine the air charge then follows from (7.1.6)

$$\begin{aligned} M_{\mathrm{i}} &= \frac{1}{\eta_\lambda \eta_{\mathrm{ign}}} M_{\mathrm{i,corr}} \\ M_{\mathrm{i,corr}} &= M_{\mathrm{eng}} + M_{\mathrm{drg}}. \end{aligned} \tag{6.5.57}$$

Figure 6.5.11 shows the look-up tables for the torque $M(\alpha_{\mathrm{th}}, n)$ and the inverted look-up table $\alpha_{\mathrm{th}} = f^{-1}(M, n)$ as an example.

The torque-oriented control of *diesel engines* is based on the internal torque model, the base look-up table

$$M_{\mathrm{i}} = f(u_{\mathrm{inj}}, n_{\mathrm{eng}}) \tag{6.5.58}$$

see Fig. 8.1.1. The injected fuel mass then follows from the inverted look-up table

$$m_{\mathrm{f}} \mathrel{\widehat{=}} u_{\mathrm{inj}} = f^{-1}(M_{\mathrm{i}}, n_{\mathrm{eng}}) \tag{6.5.59}$$

as shown in Fig. 8.1.3. For the base look-up table M_{i} it is assumed that all other influencing variables are optimized, such that a multi-criterion is satisfied or M_{i} reaches a maximum. Deviations from this optimal look-up table are taken into account by correction look-up tables. Figure 6.5.12 depicts the signal flow structure, which shows multiplicative correction maps for φ_{mi}, $p_{2\mathrm{i}}$, m_{air} and $T_{2\mathrm{i}}$. To obtain the torque at the clutch, the drag torque M_{drg} is subtracted, such that

$$M_{\mathrm{eng}} = M_{\mathrm{i,eff}} - M_{\mathrm{drg}} = M_{\mathrm{i}}\eta_1\eta_2\eta_3\eta_4 - M_{\mathrm{drg}}. \tag{6.5.60}$$

Now, two possibilities are considered to determine the torque at the clutch M_{eng} for different correction factors and then to determine the required injected fuel mass, see Sequenz (2013) and Sequenz et al (2012).

- *Successive variations*
  The practical approach begins with the determination of the drag torque look-up table. The effective inner torque $M_{\mathrm{i,eff}}$ is then known after measurement of M_{eng}, compare Fig. 6.5.12. For the optimized torque M_{i} it holds $\eta = 1$ and therefore $M_{\mathrm{i}} = M_{\mathrm{i,eff}}(u_{\mathrm{inj}}, n_{\mathrm{eng}})$ can be stored as base look-up table. Then, the correction look-up tables $\eta_1 = f(\varphi_{\mathrm{mi}}, n_{\mathrm{eng}})$ etc. are determined one after another by variations around the optimized grid points of the base look-up table. Then, the torque at the clutch follows with (6.5.60) to $M_{\mathrm{eng}}(u_{\mathrm{inj}}, n_{\mathrm{eng}})$.

- *Simultaneous variations*
  Alternatively all inputs

$$\mathbf{u}^{\mathrm{T}} = \left[u_{\mathrm{inj}}, n_{\mathrm{eng}}, m_{\mathrm{air}}, p_{2\mathrm{i}}, \varphi_{\mathrm{mi}}, T_{2\mathrm{i}}\right] \tag{6.5.61}$$

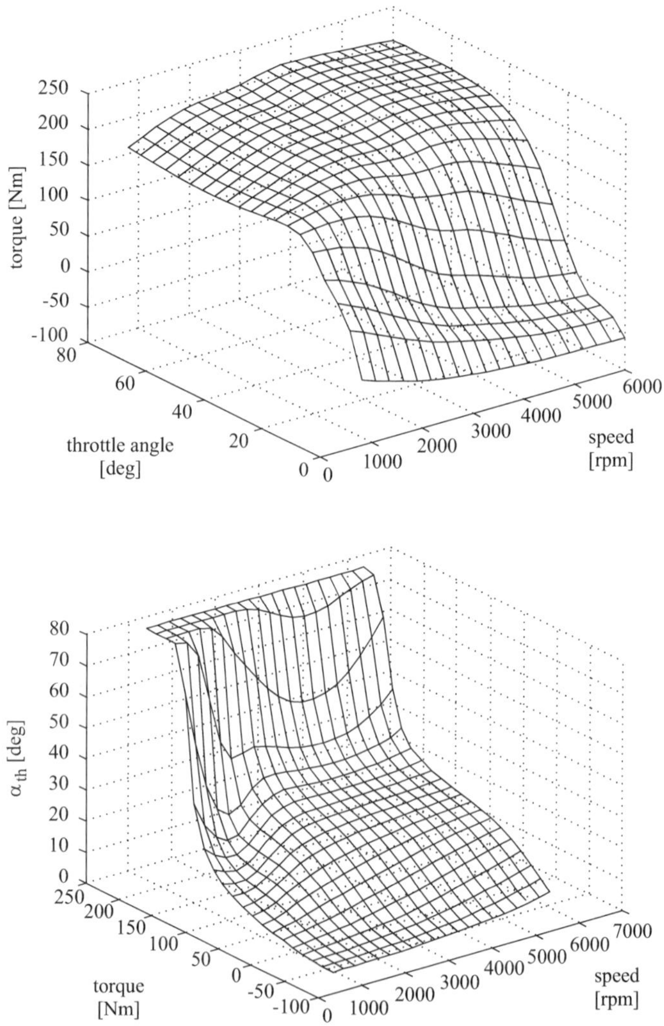

Fig. 6.5.11. Torque look-up tables of a $2.5\,\mathrm{l}$ gasoline engine. **a** measured look-up table $M = f(\alpha_{\mathrm{th}}, n)$. **b** inverted look-up table $\alpha_{\mathrm{th}} = f^{-1}(M, n)$.

can be excited simultaneously to determine $M_{\mathrm{eng}}(\mathbf{u})$. After adding the drag torque due to (6.5.60) one obtains

$$M_{\mathrm{i,eff}} = f(\mathbf{u}). \tag{6.5.62}$$

Then, based on the measurements, $M_{\mathrm{i}}^*(u_{\mathrm{inj}}, n_{\mathrm{eng}})$ is identified by parameter estimation. Using the residuals (6.5.43) the correction look-up tables η_1^* to η_4^* are identified. These efficiency or correction factors are then normalized

$$\eta_1 = \frac{\eta_1^*}{\max \eta_1^*}, \; \eta_2 = \frac{\eta_2^*}{\max \eta_2^*}, \; \ldots \tag{6.5.63}$$

to obtain values $\eta_i \leq 1$. The base look-up table is corrected due to

$$M_{\rm i}(u_{\rm inj}, n_{\rm eng}) = M_{\rm i}^*(u_{\rm inj}, n_{\rm eng})(\max \eta_1^*) \ldots (\max \eta_4^*) \tag{6.5.64}$$

and yields the base look-up table (6.5.58).
Now, the inversion of the look-up tables to determine the required injected fuel rate $u_{\rm inj}$ for a given $n_{\rm eng}$ and given torque request at the clutch is considered. This $u_{\rm inj}$ is contained in $M_{\rm i}$ and $M_{\rm drg}$. As both are monotonic functions, the inversion is possible. Then the outputs of the correction look-up tables are determined and concentrated to one value

$$\eta = \eta_1 \, \eta_2 \, \eta_3 \, \eta_4 . \tag{6.5.65}$$

This leads to the torque at the clutch

$$M_{\rm eng}(u_{\rm inj}, n_{\rm eng}) = M_{\rm i}(\mathbf{u})\eta - M_{\rm drg}(u_{\rm inj}, n_{\rm eng}). \tag{6.5.66}$$

With $n_{\rm eng}$ and η given from the current operating conditions this simplifies to a characteristic curve $M_{\rm eng} = \varphi(u_{\rm inj})$ which is inverted to $u_{\rm inj} = f^{-1}(M_{\rm eng})$ as shown in Fig. 6.5.13.

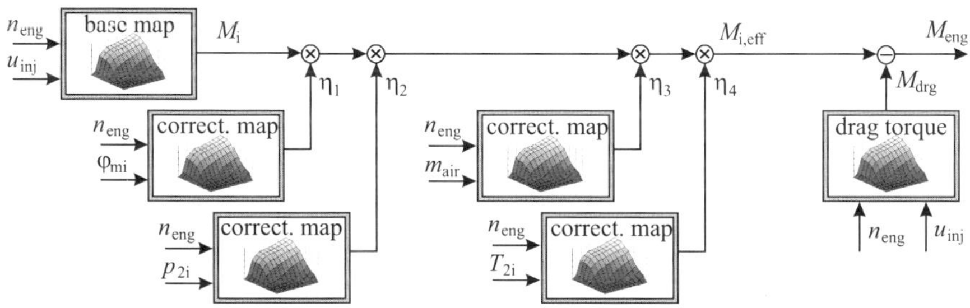

Fig. 6.5.12. Torque look-up tables for a diesel engine with 6 input variables: base map, correcting maps and drag torque map.

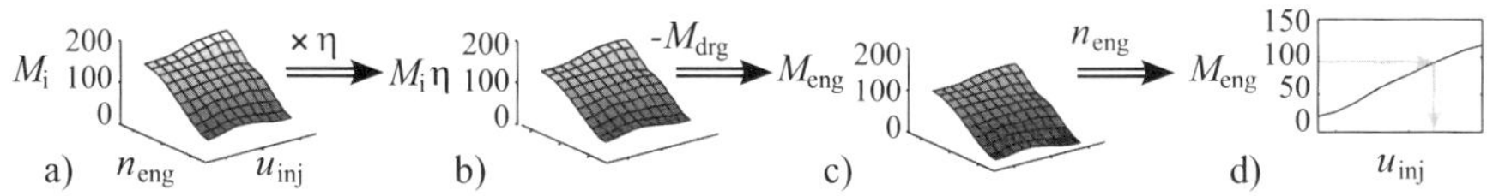

Fig. 6.5.13. Inversion of the torque look-up table of a diesel engine. **a** Basis map $M_{\rm i}(u_{\rm inj}, n_{\rm eng})$. **b** Multiplication with correcting factor η. **c** Subtraction of drag torque. **d** inverse evaluation of the torque characteristic $u_{\rm inj} = f_{\rm n}^{-1}(M_{\rm eng})$.

6.5.2 Polynomial models

An alternative to the nonparametric look-up tables is a parametric model like a polynomial which yields

$$Y = K_0 + K_1 U + K_2 U^2 + \ldots + K_q U^q = \sum_{i=0}^{q} K_i U^i \tag{6.5.67}$$

or

$$\begin{aligned} Y &= \mathbf{U}\mathbf{K} \\ \mathbf{U} &= \left[1 \; U \; U^2 \; \ldots \; U^q\right] \\ \mathbf{K}^{\mathrm{T}} &= \left[K_0 \; K_1 \; K_2 \; \ldots \; K_q\right] \end{aligned} \tag{6.5.68}$$

for one input U and one output Y.

The identification of this model from measurements was already discussed in Sect. 3.1.2. For two inputs U_1 and U_2 and one output Y the polynomial model becomes

$$\begin{aligned} Y = K_0 &+ K_1 U_1 + K_{11} U_1^2 + \ldots \\ &+ K_2 U_2 + K_{22} U_2^2 + \ldots \\ &+ K_{12} U_1 U_2 + K_{122} U_1 U_2^2 + K_{112} U_1^2 U_2 + \ldots \end{aligned} \tag{6.5.69}$$

As a control function has to invert these polynomials, direct solutions are only accessible for low orders. For $q = 2$ the solution of (6.5.67) is

$$U = \left(-K_1 \pm \sqrt{K_1^2 - 4K_2(K_0 - Y)} / \, 2K_2\right). \tag{6.5.70}$$

For higher orders q and several inputs U_i the direct use of polynomial models is not suitable. If polynomial models are used for the identification of engines, a more feasible way is to transform them into look-up tables and then to use the look-up tables for the calculation of the manipulated variables U_i as in the last section.

Another way is to use an approximation of the inverse of (6.5.67) in form of

$$U = \kappa_0 + \kappa_1 Y + \kappa_2 Y^2 + \ldots + \kappa_q Y^q \tag{6.5.71}$$

based on a new parameter estimation with the measured inputs and outputs. In the case of two inputs this becomes, e.g.

$$\begin{aligned} U_1 = \kappa_0 &+ \kappa_1 Y + \kappa_{11} Y^2 + \ldots \\ &+ \kappa_2 U_2 + \kappa_{22} U_2^2 + \ldots \\ &+ \kappa_{12} Y U_2 + \kappa_{122} Y U_2^2 + \kappa_{112} Y^2 U_2 + \ldots \end{aligned} \tag{6.5.72}$$

and similarly for U_2.

However, in general the generation of look-up tables is an easier and general applicable way.

6.5.3 Comparison of nonlinear model structures

Model-based engine-control methods require the implementation of nonlinear models. Therefore, it is of interest how the different approaches compare with regard to their accuracy and computational demand. A comparison for static models with look-up tables, LOLIMOT and LOPOMOT structures was conducted by Sequenz and Isermann (2011), Sequenz et al (2012) for an NO_x model in dependence on four inputs

$$\dot{m}_{NOx} = f(m_{air}, p_{2i}, \varphi_{mi}, T_{2i}) \quad (6.5.73)$$

for 21 fixed operating points (u_{inj}, n_{eng}).

The look-up table was identified in a nested, multiplicative form of two-dimensional look-up tables

$$\dot{m}_{NOx} = f_{bm}(m_{air}, \varphi_{mi}) \cdot f_{cor1}(m_{air}, p_{2i}) \cdot f_{cor2}(m_{air}, T_{2i}). \quad (6.5.74)$$

The axis-orthogonal partitioning of the global-local LOLIMOT approach was limited to ten. For the LOPOMOT model the maximal order of the polynomial is three and no further partitioning is required.

Table 6.5.1 represents the obtained *model accuracy* for training and verification by comparing the coefficient of determination R^2. All models attain a good approximation of the real data, as $R^2 > 0.99$ for the training data. Also the validation results with other data than for training are acceptable with $R^2 > 0.98$. The look-up tables suffer from its equidistant distribution of grid points. Only a minor part of the grid points are in the relevant region of the local model. This is a consequence of the common selection of grid points for high and low load.

To compare the required *computing time* the number of floating-point operations (FLOPs) is counted and it is distinguished between multiplications '$*$', summations '$+$' and evaluation of exponential functions 'exp'. Corresponding numbers are given in Table 6.5.2. The look-up tables require the smallest number of calculations, LOLIMOT requires about two times more and the local polynomial models LOPOMOT about the same as look-up tables.

The *memory demand* is determined by the number of parameters which have to be stored for a complete model. This is the number of unknowns which are identified at model training. Some additional parameters are required for the model structure, as the grid positions for look-up tables, the standard deviations and centers for each LOLIMOT partition model, and limitation values for LOPOMOT. Table 6.5.3 indicates that look-up tables require the highest number of stored parameters, due to the equidistant distribution of grid points, also in areas not required. LOLIMOT needs significantly less parameters to be stored. One reason is the adaptation of the model structure to possible local linear areas. However, additional parameters are necessary for the adaptation of the model structure. The smallest number of parameters is required for LOPOMOT. However, the bad extrapolation is a weak point.

In addition, Kernel methods were also compared. However, they show the highest number of operations and memory space and seem, therefore, not to be attractive candidates.

Hence, look-up tables are still advantageous for the representation of multidimensional nonlinear static functions with regard to small computation time and if the grid points are concentrated in the regions of interest. The local polynomial models with axis-orthogonal partitioning lead to small computation times and relatively small memory demand as well. The advantage of LOLIMOT models is their flexibility for many kinds of nonlinear static functions and their direct extension to dynamic models. However, the final use of the models and especially the realization and implementation of inverse models for control play a decisive role, as discussed in Sects. 6.5.1g) and 6.5.2.

Table 6.5.1. Model accuracies for different nonlinear static model structures of an NO_x model. Averaged values for 21 operating points

Model structure	Training		Validation	
	R^2	RMSE $\frac{mg}{s}$	R^2	RMSE $\frac{mg}{s}$
look-up tables	0.991	2.08	0.982	2.56
LOLIMOT	0.992	1.93	0.983	2.38
LOPOMOT	0.996	1.45	0.988	2.08

Table 6.5.2. Floating-point operations (FLOPS) for the global-local calculation of the output of an NO_x model

	Additions	Multiplications	Exponential functions	Sum of operations
look-up tables	143	164	-	307
LOLIMOT	346	347	34	727
LOPOMOT	115	216	-	331

Table 6.5.3. Memory demand for different nonlinear static model structures of an NO_x model

	Identified parameters	Additional parameters	Sum of all parameters
look-up tables	6300	40	6340
LOLIMOT	880	1408	2288
LOPOMOT	546	210	756

6.6 Nonlinear dynamic engine control

It is common practice to utilize linear controllers for processes that are nonlinear. However, this approach yields satisfactory results only as long as the process nonlinearity is weak. This means that for strongly nonlinear processes the design of the controllers have to take the process nonlinear behavior into account in a suitable way.

Basically, the nonlinear controllers then include the nonlinear process models. However, due to the many different structural possibilities of nonlinear relations between input and output signals of nonlinear processes one cannot expect to operate with only a few model classes.

In an initial approach nonlinear dynamic processes can be classified with regard to the possibility to generate derivatives within the nonlinear model. A first class belongs to processes with *continuously differentiable nonlinearities*. This property is, e.g. typical for mechanical processes with nonlinear drive and resistance characteristics, thermal processes with flow velocity dependent heat transfer or chemical processes with temperature dependent reactions. The specific kind of nonlinearity appears in the theoretical/physical based process models. To reduce the great variety of possible model structures approximations are used. Examples for gaining the models experimentally with identification methods are polynomial models, as Volterra, Hammerstein, or Wiener models, are treated in e.g. Schetzen (1980), Isermann and Münchhof (2011) and Sect. 3.2.3.

Non-continuously differentiable nonlinearities appear frequently in mechanical systems with dry friction and backlash and in electrical systems with magnetization hysteresis, Isermann (2005). Their models have been successfully applied together with parameter-estimation methods for the design of adaptive nonlinear controllers which compensate these nonlinearities at least approximately, Isermann et al (1992). These models are, e.g. used for the control of actuators and they are not further considered here.

In the case of continuously differentiable nonlinearities one may distinguish between *global nonlinear controllers* and *parameter-scheduled linear controllers*, see Fig. 6.6.1. In the first case the controller has an inherent nonlinear behavior which is based on the nonlinear process model. Because of the many different types of nonlinearities, this approach leads to a large variety of controllers. In the second case the controller is basically linear and designed for linear process models which are valid for special operating points. This approach is also known as *gain scheduling* or *parameter scheduling* and belongs to the class of linear-parameter-varying systems. One advantage is that linear control design techniques for the local linear process models can be used, altogether leading to a transparent and not too complicated procedure.

In the following two approaches which are suitable for engine control will be described briefly, local linear controllers and nonlinear controllers for process parts with nonlinear static and linear dynamic behavior.

6.6.1 Local linear control

The design of the local linear controllers is based on the LOLIMOT method of process identification in Sect. 3.1.3 where locally valid linear submodels are weighted by validity functions Φ_i, as in (3.2.54) which becomes for one input U and one output Y

$$Y = \sum_{i=1}^{M} \Phi_i(z) g_i(U) \tag{6.6.1}$$

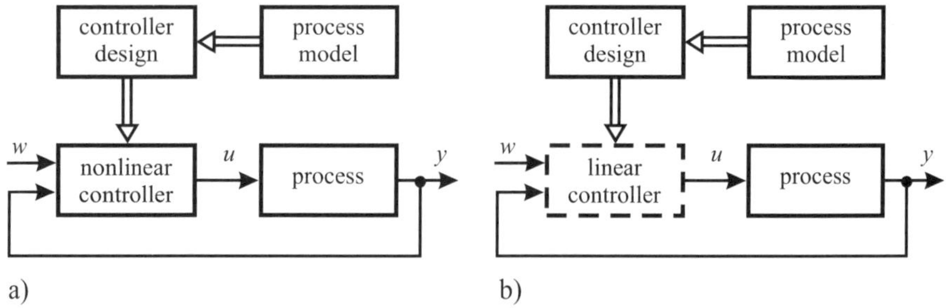

Fig. 6.6.1. Classification of nonlinear controllers. **a** global nonlinear controller. **b** parameter-scheduled linear controller.

where z describes the operating point i. Based on these local process models, local linear controllers $h_i(e)$ with the control error $e = W - Y$ can be designed. The manipulated variable then follows with interpolation of the linear controller output by

$$U = \sum_{i=1}^{M} \Phi_i(z) h_i(e) \quad (6.6.2)$$

i.e. by a weighting of several linear controllers as depicted in Fig. 6.6.2. Another approach is to weight the parameters of the local controllers, Fink (2006), Pfeil (2011). Sufficient conditions for the stability of these local linear controllers are derived in Tanaka et al (1998), Cuesta and Ollero (2004).

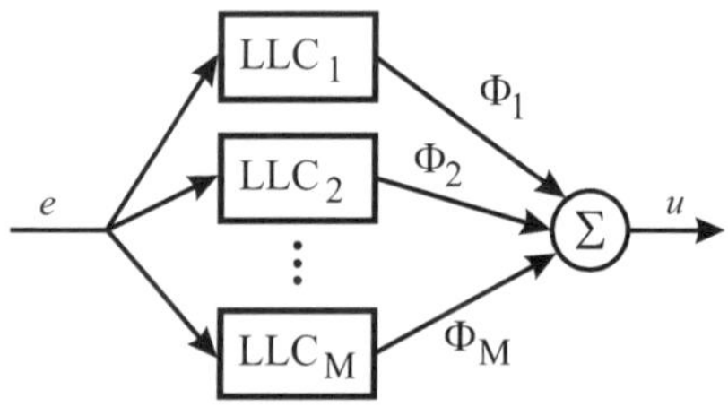

Fig. 6.6.2. Local linear controllers with interpolation of subcontroller outputs.

6.6.2 Nonlinear control for static nonlinearities

It is assumed that a nonlinear static behavior of the input, e.g. by an actuator G_a, is described by a polynomial $S_{\mathrm{u}} = f(U)$

$$S_{\mathrm{u}} = r_0 + r_1 U + r_2 U^2 \quad (6.6.3)$$

and is followed by a linear dynamic process part G_{psy}

$$a_2 \ddot{Y}(t) + a_1 \dot{Y}(t) + Y(t) = K_{\mathrm{p}} S_{\mathrm{u}}(t). \quad (6.6.4)$$

This model then has the structure of a *Hammerstein model*, see (3.2.49).

The static nonlinearity is now compensated by an inverted model $U = f^{-1}(S_u)$, e.g. by solving the quadratic equation of U or using an inverted look-up table. Then a linear controller G_c can be designed for the linear dynamic model, like a PD controller

$$S'_u(t) = K_c\left(e(t) + T_d\dot{e}(t)\right). \tag{6.6.5}$$

Figure 6.6.3 shows a corresponding signal-flow diagram. This approach of compensating static nonlinearities can be expanded to two-dimensional look-up tables if the look-up table can be inverted and a unique solution is obtained.

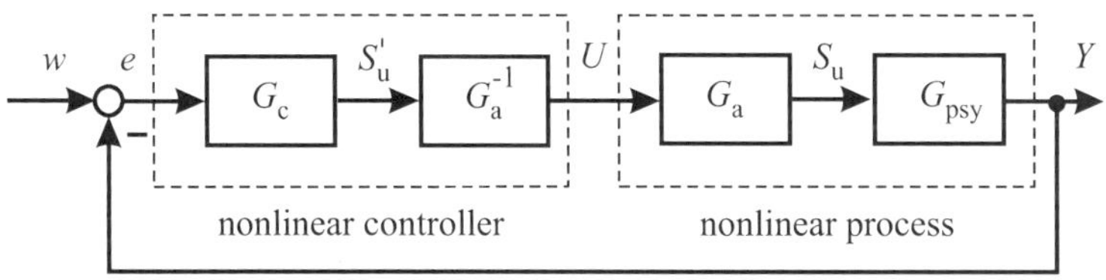

Fig. 6.6.3. Nonlinear controller for a process with nonlinear input characteristic and linear process dynamics.

6.7 Digital control

Digital computer controllers and filters operate with sampled input and output signals. Unlike continuous signals these signals have discrete amplitude values at discrete time points. Amplitude modulated pulse series emerge, for which the pulse heights are rounded up and down, according to the quantization and resolution in the A/D- and D/A-converters. If the word length is 16 bits and more, the signal amplitudes can be considered as continuous. The sampling of the signals is usually performed periodically with sampling time T_0, leading to a discrete time $k = t/T_0 = 0, 1, 2, \ldots$. If the time interval between sampling of input signals and output signals can be considered small in comparison to the time constants of the actuators and the process, synchronous sampling of input and output signals can be assumed.

The development of digital control systems requires the theory of sampled-data systems, for which special textbooks exist, e.g. Ackermann (1983), Isermann (1980), Kuo (1995), Åström and Wittenmark (1997), Franklin and Powell (2006).

6.7.1 Sampled-data processing

The signal processing of sampled data needs a special treatment because several effects arise which are different from continuous-time signals. For example, Shannon's sampling theorem shows that side spectra appear at the sampling frequencies

$\omega_0 = \nu 2\pi/T_0$, $\nu = 1, 2, \ldots$ and that therefore the sampling frequency has to be chosen

$$\omega_0 > 2\omega_{\max}$$

where $\omega_{\max}$ is the maximal frequency of interest. In the following some few basics for sampled data systems are summarized, for details, see e.g. Isermann (1989).

Usually a sampler is followed by a holding element. A holding element holds the sampled signal $x(kT_0)$ for one sampling interval such that staircase signals $m(t)$ result. If the impulse train after a sampler is described by

$$x^*(t) = \sum_{k=0}^{\infty} x(kT_0)\,\delta(t - kT_0) \tag{6.7.1}$$

where the impulse function $\delta(t)$ approximates the output of a sampler, the transfer function of a zero-order hold becomes

$$H(s) = \frac{m(s)}{x^*(s)} = \frac{1}{s}\left[1 - e^{-T_0 s}\right]. \tag{6.7.2}$$

If the sampling time T_0 is small compared to the time constants of a controlled process, the continuous time differential equations of the processes and controllers can be transformed to difference equations by *discretization*, such as

$$\begin{aligned} \tfrac{dx(t)}{dt} &\approx \tfrac{x(kT_0)-x((k-1)T_0)}{T_0} \\ \tfrac{d^2x(t)}{dt^2} &\approx \tfrac{x(kT_0)-2x((k-1)T_0)+x((k-2)T_0)}{T_0^2} \\ &\vdots \end{aligned} \tag{6.7.3}$$

However, in order to save computing time the sampling time T_0 is usually larger, so these approximations are not valid any more. Therefore, the theory of sampled data systems uses the z-transform instead of the s-transform with the variable

$$z = e^{T_0 s} = e^{T_0(\delta + i\omega)} \tag{6.7.4}$$

which leads to the z-transform of a signal $x(t)$

$$x(z) = \mathfrak{Z}\{x(kT_0)\} = \sum_{k=0}^{\infty} x(kT_0) z^{-k}. \tag{6.7.5}$$

As an example, the z-transform of a step function $x(kT_0) = 1(kT_0)$ then becomes

$$x(z) = 1 + z^{-1} + z^{-2} + \ldots = \frac{1}{1 - z^{-1}} = \frac{z}{z - 1} \tag{6.7.6}$$

and can be represented in closed form. This is also valid for many other standard signals like e^{-at} or $\sin \omega t$.

Applying the convolution sum for sampled input and output signals it holds for a linear process with impulse response $g(t)$

$$y(nT_0) = \sum_{k=0}^{\infty} u(kT_0)\, g\left((n-k)T_0\right) \tag{6.7.7}$$

which leads after Laplace transform to

$$y^*(s) = G^*(s)\, u^*(s) \tag{6.7.8}$$

and with (6.7.4) to the z-transfer function of a process

$$G(z) = \frac{y(z)}{u(z)} = \sum_{q=1}^{\infty} g(qT_0)\, z^{-q} = \mathfrak{Z}\{g(q)\}. \tag{6.7.9}$$

Making use of z-transform tables the continuous time transfer functions $G(s)$ can be transformed directly to the z-transfer function $G(z)$

$$G(z) = \mathcal{Z}\{G(s)\}. \tag{6.7.10}$$

For example a first order process leads to

$$G(z) = \mathcal{Z}\left\{\frac{K'}{a+s}\right\} = \frac{K'z}{z - e^{-aT_0}} = \frac{b_0 z}{z - a_1} = \frac{b_0}{1 + a_1 z^{-1}}. \tag{6.7.11}$$

If the sampler is followed by a hold, then one has to use for a process

$$G_{\mathrm{p}}(z) = HG(z) = \mathcal{Z}\{H(s)\, G(s)\} \tag{6.7.12}$$

and this yields

$$HG(z) = \frac{z-1}{z} \mathcal{Z}\left\{\frac{G(s)}{s}\right\}. \tag{6.7.13}$$

For a first order process it then follows

$$HG(z) = \frac{z-1}{z} \mathcal{Z}\left\{\frac{K'}{s(a+s)}\right\} = \frac{b_1 z^{-1}}{1 + a_1 z^{-1}}. \tag{6.7.14}$$

Hence, the determination of z-transfer functions of processes usually has to include a holding element and results in general form in

$$G_{\mathrm{p}}(z) = \frac{y(z)}{u(z)} = \frac{B(z^{-1})}{A(z^{-1})} = \frac{b_0 + b_1 z^{-1} + \ldots + b_m z^{-m}}{1 + a_1 z^{-1} + \ldots + a_m z^{-m}} \tag{6.7.15}$$

which corresponds to a linear difference equation

$$\begin{aligned} y(k) + a_1 y(k-1) + \ldots + a_m y(k-m) \\ = b_0 u(k) + b_1 u(k-1) + \ldots + b_m u(k-m). \end{aligned} \tag{6.7.16}$$

A dead time T_{d} can just be added by multiplication of $G(z)$ with z^{-d} where $d = T_{\mathrm{t}}/T_0$ and time shifts in the input arise, like $u(k-d)$. Contrast to the approximation of differential equation by discretization, (6.7.3), the process models (6.7.15) and (6.7.16) are valid for any size of the sampling time T_0 and are therefore the basis for the design of digital control and filtering algorithms.

6.7.2 Digital control algorithms

In a first approach the digital computer version of a PID controller with *small sampling time* T_0 can be obtained as an approximation by discretization of (6.4.8).

Applying rectangular integration for the integral term one then obtains

$$u(k) = K\left[e(k) + \frac{T_0}{T_I}\sum_{i=0}^{k-1} e(i) + \frac{T_D}{T_0}(e(k) - e(k-1))\right] \tag{6.7.17}$$

which is called "position algorithm". For real-time calculation the recursive version is used

$$u(k) = u(k-1) + q_0 e(k) + q_1 e(k-1) + q_2 e(k-2) \tag{6.7.18}$$

with the parameters

$$\begin{aligned} q_0 &= K\left(1 + \frac{T_D}{T_0}\right) \\ q_1 &= -K\left(1 + 2\frac{T_D}{T_0} - \frac{T_0}{T_I}\right) \\ q_2 &= K\frac{T_D}{T_0}. \end{aligned} \tag{6.7.19}$$

Because this control algorithm just calculates the change of the manipulated variable

$$\Delta u(k) = u(k) - u(k-1)$$

it is called "velocity algorithm". (This version is also suited for the case of actuator saturation if $u(k-1)$ is the realized (measured) actuator position, a simple form to avoid wind-up behavior.) A corresponding z-transfer function of a discrete-time controller (for any size of a sampling time T_0) is

$$G_c(z) = \frac{u(z)}{e(z)} = \frac{Q(z^{-1})}{P(z^{-1})} = \frac{q_0 + q_1 z^{-1} + \ldots + q_\nu z^{-\nu}}{p_0 + p_1 z^{-1} + \ldots + p_\mu z^{-\mu}}. \tag{6.7.20}$$

Specialization to a PID control algorithm yields

$$G_{c,PID} = \frac{q_0 + q_1 z^{-1} + q_2 z^{-2}}{1 - z^{-1}} \tag{6.7.21}$$

or a corresponding form to (6.4.8)

$$G_c(z) = K\left[1 + c_I\frac{z^{-1}}{1 - z^{-1}} + c_D(1 - z^{-1})\right] \tag{6.7.22}$$

with the parameters

$$\begin{aligned} K &= q_0 - q_2 && \text{gain} \\ c_D &= q_2/K && \text{lead coefficient} \\ c_I &= (q_0 + q_1 + q_2)/K && \text{integration coefficient.} \end{aligned} \tag{6.7.23}$$

The basic control algorithms then are

$$\begin{aligned} u(k)|_{\text{PID}} &= u(k-1) + q_0 e(k) + q_1 e(k-1) + q_2 e(k-2) \\ &= u(k-1) + K(1 + c_D)e(k) + K(c_I - 2c_D - 1)\, e(k-1) \\ &\quad + K c_D e(k-2) \end{aligned} \tag{6.7.24}$$

$$\begin{aligned} u(k)|_{\text{PI}} &= u(k-1) + q_0 e(k) + q_1 e(k-1) \\ &= u(k-1) + K e(k) + K(c_I - 1)\, e(k-1) \end{aligned} \tag{6.7.25}$$

$$\begin{aligned} u(k)|_{\text{PD}} &= u(k-1) + q_0 e(k) + q_2 e(k-1) \\ &= u(k-1) + K(1 + c_D)e(k) + K\, c_D\, e(k-1). \end{aligned} \tag{6.7.26}$$

In the case of small sampling times T_0 the tuning rules are similar to continuous time PID controllers, see Sect. 6.4.1. For larger sampling times, tuning rules based on process transient functions are given in Isermann (1989). This book also contains many other aspects for PID controller design.

The design of *internal model controllers* can either begin in s-domain and the resulting controller is then transformed into z-domain, or the process model is formulated in the z-domain and the design is performed in the z-domain. The internal model controller in the last case is

$$G_{c,\text{IMC}}(z) = \frac{u(z)}{e(z)} = \frac{1}{G_p(z)} \frac{G_{rf}(z)}{1 - G_{rf}(z)}. \tag{6.7.27}$$

As discussed in Soler (2011) the first approach is straightforward. For the second approach one has to consider especially unstable zeros by inverting the process model. (Unstable zeros may appear even for lowpass processes, if the sampling time is sufficiently small.) However, this approach is better suitable for large sampling times.

The discrete time state space model

$$\begin{aligned} \mathbf{x}(k+1) &= \mathbf{A}\mathbf{x}(k) + \mathbf{B}\mathbf{u}(k) \\ \mathbf{y}(k) &= \mathbf{C}\mathbf{x}(k) + \mathbf{D}\mathbf{u}(k) \end{aligned} \tag{6.7.28}$$

leads directly to the *state controller*

$$\mathbf{u}(k) = -\mathbf{K}\mathbf{x}(k). \tag{6.7.29}$$

For more details it is referred to the literature on digital control systems.

Linear filters are based on difference equations like (6.7.16). Depending on the selected structure and parameters low-pass, high-pass and band-pass filters can be designed as well as special filters for recursive averaging and averaging with limited or fading memory, see e.g. Isermann (1991), Cunningham (1992), Antonion (2006). The generation of derivatives is treated in Wolfram and Vogt (2002).

6.8 Conventional and model-based engine-control design and calibration

In the past only a few manipulated engine variables like injection mass and injection angle had to be controlled. Because of higher demands on low-fuel consumption, emissions and driveability, the number of main control inputs increased to 5–10 and the number of considered outputs to 6–8. Therefore, the design and calibration of the engine control has evolved into a highly demanding task. This section first describes the different emission measures and legislative limits. Then, general procedures for the design and calibration are considered for stationary operating points and driving cycles. This is followed by the calibration of the torque-oriented control for gasoline engines. If multi-variable engine models are first determined on test benches, a comprehensive optimization with different criteria enables a computer-based calibration for stationary operating points as well as for driving cycles.

6.8.1 Emission measures and limits

The optimization of the engine control system is highly influenced by the legislative emission laws. Therefore, this section discusses the various emission measures and their limits.

a) Emission measures

If the concentration of an emission is represented as c_x [g/m^3], the emission gas mass flow rate of the specific emission x becomes

$$\dot{m}_x = c_x \dot{V}_{\mathrm{exh}} = c_x \frac{\dot{m}_{\mathrm{exh}}}{\rho_{\mathrm{exh}}} \qquad [\mathrm{g/s}] \tag{6.8.1}$$

with

c_x	concentration of emission x	[g/m^3]
$\dot{V}_{\mathrm{exh}}$	exhaust gas volume flow rate at exhaust outlet	[m^3/s]
$\dot{m}_{\mathrm{exh}}$	exhaust gas mass flow rate	[g/s]
ρ_{exh}	exhaust gas density at exhaust	[g/m^3]

Frequently, the emission gas concentrations are measured in parts per million [ppm]. Then they are converted by

$$c_x[\mathrm{g/m^3}] = c'_x[\mathrm{ppm}] \cdot 10^{-3} \frac{m_{\mathrm{mol,x}}[\mathrm{g/mol}] \cdot \rho_{\mathrm{exh}}[\mathrm{kg/m^3}]}{m_{\mathrm{mol,exh}}[\mathrm{g/mol}]} \tag{6.8.2}$$

with

$m_{\mathrm{mol,x}}[\mathrm{g/mol}]$	molar mass of emission x
$m_{\mathrm{mol,exh}}[\mathrm{g/mol}]$	molar mass of exhaust ($\approx m_{\mathrm{mol,air}} = 28.9\,\mathrm{g/mol}$)

The exhaust gas density $\rho_{\text{exh,meas}}$ at the location of measurement may be different from the density at the exhaust outlet. Then the concentrations have to be corrected by

$$c_x = c_{x,\text{meas}} \frac{\rho_{\text{exh}}}{\rho_{\text{exh,meas}}}. \tag{6.8.3}$$

The molar masses of the emissions are defined as follows (EC No 692/2008):

THC (total hydro carbons)
  $m_{\text{mol,HC}} = 13.9655$ [g/mol] for diesel
  $m_{\text{mol,HC}} = 14.9413$ [g/mol] for gasoline
CO (carbon monoxide)
  $m_{\text{mol,CO}} = 28.0101$ [g/mol]
NO_x (nitro oxygen)
  $m_{\text{mol,NO}_2} = 46.0055$ [g/mol]

The concentration of particulate matter (PM) is usually measured as

c_{PM} [g/m^3].

The type approval for attaining a certification for passenger cars and light commercial vehicles ($\leq 3.8\,\text{t}$) requires exhaust-gas tests with country-specific driving cycles on vehicle chassis dynamometers. The exhaust gas over one driving cycle is collected and the pollutant masses are measured at the end. The legislative limits are given in g/km or g/mile.

For heavy commercial vehicles, stationary exhaust-gas tests are performed on a test bench, e.g. with a 13-stage-test in the EU or dynamic tests in the U.S.A. The average emissions for the 13-stage-test are determined by

$$\overline{m}_{x,\text{P}} = \frac{\sum_i w_i \dot{m}_{x,i} \Delta t_i}{\sum_i w_i P_i \Delta t_i} \qquad [\text{g/kWh}] \tag{6.8.4}$$

with

$\dot{m}_{x,i}$	mass flow rate of emission x_i at operation point i [g/h]
w_i	weighting factor of operation point i
P_i	power at operation point i [kW]
Δt_i	time interval per stage [h]
$i = 1 \ldots 13$	stages

Hence, different measures exist for the emissions, either in [g/km], [g/kWh] or [g/s]. However, the distance-related values m_{d} in [g/km] and the work-related values m_{P} in [g/kWh] are not suitable for the calibration of the engine control, because phases with idle and vehicle standstill are included, which lead to infinite or large contributions. Therefore, the mass flow rates $\dot{m}_x$ in [g/s] are preferred for the optimization of the engine control. The average time-related emission mass flow rate over one driving cycle is

$$\overline{m}_{\text{x,cyc,t}} = \frac{1}{t_{\text{cyc}}} \int_0^{t_{\text{cyc}}} \dot{m}_x(t)\, dt \qquad [\text{g/s}] \tag{6.8.5}$$

the emitted mass per cycle

$$m_{\mathrm{x,cyc}} = \int_0^{t_{\mathrm{cyc}}} \dot{m}_x(t)\, dt \qquad [\mathrm{g}] \tag{6.8.6}$$

and the distance-related emission measure can be determined by

$$\overline{m}_{x,cyc,\mathrm{d}} = \overline{\dot{m}}_{x,\mathrm{cyc},t}\, t_{\mathrm{cyc}}/d \qquad [\mathrm{g/km}], \tag{6.8.7}$$

where d is the driven distance of the cycle. The power-related measure $\overline{m}_{x,\mathrm{P}}$ [g/kWh] follows from (6.8.4).

b) Emission limits

The legislation defines permissible emission limits and testing processes, see e.g. Robert Bosch GmbH (2011). The emission-control regulations are mainly stated by CARB (California Resources Board), EPA (Environmental Protection Agency, USA), EU (European Union), and Japan. Each of the leading countries has its own test procedures, specified as type approval, serial testing and in-field monitoring. The vehicles are categorized into the classes: passenger cars, light commercial vehicles (EU: $< 3.5\,\mathrm{t}$), heavy commercial vehicles (EU: $> 3.5\,\mathrm{t}$) and off-highway vehicles. The first two classes are tested on vehicle chassis dynamometers, the last two on engine test benches.

Exhaust-gas tests for passenger cars and light commercial vehicles are performed with test cycles, defining the specific operation conditions, e.g. the NEDC (New European Driving Cycle), the FTP-75 (Federal Test Procedure), see Fig. 6.8.1 or the US 06 for USA. For heavy commercial vehicles the stationary 13-stage-test, according to Fig. 6.8.2, and the ETC (European Transient Cycle) are used in the EU, see Fig. 6.8.3a) and the transient FTP-test for USA, see Fig. 6.8.3b).

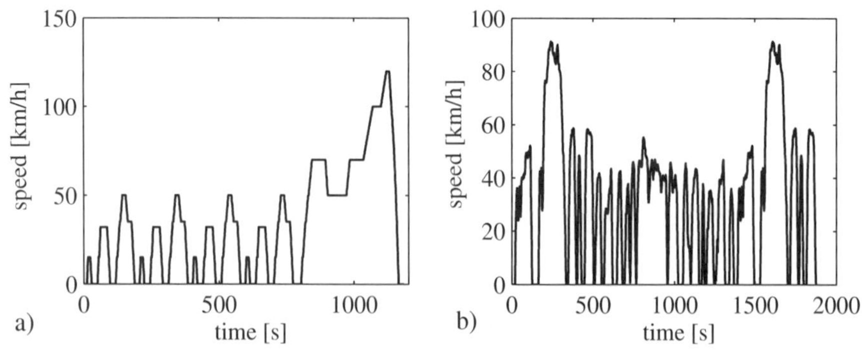

Fig. 6.8.1. Transient emission test cycles for passenger cars. **a** NEDC. **b** FTP-75.

The European ECE test cycle is originally an approximation of a city traffic cycle (UDC: Urban Driving Cycle). It consists of four identical sections of city driving of

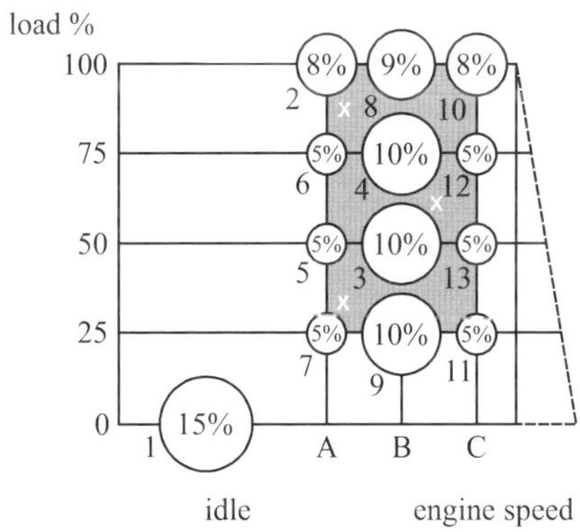

Fig. 6.8.2. Emission 13-stage-test procedure ETC for heavy commercial vehicles.

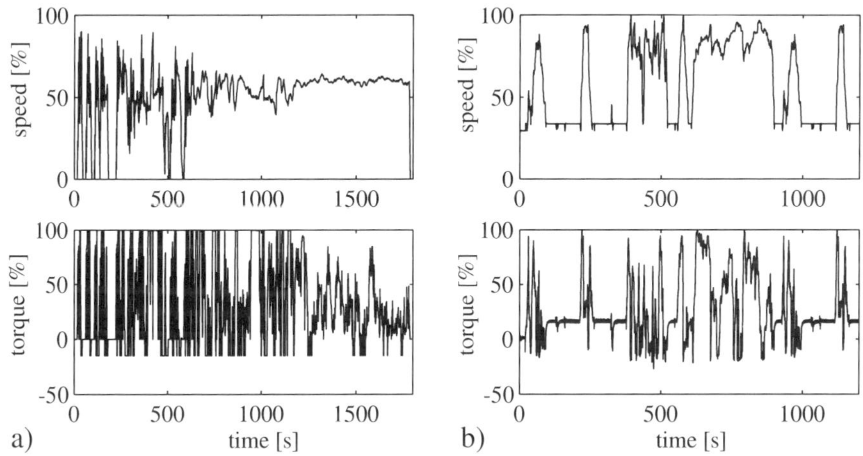

Fig. 6.8.3. Transient emission test cycles for commercial vehicles. **a** ETC (European Transient Cycle). **b** FTP (Federal Test Procedure), USA.

195 s with accelerations up to 50 km/h and decelerations with or without braking. The average speed is 18.7 km/h and the total length 4.052 km. Since 1992 it includes a highway section (EUDC: Extra-Urban Driving Cycle). This higher speed section accelerates up to 120 km/h and takes 400 s over 6.955 km. The combination of both is called NEDC (New European Driving Cycle). The overall distance of the NEDC is 11 km, the average speed 32.5 km/h and it lasts 1220 s or 20 min and 20 s including one cold start. The cold start is performed at 20 ... 30°C, lasts 40 s and is not part of the NEDC. However, since EURO 3 (2000) the cold start has been included in the MNEDC (Modified NEDC), and beginning in 2002 the starting temperature has been reduced to −7°C for gasoline engines.

The *American emission limit levels* are referred to as tiers, with Tier 1 in 1994, Tier 2 in 2004 and Tier 3 is proposed for 2014. The tiers contain different sets or bins. Manufacturers may choose between several bins for their fleet. For gasoline and diesel engines the same limits apply in the USA.

The emission limits of the *Californian CARB* are the most stringent ones. For passenger cars and light-duty trucks (LDT) they are specified in Low Emission Vehicles LEV1 for model years 1994 to 2003 with weights up to $6000\,\mathrm{lbs}$ ($2.71\,\mathrm{t}$) and since 2003 in LEV2. Manufacturers may use different exhaust-gas categories, like, in the sequence of decreasing emission limits: Tier 1, TLEV (Transitional Low-Emission Vehicle), LEV, ULEV (Ultra-Low-Emission Vehicle), SULEV (Super-Ultra-Low-Emission Vehicle), ZEV (Zero-Emission Vehicle), PZEV (Partial ZEV). LEV1 comprises Tier 1, TLEV, LEV, and ULEV, LEV2 includes LEV, ULEV and SULEV. Table 6.8.1 gives an overview of the limits.

The manufacturers have to ensure that their vehicle fleet does not exceed an average emission limit of the Non-Methane-Organic-Gases NMOG. The limits are reduced each year from $0.25\,\mathrm{g/mile}$ in 1994 to $0.035\,\mathrm{g/mile}$ in 2010.

In addition, a durability must be satisfied. This means that the manufacturers must certify that emission limits are not exceeded for 50000 miles or 5 years (intermediate useful life) or 100000 miles or 10 years (full useful life). This is checked by vehicle fleets from the production line.

The mean fuel consumption of vehicle fleets is also limited. The CAFE value (Corporate Average Fuel Economy) is 27.5 miles/gallon or 8.55l/100km since 2004 for passenger cars, and for LDT 20.3 miles/gallon or 11.6l/100km.

The *American EPA legislation* (Environment Protection Agency) covers the federal states except California and applies less strict laws than CARB. Tier 1 was used until 2004 and Tier 2 has been used since then.

The *European Union* (EU) *emission control standards* are directed by the EU commission. The EURO 1 was effective in 1992 and the emission limits are expressed in g/km. These limits must be ensured until $80\,000\,\mathrm{km}$ for EU 3 and $100\,000\,\mathrm{km}$ for EU 4, or 5 years. Table 6.8.2 shows the development of the allowed emission limits for gasoline and diesel engines.

Figure 6.8.4 shows the development of the emissions PM and NO_x + HC for diesel engines with very low values for EURO 6 in 2014.

6.8.2 Calibration for stationary operating points and driving cycles

Figure 6.8.5 depicts two schemes for the optimization of the stationary control of combustion engines at *selected stationary operating points* ($M_{\mathrm{eng,i}}, m_{\mathrm{j}}$). A conventional way is to maximize the torque $M_{\mathrm{eng}} = M$ in dependence on the speed $n_{\mathrm{eng}} = n$ for $m_{\mathrm{f,i}}$ = const.

$$\max_{\mathbf{u}} M(n, \mathbf{u}) \tag{6.8.8}$$

with regard to the manipulated variables $\mathbf{u}$ (e.g. the ignition angle and injection time of a gasoline engine) and then to store the manipulated variables $\mathbf{u}_{\mathrm{opt}}$ in form of static maps as a function of (m_{f}, n). Alternatively the specific fuel consumption can be minimized for M_{i} = const. or $m_{\mathrm{f,i}}$ = const.

$$\min_{\mathbf{u}} b_{\mathrm{sfc}}(n, \mathbf{u}). \tag{6.8.9}$$

Table 6.8.1. Emission limits for passenger cars and light-duty trucks (LDT) up to 6000 lbs for 1994-2003 and 8500 lbs since 2004 (CARB). FTP-test. [g/mile]

Model years	Emission control standard	Exhaust gas category	CO	NMOG	NO_x	PM (diesel)	durability [miles]
1994	LEV1	Tier 1	3.4	–	0.4	0.08	
		TLEV	5.5	0.20	0.9	0.10	100000
		LEV	5.5	0.13	0.5	0.10	
		ULEV	2.8	0.070	0.5	0.05	
2004	LEV2	LEV	4.2	0.090	0.07	0.01	120000
		ULEV	2.1	0.055	0.07	0.01	
		SULEV	1.0	0.010	0.02	0.01	
2014 (proposal)	LEV3	LEV		0.160		0.006	150000
		ULEV		0.125			
		SULEV		0.030			
2017 (proposal)						0.003	

Table 6.8.2. Emission limits for passenger cars and light commercial vehicles in Europe [g/km]

Engine type	Year	Emission category	CO	HC	HC + NO_x	NO_x	PM	Test cycle
gasoline	1992	EURO 1	2.72		0.97			
	1996	EURO 2	2.2		0.5			
	2000	EURO 3	2.3	0.2	–	0.15	–	MNEDC
	2005	EURO 4	1.0	0.1	–	0.08	–	MNEDC
			CO	THC	THC + NO_x	NO_x	PM	
	2009	EURO 5	1.0	0.1	–	0.06	0.005	
	2014	EURO 6	1.0	0.1	–	0.06	0.005	
diesel	1992	EURO 1	2.72		0.97		0.19	
	1996	EURO 2	1.0		0.9		0.1	
	2000	EURO 3	0.64	–	0.56	0.50	0.050	
	2005	EURO 4	0.50	–	0.30	0.25	0.025	
			CO	THC	THC + NO_x	NO_x	PM	
	2009	EURO 5	0.50	–	0.23	0.18	0.005	
	2014	EURO 6	0.50	–	0.17	0.08	0.005	

In both cases the emission masses of all operating points per driving cycle related to the distance d_{cyc} [km] or work W_{cyc} [kWh] have to be smaller than certain limits according to legislative requirements, like

$$\overline{m}_{\mathrm{CO}} \leq m_{\mathrm{CO,lim}}\,;\ \overline{m}_{\mathrm{CH}} \leq m_{\mathrm{CH,lim}}\,;\ \overline{m}_{\mathrm{NO_x}} \leq m_{\mathrm{NO_x,lim}}\,;\ \overline{m}_{\mathrm{PM}} \leq m_{\mathrm{PM,lim}} \tag{6.8.10}$$

where $\overline{m}$ stands for a distance or work related emission measure $\overline{m}_{\mathrm{x,cyc,d}}$ [g/km] or $\overline{m}_{\mathrm{x,P}}$ [g/kWh], see (6.8.7) and (6.8.4). If there are only some few manipulated variables $\mathbf{u}$, the optimal adjustments of the manipulated variables $\mathbf{u}_{\mathrm{opt}}$ can be obtained

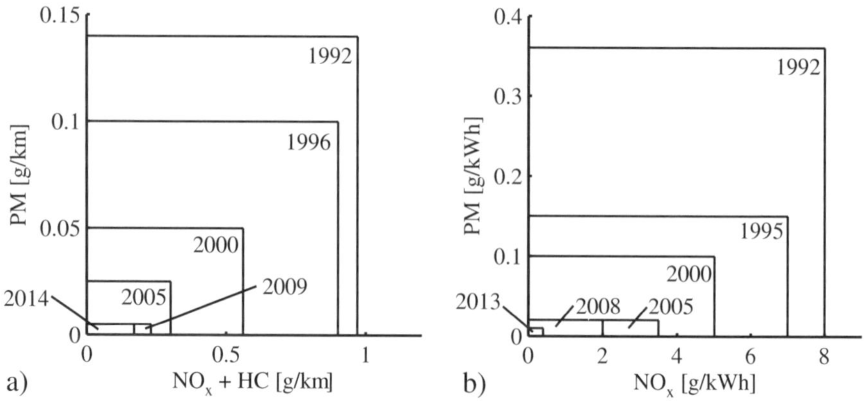

Fig. 6.8.4. European emission limits for diesel engines from EURO 1 (1992) to EURO 6 (2014) according to RL 91/441 EWG, RL 94/12/EG, RL 98/69/EG, VG(EG) 715/2007. **a** passenger cars, **b** heavy duty vehicles.

by manual optimization at the test bench and are then stored in the ECU as look-up tables in dependence on the operation points (M_i, n_j) or $(m_{f,i}, n_j)$, see Fig. 6.8.5a).

However, for the increased number of manipulated variables it is more efficient if the optimization of the calibration is not performed at the test bench, but with mathematical multi-input multi-output engine models on a computer. First, these models have to be determined at the test bench by measurement and identification methods, as described in Chap. 3. The stationary maps for the manipulated variables $\mathbf{u}_{\text{opt}}$ can then be obtained with numerical optimization methods for selected operating points, see Fig. 6.8.5b), Unland et al (1998), Hafner et al (2000), Kuder and Kruse (2000), Mitterer and Zuber-Goos (2000). In this case the engine behavior has to be measured and modeled only once and the calibration of the ECU is performed offline on development computers. The achieved results have to be validated with the engine at the test bench.

If, however, the optimization has to consider a *driving cycle*, the engines together with a simulated transmission and vehicle have to be treated using programmed settings for torque and speed (corresponding to a virtual perfect driver) and a multi criterion optimization has to be satisfied to minimize the fuel consumption for given constraints of emissions, compare Fig. 6.8.6.

This can be based on a *selection of static models* dependent on (M, n) which are determined for a driving cycle by simulating the transmission and vehicle as described in Sect. 6.8.4. A more realistic procedure is to include the dynamic engine behavior. Then the computer-based optimization with a mathematical dynamic engine model is advantageous as described in the next section. A corresponding performance criterion $J(\mathbf{u})$ is obtained by the integration of the mass flows over a driving cycle resulting in cumulated masses for fuel and emissions in [g/cycle].

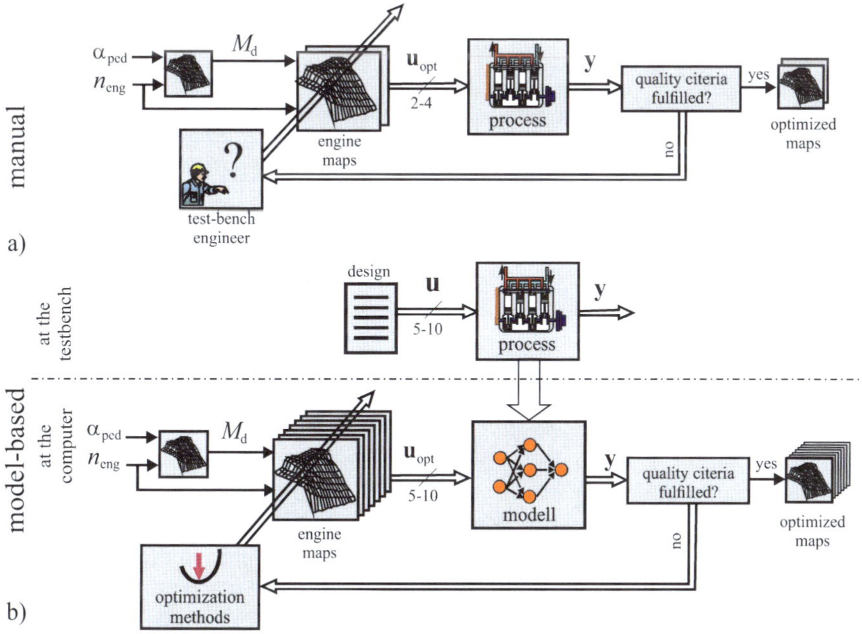

Fig. 6.8.5. Signal-flow scheme for the optimization of control functions for internal combustion engines at stationary operating points. **a** manual optimization at the test bench. **b** computer-based optimization with a mathematical multi-input multi-output engine model, Hafner et al (2000).

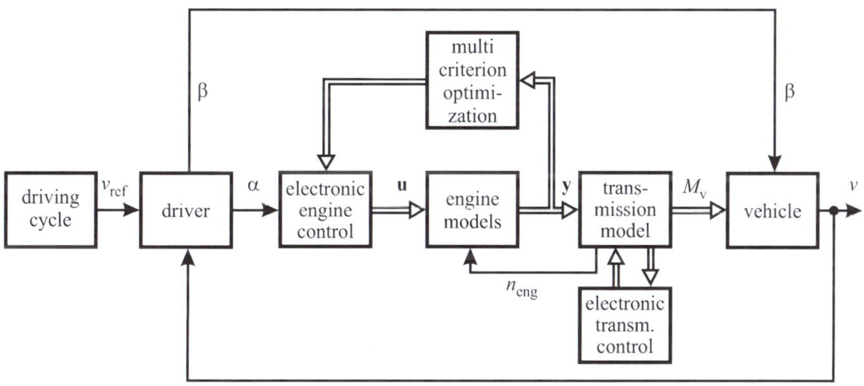

Fig. 6.8.6. Signal flow for the engine-model-based multi-criterion optimization for driving cycles, using simulated models for the automatic transmission, the vehicle and the driver. The engine models can be a selection of stationary models or dynamic models. α is the position of the acceleration pedal, β is the position of the brake pedal.

6.8.3 Calibration of the torque-oriented engine control

The measurement and calibration of the stationary control is now described for gasoline engines with homogeneous combustion as an example, according to Kuder and Kruse (2000) and Stuhler and Kruse (2003).

The development of the torque-oriented control, described in Sect. 7.1, requires the measurement of the engine torque M on a test bench for all operation points (M_i, n_j) according to the driving vehicle taking limits into account, which should not be exceeded. The adjustment of the operation point is performed by the test-bench control system. As described in Sect. 6.10, the speed of the engine is usually controlled in closed loop by the field-oriented control in the case of an AC-motor. For precise setting of the torque closed-loop control with throttle, i.e. air charge for gasoline engines or injection mass for diesel engines can be applied. However, the torque can also be feedforward controlled by appropriate adjustments of the manipulated variables.

Based on physical laws, e.g. on the thermodynamic combustion cycles in piston engines or optimal air/fuel ratio, some of the control parameters can be calculated. However, the majority of the look-up tables have to be found experimentally with dynamometer experiments because the physical knowledge, e.g. on the combustion process, fuel consumption and exhaust gas concentrations is (presently) not available in form of equations with a reasonable amount of computations.

In the past the look-up tables were adapted sequentially by manual operations on the dynamometers, requiring long adaptation times of several weeks. Since about 1997 a more systematic approach is applied based on the internal torque as a common intermediate variable and the use of different engine models. The calibration process for gasoline engines is shown in a simplified form in Fig. 6.8.7.

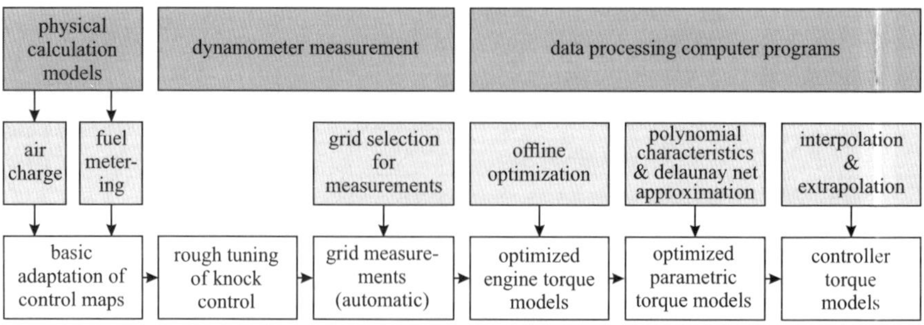

Fig. 6.8.7. Steps for the calibration of the torque-oriented control of gasoline engines, Kuder and Kruse (2000), Isermann and Hafner (2001).

Before systematic measurements with the engine on a dynamometer can be made, the engine needs a first adaption of the control parameters and maps so that the engine is able to run without being damaged. Therefore basic control functions are calculated, based on stationary air-charge models including manifold models to determine the injected fuel mass. Then, a rough tuning of the knock control is applied to protect the engine. This is followed by (automatic) grid measurements to obtain an internal torque model with fuel and also without fuel to get the drag torque models. Here, a lot of variations have to be made with regard to speed, air mass, air/fuel ratio, ignition angle, camshaft position, etc. The grid-point measurements are preselected based on certain patterns or methods for the design of experiments. Based on the measured engine maps optimal torques, ignition angles and correction factors are determined to build up the internal torque model, see Sect. 6.5.1. The optimized 3D-torque maps can then be approximated by polynomial approximators, applying a least squares approach for data smoothing. As the grid points for the required engine models are partially different from the measured ones a further approximation by Delaunay nets can be applied and used for interpolation and extrapolation resulting in final grid based engine models. These models are then the basis for designing the control parameters and maps for the ECU, see Unland et al (1998).

The measurement programs distinguish a basis calibration program and an additional calibration program, which are now considered in some more detail.

The *basis calibration measurement program* begins for the selected operating points with $n_j = \text{const}$ for air excess factor $\lambda = m_{\text{air}}/m_{\text{air,st}} = 1$ to measure the torque $M_i(\varphi_{\text{ign}})$ in dependence on the ignition angle. This is called *ignition timing cycle* and begins with the determination of the drag torque M_{drg}, see (7.1.2), with zero injection, and continues with searching the optimal torque for specified air mass m_{air} by advancing the ignition timing until a maximum torque or the knock limit is reached, see Fig. 6.8.8. Then the ignition angle is retarded stepwise until a stopping criterion like certain component temperatures, running roughness or maximal ignition angle changes, is reached.

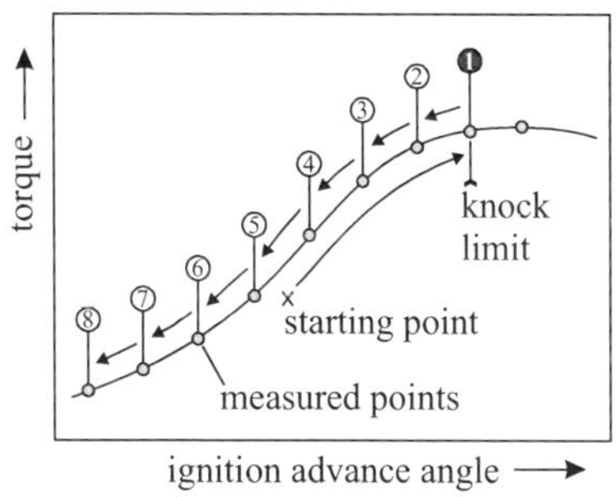

Fig. 6.8.8. Ignition timing cycle measurement.

The ignition timing cycle measurements are then repeated for different λ, resulting in about 250 cycles, which take about three min each, Kuder and Kruse (2000). The additional measurement program then includes variations of the camshaft position, the exhaust gas recirculation rate or charging pressure.

Based on the measured stationary data, the inner torque model (7.1.5)

$$M_{\text{i}}(m_{\text{air}}, n) \tag{6.8.11}$$

is calculated for the optimal ignition angle either as look-up tables or in form of approximations by polynomials, e.g. of second order.

Then, values for the torque M_{i} in dependence on additional variables, like fuel injection angle φ_{soi}, ignition angle φ_{ign}, and air excess factor λ are determined, and presented in efficiency maps as described in Sect. 7.1. Based on these models the calibration of the stationary control maps and characteristics follow as required for the gasoline control system with homogeneous combustion, Stuhler and Kruse (2003).

The development of gasoline engines with direct injection and especially with stratified combustion requires an increased amount of manipulated variables like rail pressure, multiple injections, swirl and tumble flaps and variable valve trains. Thus the number of engine input variables $\mathbf{u}$ increases to more than 10. The calibration then has to consider the requirements

$$\max_{\mathbf{u}} M(n, \mathbf{u})\,;\ \min_{\mathbf{u}} b_{\mathrm{sfc}}(n, \mathbf{u}) \tag{6.8.12}$$

under the legislative emission boundary conditions

$$\overline{m} \leq \overline{m}_{\mathrm{lim}} \tag{6.8.13}$$

where $\overline{m}$ stands for distance or work related measures of the emissions CO, HC, $\mathrm{NO_x}$ and PM.

Figure 6.8.9 shows as an example of the intersection plot for some output variables and manipulated variables that the specific fuel consumption shows minimal values for $\dot{m}_{\mathrm{egr}}$, φ_{soi} and φ_{ign}, Kuder and Kruse (2000). However the minima are at different values for $\mathrm{NO_x}$, HC and PM. Therefore appropriate compromises have to be made. This intersection plot is a feasible way to make the the relationships transparent.

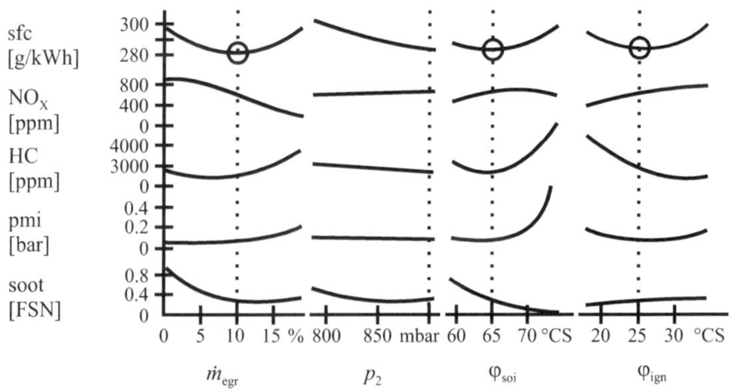

Fig. 6.8.9. Intersection plot for specific fuel consumption, emissions and engine roughness depending on EGR rate, intake pressure, start of injection and ignition angle for n = const. of a gasoline engine, Kuder and Kruse (2000).

6.8.4 Model-based calibration procedures

The general task for the optimal calibration of engine control systems can be summarized by the formulation of a multi-variable optimization criterion with different weights of the interesting output variables.

a) One operating point

For one *stationary operating point* with torque M_i and speed n_j a criterion with different weights of the output variables is

$$J(\mathbf{u}, M_i, n_j) = [w_1 b_{\text{sfc}}(\mathbf{u}) + w_2 m_{\text{CO,P}}(\mathbf{u}) + w_3 m_{\text{HC,P}}(\mathbf{u}) + w_4 m_{\text{NO}_\text{x},\text{P}}(\mathbf{u}) + w_5 m_{\text{PM,P}}(\mathbf{u})]\, P_{ij} \Delta t_{\text{ij}}\, [g] \quad (6.8.14)$$

with the power

$$P_{ij} = M_i\, \omega_j = M_i\, \pi\, n_j/30 \quad (6.8.15)$$

work-related emissions $m_{x,\text{P}}$ in [g/kWh], Δt_{ij} time interval per run, and

$$\sum_{\nu=1}^{5} w_\nu = 1 \quad (6.8.16)$$

If stationary mathematical models of the engine are available, the optimization

$$\min_{\mathbf{u}} J(\mathbf{u}, M_i, n_j) \quad (6.8.17)$$

can be performed with numerical optimization methods to find the optimal manipulated variables $\mathbf{u}_{\text{opt}}(M_i, n_j)$ for each operating point (M_i, n_j) with optimization methods described in Sect. 6.9. However, the inclusion of the emission limits requires an iterative procedure to adjust the weights w_ν.

b) Multiple operating points

If real vehicle driving is considered, multiple operating points can be selected according to their frequencies of occurence. For a certain driving cycle, representative operating points are selected for a specified transmission and vehicle. This allows to make a forecast for the fuel consumption and emissions over a driving cycle, assuming that the driving cycle can be reconstructed by a sequence of stationary operating points. An example for the calibration of four manipulated variables of a CR-diesel engine for several operating points is described in Sect. 8.3.2. A further example in Sect. 8.3.3 shows how for 200 operating points of a FTP cycle three manipulated variables are optimized by using local linear models (LOLIMOT) for the fuel consumption and two emissions.

For commercial vehicles, 13 operating points are already defined in the case of the European Steady State Test Procedure (ETC) which has to be used for type approval, see Fig. 6.8.2.

A further example for a *cycle-oriented optimization* with stationary engine models is given by Kötter (2008). Figure 6.8.10a) depicts the time history of the vehicle speed and Fig. 6.8.10b) the weight w_ν of $\nu = 10$ selected operating points according to their frequency in the driving cycle. The task is then to minimize the fuel consumption m_f for the cycle in [g/cycle] by an optimal adjustment of six manipulated variables $\mathbf{u}$ of a common-rail diesel engine, the pilot injection mass and time, charging pressure, main injection time, air mass and rail pressure. The optimization criterion then becomes

$$J_\mathrm{stat}(\mathbf{u}) = m_\mathrm{f}(\mathbf{u}) = w_1 m_\mathrm{f1}(\mathbf{u}_1) + w_2 m_\mathrm{f2}(\mathbf{u}_2) + \ldots + w_{10} m_\mathrm{f10}(\mathbf{u}_{10})\ [g/cycle] \tag{6.8.18}$$

with boundary conditions

$$\begin{aligned}
m_\mathrm{NOx}(\mathbf{u}) &= w_1 m_\mathrm{NOx1}(\mathbf{u}_1) + w_2 m_\mathrm{NOx2}(\mathbf{u}_2) + \ldots + w_{10} m_\mathrm{NOx10}(\mathbf{u}_{10}) \\
&\leq m_\mathrm{NOx,lim} \\
m_\mathrm{HC}(\mathbf{u}) &= w_1 m_\mathrm{HC1}(\mathbf{u}_1) + w_2 m_\mathrm{HC2}(\mathbf{u}_2) + \ldots + w_{10} m_\mathrm{HC10}(\mathbf{u}_{10}) \\
&\leq m_\mathrm{HC,lim} \\
m_\mathrm{CO}(\mathbf{u}) &= w_1 m_\mathrm{CO1}(\mathbf{u}_1) + w_2 m_\mathrm{CO2}(\mathbf{u}_2) + \ldots + w_{10} m_\mathrm{CO10}(\mathbf{u}_{10}) \\
&\leq m_\mathrm{CO,lim} \\
m_\mathrm{PM}(\mathbf{u}) &= w_1 m_\mathrm{PM1}(\mathbf{u}_1) + w_2 m_\mathrm{PM2}(\mathbf{u}_2) + \ldots + w_{10} m_\mathrm{PM10}(\mathbf{u}_{10}) \\
&\leq m_\mathrm{PM,lim}
\end{aligned} \tag{6.8.19}$$

where $m_\mathrm{NOx,lim}$ etc. are the legislative emission limits, according to (6.8.6).

Hence $6 \cdot 10 = 60$ values of manipulated variables $u_1, \ldots, u_6$, which are also limited like $u_{1\min} \leq u_1 \leq u_{1\max}$, have to be optimized.

The minimization of the fuel consumption $m_\mathrm{f}(\mathbf{u})$ over the selected operation points of the driving cycle as well as the boundary conditions are strongly nonlinear functions. Because several local minima may exist, hill climbing methods based on gradients may not find the global optimum. Therefore genetic algorithms are recommended. Kötter (2008) shows resulting stationary maps for the six manipulated variables.

A considerable advantage of the model-based calibration is that the engine behavior is only measured and modeled once and the calibration is performed on development computers, see Fig. 6.8.5b), and does not need expensive operation time on the test bench. This is especially attractive if the engine is calibrated for families of different transmissions and vehicles. Then, a forecast of the fuel consumption and emissions for a driving cycle and driveability can be made with the models of the engine, the vehicle and ECU functions, Schulmeister et al (2007). This is, in the context of Fig. 6.11.1, a model-in-the-loop simulation (MiL). Figure 6.8.11 shows an example how much the different operating points of an engine which is used in various vehicle variants are scattered over the operating range.

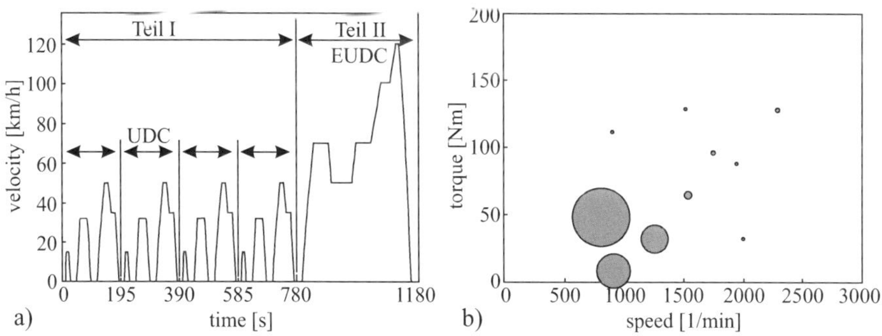

Fig. 6.8.10. **a** MNEDC cycle. **b** Frequency of 10 representative operating points in the driving cycle.

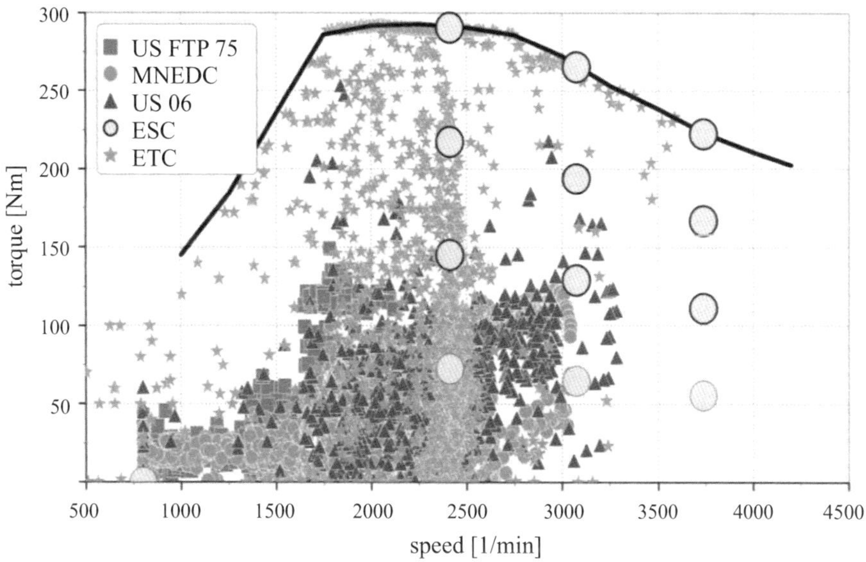

Fig. 6.8.11. Torque-speed operation points for the use of an engine in various vehicle variants and emission cycles, Schulmeister et al (2007). (ESC: in the 13-stage test procedure for commercial vehicles).

c) Calibration with dynamic engine models

The increasing requirements for the reduction of fuel consumption and emissions and for the improvement of driveability need the inclusion of the dynamic engine behavior for the control design. Dynamic delays are mainly caused by the air volume of the intake system, the moment of inertia of the turbocharger and heat storages in the combustion chambers, cylinder walls, exhaust gas channel, exhaust gas recirculation and various heat exchangers. The dynamics between combustion and crankshaft torque is influenced by the number of cylinders and can usually be neglected for the calibration of ECU (mean-value) control functions. The lag time constants of the intake system and the turbocharger are in the range of 1 to $3\,\mathrm{s}$, whereas the temperature dynamics vary from about $10\,\mathrm{s}$ to $2\,\mathrm{min}$.

For driving cycles like the FTP or NEDC the percentage of dynamic states is around $40 \ldots 50\,\%$. As the variables of the optimization criteria like fuel consumption, emissions and torque are influenced by the dynamic engine behavior, they have to be included in the optimization process, at least for high requirements and for fine-tuning.

A time-dependent version of the multi-variable optimization criterion according to (6.8.14) becomes

$$\begin{aligned} J\left(\mathbf{u}(t), M(t), n(t)\right) &= w_1 \dot{m}_{\mathrm{f}}(t) + w_2 \dot{m}_{\mathrm{CO}}(t) \\ &\quad + w_3 \dot{m}_{\mathrm{HC}}(t) + w_4 \dot{m}_{\mathrm{NOx}}(t) + w_5 \dot{m}_{\mathrm{PM}}(t) \end{aligned} \quad (6.8.20)$$

with $\dot{m}(t)$ as mass flow rates $\dot{m}$ [g/s]. However, the value of this criterion fluctuates during transient phases, as the variables depend differently on time. Hence this criterion is not suitable for calibration. Therefore its value has to be integrated over a certain dynamic phase t_{dyn}, i.e. an acceleration phase or usually a complete driving cycle with specified torques $M(t)$ and speeds $n(t)$

$$\begin{aligned} J_{\mathrm{dyn}}(\mathbf{u}(t)) = \int_0^{t_{\mathrm{dyn}}} & (w_1 \dot{m}_{\mathrm{f}}(t) + w_2 \dot{m}_{\mathrm{CO}}(t) + w_3 \dot{m}_{\mathrm{HC}}(t) \\ & + w_4 \dot{m}_{\mathrm{NO_x}}(t) + w_5 \dot{m}_{\mathrm{PM}}(t))\, dt\, [g] \end{aligned} \quad (6.8.21)$$

However, an optimization of the manipulated variables $\mathbf{u}(t)$ requires dynamic models for the integrands, which also depend on $\mathbf{u}(t)$. Hence, this task is not easy to solve.

An alternative is to reduce the optimization to the minimization of the fuel consumption over a driving cycle, compare (6.8.18) and (6.8.19),

$$J_{\mathrm{dyn}}(\mathbf{u}(t)) = m_{\mathrm{f}}(\mathbf{u}(t)) = \int_0^{t_{\mathrm{dyn}}} \dot{m}_{\mathrm{f}}(\mathbf{u}(t))\, dt\, [g] \quad (6.8.22)$$

together with the boundary conditions

$$\begin{aligned} m_{\mathrm{NOx}}(\mathbf{u}(t)) &= \int_0^{t_{\mathrm{dyn}}} \dot{m}_{\mathrm{NOx}}(\mathbf{u}(t))\, dt \leq m_{\mathrm{NOx,lim}} \\ m_{\mathrm{HC}}(\mathbf{u}(t)) &= \int_0^{t_{\mathrm{dyn}}} \dot{m}_{\mathrm{HC}}(\mathbf{u}(t))\, dt \leq m_{\mathrm{HC,lim}} \\ m_{\mathrm{CO}}(\mathbf{u}(t)) &= \int_0^{t_{\mathrm{dyn}}} \dot{m}_{\mathrm{CO}}(\mathbf{u}(t))\, dt \leq m_{\mathrm{CO,lim}} \\ m_{\mathrm{PM}}(\mathbf{u}(t)) &= \int_0^{t_{\mathrm{dyn}}} \dot{m}_{\mathrm{PM}}(\mathbf{u}(t))\, dt \leq m_{\mathrm{PM,lim}} \end{aligned} \quad (6.8.23)$$

The optimization task is then concentrated on the optimization of a single variable taking into account several constraints for the resulting emissions. The dynamic behavior can also be optimized separately from the stationary behavior, as shown in Sect.8.4.

d) Calibration of driveability control functions

The calibration of driveability control functions, like the damping of drive-train oscillations after load changes on one hand and a more sportive or comfortable tuning on the other need compromises along a Pareto curve. A software-in-the-loop simulation (SiL) with the ECU software and dynamic drive-train and vehicle models then allow a preselection of calibration candidates, Schulmeister et al (2007). Driving tests with real vehicles then are applied to finetune the corresponding control and filter functions, see e.g. Sect. 4.9 on surge damping.

6.9 Optimization methods

6.9.1 Optimization criteria

A model-based optimization of the manipulated variables $\mathbf{u}$ for the engine control requires a suitable optimization criterion or cost function $J(\mathbf{u})$. The last section has shown that different criteria can be used. These criteria are now formulated towards a general form which fits to mathematical optimization procedures.

If the *stationary behavior* has to be optimized and only one operating point (M_i, n_j) is considered, then a first possibility is to minimize a multi-objective cost function

$$J(\mathbf{w}, \mathbf{x}(\mathbf{u})) = \sum_{\nu=1}^{M} w_\nu \, x_\nu(\mathbf{u}) = \mathbf{w}^T \mathbf{x}(\mathbf{u}), \tag{6.9.1}$$

where x_ν are several output variables which depend on $\mathbf{u}$, like e.g. in (6.8.14)

$$\mathbf{x}^T = \left[x_1, \, x_2, \, \ldots, \, x_l\right] = \left[m_\mathrm{f}, \, x_\mathrm{CO}, \, x_\mathrm{CH}, \, x_\mathrm{NOx}, \, x_\mathrm{PM}\right] \tag{6.9.2}$$

and w_ν are corresponding weights

$$\mathbf{w}^T = \left[w_1, \, w_2, \, \ldots, \, w_l\right]. \tag{6.9.3}$$

The output variables $\mathbf{x}$, which have to be minimized, depend on several manipulated variables $\mathbf{u}$, which are for a gasoline engine without charging, e.g.

$$\mathbf{u}^T = \left[m_\mathrm{air}, \, m_\mathrm{f}, \, \varphi_\mathrm{inj}, \, \varphi_\mathrm{ign}, \, \dot{m}_\mathrm{egr}\right] \tag{6.9.4}$$

and for a diesel engine with turbo charger, e.g.

$$\mathbf{u}^T = \left[m_\mathrm{air}, \, u_\mathrm{vgt}, \, m_\mathrm{f}, \, \varphi_\mathrm{inj}, \, p_\mathrm{rail}, \, \dot{m}_\mathrm{egr}\right] \tag{6.9.5}$$

compare also alternatives used in Sects. 7.1 and 8.3.

In this case, the fuel consumption m_f is minimized together with the emissions x_CO, x_CH, etc., taking different weights w_ν into account. This requires that stationary engine models $\mathbf{x}(\mathbf{u})$ exist. An alternative is to minimize the fuel consumptions for L operating points (e.g. of a driving cycle)

$$J = J_\mathrm{stat}(\mathbf{w}, m_\mathrm{f}(\mathbf{u})) = m_\mathrm{f}(\mathbf{u})) = \sum_{\xi=1}^{L} w_\xi \, m_{f\xi}(\mathbf{u}) \tag{6.9.6}$$

subject to constraints or boundary conditions of the emissions

$$m_\mathrm{em}(\mathbf{u}) = \sum_{\xi=1}^{L} w_\xi \, m_{\mathrm{em}\xi} \leq m_\mathrm{em,lim} \tag{6.9.7}$$

where the index emξ stands for the emissions CO, HC, $\mathrm{NO_x}$, PM at the operating points ξ, compare (6.8.18), (6.8.19). The weights w_ξ are selected according to the relevance of the operating point ξ within a driving cycle.

The engine control optimization for the *dynamic behavior*, including acceleration and deceleration phases as for a driving cycle, needs dynamic engine models $\mathbf{x}(\mathbf{u}, t)$. Then, the minimization of a time-dependent cost function has to be performed, which becomes in the case of fuel consumption minimization as (6.8.22)

$$J = J_\mathrm{dyn}(\mathbf{u}(t)) = m_\mathrm{f}(\mathbf{u}(t)) = \int_0^{t_\mathrm{dyn}} m_\mathrm{f}(\mathbf{u}(t))dt. \tag{6.9.8}$$

The optimization goal is to find dynamic transients of the manipulated variables $\mathbf{u}(t)$, e.g. for step or ramp changes of the acceleration pedal (throttle) $\alpha(t)$ or a complete driving cycle, subject to constraints

$$g'(\mathbf{u}(t)) = \int_0^{t_\mathrm{dyn}} m_\mathrm{em}(\mathbf{u}(t))\, dt \leq m_\mathrm{em,lim} \tag{6.9.9}$$

compare (6.8.22), (6.8.23) and Hafner (2002).

Instead of optimizing the trajectories $\mathbf{u}(t)$ the parameters $\mathbf{q}$ of dynamic feedforward or feedback control algorithms can be optimized, such that

$$J(\mathbf{q}) = m_\mathrm{f}(\mathbf{q}) \tag{6.9.10}$$

$$g'(\mathbf{q}) = \int_0^{t_\mathrm{dyn}} m_\mathrm{em}(\mathbf{q})\, dt \leq m_\mathrm{em,lim}. \tag{6.9.11}$$

The optimization task can now be formulated in a general form. In all cases, one has to find optimal manipulated variables $\mathbf{u}(t)$ for the stationary behavior with $t = \mathrm{const}$ and for the dynamic behavior with variable t, so that

$$\min_{\mathbf{u}(t)} J\left(\mathbf{w}, m_\mathrm{f}(\mathbf{u}(t))\right) \tag{6.9.12}$$

In the stationary case this implies

$$\frac{dJ\left(\mathbf{w}, m_{\mathrm{f}}(\mathbf{u})\right)}{d\mathbf{u}} = \mathbf{0}. \qquad (6.9.13)$$

The emissions may be included in the cost function, like in (6.9.1), (6.9.2) or be taken into account by the constraints

$$g(\mathbf{u}(t)) = \int_0^{t_{\mathrm{dyn}}} m_{\mathrm{em}}(\mathbf{u}(t))\, dt - m_{\mathrm{em,lim}} \leq 0. \qquad (6.9.14)$$

If the controller parameters $\mathbf{q}$ are optimized in the case of dynamic behavior, it holds

$$\min_{\mathbf{q}} J(\mathbf{q}) \qquad (6.9.15)$$

$$\frac{dJ(\mathbf{q})}{d\mathbf{q}} = \mathbf{0} \qquad (6.9.16)$$

$$g(\mathbf{q}) = \int_0^{t_{\mathrm{dyn}}} m_{\mathrm{em}}(\mathbf{q})\, dt - m_{\mathrm{em,lim}} \leq 0. \qquad (6.9.17)$$

The multi-objective cost functions $J(\mathbf{w}, \mathbf{x}(\mathbf{u}))$ with included emissions or $J(\mathbf{w}, m_{\mathrm{f}}(\mathbf{u}))$ with separated boundary conditions $m_{\mathrm{em}}(\mathbf{u})$ for the case of stationary optimization depend nonlinearly on engine models $\mathbf{x}(\mathbf{u})$ or $m_{\mathrm{f}}(\mathbf{u})$ and $m_{\mathrm{em}}(\mathbf{u})$ in the form of nonparametric models, like look-up tables or complicated parametric models like polynomials of higher order or (neural) net models. As a direct optimization is not possible in that case, numerical optimization methods have to be used, for which a short overview is given in the next section.

6.9.2 Numerical optimization methods

The applicable optimization methods depend on the kind of linearity or nonlinearity of the loss functions $J(\mathbf{u})$ with regard to the quantities $\mathbf{u}$ to be optimized. Because the optimization follows from

$$\frac{dJ(\mathbf{u})}{d\mathbf{u}} = \mathbf{0} \qquad (6.9.18)$$

the type of the gradient of the loss function is relevant for selecting the optimization method. If the gradient is linear-in-the-parameters or quantities $\mathbf{u}$ and a quadratic loss function is used then linear regression methods, as the method of least squares, can be used, which leads to a direct solution $\mathbf{u}_{\mathrm{opt}}$, see Fig. 6.9.1. This can be used to estimate the parameters of polynomials for example. Then numerical search methods are not required. However, the loss functions for the optimization of the engine control are usually nonlinear in the gradients and therefore nonlinear optimization methods have to be used. As shown in the scheme of Fig. 6.9.1 one distinguishes local and global optimization methods.

The *local optimization* methods assume that there exists a unique optimum. This is always the case if the loss function gradient is linear-in-the-parameters and a direct linear regression method is used. In the case of loss function gradients which are nonlinear-in-the-parameters numerical local optimization methods start with an initial point and search in directions of neighborhood-based first and second order derivatives. This leads to a search in the vicinity of the starting point but does not necessarily result in a global optimum, if several local optima exist. However, trials with different starting points allow to find a good local optimum.

Global optimization methods take into account that several local optima exist and search for the best optimum. Therefore they search in a wider area than local optimization methods and require to cover all possible parameters or quantities $\mathbf{u}$, e.g. in form of a grid. However, as the computations increase exponentially with the number of parameters, combinations of different methods are used, e.g. a different treatment of structural and local parameters in sequential form.

The nonlinear optimization problem can be formulated in general form

$$\min_{\mathbf{u}} J(\mathbf{u}) \tag{6.9.19}$$

subject to constrains

$$\begin{aligned} g(\mathbf{u}) &\leq 0 \qquad \text{(soft boundary)} \\ h(\mathbf{u}) &= 0 \qquad \text{(hard boundary)}. \end{aligned} \tag{6.9.20}$$

For the optimization of combustion engine control the constraints are usually emissions of the exhaust gases and noise, formulated as inequalities and therefore soft boundaries.

For the optimization of control functions of combustion engines it can be assumed that a unique optimum exists in most cases. Then local optimization methods can be applied. Their principles will be described briefly in this section.

a) Local numerical optimization methods

The local optimization methods can be divided in methods with direct search and gradient-based search. Direct methods only use the values of the cost function and progress iteratively

$$\mathbf{u}(k+1) = \mathbf{u}(k) + \alpha \mathbf{s}(k) \tag{6.9.21}$$

from an initial value $\mathbf{u}(0)$ where $\mathbf{s}(k)$ is a search vector and α determines the step size. Typical representatives are section search methods, simplex methods and the Hooke-Jeeves method. However, these methods show a relatively slow convergence for large dimensions of $\mathbf{u}$. The convergence speed can be improved by inclusion of the gradient with

$$\mathbf{s}(k) = \nabla J(\mathbf{u}) \tag{6.9.22}$$

One representative of this gradient-based method is the steepest descent algorithm.

A further improvement is obtained by adding the second order derivative

$$\nabla^2 J(\mathbf{u}) = \frac{\partial^2 J(\mathbf{u})}{\partial \mathbf{u}^T \partial \mathbf{u}}. \tag{6.9.23}$$

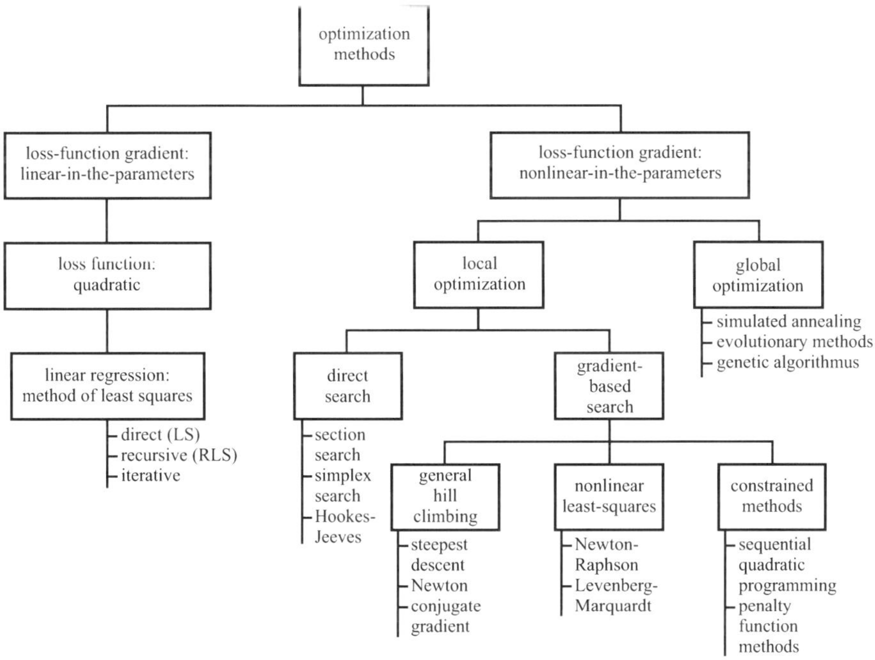

Fig. 6.9.1. Overview of direct and numerical optimization methods.

which is the Hessian matrix, to adjust the stepsize α to the nonlinear cost function. Well know methods are the Newton method and Newton-Raphson method. As the first and second order derivatives are usually not available analytically they have to be approximated numerically.

The optimization problem with *constraints* can be solved with the discussed local optimization methods by adding the constraint linearly to the cost function, like the method with Lagrange multipliers and the resulting sequential quadratic programming or by introducing special penalty functions. These basic methods of numerical optimization are e.g. treated in Gill et al (1981), Dennis and Schnabel (1996), Nocedal and Wright (2006), Nelles (2001), Isermann and Münchhof (2011).

b) Global numerical optimization methods

Global optimization methods have to search in wider areas than the local methods. As grid-based search in the parameter space takes unacceptable computing time, combinations of different methods are used, Nelles (2001). A first possibility is a *subsequent optimization*, with a global search first and then a local optimization. If the parameters can be separated, then those which can be optimized easier are searched first and then the remaining ones. This is called *nested optimization*. One example is the initial optimization of structure parameters (outer loop) and then continuation with local optimization (inner loop), like for neuronal networks or net-based models.

A *staggered optimization* is similar. First the structural parameters are optimized with a global method, keeping the other parameters fixed. Then the structure parameters are fixed and the local parameters are optimized again, etc. Hence, the idea behind these global methods is to divide the optimization in smaller and easier to handle optimizations with less parameters, instead of optimizing all parameters at once. In general, this improves the convergence speed, especially if the the parameter sets are not or only weakly coupled.

Some typical representatives for global optimization are mentioned briefly.

Simulated annealing is a stochastic method and leans on the simulation of a particle in a potential field, i.e. it uses an analogy to a physical behavior. A particle with a certain temperature has kinetic energy and moves with some randomness within the potential fields, also to higher potential energy, thus escaping local minima. The particle is annealed, i.e. its temperature decreases gradually. The particle is considered as parameter and the potential energy represents the cost function. The method begins with a wide search of the parameters and then, with decreasing temperature, concentrates on a specific region, Kirkpatrick et al (1983), Ingber (1996).

Evolutionary algorithms also have a stochastic character and they are oriented towards natural evolution. They operate with populations of individuals. The populations evolve in generations, corresponding to iterations. The individuals change by a genetic operation, like mutation or recombination, such that new individuals are generated, representing a new cost function value. Better individuals are selected by a performance index called fitness (i.e. the inverse of the cost function). Based on the selection, the population (set of parameter vectors) evolves to better individuals (parameter vectors). Hence, evolutionary algorithms operate as follows: An initial population of individuals (parameters $\mathbf{u}_0$) is chosen, the fitness $1/J(\mathbf{u})$ of all individuals of the initial population is calculated, best individuals are selected and mutated or recombined by genetic operators. This procedure is repeated iteratively until it is stopped by a termination criterion, Rechenberg (1973), Schwefel (1988).

The evolutionary algorithms can be divided into, e.g. evolution strategies or genetic algorithms. The *evolution strategies* represent the parameters in their original value, either continuously or as integers and they mostly use mutations. *Genetic algorithms* use a binary coding, i.e. the parameters are represented as a set of binary values (a bit-string), a simple analogy to the genetic information (DNA) in nature and they use recombinations.

For a detailed treatment of these methods see Rechenberg (1973), Schwefel (1988), Goldberg (1989), Davis (1991), Michalewicz et al (1992), Bäck (1996), Gen and Cheng (1997).

The application of sequential quadratic programming and evolutionary algorithms have been compared for a CR 1.9 l diesel engine with six manipulated variables and five output variables with regard to minimize the fuel consumption, Kötter (2008). A combination of evolutionary algorithms and local optimization has shown best performance, but the hill climbing method resulted in a good minimization of the fuel consumption as well. A further application with real coded genetic algorithms for the same engine to optimize six manipulated variables for ten stationary operation points (M_i, n_j) resulted in six control maps by using an interpolation method, Kötter (2008).

6.10 Test benches

Test benches are a very important facility for the experimental development of internal combustion engines. They consist of the combustion engines to be investigated and a dynamometer to convert the mechanical power of the engine into another form and to remove it in the case of positive power generation. Both, the engine and the dynamometer are provided with a rich instrumentation to measure all required variables and to automatically control some of them.

The following sections describe the test-bench types, their control and mathematical models.

6.10.1 Dynamometer types

Depending on the application, operation properties and costs, different types of dynamometer types are used, Martyr and Plint (2007). Hydrokinetic or hydraulic dynamometers consist of a cylindrical rotor with recesses. Corresponding pockets are divided by radial vanes, both in the rotor and stator. Centrifugal forces on the water result in a toroidal circulation and therefore transfer momentum from the rotor to the stator. This leads to turbulent shear in the water and dissipates heat to the water.

One distinguishes *hydrokinetic dynamometers* with constant or variable filling. In the case of constant filling the torque is varied by inserting pairs of sluice plates between rotor and stator. This is a relatively cheap solution. However, load changes are slow. For variable fill dynamometers the mass of circulating water within the casing is changed by a water outlet valve. The advantages are that they are suitable for very high power up to 130 MW, that they do not need a high power electrical connection, and that they are robust and tolerant for overload. However, they need an open water system.

Hydrostatic dynamometers consist of a fixed stroke and a variable stroke positive displacement hydraulic pump/motor system. The variable stroke pump is attached to the engine and the fixed stroke motor is the braking dynamometer. It can develop full torque at very low speeds and at stand still. However, it requires much oil and it is expensive and noisy. It is only used for special applications.

Eddy-current dynamometers have a toothed rotor with a narrow clearance to water-cooled steel loss plates. A magnetic field generated by annular coils changes the magnetic flux in the plates upon the turning rotor, resulting in eddy-currents and heat, which is transfered to cooling water flowing through the loss plates. Load is manipulated by varying the current in the coils. These eddy-current types are simple and robust and allow rapid load changes.

Most flexible are *electromotor dynamometers*, see Fig. 6.10.1. The mechanical power is converted into electrical power which is supplied to the electrical net. *Direct current motors* (DC) as used in the past are relatively easy to control by using a cascaded control system with the voltage as a manipulated variable of the major loop and the current as controlled variable in a minor loop. Then a speed or torque controller can be used as major controller. The direct current is generated through a thyristor-based AC/DC converter where the voltage is manipulated by appropriate phase triggering, synchronized to the AC net frequency. However, DC motors have a rather limited speed and high inertia. Therefore belt transmissions may be applied. In addition the commutator requires maintenance and is an electromagnetic noise source for the measurements.

Alternating current motors (AC) have a lower rotational inertia than DC motors and allow a higher speed. These AC motors are also called induction motors and consist usually of three windings per pole pair placed in stator slots which are interconnected with the individual phases of a three-voltage power supply, either in delta- or Y-connection. A rotating magnetic field with rotating frequency ω_{d} is generated where the angular velocity depends on the power-supply frequency ω_{N} and the number of pole pairs p within the stator according to $\omega_{\mathrm{d}} = \omega_{\mathrm{N}}/p$. Depending on different rotor constructions, asynchronous and synchronous induction motors can be distinguished.

Asynchronous motors have usually a squirrel-cage rotor resulting in a simple, robust and cost effective construction. The speed and torque is controlled by a field-oriented or vector control system with variable frequency and amplitude. The corresponding power electronics consist of a power grid fed converter with a net-side AC-DC converter (rectifier) which rectifies the alternating grid voltage and a motor-side three-phase DC-AC converter (inverter) that generates the three-phase voltage with variable frequency and amplitude. The power electronics is usually based on insulated gate bipolar transistors (IGBT) technology which is relatively expensive.

Based on a transformation of the three-phase voltage into a two-phase voltage system (Clarke-Park transformation) a field-oriented control can be designed. It is based on two cascaded control loops for the flux (d-current) and the torque (q-current). The flux is usually kept constant until the rated speed, but has then to be weakened. The q-current determines together with the flux the generated torque.

Synchronous motors have a similar control system. Opposite to asynchronous motors no slip between the stator field and the rotor speed ($\omega_{\mathrm{m}} = \omega_{\mathrm{d}}$) emerges. For more details, see Leonhard (1996), Lyshevski (2000), Isermann (2005).

The selection of the dynamometer also depends on the required torque-speed characteristics. Figure 6.10.2 shows the possible torque-speed quadrants. If the engine turns clockwise by looking at the flywheel, the dynamometer is in the 1st quadrant and absorbs the torque, i.e. brakes the engine. All dynamometers are designed to run in this 1st quadrant or if running counter clockwise in the 2nd quadrant. For motoring or towing, the engine and the dynamometers have to run in the 4th or 3rd quadrant. Then the choice is limited to DC or AC motors or to hydrostatic dynamometers.

The electromotor dynamometers are able to directly start the combustion engine with an appropriate, relatively small torque compared to the rated torque. The other dynamometers require either the serial engine-mounted starters or special electric motors at the other engine end of the dynamometer.

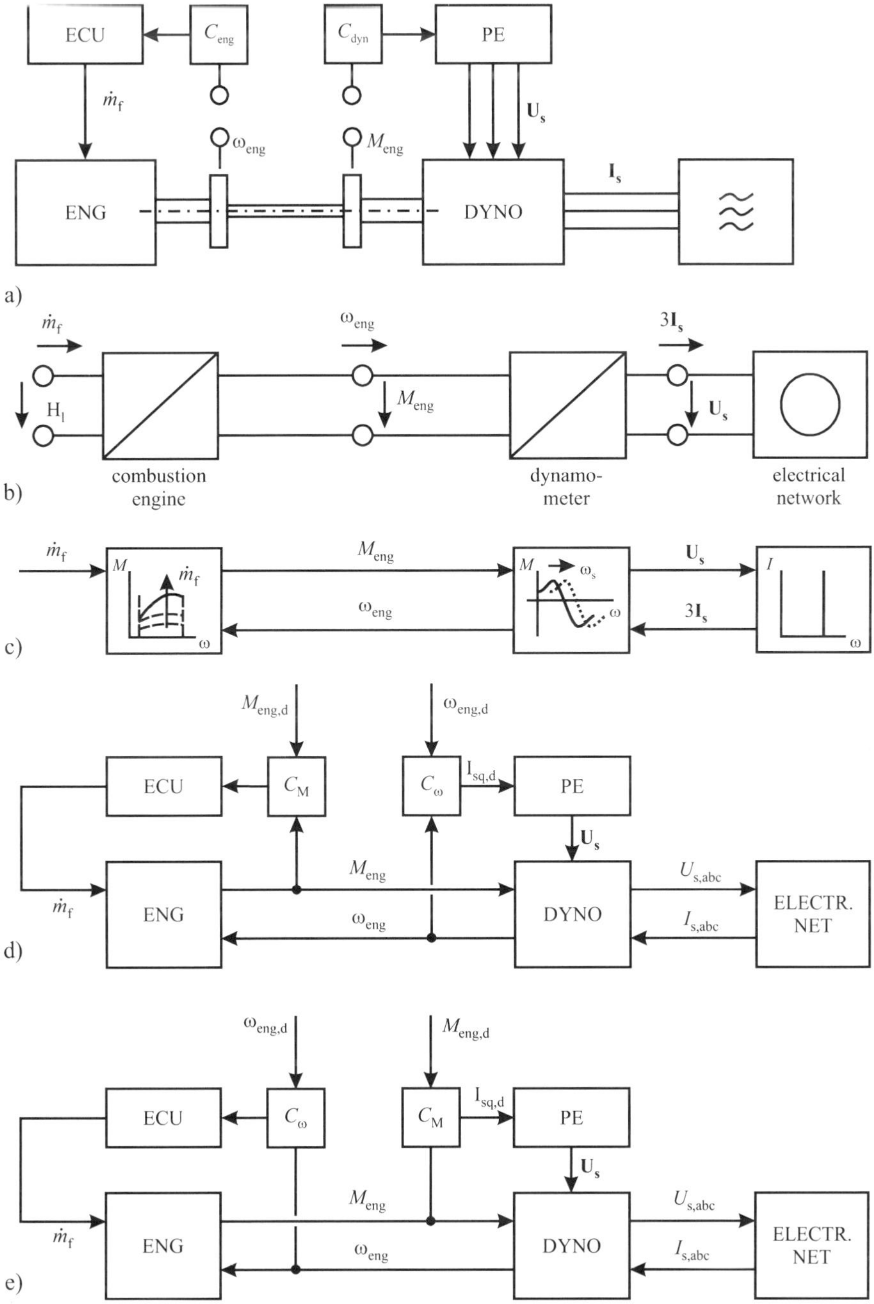

Fig. 6.10.1. Schematic representation of a combustion engine test bench with electrical dynamometer (field-oriented controlled AC-motor). ECU: Electronic Control Unit. PE: Power Electronics. **a** process scheme with controllers. **b** two-port process parts with flows and potentials. **c** two-port signal flow scheme, **d** speed control with dynamometer, torque control with engine ($n - \alpha$ control). **e** torque control with dynamometer, speed control with engine ($M - \alpha$ control).

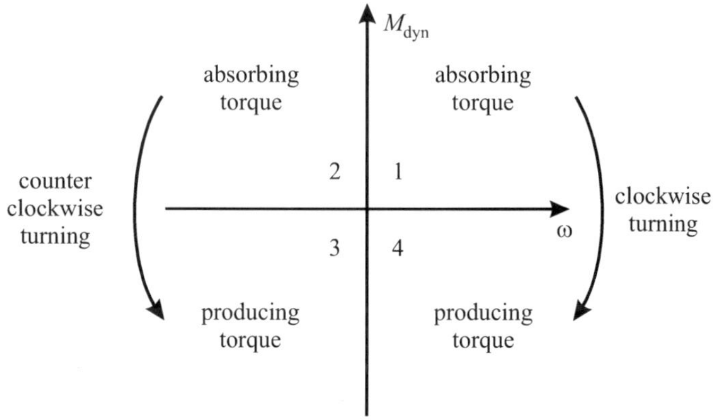

Fig. 6.10.2. Torque-speed quadrants of dynamometers.

6.10.2 Test-bench control

The main controlled variables of an internal combustion engine test bench are the speed ω_{eng} and the torque M_{eng}, besides many other variables for air and fuel conditioning, several auxiliary devices, etc. Figure 6.10.1 depicts a schematic of of a combustion engine with an electrical dynamometer, e.g. a speed controlled AC motor. The multi-port representation shows how the powers at the terminals (interfaces) $\dot{m}_{\text{f}}\, H_{\text{l}}$ for the fuel, $M_{\text{eng}}\, \omega_{\text{eng}}$ for the generated combustion engine power and the absorbed power $\text{Re}\{3\, \mathbf{U}_{\text{s}}\, \mathbf{I}_{\text{s}}\}$ of the dynamometer as a generator are linked together. It is assumed that the AC motor torque is controlled by the setpoint $I_{\text{sq,d}}$ (desired I_{sq}) respective the manipulated voltage U_{sq} of the field-oriented control, see e.g. Vas (1990), Leonhard (1996), Isermann (2005).

There are mainly the following possibilities to control the test bench

a) $n - \alpha$ control
   Speed control with dynamometer current $I = I_{\text{sq,d}}$, torque control with combustion engine throttle α
b) $M - \alpha$ control
   Torque control with dynamometer current $I = I_{\text{sq,d}}$, speed control with combustion engine throttle α
c) $n - x$ control
   Speed control with the dynamometer current $I = I_{\text{sq,d}}$, x-control with combustion engine throttle, x may be an internal engine variable, e.g. the intake pressure.

a) Test-bench models

The design of the test-bench controllers requires mathematical models of the controlled processes. Therefore, some basic models will be considered briefly.

- Elastic two-mass rotational system

It is assumed that the test bench can be represented by a two-rotational mass system with moment of inertia $J_{\rm eng} = J_1$ for the combustion engine, moment of inertia $J_{\rm dyn} = J_2$ for the dynamometer and torsional stiffness c of the shaft, damping coefficient d_s of the shaft and viscous friction coefficients $d_{\rm B1}$ and $d_{\rm B2}$ of the bearings, see Fig. 6.10.3.

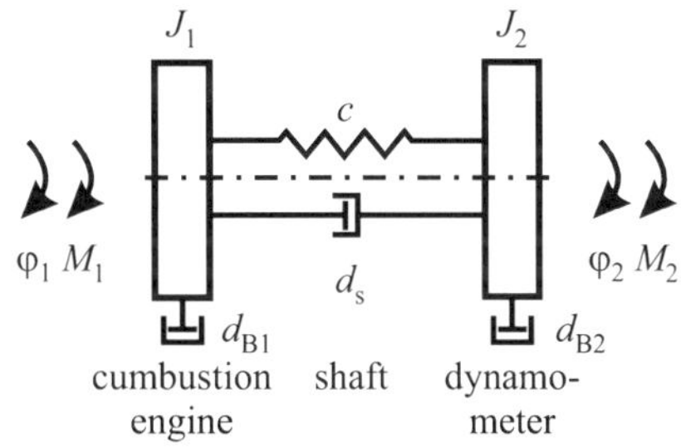

Fig. 6.10.3. Test bench as a two-mass torsional system.

For stating the differential equations, the two-mass system can be considered either as a free torsional oscillator where both rotational masses move freely, or as a fixed torsional oscillator where one of the rotational masses or the associated end of the torsion spring is fixed, see e.g. Isermann (2005).

For the free torsional oscillator the principle of angular momentum results after isolating the rotational masses and assuming $d = d_{\rm s} + d_{\rm B1} = d_{\rm s} + d_{\rm B2}$

$$J_1\ddot{\varphi}_1(t) = -c\,(\varphi_1(t) - \varphi_2(t)) - d\,(\dot{\varphi}_1(t) - \dot{\varphi}_2(t)) + M_1(t) \tag{6.10.1}$$

$$J_2\ddot{\varphi}_2(t) = -c\,(\varphi_2(t) - \varphi_1(t)) - d\,(\dot{\varphi}_2(t) - \dot{\varphi}_1(t)) + M_2(t). \tag{6.10.2}$$

A simplification can be made by introducing the torsional torque of the shaft

$$\begin{aligned} M_{\rm T} &= M_{\rm T1}(t) + M_{\rm T2}(t) \\ &= (c\,\varphi_1(t) + d\,\dot{\varphi}_1(t)) - (c\,\varphi_2(t) + d\,\dot{\varphi}_2(t)) \\ &= c\Delta\varphi^*(t) + d\Delta\dot{\varphi}^*(t) \end{aligned} \tag{6.10.3}$$

with the turning angle

$$\Delta\varphi^*(t) = \varphi_1(t) - \varphi_2(t). \tag{6.10.4}$$

A corresponding signal-flow diagram is depicted in Fig. 6.10.4

A state-space representation

$$\dot{\mathbf{x}}(t) = \mathbf{A}\mathbf{x}(t) + \mathbf{B}\mathbf{u}(t) \tag{6.10.5}$$

$$\mathbf{y}(t) = \mathbf{C}\mathbf{x}(t) \tag{6.10.6}$$

is obtained by defining the vectors

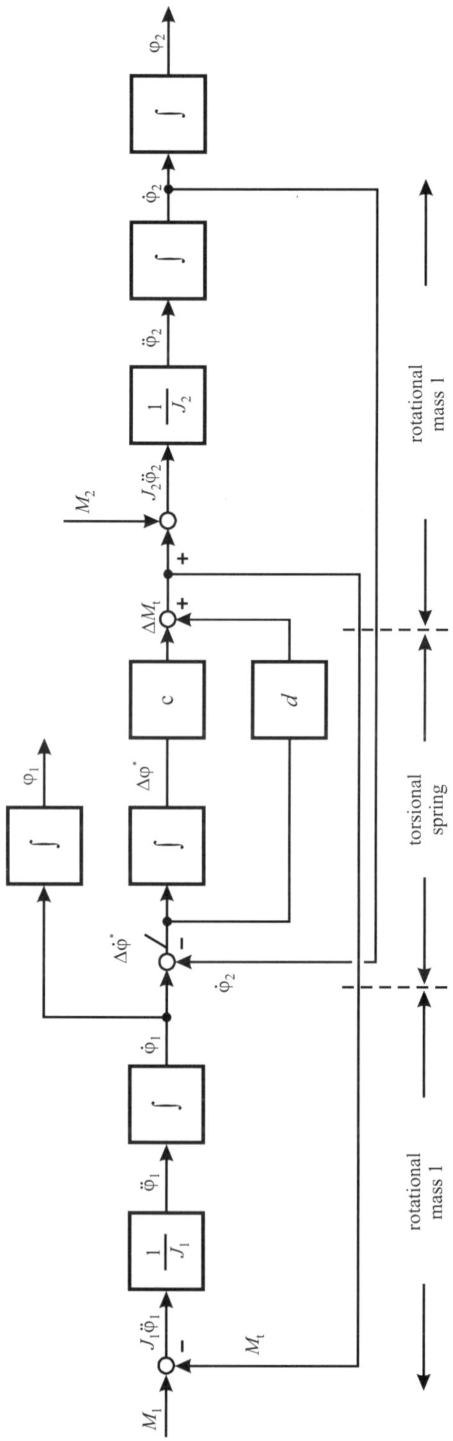

Fig. 6.10.4. Signal-flow diagram of the two-mass rotational system (test bench)

$$\mathbf{x}^T(t) = \left[\dot{\varphi}_1(t)\ \dot{\varphi}_2(t)\ \Delta\varphi^*(t)\right] \tag{6.10.7}$$

$$\mathbf{u}^T(t) = \left[M_1(t)\ M_2(t)\right] \tag{6.10.8}$$

$$\mathbf{y}^T(t) = \left[\dot{\varphi}_1(t)\ \dot{\varphi}_2(t)\right] = \left[\omega_1(t)\ \omega_2(t)\right] \tag{6.10.9}$$

with the parameters

$$\mathbf{A} = \begin{bmatrix} -\frac{d}{J_1} & \frac{d}{J_1} & -\frac{c}{J_1} \\ \frac{d}{J_2} & -\frac{d}{J_2} & \frac{c}{J_2} \\ 1 & -1 & 0 \end{bmatrix}; \quad \mathbf{B} = \begin{bmatrix} \frac{1}{J_1} & 0 \\ 0 & -\frac{1}{J_2} \\ 0 & 0 \end{bmatrix}; \quad \mathbf{C} = \begin{bmatrix} 1 & 0 & 0 \\ 0 & 1 & 0 \end{bmatrix}. \tag{6.10.10}$$

Oscillations of the free two-mass torsional system are obtained by considering the relative movements of the masses with the turning angle (6.10.4). Subtraction of the two equations (6.10.1) and (6.10.2) then yields

$$\frac{J_1 J_2}{J_1 + J_2}\Delta\ddot{\varphi}^*(t) + d\Delta\dot{\varphi}^*(t) + c\Delta\varphi^*(t) = \frac{J_2}{J_1 + J_2}M_1(t) - \frac{J_1}{J_1 + J_2}M_2(t) \tag{6.10.11}$$

with the effective rotational mass

$$J_{\text{eff}} = \frac{J_1 J_2}{J_1 + J_2} = \frac{1}{\frac{1}{J_2} + \frac{1}{J_1}} \tag{6.10.12}$$

the resulting undamped natural frequency

$$\omega_0 = \sqrt{\frac{c}{J_{\text{eff}}}} = \sqrt{\frac{c(J_1 + J_2)}{J_1 J_2}} \tag{6.10.13}$$

the damping ratio

$$\zeta = \frac{d}{2\sqrt{c}}\sqrt{\frac{J_1 + J_2}{J_1 J_2}} \tag{6.10.14}$$

the eigenfrequency

$$\omega_e = \omega_0\sqrt{1 - \zeta^2} \tag{6.10.15}$$

and the resonance frequency

$$\omega_{\text{res}} = \omega_0\sqrt{1 - 2\zeta^2}. \tag{6.10.16}$$

As the two-mass system can develop large oscillation amplitudes in the range of this resonance frequency which may brake the shaft, this frequency has to be avoided to be excited by the combustion engine. Therefore the stiffness c of the connecting shaft between the engine and the dynamometer has to be chosen appropriately. For example, if the ignition frequency is

$$f_j = \frac{\omega_{\text{eng}}}{4\pi} i_c\, j \tag{6.10.17}$$

with $\omega_{\text{eng}} = 2\pi n_{\text{eng}}$, i_c the number of cylinders and j the number of harmonics, see (4.2.8), then the resonance frequency $f_{\text{res}} = 2\pi\omega_{\text{res}}$ should be smaller than the first harmonic ignition frequency for the lowest operation speed, e.g. the idle speed

$$f_{\text{res}} < f_j. \tag{6.10.18}$$

As the undamped natural frequency $f_0 = 2\,\pi\,\omega_0$ is easier to calculate, one can require by setting $f_1 = \omega_{\text{eng}} i_c / 4\pi$ for the first harmonic

$$f_0 < \frac{n_{\text{eng}}}{2} i_c. \tag{6.10.19}$$

Example 6.5. A four cylinder engine with $i_c = 4$ has at idle with $n_{\text{eng}} = 700\,\text{rpm}$ the ignition frequency

$$f_1 = \frac{700}{2} 4 = 1400 \frac{\text{osc}}{\text{min}} \widehat{=} 23.3\,\text{Hz}.$$

The test bench yields with $c = 1000\,\text{Nm/rad}$, $J_1 = 0.13\,\text{kgm}^2$, $J_2 = 0.38\,\text{kgm}^2$ a natural undamped frequency of $f_0 = 16.2\,\text{Hz}$, which is just smaller than the ignition frequency.

□

- Stiff two-mass rotational system

If the exciting frequencies of loading the combustion engine with $\Delta M_1(t)$ are low compared to the natural angular frequency ω_0, as for driving cycle simulation, the shaft can be assumed as stiff. The two-mass system is then described by

$$(J_1 + J_2)\,\ddot{\varphi}_1(t) + (d_{\text{B1}} + d_{\text{B2}})\,\dot{\varphi}_1(t) = M_1(t) + M_2(t) \tag{6.10.20}$$

where d_{B1} and d_{B2} are bearing friction coefficients. For small changes around an equilibrium $\bar{\dot{\varphi}}_1$, $\bar{M}_1$ and $\bar{M}_2$ it follows for the process to be controlled by torque changes of the dynamometer or the combustion engine

$$(J_1 + J_2)\Delta\dot{\omega}_1(t) + d_{\text{B}}\Delta\omega_1(t) = \Delta M_1(t) + \Delta M_2(t) \tag{6.10.21}$$

with $d_{\text{B}} = d_{\text{B1}} + d_{\text{B2}}$ and the transfer function becomes

$$G_{\omega 1\text{M}}(s) = \frac{\Delta\omega_1(s)}{\Delta M_2(s)} = \frac{\Delta\omega_1(s)}{\Delta M_1(s)} = \frac{1}{(J_1 + J_2)s + d_{\text{B}}} = \frac{K_{\omega\text{M}}}{1 + T_{\omega\text{M}} s}. \tag{6.10.22}$$

Hence, the process for speed control is approximately a proportional acting process of first order with gain and time constant

$$K_{\omega\text{M}} = \frac{1}{d_{\text{B}}}; \qquad T_{\omega\text{M}} = \frac{J_1 + J_2}{d_{\text{B}}}. \tag{6.10.23}$$

If the counteracting torque of the driven load, e.g. the combustion engine, changes linearly according to

$$\Delta M_1(s) = -k_{\text{M}\omega}\,\Delta\omega_1(s) \tag{6.10.24}$$

the differential equation (6.10.20) becomes

$$(J_1 + J_2)\Delta\dot{\omega}_1(t) + (d_{\rm B} + k_{{\rm M}\omega})\Delta\omega_1(t) = \Delta M_2(t). \tag{6.10.25}$$

Gain and time constant then decrease to

$$K'_{\omega{\rm M}} = \frac{1}{d_{\rm B} + k_{{\rm M}\omega}}; \qquad T'_{\omega{\rm M}} = \frac{J_1 + J_2}{d_{\rm B} + k_{{\rm M}\omega}}. \tag{6.10.26}$$

In the case of driving cycle simulation the engine torque is applied by the vehicle simulation program. Then the engine should only accelerate its own moment of inertia $J_1 = J_{\rm eng}$ and that of the drive train and vehicle. This means that the moment of inertia $J_2 = J_{\rm dyn}$ of the dynamometer (which is not part of the vehicle) has to be accelerated by an additional compensation torque $\Delta M_{2,\rm comp}$ of the dynamometer. Simplifications of (6.10.22) by assuming $d_{\rm B} = 0$ yields

$$\Delta\omega_1(s) = \frac{1}{(J_1 + J_2)s}\left(\Delta M_1(s) + \Delta M_{2,\rm comp}(s)\right) \tag{6.10.27}$$

which has to be identical to

$$\Delta\omega_1(s) = \frac{1}{J_1 s}\Delta M_1(s). \tag{6.10.28}$$

This is satisfied by a compensation torque

$$\Delta M_{2,\rm comp} = J_2 s \Delta\omega_1 \tag{6.10.29}$$

which has to be generated by the dynamometer, Schmidt and Kessel (1999). This means that instead of an engine torque ΔM_1 the corrected torque

$$\Delta M_{1,\rm corr} = \Delta M_1 + J_2 \Delta\dot{\omega}_1 \tag{6.10.30}$$

has to be applied by the engine on the dynamometer in the case of acceleration in order to compensate for the dynamometer inertia.

b) Test-bench control

For the speed control of a dynamometer according to Fig. 6.10.1d) a PI controller can be applied, assuming $\Delta\omega_1 = \Delta\omega_2$,

$$G_c(s) = \frac{\Delta I_{\rm sq,d}(s)}{\Delta\omega_1(s)} = K_c\left(1 + \frac{1}{T_I s}\right) = \frac{K_c}{T_I s}(T_I s + 1) \tag{6.10.31}$$

which yields a closed-loop transfer function for changes of the speed-reference variable

$$G_{r\omega}(s) = \frac{\Delta\omega_1(s)}{\Delta\omega_{1d}(s)} = \frac{2 + T_I s}{\frac{T'_{\omega\rm M} T_{\rm I}}{K'_{\omega\rm M} K_{\rm c}} s^2 + \left(\frac{1}{K'_{\omega\rm M} K_c} + 1\right) T_I s + 1}. \tag{6.10.32}$$

Hence, a second order behavior results with undamped natural frequency

$$\omega_0 = \sqrt{K'_{\omega\mathrm{M}} K_{\mathrm{c}} / T'_{\omega\mathrm{M}} T_F} \tag{6.10.33}$$

and gain $G_{r\omega}(0) = 1$. If the corrected torque (6.10.30) is added, the controller (6.10.31) becomes of PID type

$$G_{\mathrm{c,corr}}(s) = K_{\mathrm{c}} \left(1 + \frac{1}{T_{\mathrm{I}} s}\right) + k_{\mathrm{d}} s \tag{6.10.34}$$

where the parameter k_{d} of the derivative function depends on J_2, the q-stator current component I_{sq} of the AC dynamometer and corresponding induction motor parameters.

Limits of the test-bench dynamics with regard to control are discussed in Grünbacher et al (2006).

6.11 Model-based control-function development with special design and simulation tools

After describing the various methods for the development of control functions and their optimization and calibration special simulation and computers are considered, which support the control function development as well as calibration and software testing. This is part of the control system integration, represented in the right branch of the V-development model in Fig. 6.1.1, and will be considered briefly in this section.

6.11.1 Model-in-the-loop simulation and control prototyping

The description of the overall procedure for engine control development in Sect. 6.1 has shown that different types of simulations are used for the development of control functions and control software, see Figs. 6.1.2 and 6.1.3.

A design of new control functions in an early development phase may be based on simulations with an engine model and an ECU model, i.e. control algorithms in a high-level language, as e.g. MATLAB-Simulink. This is called *model-in-the-loop simulation* (MiL). Both, the ECU and the engine are then represented as a model, i.e. a virtual picture of the real parts, see Fig. 6.11.1a).

If some control functions of a real development ECU can already be applied to the real engine on a test bench, because the real time functions from a former, similar engine can be used, some new control functions may be tested as prototypes with a special real-time computer in parallel to the ECU. This is called *rapid control prototyping* (RCP). Frequently the new control functions operate in a *bypass mode* and use the interfaces of the ECU to the sensors and the actuators, see Fig. 6.11.1b). The computing power for the experimental RCP computer exceeds that of the ECU and operates with a high level language. Thus, the new control functions do not have to be implemented in machine code within the limited computer power and fix point restrictions of an ECU. This may save considerable development time by trying and testing new functions directly on a higher software level with the real engine. If a development ECU is not available, a powerful real-time computer can be used if the required sensors and actuators interface are implemented. It is then called *fullpass mode*, Schäuffele and Zurawka (2005).

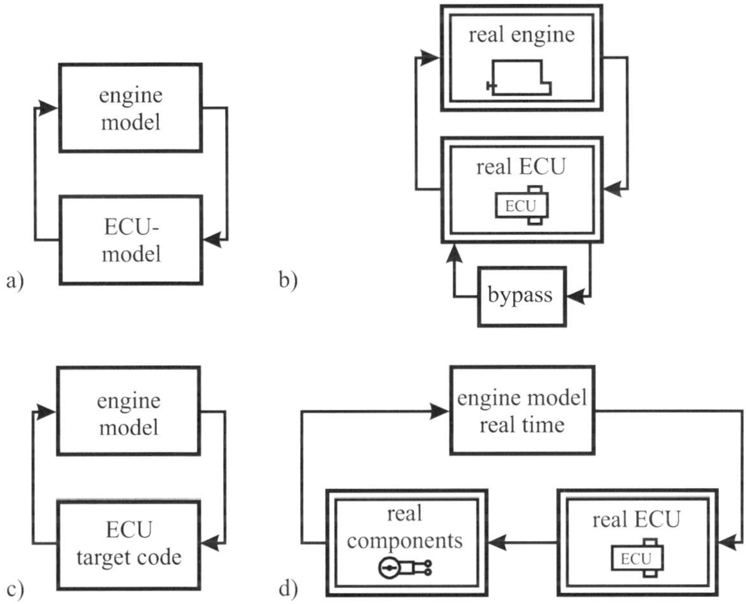

Fig. 6.11.1. Different simulation and prototyping methods for control functions and software development. **a** Model-in-the-loop (MiL) simulation. **b** Rapid control prototyping (RCP). **c** Software-in-the-loop (SiL) simulation. **d** Hardware-in-the-loop (HiL) simulation.

6.11.2 Software-in-the-loop and hardware-in-the-loop simulation

Existing engine models in high level language can be used for the validation of software functions as test candidates in an early development phase by *software-in-the-loop simulation* (SiL), see Fig. 6.11.1c). The software functions may already be implemented with fix point arithmetic and required interfaces, before they are implemented on the target ECU. Real-time behavior is not required.

For a final validation of control functions the target ECU with its interfaces has to cooperate with real signals. In order not to use real engines on expensive test benches, real time engine models are implemented in a powerful development computer. The sensor signals may be generated by special electronic modules and the output signals are frequently transferred to real actuators, like an electrical throttle, injection system or EGR-valve motor. Thus, the real ECU with implemented software operates with some real components, but with simulated real-time high performance engine models and is known as *hardware-in-the loop simulation* (HiL), see Fig. 6.11.1d). The advantages are that, e.g. software functions can be tested under real-time constraints, validation tests are reproducible and can be automated, critical boundary conditions (high speed and high load) can be realized without being dangerous, the reaction to faults and failures can be investigated, on-board diagnosis functions can be tested etc., compare Sinsel (2000), Schaffnit (2002), Zahn (2012).

This short summary of simulation and prototyping methods shows how the development of ECU control functions can be supported by using programmed dynamic engine models of different granularity and different stages of the development of the control software, see also Fig. 6.1.2. It is a basis for *virtual engine control development*. The computer-based and model-based development can also be used for a part of control calibration, as indicated in Fig. 6.1.3.

6.11.3 Examples for the HiL Simulation

a) Truck diesel engine drive train

Figure 6.11.2 shows the setup of the HiL-simulation test bench for the investigation of the engine control system of a truck diesel engine and drive train, Sinsel (2000), Isermann and Hafner (2001). It is subdivided in the following parts:

- real-time computer system including I/O modules
- periphery, consisting of the sensor and actuator interface
- PLD (pumpline nozzle injection) control unit including real actuator components, stand-alone or in combination with the real FMR-ECU
- PC with graphical user interface
- control panel

a) *The HiL-simulator*
The real-time computer system for the simulator is based on a dSpace system equipped with digital signal processors and a DEC Alpha processor. It also offers the possibility to realize all the models in MATLAB/Simulink^{TM} and to use all the benefits of a graphical simulation environment. Special I/O-modules (digital I/O-module, D/A converter, A/D converter and CAN-interface) are used for the communication with the periphery.
The coupling of the simulator and the control unit is implemented with special periphery which can be subdivided into a sensor and an actuator interface. The *sensor interface* generates the necessary sensor signals like temperatures and pressures. The pulses of the camshaft and crankshaft inductive speed sensors are generated with a board specifically designed for high speed generation. For that purpose a look-up table with the pulse-signals versus the crankshaft-angle is stored off-line in the memory. During the simulation, the signals are periodically read out, synchronous to the simulated engine speed. This realization guarantees a high flexibility in forming the pulses and adapting different gear wheels. The sensor interface also contains a relay electronic to simulate sensor faults like interruptions and short circuits.

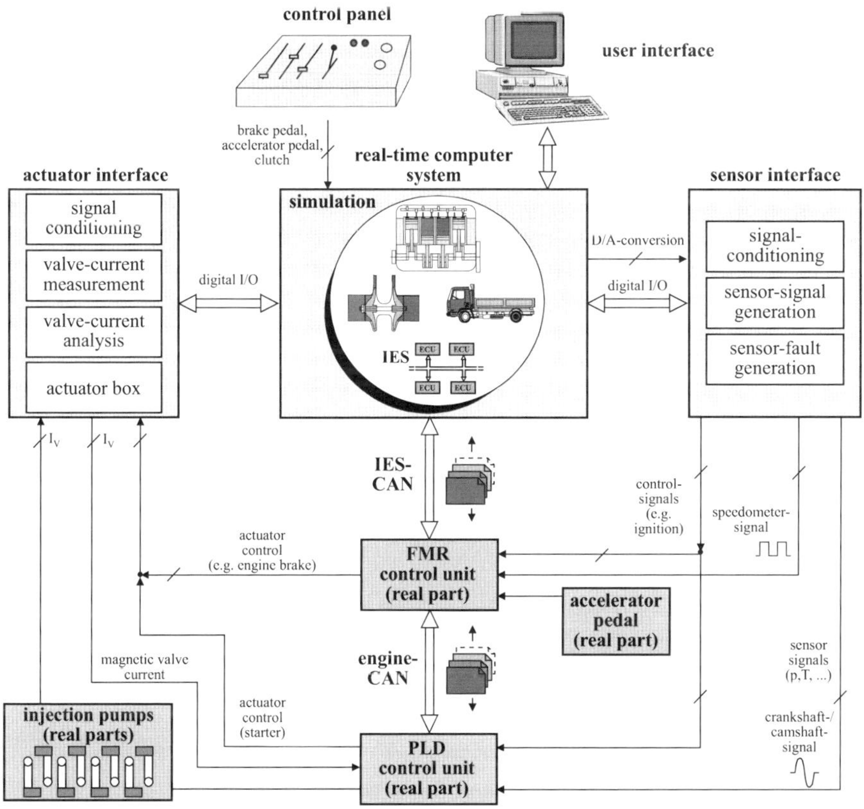

Fig. 6.11.2. Hardware-in-the-Loop test bench with simulated engine and vehicle and with real engine control unit and real injection actuators for Mercedes-Benz series 500 and 900, Sinsel (2000). FMR: vehicle management system; PLD: pump line nozzle.

The actuator interface mainly consists of the injection pumps (pump-line-nozzle injection system) which are integrated in the simulation test bench as real components, because the combination of the PLD control unit and the injection pumps represent a mechatronic unit which is difficult to model. A special electronic device measures the magnetic valve currents to reconstruct the real valve opening time and to determine the pulse width and the beginning of injection. These quantities are transferred to the real-time computer system for engine simulation. This way the real behavior of the injection pump is included.
The simulator test bench was set up with the objective of testing the PLD control unit stand alone or in combination with the FMR control unit. In the first case, the necessary FMR functions are simulated by the computer system. The data transfer is done via the engine CAN-bus. In the second case the PLD and the FMR control units are connected directly via the engine CAN-bus. The computer

system emulates other integrated electronic systems (IES) in this operation mode by transmitting the data via the IES-CAN-bus to the FMR.
For an efficient use of the simulator it is performed with a windows user interface on the host-PC which copies the functionality of a real truck-cockpit. All relevant simulation quantities can be visualized on-line or be recorded for off-line analysis. To ensure reproducible results a driver simulation is implemented which can automatically follow a given speed cycle by operating the accelerator pedal, brake, clutch and gear. As alternative, an interactive "driving" of the simulator can be performed manually with a control panel where the most important cockpit functionalities are realized.

b) *Simulation results*
In the following two HiL simulation examples for an 8-cylinder truck engine (420 kW) are represented in order to document the applicability and the performance of the simulator, using combustion models, Isermann (1999). Figure 6.11.3 demonstrates the effect of switching off a single injection pump valve. The gearbox is in neutral position and at the beginning the engine runs with idle speed. The cyclic decrease of the engine torque and the engine speed after the fuel shut off can directly be seen. The control unit gradually compensates for the missing torque of one cylinder by increasing the pulse width in order to keep the desired idle speed.
A full power acceleration of a 40 tons truck including two gear shifts (1) is depicted in Fig. 6.11.4. This figure shows the following effects:
- drive train oscillations in the drive train during acceleration (2),
- soot-limitation by limiting the A/F-ratio through reduced injection mass (3),
- speed-limitation-control at $2000\,\mathrm{rpm}$ by reducing injection mass (4)
- lagging reaction of the charging pressure because of the inertness of the (fixed) turbocharger (5).

b) Car diesel engine

Some signals for the HiL simulation of a car diesel engine in closed loop with the ECU are depicted in Fig. 6.11.5. The course of the cylinder pressure is shown for about $2\,\mathrm{s}$ and illustrates the induced oscillations of the air flow rate and some pressures. The air flow rate $\dot{m}_{\mathrm{air}}$ and the pressure p_1 at the intake inlet have a phase shift of about half a period. The pressure p_1 has a smaller amplitude than the boost pressure $p_{2\mathrm{i}}$ and a considerable phase shift, which originates from the inlet valve opening. The amplitude of the exhaust pressure p_3 is larger than that of the pressure p_4 after the turbine and shows also a small phase shift.

6.12 Control software development

According to the V-development model in Fig. 6.1.1, the first steps are the design control development and the control functions development in high-end software

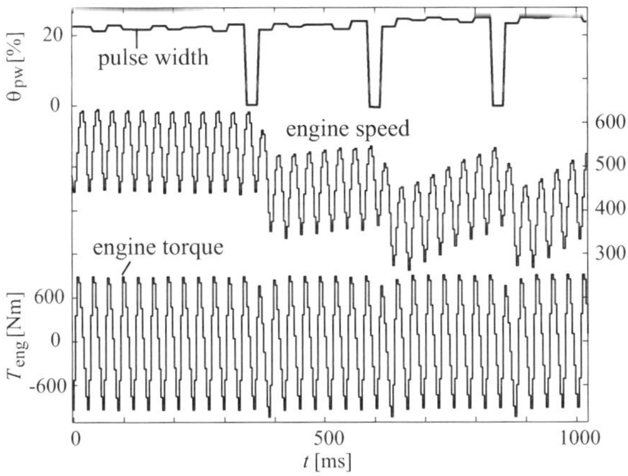

Fig. 6.11.3. HiL simulation of a single injection pump valve cut off with idle speed control.

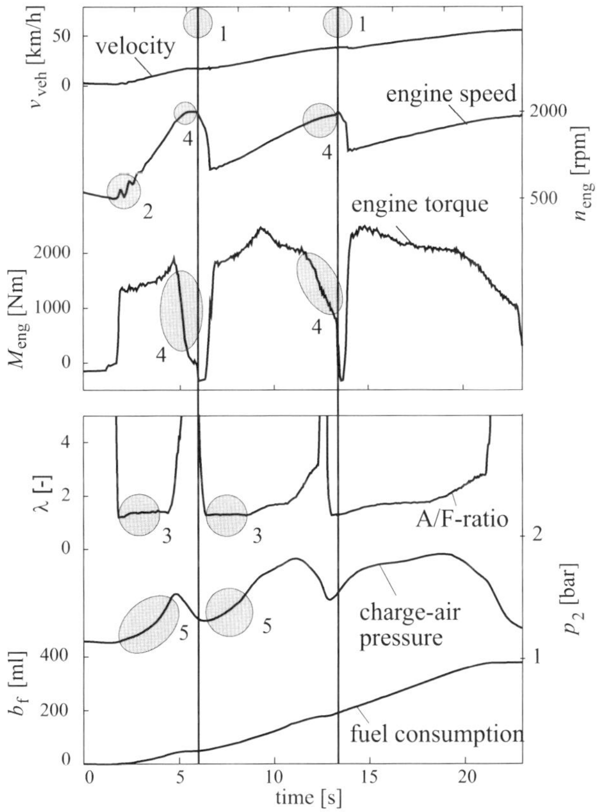

Fig. 6.11.4. HiL simulation of a full power acceleration of a 40 ton truck.

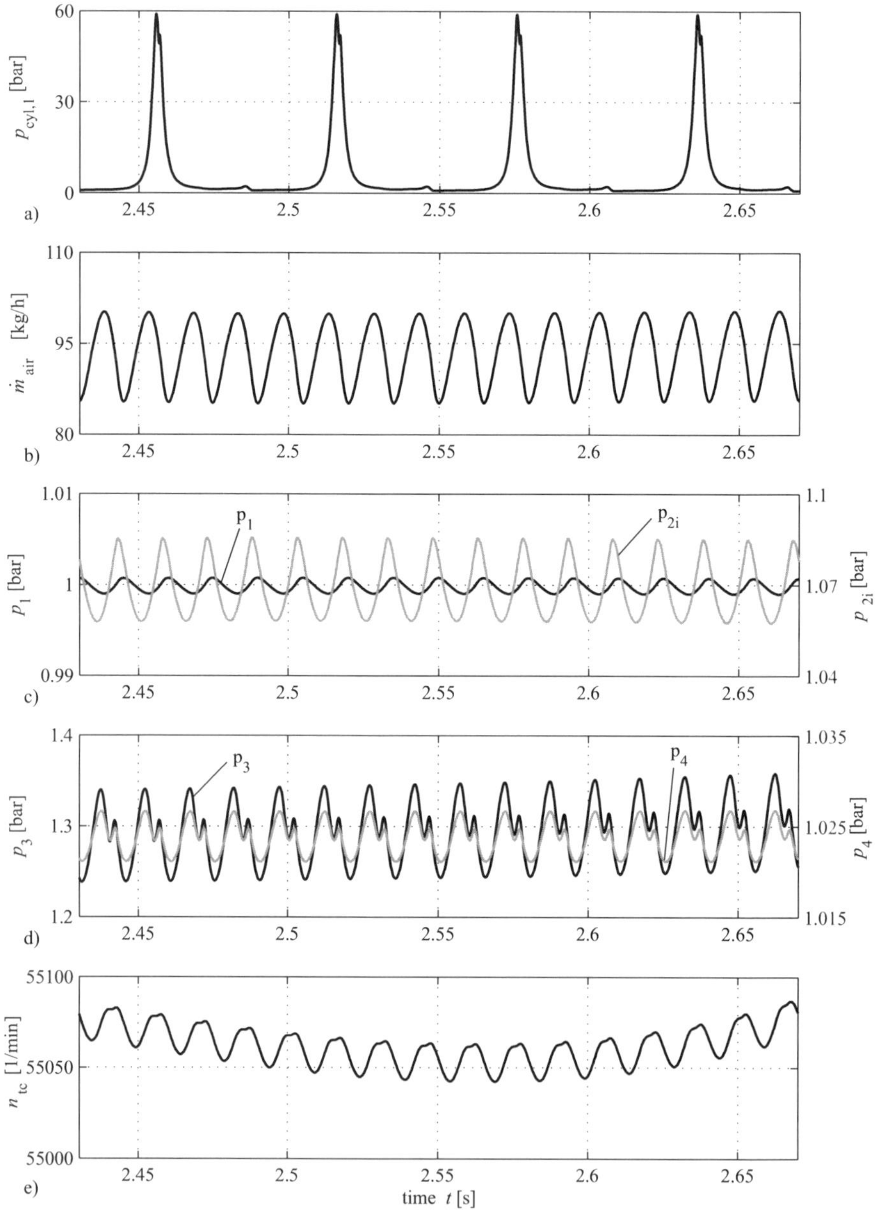

Fig. 6.11.5. HiL simulation of the stationary behavior of an 1.9l, 77 kW, 255 Nm diesel engine for $n_{\mathrm{eng}} = 2000\,\mathrm{rpm}$, $\alpha_{\mathrm{ped}} = 20\,\%$, Zahn (2012).

(e.g. MATLAB Simulink™and Stateflow™). The next steps are then the *control software development* and the *software implementation on the ECU* for series production. This is also depicted as part of the overall workflow in Fig. 6.1.2, using special software tools and computers for code generation and testing with compilers and real-time simulation. In the following, some remarks are given for the software architecture code generation and software testing.

6.12.1 Software architecture

The design of the software architecture has to consider many aspects from software development to the requirements of the target microprocessor and includes connected modules which have to be flexible with regard to continuous changes and variants. Several software layers have to be defined. The minimum is two layers, a platform software and an application software, as shown in Fig. 6.12.1, Schäuffele and Zurawka (2005). The *platform software* is oriented to the ECU and comprises the operating system, communications and network management according to OSEK/VDX (2005) standards and diagnostic protocols. OSEK stands for "Open Systems and Interfaces for Automotive Electronics" and is the result of a committee of major automotive manufacturers and component suppliers to support portability and reuseability of application software under real time constraints, started in 1993. It also contains standardized flash memory programming procedures. The standardization of the platform software is additionally advantageous during the software development with regard to software changes and parametrization. Interfaces for measurement and calibration via CAN protocols support the development phase as well, Borgeest (2008), Zimmerschied et al (2005). A hardware abstraction layer (HAL) gives access to the peripheral components of the ECU and is specified for the used microprocessors.

The *application software* can be designed by the engine or vehicle manufacturer and contains engine specific functions. Standardization takes place for control functions, ranging from look-up tables and their interpolation to dynamic control algorithms. The standardization is, e.g. treated in the MSR-MEGMA working group and ASAM (2012).

The *configuration* of standardized software components allows a specific application by using configuration tools. An automated configuration comprises, e.g. the handling of signals, messages, buses, nodes and functions. It may contain export and import interfaces with data exchange formats and a documentation interface. More details like data models for engine variants, storage in volatile (RAM) or nonvolatile memories (ROM, PROM, EPROM or Flash memory) and description files for data structure can be found, e.g. in Schäuffele and Zurawka (2005).

Activities for an open industry standard for the automotive software architecture between suppliers and manufacturers are going on in the AUTOSAR consortium (AUTomotive Open System ARchitecture) since 2003, AUTOSAR (2012), Heinecke et al (2004), ATZ extra (2013). One of the aims is an open and standardized automotive software architecture. The standard includes specifications describing software architecture components and defining their interfaces.

The AUTOSAR architecture separates the basis software from the application software and connects them by standardized interfaces, see Fig. 6.12.2. To master the complexity, several layers are defined, Wernicke and Rein (2007). The connection to the microcomputer is provided by the lowest level, the *microcontroller abstraction layer*. Here, the interfaces are defined to the memories, the I/O-drivers, their communication and to additional features which are not part of the microcontroller. The second layer is the *ECU abstraction layer*, comprising the hardware design of the ECU including driver to external components. The *service layer* at the third level provides basic software modules like the operating system, memory administration and bus communication. This layer is relatively independent on the ECU hardware. The fourth level is the *runtime environment* (RTE), which separates the basis software and application software and carries out the data exchange in both directions. Therefore, the application software components have standardized interfaces to the RTE. The RTE also integrates the application software components (SWCs), see Fig. 6.12.3. This separation and integration with standardized interfaces enables a hardware independent software development. The application software components can therefore be transferred to other ECUs and reused.

A *virtual function bus* (VFB) connects the various software components during the *design* and allows a configuration independent of a specific hardware. Thus the SWCs are runnable entities and can be linked together for development and testing. An exchange of information becomes possible through standardized software input and output ports. Thus a validation of the interaction of the SWCs and interfaces is possible before software implementation.

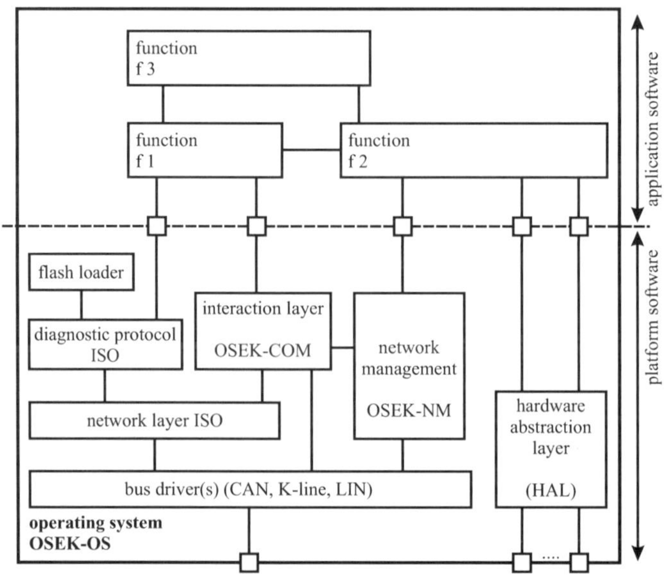

Fig. 6.12.1. Software architecture composed of standardized software components, Schäuffele and Zurawka (2005).

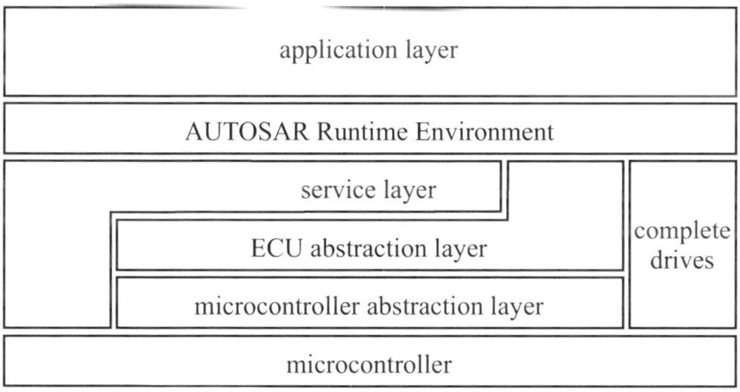

Fig. 6.12.2. AUTOSAR layer structure of automotive software, Wernicke and Rein (2007).

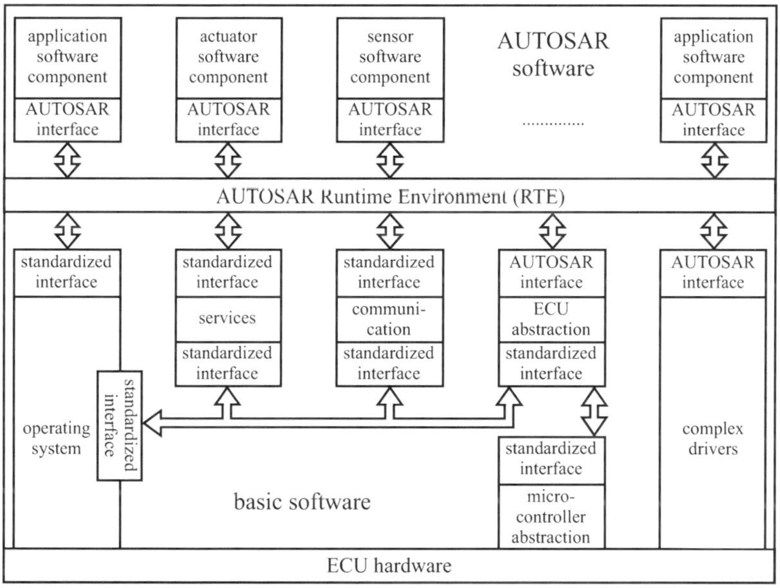

Fig. 6.12.3. AUTOSAR software architecture components and interfaces (RTE: runtime environment), Kirschke-Biller (2011).

6.12.2 Code generation

The control software development can start if the control functions are ready and available as function blocks in a high level software platform like MATLAB™ (2011), Simulink™ (2011) and Stateflow™ (2011) (The MathWorks (2011)). MATLAB™ is broadly used as an integrated function development environment for numerical calculations with a large library of mathematical design and analysis programs and special tool boxes. Simulink™ is an interactive development environment for modeling, analysis and simulation with a graphical interface and is integrated into MATLAB™. It allows the handling of models and functions with block diagrams. Stateflow™ is an expansion of Simulink™ to operate with discrete event state charts and flow diagrams.

Based on the block-oriented control functions of this control-analysis- and simulation-oriented development environment, the series production code for the ECU is developed by using special tools. The control functions are specified in graphical form and they are converted in production C-code which runs on the target processor. Thereby it is intended to reach a minimum of execution time, RAM and ROM resources and stack size, compared to human programmer's abilities.

The process of code generation is performed in following steps, Kiffmeier et al (1999), using a Software Development Tool (SDT) like Targetlink™ from dSpace or ASCET™ from ETAS or Real-Time Workshop™ from The MathWorks.

(1) The *control function block* from Simulink has to be replaced by a corresponding SDT block, taken from a block library. These blocks manage data which is used for the production code, as scaling parameters and data types.
(2) In the case of fixed-point arithmetic *scaling parameters* are required. Therefore maximal and minimal values of variables have to be known. They can be obtained from model-in-the-loop simulations to determine the range of variables, in order to prevent overflows. The scaling can be done manually or automatically by simulation. For floating-point arithmetic no scaling is needed.
(3) *Additional information* for production code generation has to be specified. This belongs to partitioning of models or algorithms for e.g. look-up tables handling, like interpolation routines.
(4) *Off-line simulations* on the host PC are performed with Simulink to detect problems with fixed-point arithmetic. The simulations are run with floating-point arithmetic.
(5) The production C code is translated for the *target microprocessor* and loaded on an evaluation board. A communication link between the evaluation board and the host PC allows to test the microcontroller code together with a Simulink simulated process model. The code is readable by humans and the development is paralleled by a documentation, which is automatically generated. Information on code size, execution times and used RAM and ROM resources are also given. The result is a general portable ANSI C-code which runs on many microcomputers. However, some specific microcomputer adaptation always has to be taken into account. If manual coding has to be applied additionally, rules given by MSRA-C have to be followed. Special toolsets allow to analyze the timing behavior of the real-time operating system application OSEK, providing run-time performance and possible modifications, before coding.

As the ECU-code must be prepared for the task of calibration, the variables are presented in standardized ASAM file format via a data dictionary. Further standards to be considered are OSEK/VDX and AUTOSAR. AUTOSAR structure elements are for example runnables, ports, and communication interfaces.

Special computers for simulation and control prototypes and software tools for control software development are compiled in the catalogues dSpace (2011) and ETAS (2011).

As modern automotive control systems are interconnected by onboard data buses like CAN, Flexray, LIN also the real-time cooperation of the engine ECU with other vehicle ECU's like for transmission control, traction control (TCS), electronic stability control (ESC), and adaptive cruise control (ACC) has to be designed and monitored.

6.12.3 Software testing

The *software testing* is part of the verification comprised in the V-model for the development, see Fig. 6.1.1. *Verification* means that the developed functions meet the specifications, which are stated at the project start by the customer, by standards or legal regulations. More general is the *validation* where the final product is checked with regard to the overall requirements, Balzert (1998), Tran (2007), see the V-model in Sect. 6.1.1. Validation ensures that the final product meets the user's needs and includes that the specifications are correct, see e.g. IEEE-STD 610.

Software testing can be understood as part of *analytical quality assurance* and contains the testing of the software. This quality assurance can be divided in analyzing methods (static) and testing methods (dynamic), Liggesmeyer (2002), Thaller (2002). The *static, analyzing methods* can be applied early because no running software is required. Methods are manual inspection, visualization and data flow analysis, usually by different persons. *Dynamic testing methods* operate with the running software and it is checked if the functions meet the specifications. One distinguishes functional tests which check the input/output behavior and consider the software program as a black-box and structural tests where the program is considered as a white-box. These structural tests can be divided in symbolic tests, diversification tests and mutation tests, Schäfer (2012).

An application of software testing during early phases of the development avoids too many iteration cycles. In order to integrate hardware components, hardware-in-the loop simulation (HiL) is used. Then the real-time software on the target ECU is tested with simulated real-time models of the engine or the vehicle and real components like actuators or injection systems in the laboratory, such avoiding tests with the real engine or vehicle, see Sect. 6.11.2.

Table 6.12.1 gives a summary of the various development test stands and tools for the control development. Engine modeling is required in the beginning and performed on-line and off-line with the engine test bench. Control function development is mainly elaborated at test benches and with MiL, HiL and RCP computers. Software development is usually done off-line with PC's and an ECU evaluation board. Fine-tuning for driveability and the emission certification are worked out with the real vehicle.

Table 6.12.1. Survey of teststands and tools for engine control development

Control development steps		Development test stands + tools								
		engine test bench			MiL	HiL	RCP	ECU	vehicle (with transmission)	
		static	dynamic		SiL	with real		evaluation	roller	vehicle
		real engine	real engine	real engine & simulated drive train	(PC)	components		board	dynamometer, simulated vehicle driving forces	on road
Engine modeling	• physical				x	x				
	• experimentally									
	– stationary	x								
	– dynamic		x	x						
control function development	• simulation				x	x				
	• development	x	x	x	x		x			
	• optimization	x	x	x			x		x	x
ECU software	• development				x					
	• testing				x	x				
	• target processor							x		
Driveability				x					x	x
Emission certification									x	x

References

Ackermann J (1983) Abtastregelung, vol 1 and 2, 2nd edn. Springer, Berlin
Antonion A (2006) Digital signal processing. McGraw Hill, New York
Åström K, Wittenmark B (1997) Computer-controlled systems – theory and design. Prentice Hall, Upper Saddle River
ATZ extra (2013) 10 years AUTOSAR. The worldwide automotive standard for E/E-systems. Special issue. Automobiltechnische Zeitschrift. Springer Vieweg/Springer Fachmedien Wiesbaden GmbH, Wiesbaden
AUTOSAR (2012) Automotive Open System Architecture. www.autosar.org
Bäck T (1996) Evolutionary Algorithms in Theory and Practice: Evolution Strategies, Evolutionary Programming, Genetic Algorithms. Oxford University Press, New York
Balzert H (ed) (1998) Lehrbuch der Software-Technik, vol 2. Spektrum Akademischer Verlag, Heidelberg
Böhm B (1979) Guidelines for verifying and validating software requirements and design specifications. Euro-IFIP, North-Holland
Borgeest K (2008) Elektronik in der Fahrzeugtechnik. Vieweg, Wiesbaden
BRD (ed) (1997) V-Modell-Entwicklungsstandard für IT-Systeme des Bundes. http://www.v-modell.iabg.de/um97.htm
BRD (ed) (2004) V-Modell-XT. http://v-modell.iabg.de
Bröhl AP (ed) (1995) Das V-Modell - Der Standard für Softwareentwicklung, 2nd edn. Oldenbourg, München
Cuesta F, Ollero A (2004) Fuzzy control of reactive navigation with stability analysis based on conicity and Lyapunov theory. Control Engineering Practice – CEP 12:625–638
Cunningham E (1992) Digital filtering – an introduction. Houghton Mifflin, Boston
Davis L (1991) Handbook of Genetic Algorithms. Van Nostrand Reinhold, New York
Dennis J, Schnabel R (1996) Numerical methods for unconstrained optimization and nonlinear equations. Classics in applied mathematics 16, Siam, Philadelphia
DIN 19226 (1994) Control technology. Terminology. Beuth Verlag, Berlin
Dorf R, Bishop R (2010) Modern control systems, 12th edn. Prentice Hall, Englewood Cliffs
Droeschel W, Wiemers M (eds) (1999) Das V-Modell 97 – Der Standard für die Entwicklung von IT-Systemen mit Anleitung für den Praxiseinsatz. Oldenburg Verlag, München
dSpace (2011) dSpace catalogue 2011. dSpace GmbH, Paderborn
ETAS (2011) ETAS catalogue 2011. ETAS GmbH, Stuttgart
Fink A (2006) Nonlinear control based on local linear neuro-fuzzy models. Dissertation Technische Universität Darmstadt. Fortschr.-Ber. VDI Reihe 8, 1096. VDI Verlag, Düsseldorf
Fisher D, Seborg D (1976) Multivariable computer control. North Holland, Amsterdam
Föllinger O (2013) Regelungstechnik, 11th edn. VDE Verlag, Offenburg

Franklin G, Powell J (2006) Feedback control of dynamic systems, 5th edn. Pearson-Prentic-Hall, Upper Saddle River

Gen M, Cheng R (1997) Genetic algorithms and engineering design

Gill PE, Murray W, Wright MH (1981) Practical optimization. Academic press, Salt Lake city, UT

Goldberg D (1989) Genetic Algorithms in Search, Optimization, and Machine Learning. Addison-Wesley Professional, Reading, MA

Grünbacher E, Langthaler P, del Re L, Kokal H, Schmidt M, Paulweber M (2006) Goal oriented experiment planning for the optimal use of dynamical engine test benches. MTZ worldwide 67:28–31

Hafner M (2002) Modellbasierte stationäre und dynamische Optimierung von Verbrennungsmotoren am Motorenprüfstand unter Verwendung neuronaler Netze. Dissertation Technische Universität Darmstadt. Fortschr.-Ber. VDI Reihe 12, 482. VDI Verlag, Düsseldorf

Hafner M, Schüler M, Nelles O, Isermann R (2000) Fast neural networks for diesel engine control design. Control Engineering Practice – CEP 8:1211–1221

Heinecke H, Schnelle KP, Fennel H, Bortolazzi J, Lundh L, Leflour J, Maté JL, Nishikawa K, Scharnhorst T (2004) AUTomotive Open System ARchitecture-an industry-wide initiative to manage the complexity of emerging automotive E/E-architectures. In: SAE 2004 Convergence, pp 325–332

Hoschek J, Lasser D (1993) Fundamentals of computer-aided geometric design. AK Peters, Wellesley, MA

Ingber L (1996) Adaptive simulated annealing (ASA): Lessons learned. Control and Cybernetics 25:33–54

Isermann R (1980) Beispiele für die Fehlerdiagnose mittels Parameterschätzung. Automatisierungstechnik – at 37:342–343, 445–447

Isermann R (1989) Digital control systems, 2nd edn. Springer, Berlin

Isermann R (1991) Digital control systems, vol 2, 2nd edn. Springer, Berlin

Isermann R (1999) Mechatronische Systeme. Springer, Berlin

Isermann R (2005) Mechatronic systems – fundamentals, 2nd edn. Springer, London

Isermann R, Hafner M (2001) Mechatronic combustion engines – from modeling to optimal control. European Journal of Control 7:220–247

Isermann R, Münchhof M (2011) Identification of dynamic systems. Springer, Berlin

Isermann R, Lachmann KH, Matko D (1992) Adaptive control systems. Prentice Hall International UK, London

Kiffmeier U, Köster L, Meyer M, Witke C (1999) Automatic prodective code generation for electronic control units. Automatisierungstechnik – at 47:295–304

Kirkpatrick S, Gelatt C, Vecchi M (1983) Optimization by simulated annealing. Science, New Series 220(4598):671–680

Kirschke-Biller F (2011) Autosar – a worldwide standard current developments, roll-out and outlook. In: 15th International VDI Congress Electronic Systems for Vehicles, Baden-Baden, Germany

Kötter H (2008) Innovative Motorvermessung. Schnelle Motorvermessung mit optimierten Testsignalen. FVV Heft 853. FVV, Frankfurt

Kuder J, Kruse T (2000) Parameteroptimierung an Ottomotoren mit Direkteinspritzung. Motortechnische Zeitschrift – MTZ 61(6)
Kuo B (ed) (1995) Digital Control Systems, 2nd edn. Oxford University Press, Oxford
Leonhard W (1996) Control of electrical drives, 2nd edn. Springer, Berlin
Liggesmeyer P (ed) (2002) Software-Qualität: Testen, Analysieren und Verifizieren von Software. Spektrum Akademischer Verlag, Heidelberg
Lyshevski S (2000) Electromechanical systems, electric machines, and applied mechatronics. CRC Press, Boca Raton, FL
Martyr A, Plint M (2007) Engine testing. Theory and practice, 3rd edn. SAE International, Warrendale
Mesarovic M (1960) The control of multivariable systems. J. Wiley, New York
Mhatre S, Brosilow C (2000) Multivariable model state feedback: computationally simple, easy-to-tome alternative to mpc. AIChE Journal 46(8):1566–1580
Michalewicz Z, Janikow C, Krawczyk J (1992) A modified genetic algorithm for optimal control problems. Computers and Mathematics with Applications 23(12):83–94
Mitterer A, Zuber-Goos F (2000) Modellgestützte Kennfeldoptimierung - ein neuer Ansatz zur Steigerung der Effizienz in der Steuergeräteapplikation. Automobiltechnische Zeitschrift – ATZ 102
Nelles O (2001) Nonlinear system identification. Springer, Heidelberg
Nelles O, Fink A (2000a) Grid-based look-up table optimization toolbox. In: 12th IFAC Symposium on System Identification (SYSID 2000), Santa Barbara, CA, USA
Nelles O, Fink A (2000b) Tool zur Optimierung von Rasterkennfeldern. atp - Automatisierungstechnische Praxis 42(5):58–66
Nocedal J, Wright S (2006) Numerical optimisation, 2nd edn. Springer, New York
Ogata K (ed) (2008) Modern control engineering, 5th edn. Prentice Hall, Upper Saddle River
Pfeil K (2011) Ladedruck- und Luftmassenregelung von aufgeladenen Dieselmotoren mit lokal linearen Modellen und Optimierung des dynamischen Emissionsverhaltens im Rauchbetrieb. Dissertation Technische Universität Darmstadt. Fortschr.-Ber. VDI Reihe 12, 744, VDI Verlag, Düsseldorf
Press W, Teukolsky W, Vetterling S, Flannery B (2007) Numerical recipes in C, 3rd edn. Cambrigde University Press, Cambrigde
Pukelsheim F (2006) Optimal design of experiments. SIAM, Philadelphia
Rechenberg I (1973) Evolutionsstrategie: Optimierung technischer Systeme nach Prinzipien der biologischen Evolution. Frommann-Holzboog, Stuttgart
Robert Bosch GmbH (ed) (2011) Automotive Handbook, 8th edn. Bentley publishers, Cambridge
Royce W (1970) Managing the development of large software projects. In: Proceed. IEEE, Wescon
Schäfer S (2012) Modellbasierte Steuerung des Kühlkreislaufes einer Brennstoffzelle mit automatisiertem Test der Software. Dissertation Technische Universität Darmstadt. Fortschr.-Ber. VDI Reihe 8, Nr. 1219 VDI Verlag, Düsseldorf

Schaffnit J (2002) Simulation und Control Prototyping zur Entwicklung von Steuergerätefunktionen für aufgeladene Nutzfahrzeug-Dieselmotoren. Dissertation Technische Universität Darmstadt. Fortschr.-Ber. VDI Reihe 12, 492. VDI Verlag, Düsseldorf

Schäuffele J, Zurawka T (2005) Automotive Software Engineering. SAE, Warrendale, PA

Schetzen M (1980) The Volterra and Wiener theories of nonlinear systems. J. Wiley, New York

Schmidt M, Kessel JA (1999) Model-based torque controller for dynamical engine test stands. In: American Control Conference, ACC 1999, San Diego, CA, USA

Schmitt M (1995) Untersuchungen zur Realisierung mehrdimensionaler lernfähiger Kennfelder in Großserien-Steuergeräten. Dissertation Technische Universität Darmstadt. Fortschr.-Ber. VDI Reihe 12, 246. VDI Verlag, Düsseldorf

Schulmeister U, Boßler M, Huber T, Johannaber M, Kruse T, Ulmer H (2007) Employment of advanced simulation methods in calibration and development process. In: 2nd International Symposium on Development Methodology, Wiesbaden, Germany

Schwefel HP (1988) Evolutionary learning optimim-seeking on parallel computer architectures. In: Sydow A, Tzafestas S, Vichnevetsky R (eds) Int. symp. on systems analysis and simulation. Akademie der Wissenschaften der DDR, Akademie Verlag, Berlin, Germany, vol 1 Theory and foundations, pp 217–225

Sequenz H (2013) Emission Modelling and Model-Based Optimisation of the Engine Control. Dissertation Technische Universität Darmstadt. Fortschr.-Ber. VDI Reihe 12, 1222. VDI Verlag, Düsseldorf

Sequenz H, Isermann R (2011) Emission model structures for an implementation on engine control units. In: Preprints of the 18th IFAC World Congress, Milano, Italy

Sequenz H, Keller K, Isermann R (2012) Zur Identifikation mehrdimensionaler Kennfelder für Verbrennungsmotoren. Automatisierungstechnik 60(6):344–351

Sinsel S (2000) Echtzeitsimulation von Nutzfahrzeug-Dieselmotoren mit Turbolader zur Entwicklung von Motormanagementsystemen. Logos, Doctoral thesis. University of Technology, Darmstadt. Berlin

Soler N (2011) Model-based thickness and tension control in multi-stand cold rolling mills. Dissertation Technische Universität Darmstadt. Fortschr.-Ber. VDI Reihe 8, 1180. VDI Verlag, Düsseldorf

STARTS Guide (1989) The STARTS purchases Handbook: software tools for application to large real-time systems, 2nd edn. National Computing Centre Publications, Manchester

Stuhler H, Kruse T (2003) Parametrierung der Motorsteuerungen. In: Isermann R (ed) Modellgestützte Steuerung, Regelung und Diagnose von Verbrennungsmotoren, Springer, Berlin, chap 3

Tanaka K, Ikeda T, Wang H (1998) Fuzzy regulators and fuzzy observers: relaxed stability conditions and LMI-based designs. IEEE Trans on Fuzzy Systems 6(2):250–265

Thaller G (ed) (2002) Software-Test-Verifikation und Validation. Heinz-Heise-Verlag, Hannover

Tikhonov A, Arsenin V (1977) Solutions of ill-posed problems. VH Winston and Sons, Washington, DC

Tran E (2007) Verification/validation/certification. In: Koopman P (ed) Topics in dependable embedded software, Carnegie Mellon University, Pittsburgh, PA

Unland S, Stuhler H, Stuber A (1998) Neue effiziente Applikationsverfahren für die physikalisch basierte Motorsteuerung ME7. Motortechnische Zeitschrift – MTZ 59

Vas P (ed) (1990) Vector control of AC machines. Clarendon Press, Oxford

VDI 2206 (2003) Design methodology for mechatronic systems. Beuth Verlag, Berlin

Wernicke M, Rein J (2007) Integration of existing ECU software in the autosar architecture. ATZ-Elektronik (1):20–25

Wolfram A, Vogt M (2002) Discrete-time filter algortihms for the computation of derivatives. Automatisierungstechnik – at 7:346–353

Wright R, Kravaris C (2006) Two-degree-of-freedom output feedback controllers for discrete-time nonlinear systems. Chemical Engineering Science 61:4676–4688

Zahn S (2012) Arbeitsspielaufgelöste Modellbildung und Hardware-in-the-Loop-Simulation von Pkw-Dieselmotoren mit Abgasturbolader. Dissertation Technische Universität Darmstadt. Fortschr.-Ber. VDI Reihe 12, 760. VDI Verlag, Düsseldorf

Zimmerschied R, Weber M, Isermann R (2005) Statische und dynamische Motorvermessung zur Auslegung von Steuerkennfeldern - eine kurze Übersicht. Automatisierungstechnik – at 53(2):87–94

7

Control of gasoline engines

The control structure of gasoline engines follows by considering the main components, the main control tasks and a resulting control-oriented block diagram of direct injection (DI) gasoline engines (or spark-ignition (SI) engines) as described in the introduction chapter, Sect. 1.3.1. Based on the block diagram Fig. 1.3.5, the software structure Fig. 5.1.5 and the hierarchical control structure of an ECU in Fig. 5.1.6, the control of gasoline engines can be subdivided in:

- Torque control
  - fuel injection control
  - air/fuel ratio control
  - ignition angle control
  - combustion control
  - knock control
- Emission control
  - exhaust recirculation control
  - catalyst control
- Idle speed control
- Coolant temperature control
- Lubrication control.

A central role for gasoline engines is the kind of *air-fuel mixture formation (A/F)*. The combustion of gasoline requires a very specific air/fuel ratio which is for stoichiometric combustion

$$m_{\mathrm{air,st}} = L_{\mathrm{st}} m_{\mathrm{f}} \tag{7.0.1}$$

with

$$L_{\mathrm{st}} = A/F = 14.7 \ldots 14.9 \approx 14.8 \text{ kg air/ kg fuel.} \tag{7.0.2}$$

A measure for the air-fuel mixture is the *air excess factor* which is defined as

$$\lambda = \frac{m_{\mathrm{air}}}{m_{\mathrm{air,st}}}. \tag{7.0.3}$$

R. Isermann, *Engine Modeling and Control*,
DOI: 10.1007/978-3-642-39934-3_7

The A/F mixture strongly influences the efficiency of the combustion and therefore the specific fuel consumption and the efficiency of the exhaust-gas treatment.

The following ranges of the excess-air factor can be distinguished:

- $\lambda = 1$: The supplied air mass corresponds to the theoretical (stoichiometric) required air mass.
- $\lambda > 1$: Usually an excess of air, called a lean mixture, is required to enable a complete combustion. Fuel consumption for intake-manifold injection is lowest for $\lambda = 1.2$ to 1.5.
- $\lambda < 1$: Too less air means a rich A/F mixture. Maximum power is at $\lambda = 0.85$ to 0.95.

For catalytic exhaust-gas treatment frequently a three-way catalyst is used, which requires $\lambda = 1$ for normal operation. This allows the catalyst to reduce all three pollutant components CO, HC and NO_x as much as possible.

The different optimal values of λ mean that an appropriate compromise has to be found. For engines with intake-manifold injection and direct injection with homogeneous combustion a range of $\lambda = 0.9$ to 1.1 is usually sufficient. However, direct injection gasoline engines with stratified combustion operate with lean combustion and $\lambda \leq 4$. Then other types of catalytic converters are required, such as NO_x accumulator-types. They need for regeneration a rich A/F ratio ($\lambda < 0.8$). In order to obtain precise A/F mixtures with different reference variables, λ-sensors are provided in the exhaust, which allow to implement a λ-feedback controller.

The gasoline air/fuel mixture formation of current gasoline engines with electronic control can be divided in external mixture formation with intake-manifold injection or internal mixture formation with direct injection. Table 7.0.1 gives a survey of the features, properties and components.

In the past only *external A/F mixture formation* was applied in the form of carburetors and and continuous, simple-point mechanical injection systems located upstream the throttle valve. However, as it is not possible to meet the high requirements for precise A/F-ratios they cannot be applied any more. They were replaced by the intake-manifold injection upstream the inlet valves, operating intermittently and for each cylinder individually. The spray of the injector is directed into the intake and on the inlet valves, allowing a fast evaporization in the turbulent airflow and only small fuel wall films. The injection has to be performed within 720°CS and an electric fuel pump in the tank providing about $p = 6$ bar is usually sufficient.

In the case of *internal A/F mixture formation* with direct injection into the cylinders, homogeneous combustion and stratified combustion are distinguished. For homogeneous combustion the fuel is injected in the induction stroke, i.e. within 180°CS. The mixture is stoichiometric with $\lambda = 1$ and distributed in the combustion chamber. The stratified combustion operates with opened throttle and full air flow and in average with a lean mixture $\lambda > 1$. The mixture is guided such that it can be ignited around the spark plug. Dependent on side or top injection and the design of the piston recess three types of injection exist, wall-directed, air-directed and jet-directed. Direct injection requires a high pressure fuel pump and very fast solenoid

Table 7.0.1. Different kind of air/fuel mixture formation for gasoline engines

Gasoline air / fuel mixture formation					
	external intake manifold injection	internal direct injection			
main features	- intermittent injection upstream inlet valves - for each cylinder	intermittent injection in cylinder head			
		- homogeneous combustion - injection in induction stroke	stratified combustion injection in compression stroke		
general properties	- injection $\leq 720°$ CS - medium injection pressure	- injection $\leq 180°$ CS - $\lambda = 1$ - high injection pressure (50-120 bar) - mixture distributed in combustion chamber	- full air flow (dethrottling) - $\lambda > 1$ (lean operation) - increased compression - high injection pressure - mixture concentrated at spark plug		
type of injection	- jets	turbulence-oriented	well-directed	air-directed	jet-directed
specific properties			- side injection - on piston recess - swirl support - condensing fuel	- side injection - on air cushion - no condensing - fuel	- top injection - next spark plug - immediate ignition - short mixture time - no condensing fuel
injectors	- solenoid coil - moving valve needle - spray formation - injection time: 20 ms	- solenoid coil with booster current - piezoelectric actuation - injection time: < 5 ms			
turbulence flap	- not required	- in front of intake valves - tumble flap or swirl flap - two-point or continuous actuation			

or piezoelectric injectors which allow injection times < 5 ms. To amplify tumble or swirl variable turbulence flaps are build in front of the intake valves.

In the next section the control structure for gasoline engines is discussed. Dependent on the A/F mixture formation combustion with different values for λ have to be considered.

On the contents of this chapter
This chapter first considers the now classical torque-oriented control structure and shows a more detailed basic control structure for gasoline engines and various signal-flow schemes. Then, simplified torque models are derived for the use in power-train control. A basic control scheme for gasoline engines is the air/fuel and catalytic converter control, also called λ-control. The ignition angle or ignition timing is usually feedforward controlled with look-up tables in dependence on various variables. The precision of the ignition-angle control can be improved by combining the feedforward look-up table with an adaptation look-up table based on the feedback of a combustion-pressure-based combustion center calculation. Thus, the ignition angles adapt to the individual cylinder properties. A further investigation shows the results of an extremum value control of the mean effective pressure by changing the ignition angle and using a sublevel combustion center control. Then knock control, idle speed control, and variable valve trains are treated.

As alternatives to the homogeneous combustion stratified combustion and homogeneous charge compression ignition (HCCI) are considered. Finally, signal-flow schemes and control systems for the coolant systems and the lubrication systems are discussed.

7.1 Gasoline engine control structure

In the past the electronic control of engines was mainly realized with two-dimensional maps between the various inputs and manipulated variables as outputs like throttle angle, fuel injection quantity, injection timing and ignition timing. However, because of the many interactions and cross couplings the control system structure without a central torque coordination was not transparent. Since about 1997 the structure has changed to a *torque-oriented scheme*, enabling a more clear physical-based structure, Gerhardt et al (1997), Gerhardt et al (1998). The underlying stationary torque model is depicted in Fig. 7.1.1. The model consists of an *inner torque* $M_{\rm i}$ in dependence on the cylinder charge (air mass) $m_{\rm air}$ and engine speed $n = n_{\rm eng}$ for stoichiometric combustion $\lambda = 1$ and optimal ignition angle. The inner torque results from the combustion and is

$$\overline{M}_{\rm i} = f_{\rm Mi}(m_{\rm air}, n) = \overline{M}_{\rm eng} + \overline{M}_{\rm drg} \tag{7.1.1}$$

where $\overline{M}_{\rm eng}$ is the mean torque at the clutch and the mean drag torque is

$$\overline{M}_{\rm drg} = \overline{M}_{\rm g,drg} + \overline{M}_{\rm f} + \overline{M}_{\rm aux1} + \overline{M}_{\rm aux2}. \tag{7.1.2}$$

The drag torque results from the gas exchange torque $M_{\rm g,drg}$ of the low pressure cycle and thermal losses during the motoring high pressure cycle, the friction torque $M_{\rm f}$

and the engine integrated auxiliary drives torque M_{aux1} and external auxiliary drives torque M_{aux2} (coolant pump, electrical generator, air conditioning compressor).

Changes from the optimal operating conditions by other air-to-fuel ratios or other ignition angles are taken into account by correction factors, called efficiencies

$$\eta_\lambda = \frac{M_{\mathrm{i}}(\lambda)}{M_{\mathrm{i}}} \;;\; \eta_{\varphi_{\mathrm{ign}}} = \frac{M_{\mathrm{i}}(\varphi_{\mathrm{ign}})}{M_{\mathrm{i}}} \tag{7.1.3}$$

such that the corrected torque becomes

$$\overline{M}_{\mathrm{icorr}}(\lambda,\, \varphi_{\mathrm{ign}}) = \eta_\lambda\, \eta_{\varphi_{\mathrm{ign}}}\, \overline{M}_{\mathrm{i}}. \tag{7.1.4}$$

The torque at the clutch is then

$$\overline{M}_{\mathrm{eng}}\,(m_{\mathrm{air}},\, n,\, \lambda,\, \varphi_{\mathrm{ign}}) = \overline{M}_{\mathrm{icorr}}(\lambda,\, \varphi_{\mathrm{ign}}) - \overline{M}_{\mathrm{drg}}(n) \tag{7.1.5}$$

compare Fig. 7.1.1.

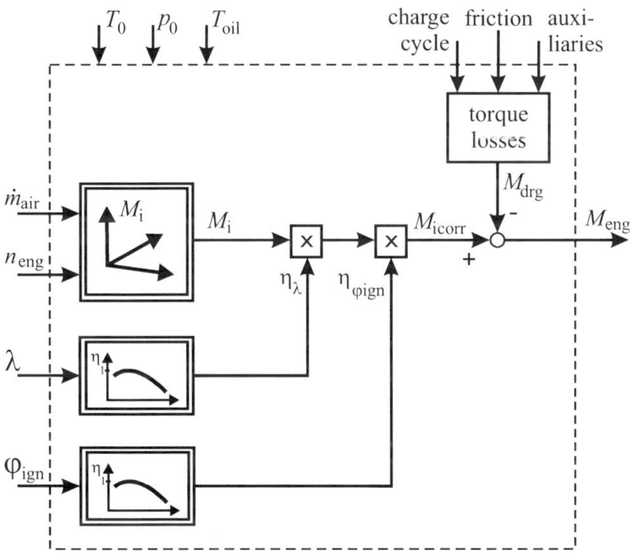

Fig. 7.1.1. Torque model for internal combustion engines, Gerhardt et al (1997). $\dot{m}_{\mathrm{air}}$: mass flow, M_{i}: internal torque, M_{eng}: crankshaft torque, M_{drg}: drag torque (torque losses).

As the torque demands from the driver, the transmission, auxiliaries, etc. play a central role in the torque-oriented control system, the torque model has to be invertible to determine the required manipulated variables, as e.g. for m_{air}

$$\begin{aligned} m_{\mathrm{air}} &= f_{\mathrm{Mi}}^{-1}(\overline{M}_{\mathrm{i}}, n) \\ \overline{M}_{\mathrm{i}} &= \frac{1}{\eta_\lambda \eta_{\varphi\mathrm{ign}}}(\overline{M}_{\mathrm{eng}} + \overline{M}_{\mathrm{drg}}) \end{aligned} \tag{7.1.6}$$

which follows from (7.1.5), (7.1.4) and (7.1.1). This becomes possible because of the two-dimensional map (7.1.1). The introduction of an optimal inner torque and corrective factors for other operation points allow to operate with two-dimensional maps and one-dimensional characteristics instead of a multi-dimensional map, see Sect. 6.5.

It turns out that the efficiency map $\eta_{\varphi\text{ign}}(\varphi_{\text{ign}})$ results in a single characteristic also for many other operating points, Gerhardt et al (1997). Thus, as a basis the engine models according to Fig. 7.1.1 have to be determined on a test bench for a specific engine. The torque model of Fig. 7.1.1 may be enhanced by further correcting factors like camshaft phasing η_{cam} or residual gases η_{res}.

The injected fuel is directly determined by the charged air mass in the cylinder and follows for a given excess-air factor from (7.0.1) to (7.0.3) to

$$m_{\text{f}}(m_{\text{air}}, \lambda) = \frac{1}{L_{\text{st}}\lambda} m_{\text{air}}. \tag{7.1.7}$$

A prerequisite for the design and implementation of the control functions of the torque-oriented control system is the existence of a *basis control calibration* of the engine, see Sect. 6.8. Therefore the steady-state behavior of the engine has to be measured on a test bench and optimal manipulated variables have to be determined for, e.g. the air charge, fuel mass, A/F ratio, injection angle, EGR mass flow, camshaft position and boost pressure in dependence on the torque M_{eng} and speed n with regard to fuel consumption, emission laws and driveability. This results in optimized steady state engine models, Stuhler and Kruse (2003):

- optimal inner torque $M_{\text{i}}(m_{\text{air}}, n)$
- optimal fuel injection $m_f(m_{\text{air}}, n)$
- optimal fuel injection angle $\varphi_f(m_{\text{air}}, n, \lambda)$
- optimal ignition angle $\varphi_{\text{ign}}(m_{\text{air}}, n, \lambda)$
- drag torque $M_{\text{drg}}(m_{\text{air}}, n, T_{\text{oil}})$
- ignition-angle efficiency $\eta_{\varphi\text{ign}}(\varphi_{\text{cs}})$
- lambda efficiency $\eta_\lambda(\lambda)$.

The torque-oriented engine control now has the task to control the engine via the given actuators according to the external or internal torque demands and limits, based on the engine models from the basis control calibration.

The torque-oriented gasoline-engine control system can be divided in different signal-flow control blocks, Robert Bosch GmbH (2004), as shown in Fig. 7.1.2.

The torque-oriented structure includes the calculation of following variables:

- *Torque demands*: determination of the required torques dependent on driving situation
- *Torque coordination*: selection of the kind of torque manipulation by appropriate commands to air charge, ignition, injection cut-off
- *Torque conversion*: determination of the desired values of the actuator outputs, the physical engine inputs, based on engine models
- *Actuator control*: determination of the desired values of the electrical actuator inputs.

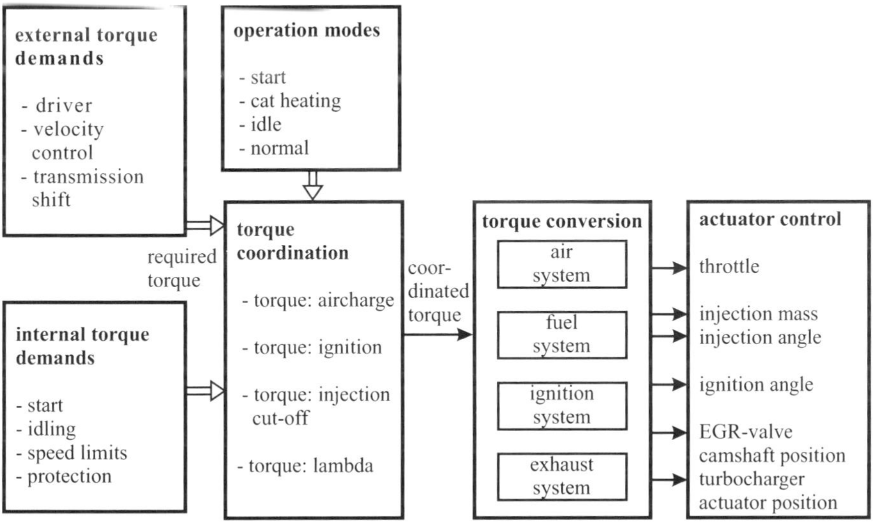

Fig. 7.1.2. Main signal-flow blocks for the control of gasoline engines.

A more detailed control structure is depicted in Fig. 7.1.3 for a gasoline engine with homogeneous combustion, which leans on Gerhardt et al (1997), Gerhardt et al (1998), Meisberger et al (1998), Hess (2003). This scheme is now further discussed as an example.

a) Torque demands

The main driving input for a vehicle is the driver's pedal position α_{ped}. According to the designed drive map (look-up table) $M(\alpha_{\text{ped}}, n)$ a desired inner torque $M_{\text{i,d}}$ is specified which for driving without gear shifting or overrun determines the load torque through air charge $M_{\text{i,ch,d}}$. Other operating conditions also require certain engine torques, like for start and idling and for additional auxiliaries. Further, the torque should never be smaller than the drag torque and not larger than protection limits given by the engine itself or the transmission. During gear shifting and control actions of drive dynamic control as, e.g. the traction control system (TCS) the engine torque should be reduced as fast as possible. Therefore, all different torque demands have to be stated.

b) Torque coordination

The torque coordination block selects, based on all torque demands, which torque has to be realized. This is based on priority selection, either by a maximum determination of the torque demands from the drive map or the accessories or the drag torque or a minimum operation with regard to the engine protection limits or the transmission demand.

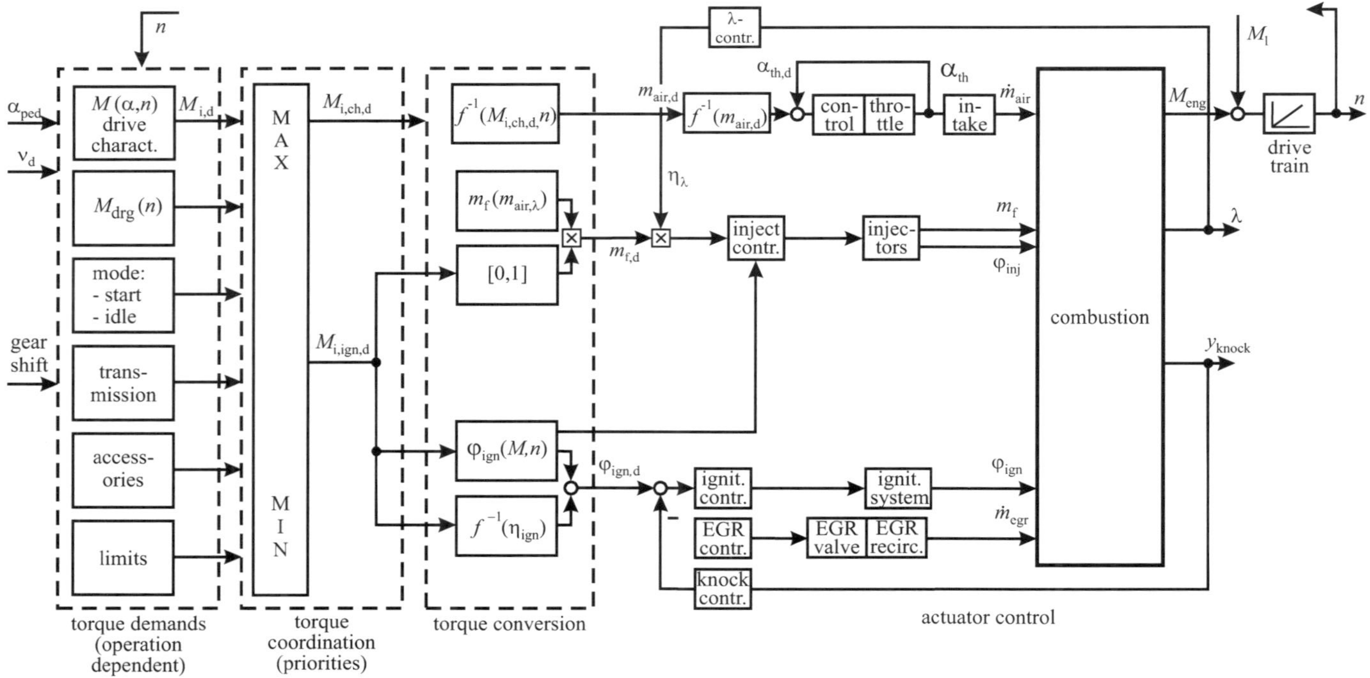

Fig. 7.1.3. Basic control structure for gasoline engines with homogeneous combustion (only main controllers, simplified, without turbocharging, without variable camshaft and valvetrain).

The torque of gasoline engines is conventionally manipulated with the throttle, the injected fuel mass or by fuel cut-off of certain cylinders or in the case of torque reduction by the ignition angle. Therefore, the kind of manipulation has to be selected for given torque demands.

The engine torque change after actuating the throttle is delayed with several $100\,\mathrm{ms}$, because of the air mass storage in the manifold and the limited speed of the throttle plate, especially for slow engine speed. The increase of the ignition spark advance angle or the reduction of the fuel injection mass can be realized crank angle synchronously and influences immediately the following combustion cycle and therefore within some $10\,\mathrm{ms}$. An increase of the spark advance from the optimum by $20°$ to $30°$CS, e.g. decreases the torque immediately by about 20 to 50 $\%$. Hence, the engine torque can be manipulated by the throttle with *slow dynamics*, determining the torque in a larger time range, and it can be reduced by ignition and/or injection timing in a *short-time range*. A fast torque reduction is required for automatic gear shifting and for drive dynamics control as for TCS. The torque coordination block, therefore, decides by a priority selection if the fast dynamics torque reduction becomes active in the sense of an auxiliary manipulating variable for the engine. The slow dynamics torque manipulation with the throttle determines the long-term torque behavior.

Figure 7.1.4 shows a detailed signal flow for the torque coordination block, divided in a slow and a fast dynamics path, according to Gerhardt et al (1997). The desired inner torque $M_{\mathrm{i,d}}$ determined from the accelerator pedal (or speed control) is the same for the slow path determining the desired inner torque through the air charge and the fast path for determining the desired inner torque through the ignition angle. (The scheme for the injection mass is similar). More details are described in Gerhardt et al (1997) and Gerhardt et al (1998), see also Meisberger et al (1998).

c) Torque conversion

The task of the torque conversion is to determine the desired physical engine inputs with the corresponding actuators by using the engine torque models. As Fig. 7.1.5 shows, the desired inner air charge torque $M_{\mathrm{i,d}}$ is corrected taking into account the efficiencies with the real λ and φ_{ign}. Then the torque map $M_{\mathrm{i}}(m_{\mathrm{air}}, n)$ for $\lambda = 1$ and the optimal calibrated ignition angle $\varphi_{\mathrm{ign,opt}}$ is inverted to calculate the required air mass m_{air}. If larger than a minimum air mass, the desired air mass $m_{\mathrm{air,d}}$ is determined.

Figure 7.1.5 depicts also the calculation of the desired ignition angle $\varphi_{\mathrm{ign,d}}$ (so called spark advance) if a fast torque reduction is required. First the desired ignition efficiency $\eta_{\mathrm{ign,d}}$ is determined based on the values of the desired torque and the possible torque according to the present (sequential) reduction step. An inversion of the ignition efficiency characteristic $\eta_{\mathrm{ign}}(\varphi_{\mathrm{ign}})$ allows then to determine the desired ignition angle $\varphi_{\mathrm{ign,d}}$, additionally taking the ignition angle change through changed λ resulting from the λ-controller into account. Thus, the fast dynamics path adds ignition angle changes $\Delta\varphi_{\mathrm{ign}}$ to the optimal calibrated $\varphi_{\mathrm{ign,opt},\lambda=1}$ if required for

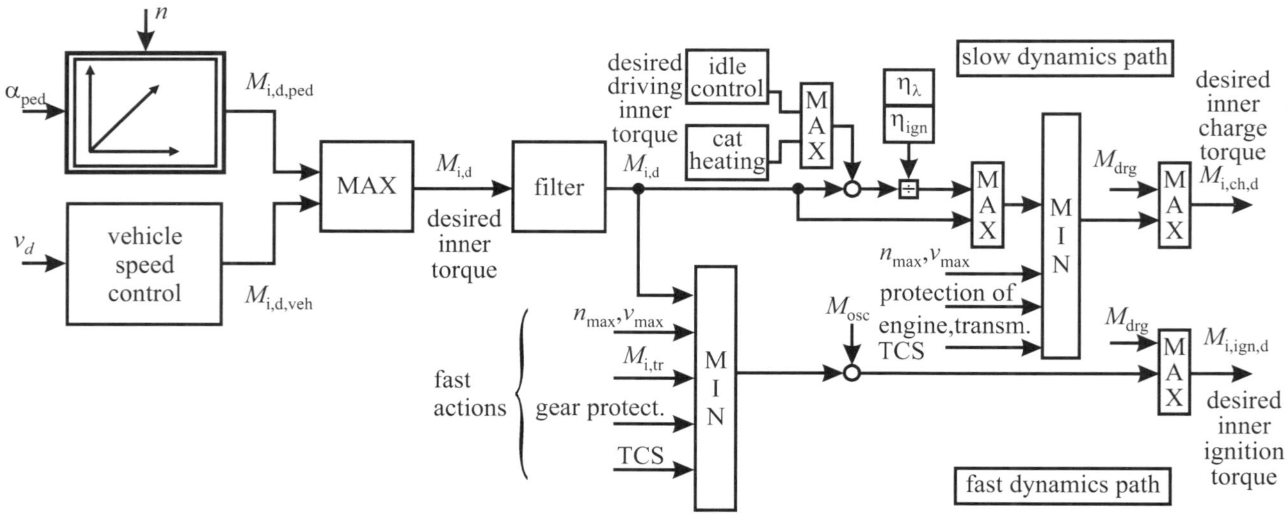

Fig. 7.1.4. Signal flow for the torque coordination of throttle-controlled gasoline engines (extract). The desired inner torque acts on the air charge (and coupled fuel mass) for slow dynamics or for fast torque dynamics on the change of ignition angle, see Fig. 7.1.5. M_{drg}: drag++ torque, M_{osc}: torque demands from surge damping of drive train oscillations. TCS: traction control system, d: desired values, Gerhardt et al (1997).

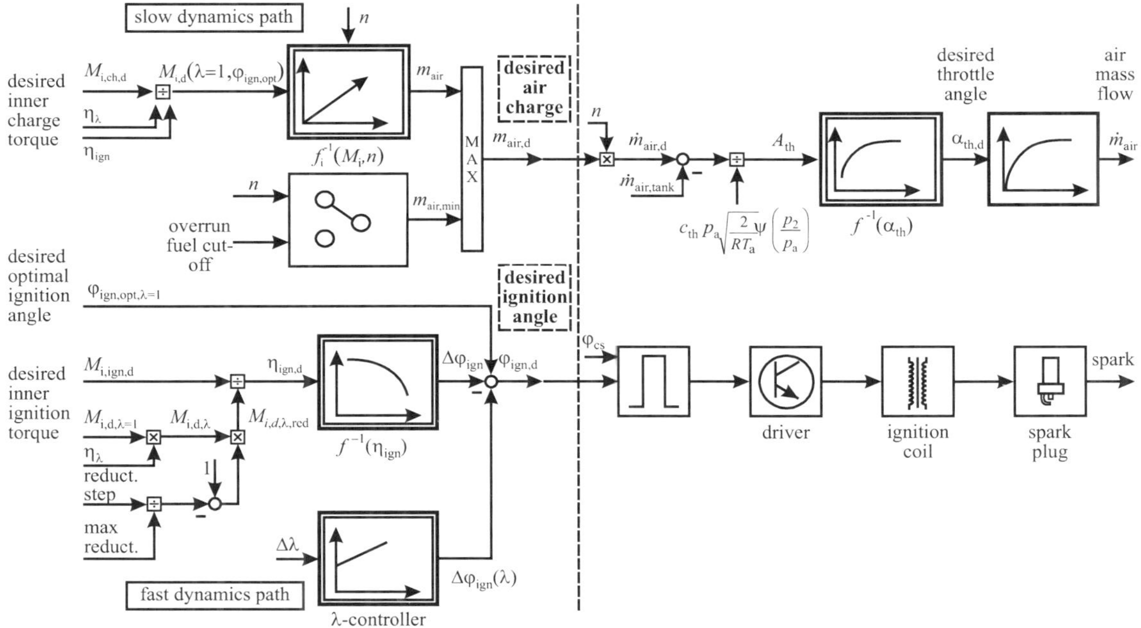

Fig. 7.1.5. Signal flow for the torque conversion and actuator control (extract). The desired actuator commands are determined from desired inner torques for the air charge and for the ignition angle.

the fast actions depicted in Fig. 7.1.4. This is in the sense of a dynamic feedforward action which passes after short time.

The injected fuel mass is calculated with (7.1.7) based on the air charge m_{air} in the cylinder. The injection angle follows the basic control calibration, see Fig. 7.1.3. In the case of fast torque reduction the fuel mass may be cut-off dependent on the sequential reduction steps, Hess (2003), where the sequence of commands to the throttle, injection mass and ignition is triggered in a cascaded manner.

d) Actuator control

The actuator control block determines the electrical inputs of the actuators for the manipulation of the physical engine inputs. Based on the desired air mass $m_{\mathrm{air,d}}$ the air mass flow rate $\dot{m}_{\mathrm{air,d}}$ is calculated, corresponding to (4.1.29), and by using the air flow rate equation (4.1.19) for the throttle and inversion of the $A_{\mathrm{th}}(\alpha_{\mathrm{th}})$ characteristic the desired throttle angle $\alpha_{\mathrm{th,d}}$ is obtained, as depicted in Fig. 7.1.5. The throttle control loop then positions the throttle to α_{th}, see Fig. 7.1.3. Delayed by the intake manifold dynamics the engine then obtains the air mass flow rate $\dot{m}_{\mathrm{air}}$ into the cylinders.

In the case of ignition-angle manipulation the desired ignition angle $\varphi_{\mathrm{ign,d}}$ is compared with the crankshaft rotating angle φ_{cs} and gives upon agreement a pulse to the electronic ignition, where the ignition driver cuts the ignition coil to generate a high voltage for the spark plug, see Sect. 5.1. The injection time and the injected fuel quantity are determined by the begin and duration of the injection and are triggered by special ASIC's, as described in Sect. 5.1.

The external exhaust gas recirculation flow rate is feedforward controlled based on the basic engine calibration, see Fig. 7.1.3.

As Fig. 7.1.3 indicates, the control functions described until now are all feedforward and based on the inner torque model and engine models from the basic control calibration. However, in order to obtain a precise stoichiometric combustion a λ-feedback controller corrects the injected fuel mass, see next section. To avoid knocking cylinders a knock feedback control manipulates the ignition angles of the individual cylinders, see Sect. 7.6, as also shown in Fig. 7.1.3.

e) Extended determination of torque demand

The torque demands according to a) and Fig. 7.1.3 apply for driving with the accelerator pedal. However, for traction control (TCS) or for vehicle speed control or adaptive cruise control (ACC) the torque at the wheels is required. Then the complete drive train has to be considered and the signal flow has to be expanded as depicted in Fig. 7.1.6.

Neglecting dynamic effects the mean crankshaft torque results according to (7.1.5) to

$$\overline{M}_{\mathrm{eng}} = M_{\mathrm{eng}} = M_{\mathrm{i,corr}} - M_{\mathrm{drg}}. \tag{7.1.8}$$

The engine output torque (after flywheel) M_{eng} acts on the clutch and according to a clutch actuation system with gliding or engaged status, the resulting torque M_{cl} is the input to the transmission. (The dynamic behavior of the drive train with the clutch in gliding or engaged is treated in Sect. 4.9.) The mean torque at the transmission output is

$$M_{\mathrm{tr}} = i_{\mathrm{tr}} M_{\mathrm{eng}} - M_{\mathrm{tr,drg}} \tag{7.1.9}$$

with the gear ratio

$$i_{\mathrm{tr}} = \frac{\omega_{\mathrm{tr,in}}}{\omega_{\mathrm{tr,out}}} = \frac{\text{transmission input speed}}{\text{transmission output speed}} \tag{7.1.10}$$

and a friction torque $M_{\mathrm{tr,drg}}$. The friction can also be taken into account with an efficiency factor η_{tr}

$$M_{\mathrm{tr}} = i_{\mathrm{tr}} \eta_{\mathrm{tr}} M_{\mathrm{eng}}. \tag{7.1.11}$$

In the case of an automatic transmission with a hydrodynamic torque converter and several planetary-gear sets the torque output is a dynamic function of the pump (impeller) torque M_{p} and the turbine torque M_{t} which both depend on their speeds ω_{p} ω_{t}, see e.g. Robert Bosch GmbH (2011). However, with a lock-up clutch the impeller and the turbine are locked together such that (7.1.9) can be applied in the engaged status and $\omega_{\mathrm{tr,in}} = \omega_{\mathrm{eng}}$.

The transmission output torque drives the propeller shaft and is converted by the differential into the differential output torque

$$M_{\mathrm{di}} = i_{\mathrm{di}} \eta_{\mathrm{di}} M_{\mathrm{tr}}. \tag{7.1.12}$$

with

$$i_{\mathrm{di}} = \frac{\omega_{\mathrm{ps}}}{\omega_{w}} = \frac{\text{propeller shaft speed}}{\text{wheel speed}} \tag{7.1.13}$$

and an efficiency η_{di} for friction losses and usually $\omega_{\mathrm{ps}} = \omega_{\mathrm{tr,out}}$. Assuming tow-driven wheels and straight driving it holds for the sum torque of the wheels

$$M_{\mathrm{w}\,\Sigma} = M_{\mathrm{di}} \eta_{\mathrm{ws}} \tag{7.1.14}$$

where η_{ws} takes the friction along the wheel shaft into account. This torque then drives the vehicle.

As indicated in Fig. 7.1.6 the engine control system has to include the signals to and from the transmission control and the clutch control.

Considering all torque losses between the engine output torque M_{eng} and the sum of the wheel torques it holds

$$M_{\mathrm{w}\,\Sigma} = \eta_{\mathrm{tr}} \eta_{\mathrm{di}} \eta_{\mathrm{ds}} M_{\mathrm{eng}}. \tag{7.1.15}$$

The required inner torque is according to (7.1.8) with now the wheel torque as input

$$M_{\mathrm{i,corr}} = \frac{1}{\eta_{\mathrm{tr}} \eta_{\mathrm{di}} \eta_{\mathrm{ds}}} M_{\mathrm{w}\,\Sigma} + M_{\mathrm{drg}}. \tag{7.1.16}$$

This torque leads with (7.1.4) to the demanded inner torque $M_{\mathrm{i,d,veh}}$

$$M_{\mathrm{i,d,veh}} = \frac{1}{\eta_{\lambda} \eta_{\mathrm{ign}}} M_{\mathrm{i,corr}}$$

e.g. as output of a vehicle speed or traction controller, Fig. 7.1.4.

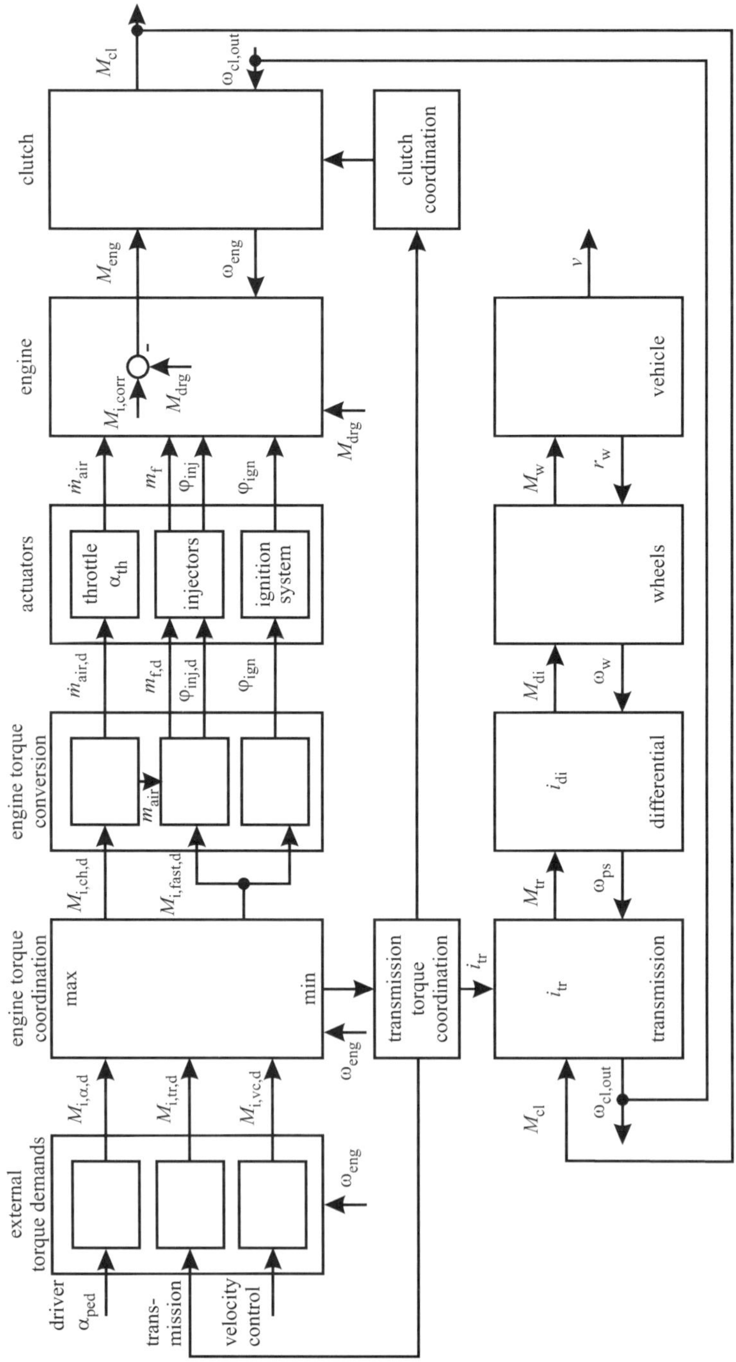

Fig. 7.1.6. Signal flow for the torque generation of a naturally aspirated gasoline engine as component within the powertrain of a vehicle.

f) Simplified gasoline engine torque models

The detailed control structures presented in Figs. 7.1.3 to 7.1.5 are required for the development and calibration of the control functions. However, for applications with the gasoline engine as part of a powertrain sometimes simplified torque models can be used.

As stated in subsection b), the demanded inner torque from external torque demands like from the driver, the transmission or the vehicle velocity control or from demanded internal torques is the input to a torque coordination, where the desired torque $M_{\mathrm{i,d}}$ is determined by priority selection. A torque conversion then calculates the required air mass $m_{\mathrm{air,d}}$ and throttle position α_{th} by an inversion of the inner torque model. The injected fuel quantity $m_{\mathrm{f}}(m_{\mathrm{air}}, \lambda)$ is directly determined from the air mass for a given excess-air factor λ. Changes of the throttle position α_{th} shows a delay in the engine torque of some 100 ms, because of the air mass storage in the manifold and result therefore in a slower torque development. The fuel injection and corresponding ignition spark advance angle can be realized crank angle synchronously and influence the torque within some 10 ms. Hence, the engine torque can be manipulated by the throttle with slower dynamics, determining the torque in a larger time range, and it can be reduced by ignition and/or injection timing in a short time range, e.g. for gear shifting. The torque conversion block combines these actuator commands, as shown previously in Fig. 7.1.3.

Figure 7.1.7 illustrates a simplified torque generation model. The main delay is caused by the intake manifold dynamics. Detailed models of the air path system are derived in Sect. 4.1. The time constant of the intake manifold is described by

$$T_{\mathrm{int}} = \frac{2V_{\mathrm{int}}}{\eta_v V_{\mathrm{D}} n_{\mathrm{eng}}} \tag{7.1.17}$$

with V_{int} the intake volume, η_v the volumetric efficiency of the cylinder group and V_{D} the displacement. Hence, T_{int} is the smaller the smaller V_{int} and the larger n_{eng}. The position control of the electrical throttle valve introduces a further delay with a time constant T_{th}. The fuel mass injection and ignition angle act only with negligible dynamics and are therefore assumed as proportional acting. A further delay may come from a dead time T_{d} which results from ECU calculation times and a cycle time, depending on the change of the injection after a command signal, which is larger for port injection than for direct injection.

The inner torque generation of a gasoline engine without supercharging can therefore be approximated by a nonlinear map

$$\overline{M}_{\mathrm{i}}(\dot{m}_{\mathrm{air}}, n_{\mathrm{eng}}) \tag{7.1.18}$$

for optimal calibrated other manipulated variables like λ and φ_{ign}. Linearization around an operating point $(\overline{M}_{\mathrm{eng}}, \overline{n}_{\mathrm{eng}})$ results for the air flow as output in the transfer function

$$G_{\mathrm{mm}}(s) = \frac{\Delta\alpha_{\mathrm{th,d}}(s)}{\Delta\dot{m}_{\mathrm{air,d}}(s)} \cdot \frac{\Delta\dot{m}_{\mathrm{air}}(s)}{\Delta\alpha_{\mathrm{th}}(s)} \approx \frac{\Delta\dot{m}_{\mathrm{air}}(s)}{\Delta\dot{m}_{\mathrm{air,d}}(s)} = \frac{K_{\mathrm{th}} K_{\mathrm{int}}}{(1+T_{\mathrm{th}}s)(1+T_{\mathrm{int}}s)} e^{-T_{\mathrm{d}}s} \tag{7.1.19}$$

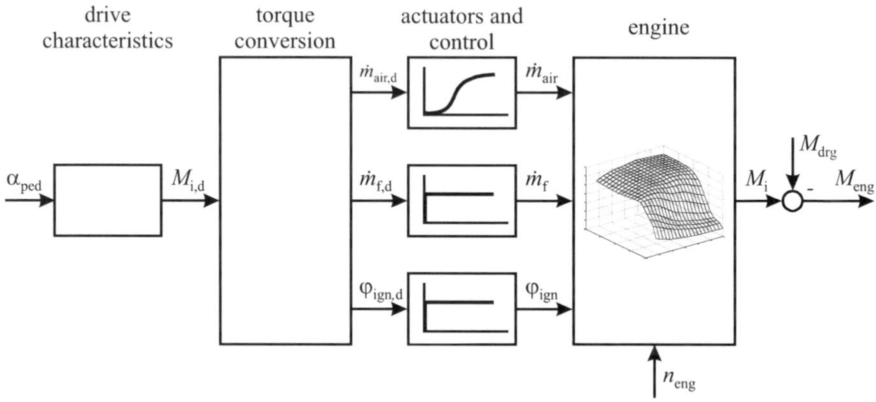

Fig. 7.1.7. Simplified torque model for a naturally aspirating gasoline engine.

and for the torque

$$G_{\mathrm{mM}}(s) = \frac{\Delta M_{\mathrm{i}}(s)}{\Delta \dot{m}_{\mathrm{air}}(s)} \cdot \frac{\Delta \dot{m}_{\mathrm{air}}(s)}{\Delta \dot{m}_{\mathrm{air,d}}(s)} = \frac{\Delta M_{\mathrm{i}}(s)}{\Delta \dot{m}_{\mathrm{air,d}}(s)} = \frac{K_{\mathrm{mM}}(\overline{M}_{\mathrm{eng}}, \overline{n}_{\mathrm{eng}})}{(1+T_{\mathrm{th}}s)(1+T_{\mathrm{int}}s)} e^{-T_{\mathrm{d}}s} \tag{7.1.20}$$

with $T_{\mathrm{th}} \approx 50 \ldots 100\,\mathrm{ms}$ and $T_{\mathrm{int}} \approx 50 \ldots 300\,\mathrm{ms}$, compare Fig. 7.1.7.

Usually, the engine torque is given as a look-up table in form of $M_{\mathrm{eng}}(\alpha_{\mathrm{th}}, n_{\mathrm{eng}})$. The inner torque then follows due to (7.1.5) to

$$M_{\mathrm{i,corr}}(\alpha_{\mathrm{th}}, n_{\mathrm{eng}}) = M_{\mathrm{eng}}(\alpha_{\mathrm{th}}, n_{\mathrm{eng}}, \lambda, \varphi_{\mathrm{inj}}) + M_{\mathrm{drg}}(n_{\mathrm{eng}}). \tag{7.1.21}$$

With α_{th} as input (7.1.19) leads to

$$G_{\mathrm{m}\alpha}(s) = \frac{\Delta \dot{m}_{\mathrm{air}}}{\Delta \alpha_{\mathrm{th}}(s)} = \frac{K_{\mathrm{m}\alpha}(\overline{M}_{\mathrm{i}}, \overline{n}_{\mathrm{eng}})}{(1+T_{\mathrm{th}}s)(1+T_{\mathrm{int}}s)} e^{-T_{\mathrm{d}}s} \tag{7.1.22}$$

and instead of (7.1.20) follows

$$G_{\alpha\mathrm{M}}(s) = \frac{\Delta M_{\mathrm{i}}(s)}{\Delta \alpha_{\mathrm{th}}(s)} = \frac{K_{\alpha\mathrm{M}}(\overline{M}_{\mathrm{i}}, \overline{n}_{\mathrm{eng}})}{(1+T_{\mathrm{th}}s)(1+T_{\mathrm{int}}s)} e^{-T_{\mathrm{d}}s}. \tag{7.1.23}$$

Hence, the torque of a naturally aspirating gasoline engine can be described by second order dynamics with a dead time.

A further simplification leads to an overall torque model for gasoline engines as depicted in Fig. 7.1.8. The nonlinear steady state behavior of the torque is described by multi-dimensional maps

$$M_{\mathrm{eng}}(\boldsymbol{u}_{\mathrm{eng}}, n_{\mathrm{eng}}, M_{\mathrm{drg}}) \tag{7.1.24}$$

where the input vector is

$$\mathbf{u}_{\mathrm{eng,gasoline}} = [\dot{m}_{\mathrm{air}}, m_{\mathrm{f}}, \varphi_{\mathrm{ign}}, \varphi_{\mathrm{inj}}]. \tag{7.1.25}$$

The desired inner torque $M_{\mathrm{i,d}}$ follows the pedal position α_{ped} according to the implemented driving characteristics and the torque conversion without delay. The main dynamic delays stem from the air path volume storage and actuator delays and can be approximated by the behavior of the delayed air flow $\dot{m}_{\mathrm{air}}$ as a representative variable. According to the dynamic model (7.1.19) it holds

$$G_{\mathrm{m}\alpha}(s) = \frac{\Delta\dot{m}_{\mathrm{air}}(s)}{\Delta\alpha_{\mathrm{ped}}(s)} = \frac{K_{\mathrm{m}\alpha}}{(1+T_{\mathrm{th}}s)(1+T_{\mathrm{int}}s)}e^{-T_{\mathrm{d}}s}. \qquad (7.1.26)$$

$\dot{m}_{\mathrm{air}}$ is then a representative dynamic input to the steady-state map (7.1.24). After linearization around an operating point $(\overline{M}_{\mathrm{eng}}, \overline{n}_{\mathrm{eng}})$ one obtains for the dynamic torque generation

$$G_{\alpha\mathrm{M}}(s) = \frac{\Delta M_{\mathrm{eng}}(s)}{\Delta\alpha_{\mathrm{ped}}(s)} = G_{\mathrm{m}\alpha}(s)K_{\mathrm{mM}} \qquad (7.1.27)$$

with

$$K_{\mathrm{mM}} = \left.\frac{\Delta M_{\mathrm{eng}}}{\Delta\dot{m}_{\mathrm{air}}}\right|_{\overline{M}_{\mathrm{eng}},\overline{n}_{\mathrm{eng}}}. \qquad (7.1.28)$$

Depending on the application the dead time T_{d} and the throttle-time constant may be neglected. $G_{\mathrm{m}\alpha}(s)$ reduces then to

$$G_{\mathrm{m}\alpha}(s) = \frac{\Delta\dot{m}_{\mathrm{air}}(s)}{\Delta\alpha_{\mathrm{ped}}(s)} = \frac{K_{\mathrm{m}\alpha}}{(1+T_{\mathrm{int}}s)}. \qquad (7.1.29)$$

If the air path time constant is much smaller than the following vehicle dynamic time constants, the torque can be represented by a proportional behavior without dynamics

$$G_{\alpha\mathrm{M}} = \frac{\Delta M_{\mathrm{eng}}}{\Delta\alpha_{\mathrm{ped}}} = K_{\mathrm{m}\alpha}K_{\mathrm{mM}} = K_{\alpha\mathrm{M}}(\overline{M}_{\mathrm{eng}}, \overline{n}_{\mathrm{eng}}). \qquad (7.1.30)$$

The torque M_{eng} acts on the crankshaft and flywheel of the engine. An angular momentum balance yields

$$J_{\mathrm{eng}}\frac{d\omega_{\mathrm{e}}(t)}{dt} = 2\pi J_{\mathrm{eng}}\frac{dn_{\mathrm{eng}}(t)}{dt} = M_{\mathrm{eng}}(n_{\mathrm{eng}},t) - M_{\mathrm{l}}(n_{\mathrm{eng}},t) \qquad (7.1.31)$$

with J_{eng} the moment of inertia of the crankshaft, pistons and flywheel and M_{l} a counteracting load torque, see Fig. 7.1.8. The load torque is either the clutch torque $M_{\mathrm{l}} = M_{\mathrm{cl,cd}}$ from a shifted transmission or the impeller torque $M_{\mathrm{l}} = M_{\mathrm{p}}$ from a hydrodynamic torque converter, see Isermann (2015). Thus, the acceleration $d\omega_{\mathrm{eng}}/dt$ of the engine is proportional to the difference between the engine and the load torque. M_{eng} acts as applied variable on the drive train and the angular engine velocity $\omega_{\mathrm{eng}} = 2\pi n_{\mathrm{eng}}$ is a dependable variable, based on the feedback of the driving resistances from the vehicle and drive train.

According to (7.1.31) and (7.1.29) the dynamic behavior of the engine speed can be directly observed by an experiment with any gasoline engine with opened clutch, i.e. in idling status. A small step input $\Delta\alpha_{\mathrm{ped}}$ on the accelerator pedal then shows for small speed of about $\overline{\omega}_{\mathrm{eng}} = 800\,\mathrm{rpm}$ a larger time constant as for higher speed of about $\overline{\omega}_{\mathrm{eng}} = 3000\,\mathrm{rpm}$.

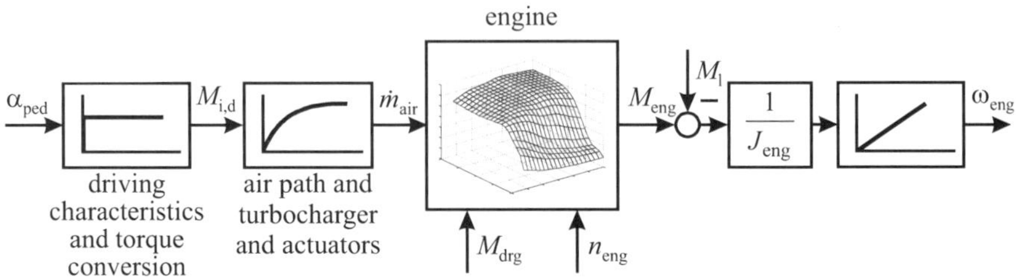

Fig. 7.1.8. Simplified overall torque model for gasoline and diesel engines with the assumption that the dominant dynamic delays can be represented by the air flow mass.

7.2 Air/fuel- and catalytic converter control

As described in Sect. 7.1 the injected fuel mass is feedforward controlled based on the measured or calculated air mass. As the combustion of gasoline engines and the *three-way catalyst* require a precise A/F ratio, the feedforward controlled fuel mass, which cannot be very precise because of measurement errors, engine tolerances, aging and not measured influences, has to be enlarged by a feedback control. Suitable sensors are λ-sensors which measure the oxygen concentration, first introduced in 1976. There are mainly two types of λ-sensors. The first type is a solid-state electrolyte made of ZrO ceramic material. At high temperatures the electrolyte becomes conductive and generates a galvanic voltage. The resulting characteristic is of jump type, see Fig. 7.2.1a). This means it can be used as a *two-point sensor* around $\lambda = 1$. (This sensor type is also called "Heated Exhaust Gas Oxygen Sensor" (HEGO).)

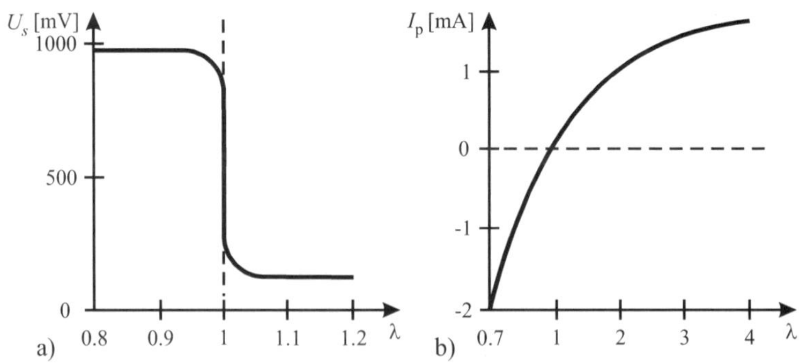

Fig. 7.2.1. Characteristics of lambda-oxygen sensors in dependence on the air excess factor λ. **a** two-point sensor: electrode-voltage U_s. **b** continuous (broad-band sensor): pump-cell current I_p.

The second type consists of multilayer ceramics, with a galvanic concentration cell (Nernst cell) and a pump cell. A voltage at the pump cell controls $\lambda = 1$ in the concentration cell by pumping oxygen away for lean exhaust gas or pumping it in for rich exhaust gas. Therefore the sign and the magnitude of the pump current becomes a nonlinear measure for λ, see Fig. 7.2.1b). This means that this *continuous lambda sensor* or broadband lambda sensor (also called "Universal Exhaust Gas Oxygen Sensor" (UEGO)) can be used for $0.7 < \lambda < 4$. A detailed description of the principles of λ-sensors is treated in Baunach et al (2006).

The exhaust gas treatment for gasoline engines with stoichiometric combustion is based on catalytic converters of different types. The *three-way catalyst* is the most used one. It is able to reduce the main pollutants CO, HC, and NO_x. CO and HC are converted to CO_2 and H_2O by oxidation. The oxygen is taken from NO_x which is converted by CO to N and O_2. The catalytic converter is designed of a sheet-metal housing, a substrate and a catalytic noble-metal coating. A ceramic or metallic substrate layer offers a large effective surface with platinum or palladium. However, the optimal conversion requires a stoichiometric combustion with $\lambda = 1$ and allows only very small deviations.

Oxygen-type catalytic converters store excess oxygen in a lean phase and give it away in a rich phase. The oxygen content may therefore oscillate to a small amount but must stay upstream accurately within $\lambda = 1 \pm 0.005$.

The NO_x *accumulator-type catalytic converter* is required for direct injection gasoline engines which operate with lean combustion. The oxygen necessary to oxidize HC and CO is taken from the excess air in the exhaust gas. This catalyst carries, e.g. barium oxide which can store NO_x. During lean operation (e.g. stratified mixture) NO_x is stored and oxidized to NO_x and is turned to e.g. barium-nitrate. If the storage is saturated, it must be regenerated by a rich combustion ($\lambda < 0.8$) in order to reduce the nitrate to N. This manipulation of lean and rich combustion requires an NO_x or λ sensor after the catalytic converter.

The starting temperature of the converters (light-off) is about $300°$C for the three-way catalyst and $200°$C for the NO_x accumulator type. The converters must be protected against too high temperatures of $> 800°$C.

A variety of *catalytic converter configurations* exist, Robert Bosch GmbH (2011). Usually a primary catalytic converter is used close to the exhaust manifold (for high temperatures and fast light-off) and a main converter under floor (for low light-off temperature), see Fig. 7.2.2 or only one main converter. For homogeneous stoichiometric combustion one or two three-way catalysts are used and for direct injection with lean mixtures a three-way catalyst followed by a NO_x accumulator-type.

The closed-loop λ-control for the catalytic converters and the combustion depends on the type of converters and the used sensors. For homogeneous stoichiometric combustion and use of *one three-way catalyst* the oxygen control variable is measured with a two-point λ-sensor upstream, i.e. in front of the catalyst. Because of the jump characteristic at $\lambda = 1$ a *two-step control loop* results. The resulting controller changes the correcting factor or efficiency η_λ which is then multiplied with the desired fuel mass $m_{f,d}$, as shown in Figs. 7.2.2 and 7.1.3. Figure 7.2.3 shows the time behavior. The manipulated variable consists of a jump and a ramp, resulting in a deviation $\Delta\lambda$ of about 0.02.

Because of the asymmetric characteristic of the λ-sensor and nonlinearities of the engine between $\Delta\eta_\lambda$ and $\Delta\lambda$ the mean value of the controlled λ has an offset. This can be compensated by making the dwell-time of the ramp signal larger for

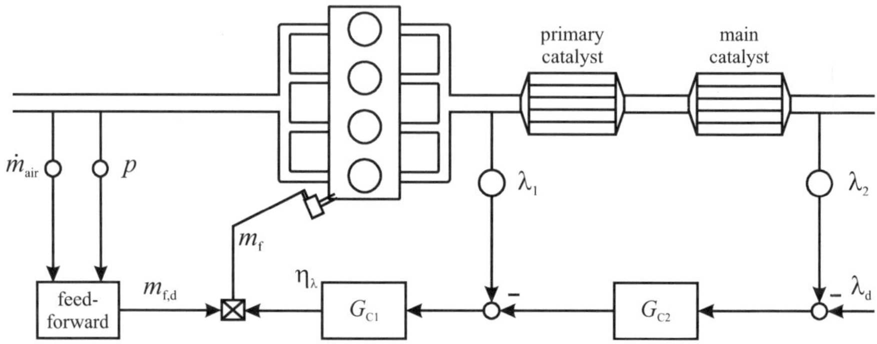

Fig. 7.2.2. Example for a λ-control system with two catalytic converters and two lambda sensors, one upstream the primary catalyst and one downstream the main catalyst.

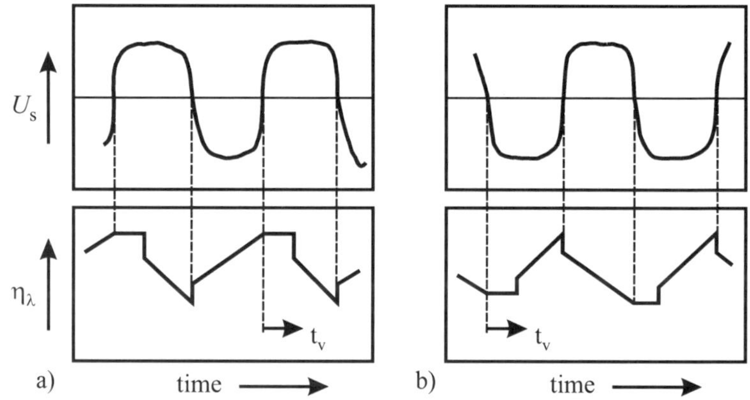

Fig. 7.2.3. Time behavior of two step λ-control with two-point λ sensor. **a** change to rich. **b** change to lean, Robert Bosch GmbH (2007). t_v: dwell time following the sensor jump.

the jump from rich to lean, see Guzzella and Onder (2010). A *steady λ-control* can be obtained by using a continuous-type λ-sensor. This may result in a better control performance and allows the control of lean or rich combustion.

For higher performance of the emission reduction *two catalytic converters* became standard with a *cascaded control system*, consisting of an inner and outer loop, as shown in Fig. 7.2.2. Then usually a continuous λ-sensor (broadband) is used in front of the primary catalyst to obtain a closed inner loop with fast dynamics. An outer loop with a second sensor downstream the main converter with two-point characteristic can then react slowly and corrects the reference variable of the inner loop. The second downstream sensor is less stressed and influenced by aging, as a sensor upstream of the primary catalyst. Hence, a cascaded control system results, with e.g. a PI controller as major controller. An additional NO_x sensor downstream of a NO_x accumulator type converter can be used to control and supervise the regeneration of this converter.

To obtain at least a rough model of the process for the λ-controller the signal flow in Fig. 7.2.4 is considered. The input of the controlled process is a change of the injected fuel mass and the output a change of the λ-sensor signal. It is assumed that the air-mass flow rate is constant.

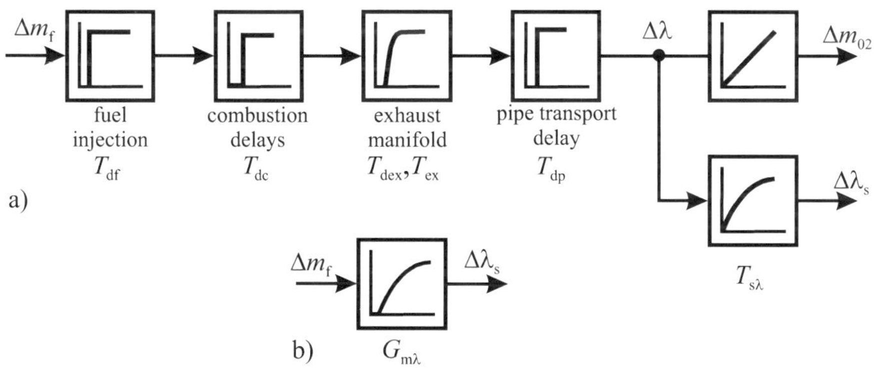

Fig. 7.2.4. Signal flow for the λ-control process of the primary catalytic converter. **a** process elements. **b** approximate model.

For intake manifold injection close to the inlet valves or direct injection in the induction stroke for homogeneous combustion it can be assumed that after a very small delay T_{df} of the ECU and the injector, the exhaust gas at the outlet valve is swept out by the pistons after about 180°to 360°CS, corresponding to an average dead time

$$T_{\mathrm{df}} = \frac{1}{\omega_{\mathrm{eng}}}(\pi + 2\pi) = \frac{3\pi}{\omega_{\mathrm{eng}}}. \qquad (7.2.1)$$

This repeats for each cylinder and because all cylinders are assumed to provide the same gas concentrations, a step function for the air excess factor λ with dead time T_{dc} results, see Fig. 7.2.4.

The gas concentrations then pass the exhaust manifold. Without mixing a pure time delay T_{dex} and with mixing a small time constant T_{ex} combined with a small time delay can be assumed. After a further transport delay T_{dp} through the exhaust pipe the oxygen concentration is measured with the λ-sensor for which data sheets give a time constant of $T_{\mathrm{s}\lambda} \approx 100\,\mathrm{ms}$.

Altogether the resulting transfer behavior may be approximated by a first order plus dead time element.

$$G_{\mathrm{m}\lambda}(s) = \frac{\Delta\lambda_{\mathrm{s}}(s)}{\Delta m_f(s)} = \frac{K_{\mathrm{m}\lambda}}{1 + T_{\mathrm{m}\lambda}s} e^{-T_{\mathrm{dm}\lambda}s} \qquad (7.2.2)$$

with a time constant

$$T_{\mathrm{m}\lambda} = T_{\mathrm{ex}} + T_{\mathrm{s}\lambda} \qquad (7.2.3)$$

and dead time

$$T_{\mathrm{dm}\lambda} = T_{\mathrm{df}} + T_{\mathrm{dc}} + T_{\mathrm{dex}} + T_{\mathrm{dp}} \approx 1.5 T_{\mathrm{dp}} \tag{7.2.4}$$

see also Alfieri et al (2009), Yildiz et al (2010).

Table 7.2.1 presents some data which was obtained for a four-cylinder diesel engine, Kimmich et al (2001). The determination of the time constant of a broadband sensor for a four-cylinder diesel engine by identification methods resulted in

$$T_{\mathrm{s}\lambda} = 116\,\mathrm{ms} - 0.07 \left[\frac{\mathrm{ms}}{\mathrm{mg}/stroke}\right] m_{\mathrm{air}} \tag{7.2.5}$$

i.e. is given in dependence on the air mass m_{air} per stroke, Kimmich et al (2001).

Table 7.2.1. Time delays and time constants of the λ-control process ($T_{\mathrm{ex}} \approx 0$). Diesel engine: 2.0 l, 4 cyl, 74kW. λ-sensor: LSU 4.2, Bosch, Kimmich et al (2001)

n	Time constant combustion	Time delay exhaust pipe	Time constant λ-sensor	Approximate model $G_{\mathrm{m}\lambda}(s)$	
[rpm]	T_{dc}[ms]	T_{dp}[ms]	$T_{\mathrm{s}\lambda}$[ms]	$T_{\mathrm{m}\lambda}$[ms]	$T_{\mathrm{dm}\lambda}$[ms]
1000	90	40	100	100	130
4000	23	15	80	80	38

The approximate model then shows larger time constants and time delays for small speed and smaller values for higher speeds. These values have about the same size as in Guzzella and Onder (2010), Rupp et al (2008). A standard controller G_{c1} for the process (7.2.2) is a PI controller. Several publications on the design of the air/fuel control exist like Ault et al (1994), Powell et al (1998), Won et al (1998), Rupp et al (2008), Yildiz et al (2010). Because of the variable dead time and gain of the controlled process, the parameters of a PI controller have to be tuned such that a robust control behavior with slow control behavior results, or the controller parameters have to be adapted to the operating points of the engine, thus leading to a so-called gain-scheduled controller.

An adaptive air/fuel ratio control is proposed by Rupp et al (2008). By applying an internal model controller to a first order low pass process with dead-time and a first order model of a wide-range lambda sensor an adaptation to changing sensor time constants is obtained with a stochastic approximation gradient method.

Yildiz et al (2010) developed model reference adaptive controllers with low-pass reference models, known dead-time and gradient-adaptation laws and obtained improvements compared to the gain-scheduled PI controller.

The oxygen storage capability of a three-way catalytic converter can be described by setting

$$\frac{d}{dt} m_{\mathrm{O_2,store}} = \dot{m}_{\mathrm{O_2}} \Delta\lambda \tag{7.2.6}$$

which leads to an integrator

$$\Delta m_{\mathrm{O_2,store}} = \frac{1}{s} \dot{m}_{\mathrm{O_2}} \Delta\lambda \tag{7.2.7}$$

see Guzzella and Onder (2010) with $\dot{m}_{\mathrm{O_2}}$ the oxygen mass flow at the inlet.

7.3 Ignition feedforward control

For gasoline engines, the compressed air-fuel mixture is ignited by an electric spark of a spark plug. The ignition system consists of an ignition circuit with an ignition coil for inductive ignition and a spark plug, where the generated high voltage generates an electric spark between the spark-plug electrodes, see Fig. 7.3.1. The ignition driven stage in the ECU switches before the moment of ignition a current through the primary winding of the ignition of the ignition coil, such that a magnetic field builds up (dwell period). At the moment of ignition the current through the primary winding is interrupted. The magnetic field energy is then discharged to the magnetic coupled secondary winding (about 1:100 turns ratio) by induction, producing a high voltage there, which generates the spark at the spark plug. The current in the primary coil needs some time, the dwell period, to build up the magnetic field with energy $E = \frac{1}{2}LI^2$ because of the induced counter voltage. If the ignition voltage at the spark pug is reached, the flashover occurs and ignition begins, compare Fig. 7.3.2.

After the flashover, the voltage at the spark plug reduces to the spark voltage. This spark voltage depends on electrode gap and flow phenomena influencing the spark plasma and is about 1kV with an ignition energy of about $30 - 100\,\mathrm{mJ}$. The spark duration varies between $100\mu\mathrm{s}$ and $2\,\mathrm{ms}$. The high-temperature plasma then starts a flame front through the air-fuel mixture.

The instant of the ignition spark has to be precisely calibrated, because it influences the combustion, torque generation, emissions and fuel consumption. It depends mainly on engine speed, air charge, excess air and engine temperature, see Fig. 7.3.3. The instant of ignition is referred to the top dead center (TDC) of the power cycle. It is called advanced if it is before top dead center, expressed as ignition advance angle

$$\varphi_{\mathrm{ign}} = -(\varphi_{\mathrm{cs,ign}} - 360°\mathrm{CS}). \tag{7.3.1}$$

The combustion of the air/fuel mixture can be divided in two phases, Kiencke and Nielsen (2000). During the *inflammation delay* Δt_{id} the cylinder pressure and temperature do not increase much. It depends on λ, p_{cyl} and T_{cyl}, is proportional to time and in the range of ns to a few ms, Stiesch (2003). The second phase depends on the flame propagation through the A/F mixture and is influenced by swirl, tumble and turbulences. It results as a combustion delay angle φ_{fd} which is almost constant for the engine operating ranges. Hence, the overall combustion delay angle can be expressed by

$$\varphi_{\mathrm{dc}} = \omega_{\mathrm{eng}}\Delta t_{\mathrm{id}} + \varphi_{\mathrm{fd}} \tag{7.3.2}$$

which corresponds to a time delay of

$$T_{\mathrm{dc}} = \Delta t_{\mathrm{id}} + \frac{\varphi_{\mathrm{fd}}}{\omega_{\mathrm{eng}}}. \tag{7.3.3}$$

Some formulas to calculate the ignition delay between the ignition and 5 % release of heat MFB_5 are summarized in Merker et al (2006), p. 169.

Figure 7.3.4 shows examples for the indicated pressure $p_{\mathrm{cyl}}(\varphi_{\mathrm{ign}})$. The largest (retarded) ignition advance angle is determined by the combustion limit and the maximum possible exhaust-gas temperature and the smallest ignition advance angle by

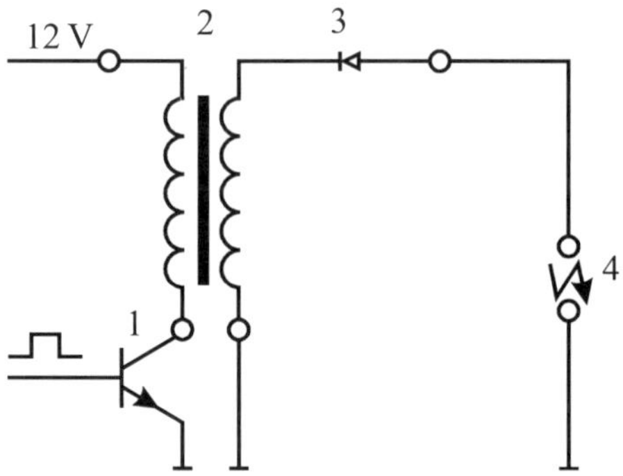

Fig. 7.3.1. Ignition circuit with single spark coils, Robert Bosch GmbH (2011). 1 ignition driver stage (ECU), 2 ignition coil, 3 activation arc diode, 4 spark plug.

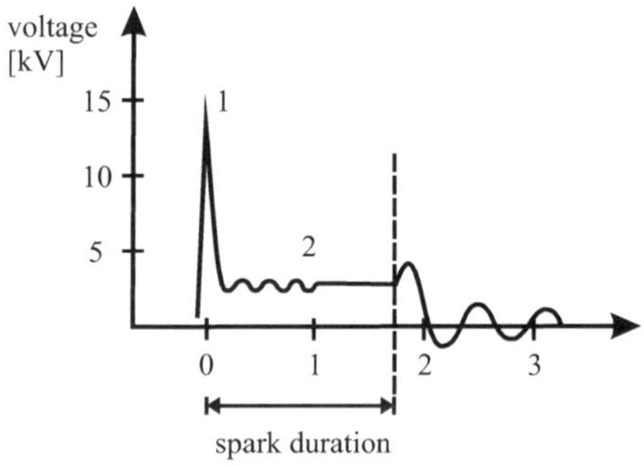

Fig. 7.3.2. Voltage at the electrodes of spark plug. 1 ignition voltage, 2 spark voltage.

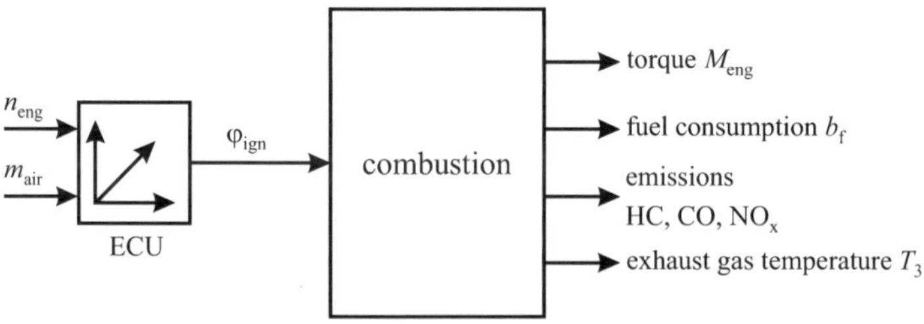

Fig. 7.3.3. Influenced variables by the ignition angle.

the knock limit, Fig. 7.3.4a). In between is an optimal ignition advance angle. This optimum is determined by the resulting torque, fuel consumption, emissions and exhaust gas temperature. A maximum torque is obtained for a high pressure curve, such that the integral over pressure $p_{\text{eff}}(\varphi)$ gets maximal, compare (4.3.14). The peak pressure should be approximately at $12 \ldots 16°$CS after TDC, Heywood (1988), p. 375, Müller (2003), or the combustion center at $8°$CS, Bargende (1995b). Generally, more spark advance increase the torque and reduce fuel consumption, but it raises HC and NO_x emissions. Retarded ignition results in a higher exhaust gas temperature.

A suitable compromise between these partially contradicting effects has to be found during the basic calibration of the engine. One begins for a certain operating point $(m_{\mathrm{air}}, n_{\mathrm{eng}})$ with an average (safe) ignition angle, then shifts to advanced angles until either the torque maximum or the knock limit is reached. Then the ignition advance angle is retarded in steps of e.g. $4.5°$CS until a stopping criterion is reached, Stuhler and Kruse (2003), compare Fig. 7.3.5a). During the measurement also the other outputs HC, $\mathrm{NO_x}$ and T_3 are measured, also for different operating points $(m_{\mathrm{air}}, n_{\mathrm{eng}})$. The ignition must usually be advanced with increasing speed, increasing air charge and lean A/F mixture. According to optimization criteria finally the optimal ignition advance angles are specified

$$\varphi_{\mathrm{ign,opt}} = f(m_{\mathrm{air}}, n_{\mathrm{eng}}, \lambda, T_{\mathrm{eng}}) \tag{7.3.4}$$

and stored in maps for the ECU, taking into account an optimal balance between b_f, T_3, HC, $\mathrm{NO_x}$, CO, and knocking limit in steady state conditions. Figure 7.3.5b) depicts an example for the measured torques in dependence on the ignition angle for different speeds and air charge. It shows the maximal torque which is not calibrated, because it is too close to the knock limit, and the optimal calibrated torque.

Figure 7.3.6a) depicts the signal flow from ignition command to torque generation. After a very short ignition delay the indicated pressure $p_{\mathrm{cyl}}(\varphi)$ of cylinder i develops depending on air charge, fuel injection and ignition angle $\varphi_{\mathrm{ign},i}$. A gas torque $M_{\mathrm{g},i}$ results, see (4.3.29) and by averaging over one stroke the average torque $\overline{M}_{\mathrm{cyl},i}$ of one cylinder is obtained. The same torque is then provided by the next cylinders with phase shifts of $180°$CS for a four cylinder and $120°$CS for a six cylinder engine. This means a step increase after an angle delay of about 10 to $40°$CS which corresponds to a dead time of

$$T_{\mathrm{dM}} = (0.06 \ldots 0.22)\frac{\pi}{\omega_{\mathrm{eng}}}. \tag{7.3.5}$$

Thus for $n_{\mathrm{eng}} = 600$ to $6000\,\mathrm{rpm}$ this dead time may vary between 0.3 and 11 ms. After subtracting the friction torque and load torque of the engine integrated auxiliaries the engine torque at clutch M_{eng} results.

The transfer behavior between a change of the ignition angle and the cylinder torque can therefore be approximated by

$$G_{\varphi\mathrm{ignM}}(s) = \frac{\Delta\overline{M}_{\mathrm{cyl}}(s)}{\Delta\varphi_{\mathrm{ign}}(s)} = K_{\varphi\mathrm{ignM}} e^{-T_{\mathrm{d}} s} \tag{7.3.6}$$

with the small dead time

$$T_{\mathrm{d}} = T_{\mathrm{dc}} + T_{\mathrm{dM}} \tag{7.3.7}$$

in the range of a few ms.

The steady state behavior is shown in Fig. 7.3.6b) and consists of the measured torque characteristic for varying ignition angle.

The ignition angle is usually feedforward controlled by using maps and correction (efficiency) factors in dependence on m_{air}, n_{eng}, λ and T_{eng} for all cylinders

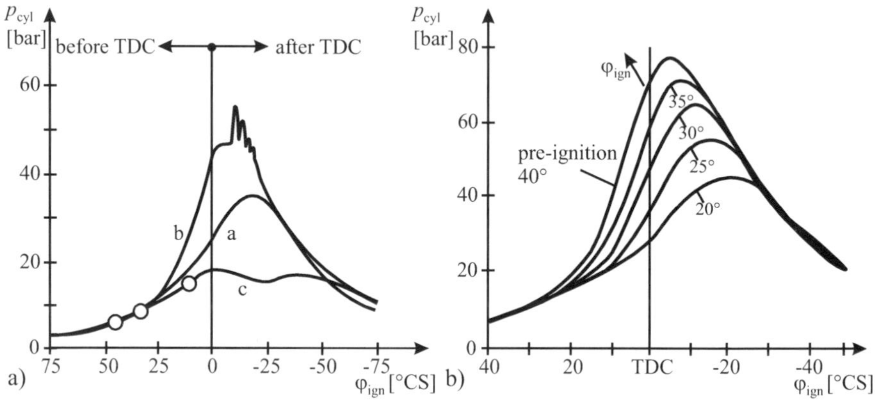

Fig. 7.3.4. Combustion pressure versus ignition advance angle φ_{ign}. **a** a: correct ignition advance, b: too much advance, c: too less advance, Robert Bosch GmbH (2011). **b** influence of the ignition angle on the indicated combustion pressure, Merker et al (2006).

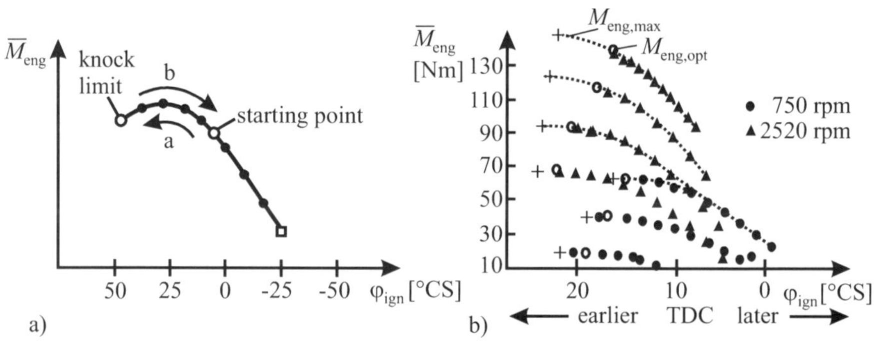

Fig. 7.3.5. Torque for homogeneous combustion in dependence on ignition angle. **a** principle of course for m_{air} and n_{eng} = constant. (ignition timing cycle). **b** maximal torque $M_{eng,max}$ and optimal calibrated torque $M_{eng,opt}$ for different speeds and air charge, BEG (2009).

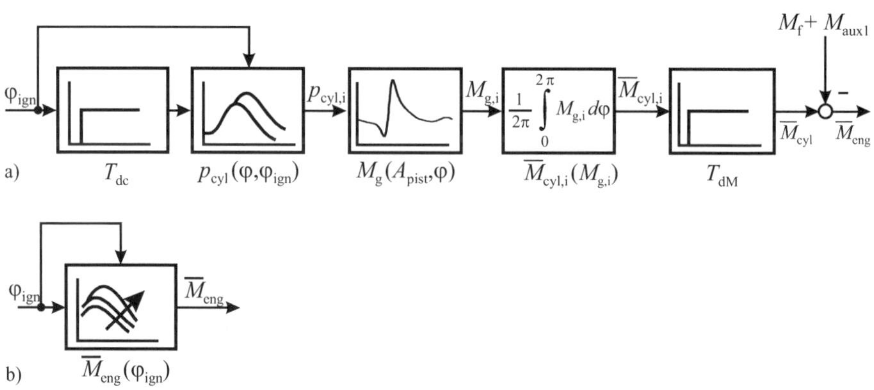

Fig. 7.3.6. Signal flows for ignition control. **a** dynamic behavior. **b** steady state behavior.

simultaneously. However, engine specific tolerances, cylinder dependent properties and engine aging are not taken into account. For example, the position of the combustion center φ_{Q50} between the cylinders of a 4.4l V8 engine show differences of $2 \ldots 8°CS$, Müller (2003). This motivates the development of closed loop ignition control systems, as described in Sects. 7.4 and 7.5.

7.4 Ignition feedback control

The objective of ignition control is to achieve optimum engine efficiency for each combustion event. General factors that influence the optimal ignition angle are engine specifications like configuration of the combustion chamber, operating conditions like engine speed, load, temperature and EGR flow rate, as well as ambient conditions such as air temperature, air pressure and humidity in the atmosphere.

Standard ignition control systems are based on feedforward control and therefore rely heavily upon the calibration of look-up tables. The database values are initially calibrated from an analysis of a nominal engine under fixed environmental conditions. However, changing environmental conditions, aging effects, different fuel quality, and manufacturing tolerances usually change an engine's characteristics and lead to a deteriorating performance. This motivates the development of closed loop systems.

However, closed loop ignition control needs a suitable control variable. An extremum value control of the measured torque was investigated by Scotson and Wellstead (1990). Ion current sensors at the spark plug need only relatively small additional effort. But the measurable amplitudes are difficult to evaluate and become small for retarded ignition angles, lean mixtures or mixtures with high residual gases and for low load, Willimowski (2003). Hence, the combustion pressure remains as an attractive variable for feedback control.

An early proposal for a combustion pressure-based ignition control by using the *peak pressure* was proposed by Glaser and Powell (1981). The control of the crank angle position of the maximal combustion pressure by changing the ignition angle was investigated by Fujii et al (1989), Zarske (1988), Rohe-Brandenburger (1986) and Heywood (1988), p. 375. The optimal position of the peak pressure was found to be $12 \ldots 16°CS$ after TDC. An ignition feedback control based on the peak pressure position was realized by Zarske (1988) and Kawamura et al (1988). However, the determination of the position of the maximal pressure requires a relatively high resolution of the crank angle measurement, at least close to TDC.

An alternative is to use a measure for the position of the *center of the combustion*, like the mass fraction burned MFB_{50} or φ_{Q50} at 50 % heat release. Optimal efficiency can be reached by placing MFB_{50} at $\varphi_{Q50} = 8 \ldots 10°CS$, Bargende (1995b), Hofmann (1984), Klimstra (1985), Matekunas (1986), or $6.5 \ldots 8.5°CS$, Hockel (1981).

An ignition control based on the center of combustion was realized by Held and Schubert (1994) and Truscott et al (2001). A linear feedback control was combined with a fixed feedforward control map. But this does not allow a fast convergence after operating point changes. The online calculation of the heat release with a special designed ASIC was realized by Hohenberg and Dolt (1993), showing that in principle it is possible to evaluate pressure-based features in real time.

Müller (2003) compared the use of the crank angle position of the peak pressure $\varphi_{\mathrm{p\,max}}$, the pressure difference maximum $\varphi_{\Delta\mathrm{p\,max}}$ and the center of combustion φ_{Q50}. Measured characteristics for a 1 l, 3 cyl gasoline engine show an approximate quadratic relation between indicated torque and crank angle position $\varphi_{\mathrm{p\,max}}$ of the *maximum combustion pressure* with a flat maximum at $10\ldots15°$CS after TDC. However, retarded combustions, especially for low loads, high exhaust gas mass flow rates or lean mixtures, lead to wrong peak pressure positions. Therefore the peak pressure maximum is not a suitable measure for the whole operating range, see also Zarske (1988).

An alternative is the use of the position $\varphi_{\Delta\mathrm{p\,max}}$ of the *combustion pressure difference* to the motored pressure

$$\Delta p_{\mathrm{cyl}}(\varphi) = p_{\mathrm{cyl}}(\varphi) - p_{\mathrm{cyl,\,mot}}(\varphi). \tag{7.4.1}$$

This way, the problem of peak pressure detection for retarded combustions is avoided. But this pressure difference alone is not a measure for the converted energy. However, as for the converted combustion energy holds

$$Q_f \sim \Delta p_{\mathrm{cyl}}(\varphi) V^n(\varphi) \tag{7.4.2}$$

for a polytropical expansion with the polytropical exponent n, the heat release function can be approximated by

$$x_{\mathrm{B},\Delta\mathrm{p}}(\varphi) = \frac{\Delta p_{\mathrm{cyl}}(\varphi) V^n(\varphi)}{\Delta p_{\mathrm{eoc}} V^n_{\mathrm{eoc}}} \tag{7.4.3}$$

with Δp_{eoc} and V_{eoc} at end of combustion.

However, this method still requires a relatively high resolution of the crank angles and the measurement of the motored cylinder pressure for all operating conditions.

The reconstruction of the *center of combustion* φ_{Q50} does not require more computations as the previously discussed methods and allows a lower resolution of the crank angle. Figure 7.4.1 shows the heat release function and the indicated torque in dependence on the combustion center position φ_{Q50}. As this feature is directly related to the converted combustion energy and retarded combustions do not lead to wrong results, its application to a closed loop ignition control will be described in the next section.

7.5 Combustion pressure-based ignition control

Cylinder pressure signals contain valuable information for closed-loop engine control. For using this information low-cost combustion pressure sensors with high long-term stability have been developed, Anastasia and Pestana (1987), Herden and Küsell (1994) and are starting to be installed into production engines, Inoue et al (1993).

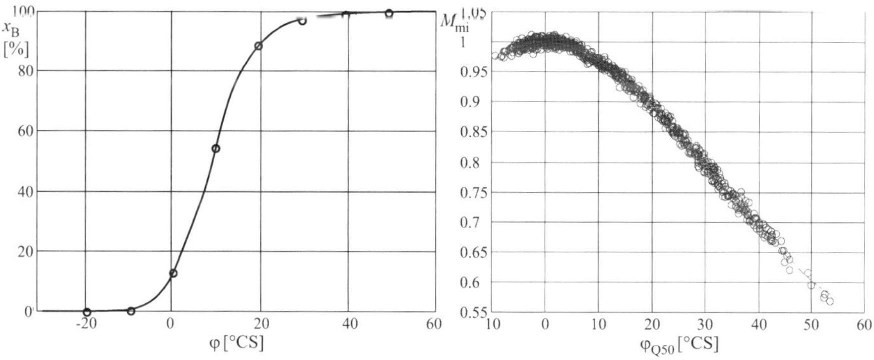

Fig. 7.4.1. **a** heat-release function for crank angle intervals of 10° CS. **b** Indicated torque in dependence on combustion center position for different ignition angles. Opel X10XE Ecotec, 1 l, 3 cyl, 40 kW, $n_{eng} = 2000$ rpm, load 35 %; $-32°\text{CS} < \varphi_{ign} < -5.25°\text{CS}$.

Real-time cylinder pressure evaluation is, however, a demanding task and requires powerful computational resources. Sampling of cylinder pressure signals is usually performed in a crankshaft synchronous manner. A typical resolution is 1, 3 or 6 crank-angle degrees (CS), which results in sampling rates up to 40, 13 or 6.7 kHz, depending on engine rotational speed. Nevertheless, increasing computational performance as well during the control design stage used rapid control prototype (RCP) systems, as for future engine control units entail an increasing interest in benefiting from combustion pressure information. A variety of engine control functions could be improved or implemented using cylinder pressure sensors, see e.g. Powell (1993), Pestana (1989), Sellnau et al (2000). The implementation of a closed loop ignition control system is described in the next Sect. 7.5.1. A further possibility to optimize the ignition angle is an extremum value control which will be described in Sect. 7.5.2.

Combustion pressure sensors allow to optimize the point of ignition of each cylinder separately. The variable which is to be controlled is calculated from the "mass fraction burned" (MFB) signal, which can be derived from cylinder pressure evaluations, see Isermann and Müller (2001). Measurements and theoretical analysis reveal, that optimal ignition, i.e. maximum torque from each combustion event, can be obtained if 50 % of fuel mass has been burned until a crank angle of about 8° after top dead center (TDC), Bargende (1995a). This crank angle is also referred to as the "*center of combustion*". The proposed approach calculates the crankshaft angle of 50 % MFB for each combustion cycle and controls it at e.g. $\varphi_{Q50} = 8°$CS after TDC. The following treatment follows Müller (2003), Vogt et al (2004), Isermann and Müller (2003).

7.5.1 Combustion pressure sensors

The combustion pressure in the combustion chambers is usually measured at engine test benches and used for basic investigations and engine development. The used sensors are mostly piezoelectric sensors which are water cooled in order to reduce temperature effects on the measurement quality. Because they are too expensive for the application in series production engines, other sensor types like piezoceramic, piezoresistive or fiber optic sensors were developed, Jost (2000), Hellemans et al (2011).

The integration of piezoelectric, piezoresistive and ionic current sensors in the glow-plug of diesel engines is reported in Borger et al (2004)), see also Moriwaki et al (2003). The piezoresistive glow-pug sensor operates with a metal diaphragm and a micro fused strain gauge Wheatstone bridge with temperature compensation. The metal diaphragm is separated from the combustion chamber by the heating rod. However, lower cost sensors show some measurement errors. These sensors usually measure differences to a zero-point. A not precisely known zero-point therefore leads to a sensor offset. The zero-point can be determined by a thermodynamic analysis, Hohenberg (1982) or with reference sensors. Changing mean temperature results in middle and long term drifts. The sensitivity (gain) of the sensors depends also on the pressure indicating a hysteresis error. The strong temperature changes during one stroke influence to measurement diaphragm by an unsteady heat flow, which shows up in short-term drift, also called thermo shock. Figure 7.5.1 shows the sensor errors for a piezoelectric sensor, modeled in Kohlhase (2011).

In the ideal case the measured voltage is proportional to the pressure

$$U_{\mathrm{s}}(\varphi) = K_{\mathrm{s}}(T,p)\, p_{\mathrm{cyl}}(\varphi) + U_{\mathrm{offset}} \tag{7.5.1}$$

and the pressure yields

$$p_{\mathrm{cyl}}(\varphi) = \frac{1}{K_{\mathrm{s}}(T,p)}\left(U_{\mathrm{s}}(\varphi) - U_{\mathrm{offset}}\right). \tag{7.5.2}$$

If for the real sensor the sensitivity and the offset is $\hat{K}_{\mathrm{s}}$ and $\hat{U}_{\mathrm{offset}}$, then the determined pressure becomes

$$\hat{p}_{\mathrm{cyl}}(\varphi) = \frac{1}{\hat{K}_{\mathrm{s}}}\left(U_{\mathrm{s}}(\varphi) - \hat{U}_{\mathrm{offset}}\right). \tag{7.5.3}$$

The measured voltage is further influenced by the mean temperature dependent offsets $U_{\mathrm{offset}}(T)$ and the short term thermo shock distortion $\Delta p_{\mathrm{ths}}(\varphi)$

$$U_{\mathrm{s}}(\varphi) = K_{\mathrm{s}}(T,p)\left(p_{\mathrm{cyl}}(\varphi) + \Delta p_{\mathrm{ths}}(\varphi)\right) + U_{\mathrm{offset}}(T). \tag{7.5.4}$$

Introducing in (7.5.3) yields the measured value with errors

$$\begin{aligned} \hat{p}_{\mathrm{cyl}}(\varphi) &= \frac{K_{\mathrm{s}}(T,p)}{\hat{K}_{\mathrm{s}}}\left(p_{\mathrm{cyl}}(\varphi) + \Delta p_{\mathrm{ths}}(\varphi)\right) + \frac{U_{\mathrm{offset}}(T) - \hat{U}_{\mathrm{offset}}}{\hat{K}_{\mathrm{s}}} \\ &= p_{\mathrm{cyl}}(\varphi) + \Delta p_{\mathrm{error}}(\varphi). \end{aligned} \tag{7.5.5}$$

Additionally, the measured voltage contains noise through the measured pressure, quantization and crank angle errors. Examples show that for a gasoline engine the typical sensor errors are as given in Table 7.5.1, Kohlhase (2011).

These sensor errors lead in the case of the optimization of the ignition angle by using the center of combustion φ_{Q50} to maximal errors of the ignition angle of about 2.5°CS and by directly using the ignition angle to about 0.5°CS, Kohlhase (2011), Kohlhase and Isermann (2009).

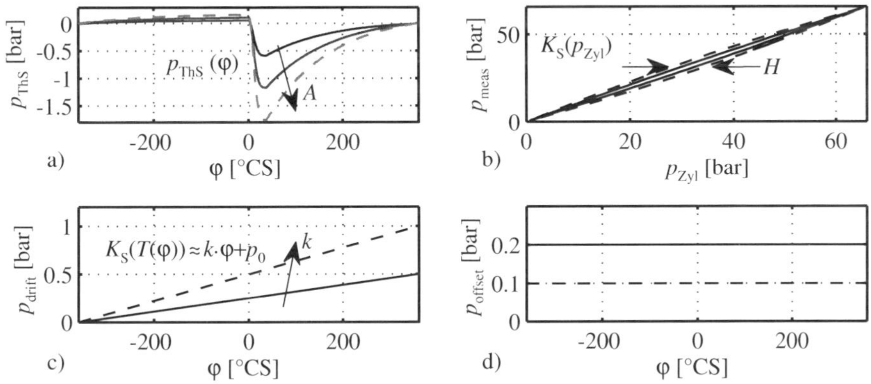

Fig. 7.5.1. Sensor errors of piezoelectric pressure sensors in dependence on the crank angle and cylinder pressure (example, simplified models), Kohlhase (2011).

Table 7.5.1. Errors for a combustion pressure piezoelectric sensor (example)

offset error	0.3 bar
hysteresis	3 % of $(p_{max} - p_{min})$
drift	0.5 bar/720°CS
thermo shock	± 1.5 % of p_{max}

7.5.2 Adaptive feedforward ignition control

a) Cylinder pressure evaluation

As described in Sect. 4.2 the mass fraction burned (MFB) $x_B(\varphi)$ from a heat release analysis is a measure for the energy conversion in dependence on the crank angle, see (4.2.65). To ease calculations, the MFB can be approximated by

$$x_B(\varphi) = \frac{p(\varphi)\,V^{\kappa}(\varphi) - p_i V_i^{\kappa}}{p_{eoc} V_{eoc}^{\kappa} - p_i {V_i}^{\kappa}} \tag{7.5.6}$$

compare (4.2.71). This approximation is valid for the assumption of a constant volume combustion cycle, compare Fig. 7.5.2, and is derived in Sect. 4.2.3c).

Figure 7.5.3 depicts the heat release for an arbitrary cycle the approximation of MFB according to 7.5.6 with the MFB calculated by thermodynamic analysis of the cylinder pressure for an arbitrary cycle. In the presented control system the crank angle location of $x_{B50}(50\,\%)$ is calculated by means of this approximation.

b) Control structure

Applying closed-loop ignition control with, for example, standard PI controllers cannot provide acceptable control performance under fast changing operating conditions since the controllers cannot be tuned for high control performance. First, this is due to the fact that the ignition angle can only be computed for the next cycle, based on measurements from the present engine cycle. Therefore a dead time of one cycle is inherent. Secondly, significant cycle fluctuations exist even under steady operating conditions. Therefore, the results of cylinder pressure evaluation of several engine cycles have to be averaged (e.g. a moving average over 10 cycles). The stochastic nature of the combustion events can be seen in Fig. 7.5.4. The upper diagram shows the measured cylinder pressure signals of the compression and the power strokes of 100 consecutive cycles of one cylinder. The dotted lines represent the reconstructed polytrope for the towed or motored pressure (without fuel injection). The lower diagram depicts the corresponding crank angle locations of 50 % MFB and shows the stochastic nature of the combustion events. As the MFB_{50} oscillates considerably from cycle to cycle and because of the dead time of one cycle the feedback controller can only be realized with small gain, thus resulting in poor control performance.

This motivates the use of Adaptive Feedforward Control (AFFC), whose general structure is shown in Fig. 7.5.5. The linear controller C is used to compensate random disturbances and to correct the adaptive feedforward controller. It does not need to have a high performance and can be designed in such a way that it provides a robust stability. Since the adaptive feedforward controller acts instead of a controller's integral term, the linear controller has preferably a simple proportional gain which may be deactivated for strong stochastic functions.

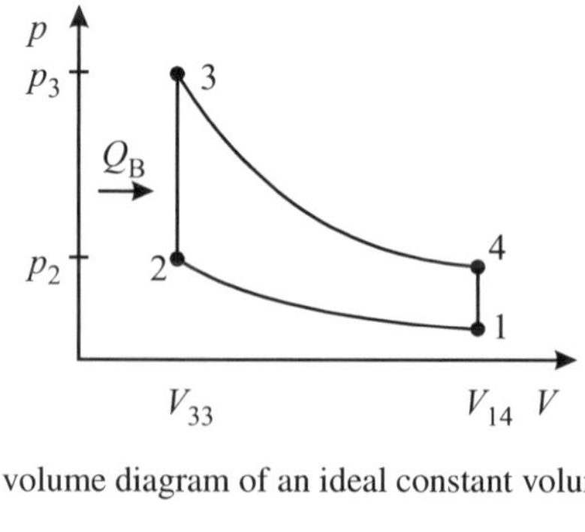

Fig. 7.5.2. Constant volume diagram of an ideal constant volume combustion cycle.

The ignition control system is divided into the conventional, fixed ignition look-up table and into the adaptive offset map as shown in Fig. 7.5.6. The operating condition is determined by the engine load (a normalized value calculated from the intake

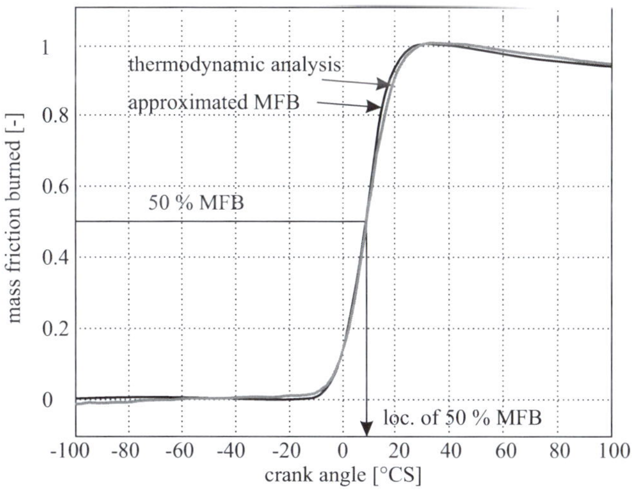

Fig. 7.5.3. Approximation of mass fraction burned (MFB) in comparison to thermodynamic heat release analysis. The location of 50 % MFB (combustion center) can be controlled by appropriate settings of the point of ignition.

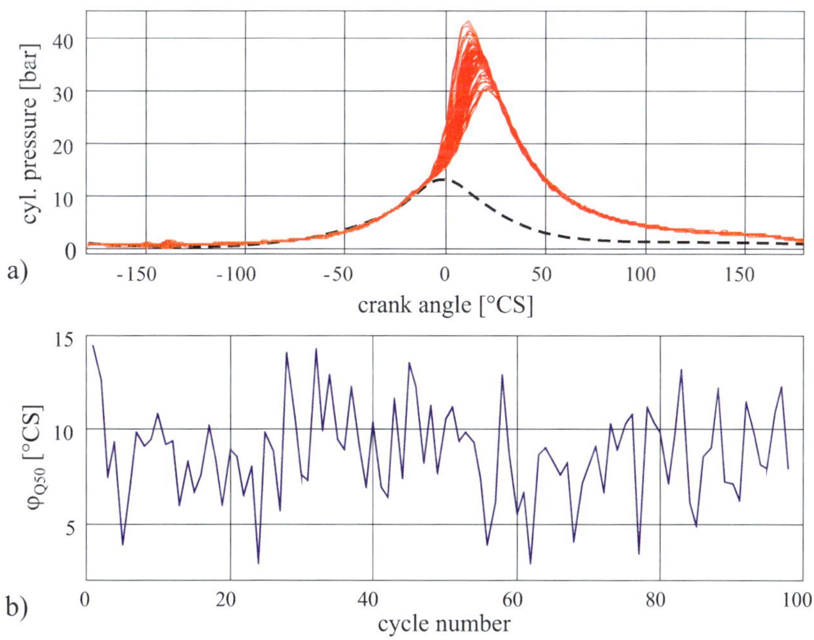

Fig. 7.5.4. a Measured cylinder pressure signals and reconstructed motored pressure signal (polytrope). **b** Calculated crank angle locations of 50 % MFB. $n_{\mathrm{eng}} = 3000\,\mathrm{rpm}$, 50 % load. Engine: Opel X10XE Ecotec, 1 l, 3 cyl —— measured - - - motored.

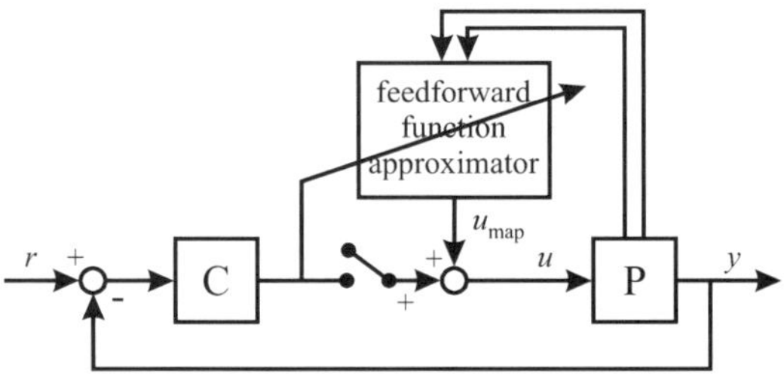

Fig. 7.5.5. General structure of an Adaptive Feedforward Controller (AFFC). The controller C is basically used for the adaptation of the feedforward map.

manifold pressure signal) and the engine's rotational speed. The conventional look-up table determines the calibrated values of the ignition angle and is valid for all cylinders, see Fig. 7.5.7a). For each cylinder the location of 50 % MFB is calculated and compared to the reference location of 8°CS. Correction values are calculated and stored in the adaptive offset look-up table of each cylinder, at the corresponding operating condition. In order to reduce noise in the controller output signal, no linear proportional controller C is used in parallel to the adaptive feedforward map. The reference map for specifying φ_{MFB50} has to be determined during basic engine calibration in dependence on load and speed. In the case of knocking or misfires the ignition angle or the reference value $\varphi_{\mathrm{MFB50,ref}}$ is retarded.

The correction ignition angle $\Delta\varphi_{\mathrm{i,adapt}}(load, n_{\mathrm{eng}})$ is determined from the control error

$$e(k) = \varphi_{\mathrm{Q50,ref}}(k) - \varphi_{\mathrm{Q50}}(k) \tag{7.5.7}$$

and stored in dependence on load and speed. The heights w_i of the map at the grid points are calculated by using the adaptive look-up table with the normalized least square estimation (NLMS), described in Sect. 3.2.

$$w_i(k+1) = w_i(k) + \beta e(k) \frac{\Phi_i\left(\mathbf{u}(k), \mathbf{c}\right)}{\sum_{j=1}^{M} \Phi_j^2\left(\mathbf{u}(k), \mathbf{c}\right)} \tag{7.5.8}$$

where w_i are the heights at the interpolation nodes, Φ_i the basis functions and β a learning rate, Müller (2003), Vogt et al (2004).

The required correction angle $\Delta\varphi_{\mathrm{ign}}$ for given load and speed is then obtained by interpolation between the heights w_i. For the experiments $\beta = 0.001$ and as map intervals $\Delta n_{\mathrm{eng}} = 500\,\mathrm{rpm}$ and 10 % of the load (air charge) was used together with an adaptation threshold $0.5°\mathrm{CS}$.

c) Experimental results

Figure 7.5.7 shows for constant engine speed of $2500\,\mathrm{rpm}$ three load changes. For the first $50\,\mathrm{s}$ only the conventional feedforward control is active. Considerable deviations of the crank angle locations of 50 % MFB from the optimal value for best engine efficiency at 8°CS after TDC are visible as shown for two cylinders. Between 50 and

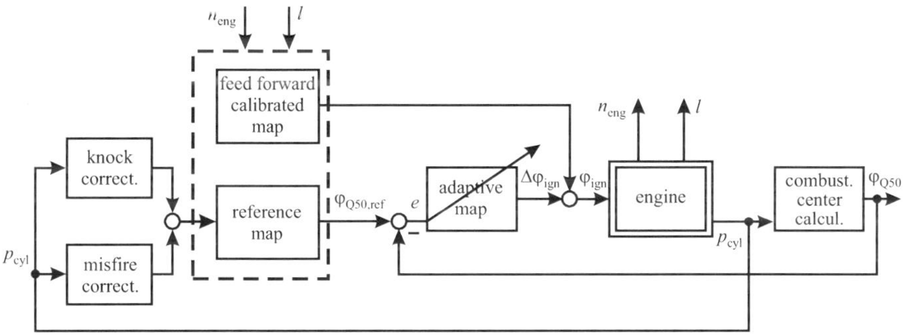

Fig. 7.5.6. Overall structure of the adaptive feedforward control system of the ignition angle, based on combustion pressure center with generation of the reference variable and superimposed corrections in case of knocking and misfire.

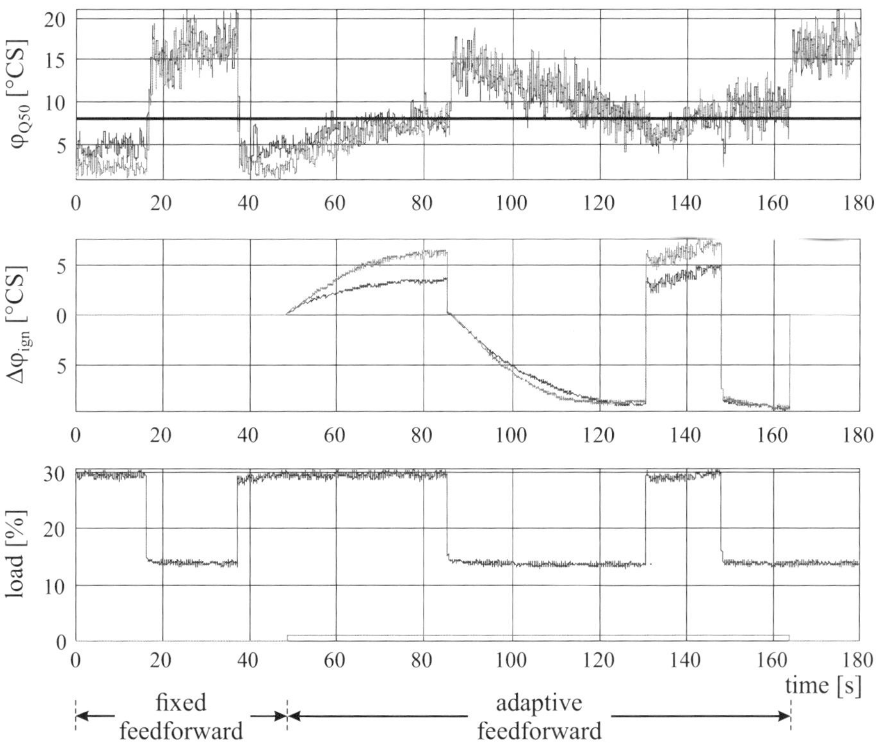

Fig. 7.5.7. Control performance of the ignition control system during three equivalent load changes. The adaptive look-up table is initialized with zero. From 50 to 160 s the adaptive feedforward control is activated, $\varphi_{Q50,d} = 8°CS$, $n_{eng} = 2500\,rpm$.

168 s the adaptive feedforward controller is active. Since the adaptive input-output map of each cylinder was initialized to zero, they first have to be adapted for both operating conditions. The control performance for the adapted feedforward control then results between 130 and 160 s, now keeping the reference value 8°CS. Then it is switched back to the conventional, fixed feedforward control. Figure 7.5.8a) shows the conventional, fixed ignition angle look-up table and b) shows the adaptive offset map of the second cylinder after a training sequence.

Despite the considerable, stochastic nature of the combustion events, the adaptive feedforward control allows the maintenance of the mean crank angle location of 50 % MFB around its optimal value at about 8°CS after TDC. The ignition angle was corrected between 0 and 5°CS dependent on the operation point. In addition, the differences between the individual cylinders showed differences of the optimal ignition angle up to 5°CS, Müller (2003).

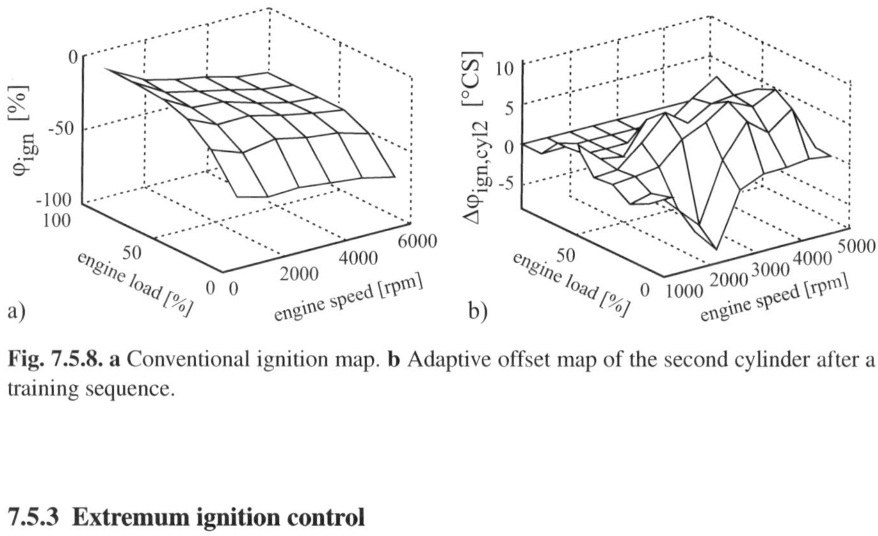

Fig. 7.5.8. a Conventional ignition map. **b** Adaptive offset map of the second cylinder after a training sequence.

7.5.3 Extremum ignition control

As described in Sect. 7.3, the ignition angle is basically feedforward controlled through calibrated maps in dependence on m_{air}, n_{eng} and corrected through efficiencies of the fast dynamic path in dependence on torque request, λ and by knock control.

Usually, all the cylinders have the same ignition advance angle φ_{ign}, optimized during the basic application of the ECU and hold for the complete lifetime of the engine. This means that an adaptation to the individual cylinder properties, wear and aging of the engine is not realized. One method to maximize the power and minimize the fuel consumption of each cylinder is to design an extremum control based on cylinder pressure measurement, Kohlhase and Isermann (2009).

The idea of extremum control was investigated by several authors. The optimization of the torque by excitation signals of the ignition angle was demonstrated by Schramm and Wurst (1977). An extremum control with parameter estimation for the ignition angle of gasoline engines is published by Wellstead and Scotson (1990), Scotson and Wellstead (1990) and Dorey and Stuart (1994). The optimization of fuel consumption for DI-gasoline engines with stratified combustion by manipulating the λ and EGR rate is proposed in Gäfvert et al (2000). In all these cases, an excitation signal is used. However, the gasoline engines were considered as an overall system, not optimizing the cylinders individually.

Figure 7.5.9 shows the relations between the mean indicated pressure p_{mi}, which is a measure for the mean torque, and the ignition angle φ_{ign} and the combustion center. φ_{Q50} can each be approximated by a quadratic function

$$V(u_j) = \theta_{0j} + \theta_{1j}u_j + \theta_{2j}u_j^2 = \boldsymbol{\psi}^T\boldsymbol{\theta} \tag{7.5.9}$$

$$\boldsymbol{\psi}^T = \left[1\ u_j\ u_j^2\right] \tag{7.5.10}$$

$$\boldsymbol{\theta}^T = \left[\theta_{0j}\ \theta_{1j}\ \theta_{2j}\right] \tag{7.5.11}$$

with $j = 1$ for the ignition angle $u_1 = \varphi_{\mathrm{ign}}$ and $j = 2$ for the combustion center $u_2 = \varphi_{\mathrm{Q50}}$ and $V(u_j) = p_{\mathrm{mi}}$. The figure shows the considerable influence of both variables on the mean indicated pressure and the strong fluctuations for the strokes. If the parameters of the quadratic polynomial (7.5.9) are estimated over many strokes in order to damp the influence of these combustion fluctuations, then the maximal value of $V(u_j)$ can be determined from

$$\frac{dV(u_j)}{du_j} = \hat{\theta}_{1j} + 2\hat{\theta}_{2j}u_j = 0 \tag{7.5.12}$$

and the optimal variable follows directly from

$$u_j\big|_{\mathrm{opt}} = -\frac{\hat{\theta}_1}{2\hat{\theta}_2}. \tag{7.5.13}$$

The parameter estimation can be performed with the recursive least squares method (RLS)

$$\hat{\boldsymbol{\theta}}(k+1) = \hat{\boldsymbol{\theta}}(k) + \boldsymbol{\gamma}(k+1)\left(y(k+1) - \boldsymbol{\psi}^T(k+1)\hat{\boldsymbol{\theta}}(k)\right) \tag{7.5.14}$$

with $y = p_{\mathrm{mi}}$ and $\boldsymbol{\gamma}(k+1)$, $\boldsymbol{P}(k+1)$ according to Sect. 3.2.2 and a forgetting factor λ, see Isermann and Münchhof (2011). To obtain a good estimation the changes of the variables u_j should be large enough for some time.

A load change influences mainly the parameter θ_{0j} but less the parameters θ_{1j} and θ_{2j}. Hence, (7.5.13) can be used also after load changes. In order to reduce the effect of load changes a standardized value for the mean effective pressure is used

$$\frac{V\left(u_j(k)\right)}{p_{\mathrm{mi,norm}}} = \theta'_{0j} + \theta'_{1j}\,u_j(k) + \theta'_{2j}\,u_j^2(k) \tag{7.5.15}$$

with $p_{\mathrm{mi,norm}} = p_{\mathrm{mi}}(k)/p_{\mathrm{mi,model}}(m_{\mathrm{air}}, n_{\mathrm{eng}})$. Based on several measurements for the strokes $k = 0, 1, 2, \ldots$ the RLS method then estimates the unknown parameters and yields an optimal value for u_j due to (7.5.13). Figure 7.5.10 shows that a load change has only a small effect on the estimates.

Figure 7.5.11 depicts the signal flow for the ignition angle as input. The relation $\varphi_{Q50}(\varphi_{ign})$ is approximately linear

$$\frac{d\varphi_{Q50}}{d\varphi_{ign}} \approx k_{Qi} \tag{7.5.16}$$

and the stationary relation $p_{mi}(\varphi_{Q50})$ is quadratic according to (7.5.9). However, due to the stochastic fluctuations of the combustions considerable noises n_1 and n_2 act on the signals, as Fig. 7.5.9 shows.

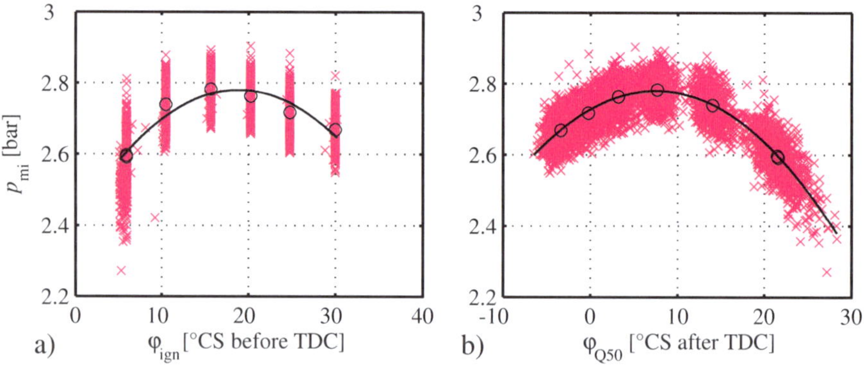

Fig. 7.5.9. Dependence of indicated pressure on **a** ignition angle, **b** combustion center for a charged direct injection gasoline engine with $n_{eng} = 2000\,\mathrm{rpm}$ and relative air charge $rl = 25\,\%$, see (4.1.9).

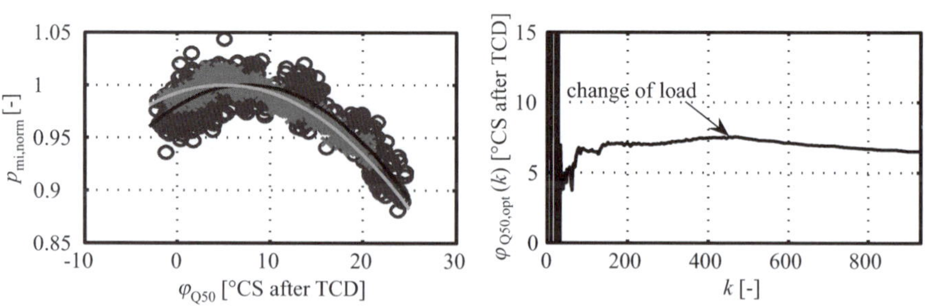

Fig. 7.5.10. Optimization of the combustion center φ_{Q50} with parameter estimation for $p_{mi}(\varphi_{Q50})$ for sufficient excitation of the ignition angle, with a load change at $k = 450$.

The online optimization of the mean effective pressure is realized as an extremum control, see Fig. 7.5.12. It operates with a sublevel combustion center control, where

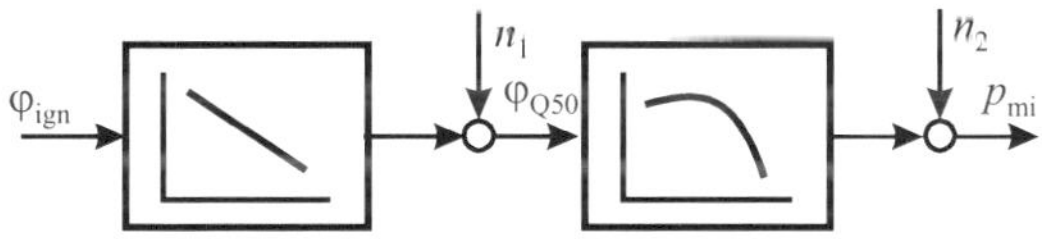

Fig. 7.5.11. Signal-flow graph for the ignition angle as input, combustion center as intermediate variable and mean effective pressure as output.

φ_{Q50} is calculated by heat release analysis based on combustion pressure measurement $p_{cyl}(\varphi)$ according to Sect. 4.2.3, with the ignition angle φ_{ign} as manipulated variable. This means that with regard to Fig. 7.5.11 the intermediate variable φ_{Q50} is closed loop controlled in order to compensate for disturbances. The changes of φ_{ign} and p_{mi} which are the result of an external excitation signal are used to estimate the parameters $\boldsymbol{\theta}'$ of the quadratic function (7.5.9). (7.5.13) yields the optimal reference value. $r_{opt} = \varphi_{Q50,opt}$ for the combustion center controller which leads to an optimal p_{mi}.

The realization of the sublevel combustion center control is depicted in Fig. 7.5.13. The ignition angle $\varphi_{ig,FF}$ is basically feedforward controlled by the look-up table obtained from the basis calibration. This ignition angle is corrected by an adaptive look-up table, which determines the corrections $\Delta\varphi_{ign}$ based on the control deviations

$$e(k) = \varphi_{Q50,ref}(k) - \varphi_{Q50}(k) \tag{7.5.17}$$

and stores them dependent on air charge and speed. This adaptive look-up table acts like an integral controller. To allow the parameter estimation for the online optimization, see Fig. 7.5.12, an excitation signal $\Delta\varphi_{Q50,ex}$ is superimposed on the reference value $\varphi_{Q50,ref}$, with a feedforward path according to (7.5.16) on the combustion center control loop.

The test signal for a persistent excitation is designed such that the averaged torque is not influenced. As Fig. 7.5.14 shows, the individual cylinders are detuned in a balanced way. This means that each two cylinders e.g. cylinders 1 and 4 change their reference value for φ_{Q50} by $+\Delta\varphi_{Q50,ex}$ and $-\Delta\varphi_{Q50,ex}$ from a starting point, and correspondingly the cylinders 2 and 3, such that

$$\overline{p}_{mi} = \frac{1}{i_C}\sum_{\nu=1}^{i_C} p_{mi,\nu}(k) = \text{const.} \tag{7.5.18}$$

For details see Kohlhase and Isermann (2009) and Kohlhase (2011).

The optimization of p_{mi} can be performed either with φ_{ign} or with φ_{Q50} as variable as shown in Fig. 7.5.9 and in (7.5.9) to (7.5.11). Taking into account different combustion pressure sensor faults discussed in Sect. 7.5.1, the optimization with the ignition angle shows smaller faults with regard to the optimal ignition angle. However, the optimization with φ_{Q50} includes a natural excitation, such that artificial test signals may not be necessary.

The practical application was realized with a four-cylinder direct injection gasoline 2l FSI engine with turbocharging and control unit Bosch MED9 with ETAS-ETK interface. The combustion pressure is measured in each cylinder with water

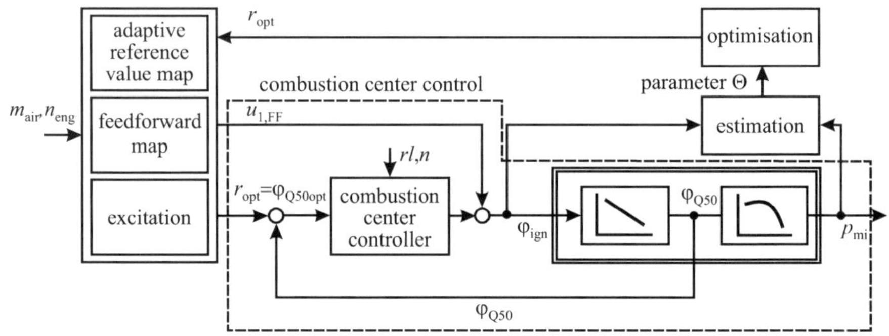

Fig. 7.5.12. Extremum control for the optimization of the mean effective pressure p_{mi}.

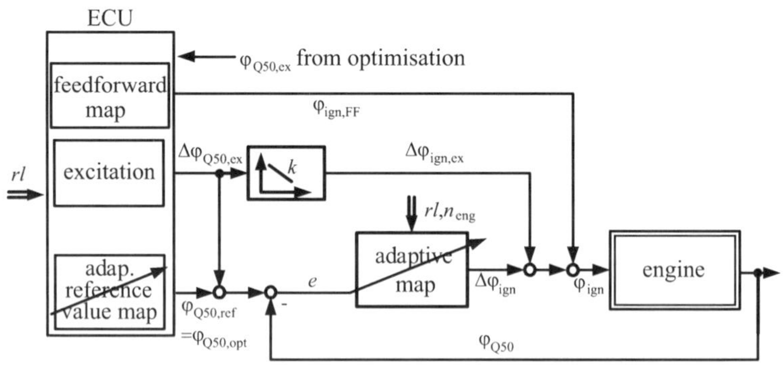

Fig. 7.5.13. Control structure of the sublevel combustion center control φ_{Q50} with feedforward control of the ignition angle φ_{ign} and correction $\Delta\varphi_{ign}$ through an adaptive look up table.

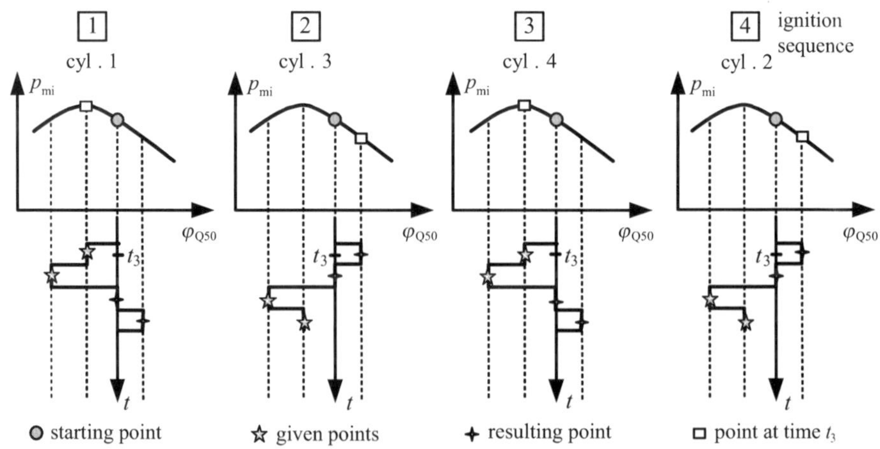

Fig. 7.5.14. Cylinder individual excitation of the combustion center position φ_{Q50} by changing the ignition angle for a 4-cylinder engine.

cooled piezoelectric sensors from Kistler. The pressure sensor signals are sampled with a self-developed Rapid-Control-Prototyping system based on dSpace components which allows a high solution sampling of several engine signals with 1°CS and real-time computations of combustion pressure-based features like the position φ_{Q50} of the combustion center based on the heat release equation (4.2.67). The manipulation of the ignition could be realized with a bypass of the ECU.

The results were obtained in wide operation areas. An extract is shown in Fig. 7.5.15. During the optimization phase the ignition angles of all four cylinders are varied in a torque-balanced way to allow the estimation of the parameters of the quadratic polynomial (7.5.15) with RLS. After about $10\,\mathrm{s}$ the estimated parameters and the optimal combustion centers $\varphi_{q50,opt}$ have converged and stay within a small tolerance band. At time $t = 42\,\mathrm{s}$ the optimal reference value is given to the combustion center controller. The resulting differences between the optimal combustion center positions are maximal 3°CS and between the ignition angle maximal 2°CS. The torque raises by about $1\,\mathrm{Nm}$ or by $4\,\%$. It can also be seen that the torque stays approximately constant during the excitation phase.

The optimization by extremum value control operates also during dynamic operating conditions, as shown by Fig. 7.5.16 where the torque is continuously increased for $t > 0\,\mathrm{s}$. After about $10\,\mathrm{s}$ the estimates have converged. This holds also for stepwise load changes.

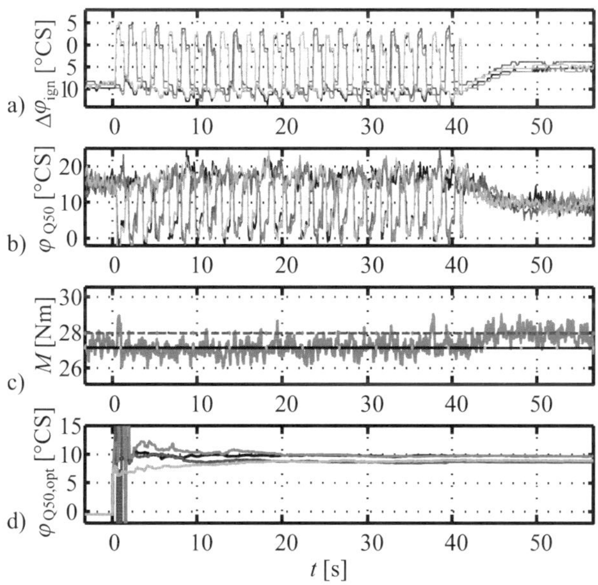

Fig. 7.5.15. Online optimization of the cylinder individual combustion center φ_{Q50}. Stationary conditions: $p_{mi} = 5\,\mathrm{bar}$, $n_{eng} = 1500\,\mathrm{rpm}$. Optimization phase: $t = 0 \ldots 42\,\mathrm{s}$ (open loop). Combustion center control $t > 42\,\mathrm{s}$ (closed loop). **a** excitation of the ignition angle by balanced detuning of the cylinders. **b** resulting time history of the combustion centers. **c** resulting measured torque. **d** Estimation of optimal combustion centers φ_{Q50}.

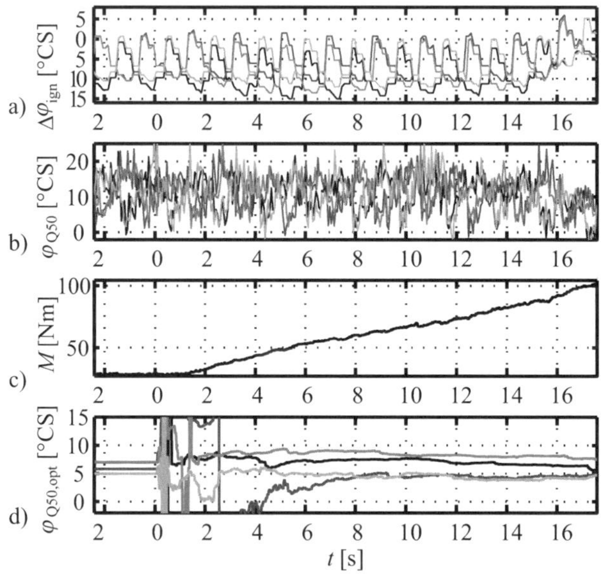

Fig. 7.5.16. Online optimization of the cylinder individual combustion center φ_{Q50}. Dynamic operation condition: increasing torque, for $n = 1500\,\mathrm{rpm}$. **a**, **b**, **c**, **d** as for Fig. 7.5.15.

The optimization results for wide operation ranges show, that the optimal combustion center positions are 4 to 6°CS for higher torque and speed and about 8°CS for low torque and speed. This means that they depend on the operation point, compare Tschöke et al (2011). The standard deviation of p_{mi} is smaller than $0.13\,\mathrm{bar}$, i.e. a small roughness is obtained, too, see Kohlhase (2011).

Another possibility for the ignition angle extremum control is the application of *correlation functions*. As the mean effective pressure p_{mi} and the combustion center both have considerable stochastic fluctuations, see Fig. 7.5.9, and their crosscorrelation sufficiently shows approximately a linear dependence, the crosscorrelation coefficient

$$\rho\left(\varphi_{Q50}, p_{mi}\right) = \frac{\sum_{i=1}^{N}\left(\varphi_{Q50}(i) - \overline{\varphi}_{Q50}\right)\left(p_{mi}(i) - \overline{p}_{mi}\right)}{\sqrt{\sum_{i=1}^{N}\left(\varphi_{Q50}(i) - \overline{\varphi}_{Q50}\right)^2}\sqrt{\sum_{i=1}^{N}\left(p_{mi}(i) - \overline{p}_{mi}\right)^2}} \qquad (7.5.19)$$

yields values in the interval $[-1, 1]$ and becomes $\rho\left(\varphi_{Q50}, p_{mi}\right) = 0$ for the optimal combustion center. An adaptive look-up table with $\rho = [-1, 1]$ as input and the ignition angle correction $\Delta\varphi_{ign}$ as output or with a cascaded structure and φ_{Q50}-feedback control as the inner loop then allows to obtain convergent values for the optimal ignition angles, Kohlhase (2011). This method does not need extra excitation signals.

7.6 Knock control

Knocking is an abnormal combustion phenomenon. It results from the autoignition of the compressed mixture of air, fuel and residual gases in front of the regular flame front or from surface ignition at hot spots. Depending on the compression, temperature and fuel the autoignition leads to a strong pressure increase and pressure oscillations, which may damage the engine. These pressure oscillations have different resonance frequencies of the combustion chamber, are transferred to the engine block and can be observed as structure-born noise with knock sensors. As the optimal ignition angle with regard to best efficiency comes close to the knocking limit or is within the range of knocking, some safety margins have to be provided, like lower compression, retarded ignition, such worsening of fuel consumption and torque. However, knock control for the individual cylinders allows making the safety margin smaller in order to improve fuel consumption and torque and to adapt to different fuel quality, manufacturing tolerances and aging.

Knock sensors are piezoelectric acceleration sensors with a seismic mass which detect the high-frequency vibrations from the cylinder pressure in the expansion part of the pressure curve, see Fig. 7.3.4a).

Broadband sensors have a frequency band of 5 to 20 kHz. Resonance sensors yield only one frequency. These sensors are mounted at appropriate positions of the engine block. For four cylinder engines one or two knock sensors are applied, six cylinder engines need two sensors. The signals of the knock sensor are band pass filtered around the expected knock resonance frequency, see Fig. 7.6.1. The signals are then evaluated during a time window shortly after TDC of each cylinder. By squaring the amplitude and integration over the time window a signal energy measure E_{k} is generated, usually in a special knock integrated circuit (IC), Kiencke and Nielsen (2000). If the signal energy passes a threshold $E_{\mathrm{k}} > E_{\mathrm{kth}}$ knocking is detected and related to special cylinders. The calibration of the signal evaluation and threshold is usually performed by measurements with combustion pressure sensors at test benches.

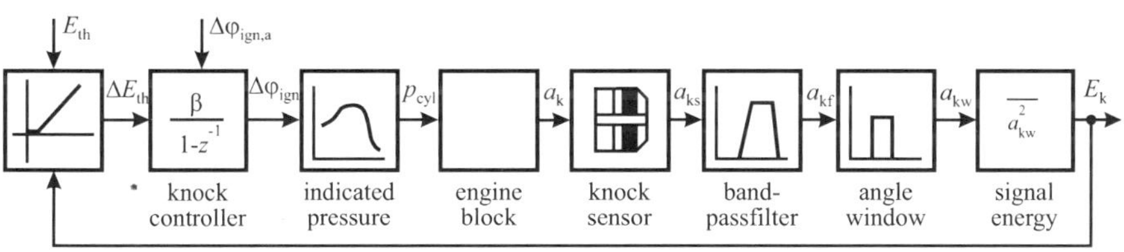

Fig. 7.6.1. Signal-flow graph of knock control.

The knock controller receives the energy difference

$$\Delta E_{\mathrm{k}} = E_{\mathrm{k}} - E_{\mathrm{kth}} \tag{7.6.1}$$

which has only positive values. Then the ignition angle is retarded

$$\varphi_{\mathrm{ign}}(\nu) = \varphi_{\mathrm{ign}}(\nu-1) + \Delta\varphi_{\mathrm{ign,a}} - \beta \Delta E_{\mathrm{k}}(\nu) \tag{7.6.2}$$

with $\nu = 1, 2, \ldots$ for the combustion cycles, Kiencke and Nielsen (2000). $\Delta\varphi_{\mathrm{ign,a}}$ is a permanent ignition angle advance and $\beta \Delta E_k$ the ignition angle retard at knocking. Thus during knocking the ignition angle is retarded with the gain β. If knocking stops, the ignition angle is advanced by $\Delta\varphi_{\mathrm{ign}}$. This way, the knocking limit is always approached again in a saw tooth manner, thus adjusting the ignition angle to changing knocking conditions, also for each cylinder.

Adaptive knock control stores the determined ignition angles in dependence on the operating point as load and speed. This allows a fast feedforward control, especially for transients. Hence, the actual ignition angle is

$$\varphi_{\mathrm{ign}}(\nu) = \varphi_{\mathrm{ign,o}} + \Delta\varphi_{\mathrm{ign}}(\nu) + \Delta\varphi_{\mathrm{ign,ad}} \tag{7.6.3}$$

with $\varphi_{\mathrm{ign,o}}$ from basic calibration, $\Delta\varphi_{\mathrm{ign}}$ from knock control, $\Delta\varphi_{\mathrm{ign,ad}}$ from an adaption map, Kiencke and Nielsen (2000). Some more details on knock control can be found in van Basshuysen and Schäfer (2004), Robert Bosch GmbH (2011).

7.7 Idle speed control

During idling the engine has to generate a torque which compensates the own combustion cycle losses, the friction of the engine, clutch and transmission and the required torque by the auxiliaries like coolant pump, oil pump, servo steering, air conditioning compressor and electric generator with board net load. The required torque therefore depends strongly on the varying loads and the engine state in the phases after start, during warming up and normal operation and especially the engine temperatures have a significant effect.

Because of the many disturbances the idling speed has to be feedback controlled. The reference value for the speed should be as low as possible in order to save fuel and emissions. However, combustion stability of single strokes, noise and vibration effects have to be considered. Especially engine stalling must be prevented under all operating conditions. In addition, the transition from and to idling should be smooth and controlled with good performance. The reference value for the idle speed controller usually depends on engine and environmental temperature and auxiliary loads. The design of the idle speed controller has to be robust to the varying behavior of the engine, tolerances between different production engines and aging.

The main control variable of the idle speed controller is the engine angular speed ω_{eng} and the main manipulated variable for gasoline engines is the air mass flow rate $\dot{m}_1$ to the manifold intake or the air charge m_{air} arriving in the cylinders. In former time, the air mass flow rate was controlled by a throttle bypass valve guarding the air flow around the closed air flow throttle. Modern gasoline engines use the electrical actuated throttle directly to manipulate the air flow or in the case of fully variable valve trains the inlet valve lift.

A signal-flow scheme of the idle speed control is depicted in Fig. 7.7.1 According to the model of the manifold Fig. 4.1.5 the change of the throttle plate position α_{th}

influences the air mass flow rate $\dot{m}_1$ according to the throttle characteristic. During idling the throttle is only slightly opened, such that overcritical pressure drop can be assumed and the intake pressure has no influence on the flow rate through the throttle. The intake manifold volume then forms a significant storage according to the mass balance (4.1.11) and the resulting nonlinear differential equation (4.1.16) for the intake pressure p_2 with the time constant T_{int} due to (4.1.17)

$$T_{\text{int}} = \frac{2V_{\text{int}}}{\eta_{\text{v}} V_D n_{\text{eng}}} \tag{7.7.1}$$

which depends on the volumetric efficiency and the engine speed. The lower the engine speed, the larger the time constant. The air mass flow rate $\dot{m}_{2,\text{gas}}$ into the cylinders is then delayed by the intake manifold pressure $p_2(t)$. This can also be called manifold filling dynamics. The changed air charge and following combustion then change the mean value engine torque proportionally after a short dead time, T_{dCD}. This dead time is an intake–to combustion stroke delay, corresponding to $180° \ldots 360°$ crank shaft angle, Hrovat and Sun (1997) and therefore

$$T_{\text{dDC}} = \frac{\pi}{\omega_{\text{eng}}} \ldots \frac{2\pi}{\omega_{\text{eng}}}. \tag{7.7.2}$$

The changed engine torque then accelerates the engine with its ratio of inertia, leading to a further first order differential equation

$$J_{\text{eng}} \frac{d\omega_{\text{eng}}}{dt} = M_{\text{eng}}(\omega) - M_{\text{aux2}}(\omega) - M_{\text{l}}(\omega) \tag{7.7.3}$$

with the engine torque $M_{\text{eng}}(\omega)$ at clutch according to (4.3.11). The torque $M_{\text{aux2}}(\omega)$ comprises the auxiliary drives like servo power steering pump, air conditioning compressor and electric generator, which cannot be considered as integrated to the engine like oil and water pump, see Sect. 4.3. The torque $M_l(\omega)$ represents operation dependent load torques like engagements of the manual or automatic transmission.

For small changes of the idling speed it is now assumed that the torques consist of a constant and a linear speed dependent term

$$M_{\text{l}\Sigma}(\omega) = M_{\text{aux2}}(\omega) + M_{\text{l}}(\omega) = M_{\text{l}\Sigma 0} + M_{\text{l}\Sigma 1}\, \omega_{\text{eng}} \tag{7.7.4}$$

$$M_{\text{eng}} = M_{\text{eng}\,0} - M_{\text{eng}1}\, \omega_{\text{eng}}. \tag{7.7.5}$$

The introduction in (7.7.2) yields for the rotational dynamics

$$J_{\text{eng}} \frac{d\omega_{\text{eng}}(t)}{dt} = M_{\text{eng}0}(\omega) - M_{\text{l}\Sigma 0} - (M_{\text{eng}1}(\omega) + M_{\text{l}\Sigma 1})\, \omega_{\text{eng}} \tag{7.7.6}$$

and with

$$\begin{aligned} M_0 &= M_{\text{eng}0} - M_{\text{l}\Sigma 0} \\ M_1 &= M_{\text{eng}1} - M_{\text{l}\Sigma 1} \end{aligned}$$

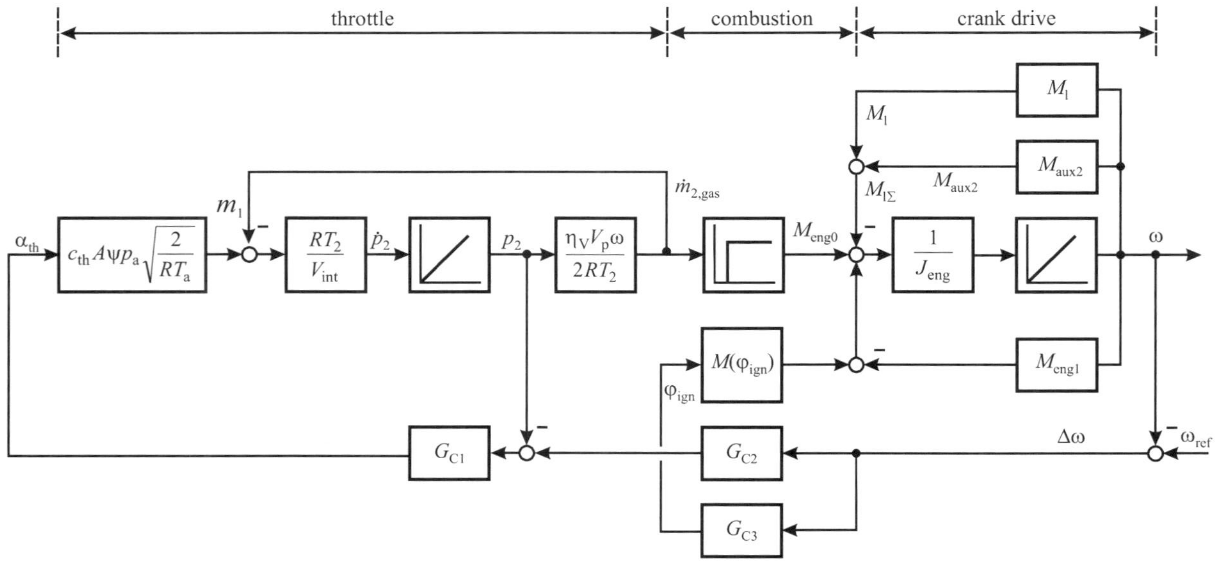

Fig. 7.7.1. Signal-flow diagram for the idle speed control of a gasoline engine with the throttle as actuator ($\omega = \omega_{eng}$).

$$\frac{J_{\text{eng}}}{M_1}\frac{d\omega_{\text{eng}}(t)}{dt} + \omega_{\text{eng}}(t) = \frac{1}{M_1}M_0(t). \tag{7.7.7}$$

A change of the throttle angle α_{th} and therefore $\dot{m}_{2,\text{gas}}$ leads to a change ΔM_0. The corresponding crank drive time constant follows as

$$T_{\text{CD}} = \frac{J_{\text{eng}}}{M_1}. \tag{7.7.8}$$

M_1 comprises all linear speed dependent feedbacks of the engine M_{eng1} treated in Sect. 4.3, and of the auxiliary drives and load $M_{\text{l}\Sigma 1}$.

The controlled process now consists of two first order delay elements and a dead time

$$G_{\alpha\text{p2}}(s) = \frac{\Delta\dot{m}_{2,\text{gas}}}{\Delta\alpha_{\text{th}}(s)} = \frac{K_{\text{int}}}{1+T_{\text{int}}s} \tag{7.7.9}$$

$$G_{\dot{\text{m}}2\omega}(s) = \frac{\Delta\omega_{\text{eng}}(s)}{\Delta\dot{m}_{2,\text{gas}}(s)} = \frac{K_{\text{CD}}}{1+T_{\text{CD}}s}e^{-T_{\text{dCD}}s} \tag{7.7.10}$$

$$G_{\alpha\omega}(s) = \frac{\Delta\omega_{\text{eng}}(s)}{\Delta\alpha_{\text{th}}(s)} = G_{\alpha\text{p2}}(s)\,G_{\dot{\text{m}}2\omega}(s) \tag{7.7.11}$$

For idling speeds of $600\ldots 800\,\text{rpm}$ the time constants T_{int} and T_{CD} are in the range of 0.5 ... 1.5 s and the dead time T_{dCD} is about 50 ... 150 ms.

Another important manipulated variable to control the idle speed is the advance ignition angle. As described in Sect. 7.1 and 7.3, and Figs. 7.1.3 to 7.1.5 and 7.3.6 advancing the ignition angle increases the engine torque immediately, because no intake filling delays are included. However, possibly a not torque optimal ignition angle has to be adjusted before, see Fig. 7.3.5. Therefore the idling speed can be increased fast according to a characteristic $M_{\text{eng}}(\varphi_{\text{ign}})$, especially to avoid undershoots with the danger of the engine stalling, see the additional feedback with G_{C3} in Fig. 7.7.1.

The idling speed controller is conventionally an outer-loop PID controller acting on the throttle. This may be added by a proportional inner-loop ignition angle feedback and several feedforward controls using auxiliary load information. A survey of published control methods is given by Hrovat and Sun (1997) and a case study with numerical values in Guzzella and Onder (2010). An improvement of the control performance can be obtained if the intake manifold pressure p_2 is feedback controlled with an inner loop controller G_{C1} and the idle speed ω_{eng} by an outer loop controller G_{C2}, thus forming a cascaded control system. The inner loop contains only the intake manifold dynamics and compensates disturbances of the air path. Usually, a P-controller is sufficient and a PI or PID controller for G_{C2}. Kiencke and Nielsen (2000) applied a linear state-space controller of second order and Loh and Lu (2001) a nonlinear state-space controller of fifth order. Alt (2010) investigated a sliding-mode controller by using a nonlinear model. Instead of the measured pressure p_2 a model-based estimate $\hat{p}_2$ can be used. In order to compensate for large load torque disturbances the ignition angle can be retarded in steady-state idling operating condition, such that an advance angle impulse via a derivative term controller G_{C3} with DT_1 behavior can be generated, which lasts only a short time period.

The required controller parameters can be partially estimated based on the parameters of the simplified models. However, they depend on the speed, engine and ambient temperature, engaged gear, etc. These nonlinear dependencies have to be represented in look-up tables which are determined by experiments.

The idle speed control is usually integrated in the torque-oriented structure of the ECU. Then the output of the controller G_{C1} is a required idle torque, see Fig. 7.1.4, which determines the desired throttle angle according to Fig. 7.1.5 and correspondingly for the ignition advance angle.

The idle speed control for *diesel engines* acts directly on the injected fuel mass and does not comprise the intake manifold as a significant storage delay. Therefore, a PI idle speed controller directly manipulates the required torque, supported by a fast acting DT_1 derivative of the control deviation $\Delta\omega$.

7.8 Variable valve trains (VVT)

Until about 1983 the opening and closing times of the engine valves were fixed by applying non-adjustable camshafts. In order to improve fuel consumption and torque, first the inlet valves and later also the outlet valves of gasoline engines have been provided with variable timing, usually by adjusting the *phase of the camshafts*. This is realized by turning the camshaft with a hydraulic or electrical actuator relatively to the crankshaft. Then only the angle phase is manipulated, without influencing the opening period and lift, Robert Bosch GmbH (2011). An example to replace the conventional hydraulic phase actuator by an electrical actuator is given by Krebber-Hortmann et al (2013).

The intake-camshaft is retarded in idle speed or for high engine speed. For idling a lower exhaust internal gas recirculation is demanded to obtain a stable running and at high speeds charging effects can be used to increase the engine power. For medium speed the camshaft is advanced, such that the intake valve closes earlier and prevents the air from flowing backwards. Also, if the exhaust-camshaft can be adjusted, advancing increases the internal exhaust gas recirculation. Hence, the conventional valve phasing of the intake and exhaust valves usually requires two adjustable camshafts.

Another recent mechanical design has two adjustable concentric camshafts instead of one camshaft. This allows to adjust the phase of two parallel valves relatively to each other and such the opening range, Schneider and Lettmann (2008).

As the valve timing has a considerable effect on the combustion, more variabilities of the valve actuation have been developed. The main goals are an increase of the torque and a reduction of fuel consumption and emissions. More specifically, flexible valve actuation allows to improve the air-fuel mixture, to reduce the charge cycle losses and to control the internal exhaust gas recirculation or the residual gases, Martin (2009), Köhler and Flierl (2012).

Especially in the case of turbocharged gasoline engines with direct injection a controlled *scavenging* can be applied by simultaneously opening the inlet- and outlet valves during the gas exchange at TDC. The air mass flow rate then increases and the efficiency of the turbocharger improves such that a higher charging pressure can be obtained. At low speeds, combined with higher injection rates, this leads to an increase of the torque and better dynamics, see e.g. Hofmann et al (2011).

Figure 7.8.1 gives a survey of the possibilities to modify the valve lift curve, from fixed timing, through phase shifting, lift adjustment and various combinations.

Valve trains with *camshaft-lobe control* alternate between two camshaft lobes which change the phase and the lift of the valves, see Fig. 7.8.1b), Robert Bosch GmbH (2011), Brüstle and Schwarzenthal (1998). The lower lobe is designed for lower and medium speeds and the higher lobe for high engine speeds. Figure 7.8.2b) shows an example. The camshaft-lobe actuation can have two selective positions (profile switching) or it can be continuously adjusted e.g. by lateral shifting of the camshaft. In this case the throttling losses from the intake throttle actuator can be decreased if the inlet valve can be adjusted to small values.

Fully variable valve trains (VVT) allow to adjust the valve lift as well as the opening angle range, Klüting et al (1999). One example for an *electro-mechanical VVT* in series production is shown in Fig. 7.8.3.

This VVT valvetronic (1st generation) is characterized by an intermediate lever, an excenter axle and a return spring. It changes the valve lift according to Fig. 7.8.3, Liebl et al (2001a), Unger et al (2008). The intermediate lever touches the excenter axle and turns around the touching point. This turning point is shifted by the excenter lobe. Therefore the touching point between the intermediate lever and the rocker arm is shifted and therefore the operating curve of the intermediate lever, moved by the camshaft lobe. This way the valve lift can be varied continuously between $0.3 - 0.5\,\mathrm{mm}$ for idling and $9 - 10\,\mathrm{mm}$ for full load. The adjustment of the excenter axle is realized with an electrical servo motor (brush commutation) with a worm gear which is controlled by a separate electronic control unit with drivers, position control and diagnosis functions. The load commands to this smart actuator are transmitted by a local CAN-Bus instead of the electrical throttle valve. The conventional electrical throttle valve is still implemented for controlling a small vacuum pressure in the intake, which is required for the tank ventilation system, and as a back-up actuator for the VVT, Liebl et al (2001b), Konrad and Krämer (2002). Based on the resulting saving of charging losses of about 30 % through early inlet closing, 12 % fuel saving could be reached.

A further development led to a second generation of valvetronic, Klaus et al (2005), with the goal to increase the speed, reduction of pumping losses due to shorter lift curves and improved combustion. The intermediate lever now has a fixed center of rotation and is supported at the top by a separate guide rail, see Fig. 7.8.3. The use of ball or needle bearings only allow a smaller friction, and low masses and high stiffness improve the dynamics. The shorter partial lift curves enable an earlier inlet valve closing, leading to a further 7 % reduction of charging losses. Different lifts for two inlet valves of the same cylinder induce swirl and tumble. The electro-mechanical VVTs require a high effort to obtain a mass production with very small tolerances and several geometry compensation techniques, Unger et al (2008).

Other solutions are *electro-hydraulic VVT*. The goal is to realize an independent manipulation of the air charge, the residual gases and the charge motion by changing

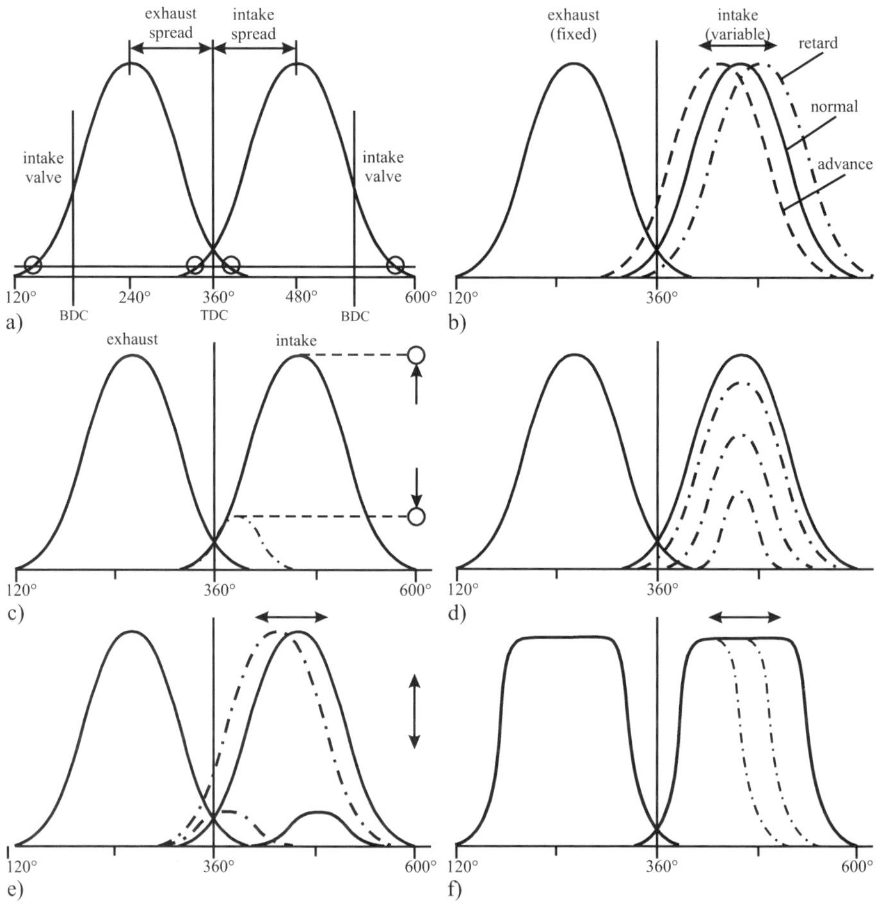

Fig. 7.8.1. Valve lift curves as a function of crankshaft angle for fixed and variable valve trains. **a** fixed timing. **b** phase shifting. **c** alternating lobes (lift and phase). **d** continuous lift adjustment (mechanical). **e** combined lift and phase adjustment (el.-mech., el.-hydr.). **f** camless valve actuation (hydraulic, electrical).

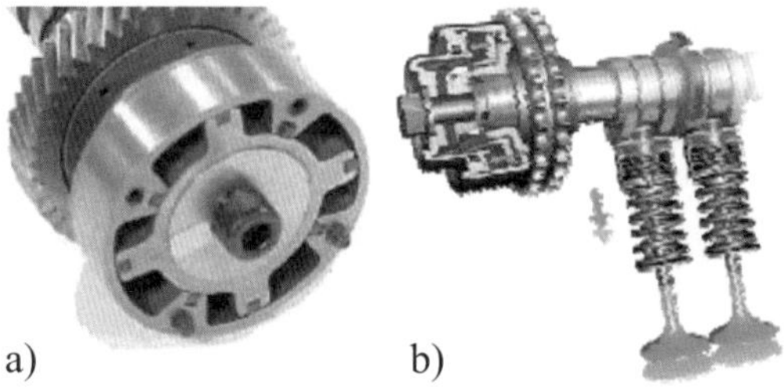

Fig. 7.8.2. Examples for mechanical adjustable valve trains. **a** hydraulic camshaft phase actuator (since 1983). **b** alternating camshaft lobe control (e.g. Porsche, 2001).

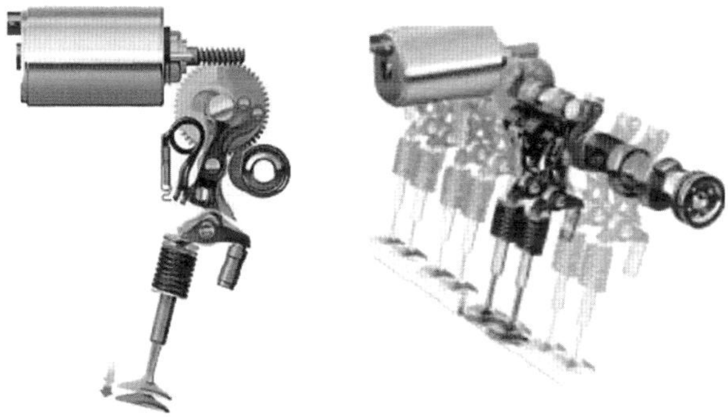

Fig. 7.8.3. Schemes of electro-mechanical continuously adjustable valve lift (2003), BMW valvetronic.

the opening and closing angles and the lift. A prototype realization EHVS with a high pressure (50 − 200 bar) and low pressure rail (5 bar) is described by Denger and Mischker (2004). It has each a 2/2 magnetic valve on the high and low pressure side acting on the two sides of a piston connected to the valve, including a brake throttle.

The *electro-hydraulic VVT* Uniair transmits the intake camshaft lift through a hydraulic chamber, see Fig. 7.8.4. The volume of this chamber can be controlled by an on/off solenoid valve. When the solenoid valve is closed, full lift of the camshaft lobe is transmitted. Controlling the opening and closing time of the solenoid valve allows a wide range of lift curves, like Fig. 7.8.1c), d) and e). Soft landing for valve closing by a spring is supported by a hydraulic brake, Bernard et al (2009). The further development, manufacturing and first series implementation in 2009 is reported in Haas and Rauch (2010).

The largest flexibility with regard to valve motions possess *electrical valve trains* without a mechanical camshaft. Each valve obtains an own drive system in the form of two electromagnets, which hold the armature in the valve shaft in their end positions, as depicted in Fig. 7.8.5. Valve opening is supported by a spring after release of the upper solenoid current. After approaching the open-end position, the lower solenoid brakes the motion by an appropriate current course with the goal to reach a "soft landing" in the valve seat. Valve closing is performed analogously, Flierl and Kairies (2006), Salber et al (2000). For zero control current the valve stays in a middle position.

A detailed modeling of the strongly nonlinear behavior of the electromagnetic spring-mass-damper valve system is treated in Straky (1996) including the design of an energy-optimal cascade control system with an inner PD-speed controller and superimposed nonlinear P-position controller.

Hence, fully variable electro-mechanical valve trains *with camshaft* allow to manipulate the air charge by changing the lift coupled with a change of the opening range by an electrical actuator. Additionally, the center of the opening range (spread) can be shifted by a conventional camshaft turning actuator. If the exhaust camshaft

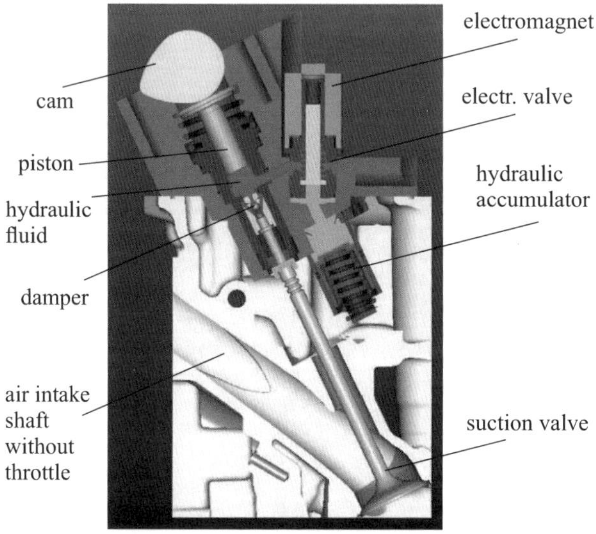

Fig. 7.8.4. Scheme of an electro-hydraulic-mechanical valve train. Fiat UNIAIR.

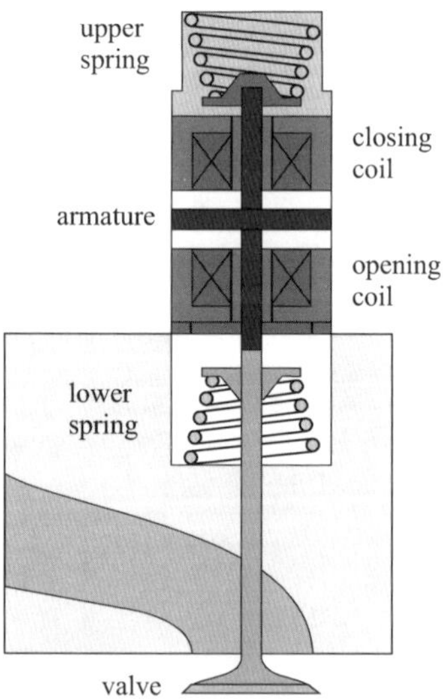

Fig. 7.8.5. Scheme of an electrical valve train.

can then be adjusted by changing the overlap of intake and exhaust valves, the internal exhaust gas recirculation can be controlled, especially with regard to improving the emissions, especially NO_x. Electro-hydraulic VVT enable lift as well as phase control and combinations of thereof.

Fully variable valve trains *without camshaft* are either hydraulically or electrically driven and are able to adjust the lift and phase of the valves as well as the actuation speed. There exist several prototypes but no serial production until now.

Table 7.8.1 presents a comparison of different variable valve trains for gasoline engines. The fuel consumption can, e.g. be reduced for mechanical phase shift by 3–4 %, for two position phase shift by 5–7 %, for continuously variable lift by 8–10 % and for electro-hydraulic and electrical VVT by 7–16 %. On the other hand the complexity, the requirements for manufacturing, the power consumption and the costs play important roles. Therefore, the mechanical and electro-mechanical solutions have been (until now) reasonable compromises, see also Schneider and Lettmann (2008) for diesel engines.

Table 7.8.1. Comparison of variable valve trains for gasoline engines, Schulz and Kulzer (2006)

VVT basic design	Mechanical		Electro-mechanical	Electro-hydraulic		Electrical
adjustment	phase shift	lift: 2 positions	lift: contin.	lift and phase	lift and phase	lift and phase
type	conv.	Porsche	BMW valvetronic	uniair	Bosch EHVV	EMVS
saving of fuel	3–4 %	5–7 %	8–10 %	7–10 %	16 %	14–16 %
HC emissions	-20 %	-20 %	-25 %	-25 %	-30 %	-25 %
torque	+3 %	+5 %	+7 %	+7 %	+11 %	+9 %
complexity	0	–	– –	– – –	– – – –	– – – –

7.9 Alternative combustion processes

The previous sections mainly considered the classical direct injection for *homogeneous operation*, where the fuel is injected in the induction stroke and A/F mixture is controlled to obtain a stoichiometric ratio ($\lambda = 1$). The mixture formation is then similar to external intake manifold injection. However, the fuel has to be injected within 180°CS and therefore requires higher fuel pressure, which leads to an improved mixture formation, compare Table 7.0.1.

7.9.1 Stratified combustion

Another operation mode is the *stratified-charge operation* where the throttle is intended to stay open in order to avoid charging losses through throttling. The fuel is then injected such that only part of the air mixes with the fuel and then the mixture is transported to the plug. The rest of the air surrounds this A/F mixture, thus forming a stratified cylinder charge. Hence, the A/F mixture is only homogeneous in a spatially restricted area, such that in an average a lean mixture $\lambda > 1$ results for the overall combustion chamber. Three types of injection are distinguished, see also Table 7.0.1, Robert Bosch GmbH (2011), Alt et al (2002), and Fig. 7.9.1.

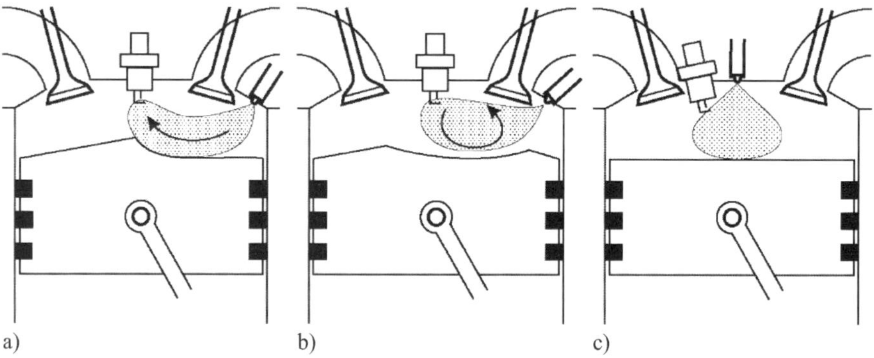

Fig. 7.9.1. Air-fuel formation for gasoline direct injection with stratified-charge combustion. **a** wall-directed. **b** air-directed. **c** jet-directed, Merker et al (2006).

Wall-directed combustion

The fuel is injected into the combustion chamber from the side. A recess in the the piston directs the spray towards the spark plug. The allowable injection time is relatively short. Fuel condensation on the cylinder wall may lead to incomplete combustion and formation of HC.

Air-directed combustion

The fuel is injected from the side, but the fuel spray is separated from the piston recess by an air cushion resulting from a forced tumble motion. However, as this is difficult to reach, a mixture of wall-directed and air-directed combustion results frequently.

Jet-directed combustion

The fuel is injected from the top and the spark plug is located close to the injector. Therefore the fuel spray is ignited shortly after injection. Because of a very short time for mixture formation, a higher injection pressure is required. However, fuel condensing on cylinder walls is avoided. This kind of injection has the highest potential for fuel saving and lowering of raw emissions.

Other operating modes
Some special operating modes result from combinations of basic modes. Combustion with gasoline is usually possible for air excess factors of $0.6 < \lambda < 1.6$. Within a transitional range between stratified and homogeneous mode an operation with *homogeneous lean mixture* and $\lambda > 1$ is possible. This reduces the fuel consumption because of dethrottling the air flow. An *homogeneous-stratified mode* is obtained with a first injection to generate a homogeneous-lean mixture during the intake stroke. A second injection then is performed during the compression stroke to reach a rich mixture around the spark plug, which also ignites the lean mixture after ignition. This operating mode can be applied during a transition between stratified and homogeneous operation.

Other operating modes with double injection are *homogeneous-knock protection* and stratified *catalytic converter heating*, see e.g. Robert Bosch GmbH (2011), Landenfeld et al (2002).

The realization of the different operating modes for direct injection gasoline engines depends on the speed and load. Figure 7.9.2 depicts the various modes in a torque-speed diagram. The stratified combustion can be performed in part load. At higher speed the injection time becomes too short to ensure a good mixture formation around the spark plug, and for high loads the mixture would become too rich and result in too high particulate emissions. Therefore, homogeneous operation is required for high load and high speed. Between stratified and homogeneous mode a region with homogeneous-lean operation becomes possible. Both the stratified and homogeneous-lean operation require an exhaust gas recirculation to reduce NO_x raw emissions.

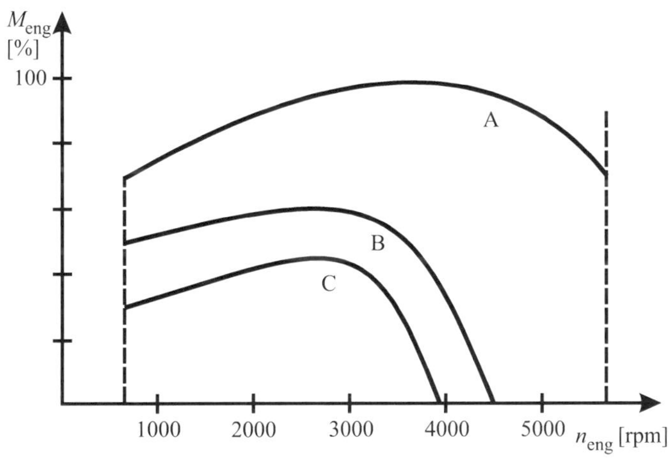

Fig. 7.9.2. Operating modes for direct injection gasoline engines. A homogeneous ($\lambda = 1$). B homogeneous-lean ($\lambda > 1$). C stratified ($\lambda > 1$).

This operation with stratified and lean combustion improves the efficiency. In addition the compression ratio can be increased as knocking is reduced by direct injection which leads to evaporation with a cooling effect (internal cooling). This also allows a larger air charge. Through these measures a fuel saving of about $10 - 15\,\%$ is reached compared to solely homogeneous combustion, Merker et al (2006), Alt et al (2002), see also Spicher (2007).

Because of the lean operation with $\lambda > 1$, the three-way catalytic converter cannot be applied for direct injection gasoline engines. Instead, a *storage-type catalytic converter* is required for the reduction of NO_x. As described in Sect. 7.2 it must be regenerated with a rich mixture ($\lambda < 1$) and an NO_x- or λ-sensor after the catalyst is required to control the regeneration. As this converter is poisoned by higher sulfur contents of the fuel, DI gasoline engines cannot be used in all markets.

The direct injection gasoline engine with different operating modes, transition between the modes and sophisticated exhaust gas treatment requires an increased effort for the electronic management and its calibration. The structure of the control of DI gasoline can be based on the control systems for homogeneous combustion. According to Fig. 7.1.2 the block operating modes are extended by the direct injection combustion modes and their commands to the torque coordination.

An addition to a jet-directed stratified direct injection engine by *turbocharging* can be a next step for further reduction of fuel consumption, Willand et al (2009). Experiments on a test bed have shown that for $p_{\mathrm{mi}} = 2\,\mathrm{bar}$ a lean stratified operation ($\lambda > 1$) with exhaust gas recirculation and for higher load up to $p_{\mathrm{mi}} = 6\,\mathrm{bar}$ a lean homogeneous operation ($\lambda > 1$) shows lowest fuel consumption and NO_x emissions. For loads higher than $8\,\mathrm{bar}$ the mode has to be switched to homogeneous stoichiometric injection ($\lambda = 1$) into the induction stroke. The advantages for the stratified operation are mainly caused by a reduction of the charging losses. However, the advantages through a higher dilution with air is compensated by a reduced turbine efficiency because of the lower exhaust gas temperature. Therefore, the potentials of downsizing concepts with increased torque for low speeds are higher than for stratified combustion with turbocharging, Willand et al (2009).

7.9.2 Homogeneous charge compression ignition for gasoline engines (HCCI)

In order to lower both fuel consumption and emissions, the homogeneous charge compression ignition (HCCI), also called controlled auto ignition (CAI), is a promising approach not only for diesel engines but also for gasoline engines. This combustion process is characterized by a homogeneous fuel/air/exhaust gas mixture which ignites itself during compression. The combustion then results at many spots without a flame front and diffusive combustion. It processes faster over about $10°$CS, compared to $30 - 40°$CS of a standard spark ignited gasoline engine, Ravi et al (2012). The lean combustion and therefore large air/fuel ratio leads to a more opened throttle (de-throttling) and therefore smaller charging losses. A lower temperature and fast heat release enable a constant volume cycle and thus an improved thermal efficiency, see Sect. 4.2.1. As the auto-ignition and following combustion start simultaneously at several locations within the combustion chamber there are no local constraints and the fluctuations around the center of heat release are smaller than for conventional gasoline engines with spark ignition, Kulzer et al (2009). However, the starting of auto-ignition depends strongly on temperature, pressure and mixture composition.

The lean and homogeneous heat release at lower burning temperatures result in a drastic reduction of NO_x emissions. In comparison to a fuel stratified combustion with peak combustion temperatures over 2200 K, a reduction of the combustion temperature with HCCI smaller than 2000 K leads to much smaller NO_x emissions, according to the nonlinear thermal Zeldovich mechanism. The HC and CO emissions can be kept at low levels, Willand et al (2008). Therefore the conventional three-way-catalytic converter can be applied without extra exhaust gas treatment. However, to reach and hold the homogeneous charge compression, ignition is very sensitive to several variables also because of a missing external combustion trigger such as a spark and therefore requires combustion pressure sensors. It is also restricted to part load. At low load HCCI is limited through an increased ignition delay. At higher load a too early begin of ignition and a too fast ignition appears because of limitations in leaning the mixture and increasing the exhaust gas rate which leads to a not acceptable mechanical stress. Hence, the HCCI combustion cannot cover the entire operation range and a transition to a spark ignition mode is required.

Gasoline has the advantage of already evaporating at lower burning temperatures. It is, however, resistant to ignition and requires an ignition temperature of about $1000 - 1200°C$. As this temperature cannot be reached by compression, the intake air has to be heated or hot exhaust gases have to be recirculated uncooled, Merker et al (2006). EGR can be realized by external or internal recirculation.

The internal EGR leads to higher temperatures and small delays. A first possibility is to use *exhaust gas retension* where the exhaust valve is closed early in order to trap a part of the burned gases and the intake valve opens late. This can be called a negative valve overlap and leads to recompression of the exhaust gases around 360 degrees, see Fig. 7.9.3, Alt et al (2008), Kulzer et al (2009) ("Exhaust recompression HCCI"). A second possibility is to open the exhaust valves a second time during the intake stroke, see Fig. 7.9.4, Willand et al (2008). This is called *exhaust-port recirculation* or "exhaust reinduced HCCI". The exhaust gases are then sucked back from the exhaust manifold at the beginning of the intake stroke, see also the HCCI versions of diesel engines in Sect. 8.4. As the HCCI operation is still within the focus of research, though first reports appeared in the 1980's, some publications are discussed in the sequel.

Investigations with fibreoptical endoscopes of a one-cylinder engine with direct injection and adjustable *exhaust gas retension* by negative valve overlap have shown that the ignition of the A/F-mixture does not occur in the whole combustion chamber at the same time. The auto-ignition begins in some regions early and in others with some delay depending on the the local mixture and temperature distribution. Then a sequential auto-ignition process follows which covers other areas, Sauter et al (2008). Experimental investigations for a single cylinder engine with fully variable electromechanical valve train show the effects of several variables, Babic et al (2010).

In the case of *internal exhaust-gas recirculation* the residual gas content directly depends on the timing of the exhaust valve closing and determines the A/F-ratio

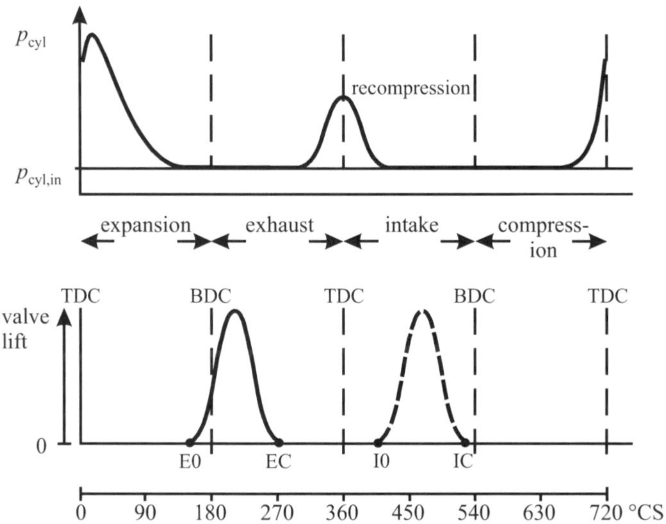

Fig. 7.9.3. Valve lift and cylinder pressure course for a HCCI gasoline engine with *exhaust gas retension*, negative valve overlap and recompression.

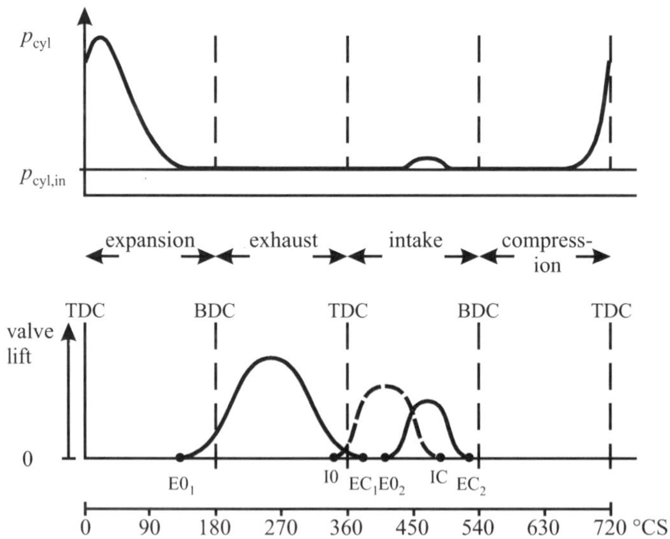

Fig. 7.9.4. Valve lift and cylinder pressure course for a HCCI gasoline engine with *exhaust-port recirculation*, positive valve overlap by a second exhaust valve opening and reduced intake valve lift.

(with throttle fully opened). A comparison of the different operating strategies has shown a fuel saving of about $15 \ldots 20\,\%$ for $2 \leq p_{\mathrm{mi}} \leq 3\,\mathrm{bar}$ and $2000 < n_{\mathrm{eng}} < 3000\,\mathrm{rpm}$ compared to conventional spark ignition combustion. However, stratified combustion gave even better results. A comparison of different charge compression ignition processes with internal and external exhaust gas recirculation is given in Bücker (2008).

The transition between auto-ignition and spark-ignition is based on phase adjustments of the valves. The spark-ignition operation is brought to a high exhaust gas rate up to an operation limit. However, auto-ignition requires higher exhaust gas rates, such that there exists a gap between the operating modes. Together with manipulating the throttle valve, intake valve timing and spark ignition angle it is then possible to allow a fast switch from spark-ignition to auto-ignition. However, a change in the other direction cannot be realized fast.

Another investigation is reported for the HCCI operation with *exhaust-gas retention* on the basis of a four-cylinder gasoline engine using a fast electrical camshaft phasing and two-point valve lift switching, Alt et al (2002). The temperature increase through exhaust gas retention is obtained by an early closing of the the exhaust valve and a late opening of the inlet valve, leading to a negative valve overlap of about 120 degrees and a successive recompression of the exhaust gases with a first fuel injection. The second injection occurs during the compression phase. The valve timing as well as fuel injection are part of a closed loop strategy with combustion pressure sensors and determination of the combustion center. HCCI operation is possible up to $40\,\%$ load at $1000\,\mathrm{rpm}$ and $25\,\%$ at $3000\,\mathrm{rpm}$.

A further prototype gasoline engine on the basis of a $1.6\,\mathrm{l}$ FSI engine was modified for *exhaust-port recirculation* with a double cam contour of the exhaust valves and reduced intake valve lift, Willand et al (2009). Low fuel consumption could be combined with low emissions. Especially extremely low $\mathrm{NO_x}$ emission could be achieved as a result of low burning temperatures. CO and HC emissions depend on the MFB50 combustion center. However, the combustion process reacts unusually sensitive to small changes in the setting variables.

A *control system* for a four cylinder gasoline engine with direct injection, variable valve train with phase manipulation and two-point lift switching for the intake and exhaust valves, internal exhaust retension and external exhaust gas recirculation is described by Kulzer et al (2009). The manipulated variables are injection time and mass, cam phases, valve lift, throttle, EGR valve and ignition time. The combustion control uses the mass fraction burned MFB50 and mean indicated pressure based on individual cylinder pressure measurement as controlled variables. A torque coordinator determines the reference variables for these control loops. Static feedforward control and a model based predictive feedforward control support fast dynamic actions and the transition from HCCI to SI operation. This transition is divided in four phases, from pre-switch to switch. Some few measurements are shown.

Modeling of HCCI gasoline engines for exhaust gas retension is reported in Ravi et al (2009) and used for control of the mean indicated pressure by the fuel mass and the combustion center MFB50 by valve timings, see also Shaver et al (2004).

Mean value models and observers for the control with exhaust-port recirculation are described in Rausen et al (2005) and Chiang et al (2007), see also Günther (2004).

A model based control of HCCI for diesel engines with experimental results is described in detail in Sect. 8.7.1.

The development of *HCCI for gasoline and diesel engines* shows several features in common. On the one hand gasoline has a good evaporation at low temperature but needs high temperatures for ignition. On the other hand diesel fuel has a low evaporation ability but ignites at lower temperatures. Thus, increasing the EGR rate rises the temperature of the gas mixture and improves the ignition ability for gasoline and the evaporation ability for diesel fuel within the homogeneous mixtures at part load. The optimization of the HCCI processes leads to compression ratios of about 11-13 and is therefore higher for spark ignition gasoline engines with typical ratios of 8-10 and lower for compression ignition diesel engines with typical values of 16-22. Roughly speaking a blending of Otto and Diesel combustion processes tends to develop. A research prototype engine with port injected gasoline and compression direct injected diesel fuel (6–25 % of fuel mass) uses the advantages of both fuel types and results in a significant reduction of NO_x and particulate emissions, Teetz et al (2012), see also the discussion of HCCI for diesel engines in Sect. 8.4.

Summing up, several research results for the physical interpretation, modeling and multi-variable control of HCCI gasoline engines for exhaust gas retention with negative valve overlap as well as for exhaust-port recirculation in part-load are published. Because of the high sensitivity of the auto-ignition process to several variables, closed loop control based on combustion pressure sensors seems to be unavoidable. However, an introduction into series production has not shown up until now.

7.10 Coolant temperature control

The goal of the coolant temperature control system is to protect the engine against overheating, especially to avoid hot spots at the cylinder walls and cylinder heads and to allow a fast warming up phase after cold start. In normal operation the cooling temperature should be high enough in order to minimize friction losses. The cooling system comprises the engine block, a cooling radiator, cooling valve with by pass, oil coolers, exhaust gas recirculation coolers, and heat exchangers for interior heating. As the design and control of the cooling system is emission and relevant to fuel saving it has received considerable attention in recent years.

The *conventional coolant temperature control* is based on a thermostatic three way or two way valve, Fig. 4.7.1. It manipulates the coolant flow into the radiator according to

$$\dot{m}_{\mathrm{cr}}(u_1) = \dot{m}_{\mathrm{c}} u_1 \tag{7.10.1}$$

where u_1 is the valve position and $\dot{m}_{\mathrm{c}}$ the coolant mass flow through the engine block. Hence the flow into the bypass becomes

$$\dot{m}_{\mathrm{cb}} = (1 - u_1)\,\dot{m}_{\mathrm{c}} \tag{7.10.2}$$

see (4.7.18) and (4.7.19). The heat flow balance (4.7.20) at the mixing point in front of the pump yields, see (4.7.22)

$$\vartheta_{\mathrm{ce,in}} = \vartheta_{\mathrm{c}} - u_1(\vartheta_{\mathrm{c}} - \vartheta_{\mathrm{cr,out}}). \tag{7.10.3}$$

Figure 7.10.1 depicts a simplified signal-flow scheme. Based on the models derived in Sect. 4.7 the coolant temperature process dynamics for small u_1 is mainly determined by the dynamics of the engine block and can be approximated by a first order differential equation plus a dead time, see (4.7.47).

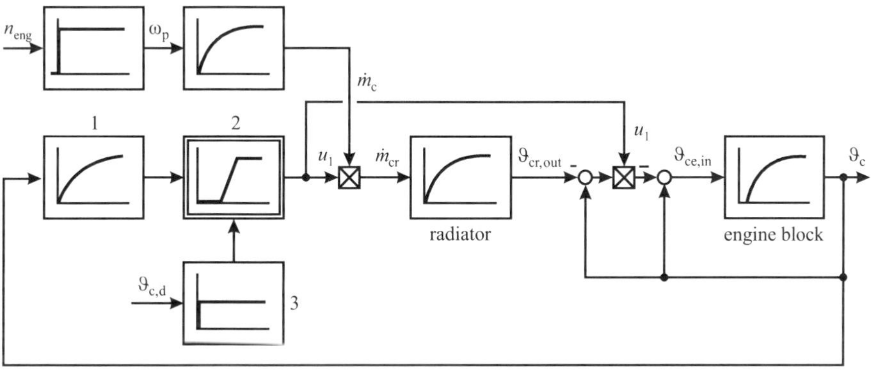

Fig. 7.10.1. Simplified signal-flow chart for the coolant temperature control with a thermostat. 1 coolant temperature sensor (wax element) 2 thermostatic valve 3 reference adjustment (heating element).

The thermostatic valve usually has a linear static characteristic in a range between, for example, 80°C<ϑ_c<98°C. Because of the proportional behavior the controlled variable varies in normal operation within this range with lower values for low loads and higher values for higher loads. However, if a map controlled thermostat is used, the coolant temperature follows a desired value more precisely, e.g. 90°C. Then at low load higher temperatures and at high load lower temperatures can be programmed in the ECU, in order to save fuel and to protect engine components.

A further task in designing the coolant control system is to enable a fast warming up after cold start. Therefore, the valve is kept closed until a certain coolant temperature is reached and then it is slowly opened according to the characteristic of the thermostat, e.g. at 80°C. However, overshoots over the desired value have to be avoided.

Figure 7.10.2 shows a more detailed signal-flow chart for the cooling system with a three-way valve manipulating the coolant flow through the radiator based on the models derived in Sect. 4.7. It includes several other influencing variables. If an electrical controllable valve is used and the engine inlet temperature $\vartheta_{\mathrm{ce,in}}$ and the coolant temperature ϑ_{c} at the valve are measured, a *cascaded electronic control system* can be realized, e.g. with PI controllers for the inner and outer controller. The inner controller controls all changes and disturbances in the radiator, control valve and oil cooler relatively quickly.

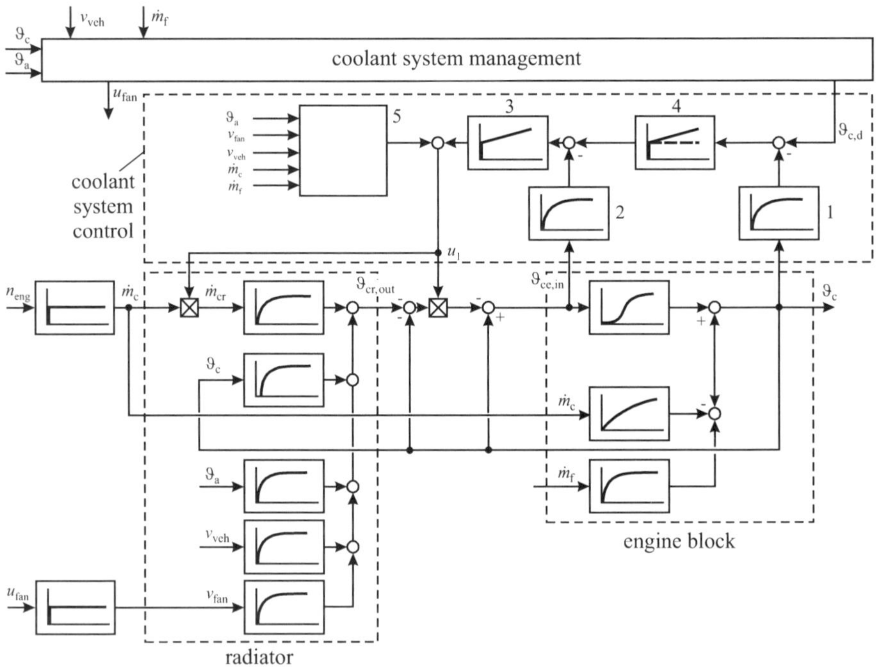

Fig. 7.10.2. Signal-flow chart for an advanced coolant system control with a coolant control valve-. 1 coolant temperature sensor at engine exit 2 coolant engine inlet temperature sensor 3 inlet temperature feedback controller 4 exit temperature feedback controller 5 feedforward controllers.

The outer controller has the task to control the remaining coolant temperature changes by the heat transfer in the engine block, changes of the coolant mass flow and interior heating. This cascade control system enables an improved control performance compared to Fig. 7.10.1. Smaller coolant temperature changes for example have the advantage that smaller thermal stress of materials results and the reference value of the coolant temperature can be moved closer to higher temperature limits.

The control performance can be further improved by *feedforward control functions* with inputs from the ECU like injected fuel mass flow $\dot{m}_{\mathrm{f}}$, coolant mass flow $\dot{m}_{\mathrm{c}}$ or engine speed n_{eng}, air speed v_{air}, vehicle speed v_{veh}, fan speed ω_{fan}, and ambient temperature ϑ_{a}, see Fig. 4.7.5. The feedforward control algorithms can be designed with proportional static or with first order dynamic behavior, thus compensating the coolant temperature dynamics, presented in Fig 4.7.5.

A further advanced coolant temperature control can be realized by manipulating the coolant circuit with an electrical (or mechanical) driven *speed-controlled pump*. In that case, the coolant mass flow is independent of the engine speed and can be adjusted to the required engine cooling power, thus saving engine power. This leads to a reduction of fuel consumption, of emissions because of a shorter warm-up phase, post cooling of the engine and its components (cylinder head, turbocharger, ...) and use of residual heat for interior heating after shut-off, Genster and Stephan (2004), Brömmel et al (2010). Also, combinations as electrically driven pumps for low load and mechanical driven ones for high load are recommended, e.g. for light duty vehicles, Keller et al (2010).

Controlling the coolant exit temperature with the pump speed and therefore by changing the coolant mass flow $\dot{m}_c$ also has control dynamic advantages, as the transfer behavior of the engine block is of low order and has almost no dead time compared to changing the engine coolant inlet temperature. However, the control behavior becomes more nonlinear and needs a comprehensive control design.

A multi-variable control scheme with the pump speed controlling the temperature difference $(\vartheta_c - \vartheta_{ce,in})$ across the engine and the coolant valve controlling the coolant temperature $\vartheta_c$is described by Guzzella and Onder (2010). Including feedforward control with measured engine input and output temperature, inversion of an external cooling model and an observer for engine block and wall temperature, a good control performance could be obtained and a fuel saving of about 3 % for an MVEG-95 test cycle, mainly due to a much reduced coolant flow compared to the mechanical driven pump, see also Cortona (2003).

The *overall coolant control system* requires a second level called the coolant system management, as shown in Fig. 7.10.2. Here, the selection of the desired coolant temperature $\vartheta_{c,d}$, the on/off switching or speed control of the radiator fan, including adjustments of the ambient temperature φ_a, vehicle speed v_{veh} and engine load, e.g. by injection mass flow $\dot{m}_f$, can be implemented.

The overall coolant control is also called *thermo-management*. This part of engine control has increasingly obtained attention in recent years and is influenced by saving fuel and emissions, optimization of the warming up phase and internal heating and defrosting comfort. Further requirements arise with start-stop systems and mechanical or electrically driven pumps and for hybrid drives with an increased frequency of starts and stops. In these cases the thermo-management has to enable an efficient cooling, avoiding damages through e.g. hot spots and avoiding an operation with too low coolant temperatures.

7.11 Oil pressure control

The oil pressure of the lubrication system described in Sect. 4.8 is conventionally controlled by a pressure-relief valve after the oil pump, see Fig. 4.8.1. The oil pressure acts on a spring-loaded valve leading to a proportional acting P-controller with a constant setpoint of about 3-6 bar. Figure 7.11.1 shows the course of oil pressure and oil temperature on the speed for a gasoline engine, Voigt (2010). For idling speed the oil pressure is about 2 bar and then increases with larger speed of 2000 rpm to 5.5 bar according to the pump characteristic, when the oil relief valve opens, see Sect. 4.8. The controlled oil pressure further increases to about 7.3 bar for 6000 rpm due to the proportional control characteristic of the mechanical oil pressure controller. The diagram also shows the oil pressure after the oil filter and in the cylinder head, indicating the corresponding pressure losses. The oil temperature increases from 80°C at idling to 130°C for maximum speed, showing the increase of heat transfer through friction and warm engine parts.

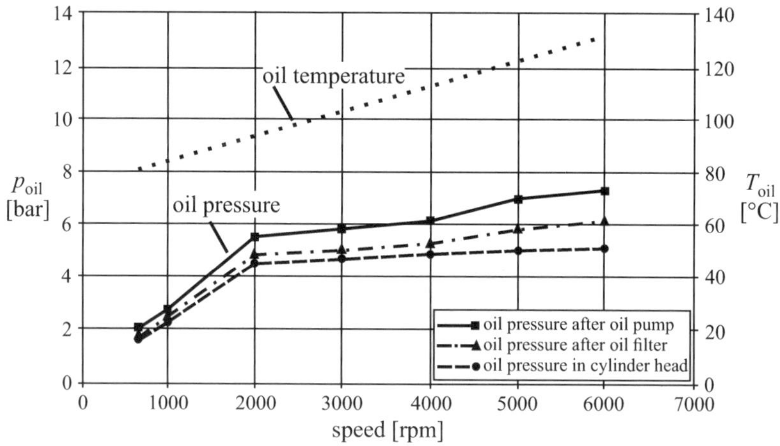

Fig. 7.11.1. Oil pressure and oil temperature of a 1.3 l, 4 cyl. gasoline engine. Oil pressure limitation at 5.5 bar, Voigt (2010).

The required oil pressure depends first on the minimal oil pressure at the different lubrication points of the engine to enable the necessary oil flow rates and cooling of the lubrication partners. Additionally, a certain minimal oil pressure is required for hydraulic actuators, for example, for camshaft phasing.

According to Fig. 4.8.3 the volume flow rate $\dot{V}_{\mathrm{eng}}(\omega_{\mathrm{eng}}, T_{\mathrm{oil}})$ increases with the engine speed, as well as the required pump power according to (4.8.2). The resulting pump power at maximum speed for passenger car engines is in the range of 1 to 2 kW. As the design of the oil supply is usually oriented to guarantee a sufficient oil pressure of about 1 bar at idling speed for maximum oil temperature of about 140°C, the resulting oil pressure and oil volume flow rate then increase for higher speed with a non-controlled *constant oil pump* to higher values than required for lubrication and cooling. This leads to throttling losses $P_{\mathrm{cv}} = \dot{V}_{\mathrm{sv}} p_{\mathrm{p}}$ where $\dot{V}_{\mathrm{cv}}$ is the flow rate through the pressure relief or control valve, see Fig. 4.8.3.

In order to save the oil pump driving power, controlled oil pumps were developed and introduced in series production in 2002, see Voigt (2003), Lamparski (2007), Wöckel and Leimbröck (2007) and Voigt (2010) The oil flow rate of these *variable oil pumps* is manipulated with electro-hydraulic valves by shifting the gear wheels of gear pumps relatively to each other or the excentricity of vane pumps, thus changing the displacement volume $V_{\mathrm{p,th}}$ per turn. This allows to control the oil pressure according to the absorbed engine oil flow, as indicated in Fig. 4.8.2d). A control scheme as in Fig. 7.11.2 results where the setpoint for the pressure control loop is adjusted by a look-up table dependent on the engine speed and the oil temperature, Wöckel and Leimbröck (2007). The oil pressure setpoint can be adjusted continuously or in one or two speed dependent steps, Voigt (2010). As the oil pressure according to (4.8.21) behaves like a proportional first order process, a PI controller with gain-scheduling is a first choice.

The oil flow rate at higher speeds is then reduced considerably and since throttling losses in the pressure release valve are avoided, fuel savings of 1 to 3 % are reported.

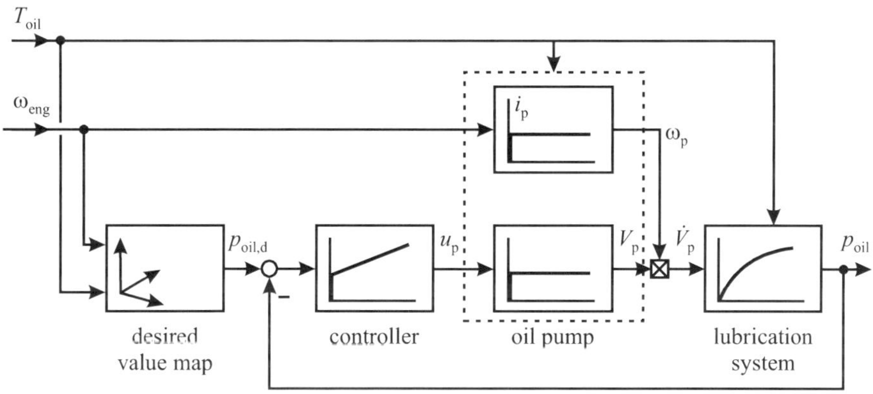

Fig. 7.11.2. Oil pressure control signal flow with variable oil pump and continuously adjusted setpoint.

The *oil temperature* depends on the friction heat of the lubrication points, the transferred heat in the engine oil channels, the cooled engine components and the water or air cooled oil cooler. The oil temperature for passenger cars is usually not controlled and varies during operation, except for the warming-up phase, where a thermostat may open a by-path of the oil cooler to quickly reach the operating temperature.

References

Alfieri E, Amstutz A, Guzzella L (2009) Gain-scheduled model-based feedback control of the air/fuel ratio in diesel engines. Control Engineering Practice – CEP 17:1417–1425

Alt B (2010) Modellbasierte Regelung ausgewählte Komponenten im Antreibsstrang eines Kraftfahrzeuges mit Ottomotor. Diss. Universität der Bundeswehr München. Fortschr.-Ber. VDI Reihe 8, 729. VDI Verlag, Düsseldorf

Alt M, Quarg J, Bargende M (2002) Gasoline direct injection - potentials of homogeneous and stratified operation (in German). In: 11th Aachener Kolloquium Fahrzeug- und Motorentechnik, Aachen, Germany

Alt M, Grebe U, Dulzo J, Chang MF (2008) Closed loop combustion control for HCCI. In: 8. Int. Stuttgarter Symposium, Stuttgart
Anastasia C, Pestana G (1987) A cylinder pressure sensor for closed-loop engine control. In: SAE Technical paper Series, Warrendale, PA, 870288
Ault B, Jones V, Powell J, Franklin G (1994) Adaptive air-fuel ratio control of a spark ignition engine. In: SAE Technical paper Series, Warrendale, PA, 940373
Babic G, Berner H, Bargende M (2010) Operating strategies on hcci combustion. In: 10. Internationales Stuttgarter Symposium, pp 543–561
Bargende M (1995a) Most optimal location of 50 percent mass fraction burned and automatic knock detection components for automatic optimization of SI engine calibrations. MTZ worldwide 56(10)
Bargende M (1995b) Schwerpunkt-Kriterium und automatische Klingelerkennung. Motortechnische Zeitschrift - MTZ 56(10):632–638
van Basshuysen R, Schäfer F (2004) Internal combustion engine handbook: basics, components, systems, and perspectives. SAE International, Warrendale
Baunach T, Schänzelin K, Diehl L (2006) Sauberes Abgas durch Keramiksensoren. Physik Journal 5(5):33–338
BEG (2009) Bosch Engineering GmbH. Training notes Momentenstruktur. ME(D)17. Robert Bosch Engineering GmbH, Abstatt, Germany
Bernard L, Ferrari A, Micelli D (2009) Electro-hydraulic valve control with MultiAir technology. ATZ autotechnology 9(6):32–37
Borger M, Haussner M, Houben H, Marto A, Pechhold F (2004) Pressure sensor glow-plug for diesel engines. MTZ worldwide 65
Brömmel A, Rombach M, Wickerath B, Weinecke T, Durand JM, Armenio G, Squarcini R, Gibat T (2010) Electrification powers pump innovations. MTZ extra pp 89–96
Brüstle C, Schwarzenthal D (1998) The two-in-one engine - Porsche's variable valve system (VVS). In: SAE Technical paper Series, Warrendale, PA, 980766
Bücker C (2008) Betriebsstrategien zur kontrollierten Selbstzündung von Ottomotoren. Dissertation. RWTH Aachen, Aachen
Chiang CJ, Stefanopoulou A, Jankovic M (2007) Nonlinear observer-based control of load transitions in homogeneous charge compression ignition engines. IEEE Trans on Control Systems Technology 15(3):438–448
Cortona K (2003) Thermomanagement for fuel consumption reduction. Doctoral thesis. ETH Zürich, Zürich
Denger D, Mischker K (2004) The electro-hydraulic valvetrain system. MTZ worldwide 65(12):10–13
Dorey R, Stuart G (1994) Self-tuning control applied to the in-vehicle calibration of a spark ignition engine. In: Proc. of the 3rd Conf. on Control Applications, Glasgow, vol 1, pp 121–126
Flierl R, Kairies D (2006) Comparison of fully variable mechanical and electromechanical valve trains. In: AUTOREG 2006, VDI, Wiesloch, Germany, vol VDI-Berichte 1931, pp 345–358

Gäfvert M, Årzén KE, Pedersen L (2000) Simple linear feedback and extremum seeking control of GDI engines. In: Proceedings of Seoul 2000 FISITA World Automotive Congress, Seoul, Korea
Genster A, Stephan W (2004) Always at the correct temperature – thermal management with electric coolant pump. MTZ worldwide 65(11):6–7
Gerhardt J, Benninger N, Hess W (1997) Torque-oriented functional structure of an electronic engine management as a new basis of drivetrain systems (in German). In: 6th Aachener Kolloquium Fahrzeug- und Motorentechnik, Aachen, Germany
Gerhardt J, Höninger H, Bischof H (1998) A new approach to functional and software structure for engine management systems - Bosch ME97. In: SAE 980801, Aachen, Germany
Glaser I, Powell JD (1981) Optimal closed-loop spark control of an automotive engine. Tech. rep.
Günther M (2004) Untersuchungen der Eigenschaften und Kontrollmöglichkeiten der homogene kompressionsgezündeten Verbrennung von Ottokraftstoff. Diss. Universität Karlsruhe. Karlsruhe
Guzzella L, Onder C (2010) Introduction to modeling and control of internal combustion engine systems, 2nd edn. Springer, Berlin
Haas M, Rauch M (2010) Electro-hydraulic fully variable valve train system. MTZ worldwide 71
Hellemans A, Landreve L, Venzel S, Walker E (2011) In-cylinder pressure sensor. MTZ worldwide 72(10):42–47
Herden W, Küsell M (1994) A new combustion prressure sensor for advanced engine management. In: SAE Technical paper Series, Warrendale, PA, 940379
Hess W (2003) Drehmomentorientierte Struktur der Motorsteuerungen. In: Isermann R (ed) Modellgestützte Steuerung, Regelung und Diagnose von Verbrennungsmotoren, Springer, Berlin
Heywood JB (1988) Internal combustion engine fundamentals. McGraw-Hill, New York
Hockel K (1981) Einfluss der zeitlichen Energieumsetzung auf den Wirkungsgrad beim Ottomotor. Automobil-Industrie 3:315–320
Hofmann D, Heinkle M, Heinrich D, Husch F, Zein T, Brandt M (2011) Scavening, a challenge for engine control (in German). In: AUTOREG 2011, VDI, Wiesloch, Germany, vol VDI-Berichte 2135, pp 75–85
Hohenberg G (1982) Der Verbrennungsverlauf. In: 4th Wiener Motorensymposium. Fortschr.-Ber. VDI Reihe 6, Nr. 103, VDI, Düsseldorf, pp 71–88
Hohenberg G, Dolt R (1993) Ein Konzept zur adaptiven Steuerung/Regelung von Verbrennungsmotoren unter Verwendung eines Online-Brennverlaufsrechners. In: Fachtagung Integrierte mechanisch-elektronische Systeme, VDI, Düsseldorf, Darmstadt, Germany, vol Fortschr.-Ber. VDI Reihe 12 Nr. 179, pp 58–69
Hrovat D, Sun J (1997) Models and control methodologies for IC engine idle speed control design. Control Engineering Practice – CEP 8(4):1093–1100
Inoue T, Matsushita S, Nakanishi K, Okano H (1993) Toyota lean combustion system - the third generation system. In: SAE Technical paper Series, Warrendale, PA, 930873

Isermann R (2015) Combustion-engine diagnosis. To be published. Springer, Berlin
Isermann R, Müller N (2001) Nonlinear identification and adaptive control of combustion engines. In: IFAC Workshop on adaptation and learning in control and signal processing, Como, Italy
Isermann R, Müller N (2003) Design of computer-controlled combustion engines. Mechatronics 13:1067–1089
Isermann R, Münchhof M (2011) Identification of dynamic systems. Springer, Berlin
Jost O (2000) Einsatz optischer Zylinderdrucksensoren zur Steuerung, Regelung und Überwachung von Pkw-Dieselmotoren mit Direkteinspritzung. In: 20. Tagung. Elektronik im Kraftfahrzeug, Haus der Technik
Kawamura A, Haneyoshi T, Hoft R (1988) Deadbeat controlled PWM inverter with parameter estimation using only voltage sensor. IEEE Trans on Power Electronics 3(2):118–125
Keller P, Wenzel W, Becker M, Roby J (2010) Hybrid coolant pumps with electrical and mechanical drive. ATZ autotechnology 10(6):52–56
Kiencke U, Nielsen L (eds) (2000) Automotive control systems. For engine, driveline and vehicle. Springer, Berlin
Kimmich F, Schwarte A, Isermann R (2001) Model-based fault detection for diesel engines. In: Aachen Colloquium, Automobile and Engine Technology, Aachen, Germany
Klaus B, Drexler G, Eder T (2005) Further development of BMW's fully-variable vavle control system Valvetronic. MTZ worldwide 66(9):10–13
Klimstra J (1985) The optimum combustion phasing angle-a convenient engine tuning criterion. In: SAE Technical paper Series, Warrendale, PA, 852090
Klüting M, Flierl R, Grudno A, Luttermann C (1999) Drosselfreie Laststeuerung mit vollvariablen Ventiltrieben. MTZ-Motortechnische Zeitschrift 60(7-8):476–485
Köhler E, Flierl R (2012) Verbrennungsmotoren, 6th edn. Vieweg, Wiesbaden
Kohlhase M (2011) Brennraumdruckbasiertes Motormanagement für Otto- und Dieselmotoren zur Verbrauchs- und Emissionsreduktion. Dissertation Technische Universität Darmstadt. Fortschr.-Ber. VDI Reihe 12, 743, VDI Verlag, Düsseldorf
Kohlhase M, Isermann R (2009) Self-tuning extremum value control based on cylincer pressure sensors for direct injection spark ignition engines (in German). at - Automatisierungstechnik 57(1):23–31
Konrad H, Krämer G (2002) The development of the control functions for the bmw valvetronic engines (in german). at - Automatisierungstechnik pp 360–367
Krebber-Hortmann K, Köster A, Tönnesmann A, Brunetti C (2013) Possibilities for fuel consumption through electrical camshaft actuation. MTZ - Motortechnische Zeitschrift 73(7-8):854–859
Kulzer A, Fischer W, Karrelmeyer R, Saver C, Wintrich T, Benninger K (2009) Homogeneous charge compression ignition on gasoline engines. MTZ worldwide 70(1):32–57
Lamparski C (2007) Bedarfsgerechte Ölversorgnung: Regelpumpen im Serieneinsatz. In: Berg M (ed) Ölkreislauf von Verbrennungsmotoren II, expert verlag, Renningen, pp 83–108

Landenfeld T, Gerhardt J, Küsell M (2002) Gasoline direct injection - an attractive concept for emission reduction (in German). In: 11th Aachener Kolloquium Fahrzeug- und Motorentechnik, Aachen, Germany, pp 177–197

Liebl J, Klüting M, Poggel J, Missy S (2001a) Der neue BMW Vierzylinder-Ottomotor mit Valvetronic. Motortechnische Zeitschrift – MTZ 62(7-8):570–579

Liebl J, Munk F, Hohenner H, Ludwig B (2001b) Die Steuerung der neuen BMW VALVETRONIC Motoren. Motortechnische Zeitschrift – MTZ 62(7-8):516–527

Loh R, Lu M (2001) Crank-angle domain modeling and optimal tracking control for automotive engine idle speed. In: IFAC Conferenc on New Technologies for Computer Control, Hong Kong, pp 247–258

Martin G (2009) Variable Ventiltriebe für Verbrennungsmotoren. ATZ extra (11):204–218

Matekunas F (1986) Engine combustion control with ignition timing by pressure ratio management. Patent US 4,622,939

Meisberger D, Albert C, Bourdon K (1998) Die neue Motorsteuerung ME7.2 von Bosch für den BMW V8-Motor. Motortechnische Zeitschrift – MTZ 59(12):826–834

Merker G, Schwarz C, Stiesch G, Otto F (2006) Simulating combustion. Springer, Berlin

Moriwaki J, Murai H, Kameshima A (2003) Glow plug with comubstion pressure sensor. In: SAE 2003 World Congress and Exhibition, Detroit, MI, 2003-01-0707

Müller N (2003) Adaptive engine control for gasoline engines using combustion pressure sensors (in German). Fortschr.-Ber. VDI Reihe 12, 545. VDI Verlag, Düsseldorf

Pestana G (1989) Engine control methods using combustion pressure feedback. In: SAE Technical paper Series, Warrendale, PA, 890758

Powell D (1993) Engine control using cylinder pressure: past, present, and future. Journal of Dynamic Systems, Measurement and Control 115:343–350

Powell J, Fekete N, Chang C (1998) Observer-based air-fuel ratio control. IEEE Control Systems Magazine 18:72–83

Rausen D, Stefanopoulou A, Kang JM, Eng J, Kuo TW (2005) A mean-value model for control of homogeneous charge compression ignition (HCCI) engines. J of Dyn Meas and Control 127:355–362

Ravi N, Roelle M, Chang C, Gerdes J (2009) Physical modeling and control of a multi-cylinder HCCI engine. In: Proceedings of the ASME 2009 Dynamic Systems and Control Conference, DSCC2009, Hollywood, California, USA

Ravi N, Liao HH, Jungkunz A, Widd A, Gerdes J (2012) Model predictive control of HCCI using variable valve actuation and fuel injection. Control Engineering Practice 20(4):421–430

Robert Bosch GmbH (2004) Ottomotor-Management. Vieweg, Wiesbaden

Robert Bosch GmbH (ed) (2007) Automotive electrics, automotive electronics. J. Wiley, Chichester, UK

Robert Bosch GmbH (ed) (2011) Automotive Handbook, 8th edn. Bentley publishers, Cambridge

Rohe-Brandenburger K (1986) Einfluss des Zündwinkels auf den Wirkungsgrad eines Mehrzylinder-Ottomotors im instationären Betrieb. Fortschr.-Ber. VDI Reihe 12, 79, VDI Verlag, Düsseldorf

Rupp D, Onder C, Guzzella L (2008) Iterative adaptive air/fuel ratio control. In: Proceedings IFAC Symposium in Advances in Automotive Control, Seascape Resort Aptos, CA, USA, pp 609–615

Salber W, Kemper H, Staay F, Esch T (2000) The electro-mechanical valve train–a system module for future powertrain concepts. MTZ worldwide 61(12):826–826

Sauter W, Hensel S, Spicher U, Schubert A, Schiessl R (2008) Controlled auto-ignition-reduction of engine emissions in gasoline engines. MTZ worldwide 69(1):40–47

Schneider F, Lettmann M (2008) Variable valve timing for any engine concept. MTZ worldwide 69(5):27–28

Schramm H, Wurst B (1977) Vorrichtung zur Extremwert-Regelung bei Kraftmaschinen. Patent DE 2739568C2. Robert Bosch GmbH

Schulz T, Kulzer A (2006) Überblick und Bewertung variabler Ventiltriebe. In: AUTOREG 2006, VDI, Wiesloch, Germany, vol VDI-Berichte 1931, pp 339–344

Scotson P, Wellstead R (1990) Self-tuning optimization of spark ignition automotive engines. In: IEEE Control Magazine, vol 10, pp 94–101

Sellnau M, Matekunas F, Battiston P, Chang CF, Lancaster D (2000) Cylinder-pressure-based engine control using pressure-ratio-management and low-cost non-intrusive cylinder pressure sensors. In: SAE Technical paper series, Warrendale, PA, 2000-01-0932

Shaver G, Gerdes J, Roelle M (2004) Physics-based closed-loop control of phasing, peak pressure and work output in HCCI engines utilizing variable valve actuation. In: American Control Conference, vol 1, pp 150–155

Spicher U (ed) (2007) Direkteinspritzung im Ottomotor III. expert Verlag, Renningen

Stiesch G (2003) Modeling Engine Spray and Combustion Processes. Springer, Berlin

Straky H (1996) Modellgestützter Funktionsentwurf für Kfz-Stellglieder; Regelung der elektromechanischen Ventiltriebaktorik und Fehlerdiagnose der Bremssystemhydraulik. Dissertation Technische Universität Darmstadt. Fortschr.-Ber. VDI Reihe 12, 546. VDI Verlag, Düsseldorf

Stuhler H, Kruse T (2003) Parametrierung der Motorsteuerungen. In: Isermann R (ed) Modellgestützte Steuerung, Regelung und Diagnose von Verbrennungsmotoren, Springer, Berlin, chap 3

Teetz C, Bergmann D, Schneemann A, Eichmeier J (2012) MTU HCCI engine with low raw emissions. MTZ worldwide 73(9):4–9

Truscott A, Noble A, Akoachere A, Müller R, Hart M, Krötz G, Richter C, Cavalloni C, Gnielka M (2001) In-cylinder pressure sensing and model-based control of a spark ignition engine. In: Proceedings. JSAE Annual Congress, 36-01, pp 5–8

Tschöke H, Schultalbers M, Gottschalk W, Huthöfer E, Jordan A (2011) Thermodynamic optimizatioon criteria for the ignition time of modern SI engines. MTZ worldwide 72(1):52–57

Unger H, Schwarz C, Schneider J, Koch F (2008) The valvetronic experience from seven years of mass production. MTZ worldwide 69(7-8):30–37

Vogt M, Müller N, Isermann R (2004) Online adaptation of grid-based look-up tables using a fast regressor technique. Transactions of the ASME 126(Dec):732–739

Voigt D (2003) Kraftstoffverbrauchsvorteile durch Regelölpumpen. Motortechnische Zeitschrift - MTZ 64(12)

Voigt D (2010) Motorschmierung bei PKW-Motoren. In: Eberan-Eberhorst C (ed) Schmierung von Verbrennungsmotoren, expert verlag, Renningen, pp 140–151

Wellstead R, Scotson P (1990) Self-tuning extremum control. In: IEE-Proceedings Control Theory and Applications, vol 137, pp 165–175

Willand J, Jelitto C, Jakobs J (2008) The GCI combustion process from Volkswagen. MTZ worldwide 69(4):56–61

Willand J, Schintzel K, Hoffmeyer H (2009) The potential of turbo-charged gasoline engines with spray guided direct injection. MTZ worldwide 70(2):12–18

Willimowski M (2003) Verbrennungsdiagnose von Ottomotoren mittels Abgasdruck und Ionenstrom. Dissertation Technische Universität Darmstadt. Shaker Verlag, Aachen

Wöckel N, Leimbröck M (2007) Optimierter Ölkreislauf der neuen Porsche V8-Motoren mit variabler Ölpumpe. In: Berg M (ed) Ölkreislauf von Verbrennungsmotoren II, expert verlag, Renningen, pp 2–22

Won M, Choi S, Hedrick J (1998) Air-to-fuel ratio control of si engines using gaussian network sliding control. IEEE Control System Technology 6(5):678–687

Yildiz Y, Annaswamy A, Yanakiev D, Kolmanovsky I (2010) Spark ignition engine fuel-to-air control: an adaptive control approach. Control Engineering Practice – CEP 18:1369–1378

Zarske R (1988) Dynamisches Steuerkonzept für Einspritzung und Zündung von Ottomotoren. Braunschweig

8

Control of diesel engines

The main components, the main control tasks and a resulting control-oriented block diagram for turbocharged diesel engines were already discussed in Sect. 1.3.1. As depicted in the block diagram Fig. 1.3.7, in the software structure in Figs. 5.1.4 and 5.1.5 and in the hierarchical control structure of an ECU in Fig. 5.1.6 the control of diesel engines can be subdivided in:

- Torque control
  - fuel injection control
    - fuel mass control
    - injection angle control
    - injection pressure control
    - multiple injection control
  - air mass control
  - charging pressure control
  - variable valve control (if available)
- Emission control
  - exhaust recirculation control
  - particulate filter or SCR control
- Smoke limitation control
- Idle speed control
- Coolant temperature control.

The diesel engine operates with internal mixture formation and compression ignition. During the compression stroke the air from the intake is compressed to $30 \ldots 55\,\mathrm{bar}$ in naturally aspirating, or $80 \ldots 110\,\mathrm{bar}$ in charged engines and therefore the air heats up to about $700 \ldots 900°\mathrm{C}$, according to compression ratios of 6:1 to 24:1. The fuel is injected shortly before the TDC and is auto-ignited by the warm air. The injected fuel spray leads to a heterogeneous combustion, based on finely atomized and easily combustible fuel droplets. The cylinder charge heats up and the cylinder pressure rises more, leading to pressure forces at the piston. The local air-fuel ratio varies between $\lambda = \infty$ for pure air and $\lambda = 0$ for pure fuel in the spray core. A combustion

R. Isermann, *Engine Modeling and Control*,
DOI: 10.1007/978-3-642-39934-3_8

arises in the range of $0.3 < \lambda < 1.5$ around the fuel droplets. Herewith the combustible mixture develops by diffusion and turbulence. The kind of the injected fuel spray and its kinetic energy have a large influence on the combustion.

In former times partial combustion started in a prechamber to obtain a good mixture in the main combustion chamber. Modern direct injection diesel engines form the mixture by the air movement and turbulence in the intake channel and valve contraction, the piston movement and the injected jet. High injection pressures up to 2000 bar, multihole injector nozzles and swirl and tumble supported by the design of the intake duct and the piston recess enable a good mixture formation.

The combustion process can be subdivided into three phases, Merker et al (2006). Firstly, the fuel jet mixes with the air and combusts very quickly after an ignition delay (pre-mixed combustion). The air/fuel mixture then continues and the combustion is determined by the mixing rate (diffusion combustion). Towards the end of combustion the burning becomes relatively slow (post-combustion). In this third phase intermediate products are further oxidized and especially previously formed soot is oxidized. This is described in more detail in Sect. 8.2.2. A pilot injection before the main injection enables to reduce combustion noise and allows an optimization of the main injection(s), which results in fuel savings and lower emissions.

Unlike gasoline engines the air flow is not throttled, such avoiding low pressure charge losses and the combustion is performed with large air excess factors of about $1.2 < \lambda < 10$. The torque is mainly determined by the injected fuel mass.

On the contents of this chapter

The chapter begins by describing the diesel-engine control structure and its various operating modes. Some simplified torque models are derived, and the measured stationary overall behavior and the dynamic behavior after step changes of four important input variables and 20 output variables is illustrated. Combustion models are treated for the pre-mixed and diffusion combustion for the assumption of one cylinder zone. Then, the optimization of the feedforward control with three main manipulated variables is considered for fixed operating points and a driving cycle by using numerical optimization methods, different criteria and model-identification methods. In order to meet the emission limits and to achieve good fuel consumption and acceleration, modern passenger car diesel engines are equipped with external exhaust gas recirculation (EGR) and with variable geometry turbochargers (VGT). The EGR mass flow is used to decrease the nitrogen oxide emissions. The turbocharger mainly increases the intake manifold pressure and, thus, increases the engine power.

A detailed presentation of the model-based design of a two-variable control system for the charging pressure and air flow rate describes how with identified local models and a combination of dynamic feedforward and feedback controls a precise control over the whole operating range can be reached.

The development of a combustion pressure heat release control opens ways for a feedback control of individual cylinder combustions. Adaptive look-up tables allow to improve the fixed feedforward look-up tables individually for all cylinders. It is then shown how a combustion-pressure-based control system can be realized to control homogeneous compression charge ignition (HCCI) combustions in a relatively large part-load range which leads to considerable reduction of NO_x and soot.

Three methods are described how smoke limitation control for strong accelerations can be designed without and with smoke models. The development of emission control systems requires overall emission models for NO_x, PM and λ. It is shown how stationary and dynamic models are identified with local polynomial models. Finally, raw emissions control with NO_x- and λ-sensors are discussed and the various exhaust aftertreatment systems and its control are outlined.

8.1 Diesel engine control structure

This section considers the basic structure of diesel-engine control, overall torque models and illustrates experimental results to obtain a first insight into the stationary and dynamic behavior.

8.1.1 Torque-oriented structure

The structure of modern diesel engine control systems followed the development for gasoline engines, Gerhardt et al (1998), Guzzella and Amstutz (1998) and is also characterized by a torque-oriented structure. The underlying stationary torque model is similar to that of gasoline engines, see Fig. 8.1.1. The inner torque M_i is mapped in dependence on the injected fuel mass m_f and engine speed n_eng for optimal combustion. Taking the torque losses M_drg according to (7.1.2) into account it holds for the inner torque in steady-state

$$\overline{M}_\mathrm{i}\left(m_\mathrm{f}, n_\mathrm{eng}\right) = f_\mathrm{i}\left(m_\mathrm{f}, n_\mathrm{eng}\right) = \overline{M}_\mathrm{eng} + \overline{M}_\mathrm{drg}. \tag{8.1.1}$$

Changes from the optimal operating conditions by other injection angles, charging pressures, or exhaust gas recirculation mass flow rates are taken into account by efficiencies like

$$\eta_\varphi = \frac{M_\mathrm{i}(\varphi_\mathrm{inj})}{M_\mathrm{i}}\,; \quad \eta_\mathrm{p} = \frac{M_\mathrm{i}(p_2)}{M_\mathrm{i}}\,; \quad \eta_\mathrm{egr} = \frac{M_\mathrm{i}(\dot{m}_\mathrm{egr})}{M_\mathrm{i}} \tag{8.1.2}$$

such that the corrected inner torque becomes

$$\overline{M}_\mathrm{i,corr} = f_\mathrm{Mi,corr}\left(m_\mathrm{f}, n_\mathrm{eng}, \dot{\varphi}_\mathrm{eng}, p_2, \dot{m}_\mathrm{egr}\right) = \eta_\varphi\, \eta_\mathrm{p}\, \eta_\mathrm{egr}\, \overline{M}_\mathrm{i}. \tag{8.1.3}$$

The torque at the clutch then follows as, see Fig. 8.1.1,

$$\overline{M}_\mathrm{eng}(m_\mathrm{f}, n_\mathrm{eng}, \varphi_\mathrm{inj}, p_2, \dot{m}_\mathrm{egr}) = \overline{M}_\mathrm{i,corr} - \overline{M}_\mathrm{drg}. \tag{8.1.4}$$

By inversion of the inner torque model the required fuel injection mass can be determined with (8.1.3) which delivers the corrected inner torque

$$m_\mathrm{f,corr} = \frac{1}{\eta_\varphi\, \eta_\mathrm{p}\, \eta_\mathrm{egr}}\, f_\mathrm{Mi}^{-1}(\overline{M}_\mathrm{i}, n_\mathrm{eng}). \tag{8.1.5}$$

This is quite similar to gasoline engines as described in Sect. 7.1. As for gasoline engines it is required for the design and implementation of the control functions that a *basic control calibration* of the diesel engines has to be carried out by measuring the steady-state behavior on a test bench to determine optimal manipulated variables. The optimization then requires to determine following steady-state models ($n = n_{\text{eng}}$):

- inner torque $M_{\text{i}}(m_{\text{f}}, n)$ (for other optimal manipulated variables)
- fuel injection angle $\varphi_{\text{inj}}(m_{\text{f}}, n)$
- air mass in the cylinder $m_{\text{air}}(m_{\text{f}}, n)$
- charging pressure $p_2(m_{\text{f}}, n)$
- exhaust gas recirculation mass flow $\dot{m}_{\text{egr}}(m_{\text{f}}, n)$
- rail pressure $p_{\text{rail}}(m_{\text{f}}, n)$
- torque $M_{\text{drg}}(n, T_{\text{oil}})$.

Instead of e.g. $\varphi_{\text{inj}}(m_{\text{f}}, n)$ also $\varphi_{\text{inj}}(M_{\text{i,d}}, n)$ can be used.

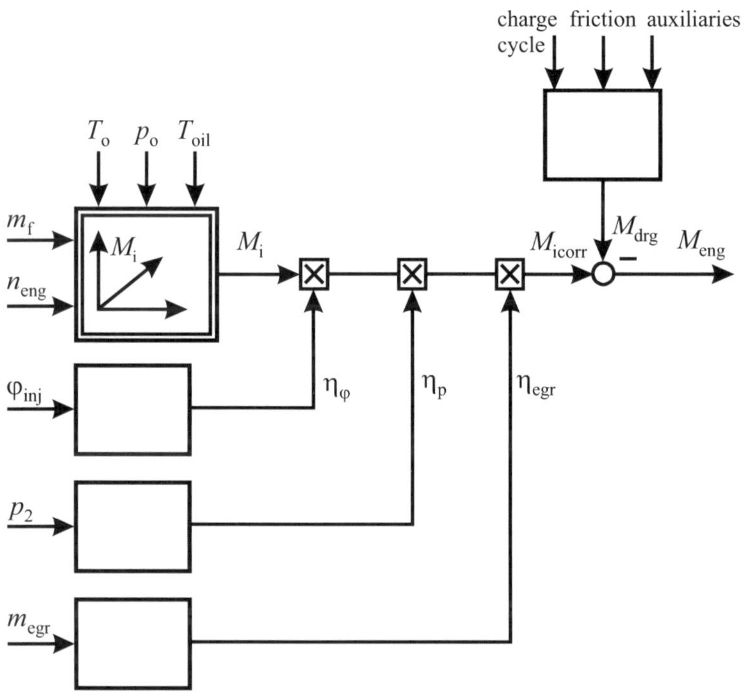

Fig. 8.1.1. Torque model of a diesel engine. M_{i}: inner torque; $M_{\text{i,corr}}$: corrected inner torque; M_{drg}: drag torque (torque losses).

Based on these look-up tables and determination of efficiencies (8.1.2) the required injected fuel mass with correction $m_{\text{f,corr}}$ is determined with (8.1.5). This basic control calibration is then represented in corresponding look-up tables and used for the design of the torque-oriented engine control with regard to internal and external torque demands.

The torque-oriented diesel engine control system can be divided in different signal-flow control blocks, as depicted in Fig. 8.1.2, correspondingly to gasoline engines,

(a) external and internal torque demands
(b) operation modes
(c) torque coordination
(d) torque conversion
(e) engine control with actuators.

The *control function blocks* comprise for example following tasks, compare Fig. 8.1.3.

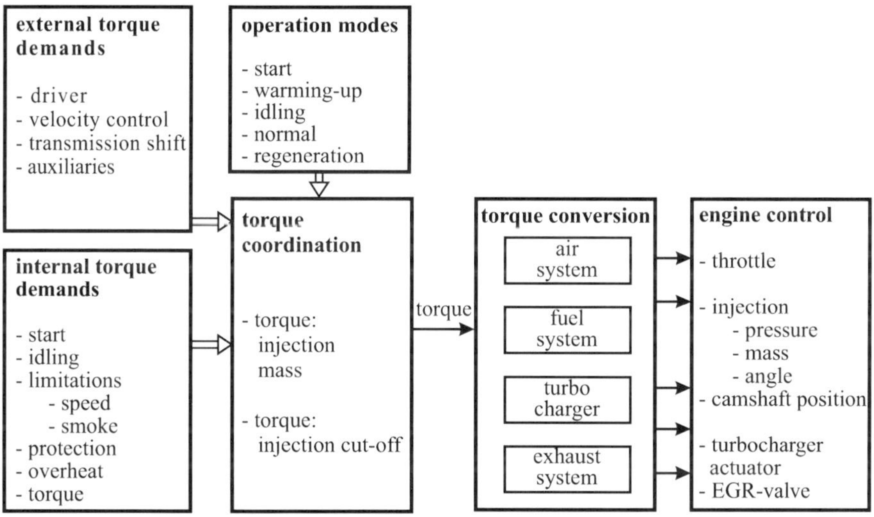

Fig. 8.1.2. Main signal-flow blocks for the control of diesel engines with common rail injection.

a) Torque demands

The pedal position α_{ped} from the driver determines the desired inner torque according to the drive map $M_{\text{i,d}}(\alpha_{\text{ped}}, n)$ for normal driving without gear shifting. Other operating conditions like starting, idling and adding of further auxiliary drives require driver independent torque demands. The inner torque should never be smaller than the drag torque (except during motoring) and not larger than protection limits by the engine itself or the transmission. Further limits of the torque for diesel engines are a limitation of the engine speed and smoke at full load.

b) Torque coordination

The torque coordination decides which of the different torque demands is given to determine the injected fuel quantity. This is realized as a priority selection with minimum and maximum operators or as addition of several torque requests. For example, the maximum of the demanded inner torque or the drag torque is selected, or the minimum of the desired torque from the drive map or the protection torque limit or the smoke torque limit.

Parallel to the selection of the torque demands a dynamic transfer element in form of a lead-lag (DT_1) element is realized in order to compensate for a part of the dynamic inertness of the turbocharger and the air flow into the cylinders in the case of a jump increase of the desired torque. This way a sudden increase of the injected fuel quantity is paralleled by an accelerated air mass flow. As the driver command is delayed by the accelerator-pedal filter and the lead path is applied to the unfiltered pedal sensor, a synchronous dynamic change of fuel mass and air mass is intended. For deceleration this lead-lag element may be circumvented.

c) Torque conversion

The torque conversion determines the desired command values to the actuators directly (feedforward) or to closed loops as reference variables or setpoints (feedback control). The torque conversion block therefore comprises the optimized steady-state models resulting from the basis control calibration.

First, the torque map of the desired inner torque $M_{\mathrm{i,d}}(m_{\mathrm{f}}, n)$ is inverted to calculate the required fuel injection mass $m_{\mathrm{f,d}}$ (further subdivided in main injection $m_{\mathrm{f,d,mi}}$ and pilot injections $m_{\mathrm{f,d,pi}}$) and desired air mass $m_{\mathrm{air,d}}$ as shown in Figs. 8.1.3 and 8.1.4. The other manipulated variables or reference variables follow from the optimized maps, like injection angles (begin of injections) $\varphi_{\mathrm{inj,d}}$, rail pressure $p_{\mathrm{rail,d}}$ and charging pressure $p_{2,\mathrm{d}}$. For strong torque reduction, injections of certain cylinders can be shut off.

d) Engine control

According to Fig. 8.1.2 the signal-flow block for engine control receives the required commands from the torque conversion block and determines the required air mass, the injection mass and injection angles and other manipulated variables for the combustion process. This is detailed further in Fig. 8.1.3 for the case that the engine runs in *normal operation*. The torque conversion block uses the steady-state models from the basic control calibration and contains maps with desired (reference) variables for the feedforward control of the fuel injection and the rail pressure and feedback control of the air mass flow and the charging pressure. Sections 8.3, 8.4 and 8.3 treat these controls more precisely. Only some remarks are given here.

The time delay of the *charged air* and *charging pressure* has a large influence on the dynamic behavior of diesel engines because of the inertness of the turbocharger and the intake volume storage. The volume of the air path is relatively large because

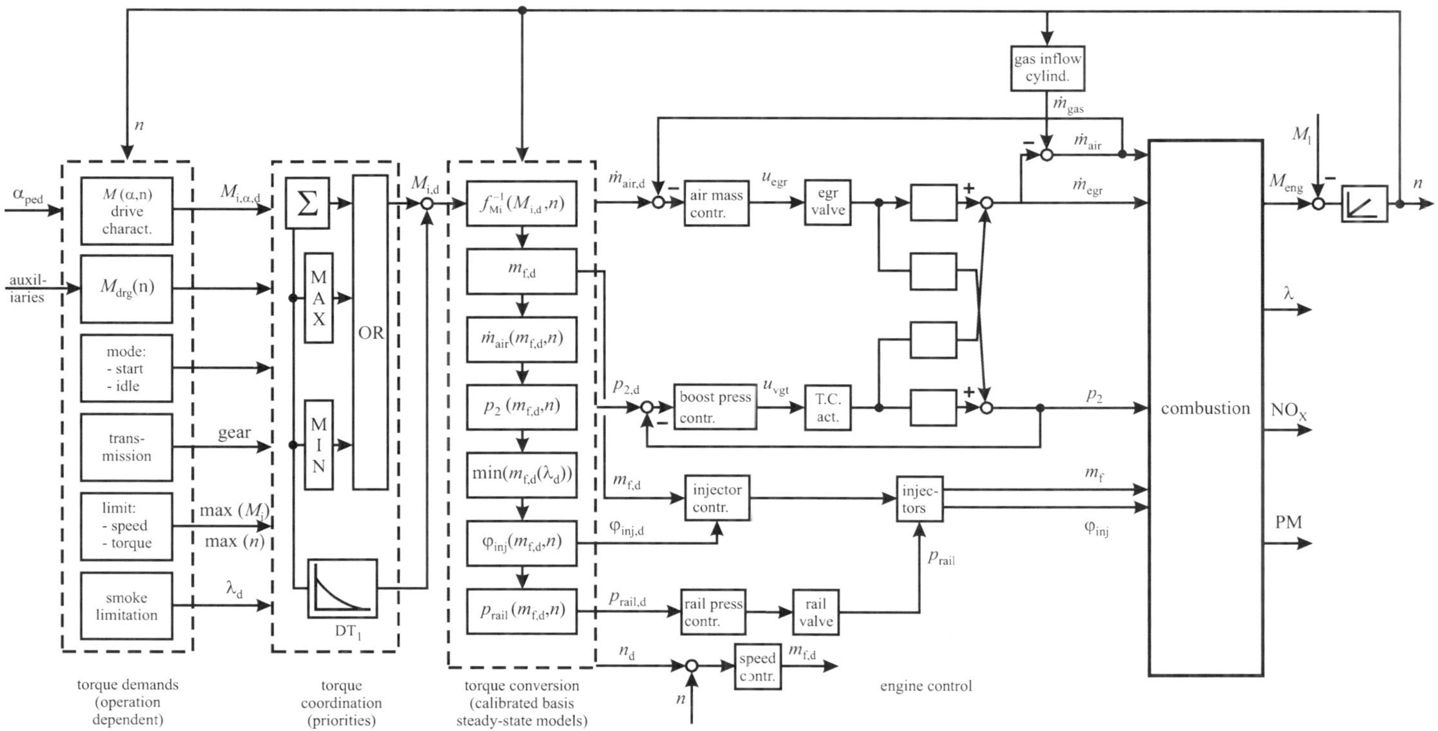

Fig. 8.1.3. Basic control structure for common-rail diesel engines with exhaust turbocharger (only main control functions for normal operation). n stands for the engine speed n_{eng}. DT_1 is a lead-lag element for compensation of dynamic delays.

the air has to be conducted to the intercooler and back to the engine. This is because the charge-air cooler must lie in a suitable air stream of the vehicle, mostly in the front part.

The air mass flow rate is measured by the air flow sensor and it follows

$$\dot{m}_{\mathrm{air}} = \dot{m}_{\mathrm{gas}}(p_2, T_2, n) \; - \; \dot{m}_{\mathrm{egr}}(u_{\mathrm{egr}}, p_3, p_2, T_2) \tag{8.1.6}$$

compare (4.1.12).

The corresponding manipulated variable for the air mass flow rate is the exhaust gas mass flow rate and is changed with the position u_{egr} of the EGR valve. On the other hand, the charging pressure p_2 is measured after the compressor and manipulates either by the wastegate of a fixed turbocharger or the blade position of a variable geometry (VGT) turbocharger u_{vgt}. Both closed loops are mutually coupled as u_{egr} influences p_2 and u_{vgt} influences the air flow $\dot{m}_{\mathrm{air}}$ as illustrated in Fig. 8.1.3. The behavior of this multi-variable control system will be treated in Sect. 8.5.

A specialty of diesel engines is a *full-load smoke limitation*. In full-load condition the air excess factor λ may approach 1 which leads to production of soot or particulates. Therefore the measurement of the air mass flow enters together with the engine speed a smoke-limitation map to determine a desired air/fuel ratio $\lambda_{\mathrm{d}}(\dot{m}_{\mathrm{air}}, n)$. Based on this and the known air mass, a maximum fuel mass $m_{\mathrm{f,max}}$ is calculated, which limits the fuel injection by a minimum operator in the torque coordination block. This feedforward limitation can be added by a feedback path if a λ-sensor in the exhaust pipe is used which corrects the maximal fuel mass dependent on the difference between the actual and the desired λ, Robert Bosch GmbH (2011). Various principles of smoke limitation control are treated in Sect. 8.8.

A further limitation is the maximum speed of the engine to avoid damage through excessive speeds (torque demand block). Therefore a *maximum speed controller* reduces the injected fuel continuously via a ramp function up to a maximum speed value (reference speed variable n_{d}). The *idle speed control* is briefly described in Sect. 7.7.

Further control systems of the engine are realized to obtain a smooth running. The *active surge damping* weakens excited oscillations of the drive train after strong engine torque changes. This is reached firstly by a low-pass filter after the pedal signal. The detection of engine-speed oscillations and modulation of the injected fuel then secondly counteract or damp oscillations, see Sect. 4.9.

Based on the detection of oscillations in the measured engine speed a *smooth-running control* or *fuel-balancing control* compensates a possible different torque of the cylinders by correcting the injected fuel mass for the individual cylinders, especially for low and medium speeds.

This chapter also considers combustion-pressure-based heat release control, which enables cylinder-individual combustion control and homogeneous charge compression ignition (HCCI), see Sect. 8.7.

8.1.2 Diesel-engine torque models

The detailed control structure as in Fig. 8.1.3 is required for the development and calibration of the diesel-engine control functions. However, with regard to treating the diesel engine within a complete power train as a torque generation unit, simplified torque models are more feasible.

As shown in Fig 8.1.4 the external torque demands determine the demanded inner torques $M_{\mathrm{i,d}}$ from the accelerator pedal, the transmission or the velocity control with a priority selection in the torque coordination block, see Figs. 8.1.2 and 8.1.3. An inversion of the inner torque look-up table in the torque conversion block then yields the desired injected fuel mass per stroke $m_{\mathrm{f,d}}$ and calibrated values for the desired values of the air mass $m_{\mathrm{air,d}}$, charging pressure $p_{2,\mathrm{d}}$, injection angle $\varphi_{\mathrm{inj,d}}$ and rail pressure $p_{\mathrm{rail,d}}$.

The air mass flow control loop with the manipulated exhaust gas recirculation flow and the charging pressure control loop with the manipulated turbocharger (either variable turbine or wastegate), according to Fig. 8.1.3, then determine the dynamics of the cylinder air charge and the coordinated injected fuel mass. Both control loops are coupled mutually. This will be considered in more detail in Sect. 8.5. The engine torque M_{eng} acts on the drive train including transmission and via the resulting torques on the movement of the vehicle. An automatic transmission control system, if present, then determines the selected gear and desired torque $M_{\mathrm{i,tr,d}}$.

Similar as for gasoline engines a simplified dynamic model for the torque generation of a diesel engine can be obtained. The generated inner torque is represented by a nonlinear map

$$\overline{M}_{\mathrm{i}}(m_{\mathrm{f}}, n_{\mathrm{eng}}). \tag{8.1.7}$$

for optimal calibrated manipulated variables, like $p_2, p_{\mathrm{rail}}, \varphi_{\mathrm{inj}}, \dot{m}_{\mathrm{egr}}$.

The torque generation after a change α_{ped} of the accelerator pedal is determined by the changes of the injected fuel mass Δm_{f} and the charge air mass Δm_{air}. Now, small changes around an operating point $(\overline{M}_{\mathrm{eng}}, \overline{n}_{\mathrm{eng}})$ are considered.

The change of the demanded inner torque $M_{\mathrm{i,d}}$ after a change $\Delta\alpha_{\mathrm{ped}}$ of the accelerator pedal is described by a constant, compare Fig. 8.1.5,

$$\frac{\Delta M_{\mathrm{i,d}}}{\Delta\alpha_{\mathrm{ped}}} = K_{\mathrm{Md},\alpha}\left(M_{\mathrm{i}}, n_{\mathrm{eng}}\right) \tag{8.1.8}$$

and a resulting change of injected fuel mass by

$$\frac{\Delta m_{\mathrm{f}}}{\Delta M_{\mathrm{i,d}}} = K_{\mathrm{mf,M}}\left(M_{\mathrm{i}}, n_{\mathrm{eng}}\right). \tag{8.1.9}$$

The inner torque reacts after a change of the injection with a time delay T_{d}

$$\frac{\Delta M_{\mathrm{i}}(s)}{\Delta m_{\mathrm{f}}(s)} = K_{\mathrm{M,mf}}\left(M_{\mathrm{i}}, n_{\mathrm{eng}}\right) e^{-T_{\mathrm{d}}s}. \tag{8.1.10}$$

The dead time T_{d} results from the time period of about one cycle, e.g. $40\,\mathrm{ms}$ for $3000\,\mathrm{rpm}$, compare Fig. 8.6.1, and ECU calculation times, which can usually be neglected.

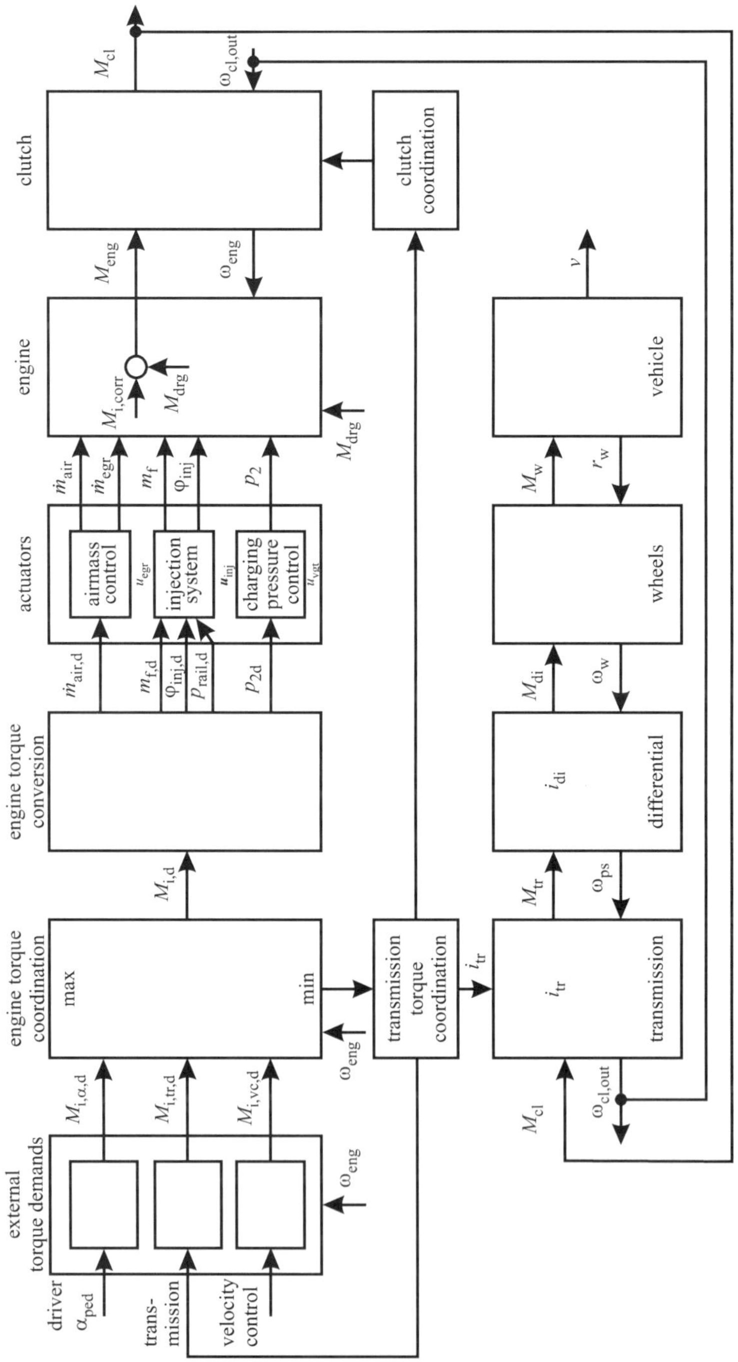

Fig. 8.1.4. Signal flow for the torque generation of a diesel engine with turbocharger as component within the power train of a vehicle.

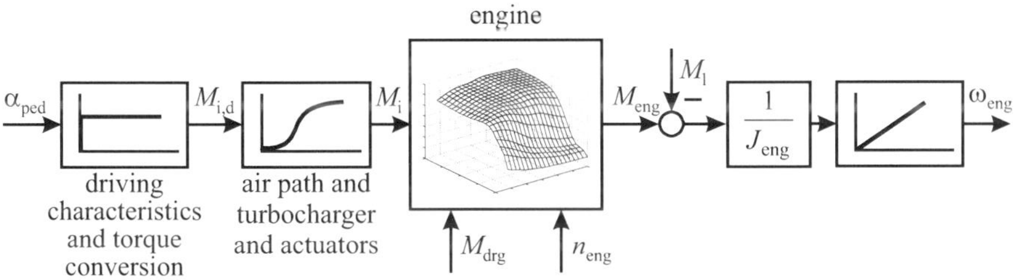

Fig. 8.1.5. Simplified torque model for a turbocharged diesel engine.

Hence, with (8.1.9) and (8.1.10) one obtains for the resulting inner torque through the fuel injection

$$G_{MM1}(s) = \frac{\Delta M_i(s)}{\Delta M_{i,d}(s)} = K_{MM1} e^{-T_d s}. \quad (8.1.11)$$

The air mass flow is delayed by the air path storage and the turbocharger including the actuator and can be approximated by first order elements, compare (4.4.100)

$$G_{mair,M}(s) = \frac{\Delta \dot{m}_{air}(s)}{\Delta M_{i,d}(s)} = \frac{K_{mair,M}}{(1 + T_{act}s)(1 + T_{M,air}s)} e^{-T_d s}. \quad (8.1.12)$$

and with

$$\frac{\Delta M_i}{\Delta \dot{m}_{air}} = K_{M,mair}(M_i, n_{eng}) \quad (8.1.13)$$

one obtains for the resulting proportion of the inner torque through the air flow

$$G_{MM2}(s) = \frac{\Delta M_i(s)}{\Delta M_{i,d}(s)} = \frac{K_{MM2}}{(1 + T_{act}s)(1 + T_{M,air}s)} e^{-T_d s}. \quad (8.1.14)$$

The overall dynamic change of the torque ΔM_i results from a superposition of the effect from Δm_f and $\Delta \dot{m}_{air}$. As usually the effect of Δm_f dominates and is faster, a first order delay and a dead time are used as an approximation

$$G'_{MM}(s) = \frac{\Delta M_i(s)}{\Delta M_{i,d}(s)} = \frac{K'_{MM}}{(1 + T_M s)} e^{-T_d s}, \quad (8.1.15)$$

and with (8.1.1) and (8.1.8) and neglected effects of drag torque changes one obtains as a simplified model

$$G_{M\alpha}(s) = \frac{\Delta M_i(s)}{\Delta \alpha_{ped}(s)} = \frac{K_{M\alpha}}{(1 + T_M s)} e^{-T_d s} \quad (8.1.16)$$

with

$$K_{M\alpha} = K_{Md,\alpha} K'_{MM}. \quad (8.1.17)$$

It will be shown in the next section that for a four-cylinder diesel engine the measured transient responses of the torque yield approximately following time characteristics: for $n_{eng} = 3000\,\text{rpm}$: $T_d \approx 100\,\text{ms}, T_M \approx 100\,\text{ms}$ and for $n_{eng} =$

$1900\,\mathrm{rpm}\;:\;T_{\mathrm{d}}\;\approx\;150\,\mathrm{ms}, T_{\mathrm{M}}\;\approx\;120\,\mathrm{ms}$. However, this includes the dynamics of the crankshaft and fly wheel and the torque measurement. After a change of the fuel injection one cylinder after another yields a change of the torque such that for a four-stroke engine all cylinders contribute with the same torque after two turns of the crankshaft or $\Delta\varphi = 720°\mathrm{CS}$. This corresponds to a time period of $\Delta t = 4\pi/\omega_{\mathrm{eng}} = 4\pi/2\pi n_{\mathrm{eng}} = 2/n_{\mathrm{eng}}$. Therefore, an averaged torque responds approximately with a time constant of $T_{\mathrm{M}} = 2/n_{\mathrm{eng}}$, which is $T_{\mathrm{M}} = 66.7\ldots 40\,\mathrm{ms}$ for $n_{\mathrm{eng}} = 1800\ldots 3000\,\mathrm{rpm}$.

If, depending on the application, the dynamics of the engine can be neglected because the following longitudinal vehicle dynamic time constants are dominating, the engine can be approximated by a proportional behavior without dynamics

$$G_{\mathrm{M}\alpha}(s) = \frac{\Delta M_{\mathrm{eng}}}{\Delta\alpha_{\mathrm{ped}}} = K_{\mathrm{M}\alpha}(\overline{M}_{\mathrm{eng}}, \overline{n}_{\mathrm{eng}}). \tag{8.1.18}$$

The torque M_{eng} acts on the crankshaft and flywheel of the engine. An angular momentum balance yields (7.1.31), see the discussion to Fig. 7.1.8.

8.1.3 Examples for the stationary and dynamic behavior of diesel engines

As the stationary and dynamic behavior of the engines plays an important role for the development of the various control functions, some results from test-bench measurements are considered first to discuss some properties of passenger car diesel engines.

a) Stationary behavior of torque and emissions

Figure 8.1.6 depicts some look-up tables (maps) of a four-cylinder passenger car diesel engine where the specific fuel consumption b_{sfc}, the $\mathrm{NO_x}$ and soot emissions and the combustion noise are given as contour lines within the diagrams of the brake mean effective pressure p_{bmep} versus the speed n_{eng}. According to (4.3.2) p_{bemp} is proportional to the torque M_{eng} at the clutch.

The specific fuel consumption has a minimum for medium speed and medium/high torque and increases strongly nonlinear for lower torque. $\mathrm{NO_x}$ increases nonlinearly with increasing torque and speed and has a minimum for low torque and low speed. Soot shows several maxima for medium torque and low to medium/high speed, and smallest values for low torque. The combustion noise increases with increasing torque and speed, i.e. with the power. Hence, all output variables change nonlinearly and rather differently within the $p_{\mathrm{bmep}}(n_{\mathrm{eng}})$ diagram and underline the difficulty to optimize the fuel consumption and emissions simultaneously.

b) Stationary look-up tables for control

Stationary and dynamic measurements are now discussed for another four-cylinder diesel engine depicted in Fig. 8.1.7.

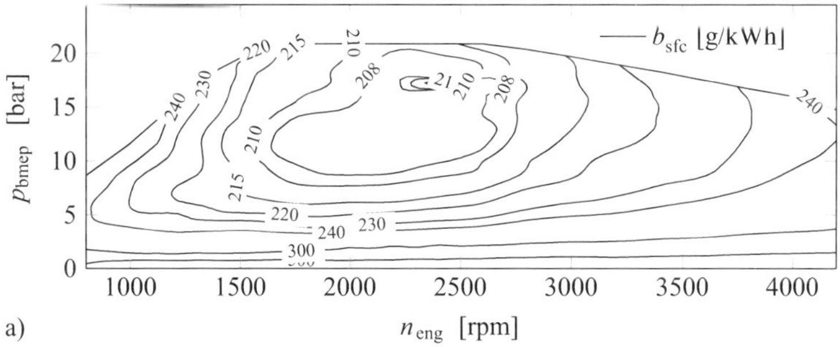

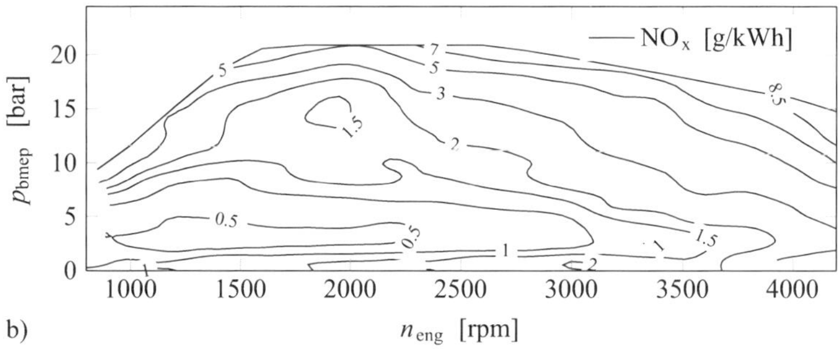

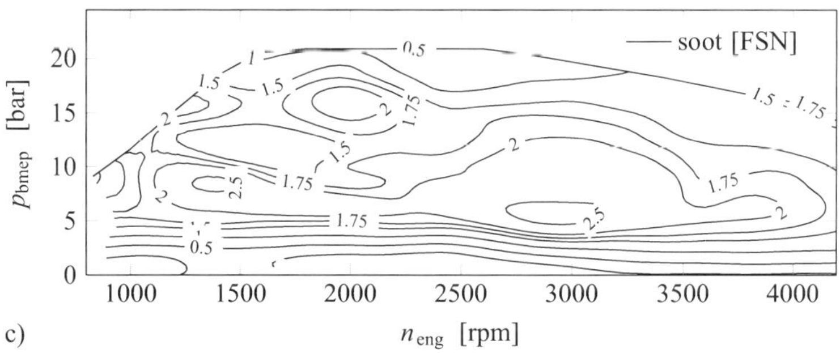

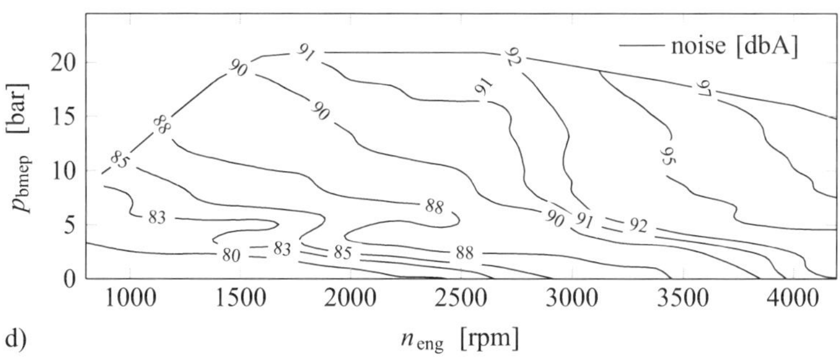

Fig. 8.1.6. Steady-state behavior of a four-cylinder, 16-valve, 110 kW diesel engines with $p_{\mathrm{rail,max}} = 2000$ bar, two main injections, particulate filter and DeNO$_x$ exhaust aftertreatment for Euro 6, fuel consumption optimized. By courtesy of Robert Bosch GmbH (2013). **a** specific fuel consumption. **b** NO$_x$ raw emissions. **c** filter smoke number. **d** combustion noise.

Figure 8.1.8 illustrates the steady-state maps for the control of the engine as measured on the test bench. The drive map for the torque in dependence on the driving pedal position α_{ped} is locally linear and for high load and low speed nonlinear (engine protection). $\dot{m}_{\mathrm{air}}$ is weakly nonlinear for high load, p_2 and p_{rail} for low load and low speed. The injection variables u_{mi} for the main injection mass and φ_{mi} for the main injection angle are relatively linear except for some areas at lower speed. Hence, the steady-state maps can be well approximated by locally linear models which vary with the operating point $[u_{\mathrm{mi}}, n_{\mathrm{eng}}]$. The areas with significant nonlinear relations are placed at different map regions.

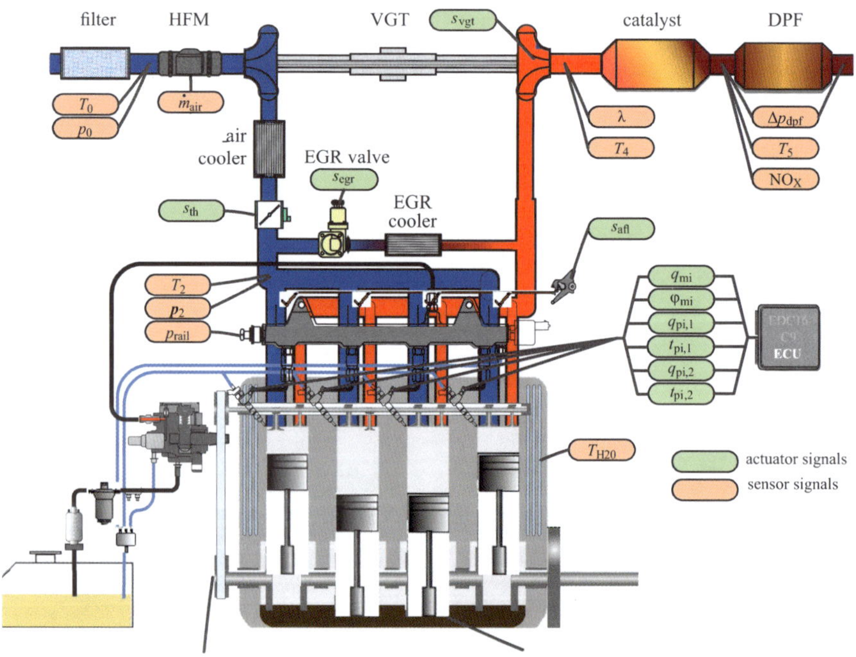

Fig. 8.1.7. Schematic of a diesel engine with turbocharger and common-rail injection on a test bench (Opel Z19 DTH, 1.9l, 315 Nm, 110kW, $\epsilon = 17.5$) and used sensor and actuator signals.

c) Step responses for the dynamic behavior

In order to give a view on the dynamic behavior of the diesel engine, various step responses or transient functions are shown for $n_{\mathrm{eng}} = 3000\,\mathrm{rpm}$ and $n_{\mathrm{eng}} = 1900\,\mathrm{rpm}$. Table 8.1.1 depicts the operating points for $n_{\mathrm{eng}} = 3000\,\mathrm{rpm}$ and the chosen step changes of each one of the manipulated variables, whereby all others have been kept constant. The engine was running with controlled speed by the test bench controller and with closed loop control for the rail pressure, positions of the EGR valve and

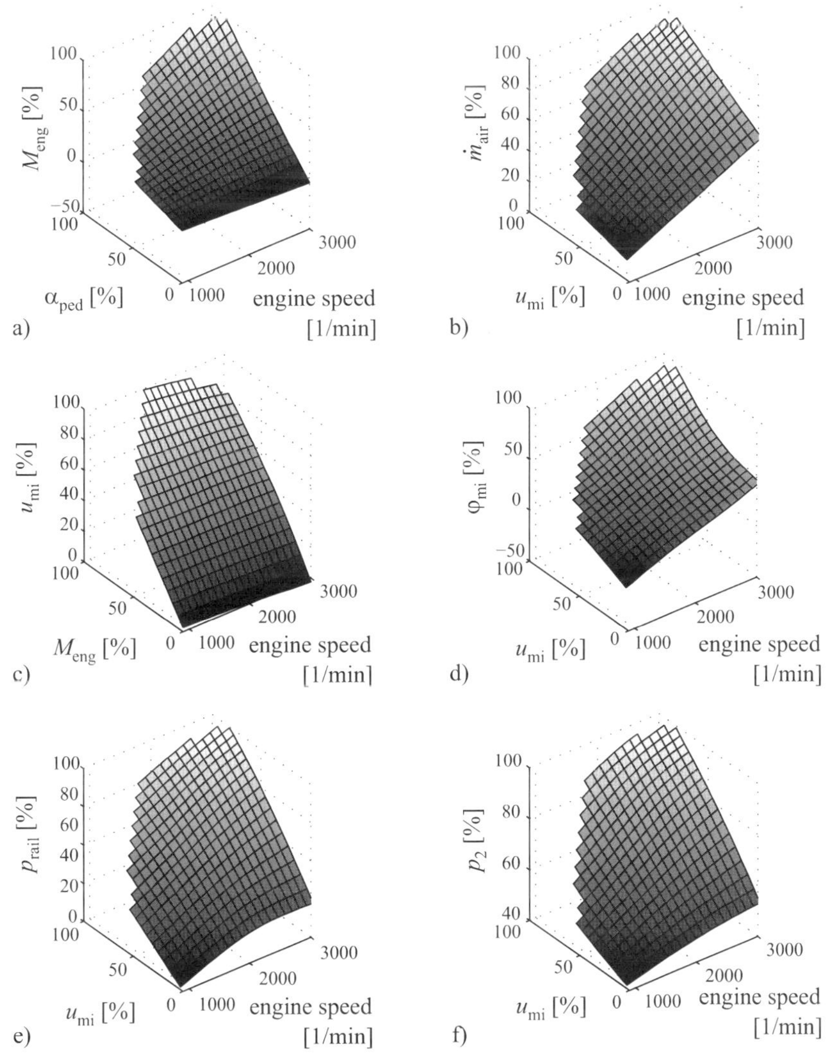

Fig. 8.1.8. Different maps for the steady-state control of the investigated 1.9 l CR diesel engine in dependence on the operating point $[u_{\mathrm{mi,d}}, n_{\mathrm{eng}}]$. The maps show measured values as determined at the test bench. **a** measured torque in dependence on pedal position and speed (driving map). **b** measured air flow rate. **c**, **d** manipulated variables main injection mass and angle. **e**, **f** measured control variables.

VGT actuator, but without control of the air-flow rate, fuel injection rate, fuel injection angle, charging pressure, and the torque, which means open loop for some main controlled and manipulated variables of the ECU. The speed $n_{\mathrm{eng}} = 3000\,\mathrm{rpm}$ or 50 rpm corresponds to 25 strokes/s/cyl or a cycle time of 40 ms for one cylinder. The variables for $p_{\mathrm{mi,hp}}$, $p_{\mathrm{mi,lp}}$ (mean effective indicated pressure for high pressure and low pressure cycle), and φ_{Q50} (combustion center crank angle) are averaged over four cycles, i.e. over 160 ms and therefore described by $\overline{p}_{\mathrm{mi,hp}}$, $\overline{p}_{\mathrm{mi,lp}}$ and $\overline{\varphi}_{\mathrm{Q50}}$. Hence, a steady-state of these averaged values is reached after 160 ms.

The following figures show 15 measured transient functions of output variables for one positive and one negative step function of one manipulated variable over a time interval of 40 s and a zoomed presentation of the initial course for 4 s, Mrosek (2014). Thus, the timely behavior of the variables can be compared to each other, also in dependence of the sign of the changed input.

Table 8.1.1. Diesel engine operation points and chosen step functions for $n_{\mathrm{eng}} \approx 3000\,\mathrm{rpm}$

step	Figure	n_{eng} [rpm]	u_{inj} [mm³/cyc]	$\varphi_{\mathrm{mi}} - 180$ [°CS]	p_{rail} [bar]	s_{egr} [-]	s_{t} [-]	q_{mi} [mm³/cyc]
φ_{mi}	8.1.11	3000	17	165 → 175	700	0.43	1	15.9
q_{mi}	8.1.10	3000	-	170	700	0.43	1	15 → 10
s_{egr}	8.1.12	3000	17	170	950	0.63 → 0.24	0.53	15.9
s_{t}	8.1.13	3000	17	170	950	0.44	0 → 1	15.9

Many of the transient functions can be approximated by a low pass behavior with dead time as shown in Fig. 8.1.9 corresponding to a transfer function

$$G(s) = \frac{y(s)}{u(s)} = \frac{K}{\prod_{i=1}^{n}(1+T_i s)}\, e^{-T_{\mathrm{d}} s}. \tag{8.1.19}$$

The dead time T_{d} is usually a transportation or calculation time and the transient function of the nth order low pass process with time constants $T_1, T_2, \ldots, T_n$ can be characterized by constructing an inflectional tangent with a lag time T_{l} and a slope time T_{s}. An overall characteristic is the 95% settling time T_{95}, see e.g. Isermann and Münchhof (2011). In the case of a first order process it holds $T_{\mathrm{s}} = T_1$. If a dead time T_{d} as well as a lag time T_{l} are included, an overall delay time $T_{\mathrm{dl}} = T_{\mathrm{d}} + T_{\mathrm{l}}$ results.

The figures include the time behavior of the calculated mean indicated pressure $p_{\mathrm{mi,hp}}$ for the high pressure cycle and $p_{\mathrm{mi,lp}}$ for the low pressure cycle according to (4.2.21). For convenience $p_{\mathrm{imep}} = p_{\mathrm{mi}}$ is set. (4.2.21) leads to

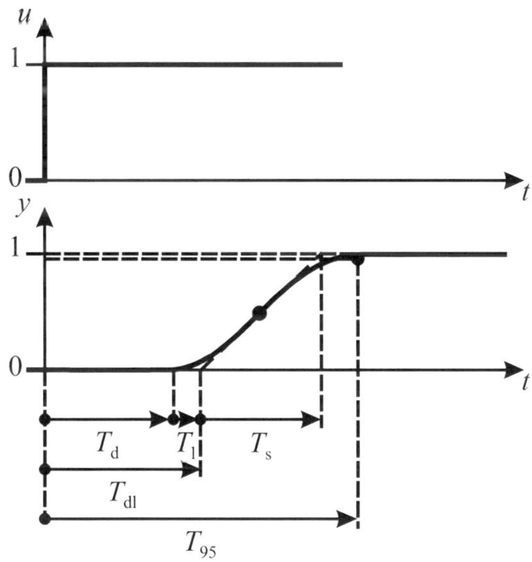

Fig. 8.1.9. Transient function of a low pass process with dead-time and characteristic values T_d: dead time (time delay); T_l: lag time; T_s: slope time; T_{95}: settling time; $T_\mathrm{dl} = T_\mathrm{d} + T_\mathrm{l}$: delay time.

$$\begin{aligned} p_\mathrm{mi,hp} &= \frac{W_\mathrm{mi,hp}}{V_\mathrm{d}} = \frac{1}{V_\mathrm{d}} \int_{\varphi=180^\circ}^{540^\circ} p_\mathrm{cyl}(\varphi)\, dV(\varphi) \\ &\approx \frac{1}{V_\mathrm{d}} \sum_{\kappa=180^\circ/\varphi_0}^{540^\circ/\varphi_0} p_\mathrm{cyl}(\kappa\varphi_0) \Delta V(\kappa\varphi_0) \end{aligned} \tag{8.1.20}$$

$$\begin{aligned} p_\mathrm{mi,lp} &= \frac{W_\mathrm{mi,lp}}{V_\mathrm{d}} = \frac{1}{V_\mathrm{d}} \int_{\varphi=540^\circ}^{180^\circ} p_\mathrm{cyl}(\varphi)\, dV(\varphi) \\ &\approx \frac{1}{V_\mathrm{d}} \sum_{\kappa=540^\circ/\varphi_0}^{180^\circ/\varphi_0} p_\mathrm{cyl}(\kappa\varphi_0) \Delta V(\kappa\varphi_0) \end{aligned} \tag{8.1.21}$$

with the discrete crank angle $\kappa = \varphi/\varphi_0 = 0, 1, 2, \ldots$ and $\varphi_0 = 1^\circ$CS. The cylinder volume follows (4.3.20) and (4.3.17).

- *Step of the main injection rate* q_mi (Fig. 8.1.10, $n_\mathrm{eng} = 3000\,\mathrm{rpm}$)
  - The mean indicated pressure of the high pressure cycle $\overline{p}_\mathrm{mi,hp}$ changes immediately and reaches a steady-state after $T_{95} \approx 160\,\mathrm{ms}$ (due to averaging over 4 cycles).
  - The measured torque M_eng at the asynchronous motor of the test bench shows a dead time $T_\mathrm{d} = 100\,\mathrm{ms}$ and reaches steady-state after $T_{95} \approx 300\,\mathrm{ms}$ after some oscillations which originate from the test bench dynamics.

- The calculated combustion center φ_{Q50} changes immediately with a steady-state after $T_{95} \approx 160\,\text{ms}$ (due to averaging).
- The exhaust pressure p_3 in the exhaust manifold first shows a small step which is proportional and delayed to $\overline{p}_{\text{mi,hp}}(t)$ and $\overline{\varphi}_{Q50}(t)$ by a dead time $T_\text{d} \approx 100\,\text{ms}$. Then a slower transient with $T_{95} \approx 4\,\text{s}$ to steady-state follows.
- Correspondingly to p_3 the exhaust gas flow rate $\dot{m}_\text{egr}$ which depends on the pressure difference

$$\Delta p_\text{egr}(t) = p_3(t) - p_2(t)$$

over the EGR valve develops and also the air-flow rate $\dot{m}_\text{air}$ according to (8.1.6).
- The rotational speed n_tc of the turbine first shows a delay time of $T_\text{dl} \approx 150\,\text{ms}$ and then a transient function with a time constant of $T_1 \approx 0.9\,\text{s}$ and a settling time of $T_{95} \approx 3.5\,\text{s}$. The transient response is a bit slower for an increase of the fuel injection than for a decrease.
- The charging pressure p_{2c} has about the same dynamics as the turbocharger speed n_tc.
- The exhaust temperatures T_3 and T_4 have the slowest responses, with a delay time of $T_\text{dl} \approx 200\,\text{ms}$ and a settling time of $T_{95} \approx 10$ to $20\,\text{s}$ for T_3 and 20 to 40 s for T_4. This is due to the pressure dynamics, the heat exchange with the heat capacities of the cylinder, cylinder head, exhaust pipes, turbocharger and sensor dynamics. T_{2c} is more retarded with $T_\text{dl} \approx 1\,\text{s}$ and $T_{95} \approx 30\,\text{s}$.
- The oxygen content of the exhaust gases measured by the λ-sensor has a delay time of $T_\text{dl} \approx 150\,\text{ms}$, a small overshoot and settles after about $1\,\text{s}$. The initial course of $\lambda(t)$ is directly related to the fast changes of $\overline{p}_{\text{mi,hp}}(t)$ and $\overline{\varphi}_{Q50}(t)$ and then influenced more slowly by the air-flow rate $\dot{m}_\text{air}(t)$. However, the included lag time of the λ-sensor has a dominating effect with a time constant of $T_1 \approx 80\,\text{ms}$, see Table 7.2.1.
- The emission concentration c_{NOx} shows a dead time plus lag time of $T_\text{dl} \approx 400\,\text{ms}$ and settles after $T_{95} \approx 2$ to $3\,\text{s}$. This time behavior is to a good part caused by the sensor dynamics. Table 8.9.1 shows that the NO_x sensor has a dead time of $T_\text{d} \approx 200\,\text{ms}$ and the sum of the time constants is $T_\Sigma \approx 500\,\text{ms}$.
- The opacity c_op is a measure for the particulates and shows a dead time of $T_\text{d} \approx 400\,\text{ms}$ and a small time constant of $T_1 \approx 100\,\text{ms}$, which corresponds to the sensor dynamics in Table 8.9.1. Hence, it can be assumed that both, NO_x and opacity in the exhaust manifold immediately follow changes in the combustion, indicated by the time behavior of $\overline{p}_{\text{mi,hp}}(t)$ and $\overline{\varphi}_{Q50}(t)$. Table 8.1.2 summarizes the characteristics for the changes of the main injection.

- *Step of the injection angle* φ_mi (Fig. 8.1.11)
  - $\overline{p}_{\text{mi,hp}}$, M_eng, $\overline{\varphi}_{Q50}$ and λ change immediately within 160 ms, similarly as for Δq_mi.
  - p_3, p_{2c}, n_tc behave similarly delayed as for the change of the fuel injection volume.

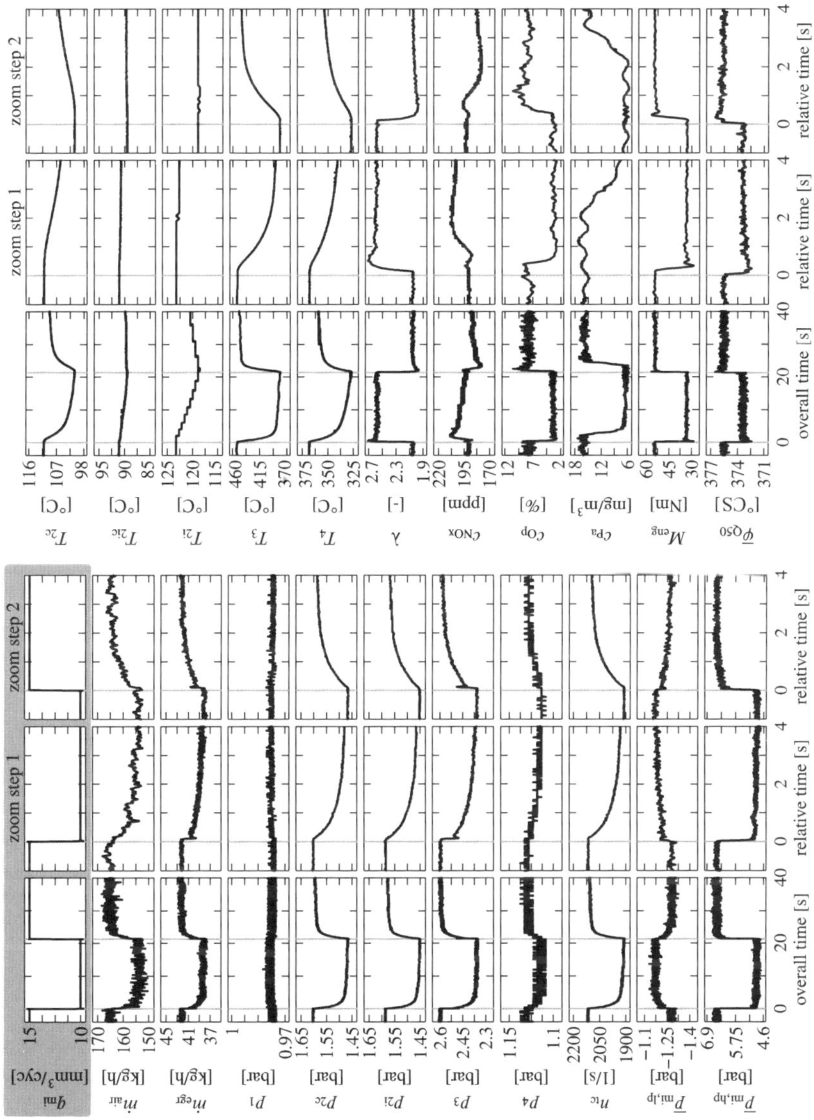

Fig. 8.1.10. Step responses of a step in q_{mi} at $n_{\mathrm{eng}} \approx 3000\,\mathrm{rpm}$. $p_{\mathrm{mi}} = p_{\mathrm{imep}}$.

Table 8.1.2. Step response characteristics for a change of the injection volume q_{mi} at $n_{\mathrm{eng}} = 3000\,\mathrm{rpm}$ in [ms] or [s]

	fast dynamics				medium dynamics							slow dynamics		
	$\overline{p}_{\mathrm{mi,hp}}$	$\overline{\varphi}_{\mathrm{Q50}}$	M_{eng}	p_3	n_{tc}	$p_{2\mathrm{c}}$	$\dot{m}_{\mathrm{egr}}$	$\dot{m}_{\mathrm{air}}$	λ	c_{NOx}	c_{op}	$T_{2\mathrm{c}}$	T_3	T_4
T_{d}			100	100							400			
T_{dl}					150	150			150	400		1 s	200	200
T_{s}			100		900	900					100			
T_{95}	150	150	300	4 s	3.5 s	3.5 s	4 s	4 s	1 s	2.5 s		30 s	15 s	30 s

- The emission concentrations c_{NOx} and c_{op} and temperatures T_3 and T_4 show a strong change and similar dynamics to Fig. 8.1.10. Hence, the injection angle has a significant influence on the emissions.

- *Step of the EGR valve position* s_{egr} (Fig. 8.1.12)
  - The EGR flow rate $\dot{m}_{\mathrm{egr}}$ changes immediately and reaches a steady-state within $T_{95} \approx 100\,\mathrm{ms}$.
  - The air flow rate $\dot{m}_{\mathrm{air}}$ shows a time constant of $T_i \approx 300\,\mathrm{ms}$ for an increase and 200 ms for a decrease, i.e. lags the $\dot{m}_{\mathrm{egr}}$ (because of the intake volume).
  - $\overline{p}_{\mathrm{mi,hp}}$ and $\overline{\varphi}_{\mathrm{Q50}}$ change with a time constant of $T_i \approx 1\,\mathrm{s}$, with only a very small change of the torque M_{eng}.
  - p_3 changes fast initially and reaches a steady-state after $T_{95} \approx 1$ to 2 s.
  - $p_{2\mathrm{c}}$ and n_{tc} change with a time constant of $T_i = 1\,\mathrm{s}$ (and $p_{2\mathrm{c}}$ with a small initial undershoot, i.e. all pass behavior, because of the fast rising p_3).
  - The emissions c_{NOx} and c_{op} show large changes in opposite directions, with delay times of $T_{\mathrm{dl}} \approx 200$ to 300 ms and settling times of $T_{95} \approx 2$ to 3 s. λ changes faster and with a settling time of $T_{95} \approx 1.5\,\mathrm{s}$. (These values include the sensor dynamics.)
  - The temperatures T_3, T_4 and $T_{2\mathrm{c}}$ have delay times of $T_{\mathrm{dl}} \approx 600$ to 800 ms and settling times of $T_{95} \approx 20$ to 40 s. The slowest response is measured for $T_{2\mathrm{c}}$.
  - Hence, changing the EGR valve position has a strong effect on $\dot{m}_{\mathrm{egr}}$, $\dot{m}_{\mathrm{air}}$, n_{tc}, c_{NOx} and c_{op}. Also the pressures and temperatures change significantly, but more slowly.

- *Step change of the VGT actuator position* s_{t} (Fig. 8.1.13)
  - $p_{2\mathrm{c}}$ and n_{tc} change after a small lag time of $T_{\mathrm{l}} \approx 100\,\mathrm{ms}$ with a time constant of $T_i \approx 1.3\,\mathrm{s}$ for increasing and about 0.8 s for decreasing values.
  - p_3 and $\dot{m}_{\mathrm{egr}}$ change with a time constant of $T_i \approx 0.6\,\mathrm{s}$ in positive direction and of about 0.15 s in negative direction, i.e. they are significantly direction dependent.
  - $\dot{m}_{\mathrm{air}}$ increases slowly with a lag time of $T_{\mathrm{l}} \approx 0.6\,\mathrm{s}$ and a time constant of $T_i \approx 1.6\,\mathrm{s}$ because of the increasing $p_{2\mathrm{c}}$. In negative direction it shows a strong undershoot because of the fast decreasing $\dot{m}_{\mathrm{egr}}$. A part of this direc-

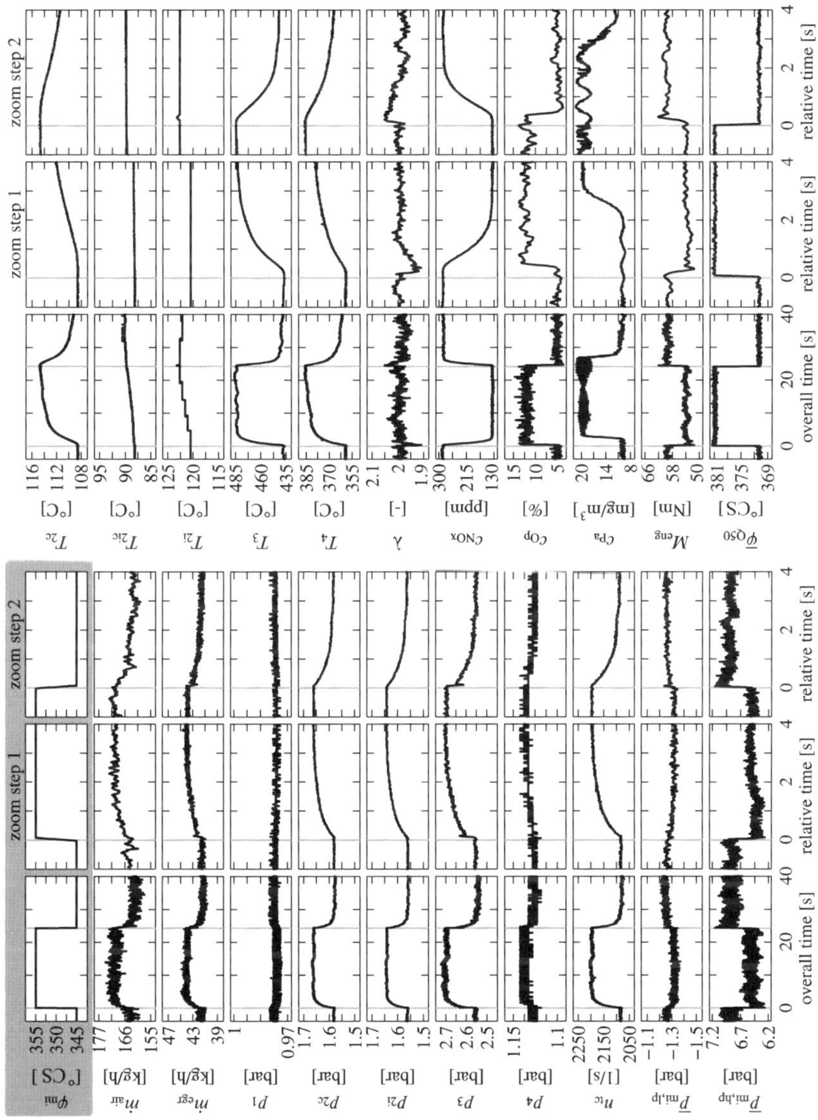

Fig. 8.1.11. Step responses of a step in φ_{mi} at $n_{eng} \approx 3000\,\text{rpm}$.

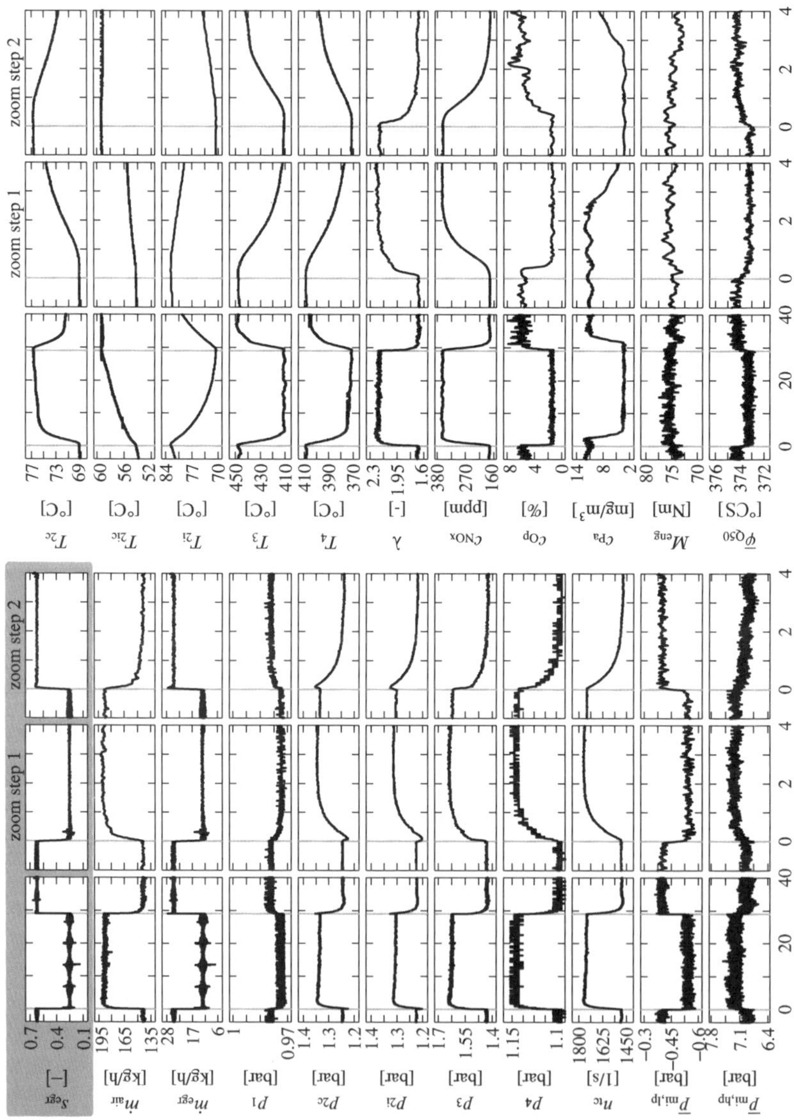

Fig. 8.1.12. Step responses of a step in s_{egr} at $n_{eng} \approx 3000$ rpm.

tion dependent behavior is caused by the different dynamic behaviors of the pressure difference $\Delta p_{\rm v} = p_3(t) - p_{2c}(t)$ over the EGR valve.
- $\overline{p}_{\rm mi,hp}$ changes according to p_{2c} and $\overline{\varphi}_{\rm Q50}$ changes only very little.
- The torque $M_{\rm eng}$ decreases because $\overline{p}_{\rm mi}$ decreases as $\overline{p}_{\rm mi,lp}$ increases significantly, following the direction dependent course of p_3.
- The emissions $c_{\rm NOx}$ and $c_{\rm op}$ show significant changes in opposite directions and have a delay time of $T_{\rm dl} \approx 0.6\,{\rm s}$ an a settling time of $T_{95} \approx 2\,{\rm s}$. λ shows a passing deviation, following $\dot{m}_{\rm air}(t)$.
- The temperatures T_3, T_4 and T_{2c} have delay times of $T_{\rm dl} \approx 1\,{\rm s}$ and settling times of $T_{95} \approx 10$ to $20\,{\rm s}$.

Summing up, the dynamics of the measured output variables for different steps of manipulated variables and $n_{\rm eng} = 3000\,{\rm rpm}$ can be subdivided into three groups:

(1) *Fast dynamics* ($T_{95} < 200\,{\rm ms}$)
Mean effective pressure $\overline{p}_{\rm mi,hp}$, combustion center $\overline{\varphi}_{\rm Q50}$, torque $M_{\rm eng}$, initial course of exhaust manifold pressure p_3 react immediately for changes of $q_{\rm mi}$ and $\varphi_{\rm mi}$, as well as $\dot{m}_{\rm egr}$ for $\Delta s_{\rm egr}$.

(2) *Medium retarded dynamics* ($200\,{\rm ms} < T_{95} < 4\,{\rm s}$)
Turbocharger speed $n_{\rm tc}$, charging pressure p_{2c}, final course of exhaust manifold pressure p_3, exhaust recirculation mass flow rate $\dot{m}_{\rm egr}$, air flow rate $\dot{m}_{\rm air}$, emission concentrations λ_3, $c_{\rm NOx}$, $c_{\rm op}$ react with medium delay times for changes of $q_{\rm mi}$, $\varphi_{\rm mi}$, $s_{\rm egr}$ and $s_{\rm t}$.

(3) *Slow dynamics* ($4\,{\rm s} < T_{95} < 40\,{\rm s}$)
Exhaust temperatures T_{2c}, T_3, T_4 exhibit significant lag times and settling times to all changes of the manipulated variables.

- *Step change of the main injection rate* $q_{\rm mi}$ *at* $n_{\rm eng} = 1900\,{\rm rpm}$ (Fig. 8.1.14)
Figure 8.1.14 illustrates the transient functions for a lower speed $n_{\rm eng} = 1900\,{\rm rpm}$ and therefore a longer settling time $T_{95} \approx 250\,{\rm ms}$ for the combustions of all four cylinders. One observes:
  - $\overline{p}_{\rm mi,hp}$ and $\overline{\varphi}_{\rm Q50}$ change immediately. The torque $M_{\rm eng}$ shows a settling time of $T_{95} \approx 300\,{\rm ms}$.
  - $n_{\rm tc}$ and p_2 have a delay time of $T_{\rm dl} \approx 150\,{\rm ms}$, a time constant of $T_i \approx 1.5\,{\rm s}$ and a settling time of $T_{95} \approx 2.5$ to $3\,{\rm s}$.
  - The dynamics of the emission concentrations λ, $c_{\rm NOx}$ and $c_{\rm op}$, as well as the temperatures T_{2c}, T_3, T_4 are similar to those of $n_{\rm eng} = 3000\,{\rm rpm}$.
  - $\dot{m}_{\rm air}$ and $\dot{m}_{\rm egr}$ show no significant changes.

  Hence, for lower speed mainly the dynamics of the turbocharger are slower. The other dynamics are similar to those at $n_{\rm eng} = 3000\,{\rm rpm}$. Because of the small values for p_{2c} and small differences $p_3 - p_{2c}$, the air mass flow $\dot{m}_{\rm air}$ and $\dot{m}_{\rm egr}$ are not significantly affected.

- *Step change of the rail pressure* $p_{\rm rail}$ *at* $n_{\rm eng} = 1900\,{\rm rpm}$ (Fig. 8.1.15)
Figure 8.1.15 shows the transient responses for a change of the rail pressure $p_{\rm rail}$.

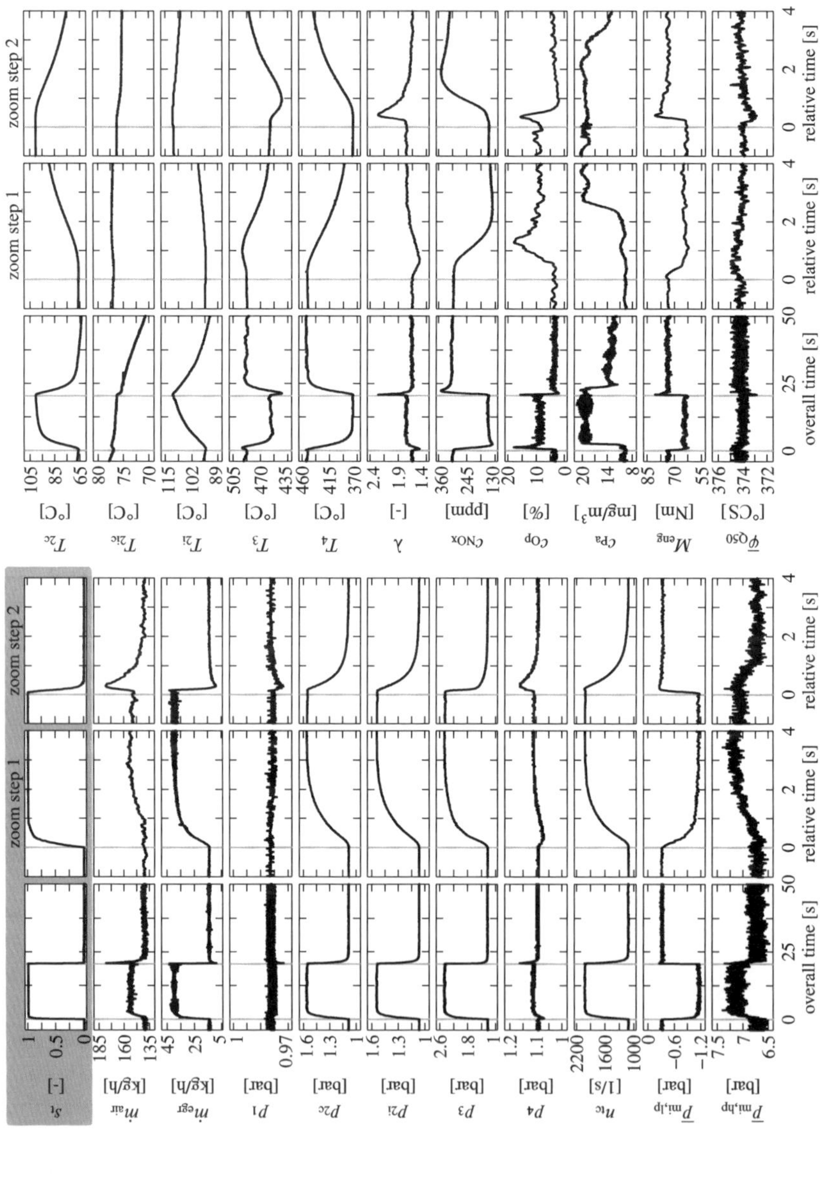

Fig. 8.1.13. Step responses of a step in s_t at $n_{eng} \approx 3000$ rpm, positive coupling to $\dot{m}_{air}$.

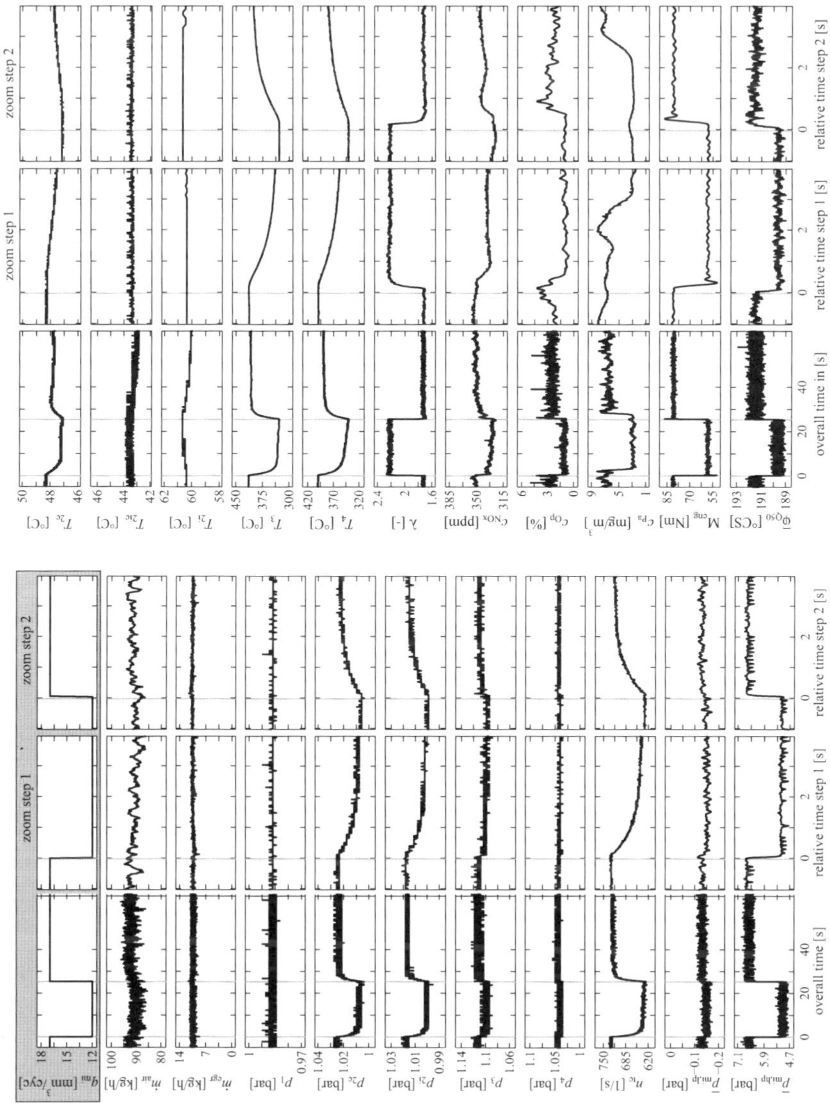

Fig. 8.1.14. Step responses of a step in q_{mi} at $n_{eng} \approx 1900$ rpm.

- $\overline{p}_{\mathrm{mi,hp}}$, $\overline{\varphi}_{\mathrm{Q50}}$ reach a steady-state almost immediately with about the same dynamic delay as the rail pressure itself. M_{eng} is slightly more delayed, by about 100 ms.
- The emission concentrations c_{NOx} and c_{op} are significantly influenced in opposite directions with a delay time of $T_{\mathrm{dl}} \approx 500\,\mathrm{ms}$ and $T_{95} \approx 2\,\mathrm{s}$ for c_{NOx} and $T_{95} \approx 1.2\,\mathrm{s}$ for c_{op}. λ changes little.
- n_{tc}, T_3 and T_4 show very small changes. All other variables remain constant.

Hence, the rail pressure p_{rail} has a significant influence on the combustion and therefore on the torque and the emissions.

- *Summary*
  These *measured transient functions* for the main manipulated variables q_{mi}, φ_{mi}, s_{egr}, s_{t} and p_{rail} of a diesel engine demonstrate that not only the main controlled variables M_{eng}, $\dot{m}_{\mathrm{air}}$ and p_2 are influenced, but almost all pressures, temperatures and emission concentrations with different dynamics. This underlines the multi-variable behavior of combustion engines and the significance of the many crosscouplings. In the following sections the stationary and dynamic behavior will be modeled in detail to be used for the design of engine control systems.

8.2 Combustion models for diesel engines

This section considers the special combustion phases for diesel engines and simplified combustion models, in addition to the general high and low combustion cycle models for four-stroke engines discussed in Sect. 4.2. In contrast to gasoline engines the processes for diesel engines do not run sequentially but mainly in parallel whereby they influence each other mutually.

8.2.1 Combustion phases and combustion modeling

The fuel of modern diesel engines is injected directly into the combustion chamber with high pressure usually before the top dead center. A resulting spray expansion leads to a formation and evaporation of fuel drops thereby mixing with the compressed and heated air. Then auto-ignition starts with some delay followed by various phases of combustion exhaust gas development and expansion with transfer of the thermal work into mechanical work at the crankshaft. Figure 8.2.1 depicts some phases of the diesel combustion process, compare Merker et al (2006).

a) Fuel injection

In the following, common-rail injection systems are assumed where the injection pressure can be adjusted independently of engine speed, fuel quantity, and crankshaft angle. Therefore, much more freedom for selecting optimal injection variables is possible, compared to cam driven in-line injection pumps, distribution injection

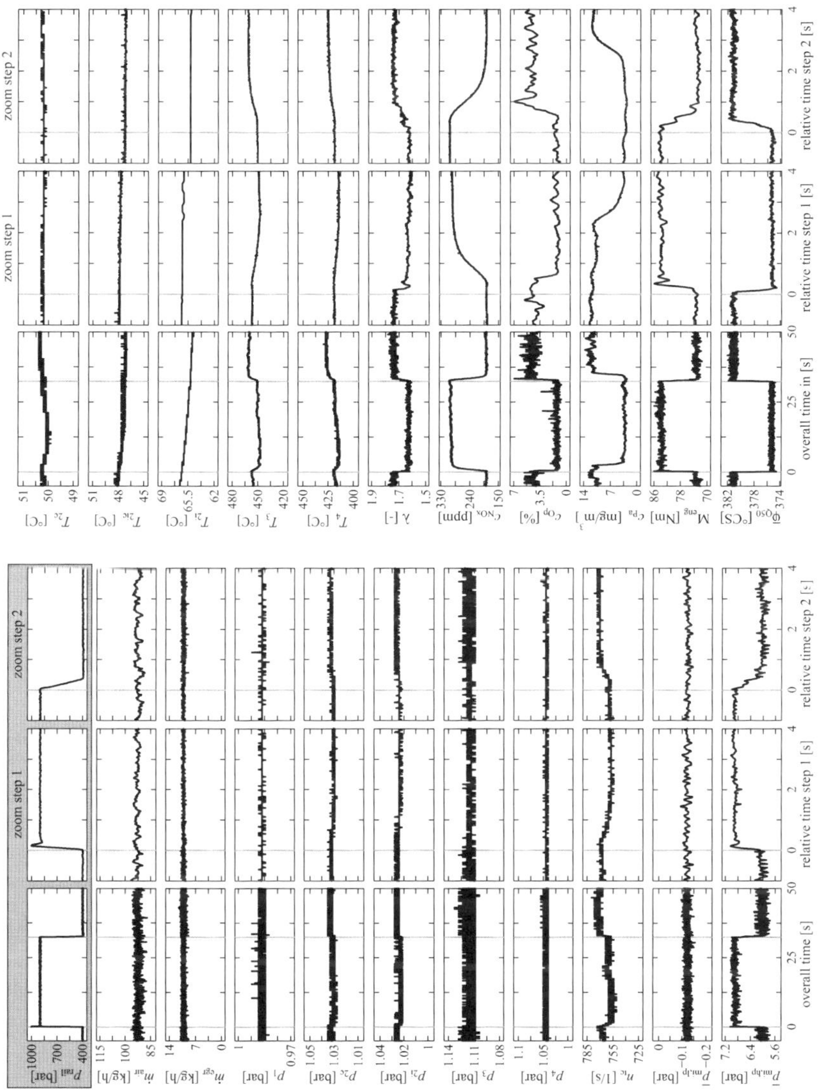

Fig. 8.1.15. Step responses of a step in p_{rail} at $n_{\text{eng}} \approx 1900$ rpm.

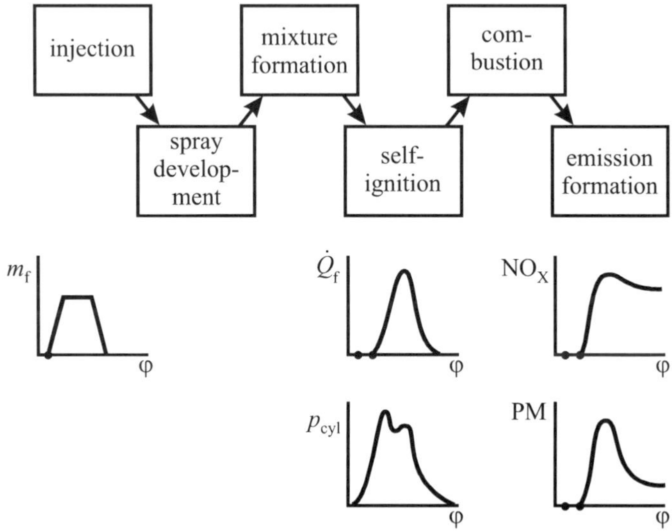

Fig. 8.2.1. Phases from injection to emission formation for diesel fuel combustion.

pumps or unit injector systems, see e.g. Robert Bosch GmbH (2011). The common-rail (CR) serves as accumulator and allows a separation of pressure generation and fuel injection. Presently CR pressures of $1600 - 2000$ bar are used for passenger cars and 1400 bar for commercial vehicles. The direct injection into the cylinder is performed by a solenoid-valve injector or piezo-injector, allowing multiple injections through throttling-pintle nozzles or hole-type nozzles.

b) Mixture formation

During the compression stroke the intake gas (air plus recirculated exhaust gas) is compressed to $30 \ldots 55$ bar for naturally aspirated engines or $80 \ldots 110$ bar for charged engines. The temperature thereby rises to $700 \ldots 900°$C and allows self-ignition of the fuel/gas mixture before reaching the top dead center (TDC). The injected spray consists of many small fuel droplets, where the local air/fuel ratio λ varies from pure air in the periphery ($\lambda = \infty$) to pure fuel ($\lambda = 0$) in the core of the spray such forming a heterogeneous mixture which is influenced by turbulence and diffusion. The combustion arises only between an air/fuel ratio of $0.3 < \lambda < 1.5$.

The kinetic energy of the spray and the spray pattern influence the resulting turbulence, the droplet sizes and the interaction of fuel and air. The generation of turbulence is supported by the shape of the combustion chamber, the piston and its recess, by special induction track geometries and swirl-flaps and by the piston movement with resulting squish.

The liquid fuel is atomized during the injection process in up to 10^8 droplets with diameters in the range of $10\,\mu$m, Stiesch (2003) leading to a fast evaporation. One distinguishes a primary spray decay in the vicinity of the injector, forming larger droplets and a secondary spray decay downstream with microdroplets by mixing with the turbulent air. The heated air and the cylinder wall temperatures then vaporize the droplets.

c) Combustion process

The start of ignition follows the start of injection and mixture formation after a certain period of time, the ignition delay. It depends on a physical part, which comprises the spray development, mixture formation and fuel evaporation, and a chemical part where reactive radicals like OH are formed and lead to the ignition.

The ignition delay depends on the ignition quality of the fuel, expressed by the cetane number ($CN = 45 \ldots 50$), the compression pressure and compression temperature.

The combustion process can be divided into three phases, see Fig. 8.2.2, Merker et al (2006). During the *pre-mixed flame combustion* the injected fuel and air mixture combusts very rapidly with high heat release and leads to a fast pressure rise. This phase is responsible for the hard noise and forming of NO_x because of the large air excess. The following *diffusion flame combustion* is based on the heterogeneous mixture with strongly varying air excess for rich and lean mixture parts. The rich mixture parts are responsible for soot generation. The burning rate is slower than the pre-mixed combustion, because of the physically-based turbulent mixture process. The third phase is the *post-combustion* and is characterized by slower chemical reactions because of the decreasing pressure, temperature and turbulence. In this phase the remaining fuel is burned, intermediate combustion products and soot are oxidized and the final emissions are determined. The division of the fuel injection in pre-injection(s) and main-injection(s) allows to reduce the combustion noise and to obtain an early main-injection.

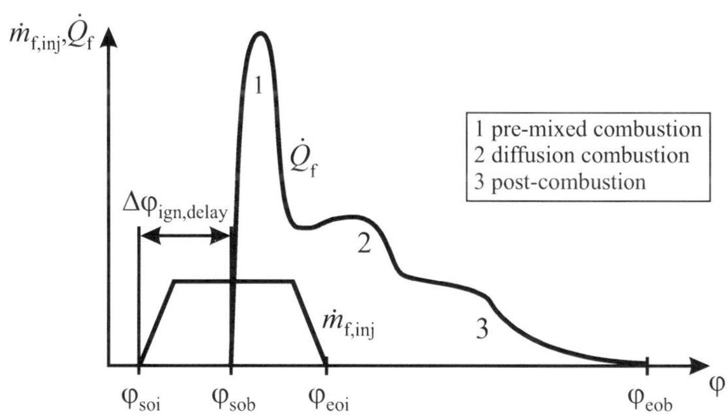

Fig. 8.2.2. Course of fuel injection and heat release of the diesel combustion (qualitative) for one main injection, Merker et al (2006).

8.2.2 Combustion models

For calculation and simulation of the combustion process different methods were developed. They can be divided into empirical, phenomenological and three-dimensional models, Stiesch (2003) and Merker et al (2006). The *empirical models* assume one or two zones of the combustion chamber and describe the state variables pressure, temperature and gas composition in dependence on the crank angle or time. One example of this type with one zone is the approximation of the gross heat release with the Vibe function (4.2.40). *Phenomenological models* determine the heat release by assuming a local distribution of combustion chamber zones and include, e.g. spray development, mixture formation and chemical reactions. One input is the profile of the injection flow rate. *Three-dimensional models* (3D) include the spatial distribution of the flows on a gridnet by using computational fluid dynamic simulation programs with consideration of chemical reactions and emission formation.

The required calculation time for one stroke ranges between less than one ms for empirical, less than one ms to some min for phenomenological and some min to hours for 3D-models. For fast calculations in connection with use for control functions and for HiL real-time simulation therefore one-zone phenomenological models according to Zahn (2012) are considered subsequently, in addition to the empirical models described in Sect. 4.2.

One-zone cylinder model (according to Zahn (2012))
Assuming an ideally mixed combustion chamber, see Fig. 8.2.3, the mass balance for one cylinder during one stroke yields

$$\frac{dm_{\mathrm{cyl}}}{dt} = \dot{m}_{\mathrm{cyl,in}} + \dot{m}_{\mathrm{f}} - \dot{m}_{\mathrm{cyl,out}} \tag{8.2.1}$$

neglecting leakage losses at the piston. Correspondingly, the energy balance becomes (compare (4.2.24))

$$\frac{dU_{\mathrm{cyl}}}{dt} = \dot{Q}_{\mathrm{f}} + \dot{Q}_{\mathrm{w}} + P_{\mathrm{cyl}} + h_{\mathrm{cyl,in}}\dot{m}_{\mathrm{cyl,in}} - h_{\mathrm{cyl,out}}\dot{m}_{\mathrm{cyl,out}} \tag{8.2.2}$$

where $\dot{Q}_{\mathrm{f}}$ is the heat release rate, $\dot{Q}_{\mathrm{w}}$ the heat flow from the wall to the gas and P_{cyl} the mechanical power output at the piston. The enthalpies of the flow rates follow from a charge cycle model. For the mechanical power holds

$$P_{\mathrm{cyl}} = -p_{\mathrm{cyl}}\frac{dV_{\mathrm{d}}}{dt} = -p_{\mathrm{cyl}}\,\omega_{\mathrm{eng}}\frac{dV_{\mathrm{d}}}{d\varphi} \tag{8.2.3}$$

and for the specific enthalpy

$$h_{\mathrm{cyl,out}} = u_{\mathrm{cyl}}(T_{\mathrm{cyl}}, \lambda_{\mathrm{cyl}}) + p_{\mathrm{cyl}}\,v_{\mathrm{cyl}} = u_{\mathrm{cyl}}(T_{\mathrm{cyl}}, \lambda_{\mathrm{cyl}}) + R_{\mathrm{g}}\,T_{\mathrm{cyl}}. \tag{8.2.4}$$

For the internal energy in (8.2.2) follows

$$\frac{dU_{\mathrm{cyl}}}{dt} = \frac{d}{dt}(m_{\mathrm{cyl}}u_{\mathrm{cyl}}) = m_{\mathrm{cyl}}\frac{du_{\mathrm{cyl}}}{dt} + u_{\mathrm{cyl}}\frac{dm_{\mathrm{cyl}}}{dt}. \tag{8.2.5}$$

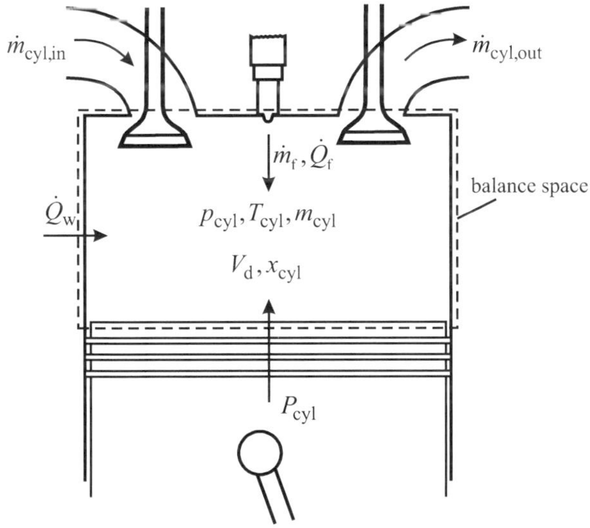

Fig. 8.2.3. Combustion chamber and variables of the one-zone model.

The specific internal energy u_{cyl} is a function of the temperature, pressure and gas composition. As the dependence on the pressure is small, it can be neglected for temperatures smaller than 1800 K, Pischinger et al (2009). Then it holds

$$\frac{du_{\text{cyl}}}{dt} = \left.\frac{\partial u_{\text{cyl}}}{\partial T_{\text{cyl}}}\right|_{\lambda_{\text{cyl}}} \frac{dT_{\text{cyl}}}{dt} + \left.\frac{\partial u_{\text{cyl}}}{\partial \lambda_{\text{cyl}}}\right|_{T_{\text{cyl}}} \frac{d\lambda_{\text{cyl}}}{dt} \tag{8.2.6}$$

where polynomials from Justi (1938) are applied for an approximation of (8.2.6).

The heat release rate $\dot{Q}_{\text{f}}$ in (8.2.2) follows from the combustion models described in Sect. 8.2.3. A rough overall measure is

$$\dot{Q}_{\text{f}} = \dot{m}_{\text{f}} H_{\text{l}}$$

where H_{l} is the lower heating value ($H_{\text{l}} = 42500\,\text{kJ/kg}$).

The air content in the cylinder is defined by

$$x_{\text{cyl}} = m_{\text{air}}/m_{\text{cyl}} \tag{8.2.7}$$

and its timely behavior is, assuming lumped parameters,

$$\frac{dx_{\text{cyl}}}{dt} = \frac{1}{m_{\text{cyl}}} \left[(x_{\text{cyl,in}} - x_{\text{cyl}})\dot{m}_{\text{cyl,in}} - x_{\text{cyl}}\,\dot{m}_{\text{f}} \right]. \tag{8.2.8}$$

The differential equations (8.2.1), (8.2.2) and (8.2.8) are solved for T_{cyl}. The cylinder pressure follows from the gas equation

$$p_{\text{cyl}} = \frac{1}{V_{\text{d}}} R_{\text{g}}\, m_{\text{cyl}}\, T_{\text{cyl}}. \tag{8.2.9}$$

This basic one-zone model requires the determination of in- and out-flowing gas mass flow rates $\dot{m}_{\mathrm{cyl,in}}$ and $\dot{m}_{\mathrm{cyl,out}}$. These have to be calculated by charge cycle models, where the mass flows are determined by the flow laws of compressible fluids through cam position dependent valve orifices, as shown by Zahn (2012). This may include forward flows as well as backward flows through the valves, depending on the pressure differences over the valves. Figure 8.2.4 summarizes the calculation of the cylinder pressure by using (8.2.1) to (8.2.9).

8.2.3 Pre-mixed and diffusion combustion models

a) Pre-mixed combustion

The pre-mixed combustion during pre-combustion and beginning of main-combustion is mainly chemical based and comprises several partial reactions which can be summarized for an ideal combustion by the gross reaction equation

$$\nu_{\mathrm{f}}\,\mathrm{C}_x\,\mathrm{H}_x + \nu_{\mathrm{O2}}\,\mathrm{O}_2 = \nu_{\mathrm{CO2}}\,\mathrm{CO}_2 + \nu_{\mathrm{H2O}}\,\mathrm{H}_2\mathrm{O} + \Delta\,\mathrm{H}_{\mathrm{c}} \tag{8.2.10}$$

where ν_i are the stoichiometric reaction coefficients and $\Delta\,\mathrm{H}_{\mathrm{c}}$ the generated reaction enthalpy. Since the combustion is not ideal in reality, further components result like $\mathrm{NO_x}$ and particulates.

The following description of phenomenological models is based on Chmela and Orthaber (1999), Chmela et al (2006) and follows Zahn (2012).

The reaction velocity of the gross reaction according to (8.2.10) can be expressed by

$$v_{\mathrm{r}} = k_{\mathrm{n}} c_{\mathrm{f}} c_{\mathrm{O2}} = k_{\mathrm{n}} \frac{\beta_{\mathrm{f}}}{M_{\mathrm{f}}} \frac{\beta_{\mathrm{O2}}}{M_{\mathrm{O2}}} \quad \left[\frac{\mathrm{mol}}{\mathrm{m^3 s}}\right] \tag{8.2.11}$$

with c_{f} and c_{O2} the component concentrations, $\beta_{\mathrm{f}} = m_{\mathrm{f}}/V_{\mathrm{d}}$ and $\beta_{\mathrm{O2}} = m_{\mathrm{O2}}/V_{\mathrm{d}}$ the mass concentrations and M_{f} and M_{O2} the molar masses of the fuel and oxygen. The reaction velocity coefficient follows an Arrhenius equation

$$k_{\mathrm{n}} = A p^{\mathrm{n}} \exp\left(\frac{-E_{\mathrm{a}}}{R_{\mathrm{m}} T}\right) \approx A p_{\mathrm{cyl}}^{\mathrm{n}} \exp\left(\frac{-T_{\mathrm{a}}}{T_{\mathrm{cyl}}}\right) \tag{8.2.12}$$

with E_{a} the activation energy, R_{m} the universal gas constant and the activation temperature $T_{\mathrm{a}} = E_{\mathrm{a}}/R_{\mathrm{m}}$. The coefficients A and n have to be determined experimentally.

Multiplying (8.2.11) with the molar mass M_{f} and inserting (8.2.12) yields for the reaction rate

$$r_{\mathrm{r}} = C'_{\mathrm{arr}}\, \beta_{\mathrm{f}}\, \beta_{\mathrm{O2}}\, p_{\mathrm{cyl}}^{\mathrm{n}} \exp\left(\frac{-T_{\mathrm{a}}}{T_{\mathrm{cyl}}}\right) \quad \left[\frac{\mathrm{kg}}{\mathrm{m^3 s}}\right] \tag{8.2.13}$$

whereby $C'_{\mathrm{arr}} = A/M_{\mathrm{O2}}$. The resulting heat release follows with H_{l} the heat value of the fuel from

$$\dot{Q}_{\mathrm{f,pre}} = r_{\mathrm{r}}\, V_{\mathrm{r}}\, H_{\mathrm{l}}. \tag{8.2.14}$$

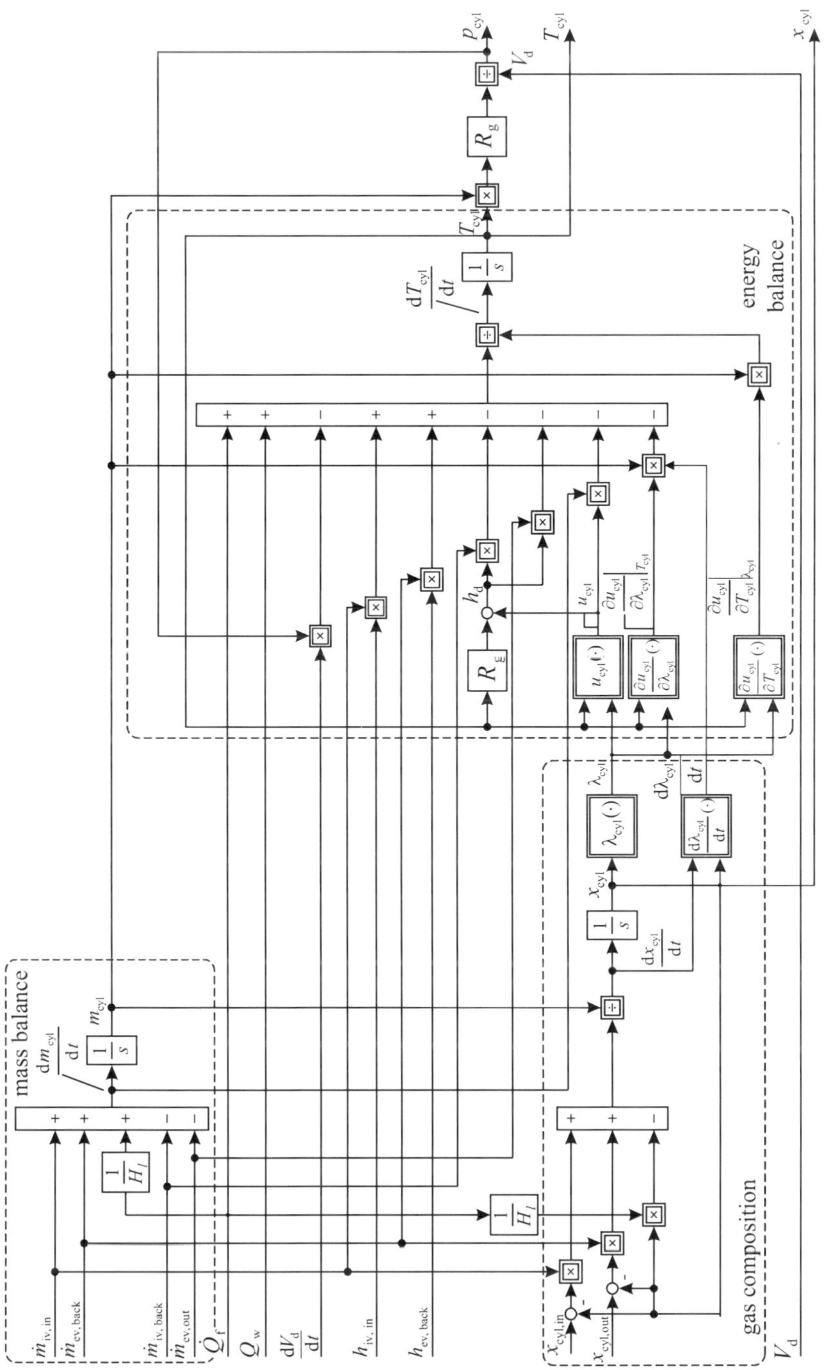

Fig. 8.2.4. One zone combustion model for the cylinder pressure and temperature, including in-flowing gas mass valve flows $\dot{m}_{\mathrm{iv,in}}$ and $\dot{m}_{\mathrm{ev,back}}$ and out-flowing gas mass flows $\dot{m}_{\mathrm{ev,out}}$ and $\dot{m}_{\mathrm{iv,back}}$, Zahn (2012).

The reaction volume is assumed as a half-spherical section of a flame front with width s_{ff}

$$V_{\mathrm{r}} = V_{\mathrm{ff}} \approx 2\pi s_{\mathrm{ff}} (t - t_{\mathrm{soc}})^2. \tag{8.2.15}$$

Hence, the heat release becomes

$$\dot{Q}_{\mathrm{f,pre}}(t) = C_{\mathrm{arr}}(t)\, H_{\mathrm{l}}\, \beta_{\mathrm{f}}\, \beta_{\mathrm{O2}}\, p_{\mathrm{cyl}}^{\mathrm{n}} \exp\left(\frac{-T_{\mathrm{a}}}{T_{\mathrm{cyl}}}\right) (t - t_{\mathrm{soc}})^2. \tag{8.2.16}$$

The available fuel mass to determine β_{f} follows from

$$m_{\mathrm{f,pre}}(t) = \chi_{\mathrm{pre}} \int_{t_{\mathrm{soi}}}^{t} [\sigma(\tau - t_{\mathrm{soi}}) - \sigma(\tau - t_{\mathrm{sob}})]\, \dot{m}_{\mathrm{f}}(\tau) d\tau - \frac{Q_{\mathrm{f,pre}}(t)}{H_{\mathrm{l}}} \tag{8.2.17}$$

with $\sigma = \{0, 1\}$ for unit step functions and $0 < \chi_{\mathrm{pre}} \leq 1$ a factor for the part of fuel burned in the pre-mix phase. The current oxygen mass to determine β_{O2} follows from

$$m_{\mathrm{O2}}(t) = m_{\mathrm{O2,ic}} - O_{2,\mathrm{st}} \frac{Q_{\mathrm{f}}(t)}{H_{\mathrm{l}}} \tag{8.2.18}$$

with $O_{2,\mathrm{st}} \approx 3.36\,\mathrm{kg\,O_2/kg\,fuel}$ the stoichiometric oxygen demand. For the oxygen content at inlet closes holds

$$m_{\mathrm{O2,ic}} = \xi_{\mathrm{O2}} \left[(1 + L_{\mathrm{st}}) x_{\mathrm{cyl,ic}} - L_{\mathrm{st}}\right] m_{\mathrm{cyl,ic}}. \tag{8.2.19}$$

ξ_{O2} is the mass fraction of oxygen in the air ($\xi_{\mathrm{O2}} = 0.232\,\mathrm{kg\,O_2/kg\,air}$).

b) Diffusion combustion

The combustion process of the diffusion phase is determined by chemical reactions during the heterogeneous mixture of the spray with the air. Therefore its burning rate is mainly influenced by the mixing process of the fuel with the air and depends on the turbulence generated by the injection and geometrical induced swirl and squish.

According to Magnussen and Hjertager (1977) the combustion reaction rate of turbulent not pre-mixed flames can be expressed by

$$r_{\mathrm{D}} = C'_{\mathrm{mag}} \beta_{\mathrm{f}} \left(\frac{\epsilon}{k}\right) \quad \left[\frac{\mathrm{kg}}{\mathrm{m^3 s}}\right] \tag{8.2.20}$$

with C'_{mag} a constant, $\beta_{\mathrm{f}} = m_{\mathrm{f,diff}}/V_{\mathrm{d}}$ the local mass concentration of the fuel, k the turbulent kinetic energy density and ϵ the dissipation velocity of the kinetic energy density, Stiesch (2003). $\epsilon/k\,[1/\mathrm{s}]$ can be interpreted as the mixing frequency $f_{\mathrm{mix}} = 1/\tau_{\mathrm{mix}}$. The turbulent kinetic energy k follows from the differential equation

$$\frac{dk}{dt} = \frac{dk_{\mathrm{prod}}}{dt} - \epsilon. \tag{8.2.21}$$

For the dissipation rate holds

$$\epsilon = C''_{\text{diss}} \frac{k^{3/2}}{l_{\text{I}}}. \tag{8.2.22}$$

l_{I} is a turbulent length measure

$$l_{\text{I}} = \sqrt[3]{\frac{6V_{\text{d}}}{\pi}}. \tag{8.2.23}$$

Assuming that the turbulence is mainly generated by the injection spray the generation term dk_{prod}/dt can be determined from the energy rate of the injection jet with the velocity $c_{\text{f,inj}}$

$$\frac{dE_{\text{inj}}}{dt} = 0.5\, \dot{m}_{\text{f,inj}}\, c^2_{\text{f,inj}} = 0.5 \frac{m^2_{\text{f,inj}}}{\rho^2_{\text{f}} A^2_{\text{inj}}}. \tag{8.2.24}$$

The kinetic energy density of the injection jet is approximated by

$$\frac{dk_{\text{inj}}}{dt} = \frac{1}{m_{\text{cyl}}} \frac{dE_{\text{inj}}}{dt}. \tag{8.2.25}$$

Hence, the turbulent kinetic energy rate becomes

$$\frac{dk}{dt} = \frac{dk_{\text{prod}}}{dt} - \epsilon = 0.5 \frac{\dot{m}^3_{\text{f,inj}}}{\rho^2_{\text{f}}\, A^2_{\text{inj}}\, m_{\text{cyl}}} - C'_{\text{diss}} \frac{k^{3/2}}{\sqrt[3]{V_{\text{d}}}}. \tag{8.2.26}$$

Inserting (8.2.22) in (8.2.20) yields

$$r_{\text{D}} = C'_{\text{mag}}\, \beta_{\text{f}} \frac{\sqrt{k}}{\sqrt[3]{V_{\text{d}}}}. \tag{8.2.27}$$

Thus, the heat release of the diffusion combustion becomes with $\beta_f = \frac{m_{\text{f,diff}}}{V_{\text{r}}}$

$$\dot{Q}_{\text{f,diff}} = r_{\text{D}}\, V_{\text{r}}\, H_{\text{l}} = C'_{\text{mag}}\, H_{\text{l}} m_{\text{f,diff}} \frac{\sqrt{k}}{\sqrt[3]{V_{\text{d}}}}. \tag{8.2.28}$$

Hence, the heat release during the diffusion combustion is mainly determined by the available fuel mass and the turbulent kinetic energy density, Zahn (2012). The available fuel mass is calculated by

$$m_{\text{f,diff}}(t) = \int_{\text{t}_{\text{soi}}}^{t} \dot{m}_{\text{inj}}(\tau) d\tau - \frac{1}{H_{\text{l}}} \left(\dot{Q}_{\text{f,pre}}(t) + \dot{Q}_{\text{f,diff}}(t) \right). \tag{8.2.29}$$

c) Overall combustion model

The overall heat release for one stroke and for one pre- and one main injection results from the summation of

- pre-combustion as pre-mixed combustion

$$\dot{Q}_{\text{f,pre-comb}} = \dot{Q}_{\text{f,pre,pc}} \tag{8.2.30}$$

according to (8.2.16) and (8.2.17), and

- main-combustion with pre-mixed combustion and diffusion combustion

$$\dot{Q}_{\mathrm{f,main-comb}} = \dot{Q}_{\mathrm{f,pre,mc}} + \gamma_{\mathrm{pre}} \dot{Q}_{\mathrm{f,diff,mc}} \tag{8.2.31}$$

  using (8.2.16), (8.2.17) and (8.2.29)

The main combustion begins with pre-mixed combustion and continues with the diffusion combustion. The coefficient

$$\gamma_{\mathrm{pre}} = \left(\frac{\dot{Q}_{\mathrm{f,pre,mc}}}{m_{\mathrm{f,pre,mc}} H_{\mathrm{l}}} \right)^{l} \tag{8.2.32}$$

serves to model the delay of the diffusion combustion through the pre-mixed combustion with $l = 4$, Barba (2001). $m_{\mathrm{f,pre,mc}}$ is the injected and with $\chi_{\mathrm{pre,mc}}$ weighted fuel mass according to (8.2.16).

Hence, the total heat release becomes in dependence on time

$$\dot{Q}_f(t) = \dot{Q}_{\mathrm{f,pre,pc}}(t) + \dot{Q}_{\mathrm{f,pre,mc}}(t) + \gamma_{\mathrm{pre}} \, \dot{Q}_{\mathrm{f,diff,mc}}(t) \tag{8.2.33}$$

and by using

$$\frac{dQ_{\mathrm{f}}(\varphi)}{d\varphi} = \frac{dQ_{\mathrm{f}}}{d\varphi} \frac{d\varphi}{dt} \frac{1}{\omega} = \frac{1}{\omega} \frac{dQ_{\mathrm{f}}(\varphi)}{dt} = \frac{1}{\omega} \dot{Q}_{\mathrm{f}}(t) \tag{8.2.34}$$

one obtains for constant speed ω

$$\frac{dQ_{\mathrm{f}}}{d\varphi} = \frac{dQ_{\mathrm{f,pre,pc}}(\varphi)}{d\varphi} + \frac{dQ_{\mathrm{f,pre,mc}}(\varphi)}{d\varphi} + \gamma_{\mathrm{pre}} \frac{dQ_{\mathrm{f,diff,mc}}(\varphi)}{d\varphi} \tag{8.2.35}$$

the *heat release in dependence on the crank angle* φ.

8.2.4 Ignition delay model

The start of combustion (soc) for the pre-mixed and diffusion combustion follows the start of injection (soi) with a certain time interval, which is called ignition delay. The physical processes during this interval are determined by fuel atomization, evaporation and air/fuel mixing until an ignitable mixture with local air/fuel ratios within $0.5 < \lambda < 0.7$ are reached to start the chemical processes, Merker et al (2006). The oxidation of C_xH_y fuels is a branched propagation progress with hundreds of intermediate species. Many investigations have shown that the ignition delay

$$\Delta t_{\mathrm{ic}} = t_{\mathrm{soc}} - t_{\mathrm{soi}} \tag{8.2.36}$$

is for charged diesel engines in the range of 0.3 to $0.8\,\mathrm{ms}$ which corresponds to about 2.5 to $7.2°\mathrm{CS}$ for $1500\,\mathrm{rpm}$, see e.g. Mollenhauer and Tschöke (2010). The chemical ignition delay can be described by an Arrhenius equation, compare (8.2.11)

$$\Delta t_{\mathrm{ic,chem}} = \frac{1}{k_{\mathrm{n}}} = A \frac{\lambda}{p_{\mathrm{cyl}}^2} \exp\left(\frac{E_{\mathrm{a}}}{R_{\mathrm{m}} T_{\mathrm{cyl}}} \right), \tag{8.2.37}$$

Merker et al (2006), where A is a constant. Zahn (2012) approximates the ignition delay by

$$\Delta t_{\text{ic}} = \Delta t_{\text{ic,phys}} + \Delta t_{\text{ic,chem}} = k_{\text{ic},1} + k_{\text{ic},2}\, p_{\text{cyl,soi}}^{-k_{\text{ic},3}} \exp\left(\frac{k_{\text{ic},4}}{T_{\text{cyl,soi}}}\right) \tag{8.2.38}$$

where the constants $k_{\text{ic},1} \ldots k_{\text{ic},4}$ are determined by least-squares parameter-estimation based on the measured data of a pressure-based net heat release analysis as described in Sect. 4.2.2. The obtained ignition delays are in the range of 4 to 12°CS for pre-combustion and between 3 and 14°CS for the main combustion for $1000 \leq n_{\text{eng}} \leq 3000\,\text{rpm}$. The accuracy of the model (8.2.38) is about ±2°CS.

8.2.5 Adaptation of the combustion models

The phenomenological combustion models carry several parameters which have to be adapted to a given engine. These parameters can be integrated in the parameter vectors

$$\boldsymbol{\theta}_{\text{pre-comb}} = \left[C'_{\text{arr,pc}},\ n_{\text{pc}},\ T_{\text{a,pc}}.\ \chi_{\text{pre,pc}}\right]^{\text{T}} \tag{8.2.39}$$

$$\boldsymbol{\theta}_{\text{main-comb,pre}} = \left[C'_{\text{arr,mc}},\ n_{\text{mc}},\ T_{\text{a,mc}},\ \chi_{\text{pre,mc}}\right]^{\text{T}} \tag{8.2.40}$$

$$\boldsymbol{\theta}_{\text{main-comb,diff}} = \left[C'_{\text{mag,mc}},\ C'_{\text{diss,mc}}\right]^{\text{T}}. \tag{8.2.41}$$

Based on the measured cylinder pressure $p_{\text{cyl}}(\varphi)$ and the calculated net heat release $dQ_{\text{h}}(\varphi)/d\varphi$, similar to (4.2.33), these parameters are estimated with nonlinear least-squares estimation, as described with much more detail in Zahn (2012). This also includes the estimation of the ignition delay. Figure 8.2.5 depicts the cylinder pressure based heat release rate in comparison to the results obtained with the phenomenological combustion models. The pre-combustion as well as the main combustion show a good agreement. The main combustion in Fig. 8.2.5a) is determined to a large extent by the pre-mixed combustion, and in Fig. 8.2.5b) with the larger injection mass it is determined mostly by the diffusion combustion.

The parameter vectors (8.2.39) to (8.2.41) have to be determined for each operating point $(n_{\text{eng}}, M_{\text{eng}}, m_{\text{air}})$. However, if they are determined for one reference operating point, they can be transformed to other operating points using multiplicative correcting factors given by Zahn (2012).

8.2.6 Emission models

Beside the ideal combustion to CO_2 and H_2O according to (8.2.10) the exhaust gases contain the nitrogen (N_2) of the air and unused, residual oxygen (O_2). Because of the non-ideal combustion other components develop like carbon monoxide (CO), unburned hydrocarbons (HC), nitrogen oxide (NO_x), and particulates (PM). About 90% of the emissions CO and HC are converted in an oxidation catalyst to CO_2 and H_2. Therefore, especially NO_x and particulates have to be reduced as much as

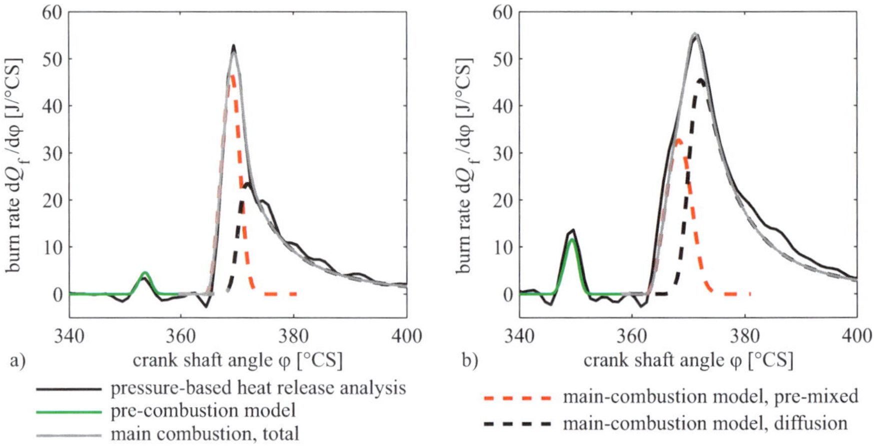

Fig. 8.2.5. Heat release rate determined by the cylinder pressure based heat release analysis and the various phenomenological combustion models (Opel 1.9 l, 255 Nm, 77 kW), Zahn (2012). **a** $n_{\text{eng}} = 1580\,\text{rpm}$, $m_{\text{f}} = 12.2\,\text{mg/inj}$. **b** $n_{\text{eng}} = 2020\,\text{rpm}$, $m_{\text{f}} = 19.3\,\text{mg/inj}$.

possible during the combustion process. The term NO_x summarizes the species *nitrogen monoxide* (NO) and *nitrogen dioxide* (NO_2). Usually NO is dominant which results from several chained reactions. Thermal NO is formed at high temperatures $T_{\text{cyl}} > 2000\,\text{K}$ from the atmospheric nitrogen N_2 and oxygen O_2 via the Zeldovich mechanism behind the flame front. Additionally, prompt NO develops in the flame front as a reaction of CH radicals and N_2, according to the Fenimore mechanism for temperatures $> 1000\,\text{K}$. About 90-95% of NO originates as thermal NO. In the atmosphere NO converts after some time in NO_2. Figure 8.2.6 shows the development of NO during the combustion. It requires air excess and is generated in lean areas, mainly during the pre-mixed combustion. During the expansion phase a part of the NO is reduced. However, it then remains constant and appears in the exhaust gases.

Particulate matter is formed at high temperatures, especially in the ranges of $1500\,\text{K} < T_{\text{cyl}} < 1900\,\text{K}$ and lack of oxygen with $\lambda < 0.6$. Diesel particulates consist of about 95% organic components, like polycyclic aromatic hydrocarbons and carbon (soot) and 5% inorganic components like ashes of oil. The main part of particulates is formed at the beginning of the combustion during local rich areas. However, it is then oxidized during the main and post-combustion phases, as shown in Fig. 8.2.6, and only a small fraction is then measurable in the exhaust gases.

Mathematical modeling of the emissions is rather complicated and generally requires multi-zone models because of the dependence on local temperatures and air/fuel ratios. A survey of the state-of-art is, for example, given in Stiesch (2003) and Merker et al (2006). Hence, usually only measured data based empirical mean-value models for the exhaust gases are available, e.g. based on polynomial or neural net type models. Analytically derived emission models are in the status of research, Guzzella et al (2012).

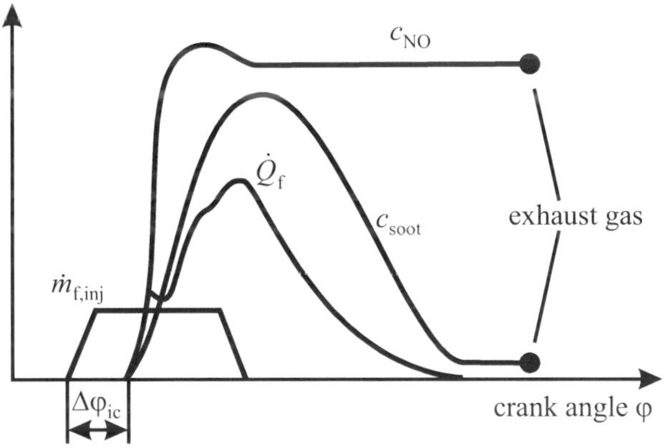

Fig. 8.2.6. Course of NO and particulates during the combustion and expansion phase, Mollenhauer and Tschöke (2010).

8.3 Optimization of steady-state feedforward control of diesel engines

As mentioned in Sect. 6.8 the optimization of control functions in dependence on the operating point and environmental variables is a complex task. It requires the treatment of a strongly coupled nonlinear multiple-input multiple-output system with optimization methods.

Sections 8.3.1 and 8.3.2 consider the optimization of the static behavior of diesel engines for a selection of stationary operation points. This is followed by an optimization for a driving cycle, using a sequence of stationary engine models. In order to save measurement time it is based on a combination of dynamic and static measurements. Dynamic local linear net models LOLIMOT are used as an identification method, where the steady-state behavior is extracted as a special case, Schüler et al (2000), Schüler (2001).

8.3.1 Static engine models from dynamic measurements at fixed operating points

In the following a four-cylinder direct injection diesel engine VW 1.9 l TDI, 81 kW, with VGT turbocharger and cooled exhaust gas recirculation is considered as an example. The optimization is restricted to the outputs fuel consumption b_{sfc}, NO_x and particulates, measured as opacity c_{op}. Instead of measuring only steady-state values and omitting the transients between these steady-states a sequence of input step functions in short time intervals is applied for a constant operating point (M_i, n_j), where all measured data are stored. This allows saving measurement time at the test bench.

First, the engine runs at the specified speed and load until stationary conditions are reached. Then, the manipulated variables (φ_{soi}, u_{vgt}, u_{egr}) are changed in steps in order to extract the relevant process dynamics from the step responses and the estimation of the stationary behavior. A fast torque controller compensates for the influence of these changes on the torque by adjusting the fuel mass. Short steps of 5–10 seconds duration allow to adjust the three different inputs in the relevant range of the input variables and still lead to reasonable measurement times. This is the main difference between this concept and those based on the design of experiments (DoE) for the stationary behavior.

The restriction on one specific operation point per measurement sequence guarantees an approximately constant engine temperature and thus reduces drift effects on emission values. Consequently, the derived models are local models and cannot be used to calculate the engine behavior during driving cycles because the training data does not contain any load-/speed changes.

Figure 8.3.1 shows the measurements with step function inputs. The step function sequence of each input is varied, such that for a certain period only one input varies more frequently with a time interval of about 5–10 s, in order to scan the interesting operating range. The measurement time for one operating point was about $10\,\mathrm{min}$ and is therefore much shorter than for stationary measurements.

The used dynamic discrete time models are of second order with dead time and sampling time $T_0 = 100\,\mathrm{ms}$, as e.g.

$$\begin{aligned} x(k) = f_{\text{LOLIMOT}}\big[&u_{\text{egr}}(k-1-d_1),\ u_{\text{egr}}(k-2-d_1),\\ &\varphi_{\text{soi}}(k-1-d_2),\ \varphi_{\text{soi}}(k-2-d_2),\\ &u_{\text{vgt}}(k-1-d_3),\ u_{\text{vgt}}(k-2-d_3),\\ &x(k-1),\ x(k-2)\big]. \end{aligned} \tag{8.3.1}$$

The outputs of the identified dynamic models indicate a good agreement with the measured values, also for generalization measurements, as shown in Schüler (2001).

The stationary models are then determined by $\lim_{k\to\infty} x(k)$ for fixed values of the three inputs, leading to the characteristic maps shown in Fig. 8.3.2 for one operating point. Finally, a steady-state overall model in dependence on the operation point according to Fig. 8.3.3 could be obtained and compared with direct stationary measurements for the series production calibration, with mean absolute errors of $\Delta\dot{m}_\text{f} = 0.05\,\text{g/s}$, $\Delta\,\text{NO}_x = 0.003\,\text{g/s}$, $\Delta\dot{m}_\text{air} = 2.3\,\text{g/s}$, $\Delta p_2 = 0.03\,\text{bar}$.

8.3.2 Optimization of the stationary feedforward control for selected operating points (STATOP-LOC)

The goal of the optimization of the stationary (static) engine control is to minimize the fuel consumption within given emission limits over a certain driving cycle, see Sect. 6.8. Usually, some of the goals oppose one another. With the prerequisite that the process can be accurately modeled, optimized manipulated variables $\mathbf{U}$ can be determined by an offline optimization as depicted in Fig. 6.8.5b).

In the following, static engine models will be used for the *local optimization* of individual operating points and applied for a specific driving cycle. The basis

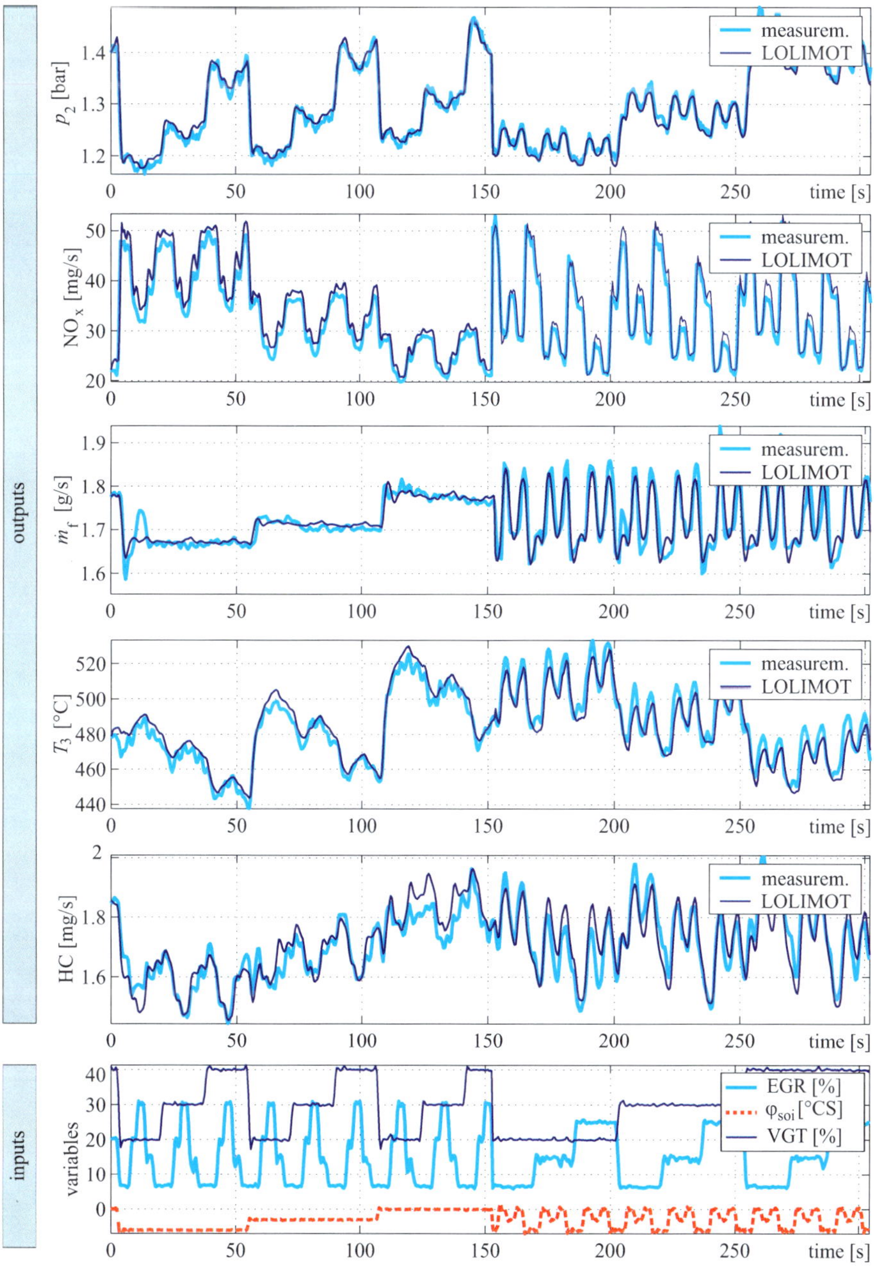

Fig. 8.3.1. Dynamic step function measurements for a diesel engine with three manipulated variables and five output variables, $M_{\mathrm{eng}} = 90\,\mathrm{Nm}$, $n_{\mathrm{eng}} = 3000\,\mathrm{rpm}$, Schüler (2001). Engine: VW, $1.9\,\mathrm{l}$ TDI with cooled EGR and VGT turbocharger, $81\,\mathrm{kW}$, $235\,\mathrm{Nm}$, distributor fuel pump VP37.

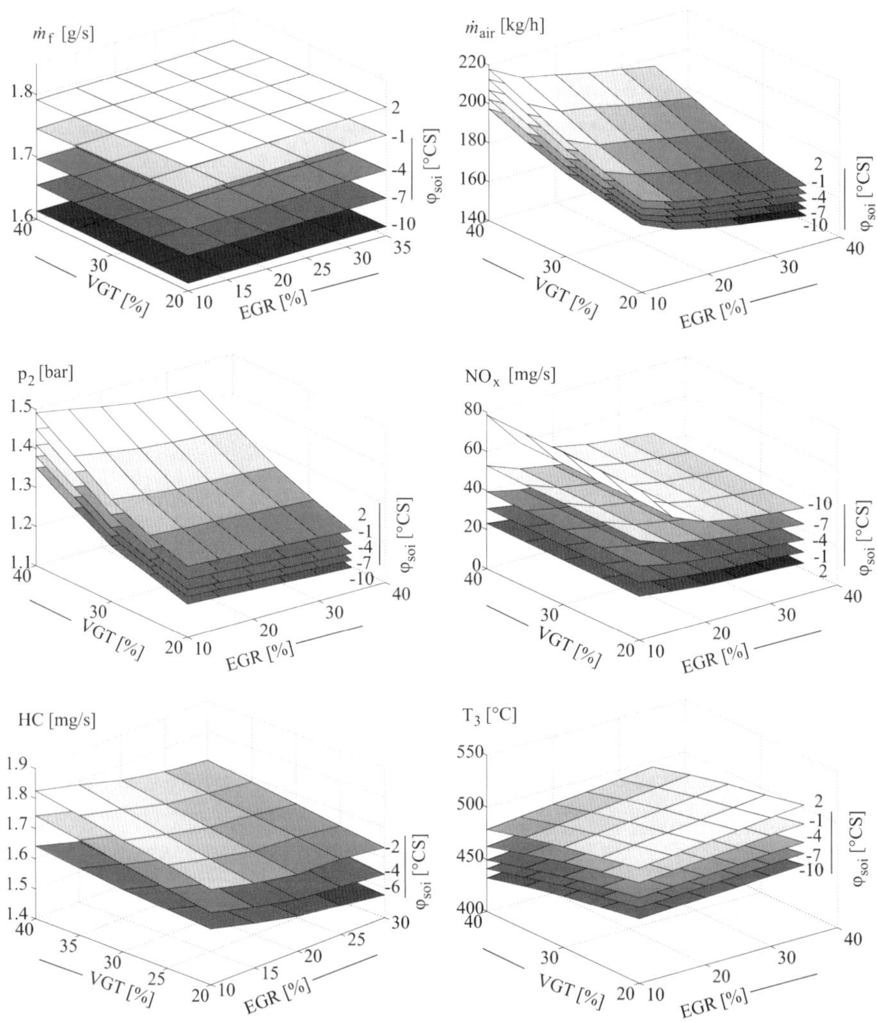

Fig. 8.3.2. Characteristic maps for the steady-state behavior based on dynamic step measurements for $M_{\mathrm{eng}} = 90\,\mathrm{Nm}$, $n_{\mathrm{eng}} = 3000\,\mathrm{rpm}$.

of the optimization is the loss function J, which represents, depending on the engine settings to be optimized, a scalar, which has then to be minimized by efficient mathematical optimization algorithms as described in Sect. 6.8.4. It should be noted that there are no specific emission constraints in the individual operating points because no specific driving cycle is considered. The aim is to find a good compromise between consumption and emissions in each operating point with the constraint to guarantee a specific engine torque or other constraints like charging pressure, cylinder peak pressure, etc.

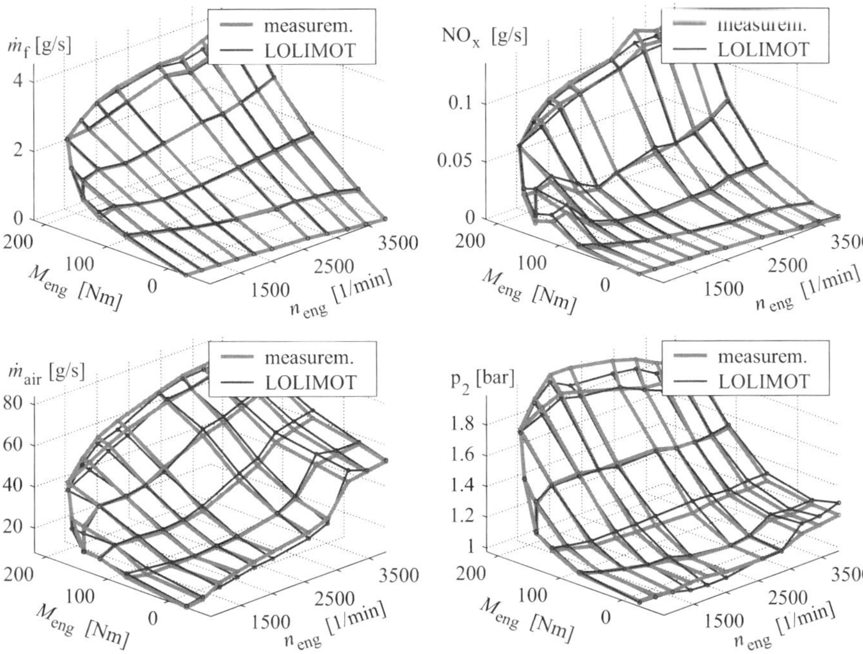

Fig. 8.3.3. Characteristic maps for four output variables in dependence on the operating point (M_{eng}, n_{eng}). Identified behavior and directly measured behavior for the series calibration. VW 1.9 l TDI.

A simultaneous improvement of both consumption and emissions is usually only possible to a certain extent. Therefore, the components of the loss functions (consumption and emissions) have to be weighted one against the other. Confining oneself on the most critical outputs, namely the specific fuel consumption b_{sfc}, and the emissions c_{NOx} and c_{op} (opacity as an equivalent of the particulate matter), one comes to the loss function in the operating point (α, n) = (gas pedal value, engine speed). The specific components of the loss function depend on the engine inputs to be optimized and are represented by three net models (LOLIMOT).

$$\begin{aligned} b_{\text{sfc}} &= f_{\text{sfc}}\,(u_{\text{egr}}, u_{\text{vgt}}, \varphi_{\text{soi}}, m_{\text{inj}}, n_{\text{eng}}) \\ c_{\text{NOx}} &= f_{\text{NOx}}(u_{\text{egr}}, u_{\text{vgt}}, \varphi_{\text{soi}}, m_{\text{inj}}, n_{\text{eng}}) \\ c_{\text{op}} &= f_{\text{op}}\,(u_{\text{egr}}, u_{\text{vgt}}, \varphi_{\text{soi}}, m_{\text{inj}}, n_{\text{eng}}). \end{aligned} \tag{8.3.2}$$

The multi-variable optimization criterion used is

$$J(M_{\text{eng}}, n_{\text{eng}}) = w_1 \frac{b_{\text{sfc}}(\mathbf{u})}{b_{\text{sfc0}}} + w_2 \frac{c_{\text{NOx}}(\mathbf{u})}{c_{\text{NOx0}}} + w_3 \frac{c_{\text{Op}}(\mathbf{u})}{c_{\text{Op0}}} \tag{8.3.3}$$

and its minimization requires

$$\left. \frac{dJ(M_{\text{eng}}, n_{\text{eng}})}{d\mathbf{u}} \right|_{\mathbf{u}=\mathbf{u}_{\text{opt}}} = \mathbf{0} \tag{8.3.4}$$

for each operating point with regard to the manipulated variables

$$\mathbf{u}^T = \left[m_{\text{inj}}\ \varphi_{\text{soi}}\ u_{\text{vgt}}\ u_{\text{egr}}\right] . \tag{8.3.5}$$

The components of the performance criterion have to be normalized in order to get a comparable sensitivity of the loss function with regard to each component. In this case, the process behavior for standard ECU settings was used for the normalization. The weighting factors have to be determined beforehand. Different combinations which seem adequate are used for the optimization leading to a set of optimal control maps (once slightly better for some emissions, once for the consumption etc.). Afterwards, the best result which fulfills the limits for a driving cycle with minimum consumption can be implemented as the default setting for the car.

Results for the application of this STATOP-LOC optimization procedure are shown by Hafner (2002) for a 2 l, 74 kW diesel engine with wastegate turbocharger for different weighting factors w_i.

Figure 8.3.4 depicts as a result the optimized maps of the manipulated variables and output variables after smoothing. The weighting factors for the optimization criterion (8.3.3) have been systematically varied for an FTP driving cycle from $w_1 = 1$, $w_2 = 1$, $w_3 = 1$ to $w_1 = 4.2$, $w_2 = 0.75$, $w_3 = 0.7$, until the emission limits were satisfied.

8.3.3 Optimization of the stationary feedforward control for a driving cycle (STATOP-CYC)

The optimization of the manipulated variables $\mathbf{u}$ for a complete driving cycle with stationary engine models requires the consideration of many operating points ($M_{\text{eng,i}}$, $n_{\text{eng,j}}$) which are used in the driving cycle. However, dynamic effects are then not taken into account. An example will now be considered for the VW 1.9 l TDI used in Sect. 8.3.1. The goal is to find the optimal manipulated variables

$$\mathbf{u}^T = \left[\varphi_{\text{soi}}\ u_{\text{vgt}}\ u_{\text{egr}}\right] . \tag{8.3.6}$$

The injected fuel mass $m_{\text{inj}}(\alpha, n_{\text{eng}})$ is determined beforehand to satisfy the demanded torque.

For a stationary optimization the cycle is regarded as a sequence of independent stationary operating points. By clustering similar operating points, the task can be divided in the minimization of the consumption over N operating points, which are weighted by factors k_i according to their frequency in the cycle. Hence, the criterion

$$J_{\text{cyc}} = \sum_{i=1}^{N} w_i m_{\text{inj}}(M_{\text{eng}}, n_{\text{eng}}, \varphi_{\text{soi}}, u_{\text{vgt}}, u_{\text{egr}}) \tag{8.3.7}$$

has to be minimized with emission constraints

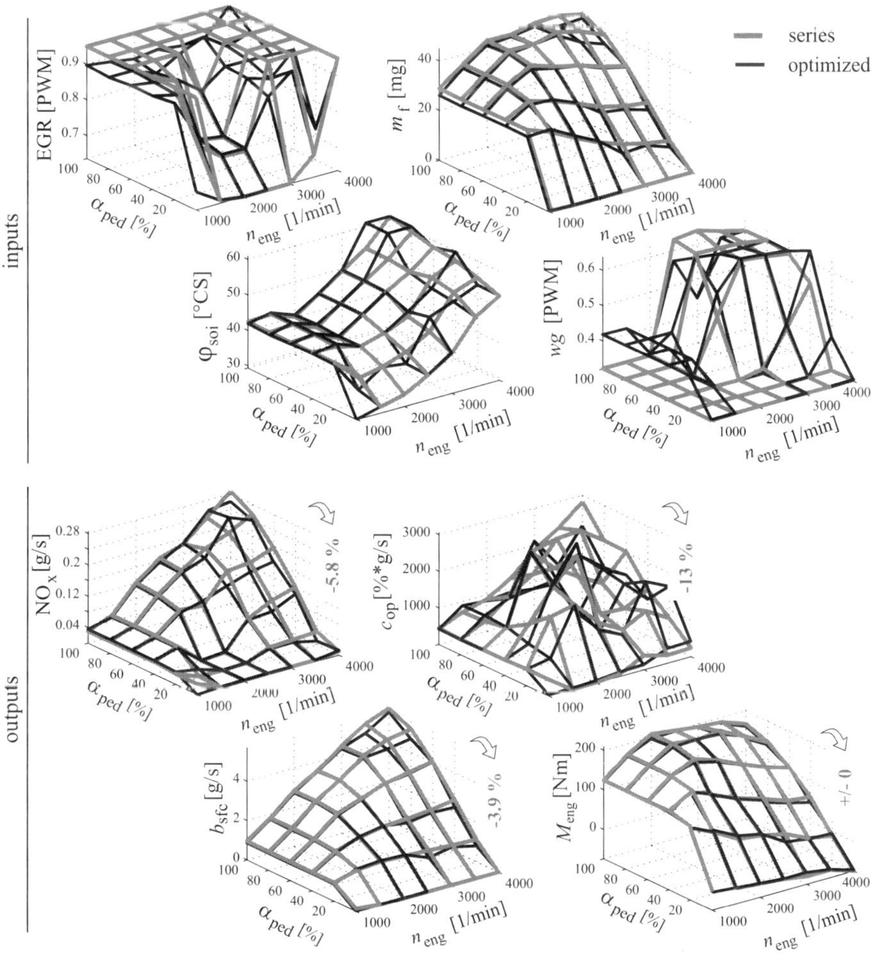

Fig. 8.3.4. Optimized maps for the stationary behavior of an Opel, 2 l DTI, 4 cyl., 74 kW, 205 Nm diesel engine in dependence on the operating point. Black: series; grey: steady-state optimized for weighting factors $w_1 = 4.2$, $w_s = 0.75$, $w_3 = 0.7$, Hafner (2002).

$$\sum_{i=1}^{N} w_i \, \mathrm{NO_x}(M_{\mathrm{eng}}, n_{\mathrm{eng}}, \varphi_{\mathrm{soi}}, u_{\mathrm{vgt}}, u_{\mathrm{egr}}) \leq c_{\mathrm{NOx,cycle}} \tag{8.3.8}$$

$$\sum_{i=1}^{N} w_i \, \mathrm{PM}\,(M_{\mathrm{eng}}, n_{\mathrm{eng}}, \varphi_{\mathrm{soi}}, u_{\mathrm{vgt}}, u_{\mathrm{egr}}) \leq c_{\mathrm{PM,cycle}} \tag{8.3.9}$$

concentrating on these main emission components, see Sect. 6.8. In addition, several other constraints have to be taken into account, like peak pressure, exhaust gas temperature and noise.

A straightforward way is to optimize the manipulated variables $\mathbf{u}$ directly by

$$\left. \frac{dJ_{\text{cyc}}}{d\mathbf{u}} \right|_{\mathbf{u}=\mathbf{u}_{\text{opt}}} = \mathbf{0}. \tag{8.3.10}$$

However, this leads to a large computational effort. Additionally, the values at neighboring operating points may vary strongly, resulting in discontinuous control maps. An alternative is to model the manipulated variables with a local linear network (LOLIMOT), dependent on the operating point

$$\begin{aligned} u_{\text{vgt}} &= f_{\text{LOLIMOT}}(M_{\text{eng}}, n_{\text{eng}}) \\ &= \sum_{j=1}^{M} u_{\text{vtg},j} = \sum_{j=1}^{M} \phi_j(M_{\text{eng}}, n_{\text{eng}}) \end{aligned} \tag{8.3.11}$$

and linear local models $j = 1 \ldots M$

$$\begin{aligned} u_{\text{vgt},j} &= a_{0,\text{vgt},j} + a_{1,\text{vgt},j} n_{\text{eng}} + a_{2,\text{vgt},j} M_{\text{eng}} \\ &= \mathbf{a}_{\text{vgt,j}}^T \left[1 \; n_{\text{eng}} \; M_{\text{eng}} \right] \end{aligned} \tag{8.3.12}$$

and correspondingly for $u_{\text{egr},j}$ and $\varphi_{\text{soi},j}$. This means that a *parametric manipulation model* is used as an intermediate model.

The optimization then has to find

$$\left. \frac{dJ_{\text{cyc}}}{d\mathbf{a}} \right|_{\mathbf{a}=\mathbf{a}_{\text{opt}}} = \mathbf{0} \tag{8.3.13}$$

with

$$\mathbf{a}^T = \left[\mathbf{a}_{\text{vgt},1}^T \; \mathbf{a}_{\text{egr},1}^T \; \mathbf{a}_{\text{soi},1}^T \; \ldots \; \mathbf{a}_{\text{vgt,M}}^T \; \mathbf{a}_{\text{egr,M}}^T \; \mathbf{a}_{\text{soi,M}}^T \right]. \tag{8.3.14}$$

Then, the amount of parameters to be determined is only a function of the structure of the used network model and does not depend on the number of discrete operating points. This allows to include many operating points of the cycle. The structure of the applied LOLIMOT network can even be based on a similar engine.

For the considered example it was sufficient to choose $M = 3$ local models with three parameters in each model, namely one offset and one gain for each of the two model inputs engine speed and torque, as (8.3.12). Hence, 27 parameters have to be determined altogether.

The structure of the proposed method is depicted graphically in Schüler (2001). The first step is to transfer the transient test cycle into static operating points (speed-load) and then to weight them according to their occurrence in the cycle. The manipulated variables for these points are then calculated with the identified engine models. These manipulated variables serve as model inputs for the emission/consumption models. A nonlinear optimization method determines the parameters $\mathbf{a}$ of the engine manipulated variable networks (8.3.12) in a way that the consumption is minimized under the constraint of the emission limits, (8.3.7) to (8.3.9).

200 different operating points were extracted from a FTP cycle for the optimization. As a result, a consumption improvement of 4% could be achieved for about the same emissions.

Figure 8.3.5 compares the standard engine settings with the optimized settings by means of 3-D maps. The EGR was obviously enlarged by the optimization in correspondence with lower VGT settings (lower charging pressure) in areas which occur more often in the cycle. For low speed conditions, the optimization also delays the injection angle to later values in order to lower the NO_x-emissions there, whereas the injection is set to earlier values for high speeds in order to realize a lower consumption. Those areas of the control maps which are not relevant for the cycle were not adapted.

Two main advantages of the proposed method are the possible integration of many operating points and the avoidance of a discontinuous engine setting behavior. It would also be possible to choose the weighting factors in a way that puts more emphasis on low consumption in those areas that are not covered by the emission cycle.

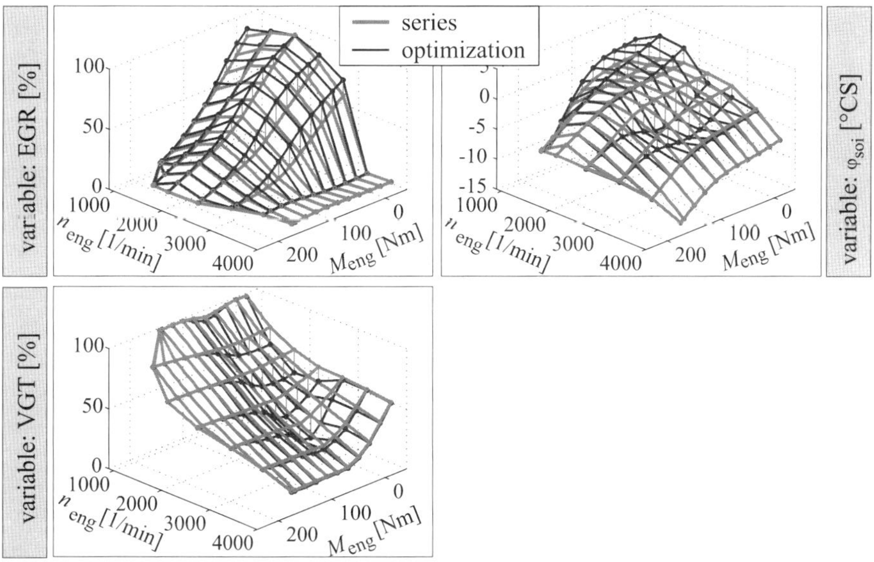

Fig. 8.3.5. Maps for the manipulated variables: Comparison of optimized and standard ECU values, VW 1.9 l, 81 kW.

8.4 Optimization of dynamic feedforward control

As discussed in Sect. 6.8.4, the increased requirements for the reduction of fuel consumption, emissions and driveability need the inclusion of the dynamic engine behavior.

For diesel engines the stationary controlled fuel mass injection and air mass leads to emission peaks during transients, because of too low air/fuel ratio. The charging pressure p_2 rises during an acceleration only within certain time delays after manipulation of the guide blade angles or wastegate valve due to inertia effects and gas flow delays. The changing pressure condition before the turbine and behind the compressor will also change the pressure gradient over the EGR channel and influences the EGR rate. On the other hand, new EGR valve settings influence the pressure characteristics of the turbocharger and the charging pressure itself. A mere static open loop control of the manipulated variables does not consider these dynamic effects. Therefore, dynamic predictive control functions for EGR and VGT (and also for the injection angle) are required in order to harmonize unintended effects during transient engine conditions.

In the following it will be shown how dynamic engine models for the turbocharger and exhaust gas recirculation are obtained through dynamic identification methods and how they can be used for the design of the coupled charging pressure and exhaust recirculation mass flow control, Hafner (2002), Schüler et al (2000).

8.4.1 Dynamic engine models with APRBS excitation (DYNMET)

In order to measure the transient engine behavior for all operating conditions, it is necessary to excite all relevant input variables in the whole operating regime. In addition, all frequencies and amplitudes of the process have to be excited well enough, see Chap. 3. This can be achieved by so called amplitude-modulated pseudo random binary signals (APRBS) which cover a broad range of frequencies and amplitudes.

In order to avoid measurements in areas which could harm the engine (e.g. too high peak pressures) or which the engine would never be run in (e.g. too small lambda values producing too much soot), the APRBS signal is modulated with a specific maximum amplitude on the manipulated variables designed earlier for a similar engine. If no corresponding engine is available, the limits could be very roughly measured in some short prior measurements. Figure 8.4.1 depicts the procedure if the standard values of a similar engine are already known and the APRBS signals are superimposed on the standard manipulated variables of the series ECU. It is obvious that high EGR settings are only realized at low load conditions in order to avoid areas where high soot emissions are most likely to occur. A complete dynamic engine measurement coping with five inputs and the APRBS method could be performed in less than one hour.

The identification was performed with the local linear model network LOLIMOT, described in Sect. 3.2.3. Depending on the inputs and outputs, discrete time models with first and second order and dead time according to (8.3.1) were used with 5–10 local models each. The operating point $(M_{\mathrm{eng},i}, n_{\mathrm{eng},j})$ was kept constant during dynamic excitation of the three inputs u_{egr}, u_{vgt}, φ_{soi} for 10 min, see Fig. 8.4.1. The considered outputs were b_{sfc}, $\mathrm{NO_x}$, c_{op} and p_2, using a sampling time of $T_0 = 100\,\mathrm{ms}$. Figure 8.4.2 depicts the generalization for one operating point. It shows a relatively good agreement with mean square errors of about 2–5%. Also the determination of the steady-state behavior from the dynamic models led to good results, Schüler et al (2000).

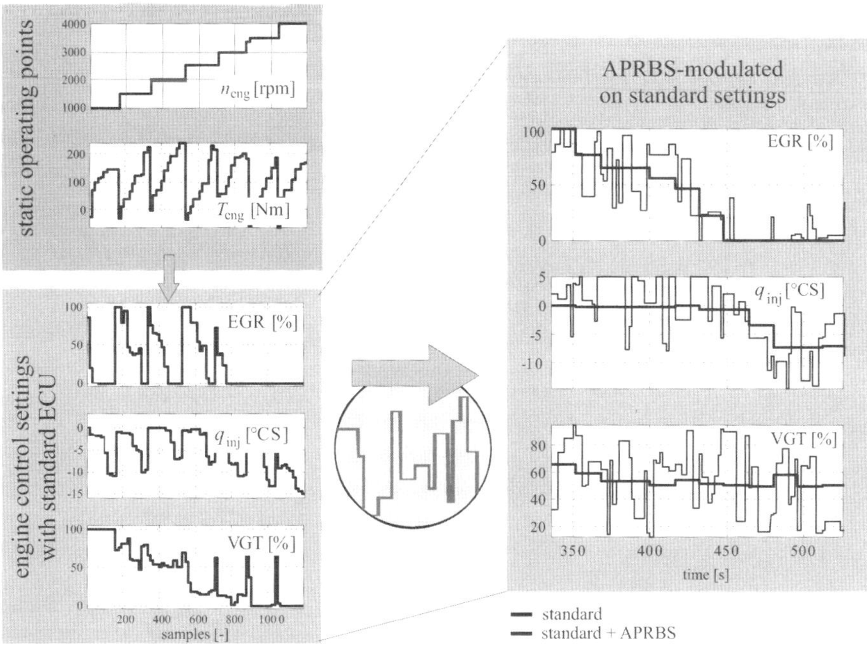

Fig. 8.4.1. Dynamic measurement of the engine behavior with APRBS signals, superimposed on the series ECU values.

8.4.2 Optimization of the dynamic feedforward control of the charging pressure and exhaust-recirculation flow rate

Figure 8.4.3 shows the signal-flow chart for the considered two-variable turbocharger and exhaust recirculation system. A static feedforward control of VGT and EGR results in the described delays compared to the injection mass m_{inj} and because of a too low air excess leads to a soot peak. This effect can be reduced by a derivative control term according to a proportional lead-lag element PDT_1 with a transient function

$$u_2(t) = u_{10} + (u_{20} - u_{10})\left(1 - \left(1 - \frac{T_D}{T_1}\right) e^{-\frac{t}{T_1}}\right). \tag{8.4.1}$$

(Such a lead-lag feedforward control was already addressed in Sect. 6.3.) The steady-state values u_{10} and u_{20} follow directly from the optimized stationary maps. T_1 is adapted to the process dynamics, with $T_1 = 0.25 \dots 1\,\mathrm{s}$, and is kept constant. $T_{D,egr}$, $T_{D,vgt}$ and T_D, φ_{egr} are determined by minimizing a loss function which integrates the weighted NO_x, c_{op}, and M_{eng} over a time window, Hafner (2002). This way, maps for the derivative times $T_{Di} = -1\,\mathrm{s} \dots +1\,\mathrm{s}$ are generated, depending on step size $\Delta\alpha$ of the pedal and engine speed n_{eng}. In addition the injection mass was delayed with a low pass filter with time constant $T_1 = T_D/2 = 100 \dots 200\,\mathrm{ms}$.

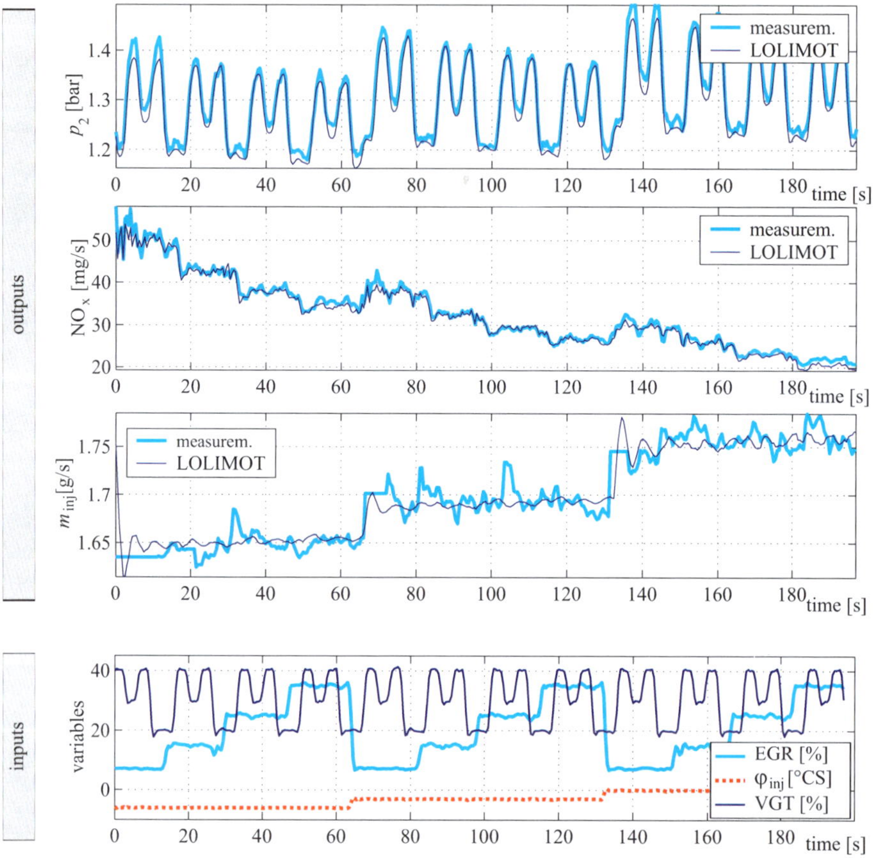

Fig. 8.4.2. Generalization measurement for comparison of the identified dynamic engine models with measurements. $M_{eng} = 90\,\mathrm{Nm}$, $n_{eng} = 3000\,\mathrm{rpm}$

These dynamic elements then lead to a reduction of the soot peaks, as depicted in Fig. 8.4.4. For a FTP driving cycle this control resulted in improvements of the opacity by 35% and a small increase of 5% for NO_x, Hafner (2002).

The combination of a dynamic PDT_1 feedforward control with feedback control of p_2 and $\dot{m}_{air}$ is described in Sect. 8.5.

8.5 Air flow and charging pressure control with exhaust gas recirculation – an in-depth case study[1]

In order to develop a model-based nonlinear multi-variable control system of the airflow and charging pressure a passenger car diesel engine according to Fig. 8.5.1 is

[1] compiled by Karl von Pfeil

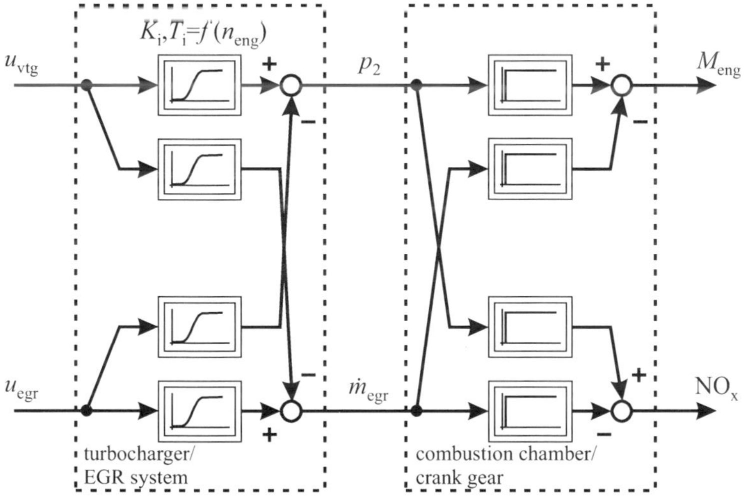

Fig. 8.4.3. Signal-flow chart for the coupled nonlinear two-variable turbocharger and exhaust recirculation system. Inputs are the manipulated variables for the turbocharger and the EGR valve and outputs are torque and NOx.

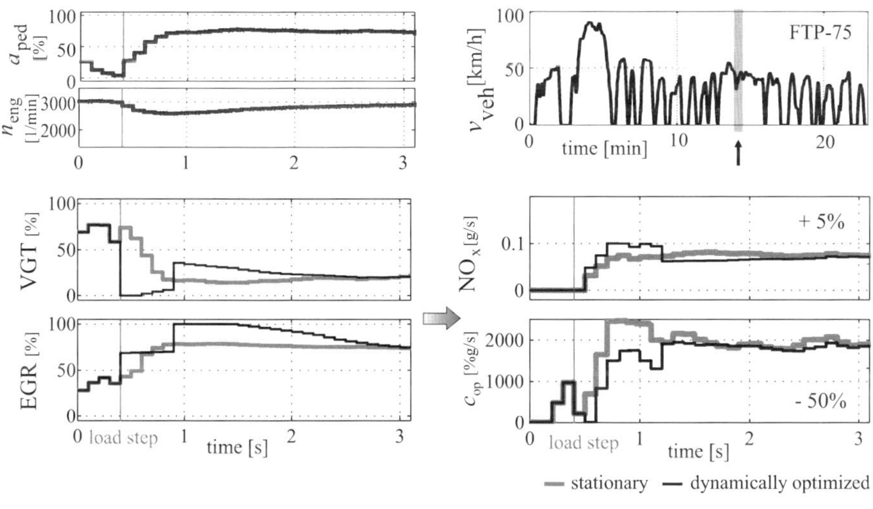

Fig. 8.4.4. Dynamic behavior of soot for static and dynamic changes of the manipulated variables u_{vgt} of the turbocharger and u_{egr} of the EGR rate (feedforward control) for an acceleration within a FTP cycle (simulation), Hafner (2002).

considered. The engine is equipped with an external exhaust gas recirculation (EGR) and with a variable geometry turbocharger (VGT). The EGR mass flow is used to decrease the nitrogen oxide emissions. The turbocharger mainly increases the intake manifold pressure and, thus, increases the engine power. In the following, an in-depth case study is described, based on Pfeil et al (2007a), Pfeil et al (2007b), Pfeil (2011).

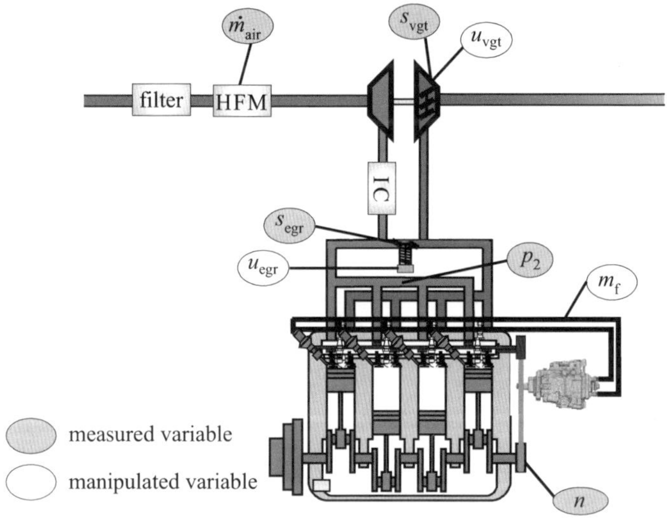

Fig. 8.5.1. Scheme of the investigated diesel engine (Opel JTD, 1.9 L, CR, 77 kW, 255 Nm). u_{egr}: control input of the EGR valve, u_{vgt}: control input of the VGT, s_{egr}: position of the EGR valve, s_{vgt}: position of the VGT, n: engine speed, m_{f}: injected fuel mass, p_2: intake manifold (charging) pressure, $\dot{m}_{\mathrm{air}}$: air-mass flow.

Figure 8.5.2 depicts the corresponding signal flow of the air and EGR system. The goal of its control is to precisely supply the cylinders with a gas mixture charge, consisting of air mass m_{air} and exhaust gas mass m_{egr} with charging pressure p_2.

The mass-flow balance at the cylinder inlet yields

$$\dot{m}_{\mathrm{air}} = \dot{m}_{\mathrm{cyl,in}} - \dot{m}_{\mathrm{egr}} \tag{8.5.1}$$

where for the cylinder gas mass inflow holds

$$\dot{m}_{\mathrm{cyl,in}} = \eta_{\mathrm{v}}(n_{\mathrm{eng}}, \rho_2)\, \frac{p_{2\mathrm{i}}(r_{\mathrm{egr}})}{2RT_{2\mathrm{i}}} V_{\mathrm{D}} n_{\mathrm{eng}} \tag{8.5.2}$$

see (4.1.2) to (4.1.8). For the EGR rate, see (8.7.5). The gas mass flowing through the inlet valves thus depends mainly on the volumetric efficiency η_{v}, gas state variables and the speed of the engine. If the air mass per stroke is $m_{\mathrm{air,cyl}}$, the corresponding air mass flow rate of one cylinder is

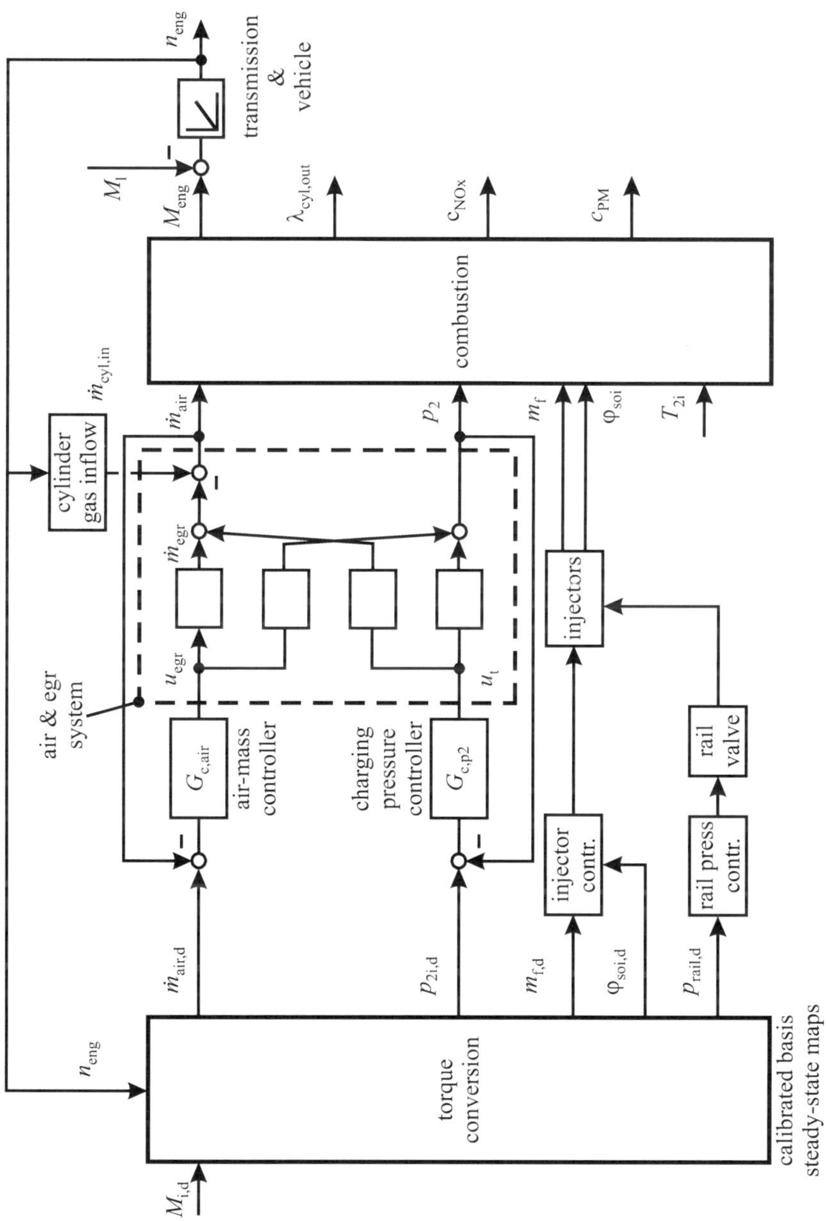

Fig. 8.5.2. Signal flow for the air flow and charging pressure control with exhaust gas recirculation (EGR) and combustion outputs.

$$\dot{m}_{\mathrm{air,cyl}} = \frac{dm_{\mathrm{air,cyl}}}{dt} = m_{\mathrm{air,cyl}}\frac{n_{\mathrm{eng}}}{2} \tag{8.5.3}$$

and for i cylinders holds

$$\dot{m}_{\mathrm{air}} = i\, m_{\mathrm{air,cyl}}\frac{n_{\mathrm{eng}}}{2}. \tag{8.5.4}$$

According to the mass-balance equation (8.5.1) the air mass flow can be controlled by manipulating the exhaust gas flow rate $\dot{m}_{\mathrm{egr}}$.

As shown in Fig. 8.5.2 the air mass controller $G_{\mathrm{c,air}}$ controls the air flow rate $\dot{m}_{\mathrm{air}}$ by manipulating the exhaust gas flow valve and the charging pressure p_2 by manipulating the actuator of a variable geometry turbocharger or a turbocharger with wastegate. Alternatives of this frequently applied control system is the direct control of the EGR mass flow, Rückert et al (2004) or the EGR rate, M.J. van Nieuwstadt et al (2000), Weirich et al (2006).

A turbine adjustment affects not only the charging pressure, but also the EGR mass flow, (because of the change of the EGR pressure drop). On the other hand, a change of the EGR mass flow affects the turbine-mass flow, as well as the charging pressure. These mutual couplings of both control loops require that the design of the control has to treat the process as a coupled two-input/two-output (TITO) system, which is strongly nonlinear.

Constantly rising requirements of the reduction of the pollutant emissions and increased power of charged diesel engines put new demands on the control performance of this intake manifold pressure/air-mass flow controller. The direct parameterization of this control system at the test bench is therefore more time-consuming due to the necessarily rising complexity (nonlinear behavior, coupling, etc.). Employment of model-based control design allows to develop and calibrate new controller structures in a systematic manner requiring less time at the test bench.

8.5.1 Physical model of the air and exhaust recirculation path

Figure 8.5.3 depicts the process scheme with the compressor and turbine, a charge-air cooler, an intake and exhaust manifold and an exhaust cooler and exhaust gas recirculation valve.

A signal-flow chart for the dynamic behavior of the air and exhaust path is shown in Fig. 8.5.4. It consists of nonlinear models of the VGT actuator and EGR valve, dynamic models of the intake and exhaust manifold, and a nonlinear turbine and compressor model, including rotor dynamics.

The mass-balance equation for the intake and exhaust manifold gives, compare (4.1.56)

$$\frac{dp_{2\mathrm{i}}}{dt} = \frac{RT_{2\mathrm{i}}}{V_2}\left(\dot{m}_{2\mathrm{c}} + \dot{m}_{\mathrm{egr}} - \dot{m}_{2\mathrm{i}}\right) \tag{8.5.5}$$

$$\frac{dp_3}{dt} = \frac{RT_3}{V_3}\left(\dot{m}_{3,\mathrm{cyl}} - \dot{m}_{\mathrm{egr}} - \dot{m}_{\mathrm{t}}\right) \tag{8.5.6}$$

The EGR valve is modeled by, compare (4.1.57)

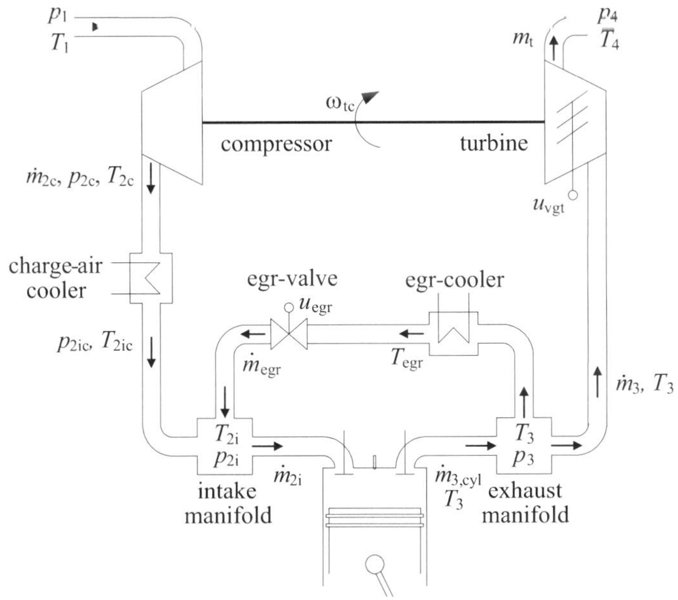

Fig. 8.5.3. Scheme of the air path and exhaust path of a turbo-charged diesel engine.

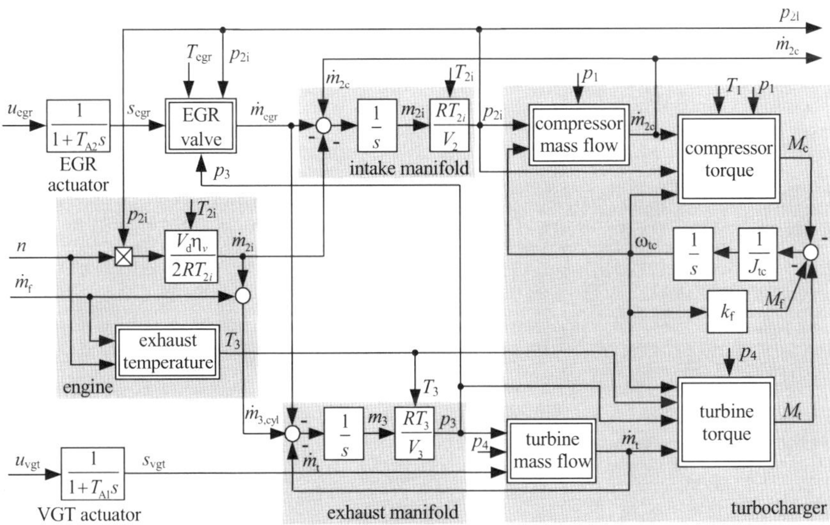

Fig. 8.5.4. Signal-flow chart for the air path and exhaust path with turbocharger and exhaust gas recirculation, without temperature models, Pfeil (2011).

$$\dot{m}_{\mathrm{egr}} = c_{\mathrm{d,egr}}\, A_{\mathrm{egr}} \sqrt{\frac{2}{R\, T_{\mathrm{egr}}}}\; \psi\!\left(\tfrac{p_{2\mathrm{i}}}{p_3}\right) p_3 \tag{8.5.7}$$

and the intake gas flow of the engine with volumetric efficiency η_{v} according to (4.1.5)

$$\dot{m}_{2\mathrm{i}} = \frac{\eta_{\mathrm{v}}\, V_{\mathrm{D}}\, \omega_{\mathrm{eng}}}{4\pi\, R\, T_{2\mathrm{i}}}\, p_{2\mathrm{i}} \tag{8.5.8}$$

For the rotor dynamics of the turbocharger holds

$$J_{\mathrm{tc}} \frac{d\omega_{\mathrm{tc}}}{dt} = M_{\mathrm{t}} - M_{\mathrm{c}} - M_{\mathrm{f}} \tag{8.5.9}$$

with the friction torque, compare (4.4.45)

$$M_{\mathrm{f}} = c_{\mathrm{f,tc}}\, \omega_{\mathrm{tc}} \tag{8.5.10}$$

The torque for the turbine follows by thermodynamic modeling from (4.4.27)

$$M_{\mathrm{t}} = \dot{m}_{\mathrm{t}} \frac{c_{\mathrm{p}}\, T_3^*\, \eta_{\mathrm{t,is}}}{\omega_{\mathrm{tc}}} \left(1 - \left(\frac{p_4}{p_3}\right)^{\frac{\kappa-1}{\kappa}}\right) \tag{8.5.11}$$

where the isentropic efficiency $\eta_{\mathrm{t,is}}$ is represented as a look-up table, see Sect. 4.4. The turbine mass flow results from the flow equation through a contraction

$$\dot{m}_{\mathrm{t}} = A_{\mathrm{t,eff}}\, \frac{p_3}{\sqrt{R T_3^*}}\, \psi\!\left(\tfrac{p_4}{p_3}\right) \tag{8.5.12}$$

see (4.4.39).

The torque of the compressor, based on the thermodynamic model (4.4.15), becomes

$$M_{\mathrm{c}} = \dot{m}_{2\mathrm{c}}\, \frac{c_{\mathrm{p}}\, T_1^*}{\omega_{\mathrm{tc}}\, \eta_{\mathrm{c,is}}} \left(\left(\frac{p_2}{p_1}\right)^{\frac{\kappa-1}{\kappa}} - 1\right) \tag{8.5.13}$$

The isentropic efficiency of the compressor depends on the mass flow rate and rotor speed

$$\eta_{\mathrm{c,is}} = f(\dot{m}_{2\mathrm{c}}, \omega_{\mathrm{tc}}) \tag{8.5.14}$$

and is, based on measurements, represented as a map. The mass flow rate $\dot{m}_{2\mathrm{c}} = \dot{m}_{\mathrm{c}}$ is usually obtained from the manufacturer and represented also in a map

$$\dot{m}_{\mathrm{c}} = f\left(\tfrac{p_2}{p_1}, \omega_{\mathrm{tc}}\right) \tag{8.5.15}$$

or, if corrected variables are used, according to (4.4.17). Instead of operating with parametric models of a map, the polynomials of (4.4.19) and (4.4.20) can be used as approximations.

The models for the coolers are not included in Fig. 8.5.4. The temperatures $T_{2\mathrm{ic}} = f(\dot{m}_{2\mathrm{c}}, T_{2\mathrm{c}}, T_{\mathrm{a}})$ of the charge-air cooler and $T_{\mathrm{egr}} = (\dot{m}_{\mathrm{egr}}, T_3, T_1)$ can be based on the models for heat exchangers described in Sect. 4.6, e.g. by the lumped parameter model of Sect. 4.6.3 or the distributed parameter model of Sect. 4.6.4.

Simplified stationary models for the charge-air cooler (intercooler) are

$$T_{2\text{ic}} = T_{2\text{c}} - \epsilon_{\text{ic}}(T_{2\text{c}} - T_1) \tag{8.5.16}$$

and the EGR cooler

$$T_{\text{egr}} = T_3 - \epsilon_{\text{egr}}(T_3 - T_{\text{c}}) \tag{8.5.17}$$

where the efficiencies $\epsilon_{\text{egr}} = f(\dot{m}_{\text{egr}}, n_{\text{eng}})$ and $\epsilon_{\text{ic}} = f(\dot{m}_{2\text{c}}, v_{\text{veh}})$ are described by maps.

The signal-flow chart of Fig. 8.5.4 depicts the many nonlinearities and interconnections between the process elements. It shows that the process for controlling the charging pressure p_2 and the air mass $\dot{m}_{2\text{c}} = \dot{m}_{\text{air}}$ is a strongly coupled, nonlinear, two-variable process and needs many parameters for modeling and simulation.

The dynamic behavior is caused by the turbocharger inertia J_{tc} and the air path volume V_2 and leads to a second order process. Hence, in order to obtain small time lags, the turbocharger ratio of inertia and the intake volume have to be small.

Some comparisons between the physical derived and experimentally identified models of a turbocharger show a relatively good agreement, as shown in Sect. 4.4.

With regard to the design of the control system for the air path with exhaust gas recirculation, a signal-flow scheme of the two-variable process can be used, see Fig. 8.5.5. It shows the direct paths from the manipulated variables to the controlled variables and the cross couplings in a so called P-canonical structure of multivariable controlled processes, see e.g. Isermann (1991).

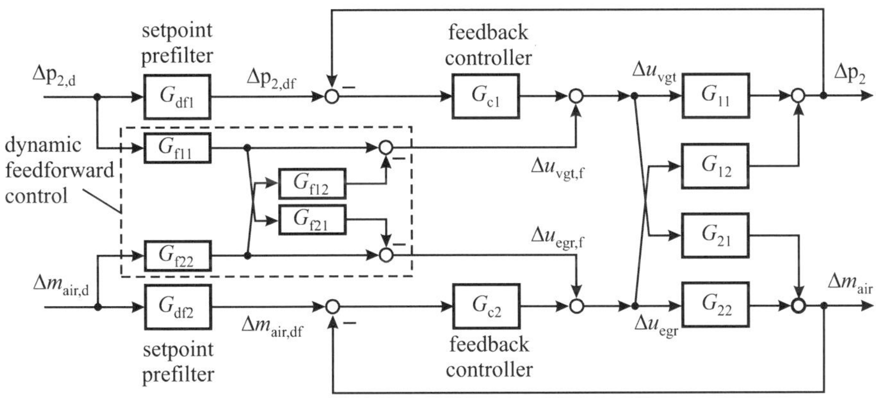

Fig. 8.5.5. Charging pressure and air mass flow rate control as a coupled linearized two-variable process in P-canonical structure and decentralized controllers with decoupling elements for feedforward control.

8.5.2 Experimental modeling (identification) of the air- and exhaust recirculation path

Models of the air- and exhaust manifold of turbocharged diesel engines based on physical laws as described by Nieuwstadt and Upadjyay (2000), Wellers and Elicker (2004) have the advantage of good interpretability. These "white-box" models are strongly nonlinear and the number of states varies between 3 and 8, depending on the level of detail. The controller design based on these nonlinear models is quite extensive. Therefore, simplified models, like linear MIMO processes of first order, Rückert et al (2004), or Hammerstein models of second order, Wellers and Elicker (2004), may be considered. These models are frequently valid for one operating point represented by the engine speed n and the engine torque M.

Contrary to the theoretical modeling, data-based modeling (*identification*) of the air- and exhaust manifold results in so called "*black-box*" models which just describe the input/output behavior. A compromise is to develop the model structure from physical modeling and to determine the unknown parameters from experiments with parameter estimation. Then "gray-box" or "semi-physical" models are obtained. This procedure is described in the following by applying local linear models in the frame of LOLIMOT identification, see Sect. 3.4 and 3.5. The advantage is that linear controller design methods can be utilized.

The operating point dependent local linear models describe the nonlinear behavior by weighted superposition.

a) Local linear model structures

In order to cope with the nonlinear behavior a local linear model structure is applied utilizing the LOLIMOT method described in Sects. 3.1.3e) and 3.2.3c). The LOLIMOT construction algorithm decomposes the input space in local subregions by using axis-orthogonal partitioning and inserts submodels step by step in more nonlinear areas. It uses an extended parameter varying model developed by Zimmerschied (2008). This model structure exhibits a delayed parameter changing and an offset with numerator dynamics which result in better identification results.

b) Identification of a diesel engine

In order to conduct the identification of the diesel engine, the EGR valve and the VGT are simultaneously excited with APRB signals (Amplitude modulated Pseudo Random Binary Signal) at different engine operating points ($\alpha_{\mathrm{ped},i}$, n_j). The APRB signal was specifically designed for the identification of nonlinear processes. It consists of multiple steps with different amplitudes and pulse intervals or clock time. It excites the process within a wide-frequency spectrum and equally distributed amplitudes, Isermann and Münchhof (2011). The amplitudes are fitted to each engine operating point in order to avoid the violation of engine limits, like the maximum combustion pressure. Choosing a minimum pulse time of 0.8 s for the APRB signals, a sampling time of $T_0 = 0.01\,\mathrm{s}$, a total measurement time of about 60 min is required for the 25 operating points. A minimum pulse time of 0.8 s is selected, since the dominant time constants of the air and exhaust manifold relevant for the control system are about of this size. An extract of the excitation signals is illustrated in Fig. 8.5.6.

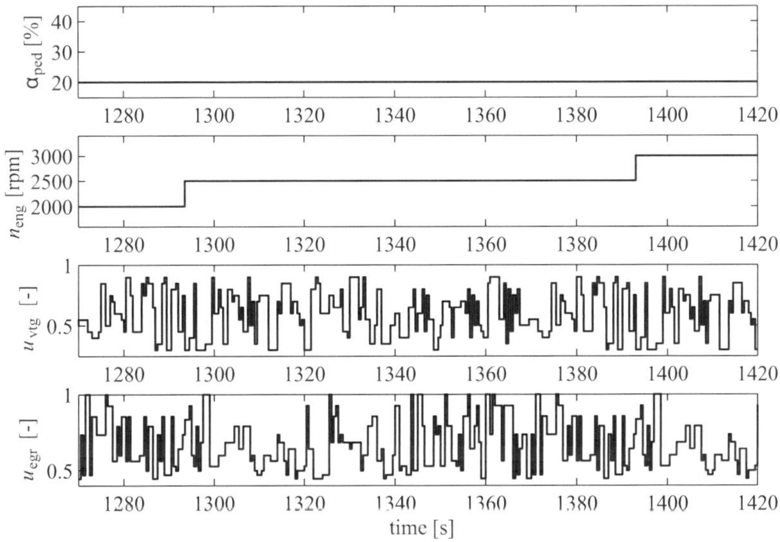

Fig. 8.5.6. Extract of the identification signals. The EGR valve and VGT are simultaneously excited with APRB signals, whereas the operation point (α_{ped}, n) is changed stepwise.

The engine measurements are conducted at a dynamic engine test bench with a common-rail diesel engine (OPEL/Fiat JTD 1.9L, EDC16 control unit), as depicted in Fig. 8.5.1. The electric EGR valve and the pneumatically actuated VGT are equipped with position sensors. A position controller is implemented for the EGR valve. Both actuators are controlled by a rapid control prototyping system. The controllers are implemented in MATLAB/SIMULINK.

A PDT_2 structure is chosen both for the model of the intake manifold pressure and the air mass. The number of local models is set to $L = 20$. The model inputs and the z-regressors are the engine speed n, the pedal position α_{ped}, the EGR valve position s_{egr} and the VGT position s_{vgt}. In order to enable a global NOE estimation (nonlinear output error), the number of measurement points is reduced by carrying out the identification with a sampling time $T_0 = 0.02$ s. Since the controller shall operate with a sampling time $T_0 = 0.01$ s, the identified models are afterwards converted to this sampling time $T_0 = 0.02$ s, Nelles (2001).

The parameter estimation is conducted with the extended local linear model structure proposed by Zimmerschied (2008). Identification studies with a simulated physical diesel engine model according to (8.5.5) to (8.5.17) show that the usage of the extended model structure significantly increases the model quality. Furthermore, it can be shown that only the parameters of the extended structure match the linearized process dynamics, Zimmerschied (2008). The identified models are obtained as discrete time models. They correspond to a continuous-time model

$$\begin{pmatrix} \Delta p_2 \\ \Delta m_{\mathrm{air}} \end{pmatrix} = \begin{pmatrix} \frac{K_{11}(T_{\mathrm{D}11}s+1)}{(T_{11}s+1)(T_{12}s+1)(T_{\mathrm{A}1}s+1)} & \frac{K_{12}(T_{\mathrm{D}12}s+1)}{(T_{11}s+1)(T_{12}s+1)(T_{\mathrm{A}2}s+1)} \\ \frac{K_{21}(T_{\mathrm{D}21}s+1)}{(T_{21}s+1)(T_{22}s+1)(T_{\mathrm{A}1}s+1)} & \frac{K_{22}(T_{\mathrm{D}22}s+1)}{(T_{21}s+1)(T_{22}s+1)(T_{\mathrm{A}2}s+1)} \end{pmatrix} \begin{pmatrix} \Delta u_{\mathrm{vgt}} \\ \Delta u_{\mathrm{egr}} \end{pmatrix} . \tag{8.5.18}$$

with $T_{\mathrm{A}1}, T_{\mathrm{A}2}$ as actuator time constants.

Figure 8.5.7 shows the gains of this two-input two-output process. The gain $K_{11} = \Delta p_2/\Delta u_{\mathrm{vgt}}$ for the charging process indicates a strong dependence on the operation point (load and speed) in a range of 1:14. The gain $K_{22} = \Delta m_{\mathrm{air}}/\Delta u_{\mathrm{egr}}$ has also a dependence on the operation point, but with a smaller ratio of 1:1.5. Also the crosscoupling gains K_{12} and K_{21} vary strongly with the operation point. The gain $K_{21} = \Delta m_{\mathrm{air}}/\Delta u_{\mathrm{vgt}}$ even changes the sign. It is negative for low loads because the EGR valve is wide open and becomes positive if the EGR valves closes for high loads. Also the time constants depend strongly on the operation point. The dominant time constant T_{12} for the charging pressure is in the range of 0.7 to 1.4 s, and that of the air mass in a range of 0.2 to 0.8 s. The crosscoupling path G_{12} between Δu_{egr} and Δp_2 shows non-minimum-phase behavior.

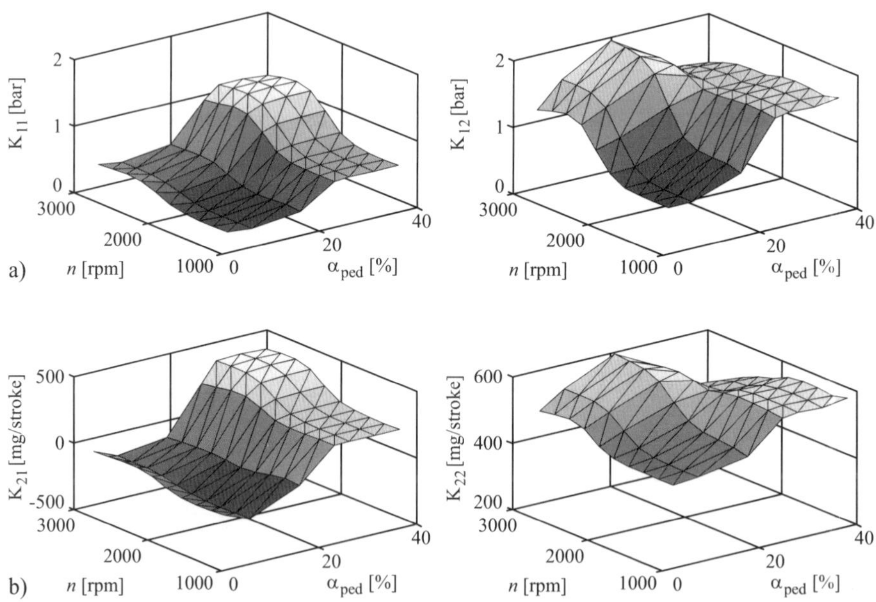

Fig. 8.5.7. Gains K_{ij} of the airpath and exhaust recirculation process: a) charging pressure p_2 as output and inputs: $K_{11} = \Delta p_2/\Delta u_{\mathrm{vgt}}$ [bar/−], $K_{12} = \Delta p_2/\Delta u_{\mathrm{egr}}$ [bar/−], b) air mass m_{air} as output and inputs: $K_{22} = \Delta m_{\mathrm{air}}/\Delta u_{\mathrm{egr}}$ [mg/stroke/−], $K_{21} = \Delta m_{\mathrm{air}}/\Delta u_{\mathrm{vgt}}$ [mg/stroke/−]. The gains are related to the dimensionless relative changes of the actuator positions $u_{\mathrm{vgt}} = 0 \ldots 1$ and $u_{\mathrm{egr}} = 0 \ldots 1$, Pfeil (2011).

An extract of the generalization data is depicted in Fig. 8.5.8. The excitation possesses the same properties as the training data. Generally, the simulation data fit the measured data well. The model errors of the intake manifold pressure are slightly higher than those of the air mass. This leads back to hysteresis effects of the nozzle guide vane of the VGT which are not captured by the model.

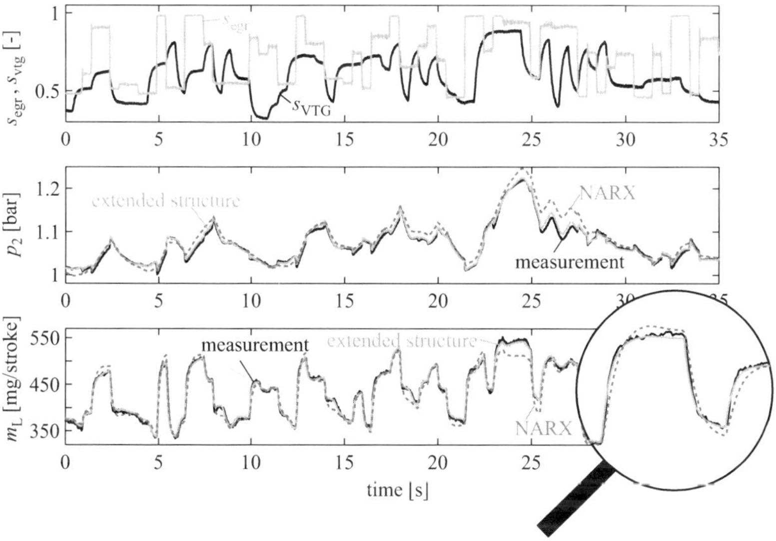

Fig. 8.5.8. Extract of the generalization data for the extended local linear model structure. (NARX stands for Nonlinear AutoRegressive with eXogenous input and is described in Zimmerschied (2008) and Sect. 8.7.2).

Figure 8.5.9 depicts measured and simulated EGR and VGT step responses at the constant engine operating point $n = 2000$ rpm and $\alpha_{\text{ped}} = 20\%$. The graph visualizes the plant dynamics and allows a comparison with the model dynamics. Figure 8.5.9a) shows VGT step responses for three different EGR valve positions and, vice versa, Fig. 8.5.9b) EGR step responses for three different VGT positions. The electrical EGR actuator is significantly faster than the pneumatic VGT actuator. Since the VGT actuator is operated in open loop (no position control), there is a steady-state error between the VGT control signal and the VGT position. By opening the EGR valve, the EGR mass flow is almost instantaneously increased. Hence, the intake manifold is filled and the charging pressure is raised. At the same time the turbine mass flow and the turbine power is reduced and, as a result, also the compressor mass flow and the charging pressure is decreased. However, this decrease is delayed due to the inertia of the turbocharger. In summary, this leads to a non-minimum phase behavior between the EGR valve position and the intake manifold pressure. A comparison between the simulated and measured step responses shows that the models describe the process dynamics very well. The identified models correctly reproduce the non-minimum phase behavior and the sign reversal of the process gain between the VGT position and the air mass caused by different EGR positions. The settling time T_{95} for the intake manifold pressure p_2 after step changes of the VGT actuator u_{vgt} is about 2 to 3 s for the considered operation point and that of the air mass m_{air} with EGR actuator about 0.2 s.

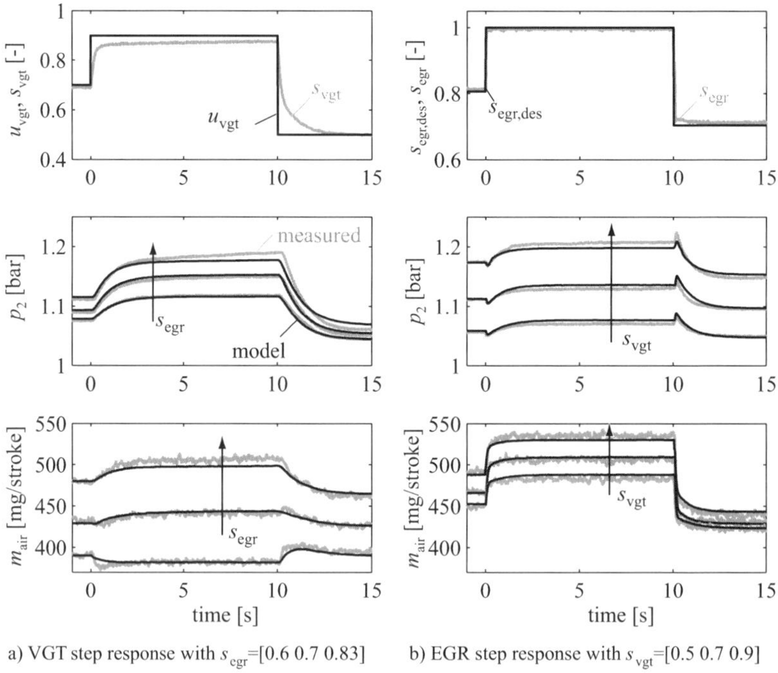

Fig. 8.5.9. VGT and EGR step responses of the identified local linear models (black) and in comparison the measured step responses (gray) at the engine operating point $n = 2000\,\mathrm{rpm}$ and $\alpha_{\mathrm{ped}} = 20\,\%$.

The identified models describe the behavior between the actuator positions and the controlled variables. They include the models of the actuators which can be approximated by first order lag elements. The identified models for the intake manifold pressure and the air mass according to the extended model structure, Zimmerschied (2008), can be transformed into a state-space representation in observer canonical form

$$\mathbf{x}(k+1) = \mathbf{A}(\mathbf{z})\mathbf{x}(k) + \mathbf{B}(\mathbf{z})\mathbf{u}(k) + \mathbf{B}_{\mathrm{d}}(\mathbf{z})\mathbf{d}(k) \tag{8.5.19a}$$

$$\mathbf{y}(k) = \mathbf{C}(\mathbf{z})\mathbf{x}(k). \tag{8.5.19b}$$

In this equation, the vector $\mathbf{u} = [u_{\mathrm{vgt}}, u_{\mathrm{egr}}]^{\mathrm{T}}$ denotes the manipulated variables and $\mathbf{y} = [p_2, m_{\mathrm{air}}]^{\mathrm{T}}$ the controlled variables. The engine speed, the pedal position and the offset are combined into the disturbance vector $\mathbf{d} = [n, \alpha_{\mathrm{ped}}, 1]^{\mathrm{T}}$.

The operating point $\mathbf{z}$ of the model contains the engine speed, the pedal position, the EGR valve position and the VGT position ($\mathbf{z} = [n, \alpha_{\text{ped}}, s_{\text{egr}}, s_{\text{vgt}}]$). A local linearization for a given engine operating point (n, α_{ped}) is obtained by first calculating the corresponding steady-state operating point $\mathbf{z}_{\text{ss}}$ and inserting it into (8.5.19). As a result, the linearized and into continuous time converted transfer behavior between the input signals and the controlled variables is given by (8.5.18). The parameters T_{A1} and T_{A2} represent the time constants of the actuators. The other parameters characterize the behavior of the intake and exhaust system including the turbocharger. (Even though the model is in discrete time, for simplicity it is shown here in continuous time). If the model parameters are plotted versus the engine operating points, smooth curves are obtained, which visualize the nonlinearities of the process as already shown in Fig. 8.5.7.

An analysis of the local MIMO submodels shows, that each model possesses two invariant zeros ($z_1 = -1/T_{\text{IN1}}$, $z_2 = -1/T_{\text{IN2}}$). For most submodels one of them is unstable, that is, the models exhibit a non-minimum phase behavior. (For MIMO systems there is no simple relationship between the invariant zeros and the zeros of the corresponding transfer functions G_{ij}. The invariant zeros are calculated by solving the generalized eigenvalue problem). In the next section it will be shown that this system property has a significant impact on the controller design.

8.5.3 EGR/VGT control

Figure 8.5.10 shows the control structure of a multi-variable process, which comprises a feedback and feedforward controller. Using this "‘two-degree-of-freedom"’ structure, the response to setpoint changes and disturbances can be provided independently. Furthermore, there is the possibility to deactivate the feedback controller, e.g. in the case of a sensor failure or at idle engine speed.

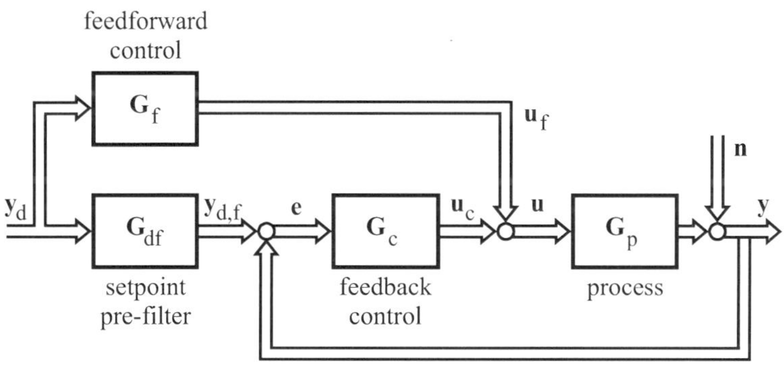

Fig. 8.5.10. Feedforward and feedback control structure for MIMO processes (Two degrees of freedom control).

The aim of controlling the intake and exhaust manifold consists of the precise regulation of the amount and composition of the cylinder charge and, as a result, the control of the exhaust emissions. Usually the air-mass flow and the charging pressure are measured. As an alternative to the air-mass, the EGR mass flow or the NO_x concentration can be measured as well. In this section, the charging pressure and the air mass are chosen as the controlled variables. Nevertheless, the proposed control design methods can also be applied to other controlled variables like the EGR rate, Kohlhase and Isermann (2009). In the following, first the design of a feedforward controller is discussed and then the design of the feedback controller.

a) Feedforward control

State of the art for the feedforward control of combustion engines are *static look-up tables*, see Sect. 6.5. The manipulated variable u_{FF} is calculated by a look-up table and is subsequently corrected based on the environmental conditions. As an option, the resulting signal is filtered. The parameters of the filter are stored in look-up tables as a function of the operating point. The filter can, for instance, be a PDT_1 element. This control structure based on simplified look-up tables without any environmental corrections is shown in Fig. 8.5.11a). The desired values for the feedback control are calculated in the same manner, see Fig. 8.5.11b). The desired values can be filtered with either low-pass or high-pass filters. For the feedback control of the charging pressure, usually a first or second order lag element is utilized, which describes the open-loop plant behavior. This approach ensures that the control deviation remains small. The feedback controller only needs to correct the remaining deviations of the feedforward controller. Without a filtering of the desired value, the feedforward and feedback controller are both active, which leads to an overshoot of the controlled variable. A characteristic of the control structure in Fig. 8.5.11 is that the feedforward control variables u_{ff} are calculated independently of the desired values.

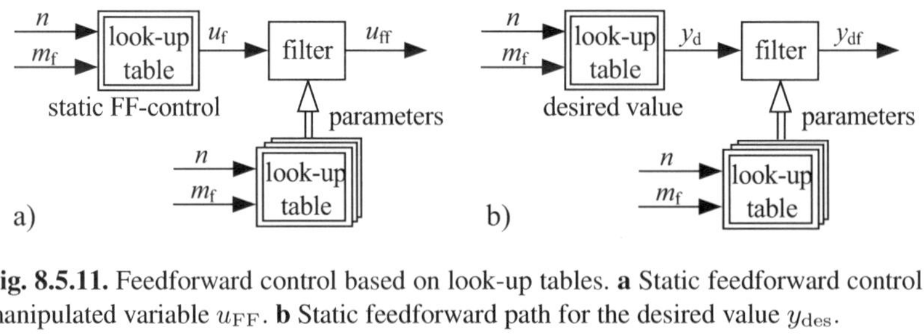

Fig. 8.5.11. Feedforward control based on look-up tables. **a** Static feedforward control of the manipulated variable u_{FF}. **b** Static feedforward path for the desired value y_{des}.

Model-based dynamic feedforward control

The *static feedforward control* with look-up tables does not take the process dynamics into account. An improvement is therefore obtained by *dynamic feedforward control*. A dynamic model-based feedforward control structure calculates the control variable u_{ff} based on a process model and the desired value y_{des}. Figures 6.3.1

and 6.3.2 depict two possible control structures for linear systems with input saturations. The approach shown in Fig. 6.3.1a) can be based on a direct inversion of the process model and a specification of the transfer function G_{rf} of the feedforward controller (6.3.11). Herein, this approach is referred to as the classical feedforward control structure. The transfer function G_{rf} needs to be chosen so that the transfer function of the feedforward controller

$$G_{\mathrm{cwu}}(s) = G_{\mathrm{p}}^{-1}(s) G_{\mathrm{rf}}(s) \tag{8.5.20}$$

is realizable, see Sect. 6.3.

Instead of explicitly inverting the model, the inversion can also be done implicitly by applying feedback control to the plant model. Figure 6.3.2b) depicts this control structure that consists of a process model, a constant state-space controller $\mathbf{k}^T$ and a constant prefilter $\mathbf{f}^T$. This control structure is also called a *reference shaping filter* Hippe (2006). The inputs to the plant model are limited by a model of the input saturation, compare the simulations shown in Fig. 6.3.3.

Local linear feedforward control with partial input-output decoupling

In order to apply the reference shaping filter to the nonlinear intake and exhaust manifold, a controller matrix $\mathbf{K}_i$ is designed offline for each local submodel. Following the local linear approach, the state-space controller is then applied in each time step by superimposing the local controller matrices weighted by the activation functions of the local submodels. Therefore, the feedforward control signal (without considering the input saturation) is given by:

$$\mathbf{u}_{\mathrm{f}}(k) = -\sum_{i=1}^{L} \mathbf{K}_i \Phi_i(\mathbf{z}(k)) \mathbf{x}_{\mathrm{M}}(k) + \mathbf{v}(k) = -\mathbf{K}(\mathbf{z}) \mathbf{x}_{\mathrm{M}}(k) + \mathbf{v}(k). \tag{8.5.21}$$

The vector $\mathbf{v}(k)$ is calculated such that the feedforward controller has no steady-state error with respect to the plant model. To meet this requirement, the steady state $\mathbf{x}_{\mathrm{ss}}$ must satisfy the equation

$$\mathbf{C}(\mathbf{z}) \mathbf{x}_{\mathrm{ss}}(k) = \mathbf{y}_{\mathrm{des}}(k). \tag{8.5.22}$$

If the steady-state of the plant model is calculated from (8.5.19), then (with (8.5.21) and (8.5.22) the vector $\mathbf{v}$ is given by

$$\begin{aligned} \mathbf{v}(k) = {} & \underbrace{\big(\mathbf{C}(\mathbf{z})\mathbf{M}(\mathbf{z})\mathbf{B}(\mathbf{z})\big)^{-1}}_{\mathbf{F}(\mathbf{z})} \mathbf{y}_{\mathrm{des}}(k) \\ & - \underbrace{\big(\mathbf{C}(\mathbf{z})\mathbf{M}(\mathbf{z})\mathbf{B}(\mathbf{z})\big)^{-1} \mathbf{C}(\mathbf{z})\mathbf{M}(\mathbf{z})\mathbf{B}_{\mathrm{d}}(\mathbf{z})}_{\mathbf{F}_{\mathrm{d}}(\mathbf{z})} \, \mathbf{d}(k). \end{aligned} \tag{8.5.23}$$

with

$$\mathbf{M}(\mathbf{z}) = \big(\mathbf{I} - \mathbf{A}(\mathbf{z}) + \mathbf{B}(\mathbf{z})\mathbf{K}(\mathbf{z})\big)^{-1}. \tag{8.5.24}$$

The matrices $\mathbf{F}(\mathbf{z})$ and $\mathbf{F}_{\mathrm{d}}(\mathbf{z})$ of this reference shaping filter can be interpreted as setpoint and disturbance filters. The vector $\mathbf{d}$ represents the operating point or disturbance, see (8.5.19). In order to avoid a steady-state error, these matrices cannot be calculated via a weighted superposition of local matrices. Instead, they need to be determined in each time step according to (8.5.23) and (8.5.24). The most important step during the feedforward controller design consists of the calculation of the local controller matrices $\mathbf{K}_i$. In principle, any controller design method can be applied. Nevertheless, an input-output decoupling design is particularly suitable. If constant controller matrices are used, a complete decoupling is only possible for minimal phase systems. For non-minimal phase systems, a so-called partial input-output decoupling can be achieved, Lohmann (1991a), Lohmann (1991b). If this method is applied to the control of the intake and exhaust manifold, the local controller matrices are calculated such that the behavior between the desired values and the controlled variables

$$\begin{pmatrix} \Delta p_2 \\ \Delta m_{\text{Air}} \end{pmatrix} = \mathbf{G}_{\text{w}} \cdot \begin{pmatrix} \Delta p_{2,\text{des}} \\ \Delta m_{\text{Air,des}} \end{pmatrix} \tag{8.5.25}$$

is given by the discrete-time equivalent of the continuous transfer matrix

$$\mathbf{G}_{\text{w}} = \begin{pmatrix} \frac{(T_{\text{IN1}}s+1)}{(T_{11}s+1)(\alpha_1 T_{12}s+1)(\alpha_1 T_{\text{A1}}s+1)} \\ 0 \end{pmatrix} \begin{pmatrix} \frac{s f_{12}(s,s^2,a_{21},a_{22})}{(T_{11}s+1)(\alpha_1 T_{12}s+1)(\alpha_1 T_{\text{A1}}s+1)(\alpha_2 T_{21}s+1)(\alpha_2 T_{\text{A2}}s+1)} \\ \frac{1}{(\alpha_2 T_{21}s+1)(\alpha_2 T_{\text{A2}}s+1)} \end{pmatrix} . \tag{8.5.26}$$

The decoupling is not complete, since there is a coupling between the desired air mass and the intake manifold pressure ($G_{\text{w12}} \neq 0$). The time constants of this transfer matrix correspond to those of the process model (8.5.18). The time constant T_{IN1} belongs to the unstable invariant zero of the plant model. It cannot be shifted by the state-space controller. The tuning parameters α_1 and α_2 are used to shift the dominant time constants of the process. The fast time constant of the intake manifold pressure model is so small, that it does not need to be shifted. The parameters a_{21} und a_{22} in the function $f_{12}(s, s^2, a_{21}, a_{22})$ significantly influence the input-output coupling. Since they can be arbitrarily chosen, they are used to minimize the coupling term G_{w12}. Detailed information about the controller design via partial decoupling can be found in Lohmann (1991a), Lohmann (1991b).

The advantages of the controller design via partial decoupling are a good setpoint behavior and a simple calibration. There are only two tuning parameters α_1 and α_2, that describe the factors by which the system is accelerated by the feedforward control.

Simplified dynamic feedforward control with SISO elements

Based on the p-canonical two-input two-output structure of Fig. 8.5.2 the feedforward control structure of Fig. 8.5.5 follows. For the charging pressure control the feedforward G_{f11} from the desired value $\Delta p_{2,\text{d}}$ (setpoint) acts directly on the manipulated variable u_{vgt}, by passing the feedback controller G_{c1}. A corresponding structure holds for the air charge control. In addition, the process cross coupling G_{12} is compensated by G_{f12} and that of G_{21} by G_{f21}. Then it yields for the feedforward part

$$\begin{bmatrix} \Delta u_{\mathrm{vgt,f}}(s) \\ \Delta u_{\mathrm{egr,f}}(s) \end{bmatrix} = \begin{bmatrix} G_{\mathrm{f11}}(s) & -G_{\mathrm{f22}}(s)G_{\mathrm{f12}}(s) \\ -G_{\mathrm{f11}}(s)G_{\mathrm{f21}}(s) & G_{\mathrm{f22}}(s) \end{bmatrix} \begin{bmatrix} \Delta p_{2,\mathrm{d}}(s) \\ \Delta \dot{m}_{\mathrm{air,d}}(s) \end{bmatrix} . \quad (8.5.27)$$

This structure has the advantage that a complete decoupling of the two-variable control loop can be obtained with

$$G_{\mathrm{f12}}(s) = \frac{G_{12}(s)}{G_{11}(s)} = \frac{K_{12}\,(T_{\mathrm{D12}}s+1)\,(T_{\mathrm{A1}}s+1)}{K_{11}\,(T_{\mathrm{D11}}s+1)\,(T_{\mathrm{A2}}s+1)} \quad (8.5.28)$$

$$G_{\mathrm{f21}}(s) = \frac{G_{21}(s)}{G_{22}(s)} = \frac{K_{21}\,(T_{\mathrm{D21}}s+1)\,(T_{\mathrm{A2}}s+1)}{K_{22}\,(T_{\mathrm{D22}}s+1)\,(T_{\mathrm{A1}}s+1)} \quad (8.5.29)$$

by using (8.5.18) and thus depends only on process transfer functions, Pfeil (2011). These coupling elements can be approximated by

$$\tilde{G}_{\mathrm{f12}}(s) = \frac{K_{12}}{K_{11}} \quad (8.5.30)$$

$$\tilde{G}_{\mathrm{f21}}(s) = \frac{K_{21}}{K_{22}\,(T_{\mathrm{A1}}-T_{\mathrm{A2}})\,s+1} . \quad (8.5.31)$$

The feedforward controllers of the main paths follow from

$$G_{\mathrm{f11}}(s) = \left(G_{\tilde{\mathrm{M}}1}\right)^{-1} G_{\mathrm{F1}} \quad (8.5.32)$$

$$G_{\mathrm{f22}}(s) = \left(G_{\tilde{\mathrm{M}}2}\right)^{-1} G_{\mathrm{F2}} \quad (8.5.33)$$

where, e.g. $G_{\tilde{\mathrm{M}}1}$ is the invertable stable model of $G_{\mathrm{M1}}(s) = G_{11}(s)$ and low-pass filters are

$$G_{\mathrm{F1}}(s) = \frac{1}{(\alpha_1 T_{12}s+1)\,(\alpha_1 T_{\mathrm{A1}}s+1)} \quad (8.5.34)$$

$$G_{\mathrm{F2}}(s) = \frac{1}{(\alpha_2 T_{21}s+1)\,(\alpha_2 T_{\mathrm{A2}}s+1)} . \quad (8.5.35)$$

Applying order reduction leads to, see Pfeil (2011),

$$G_{\mathrm{f11}}(s) = K_{\mathrm{f11}} \frac{T_{\mathrm{D,f11}}s+1}{T_{\mathrm{f11}}s+1} \quad (8.5.36)$$

$$G_{\mathrm{f22}}(s) = K_{\mathrm{f22}} \frac{T_{\mathrm{D,f22}}s+1}{T_{\mathrm{f22}}s+1} . \quad (8.5.37)$$

For the operating point $n = 2000\,\mathrm{rpm}$ and $\alpha_{\mathrm{ped}} = 20\,\%$ one obtains with $\alpha_1 = \alpha_2 = 0.6$, $T_{\mathrm{A1}} = 0.2\,\mathrm{s}$, $T_{\mathrm{A2}} = 0.05\,\mathrm{s}$

$$G_{\mathrm{f11}}(s) = 3.9 \frac{1}{\mathrm{bar}} \frac{0.87s+1}{0.48s+1} \quad (8.5.38)$$

$$G_{\mathrm{f22}}(s) = 1.53 10^{-3} \frac{1}{\mathrm{mg/stroke}} \frac{0.37s+1}{0.15s+1} \quad (8.5.39)$$

$$G_{\mathrm{f12}}(s) = 1.28 \quad (8.5.40)$$

$$G_{f21}(s) = -0.33 \frac{0.061s + 1}{0.23s + 1} \tag{8.5.41}$$

where the time constants are in seconds.

Hence, the multi-variable feedforward controller compensates approximately not only for the forward paths of the process but also the crosscoupling paths. Because the required control performance is then already reached to a considerable part by these feedforward actions, simple decentralized PI(D) controllers $G_{c1} = G_{c,air}$ and $G_{c2} = G_{c,p2}$ designed with a pole cancellation method are sufficient, according to Fig. 8.5.5 and the next subsection. The setpoint-prefilters are selected with $G_{df1} = G_{F1}$ and $G_{df2} = G_{F2}$.

The described procedure is valid for the local linear process models. The locally designed controller parameters are then stored in look-up tables depending on the operating point $[\alpha_{ped}, n]$, by using the identified LOLIMOT models. By this way an overall parameter-variable control results.

A further improvement of the feedforward control is obtained by taking the actuator restrictions (position saturation) into account, in using a feedback to the process model with the saturation characteristic instead of the process-model inversion.

b) Feedback controllers

The state of the art for the feedback control of the intake and exhaust path are gain-scheduled PID controllers Nieuwstadt and Upadjyay (2000). In addition to these decentralized controllers, multi-variable control concepts are published, e.g. model predictive control Rückert et al (2004), Wellers and Elicker (2004) or robust H_∞-control based on LPV models Jung (2003). However, since the reference shaping filter significantly improves the system behavior with regard to setpoint changes, the feedback controllers only need to correct small control errors. Therefore, a simple local linear, but operating point dependent feedback control is sufficient.

Design of gain-scheduled PI(D) controllers

The air mass is controlled via the EGR valve, the charging pressure via the VGT actuator, see Fig. 8.5.12. A reverse mapping of the control variables is not feasible, due to the sign reversal of the gain between the VGT control variable and the air mass, see Fig. 8.5.9a). A PID controller is used for controlling the charging pressure, a PI controller is sufficient for the air mass. The strong plant nonlinearities are taken into account by storing the controller parameters in look-up tables as a function of the engine operating point, see Fig. 8.5.12.

The parameterization of the PI(D) controllers are based on a local linearization of the local linear plant model at different engine operating points. Then, the linearized model is used to design the controllers by sequentially closing both control loops Isermann (1987). With this approach the controller design of the MIMO system is simplified to the design of two SISO control loops.

The SISO controllers are parameterized with the cancelation-compensation method that is based on the specification of the closed-loop transfer function, Isermann (1989). The resulting controllers exhibit a high dynamic order. Therefore, they need

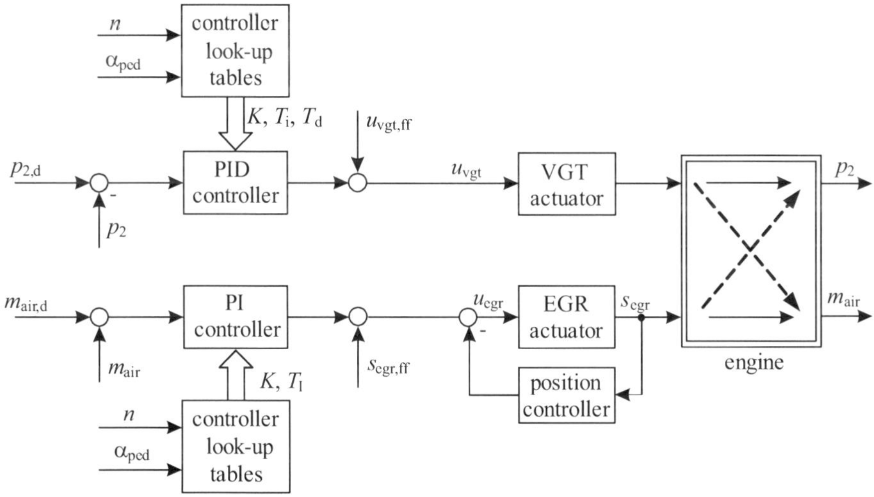

Fig. 8.5.12. Signal-flow diagram of the gain-scheduled PI(D) controllers and the process.

to be reduced to the PID structure. The utilized order reduction is based on a weighted LS approximation of the controller transfer function in the frequency domain. The frequency weighting is chosen so that the mean square difference between the output of the system controlled via the high order controller and the PID controller is minimized, Lee and Edgar (2004). Thus the controller design is based on algebraic equations and a linear LS estimation.

The calibration of the PI(D) controllers is equivalent to the reference shaping filter. It is essentially based on the specification of the parameters α_1 and α_2 of (8.5.26) that define the factors. The dominant time constants are shifted by the feedback controllers. Executing this procedure, smooth look-up tables are obtained for the controller parameters. The proportional controller gains are roughly inverse to the plant gains K_{11} and K_{22}.

Internal Model Control in state-space representation
In the previous section a two-degree-of-freedom structure was discussed, that includes a feedforward and feedback controller. A one-degree-of-freedom structure, like Internal Model Control (IMC), possesses advantages in terms of the calibration effort and the computational and memory consumption. Furthermore, IMC fulfills the requirement to allow a switching between feedback and feedforward control. There are different implementations of IMC. Figure 8.5.13a) shows the classical IMC structure with the filter $G_{\mathrm{f,b}}$ and the process model. The transfer function of the IMC filter is designed in the same manner as the classical feedforward controller. Therefore it is based on a direct inversion of the process model. Hence, this structure possesses similar properties with regard to input saturations.

In comparison, Fig. 8.5.13b) shows an IMC structure based on a state-space representation, which is also referred to as model-state feedback control, Wright and

Kravaris (2005). In this structure, the IMC filter is a reference shaping filter in state-space representation. Hence, the IMC filter consists of the controller matrix $\mathbf{k}^{\mathrm{T}}$, the prefilter F, the process model and a model of the input saturation. Since the input saturation is incorporated into the IMC filter, it possesses the same advantages as the reference shaping filter, compare Sect. 6.4.2.

The state-space IMC can be extended to local linear models in the same way as the reference shaping filter. Taking the disturbance vector $\mathbf{d}$ into account (compare (8.5.19)), a local linear IMC scheme in state-space representation is obtained, see Fig. 8.5.14 for the case of a MIMO system. The reference shaping filter avoids a steady-state error with respect to the process model. This property prevents a steady-state error of the IMC structure also with respect to the real process. A switching between feedback and feedforward control is achieved by disconnecting the feedback $\mathbf{y}(k) - \mathbf{y}_{\mathrm{m}}(k)$, see Fig. 8.5.14. The controller matrix $\mathbf{K}$ is designed based on a partial input-output-decoupling. Hence, the IMC structure differs from the reference shaping filter $G_{\mathrm{f,b}}$ only by the feedback signal $\mathbf{y}(k) - \mathbf{y}_{\mathrm{m}}(k)$.

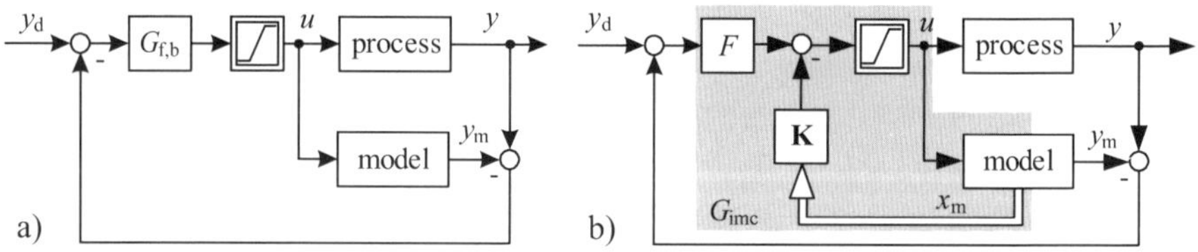

Fig. 8.5.13. Different Internal Model Control (IMC) structures. **a** classical IMC. **b** IMC in state-space representation.

Experimental results

The considered control structures have been tested on the engine test bench with the 1.9L CR diesel engine shown in Fig. 8.5.1. They are implemented on a dSPACE rapid-control-prototyping system in Simulink. The local linear state-space structures are implemented in S-functions that can generally be applied to local linear state-space models of arbitrary order.

The action of the reference shaping filter is depicted in Fig. 8.5.15 for the case of measured setpoint step-changes. Gain-scheduled PI(D) controllers are used to compensate steady-state errors. The desired values $p_{2\mathrm{des}}$ and $m_{\mathrm{air,des}}$ are inputs to the reference shaping filter. It outputs the filtered desired values $p_{2\mathrm{des,f}}$ and $m_{\mathrm{air,des,f}}$. Figure 8.5.15 a) and b) illustrate that the controlled variables follow their filtered setpoints $p_{2\mathrm{des,f}}$ and $m_{\mathrm{air,des,f}}$ very well. A comparison of the manipulated variables u_{vgt} and u_{egr} with the feedforward manipulated variables ($u_{\mathrm{vgt,FF}}$ and $u_{\mathrm{egr,FF}}$) shows that the feedback controllers only take little corrective actions.

The decoupling property of the reference shaping filter can be particularly observed during the setpoint changes of the air mass, see Fig. 8.5.15 b). Here, the reduction of the air mass is achieved by opening the EGR valve. The feedforward controller simultaneously closes the VGT in order to maintain the intake manifold pressure. The temporary increase of the intake manifold pressure caused by the open-

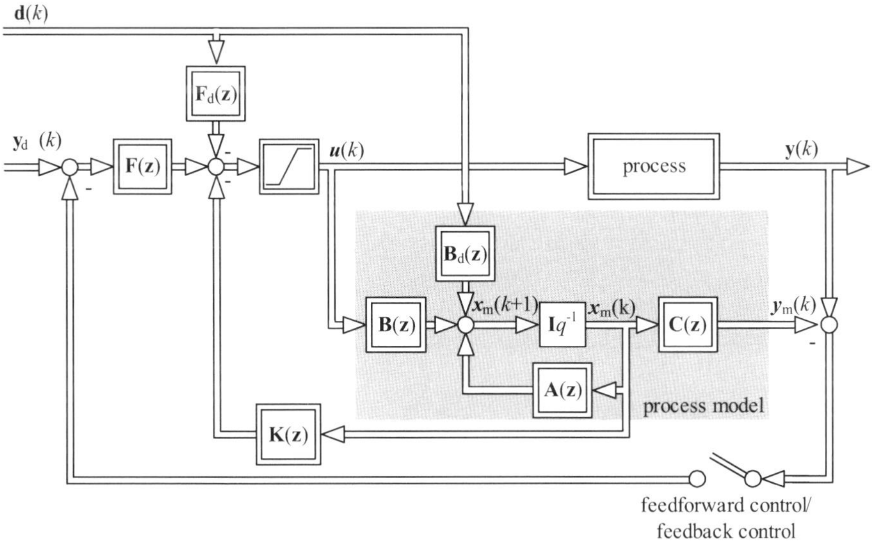

Fig. 8.5.14. Local linear IMC in state-space representation (MIMO).

ing of the EGR valve (non-minimal phase behavior between s_{egr} and p_2) is not fully compensated by the feedforward controller. This is due to the fact that the reference shaping filter is only designed to achieve a partial input-output decoupling ($G_{\mathrm{w12}} \neq 0$ in (8.5.26)). Even though the desired intake manifold pressure p_{2des} remains constant, its filtered value $p_{\mathrm{2des,f}}$ is adjusted to the non-minimal phase behavior of the plant. This property prevents the PID controller to open the VGT, which would lead to a drop of the intake manifold pressure after its temporary increase.

Figure 8.5.16 illustrates the impact of the feedforward control on the control performance. It compares the behavior of PI(D) controllers with and without a reference shaping filter as well as the performance of the IMC structure. All control schemes are stable. The PI(D) controllers without feedforward control exhibit a large and long lasting control deviation of the intake-manifold pressure for step-changes of the air mass setpoint. This deviation can be significantly reduced with the reference shaping filter. Additionally, the feedforward controller improves the response to negative setpoint changes of the air mass due to the immediate closing of the VGT and the increased pressure ratio over the EGR valve. Furthermore, the response to setpoint changes of the charging pressure is improved. In addition, Fig. 8.5.16 depicts that there are only minor differences between the performance of the PID controllers with reference shaping filter and the IMC scheme.

All control structures were successfully tested within the New European Driving Cycle (NEDC). In Fig. 8.5.17 the performance of the PI(D) controllers with static feedforward control is compared to the IMC scheme in the extra-urban part of the NEDC. (The constant sectors of the cycle are not shown in favor of a better illustration of the dynamic parts). During the dynamic operating conditions the IMC scheme exhibits a better control performance than the PID controllers. If the static feedforward controller is replaced by a reference shaping filter, the control performance is very similar to the IMC structure (not shown in Fig. 8.5.17). Consequently, the proposed control structures exhibit a good control performance within the entire engine operating range.

In comparison to the control structures based on look-up tables, both the reference shaping filter and the IMC scheme require more memory and computational resources. Nevertheless, these structures are suited for production control units with limited resources. In comparison to other proposed control structures in the literature (e.g. nonlinear model predictive control) the computational effort of the local linear structures are rather low.

Another approach to the air charge and charging pressure control is published in Wang et al (2011). Identification of linear models is performed by using PRBS and feedforward control is realized with steady-state maps. A decoupling gain-scheduled PID control structure with linear models and gain-phase-margin design specification then shows the (moderate) effect of the decoupling control. It concludes that the air-path control quality is important to control the emissions.

Other multi-variable control concepts, where the emphasis is more on feedback control, are using predictive control, Rückert et al (2004), Wellers and Elicker (2004) or robust H_∞-control, Jung (2003).

8.5.4 Summary

The identification of the dynamic behavior of the air and exhaust manifold of turbocharged diesel engines is performed with local linear models. The quality of the identification can be significantly increased by using an extended local linear model. The parameter estimation is carried out in output error configuration, since the model cannot be transformed into a pseudo-linear form. The estimated model in I/O representation is transformed into a local linear state-space representation in observer canonical form.

A local linear reference shaping filter in state-space representation is first proposed for the feedforward controller. In the presence of input saturations, this structure possesses advantages over a classical feedforward scheme. The reference shaping filter is designed based on a partial input-output decoupling. The feedback control is implemented with gain-scheduled PI(D) controllers. They are designed via a sequential closing of the feedback loops and parameterized with the compensation method. As an alternative to this two-degree-of-freedom-structure, a local linear IMC scheme in state-space representation is proposed.

After software implementation, the described structures feature a simple and intuitive calibration. Results on the test bench show that both the reference shaping filter and the IMC scheme improve the control performance of conventional control structure of combustion engines (static feedforward control with gain-scheduled PI(D) controllers). For more details see Pfeil (2011).

The combined dynamic feedforward and feedback structure can be approximated with simplified single-input single output (SISO) elements according to Fig. 8.5.5

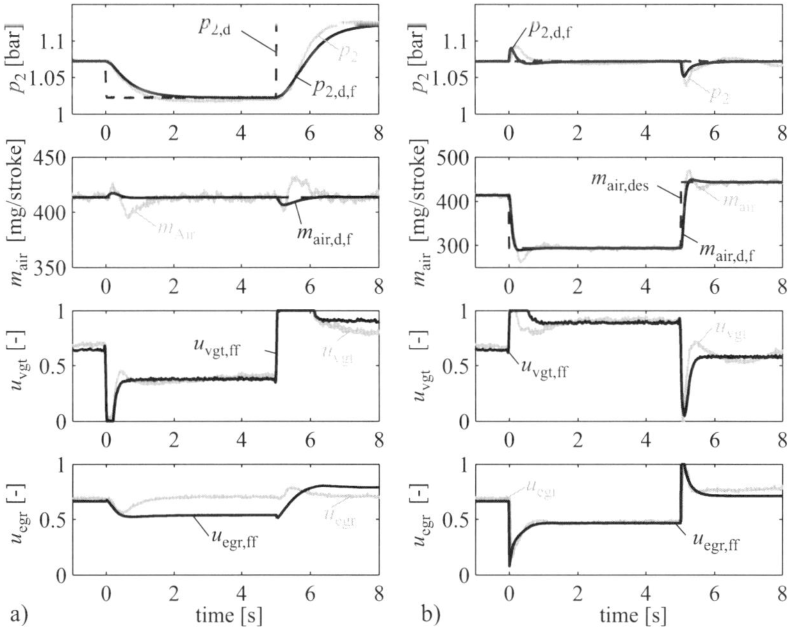

Fig. 8.5.15. Control performance of the local linear reference shaping filter for feedforward control in combination with gain-scheduled PI(D) controller at the engine test bench (engine operating point: $n = 2000\,\text{rpm}$, $\alpha_{\text{ped}} = 20\,\%$.). **a** setpoint step-change of the charging pressure. **b** setpoint step-change of the air mass. Black: feedforward control Light gray: feedback corrections.

and transfer elements (8.5.30) to (8.5.37). This leads to an especially transparent control system. However, it has not been tested experimentally.

As a general conclusion it is stated that a model-based charging pressure and air-charge control, designed to work in all operating points for normal operating mode, is a key element to reach a good control performance over the whole operating range. By putting the emphasis on a precise feedforward control relatively simple standard feedback controllers can be applied. The results also show that it is important to design the characteristics of the VGT turbocharger, the air path and exhaust recirculation path and its actuators such that a good stationary and dynamic behavior is reached to enable a good control performance. Precise dynamic control of the air mass and charging pressure is a pre-requisite to satisfying the requirements for low fuel consumption, low emissions and good driveability with regard to legislative driving cycles and real-world driving.

8.6 Combustion-pressure-based heat release control

If the combustion pressure in each cylinder is measured, different features can be calculated during one stroke. To control well-balanced combustions of each cylinder,

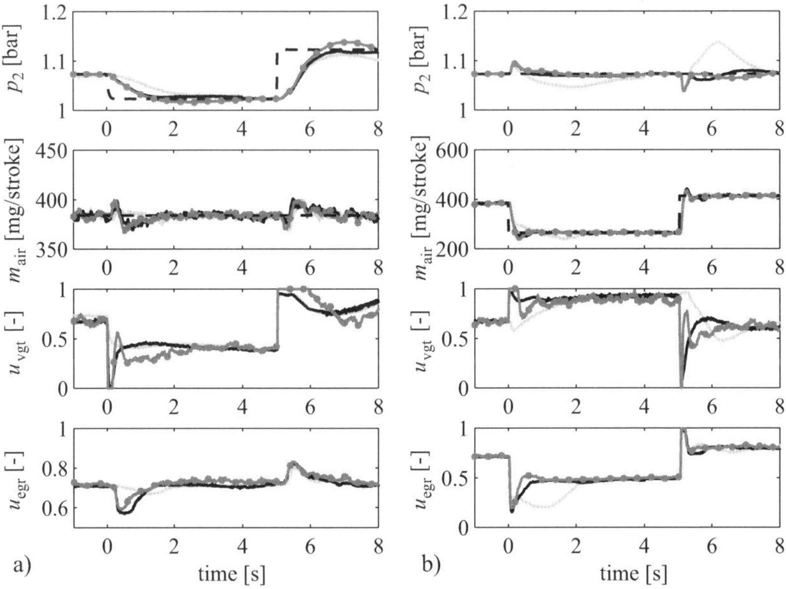

Fig. 8.5.16. Performance of different control structures measured on the engine test bench for the engine operation point $n = 2000\,\text{rpm}$ and $\alpha_{\text{ped}} = 20$. **a** setpoint step-change of the charging pressure. **b** setpoint step-change of the air mass. Light gray: PI(D) controller without feedforward control. Dark gray dash-poted: PI(D) controller with feedforward reference shaping filter. Black: IMC in state-space representation.

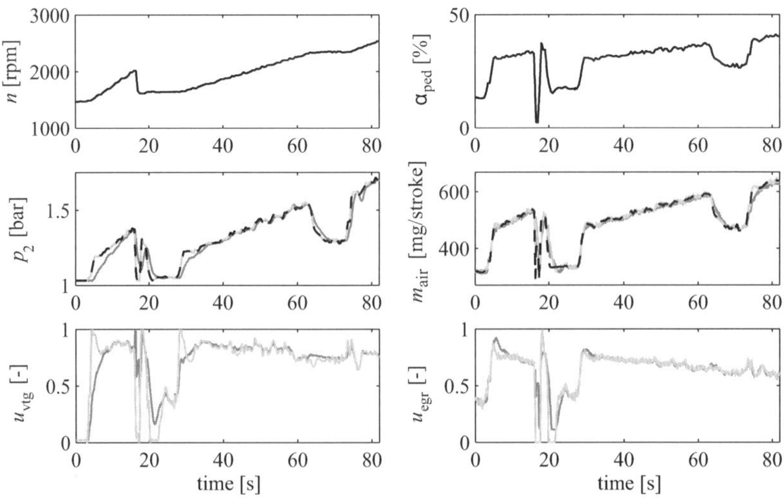

Fig. 8.5.17. Control performance during the extra-urban part of the NEDC measured on the engine test bench (constant parts of the cycle are not shown in favor of a better illustration of the dynamic parts). Dark gray: PI(D) controllers with static feedforward control. Light gray: IMC in state-space representation. Black dashed: desired values.

for example the mean indicated pressure p_{mi}, the pressure gradient, or the net heat release $Q_{\mathrm{h}}(\varphi)$ can be controlled by using different manipulated variables.

The possibilities of using the combustion pressure for the control of diesel engines was investigated in several research projects. Jeschke (2002) focused on the control of the mean indicated pressure by manipulating the injection duration and the beginning of combustion. The use of combustion pressure features for the determination of the cylinder gas mass allows the control of the cylinder charge and exhaust gas recirculation rate, Larink (2005). A control of the gas mass and the oxygen content in the combustion chamber was developed by Knippschild (2011). Together with an electromechanic valve train the raw emissions could be reduced significantly for both, conventional combustion and partially homogeneous combustion. A combustion control concept with cylinder pressure measurement and low-pressure exhaust recirculation for a series diesel engine is described by Hadler et al (2008). In all cases the NO_x and particulate raw emissions and their variances could be reduced significantly.

In the following the development of an adaptive control system is described for the net heat release, and later also for other variables.

The *net heat release* $Q_{\mathrm{h}}(\varphi)$ is a suitable measure for the converted energy. Therefore as a control variable the *combustion center crankshaft angle* φ_{Q50} for 50% released heat is a suitable choice. The incremental heat release is calculated by, see (4.2.67)

$$\Delta Q_{\mathrm{h}}(\varphi_i) = \frac{c_{\mathrm{v}}}{R} V_{\mathrm{cyl}}(\varphi_i) \Delta p_{\mathrm{cyl}}(\varphi_i) + \left(1 + \frac{c_{\mathrm{v}}}{R}\right) p_{\mathrm{cyl}}(\varphi_i) \Delta V_{\mathrm{cyl}}(\varphi_i) \tag{8.6.1}$$

for each cylinder i and its summation yields the integrated heat release

$$Q_{\mathrm{h}\Sigma i} = \sum_{\varphi=\varphi_{\mathrm{soi}}}^{\varphi_{\mathrm{E0}}} \Delta Q_{\mathrm{h}}(\varphi_i) \tag{8.6.2}$$

The 50% heat release or MFB50 is then given by

$$Q_{\mathrm{h}i}(\varphi_{\mathrm{Q50}}) = 0.5 \left(\max Q_{\mathrm{h}\Sigma i} - \min Q_{\mathrm{h}\Sigma i}\right) + \min Q_{\mathrm{h}\Sigma i} \tag{8.6.3}$$

and by comparing with (8.6.2) the crank angle φ_{Q50} is determined. The combustion pressure $p_{\mathrm{cyl}}(\varphi_i)$ and the geometrical determination of the the volume displacement $V_{\mathrm{cyl}}(\varphi_i)$ may be sampled with $\Delta\varphi_i = 1 \ldots 6°\mathrm{CS}$. The combustion center φ_{Q50} is controlled with the start of injection φ_{soi} as manipulated variable. To investigate the transfer behavior, Fig. 8.6.1 depicts step responses of all cylinders. It shows that the controlled process has proportional behavior with a dead time of about one cycle, Fig. 8.6.1a). A proportional behavior also holds between the main injection mass and the indicated pressure. Also a proportional behavior can be observed between a change of *pilot* injection quantity and maximum pressure gradient $d\Delta p_{\mathrm{max}}/d\varphi$. Hence, for small changes these processes can be presented by the transfer function

$$G_{\varphi_{\mathrm{soi}}/\varphi_{\mathrm{Q50}}}(s) = K_{\varphi\varphi} e^{-T_{\mathrm{t}} s} \tag{8.6.4}$$

where the parameters depend on the operation point.

As the combustion center values oscillate from cycle to cycle, the same control structure as described in Sect. 7.5 for ignition control of gasoline engines is applied. The resulting adaptive control system is shown in Fig. 8.6.2. It consists of a regular look-up table for proportional feedforward control and an adaptive look-up table for feedback control, which corrects the output of the fixed look-up table and stores the correction values in dependence on the operating point. The control error e is like (7.5.7), used for the normalized least squares estimation of the weights of the adaptive maps, corresponding to (7.5.8). The adaptive look-up table finally acts as an integral action controller, Kohlhase and Isermann (2009), Kohlhase (2011). The reference value for the combustion center $\varphi_{\mathrm{Q50,d}}$ is usually preselected in the range between 6 and $10°$CS after TDC. The described control system can be used for both heterogeneous and homogeneous combustions.

Experimental results of the adaptive control system are shown in Fig. 8.6.3 for three different control variables where the loops are closed sequentially. First the φ_{Q50}-control is closed for constant driver pedal position, then the p_{mi}-control and then $d\Delta p_{\mathrm{max}}/d\varphi$-control. In all cases a fast settling to the reference values can be observed, with partially significant differences of the manipulated variables for the single cylinders. The step changes of the pedal indicate, that after the second change in the same direction the control deviations are considerably smaller, due to the adaptation or "learning" effect.

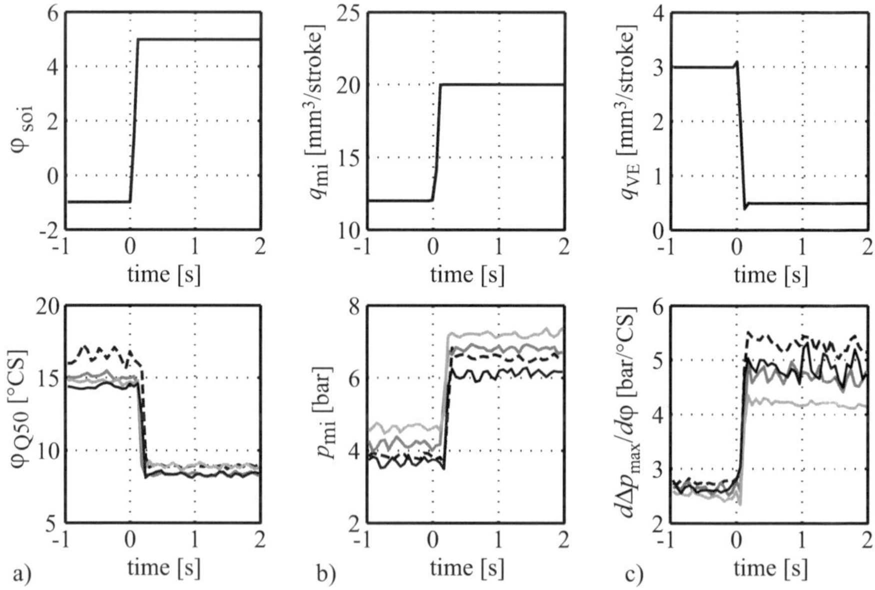

Fig. 8.6.1. Step responses for all four-cylinders of a conventional diesel combustion engine. **a** from start of injection to combustion center. **b** from main injection quantity to mean indicated pressure. **c** from pilot injection quantity to maximum pressure gradient. $n_{\mathrm{eng}} = 1500\,\mathrm{rpm}$.

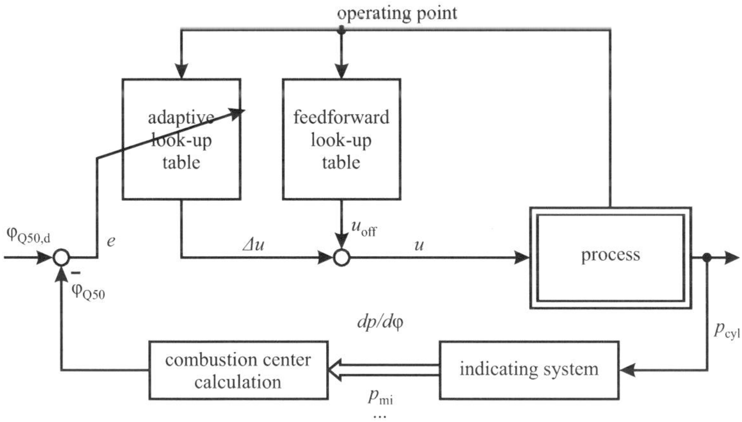

Fig. 8.6.2. Adaptive control system for the combustion center, consisting of a fixed feedforward look-up table and an adaptive correction look-up table as feedback from measured combustion pressure and calculated features like center of combustion φ_{Q50}, Kohlhase (2011).

8.7 Alternative combustion processes (HCCI) with pressure-based control

8.7.1 Homogeneous charge compression ignition (HCCI)

The conventional diesel engine combustion is based on an injection-jet-oriented heterogeneous air/fuel mixture formation. This leads to a trade-off between particulate matters (PM) and NO_x, see Fig. 8.7.1a), and fuel consumption and NO_x, because of a combustion with rich zones. Around the jet spray different zones exist. The external jet zone has a high air excess factor lambda (lean burning) which leads to a good ignition and burning with high temperatures over 2000 K and large NO_x formation. However the internal zone of the jet droplets have too little air (rich burning) so that soot is formed. Therefore always both NO_x and PM appear, which may be varied along hyperbolic curves according to the combustion influencing variables such as spray formation, multiple injection, rail pressure, turbulence, etc.

The formation of NO_x can primarily be reduced by lower combustion temperatures which can be reached by e.g. cooled exhaust gas recirculation (EGR) or late injection. A better oxidation of soot is possible by higher injection pressure, post-injection and high charging pressure. The fuel itself also has a considerable influence, Mollenhauer and Tschöke (2010), Pöttker et al (2005). This means that compromises have to be found for the conventional diesel combustion.

Alternative combustion methods for diesel engines now try to avoid too high combustion temperatures and too low air-excess areas, with the goal to operate more lean and homogeneous at lower combustion temperatures. This leads to the development of different kinds of homogeneous ignition processes, Suzuki et al (1997), Stanglmaier and Roberts (1999), Weißbäck et al (2003).

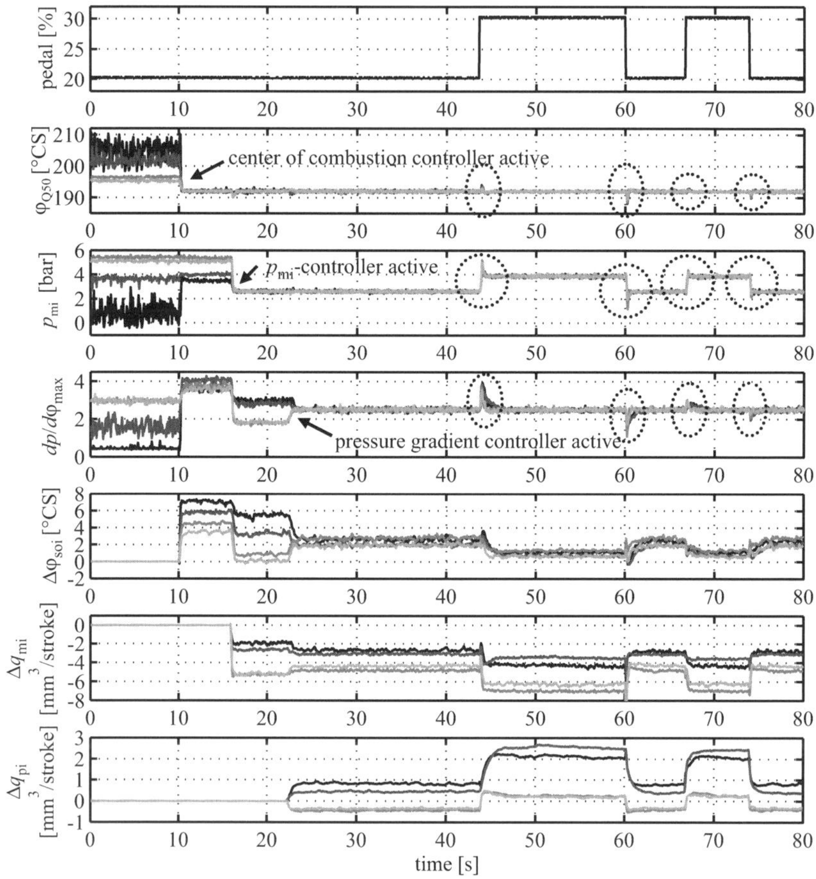

Fig. 8.6.3. Adaptive control of combustion center φ_{Q50} with start of injection φ_{soi}, mean indicated pressure p_{mi} with main injection mass q_{mi}, and maximum difference pressure gradient max $dp/d\varphi$ with pilot injection mass q_{mi}. The loops are closed sequentially, Kohlhase (2011)

In the case of the homogeneous charge compression ignition process (HCCI) the injection is performed in a very early phase of the compression cycle at about $90 - 140°$CS before TDC. The mixture then has more time to homogenize. Consequently, the combustion arises at many spots with relatively good air/fuel ratios. Additionally high EGR rates of about >40% and strong cooling lead to lower combustion temperatures, such that according to Fig. 8.7.1b) the areas for the heterogeneous combustion are avoided. Also the compression ratio has to be lowered to about 12:1 to 14:1 to prevent too early ignition. However, a problem may be the dilution of the lubrication oil by liquid fuel parts, Weßlau et al (2006).

Another slightly modified HCCI combustion is the homogeneous charge late injection process (HCLI). In order to prolong the time for a homogeneous mixture formation, the injection is more advanced compared to conventional combustions.

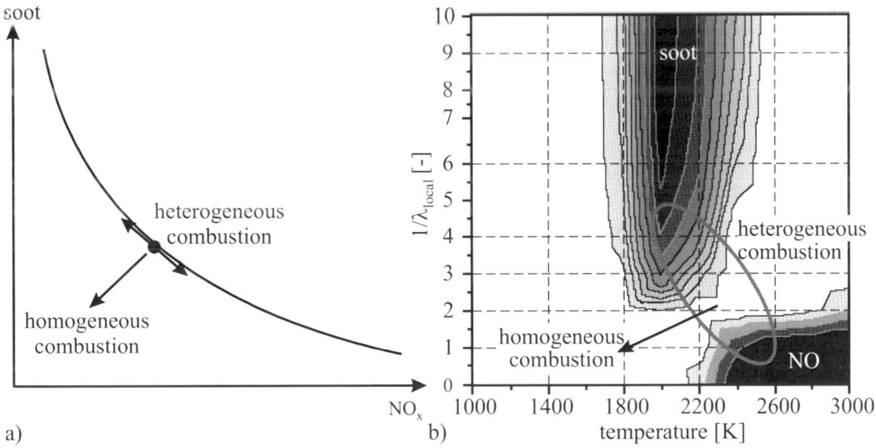

Fig. 8.7.1. Particulate (PM) and NO_x formation for heterogeneous and homogeneous diesel combustion. **a** PM-NO_x. **b** $1/\lambda$ and combustion temperature, Akihama et al (2001).

Together with high EGR rates >50%, the ignition is delayed. This combustion process comes relatively close to the conventional diesel combustion. The homogeneous combustion processes are restricted to part load. For the high load operation one has to switch to the conventional heterogeneous combustion with lower EGR rates and relatively late injection.

The developments also show other approaches such as the highly premixed late injection process (HPLI) with a long ignition delay by injection after TDC and low EGR rates, or the dilution-controlled combustion (DCCS) with very high EGR-rates >80%, see Landsmann (2010), Mollenhauer and Tschöke (2010), and Rether et al (2012). The relatively large exhaust gas recirculation rates can be obtained by external or internal EGR. The external EGR usually requires a throttle valve in the air path after the compressor in order to increase the pressure drop between the exhaust manifold and the intake manifold. This may be called "residual affected HCCI". An alternative is to close the exhaust valve late to reinduct some exhaust gases from the exhaust manifold, called "exhaust reinducted HCCI", Law et al (2001), Caton et al (2003). Further, the exhaust valve train can be closed early in order to trap the burned gases within the cylinder and to recompress them in the same upwards stroke. This is called "exhaust recompression HCCI", see e.g. Ravi et al (2012).

Since the modified HCCI/HCLI seems to be a reasonable compromise which can be realized with series production diesel engines, the development of a control system will be described for this combustion process according to Kohlhase et al (2008), Kohlhase (2011). Because of its flexibility it can also be adapted to the other HCCI processes.

A scheme of the considered diesel engine is depicted in Fig. 8.7.2. It is a 1.9 l diesel engine with common-rail injection and VGT turbocharger. The throttle valve in the intake after the compressor and the intercooler is required to adjust large EGR rates to enable HCCI.

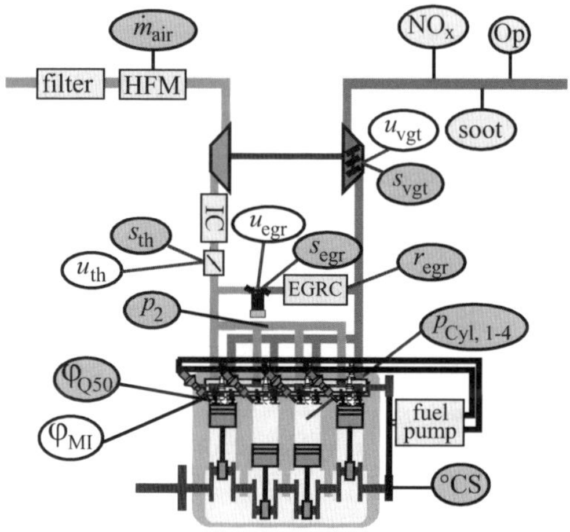

Fig. 8.7.2. Considered diesel engine Opel Z19DTH, 1.9 l, 315 Nm, 110 kW (Euro 4) with high-pressure common-rail and VGT turbocharger. *Manipulated variables* for the HCCI combustion are: main injection angle φ_{mi}, throttle valve u_{th}, EGR valve u_{egr}, VGT actuator u_{vgt} (s_j are actuator positions). *Controlled variables* are: combustion center φ_{Q50}, air mass m_{air}, EGR rate r_{egr}.

Figure 8.7.3 shows a comparison of a conventional and a homogeneous combustion for the considered diesel engine. The heterogeneous combustion has a pilot injection of duration $\Delta\varphi_{pi,conv}$ at $-19°$CS and a main injection $\Delta\varphi_{mi,conv}$ around TDC. For the homogeneous combustion there is only a main injection at $-19°$CS. Then the combustion begins later, allowing about $\Delta\varphi = 15°$CS ignition delay for homogeneous mixture formation. Hence, fuel injection and combustion are separated, Brengtsson (2004), Stein et al (2004).

Following control and manipulated variables are required to control the HCCI / HCLI process:

$$\varphi_{Q50}(\varphi_{soi}); \quad m_{air}(u_{th}); \quad r_{egr}(u_{egr}); \quad p_2(u_{th}).$$

Figure 8.7.4 depicts the multi-variable process as a signal-flow block, also indicating the many cross couplings.

The homogeneous combustion is very sensitive to deviations from an optimal state and depends especially on the large EGR rates which are different for the individual cylinders. Therefore, a combustion-pressure-based control of the combustion center φ_{Q50} by manipulating the start of injections φ_{soi} is required as a basic sub-level control for all cylinders, as described in the last Sect. 8.6. It is marked with I in Fig. 8.7.4. The charging pressure p_2 is feedforward controlled, because the EGR rate is very large in part load and the turbocharger does not have much power. Therefore, the next section describes the remaining strongly coupled air mass and EGR rate control within the box II of Fig. 8.7.4.

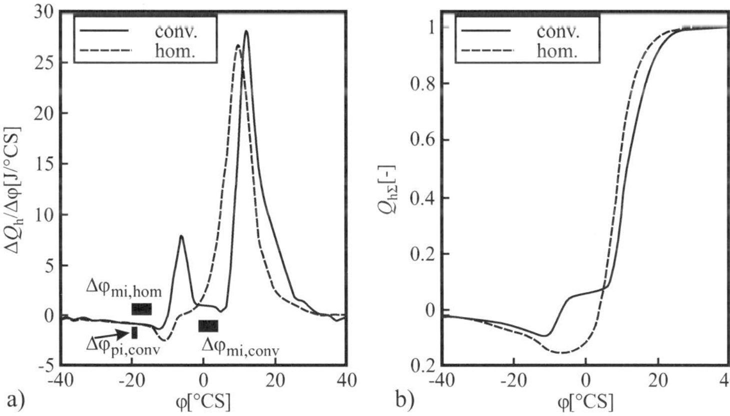

Fig. 8.7.3. Comparison of the conventional and homogeneous diesel combustion for a 1.9 l, 110 kW, common-rail DI diesel engine. **a** rate of net heat release $\Delta Q_h/\Delta\varphi$. **b** normalized integrated heat release $Q_{h\Sigma}$.

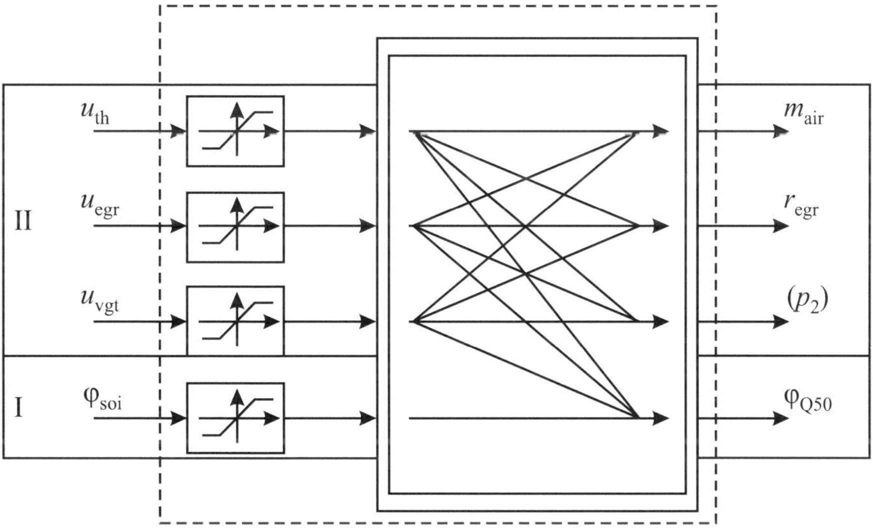

Fig. 8.7.4. Controlled and manipulated variables for the homogeneous diesel combustion. φ_{Q50}: combustion center; φ_{soi}: start of injection; $\dot{m}_{air}$: air mass flow rate; u_{th}: throttle position; r_{egr}: exhaust gas recirculation rate; u_{egr}: EGR valve position; p_2: charging pressure; u_{vgt}: blade position VGT.

8.7.2 Air mass flow and EGR rate control

Air flow rate and exhaust gas recirculation control are already described in Sect. 8.5, including physical-based process models. For the coupled control of air flow and charging pressure a special control scheme is discussed in Sect. 8.5 which is based on nonlinear identified models. This method uses local linear models and will now be transferred to the design of a two variable controller of the air mass and the EGR rate.

a) Model structure

Based on a physical model structure a local affine (local linear) state model with varying parameters is applied, see Fig. 8.7.5. It is obtained by a weighted superposition of linear state-space models that are additively extended by a vectorial offset $\boldsymbol{\xi}(\mathbf{z}(k))$, see Sect. 8.5. The vector $\mathbf{z}(k)$ describes the operating point, e.g. the engine speed n_{eng} and accelerator pedal position α_{ped}. Since the process is nonlinear in its inputs, $\mathbf{z}^T(k) = [\mathbf{u}(k), n_{\text{eng}}(k), \alpha_{\text{ped}}(k)]$ is chosen. The local affine state-space model is given by

$$\begin{aligned}\mathbf{x}(k+1) &= \mathbf{A}(\mathbf{z})\mathbf{x}(k) + \mathbf{B}(\mathbf{z})\mathbf{u}(k) + \boldsymbol{\xi}(\mathbf{z}(k))\\ \mathbf{y}(k) &= \mathbf{C}(\mathbf{z})\mathbf{x}(k) + \mathbf{D}(\mathbf{z})\mathbf{u}(k),\end{aligned} \tag{8.7.1}$$

with

$$\mathbf{u} = \begin{bmatrix} u_{\text{th}} \\ u_{\text{egr}} \end{bmatrix}; \qquad \mathbf{y} = \begin{bmatrix} m_{\text{air}} \\ r_{\text{egr}} \end{bmatrix}$$

with the initial conditions $\mathbf{x}(0) = \mathbf{x}_0$ and the operating point dependent matrix

$$\mathbf{A}(\mathbf{z}) = \sum_{j=1}^{M} \Phi_j(\mathbf{z}(k))\mathbf{A}_j. \tag{8.7.2}$$

M represents the number of local models and $\Phi_j(\mathbf{z}(k))$ is the activation function of the j-th local model. The activation functions are normalized Gaussians, that specify the validity of each local model. The matrices $\mathbf{B}(\mathbf{z}(k))$, $\mathbf{C}(\mathbf{z}(k))$, $\mathbf{D}(\mathbf{z}(k))$ and the vector $\boldsymbol{\xi}(\mathbf{z}(k))$ are defined analogously.

The advantage of this structure is that the system matrices can either be determined by means of measured data or by a linearization of a nonlinear physically-based model.

In the following, a structure with $\mathbf{D}(\mathbf{z}(k))=\mathbf{0}$ is considered. If the local affine state-space model is used in observable canonical form, the state-space representation with simultaneous parameter change can be transformed into the affine parameter-varying difference equation

$$\begin{aligned} y(k) = &\sum_{i=1}^{n} \widetilde{b}_i\left(\mathbf{z}(k-i)\right) u(k-i)\\ &- \sum_{i=1}^{n} \widetilde{a}_i\left(\mathbf{z}(k-i)\right) y(k-i) + \sum_{i=1}^{n} \widetilde{\xi}_i\left(\mathbf{z}(k-i)\right)\end{aligned} \tag{8.7.3}$$

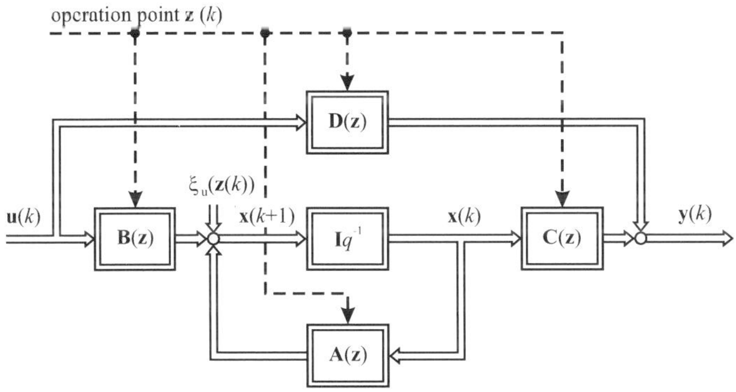

Fig. 8.7.5. Local affine state-space model for the air flow and EGR rate. $\mathbf{u}^T = [u_{\mathrm{th}} \;\; u_{\mathrm{egr}}]$, $\mathbf{y}^T = [m_{\mathrm{air}} \;\; r_{\mathrm{egr}}]$, $\mathbf{z}^T = [u_{\mathrm{th}} \;\; u_{\mathrm{egr}} \;\; n_{\mathrm{eng}} \;\; \alpha_{\mathrm{ped}}]$.

(shown here for simplicity only for the single input single output case). The parameter n represents the order of the system. In contrast to (8.7.1), the parameters

$$\widetilde{b}_i\left(\mathbf{z}(k-i)\right) = \sum_{j=1}^{M} b_{ij} \cdot \Phi_j(\mathbf{z}(k-j)) \qquad (8.7.4)$$

as well as $\widetilde{a}_i\left(\mathbf{z}(k-i)\right)$ and $\widetilde{\xi}'_i\left(\mathbf{z}(k-i)\right)$ change depending on the operating point $\mathbf{z}(k-i)$ delayed by i. The fact that the state-space model in observable canonical form can be transformed into an affine parameter-varying difference equation is used for the data-based identification of the air path.

The main controlled variable is the air mass m_{air}, determined from the measured air-flow rate $\dot{m}_{\mathrm{air}}$, and the EGR rate

$$r_{\mathrm{egr}} = \frac{\dot{m}_{\mathrm{egr}}}{\dot{m}_{\mathrm{egr}} + \dot{m}_{\mathrm{air}}} = \frac{\dot{m}_{\mathrm{cyl}} - \dot{m}_{\mathrm{air}}}{\dot{m}_{\mathrm{cyl}}}. \qquad (8.7.5)$$

The gas mass in the cylinder follows from the gas state equation

$$m_{\mathrm{cyl}} = \frac{p\left(\varphi_{\mathrm{ic}}\right)V\left(\varphi_{\mathrm{ic}}\right)}{RT(\varphi_{\mathrm{ic}})} \qquad (8.7.6)$$

with values for inlet valve closes (ic). Assuming a polytropic state change it holds

$$m_{\mathrm{cyl}} = \frac{1}{RV^{n-1}(\varphi_{\mathrm{ic}})T(\varphi_{\mathrm{ic}})}\frac{1}{N}\sum_{i=1}^{N} p_{\mathrm{cyl}}(i)V^n(\varphi_{\mathrm{ic}}) \qquad (8.7.7)$$

with averaging over N pressure measurements around TDC. The gas flow rate into i_c cylinders is then

$$\dot{m}_{\mathrm{cyl}} = \frac{1}{2} m_{\mathrm{cyl}} n_{\mathrm{eng}} i_{\mathrm{c}}. \qquad (8.7.8)$$

This yields with (8.7.5) the EGR rate.

b) Identification of the air / EGR process

The input variables for the identification are the actuator positions of the throttle s_{th} and of the EGR valve s_{egr}, and the output variables are m_{air} and r_{egr}.

For the parameter estimation the equation error and the output error approach are possible. In the nonlinear case the first one is named NARX (*Nonlinear AutoRegressive with eXogenous input*) model. This estimation has the advantage that the optimization linearly depends on the parameters, if the nonlinear model is also linear in the parameters. This is the case for (8.7.3). However, the disadvantage of the NARX structure is that a systematic parameter error (bias) occurs caused by noisy measurement data. If the nonlinear output error (NOE) approach is used, the error is nonlinear in the parameters which leads to a nonlinear optimization. Therefore first, the linear least squares method is used to calculate the parameters quickly in one step. These estimated parameters are used as an initialization for the nonlinear optimization of the NOE model. This two-step-identification leads to better results, since the NOE structure has no systematic bias, Zimmerschied (2008). After the estimation of the parameters the model of (8.7.3) is transformed into a state-space model in observable canonical form. The operating point $\mathbf{z}^T(k) = [\mathbf{u}(k), n_{\mathrm{eng}}(k), \alpha_{\mathrm{ped}}(k)]$ for the identification of the data-based model depends on the input signals $\mathbf{u}^T = [s_{\mathrm{egr}}, s_{\mathrm{th}}]$, the engine speed n_{eng} and the accelerator pedal α_{ped}. For the identification, the position-controlled EGR valve and throttle valve ($s_{\mathrm{egr,d}}$, $s_{\mathrm{th,d}}$) are excited with an APRB (*Amplitude modulated Pseudo Random Binary*) signal at different operating points (n_{eng}, α_{ped}). The APRB signal consists of multiple steps with different amplitudes and hold times in order to excite the process within a wide frequency range and equally distributed amplitudes. It is specifically designed for the nonlinear identification.

Figure 8.7.6 shows an extract of the excitation signals $s_{\mathrm{egr,d}}$ and $s_{\mathrm{th,d}}$ and the generalization data m_{air} and r_{egr}. The simulated model output fits the measured data very well.

The DC gains of the plant given by

$$\begin{pmatrix} m_{\mathrm{air}} \\ r_{\mathrm{egr}} \end{pmatrix} = \begin{pmatrix} K_{11} & K_{12} \\ K_{21} & K_{22} \end{pmatrix} \cdot \begin{pmatrix} s_{\mathrm{th,d}} \\ s_{\mathrm{egr,d}} \end{pmatrix} \tag{8.7.9}$$

are important for the controller design. In Fig. 8.7.7 the static behavior is plotted against the input signals $\mathbf{u} = [s_{\mathrm{egr,d}}, s_{\mathrm{th,d}}]$ for a constant engine speed n_{eng} and a constant accelerator pedal position α_{ped}. Since the DC gains vary strongly with the operating point, the controller design has to consider this nonlinear behavior. Figure 8.7.7 also depicts the coupling K_{12} and K_{21} between the inputs and outputs.

Beside the static behavior and the input/output coupling, the dynamic behavior is also important for the controller design. The step responses of each input are represented in Fig. 8.7.8. They show a low pass behavior of appropriately first order with small time constants of about 50 ms to 150 ms, corresponding to 1 to 3 strokes.

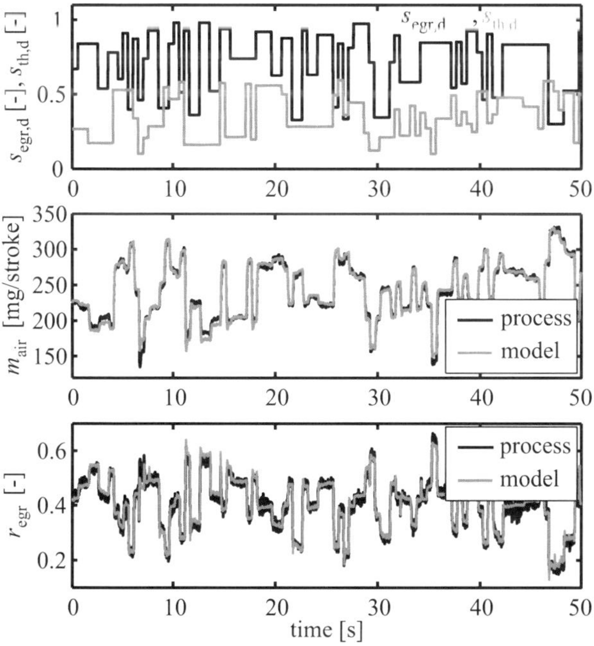

Fig. 8.7.6. Extract of the measured signals for the identification of the two variable process. Shown are generalization data with simultaneous APRBS excitation of the two inputs for constant $n_{\mathrm{eng}} = 2000\,\mathrm{rpm}$ and $\alpha_{\mathrm{ped}} = 23\%$.

c) Design of the control system

The controlled variables for the homogeneous diesel combustion are the air mass m_{air} and the EGR rate r_{egr}, belonging to part II in Fig. 8.7.4, to enable a HCCI process, and the combustion center φ_{Q50}, belonging to part I, to keep the combustions in a stable state. First, the control of the air mass and the EGR rate is considered. The control structure is shown in Fig. 8.7.9.

This structure consists of a nonlinear reference filter which generates the filtered setpoints for feedback control $\mathbf{y}_{\mathrm{d,f}}$ and the manipulated variables for feedforward control $\mathbf{u}_{\mathrm{FF}}$. This structure has the advantage that the nonlinear reference shaping filter and the feedback controller can be designed separately. The nonlinear reference shaping filter is aimed for setpoint tracking and the feedback controller for disturbance rejection. This structure is also named a two-degrees-of-freedom-structure.

Feedforward control

The nonlinear reference shaping filter for feedforward control is shown in Fig. 8.7.10. This filter is based on an identified local affine (local linear) state-space model. The manipulated variables

$$\mathbf{u}_{\mathrm{f}} = -\mathbf{K}\mathbf{x}_{\mathrm{M}} + \mathbf{v} \tag{8.7.10}$$

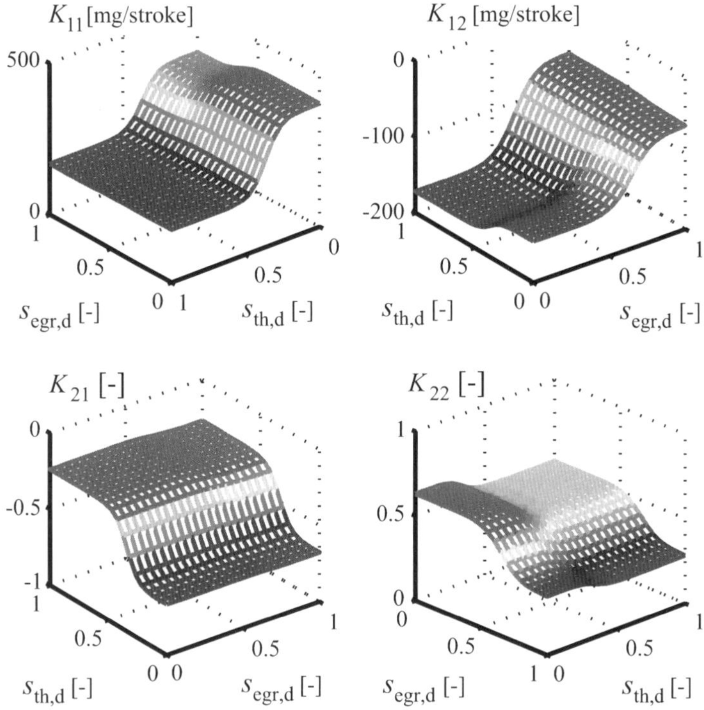

Fig. 8.7.7. Gains (stationary behavior) of the two variable process depending on the manipulated variables for $n_{\mathrm{eng}} = 2000\,\mathrm{rpm}$ and $\alpha_{\mathrm{ped}} = 20\%$.

for feedforward control are calculated with the model-state feedback $\mathbf{K}$ and the signal $\mathbf{v}(k) = \mathbf{V}\mathbf{y}_{\mathrm{d}}(k) - \mathbf{F}\boldsymbol{\xi}(\mathbf{z}(k))$, which depends on the setpoint $\mathbf{y}_{\mathrm{d}}$ and the offset $\boldsymbol{\xi}$. The manipulated variables are input variables of the model with constraint (saturation) and also of the real plant. The output variables of the model $\mathbf{y}_{\mathrm{d,f}}$ are used as setpoints for the feedback controller. If the model exactly matches the real plant, the control deviation $\mathbf{e} = \mathbf{y}_{\mathrm{d,f}} - \mathbf{y}$ is zero and the feedback controller only responds to disturbances. Additionally, model uncertainties affect the feedback controller. Therefore, a high model quality is crucial for a high control performance.

The model-based setpoint-following-control is based on state feedback. A state feedback matrix $\mathbf{K}_j$ is calculated offline for each local state-space model ($\mathbf{A}_j$, $\mathbf{B}_j$, $\mathbf{C}_j$, $\mathbf{D}_j$). Any linear state-space control design method can be exploited, e.g. pole placement, decoupling controller design etc. In this contribution an optimal controller $\mathbf{K}_j$ is calculated by solving the matrix Riccati equation for each local state-space model for the given weighting matrices ($\mathbf{Q}_j$, $\mathbf{R}_j$). The overall state feedback controller

$$\mathbf{K}(\mathbf{z}(k)) = \sum_{j=1}^{M} \Phi_j(\mathbf{z}(k))\mathbf{K}_j, \tag{8.7.11}$$

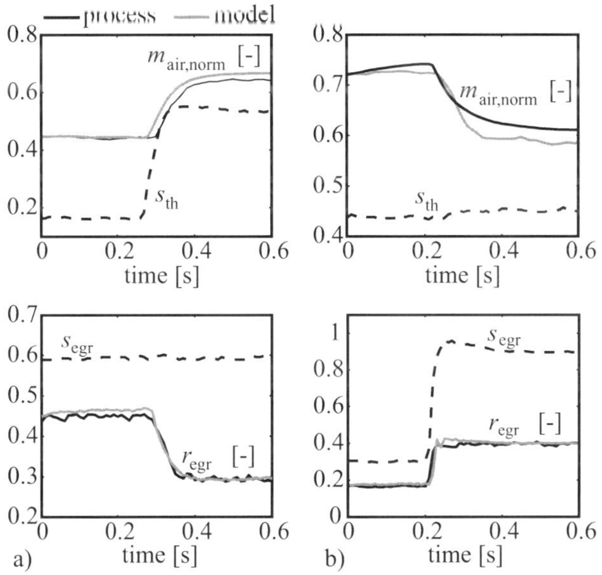

Fig. 8.7.8. Measured step responses for step changes of **a** throttle s_{th}, **b** EGR valve s_{egr} ($n_{eng} = 2000\,\text{rpm}$ and $\alpha_{ped} = 20\%$).

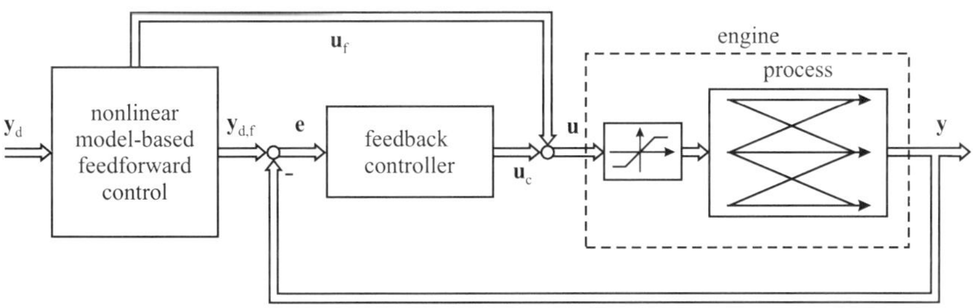

Fig. 8.7.9. Structure for the control of the airmass and the EGR rate. $\mathbf{u}^T = [u_{th} \;\; u_{egr}]$; $\mathbf{y}^T = [m_{air} \;\; r_{egr}]$.

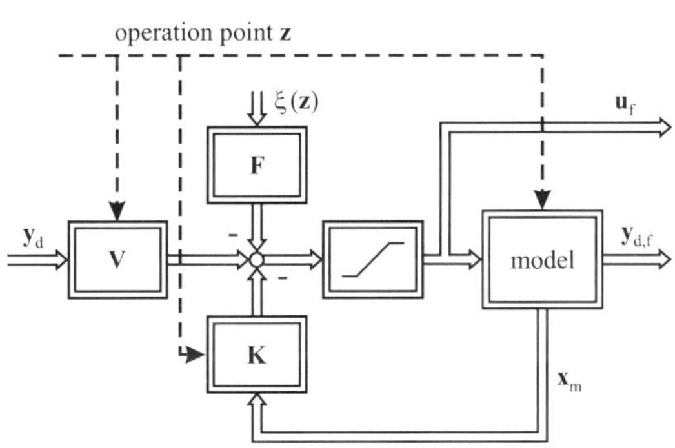

Fig. 8.7.10. Reference filter with actuator constraints.

is calculated by a superposition of the local subcontrollers $\mathbf{K}_j$ weighted with the activation functions $\Phi_j(\mathbf{z}(k))$. These normalized Gaussians are equivalent to those of the state-space model. As in the linear case, setpoint following is achieved with the setpoint filter $\mathbf{V}$ and disturbance filter $\mathbf{F}$. For this purpose the vector

$$\mathbf{v}(k) = \left(\mathbf{C}(\mathbf{z}(k))\tilde{\mathbf{V}}(\mathbf{z}(k))\mathbf{B}(\mathbf{z}(k))\right)^{-1} \quad (8.7.12)$$

$$\cdot\left(\mathbf{y}_\mathrm{d}(k) - \mathbf{C}(\mathbf{z}(k))\tilde{\mathbf{V}}(\mathbf{z}(k))\boldsymbol{\xi}(\mathbf{z}(k))\right) \quad (8.7.13)$$

$$= \mathbf{V}\mathbf{y}_\mathrm{d}(k) - \mathbf{F}\boldsymbol{\xi}(\mathbf{z}(k)) \quad (8.7.14)$$

is calculated *online* and added to the model state feedback $\mathbf{u}_\mathrm{f} = -\mathbf{K}\mathbf{x}_\mathrm{M} + \mathbf{v}$. This vector $\mathbf{v}$ is computed with the system matrices ($\mathbf{A}$, $\mathbf{B}$, $\mathbf{C}$), the overall state feedback $\mathbf{K}$, the offset vector $\boldsymbol{\xi}$ and with the desired values $\mathbf{y}_\mathrm{d}$ in each time step k, see Kroll et al (2000). In (8.7.13) the matrix $\tilde{\mathbf{V}}$ is defined as:

$$\tilde{\mathbf{V}}(\mathbf{z}(k)) = (\mathbf{I} - \mathbf{A}(\mathbf{z}(k)) + \mathbf{B}(\mathbf{z}(k))\mathbf{K}(\mathbf{z}(k)))^{-1} . \quad (8.7.15)$$

The advantage of the nonlinear reference filter is the handling of the input constraints. The manipulated variables $\mathbf{u}_\mathrm{f}$ are calculated based on the model states $\mathbf{x}_\mathrm{M}$ which depend on the *constrained* input signals (compare Fig. 8.7.10). Hence, the reference shaping filter is "aware" of the input constraint and "behaves" in a proper manner, Hippe (2006). On the other hand, the classical feedforward control concepts based on the inversion of linear models just clips the input signals. In the case of active input saturations, this can lead to reduced control performance.

Feedback control

The feedback controller can have the same structure as the nonlinear reference filter. However, an observer has to be included since not all states are measured. Furthermore, integral action is necessary to avoid a steady-state error. Due to the high model quality and therefore the high quality of the nonlinear reference filter, decentralized PI controllers generate the demanded control performance, see Fig. 8.7.11. They are tuned sequentially to guarantee stability. The EGR controller is closed first because of its fast dynamics.

Overall control

Figure 8.7.12 shows the complete control structure with the applied nonlinear reference filter, the decentralized PI controllers, the feedforward control of the VGT turbocharger and the control structure for the combustion center φ_Q50 as controlled variable. The structure of the combustion center control is based on a regular look-up table with the input signals engine speed n_eng and acceleration pedal position α_ped, see Sect. 8.6. The resulting injection angle φ_soi is corrected by the measured air mass flow $\dot{m}_\mathrm{air}$. This static correction acts as a disturbance rejection. Setpoint following of the controlled combustion center angle φ_Q50 is achieved by adaptive look-up tables.

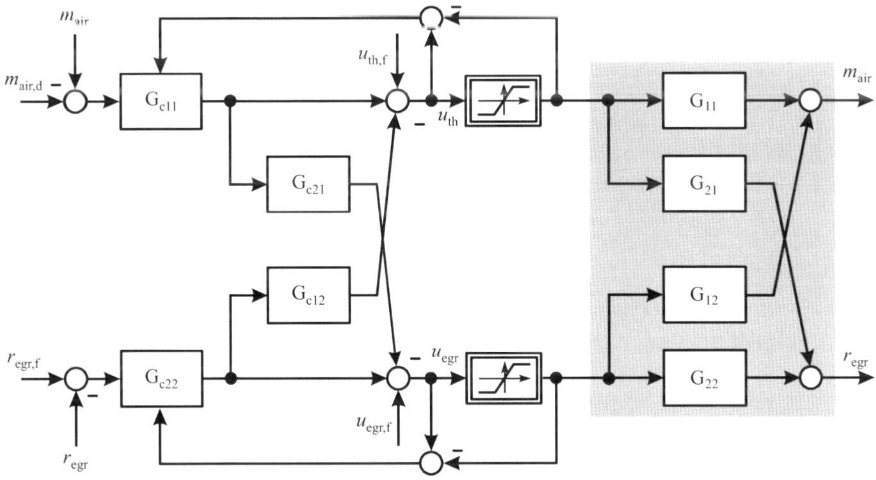

Fig. 8.7.11. Two-variable control system for the air and EGR system with decentralized controllers G_{c11} and G_{c22}, decoupling with G_{c12} and G_{c21} and anti-windup.

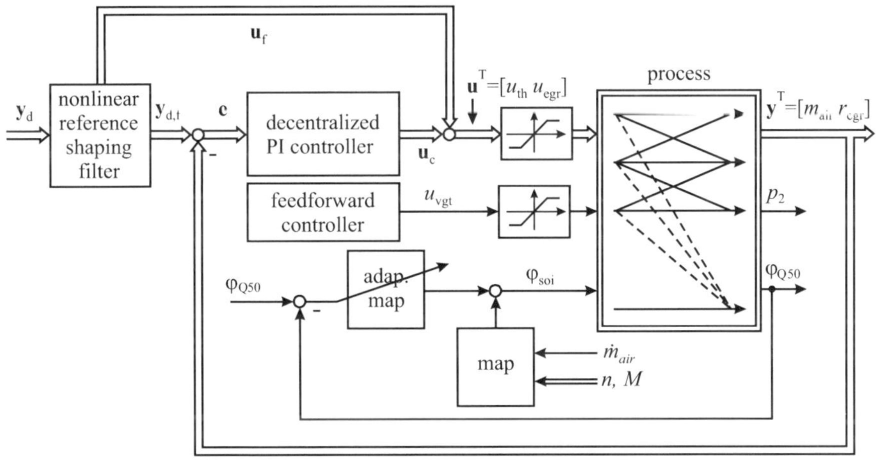

Fig. 8.7.12. Overall control scheme for the homogeneous diesel combustion. part I: combustion center control; part II: two-variable airmass/EGR control. The charging pressure p_2 is feedforward controlled.

Experimental results

The test bench used is equipped with a fully indicated common rail diesel engine (OPEL/Fiat JTD 1.9L) with an EGR valve, a throttle valve and a VGT turbocharger. Piezoelectric cylinder-pressure sensors (Kistler) were used, which were placed into the combustion chamber with glow plug adapters. The cylinder-pressure signals were measured by an in-house indicating system. The indicating system is implemented on a rapid control prototyping (RCP) system (dSpace). Cylinder-pressure characteristics are computed by the indicating system in real time and are sent to another RCP system via CAN. This system receives the CAN-messages for the cylinder-pressure-based control in bypass mode. The cylinder pressure based control algorithms were implemented on an ETAS ES1000 system. Another RCP system (dSpace) is used to control the air mass and the EGR rate.

Figure 8.7.13 depicts the transition from conventional to homogeneous operation and back. As well the air mass as the EGR rate follow the reference values, which are relatively well feedforward controlled by the reference filter. The feedforward action u_f uses the restricted manipulation range almost completely. The feedback controller reacts only to disturbances and model inaccuracies. The figure shows both, the calculated manipulation variables u_i and the actuator positions s_i.

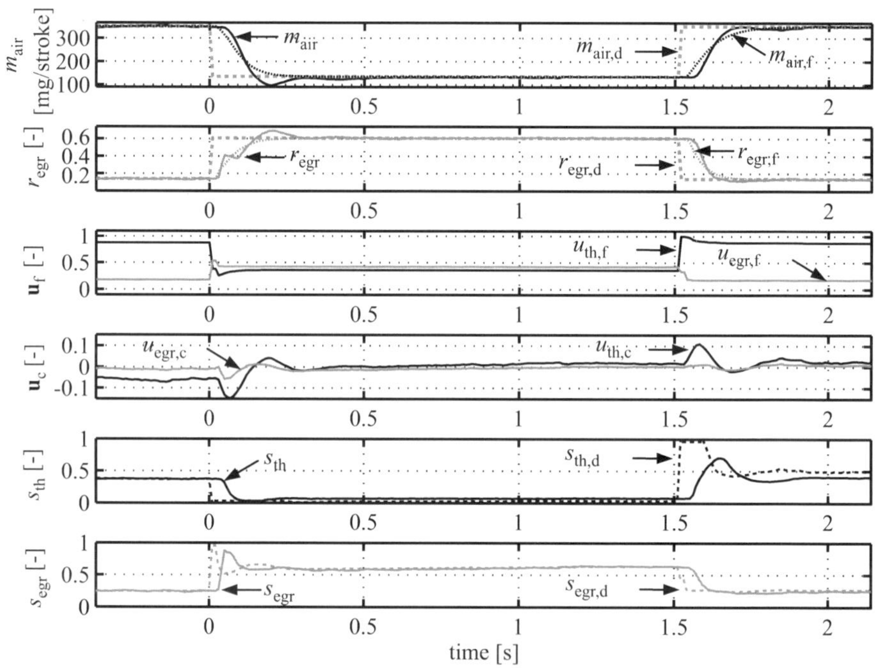

Fig. 8.7.13. Experimental results of the model-based air mass and EGR-rate control for the transition from conventional to homogeneous combustion at $t = 0$ and back at $t = 1.5\,\mathrm{s}$. $n = 2000\,\mathrm{rpm}$, $q_f = 10\,\mathrm{mm}^3/\mathrm{stroke}$. f: feedforward controlled, c: feedback controlled.

The behavior of the control system for changes of the operating point is shown in Fig. 8.7.14. For the speed change (with a torque increase of the dynamometer) the feedforward controlled manipulated variables react immediately. The overall control systems keep the air mass and EGR-rate within small tolerances. Good results are also obtained for a load change through the pedal.

Another experiment shows Fig. 8.7.15. During the transition from conventional to homogeneous combustion the center of combustion deviates temporarily. With-

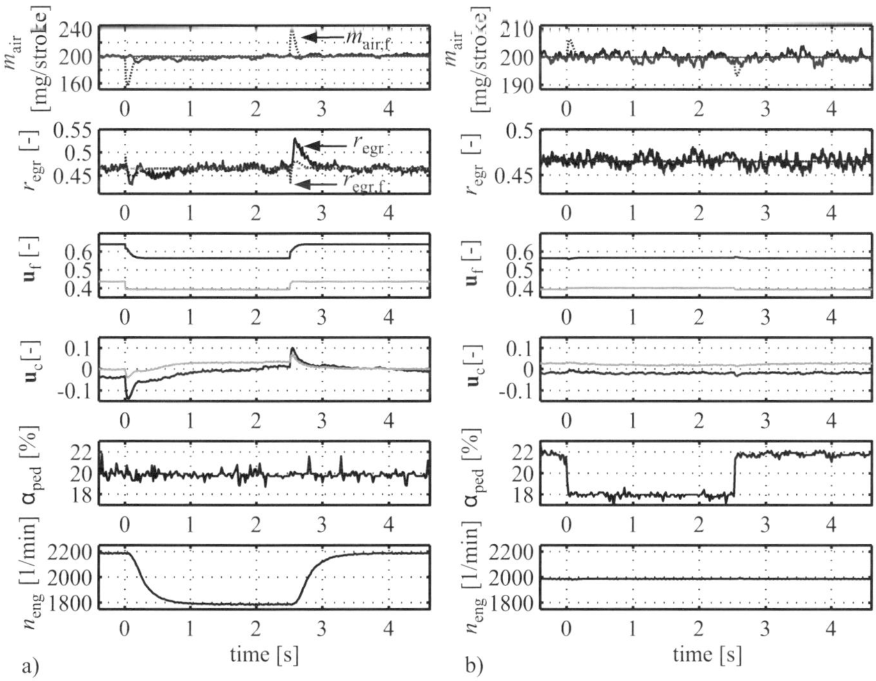

Fig. 8.7.14. Experimental results of the model-based air mass and EGR-rate control for changes of the operating point in homogeneous operation. **a** change of speed. **b** change of pedal ($n_{\mathrm{eng}} = 2000\,\mathrm{rpm}$).

out the adaptive look-up table the steady-state values show an offset, Fig. 8.7.15b). However, applying the adaptation the offsets vanish and a correct following to the setpoints can be observed, Fig. 8.7.15c). This shows that the large EGR rate of about 45% has a considerable effect on the combustions in the single cylinders and that differences can be compensated with the combustion center control by changing the start of injection. During the transition the air excess factor decreases from 2.7 to 1.2, i.e. close to stoichiometric combustion.

e) Conclusions for the control of the HCCI operation

Steady-state measurements with the investigated diesel engine resulted in diagrams of Fig. 8.7.16. By increasing the EGR-rate NO_x decreases but soot increases first. However, for very high EGR rates the soot decreases considerably if homogeneous combustion begins.

In the case of a constant operation point $n_{\mathrm{eng}} = 2000\,\mathrm{rpm}$ and $q_f = 10\,\mathrm{mm}^3/\mathrm{stroke}$ soot could be reduced by 80% and NO_x by 82% for constant torque, compared to conventional combustion. Measurements with the NEDC dynamic driving cycle and mixed conventional/homogeneous operation resulted in a reduction of soot by 65% and for NO_x by 15%. However, the HC emission increased.

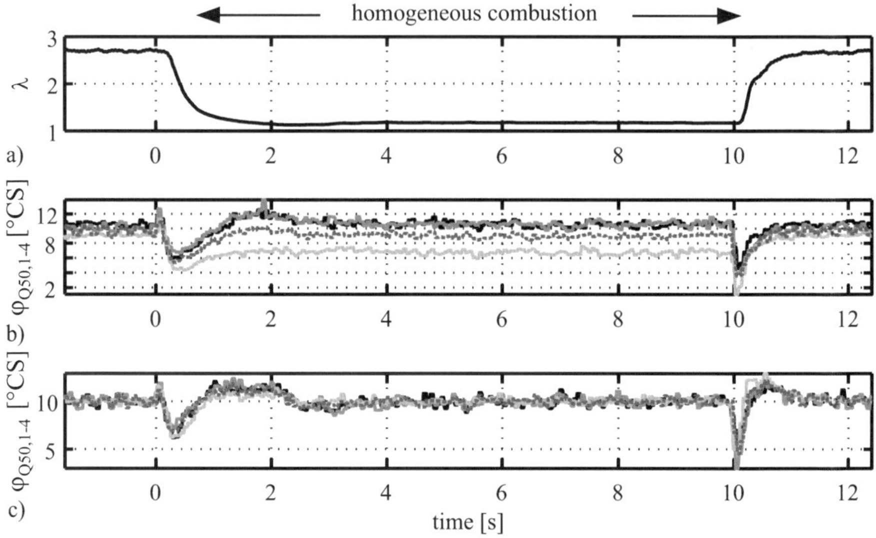

Fig. 8.7.15. Experimental results of the control of the combustion center during transition from conventional to homogeneous combustion. **a** air excess factor. **b** without adaptive look-up table. **c** with adaptive look-up table.

Hence, the HCCI combustion offers an interesting reduction of soot and emissions. After optimization of the basic design of the engine for homogeneous combustion and with variable valve trains even more improvements can be be expected. However, only a precise model-based nonlinear control can manage the sensitive HCCI combustion. Further literature on the control of HCCI combustion for gasoline engines is Rausen et al (2005), Chiang et al (2007) and for diesel engines Catanese et al (2009).

8.8 Smoke limitation control

During large load changes with increased injected fuel mass the cylinders do not get the required excess air mass because of the inert dynamic behavior of the turbocharger and the intake gas volume storage. Therefore the air/fuel ratio or excess air factor

$$\lambda = \frac{m_{\text{air}}}{m_{\text{air,st}}} = \frac{m_{\text{air}}}{L_{\text{st}}\, m_{\text{f}}} \tag{8.8.1}$$

may drop below λ_{min} such that considerable soot is generated. To avoid this temporary soot peak the fuel injection has to be limited, Nitzke et al (1999), Birk et al (2003), Nitsche et al (2004). Based on the air mass m_{air} in the cylinder the maximal allowed injected fuel mass is determined by

$$m_{\text{f,max}}(t) = \frac{m_{\text{air}}(t)}{L_{\text{st}}\, \lambda_{\text{min}}} \tag{8.8.2}$$

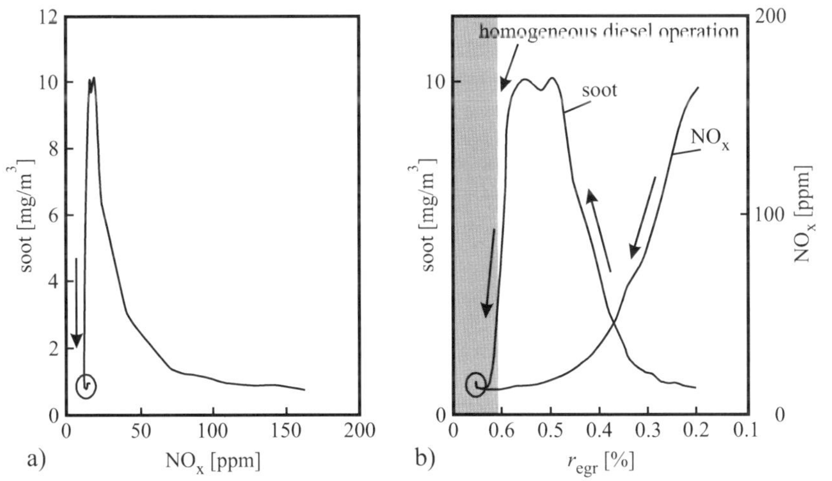

Fig. 8.7.16. Homogeneous diesel combustion. **a** trade-off between NO_x and soot. **b** dependency of NO_x and soot on the EGR rate. ($n_{eng} = 2000\,\text{rpm}$, $\alpha_{ped} = 25\%$).

assuming ideal combustion. This results in a *smoke limitation* or *smoke operation*. However, the fuel limitation during acceleration leads to a loss in dynamic torque generation. Hence, good compromises have to be found, Renninger et al (2005), Pfeil (2011).

The smoke limitation has a considerable influence on the total particulates for a driving cycle, as about 50-60% of the emissions result from dynamic acceleration phases. An application of smoke limitation control for a truck engine during a FTP-cycle shows that about 35% of particulates and 7.5% of NO_x are generated during the phases where the limitation becomes active, Pfeil (2011). Hence, the optimization of the smoke limitation control is an important part for emission reduction.

8.8.1 Smoke limitation with minimal air/fuel ratio

A direct way for smoke limitation during acceleration of the engine is to limit the injected fuel according to (8.8.2) and Fig. 8.8.1 based on the measured air mass flow rate $\dot{m}_{air}(t)$ or the calculated charged air mass $m_{air}(t)$ in the cylinder. This requires the determination of the minimal allowable air/fuel ratio λ_{min}, which is usually in the range of $\lambda_{min} \approx 1.1 \ldots 1.3$ and to store it as look-up table, see Fig. 8.8.1. It depends on the kind of the combustion with regard to soot generation and therefore also on other variables such as, e.g. p_{2i}, T_{2i}, φ_{mi} and the environmental variables p_a, T_a, see Sect. 8.9. Therefore λ_{min} has to be determined experimentally dependent on the operation point M_{eng}, n_{eng} and various engine control settings for normal, stationary operation. To simplify the procedure also a reasonable large value as $\lambda_{min} = 1.3 \ldots 1.5$ can be assumed. However, this may limit the dynamic torque development too much.

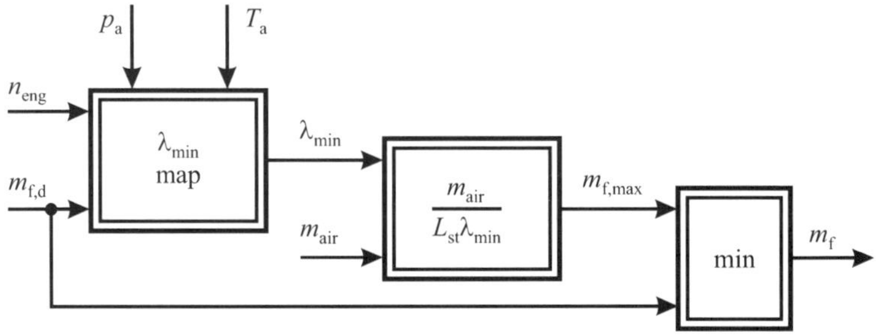

Fig. 8.8.1. Smoke fuel limitation control with minimal air/fuel ratio.

8.8.2 Smoke limitation with a soot model

Another way is to apply a soot model as described in Sect. 8.9.1 on raw emission models. For example the experimentally determined stationary local model according to (8.9.2)

$$c_{PM,j,k} = f_{j,k}(\dot{m}_{air}, p_{2i}, T_{2i}, \varphi_{Q50}) \qquad j = u_{inj}\,;\; k = n_{eng} \tag{8.8.3}$$

can be applied. See also Figs. 8.9.7, 8.9.8, and 8.9.10.

An alternative soot model is

$$c_{PM,j,k} = f_{j,k}(x_{eng,out}, p_{2i}, T_{2i}, \varphi_{Q50}) \tag{8.8.4}$$

where the gas composition $x_{eng,out}(m_{air}, m_{eng}, m_f) = (\lambda - 1)L_{st}/(\lambda L_{st} - 1)$ after combustion is used according to (8.9.8) instead of $\dot{m}_{air}$, which yields more precise soot models. A soot limit can then be given directly, e.g. as a concentration $c_{PM,min}$ in $[g/m^3]$, see Sect. 6.8.1. As the gas composition $x_{eng,out}$ contains the injected fuel mass, the corresponding fuel mass $m_{f,max}$ can then be determined by an inversion of the soot model with an iterative search procedure, Sequenz and Isermann (2011). Compare the intersection plots of the soot model in Fig. 8.9.8b).

Figure 8.8.2 depicts some variables for one step of the injection mass and for constant speed without and with smoke limitation by using (8.8.4). A soot limit $c_{PM} = 15\,g/m^3$ corresponds to an opacity value of 7%. The time histories show that after increasing the injected fuel mass the soot exceeds the limit value. Hence an operation with too much soot arises. After switching to soot limitation control at $t = 30\,s$ the injected fuel mass is reduced and therefore λ becomes larger and opacity becomes smaller. However, this method requires the measurement of the cylinder pressure to determine the combustion center φ_{Q50}. Alternatively φ_{mi} can be used, see Sequenz (2013).

Figure 8.8.3 shows the influence of the smoke-limitation control for transient states. Without smoke-limitation control a large soot peak develops after increase of fuel injection at $t = 6.1$ s. This peak is reduced significantly with smoke-limitation control at $t = 15$ s by a limitation of the fuel mass and thus reduction of the torque for about 1 s, Sequenz (2013).

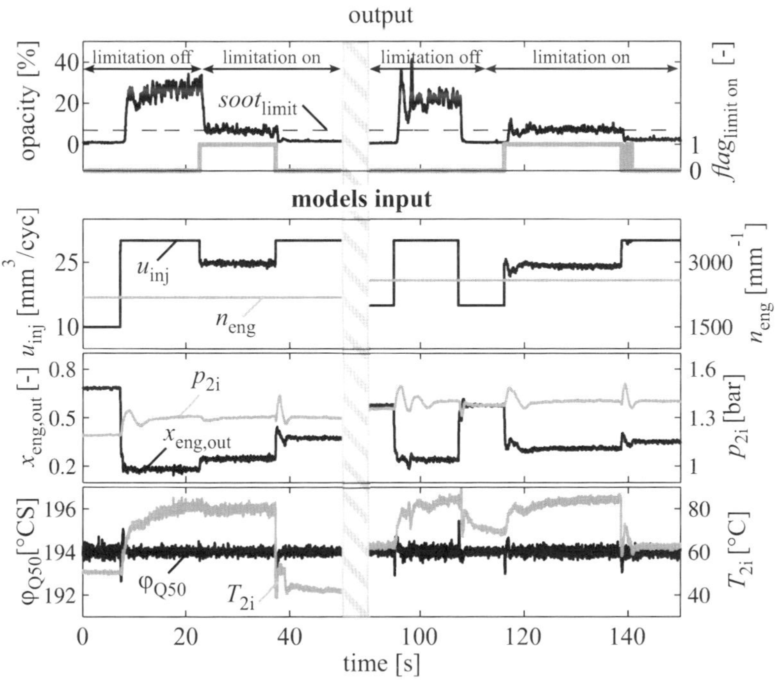

Fig. 8.8.2. Results of the smoke-limitation control by applying a soot model for step changes of the fuel injection and stationary condition with constant speed $n_{\mathrm{eng}} = 2200\,\mathrm{rpm}$ (measurements). $u_{\mathrm{inj}} = 10 \ldots 30\,\mathrm{mm}^3/\mathrm{cyc}$, $m_{\mathrm{air}} = 410\,\mathrm{mg/cyc}$, $p_{2\mathrm{i}} = 1.3\,\mathrm{bar}$, $\varphi_{\mathrm{Q50}} = 194°\mathrm{CS}$, Sequenz (2013).

8.8.3 Smoke limitation with multiple smoke operation models

The limitation of the injected fuel mass is the primary action to avoid soot peaks during acceleration, if the air/fuel ratio reaches $\lambda_{\min}$. As the soot generation and also NO_x generation in this phase can also be reduced by simultaneously changing other manipulated variables as the exhaust gas recirculation mass $\dot{m}_{\mathrm{egr}}$, the beginning of main injection φ_{mi} and the rail pressure p_{rail}, additional smoke operation models between these inputs and the dynamic particulate mass flow rate $\dot{m}_{\mathrm{PM}}(t)$, nitrogen oxide mass flow rate $\dot{m}_{\mathrm{NOx}}(t)$ and torque development $M_{\mathrm{eng}}(t)$ during acceleration may be used to optimize the smoke limitation control.

In addition to the fuel limitation according to Fig. 8.8.1 a smoke operation index r_{smoke} is introduced to determine the smoke operation control. The smoke operation index calculation is based on $\lambda_{\min}$ and m_{air} and depends on the operation point $[m_{\mathrm{f}}, n]$. During normal operation for $\lambda > \lambda_{\min}$ the smoke operation index is $r_{\mathrm{smoke}} = 0$ and the engine runs with normal operation control. If the air/fuel ratio becomes $\lambda < \lambda_{\min}$, the smoke operation index $r_{\mathrm{smoke}} = 1$ is set. Then the smoke operation control becomes active and the manipulated variables as for example

$$\mathbf{u}_{\mathrm{smoke}}^{\mathrm{T}} = \left[u_{\mathrm{egr}}, u_{\mathrm{vgt}}, \varphi_{\mathrm{mi}}, p_{\mathrm{rail}}\right] \tag{8.8.5}$$

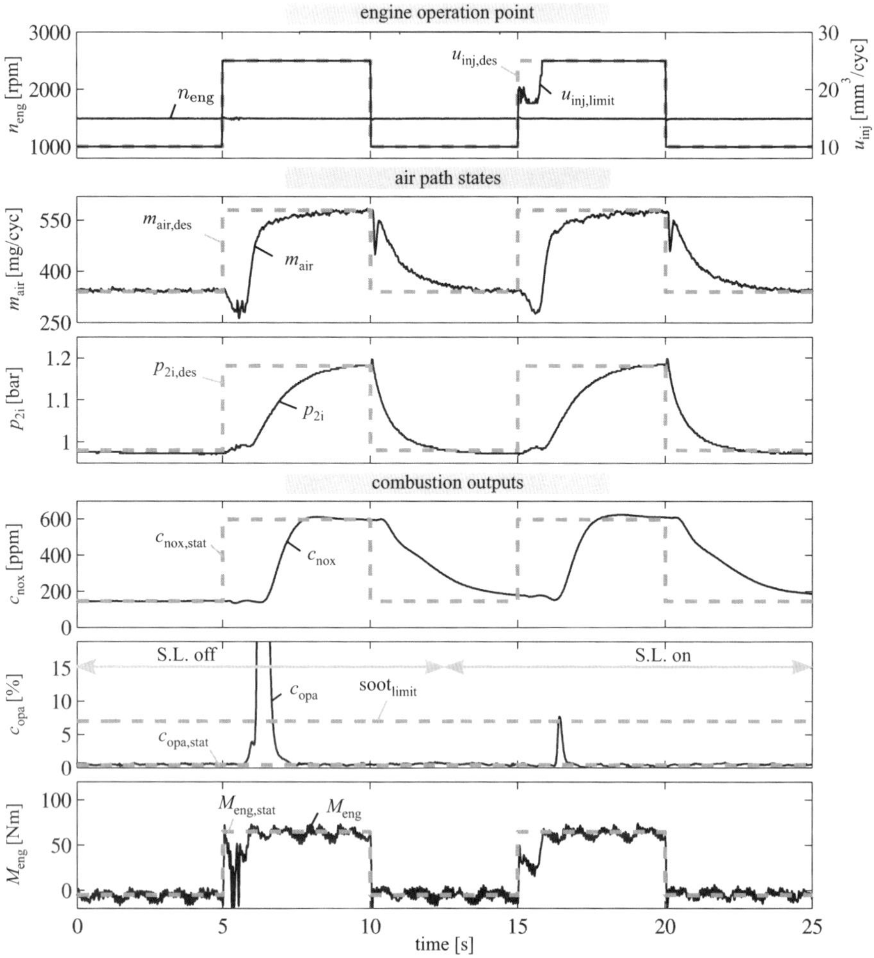

Fig. 8.8.3. Results of the smoke-limitation control by applying a soot model for constant speed and passing smoke generation. $u_{inj} = 10 \rightarrow 25\,\mathrm{mm^3/cyc}$ with $n_{eng} = 1500\,\mathrm{rpm}$. Desired values are plotted as grey dashed lines. $\varphi_{Q50} = 15°\mathrm{CS}$ is controlled by a closed-loop control. The smoke limitation is switched on at $t = 12.5\,\mathrm{s}$.

are calculated with special smoke models, Renninger et al (2006), Pfeil (2011), as depicted in Fig. 8.8.4.

In the sequel the identification and smoke control optimization is described for the last case and follows, Renninger et al (2006), Pfeil (2011). The considered engine is a Daimler 900 series, 6-cylinder truck diesel engine with unit pump injection $p_{max} = 2000\,\mathrm{bar}$, a fixed turbocharger, controlled EGR and particulate filter, Renninger et al (2006).

As the smoke limitation has to be designed for large acceleration pedal changes, models of the dynamic behavior for the output variables $M_{eng}(t)$, PM(t) and NO$_x(t)$

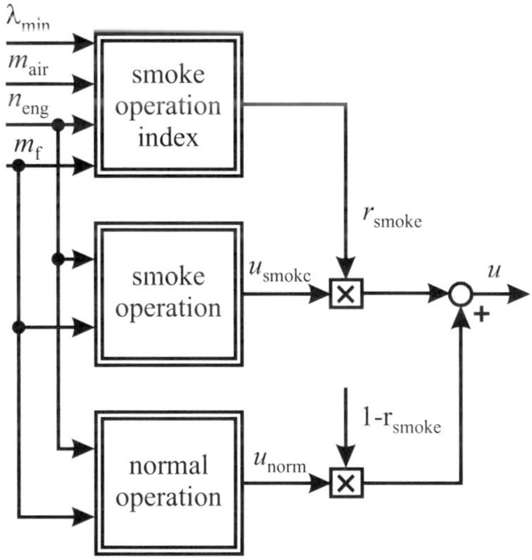

Fig. 8.8.4. Feedforward control scheme for smoke limitation with special smoke operation models $u_i(m_f, n_{eng})$. $r_{smoke} = 0$: normal operation. $r_{smoke} = 1$: smoke operation.

are required. Therefore experiments are performed for step function changes of the pedal between idle and full load at constant speed n_{eng} and for variations of u_{egr}, φ_{mi}, p_{rail} and λ_{min}. Then the averaged values of the two emissions are calculated, according to

$$\bar{m}_{NOx} = \overline{NO}_x = \frac{1}{\Delta t} \int_0^{\Delta t} c_{NOx}(t - T_{t,NOx})\, \dot{m}_{exh}(t)\, dt \qquad (8.8.6)$$

with Δt the evaluation time for the transient and $T_{t,NOx}$ the dead time of the measurement device. The transient of the torque is described e.g. by T_{90}, the 90% settling time. This measure for the dynamic torque behavior is the larger the larger λ_{min}, because the injected fuel mass is limited if λ_{min} is reached and therefore the torque reaches its steady-state value later during smoke-limitation control. These characteristic values of the outputs are determined for eight different speeds within $800\,\mathrm{rpm} < n_{eng} < 2200\,\mathrm{rpm}$ and for $5\,\mathrm{s}$ idling and $7\,\mathrm{s}$ full load. The values of the four variables are selected with a D-optimal design within a definite allowable operation space and a forward and backward sequence to eliminate eventual drift effects.

The three outputs are approximated with second order polynomials, such as

$$\overline{NO}_x(\mathbf{u}') = c_0 + \sum_{i=1}^{4} c_i\, u_i + \sum_{i=1}^{4} \sum_{j=i}^{4} c_{ij}\, u_i\, u_j \qquad (8.8.7)$$

with

$$\mathbf{u}' = \left[u_{egr}, \varphi_{mi}, p_{rail}, \lambda_{min}\right] \qquad (8.8.8)$$

taking all mutual relations of the four inputs into account and their parameters are determined with least squares parameter estimation. Figure 8.8.5 depicts intersection plots and indicates that the major influence on opacity and torque stems from u_{egr} and λ_{min}. It also shows the opposing behavior of OP and $\mathrm{NO_x}$ and the tradeoff between $\overline{\mathrm{M}}$ and $\overline{\mathrm{OP}}$ for λ_{min}. To improve the torque the EGR valve should be partially closed.

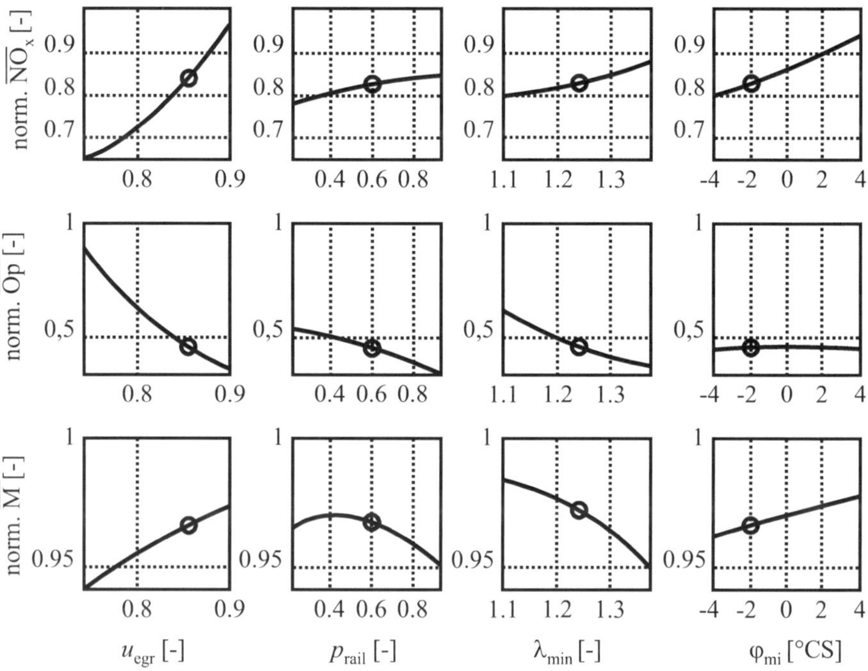

Fig. 8.8.5. Intersection plots of the normalized outputs $\overline{\mathrm{NO}}_x$, $\overline{\mathrm{OP}}$ and $\overline{\mathrm{M}}$ for pedal changes from idle to full control. $n_{\mathrm{eng}} = \mathrm{const.}$

Based on the polynomial models for $\overline{M}_{\mathrm{eng}}(\mathbf{u})$, $\overline{\mathrm{NO}}_x(\mathbf{u})$ and $\overline{\mathrm{PM}}(\mathbf{u})$, which are each valid for one speed $n_{\mathrm{eng},j}$ the emissions can be minimized by using a performance criterion

$$J_j(\mathbf{u}_{\mathrm{smoke}}) = w_1\, \overline{m}_{\mathrm{NOx}} + (1 - w_1)\, \overline{m}_{\mathrm{PM}} \tag{8.8.9}$$

with the boundary condition for the 90 % torque settling time

$$T_{90,j} < T_{90,j,\mathrm{max}}. \tag{8.8.10}$$

Using this criterion for each speed local optimization results are obtained for different weightings w_1 of the two emissions for a given dynamic torque limitation. Instead of applying the settling time $T_{90,j}$ the averaged torque $\overline{M}_j$ can also be used.

The results of this optimization are depicted in Fig. 8.8.6a). The diagrams show the known tradeoff between $\overline{\mathrm{NO}}_x$ and $\overline{\mathrm{OP}}$ (opacity). Larger values of dynamic torque settling times T_{90} yield better emission values, and vice versa. By shifting the weights w_1 from 0 to 1 one moves along a Pareto-curve. Depending on a selected $\overline{\mathrm{NO}_x}$ or w_1 the manipulated variables u_i and $\lambda_{\min}$ follow. Smaller values of $\overline{\mathrm{OP}}$ (and therefore larger values of NO_x result for $\lambda_{\min} > 1.1$, partially closing the EGR valve and retarding the start of injection. The injected fuel mass $m_{\mathrm{f,max}}$ follows from $\lambda_{\min}$ according to Fig. 8.8.1.

If the optimization with the performance criterion (8.8.9) is made for a selected range of engine speeds, and the same weight w_1 is used, then one obtains the optimal maps of the manipulated variables u_i for the feedforward control of the smoke operation in the complete operation range, compare Fig. 8.8.6b). Figure 8.8.7 shows the complete signal flow for smoke limitation control, which results from combining Figs. 8.8.1 and 8.8.4.

Pfeil (2011) has also treated a global optimization and has shown that the optimized smoke operation resulted in a 40% improvement of NO_x and soot emissions during a FTP driving cycle. Especially the soot peaks during the accelerations could be considerably reduced.

Summarizing, the simplest form of soot limitation control during acceleration is obtained by limiting the injected fuel mass $m_{\mathrm{f,max}}$ with an appropriate air/fuel ratio $\lambda_{\min}$, Fig. 8.8.1. If a soot model can be determined the modeled soot output can be limited and a limit value for the injected fuel mass can be calculated. A more general method is to use special smoke operation models for dynamic soot, NO_x and torque development and to optimize a set of combined manipulated variables, for smoke operation like $m_{\mathrm{f,max}}(\lambda_{\min})$, u_{egr}, φ_{mi} and p_{rail}, see Fig. 8.8.7.

8.9 Emission control

The exhaust gas emissions during combustion consist of harmless and unavoidable components like CO_2 and H_2O and harmful components like carbon monoxide CO, unburned carbons HC, nitrogen oxides NO_x (NO, NO_2) and particulates (solids, soot). As the legal emission values decrease strongly, see Sect. 6.8.1, the emissions have to be reduced by the combustion process itself and by exhaust gas aftertreatment. Measures to lower the raw emissions are e.g. better air/fuel mixture, better injection spray formation and exhaust gas recirculation with cooling. Exhaust gas aftertreatment systems consist of catalytic converters to reduce NO_x, CO and HC to CO_2 and H_2O and particulate filters to reduce soot. Remaining particulates have to be reduced in an aftertreatment with particulate filters (PF) or by catalytic reduction (SCR, NSC) to meet the legislative limits. However, especially particulate filters form gas dynamic resistances such that the exhaust back pressure increases as well as the fuel consumption. Therefore the flow losses have to be kept small.

This section describes some approaches to model the raw emissions especially by experiments with identification methods. Because the dynamic behavior of emission measurement sensors and devices play a considerable role, corresponding models are derived. This is followed by a discussion on raw emission control. Then various

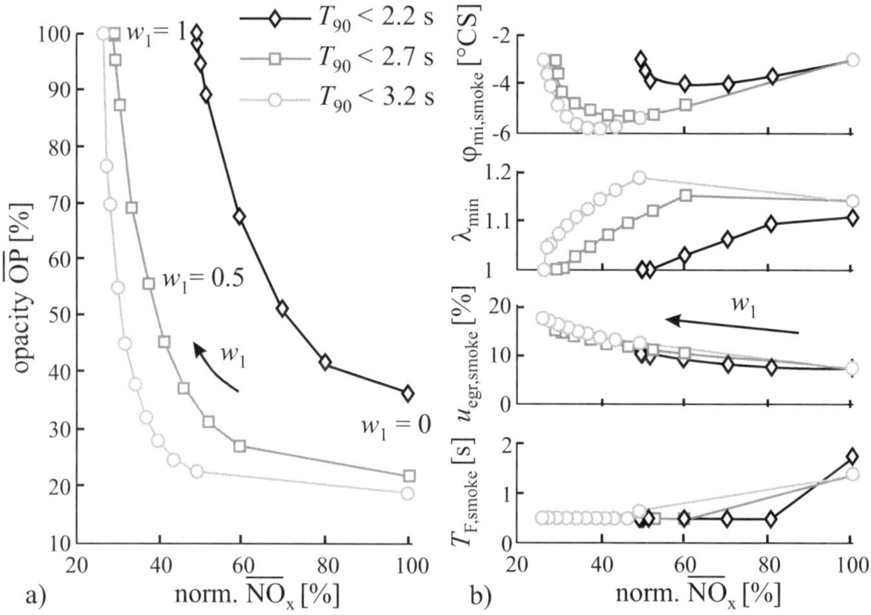

Fig. 8.8.6. Weighted optimization of the emissions NO_x and opacity (normalized) for different weighting factors $w_1 = 0 \ldots 1$ and different torque settling times $T_{90,\mathrm{min}} < 2.2\,\mathrm{s}$, $2.7\,\mathrm{s}$ and $3.2\,\mathrm{s}$ for one engine speed. $T_{\mathrm{F,smoke}}$ is a filter time constant for the switch from smoke operation to normal operation. **a** opacity vs. NO_x tradeoff for different torque dynamics. **b** resulting manipulated variables and λ_{min}.

exhaust gas after treatment components like special catalyst and particulate filters are considered, including their control.

8.9.1 Raw emission models

a) On emission modeling

Mathematical modeling of the generation of the emissions during the combustion is very complicated and generally needs a lot of computations, see Stiesch (2003), Merker et al (2006), Kirchen and Boulouchos (2008). Usually multi-zone models are required because the local concentration of fuel, air and temperature are important. One may use the observation that most of NO_x is produced as thermal NO in lean areas at high temperatures during an early phase of the pre-mixed combustion, see Fig. 8.2.6. Therefore, e.g. the brake mean effective pressure, maximum cylinder pressure, maximum temperature and the combustion center φ_{Q50} may give feasible correlations. The particulates are formed at high temperatures in local rich areas, but are then reduced in the post-combustion phase, such that several variables like fuel/air ratio, maximum temperature and the combustion center φ_{Q50} may be especially correlated. Therefore correlations of NO_x- and PM-emission concentrations

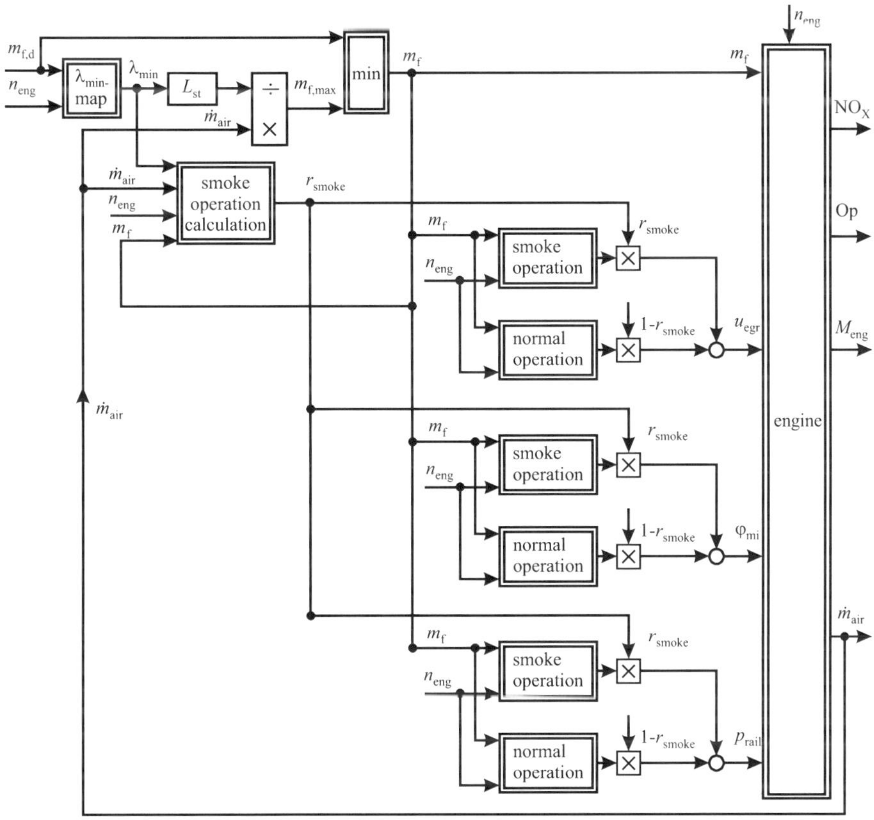

Fig. 8.8.7. Smoke limitation control with smoke operation models for the minimum air/fuel ratio λ_{min} and resulting $m_{f,max}$, and with the additional manipulated variables m_{egr}, φ_{mi} and p_{rail}. Commercial diesel engine with fixed turbocharger.

and the heat releases of the three combustion phases with pre-mixed and diffusion combustion, see Sect. 8.2.3, may be a way to build phenomenological emission models. Because of the highly dynamic processes that last only a few ms the models of the fuel injection, combustion and raw emissions can be considered as static. The main dynamics are caused by the intake and EGR system, including turbocharger and coolers.

Figure 8.9.1 shows a signal-flow scheme for the development of emission models. Emission models may be built with stationary and dynamic models of the intake and EGR system, determining the air and exhaust charge in the cylinder, the fuel-injection model, describing injected mass, rail pressure and begin of injection for pilot and main injection, and different kind of combustion models. The combustion models yield outputs as inputs of the emission models. In the case of *empirical combustion models* their outputs are, e.g. basic thermodynamic variables from combustion cycles like pressure, temperature, air/exhaust ratio, heat release and derived characteristic quantities like $p_{\mathrm{cyl,max}}$ or φ_{Q50}, see Sect. 4.2. *Phenomenological combustion models* take more details of the combustion process into account, like mixture formation, chemical reactions, and laws for pre-mixed and diffusion combustion as discussed in Sect. 8.2.3.

In all cases at least parts of the models have to be determined experimentally. If the relations between the outputs of the intake and EGR system and fuel injection models are used to identify the emission models this is a black-box approach, e.g. by adapting polynomial models or neural nets, see e.g. Christen et al (2001), Hafner et al (2000), Mrosek et al (2010). A further step is to determine empirical or phenomenological combustion models first, by applying physical-oriented relations and then to identify raw emission models with their outputs. These models are considered to be semi-physical or gray-box models, as described in Traver et al (1999), Gärtner (2001), Benz et al (2010).

A comprehensive investigation is given by Benz (2010), Benz et al (2010). For NO_x and PM a static model with piston speed and fuel mean effective pressure serves as basis. Then a deviation model is identified with additional selected inputs, like cylinder air mass, air fraction, combustion center for NO_x and air mass, rail pressure and φ_{Q90} as inputs for PM, applying a symbolic regression method.

In all cases measured data has to be used to find the emission models, which express the relations between the emission concentration and the possible manipulated variables of diesel engines.

b) Local and global emission models

In the following experimentally obtained emission models are considered, which use the variables

$\dot{m}_{\mathrm{air}}$:	air mass flow rate
p_{2i} :	intake charging pressure
T_{2i} :	intake charging temperature
φ_{Q50} :	combustion center

as inputs for the operating points

u_{inj} :	injection mass
n_{eng} :	engine speed.

The exhaust mass flow rate follows, according to (4.1.2) to (4.1.8), from

$$\dot{m}_{\mathrm{egr}} = \dot{m}_{\mathrm{eng}} - \dot{m}_{\mathrm{air}} = \eta_{\mathrm{v}}(n_{\mathrm{eng}}, p_2)\,\frac{p_{\mathrm{2i}}(r_{\mathrm{egr}})}{2RT_{\mathrm{2i}}}\,V_{\mathrm{D}}\,n_{\mathrm{eng}} - \dot{m}_{\mathrm{air}}. \tag{8.9.1}$$

If the inputs $\dot{m}_{\mathrm{air}}$, p_2 and T_{2i} are used for emission modeling, the flow rate $\dot{m}_{\mathrm{egr}}$ is implicitly taken into account.

Engine model structures can generally be divided into local and global approaches. The *local models* are determined for fixed operating points (u_{inj}, n_{eng}),

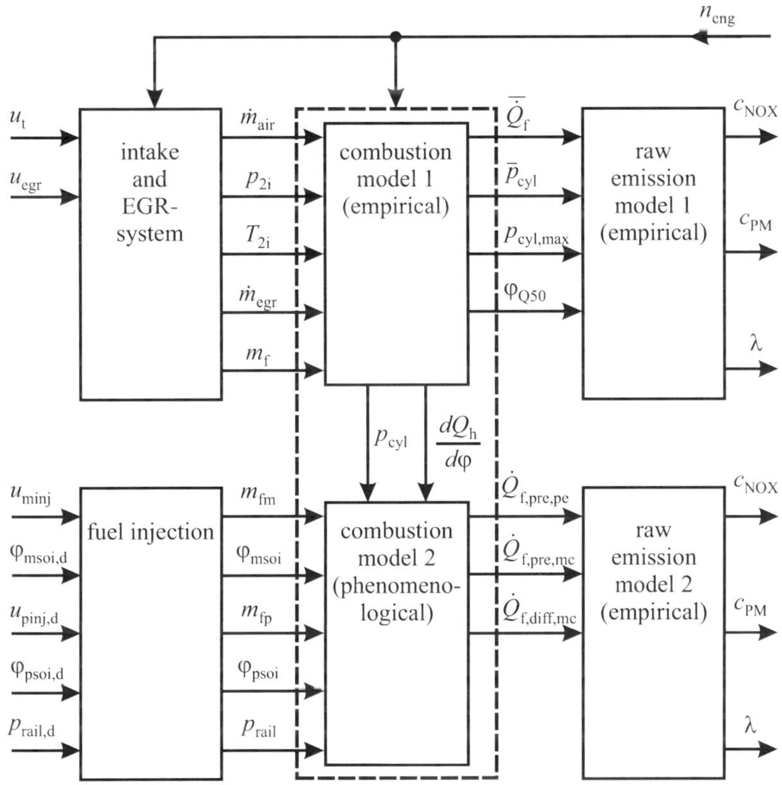

Fig. 8.9.1. Overall signal flow to generate semi-physical empirical and phenomenological raw emission models for diesel engines. (Some symbols are explained in Sect. 8.2.3.)

such that for certain operating points output models of the form

$$\hat{y}_{jk} = f_{jk}(\dot{m}_{air}, p_{2i}, T_{2i}, \varphi_{Q50}) \qquad (8.9.2)$$
$$j = u_{inj}\,; \quad k = n_{eng}$$

are determined, where $\hat{y}$ stands for c_{NOx} or c_{PM}. *Global models* include the operating points explicitly

$$\hat{y} = f(\dot{m}_{air}, p_{2i}, T_{2i}, \varphi_{Q50}, u_{inj}, n_{eng}). \qquad (8.9.3)$$

Global-local models combine the local models to generate a global output by weighting functions

$$\hat{y} = \sum_j \sum_k \Phi_{jk}(u_{inj}, n_{eng})\, f_{jk}(\dot{m}_{air}, p_{2i}, T_{2i}, \varphi_{Q50}). \qquad (8.9.4)$$

The weighting function is usually a linear interpolation between the operating points, see Fig. 8.9.5.

c) Identification of raw emission models

A scheme of the investigated CR diesel engine with VGT, high pressure and low pressure exhaust gas recirculation is depicted in Fig. 8.9.2. The goal of the identification is to obtain a mean value model for the emissions nitro oxygen NO_x and particulates PM, both for steady-state and dynamic behavior, by using relevant inputs from the intake system and the combustion, as described by Sequenz and Isermann (2011).

Figure 8.9.3 shows the underlying model structure. To obtain a mean value model of the emissions the combustion can be considered as a batch process, in this case characterized by a steady-state value of the combustion center φ_{Q50}, determined from cylinder pressure measurement $p_{cyl}(\varphi)$ or from a corresponding model. However, as the major dynamics of the emissions are determined by the dynamics of the air and exhaust gas recirculation system, the states of the gas mixture as inputs m_{air}, p_{2i}, T_{2i} are measured dynamically, assuming that they can be measured without considerable sensor time delays or simulated. The emission model then uses m_{air}, p_{2i}, T_{2i} and φ_{Q50} as inputs for local models as (8.9.2) or additionally with u_{inj}, n_{eng} for global models as (8.9.3).

The emission models then result as stationary models and describe the stationary and dynamic behavior of the emissions by using the dynamic inputs from the air and exhaust system. With reference to Fig. 8.9.1 an empirical raw emission model 1 with an empirical or analytical combustion model is obtained.

The determination of the manifold temperature T_{2i} is based on the measurement of the temperature T_{ic} after the charge air cooler and the temperature T_{hp-egr} after the exhaust air cooler, see Fig. 8.9.2, by calculating a mixture temperature

$$T_{2i} = \frac{(\dot{m}_{air} - \dot{m}_{lp-egr})T_{ic} + \dot{m}_{hp-egr}T_{hp-egr}}{\dot{m}_{air} + \dot{m}_{lp-egr} + \dot{m}_{hp-egr}}. \qquad (8.9.5)$$

The EGR mass flow rates are determined based on the adiabatic flow equation for valves

$$\dot{m}_{egr} = c_{th}\, A(s_{egr})\, \frac{p_{in}}{\sqrt{R T_{in}}}\, \psi\left(\frac{p_{out}}{p_{in}}\right) \qquad (8.9.6)$$

according to (4.1.19). Thus, for the determination of $\dot{m}_{egr-hp}$ measurements of $p_{in} = p_3$ and $p_{out} = p_{2i}$ and for $\dot{m}_{egr-lp}$ measurements of $p_{in} = p_4 - \Delta p_{PF}$ and $p_{out} \approx p_a$ are required (for the pressure drop in the particulate filter Δp_{PF} see Sect. 8.9.4). However, if $\dot{m}_{egr-lp} = 0$ then $\dot{m}_{egr}$ can be calculated with (8.9.1), without additional sensors.

For the identification a model network LOPOMOT (local polynomial model tree) as shown in Fig. 8.9.4 is applied. Local polynomial models are identified for fixed operating points, forming the z-regressors u_{inj}, n_{eng}. The local nonlinear models use m_{air}, p_{2i}, T_{2i}, φ_{Q50} as inputs, forming the x-regressors. By superposition with weighting functions Φ_{jk} the outputs c_{NOx}, c_{PM} and M_{eng} are

$$y = \sum_j \sum_k \Phi_{jk}(u_{inj}, n_{eng})(w_{0,jk} + w_{1,jk}\, x_1 + w_{2,jk}\, x_1 x_2 + w_{3,jk}\, x_1^2 + \ldots) \qquad (8.9.7)$$

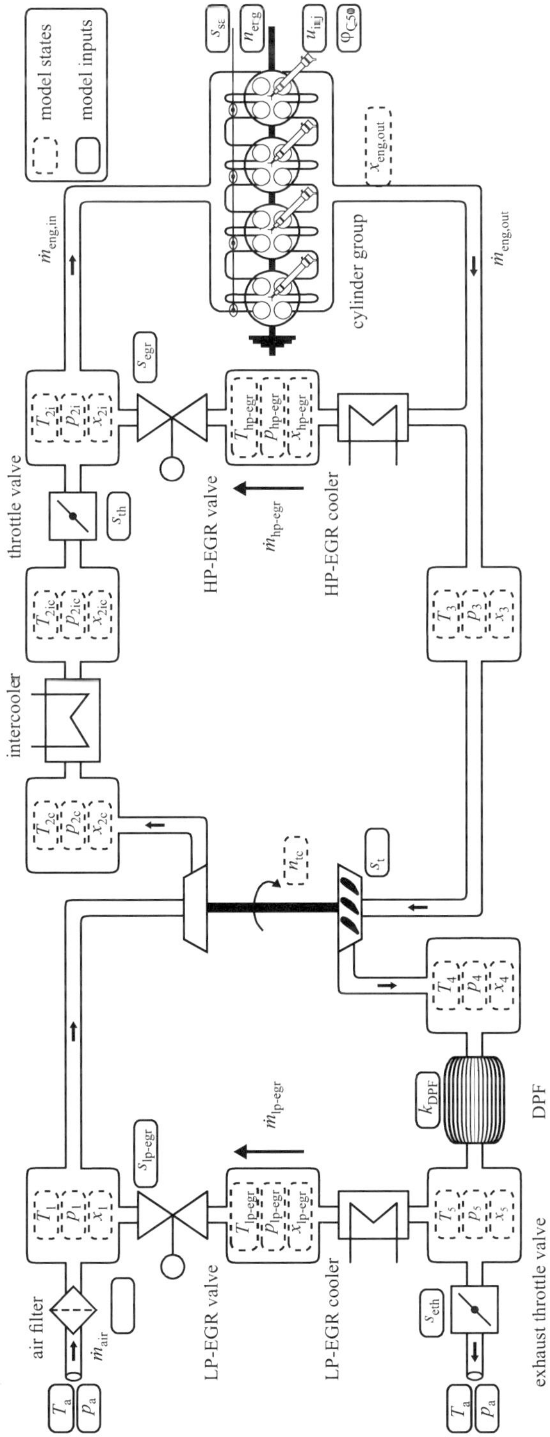

Fig. 8.9.2. Scheme of the CR diesel engine with VGT, high pressure (hp) and low pressure (lp) exhaust gas recirculation. Opel Z 19 DTH, 1.9 l, 315 Nm, 110 kW.

where Φ_{jk} are the coefficients of a linear interpolation between at least four local models according to (6.5.13) and w_ν are the estimated parameters.

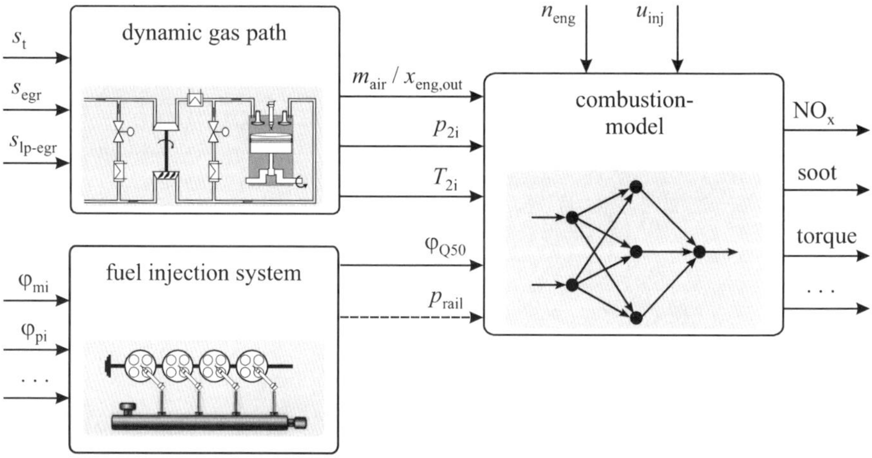

Fig. 8.9.3. Signal-flow diagram for the identification of emission models, consisting of a dynamic gas path model and stationary combustion and emission models.

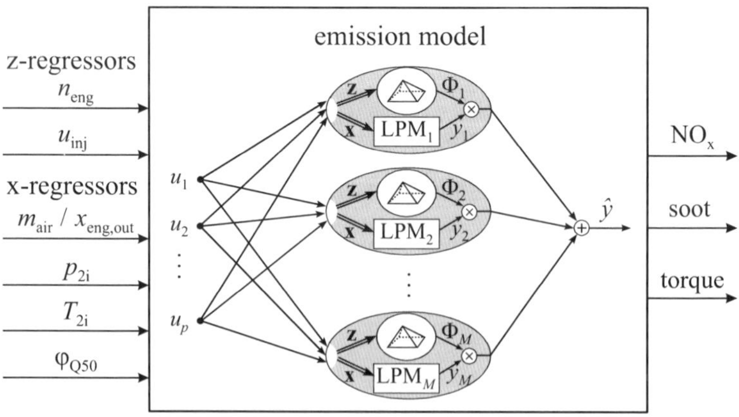

Fig. 8.9.4. Structure of the stationary emission model with local polynomial models (LPM). The local polynomial models are composed by linear interpolation to a global model output (LOPOMOT).

The distribution of the stationary engine operating points is equidistantly selected within the emission-cycle-relevant operating boundaries, as shown in Fig. 8.9.5. Validation measurements are taken in between the operating points. Measurements of

NO_x are made with an NGK sensor and measurements of the the soot are made with an AVL microsoot sensor.

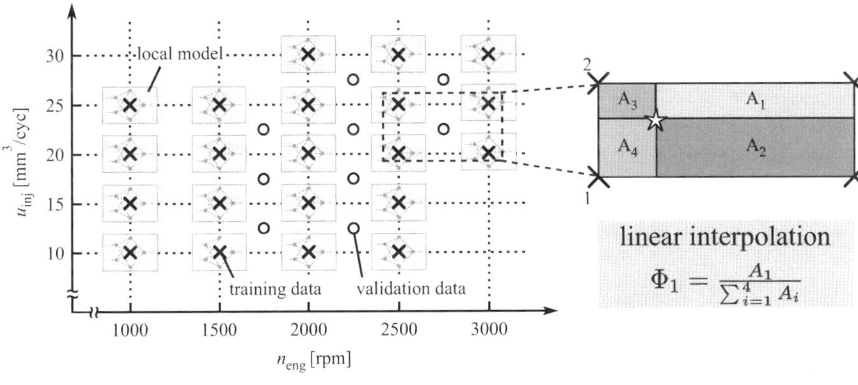

Fig. 8.9.5. Distribution of local models, training (×) and validation data (○) in the engine operation range.

Because the microsoot measurement shows a linear increase of the measurement error with the output value a multiplicative measurement error can be supposed. Therefore a logarithmic transformation of the output is made, in order to perform the identification with an additive measurement error, Sequenz and Isermann (2011).

The structure search of the LOPOMOT identification results in 21 local models of third order with altogether 421 parameters for NO_x and 448 parameters for soot. Figure 8.9.6 depicts the measured versus predicted plots, also indicating the coefficient of determination R^2 and the root mean square error (RMSE) as criteria of fit. The precision of the NO_x model is higher than that of the soot model. One reason is that the soot measurements themselves have a high variance.

A comparison of the simulated dynamic behavior with the measured variables for the extra urban part of NEDC driving cycle is presented in Fig. 8.9.7. The simulated emissions are shown as model outputs and with transport delay to the measurement point and sensor dynamics of first order. The test cycle includes several accelerations and a gear shift at $t = 1050\,\mathrm{s}$. Measured and simulated emissions agree well, especially after inclusion of the measurement dynamics in the simulations, see also the zoomed boxes.

The described emission models use the variables m_{air}, $p_{2\mathrm{i}}$, $T_{2\mathrm{i}}$, $\varphi_{\mathrm{Q}50}$ as inputs. An alternative is to use m_{air}, m_{egr}, $T_{2\mathrm{i}}$, $\varphi_{\mathrm{Q}50}$ by using (8.9.1). Applying the gas composition after combustion

$$x_{\mathrm{eng,out}} = \frac{m_{\mathrm{air}} - m_{\mathrm{air,burnt}}}{m_{\mathrm{gas,tot}}} = \frac{m_{\mathrm{air}} - m_{\mathrm{f,inj}} L_{\mathrm{st}}}{m_{\mathrm{gas}} - m_{\mathrm{f,inj}}} = \frac{(\lambda - 1) L_{\mathrm{st}}}{\lambda L_{\mathrm{st}} - 1} \tag{8.9.8}$$

instead of the air mass m_{air} yields more precise soot models, because the oxygen content in the expansion phase is relevant for the soot in the exhaust gas, Sequenz and Isermann (2011).

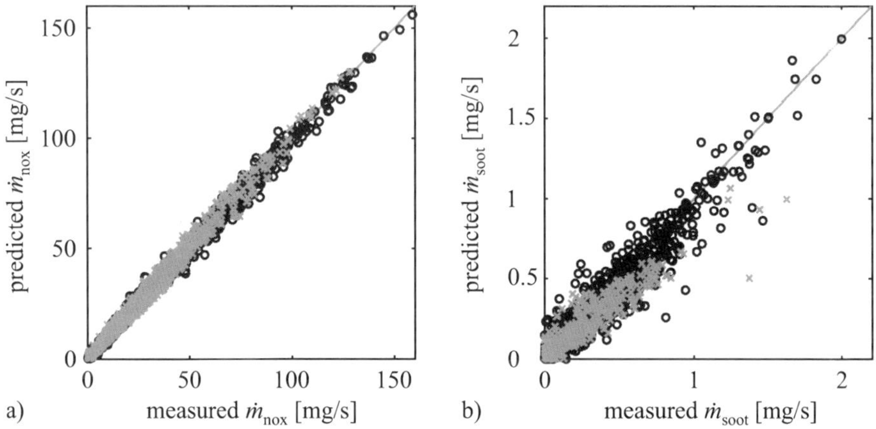

Fig. 8.9.6. a Measured vs. predicted plot for the stationary NO_x model with identification data (black circles) ($R^2 = 0.994$, RMSE = 17.5 ppm) and validation data (grey crosses) ($R^2 = 0.984$, RMSE = 24.6 ppm). **b** measured vs. predicted plot for the soot model with identification data ($R^2 = 0.970$, RMSE = 0.43 mg/m^3) and validation data ($R^2 = 0.932$, RMSE = 0.61 mg/m^3), Sequenz (2013).

To illustrate the influence of individual input variables on the outputs of these nonlinear multi-input single-output (MISO) models, intersection plots can be generated. The multidimensional model is then simulated and all inputs except one are kept constant, Sequenz and Isermann (2011). If the output is plotted in dependence on one input, a two-dimensional characteristic results. In the case of two outputs, a three-dimensional map is obtained. Figure 8.9.8 shows these two plots for the case that the gas composition $x_{eng,out}$ according to (8.9.8) is used as input instead of m_{air} to model the soot behavior. Figure 8.9.8a) depicts the strong influence of the exhaust gas composition $x_{eng,out}$ on the soot formation. The two-dimensional intersection plots in Fig. 8.9.8b) illustrate the dependence for the two outputs soot and NO_x on $x_{eng,out}$ and the charging pressure p_{2i} with their reverse behavior, the so called NO_x-soot trade-off.

d) Dynamic models of emission measurement

For measurement of emissions at engine dynamometers or chassis dynamometers, usually emission test benches or gas analyzers are used. They are usually designed for stationary measurements. However, emission control as well as the determination of dynamic engine models require a relatively fast dynamic measurement of the emissions, like oxygen content, NO_x and particulates. This can be provided by some emission measurement devices. In order to investigate their dynamic behavior experimental investigations are described for a 4-cyl. CR-diesel engine with an NO_x-sensor (NGK), an opacimeter (AVL 439) and a microsoot sensor (AVL 483).

Sensors for measurement of *nitrogen oxides* NO_x, which consist of nitrogen monoxide NO and nitrogen dioxide NO_2 are based on the reduction of NO into N_2

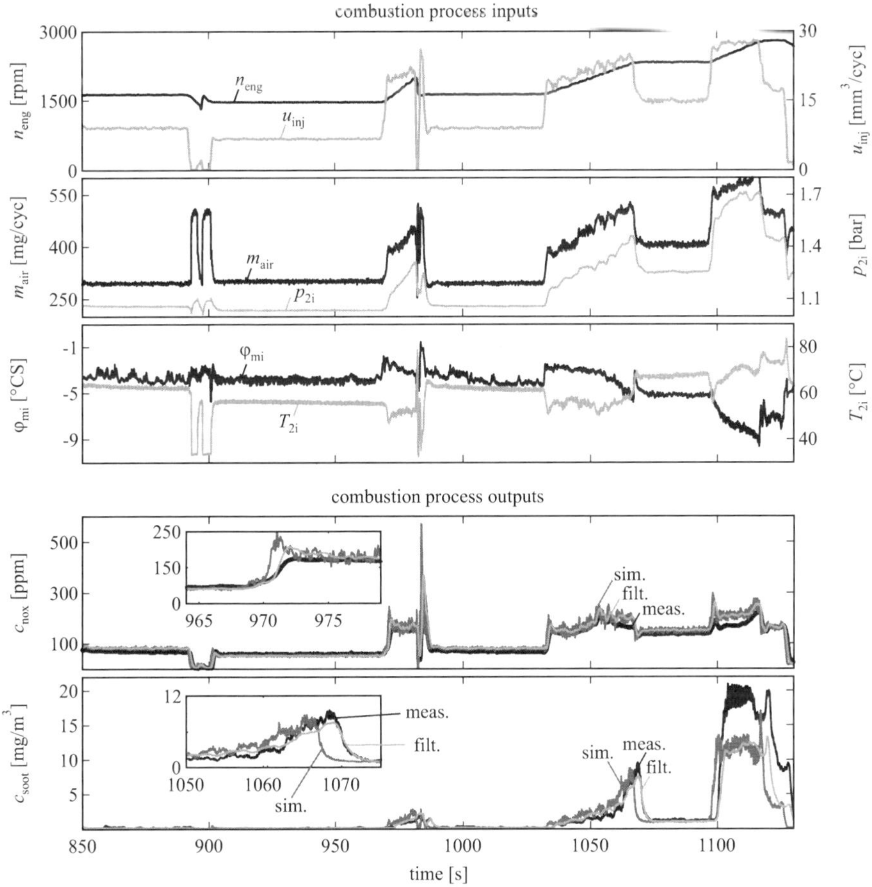

Fig. 8.9.7. Dynamic measurements and simulation with the identified global emission model for the extra urban part of the NEDC. Calibrated ECU with high-pressure EGR only. The emissions are shown as simulated model outputs (sim) without measurement delay and with measurement delay (filt) for $T_{\mathrm{d,NOx}} = 0.4\,\mathrm{s}$, $T_{\mathrm{s1,NOx}} = 1.0\,\mathrm{s}$; $T_{\mathrm{d,soot}} = 1.65\,\mathrm{s}$, $T_{\mathrm{s1,soot}} = 0.9\,\mathrm{s}$.

and O_2 by using pump electrodes for oxygen and measurement of the O_2 concentrations, see e.g. Kato et al (1997).

Microsoot devices (MSS) detect particulates for low levels in the range of $5\,\mu\mathrm{g/m}^3$, use a laser and are based on photo-acoustic principles. The modulated laser beam heats the black particulates so that the surrounding gas generates an acoustic sound in a resonance cell which is measured by a microphone. A signal processor analyzes the dominating oscillation.

Opacimeters (OPA) measure the opacity of undiluted exhaust gases by the degree of obscuration of a light beam. The measurement value is the percentage of blocked light. Therefore 0% is clean and 100% is thick black smoke. A part of the exhaust gas

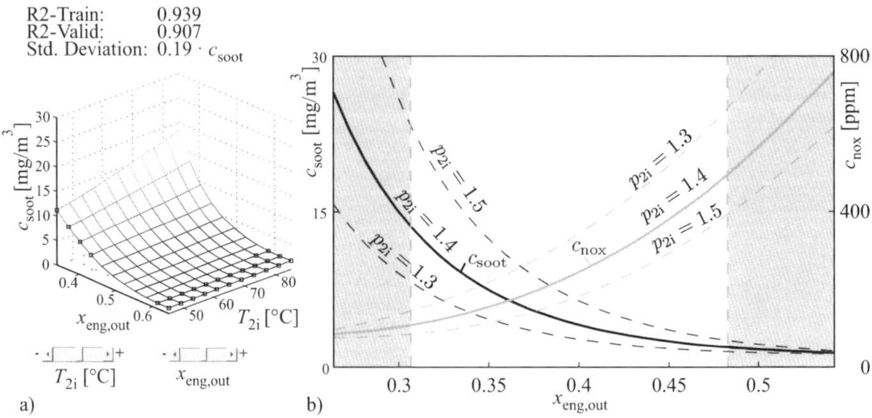

Fig. 8.9.8. Intersection plots of emission models. **a** Soot in dependence on manifold exhaust gas composition and intake temperature (with extrapolation). **b** Soot (black) and NO_x (gray) in dependence on manifold exhaust gas composition for different charging pressures. Other inputs are constant: $n_{eng} = 2200\,\text{rpm}$; $u_{inj} = 24\,\text{mm/cyc}$; $T_{2i} = 70°\text{C}$; $\varphi_{Q50} = 194°\text{CS}$. The shaded areas indicate extrapolated regions, Sequenz (2013).

is pumped by diaphragm pumps through hoses to the sensor with controlled constant volume flow rate. The transportation time therefore generates a certain dead time.

Fig. 8.9.9 shows measured transient functions for a ramp change of the rail pressure p_{rail}. The fastest response is obtained with the opacimeter, Fig. 8.9.9a). The NO_x sensor has a dead time of about 200 ms and low pass behavior. The microsoot sensor is the slowest with a dead time of about 2.5 s and low pass behavior. Hence, the dynamics of the sensors are different and depend on the flow speed of the exhaust gases, or engine speed, Fig. 8.9.9b).

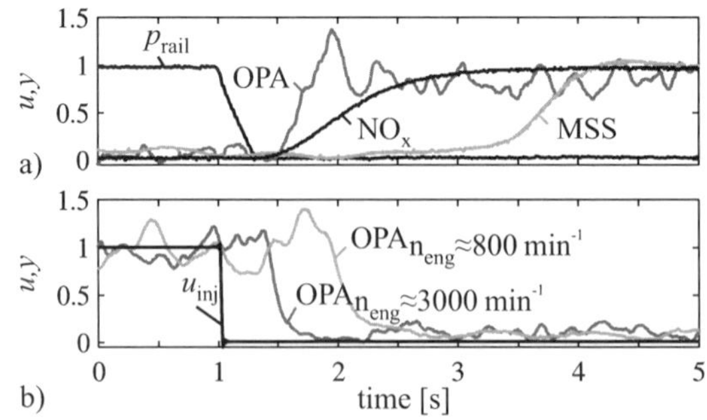

Fig. 8.9.9. Normalized emission sensor responses. **a** change of rail pressure p_{rail}; $n_{eng} = 1900\,\text{rpm}$. **b** only opacimeter. $n_{eng} = 800\,\text{rpm}$ and $3000\,\text{rpm}$ for change of injection mass u_{inj}.

Based on the measured transient functions, assuming linear dynamics, the behavior of the emission measurements can be approximated by the transfer function

$$G_s(s) = \frac{K_s}{(1 + T_{s1}s)(1 + T_{s2}s)} e^{-T_d s} \tag{8.9.9}$$

compare the model for a λ-sensor in Sect. 7.2. The time constants T_{s1} and T_{s2} are due to storage effects of the sensors and pipes and the dead time T_d is caused by several transportation times

$$T_d = T_{dc} + T_{d,V} + T_{d,s} = T_{d,eng} + T'_d \tag{8.9.10}$$

with

$$\begin{aligned} T_{dc} &= \pi/\omega_{eng} \\ T_{d,V} &= V/\dot{V}_j \qquad j = 3, 4 \\ T_{d,s} &= \tfrac{l}{v_s}. \end{aligned}$$

For a 4-stroke engine, T_{dc} is the cycle time between intake valve closing and exhaust valve opening, i.e. about one crankshaft revolution. For the NO_x sensor, which is placed after the turbocharger turbine, the travel time through the exhaust manifold is $T_{d,V} = V/\dot{V}_3$ and for the opacity and microsoot sensors, which are placed further downstream, it holds $T_{d,V} = V/\dot{V}_4$, compare Fig. 8.9.10. $T_{d,s}$ expresses the travel time through the connecting hoses and the sensor. It can be relatively large if the sensors are connected via a long hose of length l and a controlled gas velocity v_s.

The transportation times to a λ-sensor placed after the manifold range between 15 and 70 ms for a passenger car engine, Kimmich (2004), see also (7.2.5).

In order to excite only the dynamic behavior of the emission measurements (sensors and possibly hoses) a test signal should be applied, which only influences the emission generation of the combustion and not the air intake and EGR system. Therefore changes of the rail pressure p_{rail} are used as an excitation signal. A resulting change of the combustion is compensated for by a change of the start of main injection $\varphi_{mi} = f(p_{rail})$, Mrosek et al (2011). Figure 8.9.11 depicts the resulting experiment and shows that p_{2i}, $\dot{m}_{air}$, φ_{Q50} and T_3 remain constant.

A square wave of eight steps was then applied for 60 operating points varying from $n_{eng} = 790$ to $3000\,\mathrm{rpm}$. The used sampling time was $T_0 = 10\,\mathrm{ms}$. The dead times were directly determined from the recorded data.

The transient functions of Fig. 8.9.9 show that the NO_x-sensor has the behavior of a second order system. Its time constant T_{s2} can be fixed with $T_{s2} = 450\,\mathrm{ms}$. The sensors for the particulates show first order dynamics. Therefore, in all three cases the time constant T_{s1} was determined by least squares parameter estimation.

The resulting dead times and time constants T_{s1} are presented in Figs. 8.9.12 and 8.9.13 in dependence on the volume flow rates. The dead times of the sensors become smaller with increasing $\dot{V}$ and approach constant values for $\dot{V} > 1\,\mathrm{m}^3/\mathrm{s}$. Its value originate mainly from the sensor itself. The time constants decrease approximately linearly with increasing volume flow rate

$$T_{s1,j} = T_{s1,j0} - w_j \dot{V}_{3\,\mathrm{or}\,4}\,; \quad j = \text{mss, opa, nox} \tag{8.9.11}$$

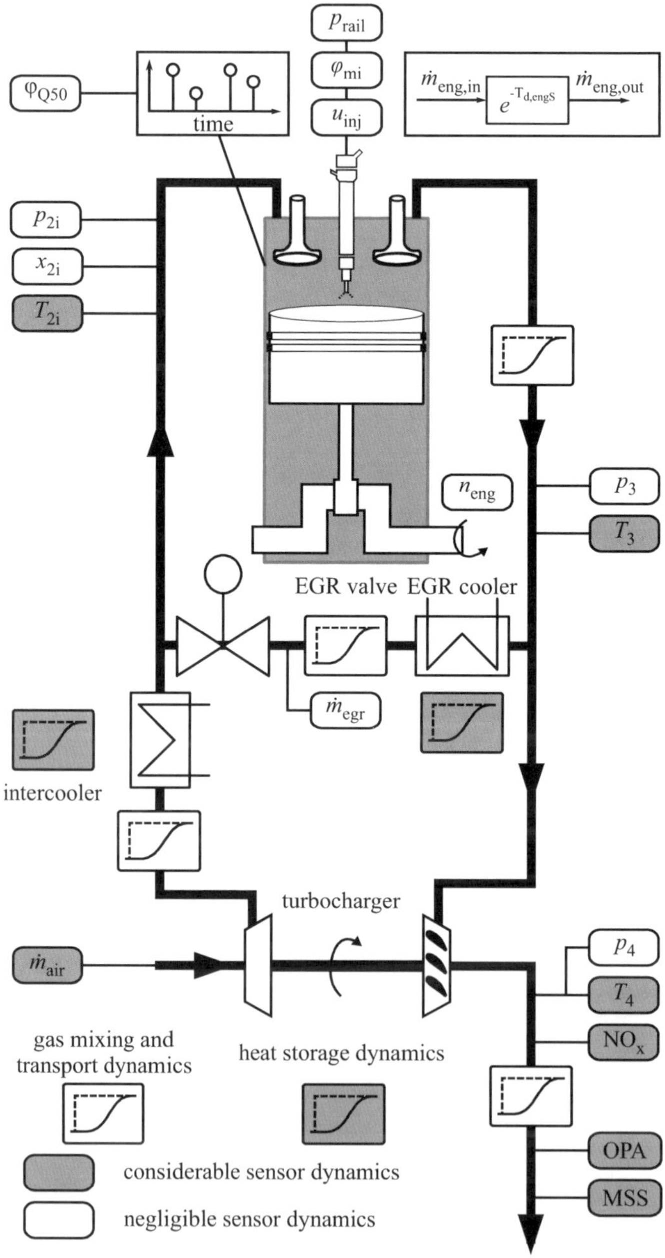

Fig. 8.9.10. Placement of emission sensors and schematic for determining the transportation dead times.

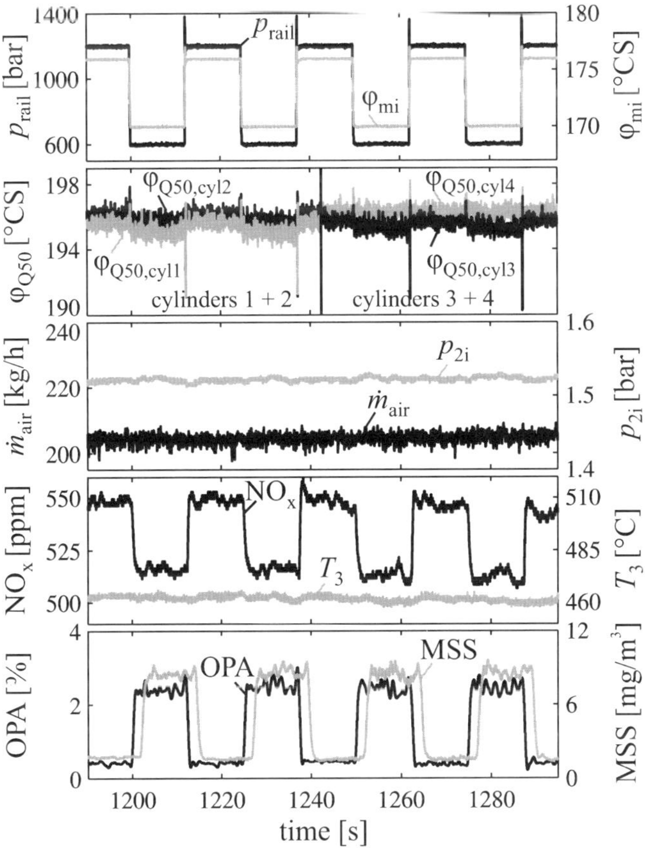

Fig. 8.9.11. Simultaneous excitation of p_{rail} and φ_{mi} to identify the emission sensor dynamics. $t < 1245\,\mathrm{s}$: cylinder 1 and 3, $t > 1245\,\mathrm{s}$: cylinders 3 and 4.

A comparison with the dynamic models of a λ-sensor considered in Sect. 7.2 shows the same dependency on the exhaust gas flow rate, with time constants of 80 to 100 ms, see Table 7.2.1. Table 8.9.1 summarizes the identified parameters.

Table 8.9.1. Identified dead times and time constants of emission sensors

$\dot{V}_{\mathrm{exh}}$	microsoot		opacimeter		NO_x		
	T'_{d}	T_{s1}	T'_{d}	T_{s1}	T'_{d}	T_{s1}	T_{s2}
$\left[\frac{\mathrm{m}^3}{\mathrm{s}}\right]$	[ms]	[ms]	[ms]	[ms]	[ms]	[ms]	[ms]
0.02	3300	950	750	80	320	190	450
0.06	2200	800	450	70	240	120	450
0.1	1950	600	300	55	220	80	450

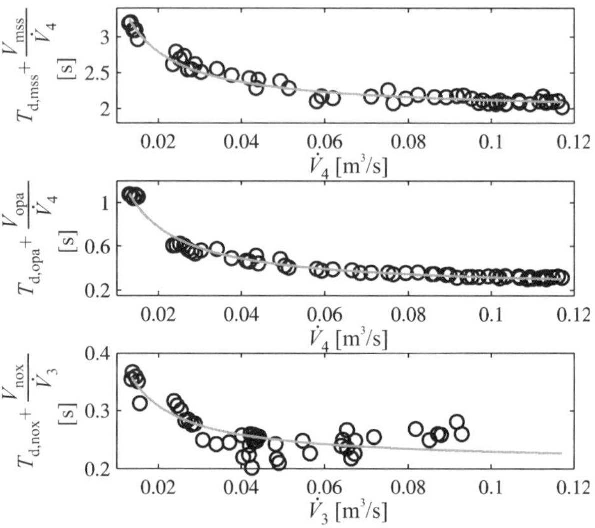

Fig. 8.9.12. Identified dead times for the emission measurement devices.

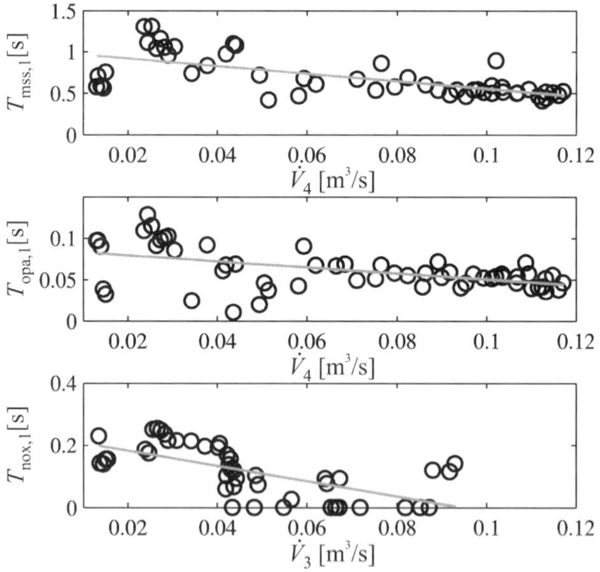

Fig. 8.9.13. Identified time constants of the emission measurement devices.

The microsoot sensor has the largest dead time and time constant. These delays are caused mainly through the connection line length. The opacimeter is faster, has a considerable dead time but a small time constant. The NO_x sensor has dead times between 200 and 300 ms and a dominating time constant of 450 ms. A comparison between modeled and measured behavior is shown in Fig. 8.9.14 and shows good agreement. Taking average values for the time constants, the corner frequencies of the frequency responses are $\omega_c = 1/T_{s1} = 1.25;\ 14.2;\ 2.2\ 1/s$ or $f_c = \omega_c/2\pi = 0.19;\ 2.76;\ 0.35\,Hz$. This means that the dynamic measurements of combustion engines have to be taken into account for frequencies larger than about 0.2 Hz (microsoot sensor), 2.5 Hz (opacimeter), and 0.35 Hz (NO_x-sensor).

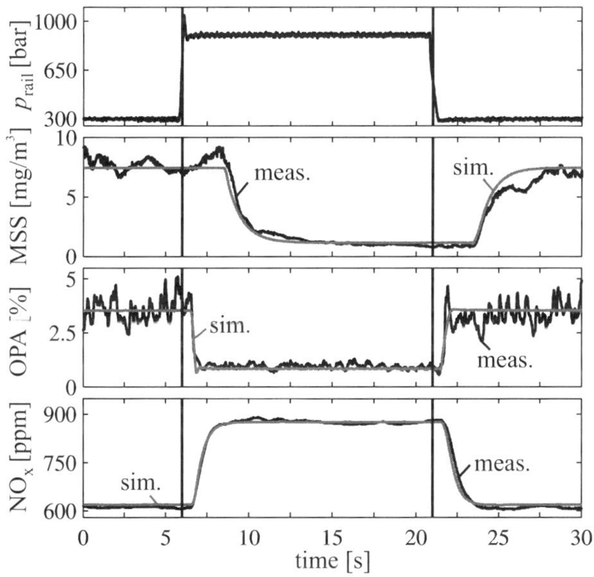

Fig. 8.9.14. Measured and simulated transient functions of the emission sensors. $n_{eng} = 913\,rpm$, $u_{inj} = 23.5\,mm^3/cyc$; $\dot{m}_{air} = 52.5\,kg/h$.

8.9.2 Raw emission control

The legislative requirements of the exhaust emissions of diesel engines are becoming very restrictive especially for NO_x and particulates (PM), as described in Sect. 6.8.1. The generation of the raw emissions is reduced on the one hand by improving the injection and combustion process and on the other by optimal calibration of the electronic control settings, see Sect. 6.8. NO_x can especially be reduced by the recirculation of cooled exhaust gas (EGR), thus leading to smaller air/fuel ratios (AFR) and gas temperature. However, with increasing EGR the formation of particulates increases. Therefore a compromise of this NO_x-PM tradeoff has to be found, see Fig. 8.9.15.

Usually, the EGR valve is used to control the air flow and therefore the air/fuel ratio and the turbine actuator to control the charging pressure. This results in a mutually coupled two-variable control system, depicted in Fig. 8.5.2. The reference values for these closed loops are optimized during the control calibration with regard to torque generation, fuel consumption, driveability and emissions. This way the preconditions for the combustion, like the air/fuel ratio and charging pressure are feedback controlled. However, the combustion process is then feedforward controlled if the combustion pressure is not measured and used for closed loop control. Changes in the features influencing the combustion such as deposits, coking or widening of injector holes, wear and aging of valves and actuators and faults of sensors then lead to changes of the emissions. Figure 8.9.15 depicts the NO_x concentration in dependence on the particulates PM with regard to the legislative limits. Usually the design and calibration has to be placed with some safety margins to the limits, in order to cope with production tolerances and aging. If the emissions can be measured and feedback controlled, the tolerances may be significantly smaller.

Production sensors exist for NO_x and λ, but until now not for soot. Then the oxygen content λ of the raw exhaust gases can be controlled by manipulation of the EGR valve and the NO_x by manipulating the start of injection φ_{soi}, as proposed by Alfieri et al (2009). This two-variable control system then forms a feedback from the states after the combustion and compensates several effects of the combustion process and charging combustion conditions. By appropriately selecting the reference variables for the λ-control the particulates can be influenced indirectly, as the oxygen content of the exhaust gases during and at the end of the expansion phase is relevant for the oxidation of the particulates, see Sect. 8.2.6.

The design of this control system for the air/fuel ratio (AFR) in the exhaust gas $\lambda_{cyl,out}$ by manipulation of the EGR valve is treated in Alfieri et al (2009). Approximating the process transfer behavior with a linear first order and dead time model a linear internal model controller follows directly, where the parameters are adapted to the operating point. Static feedforward control of the EGR-valve from the desired reference variable $\lambda_{cyl,ref}$ and from the injected fuel mass flow rate $\dot{m}_f$ improve the dynamic control performance. Hence, this exhaust gas λ-control system may replace the air mass controller in Fig. 8.5.2 and instead of the steady-state optimized reference value $\dot{m}_{air,d}$ the reference value $\lambda_{cyl,out,d}$ has to be optimized.

Together with a NO_x controller manipulating the start of injection angle φ_{soi}, it was shown that under the influence of drift in the air flow and charging pressure sensor the deviations of NO_x and PM are smaller compared to standard control, thus keeping the narrower tolerances in Fig. 8.9.15.

The shift of φ_{soi} by retarding to values closer to TDC or even after TDC may reduce the NO_x concentration considerably, see e.g. Motz et al (2011). Results from the realization of a NO_x-control system for a truck diesel engine confirm the smaller tolerances of Fig. 8.9.15 and allow a reduction of fuel consumption through larger reference values of NO_x because of the smaller variances, Eijk et al (2011).

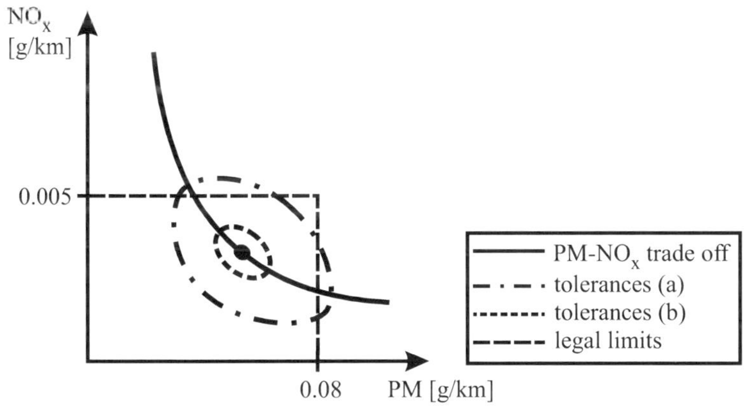

Fig. 8.9.15. Legal limits and tolerances for the nitro oxygen and particulate matter for a driving cycle. **a** tolerances through production tolerances, aging and wear. **b** smaller tolerances with emission feedback control.

8.9.3 Exhaust gas after-treatment and its control

The exhaust gas after-treatment has the task of converting the unwanted emissions of the raw exhaust gases which cannot be avoided by the optimization of the combustion process into harmless components as much as possible. For diesel engines this is realized with different types of catalytic converters and particulate filters.

a) Exhaust gas oxidation

Diesel oxidation catalysts (DOC) have the task to oxidize carbon monoxide (CO) and unburned hydrocarbon (HC) emissions with the remaining oxygen (O_2) of the exhaust gas into water vapor (H_2O) and carbon dioxide (CO_2). The DOC converter is built as a ceramic or metallic honeycomb structure where the exhaust gases are forced to flow through thin channels. The channel walls consist of a ceramic or metal substrate which is coated by a catalyst from platinum or palladium. The substrate is a washcoat consisting of aluminum, cerium or zirconium oxide and offers a large surface area. The exhaust gases diffuse into the catalyst coat and are oxidized, supported by the catalytic reaction of the washcoat. The DOC's are placed close to the exhaust gas manifold to reach the required light-off temperature at 150-200°C as quickly as possible.

b) Reduction of NO_x

A first way to reduce NO_x is to use a NO_x *storage catalyst* (NSC), also called lean NO_x trap (LNT), see Fig. 8.9.16a). The catalyst has a coating with alkali or barium oxides and stores NO_x at temperatures between 250 and 450 °C. Two different operating modes can be distinguished. In normal lean burn operation $\lambda > 1$ NO is first oxidized to NO_2 and then adheres in the form of nitrate (NO_3) to a metal oxide in a loading phase. Then a regeneration phase follows with rich exhaust gas $\lambda < 1$, by retarded injection and air throttling. The loading phase lasts 30 to 300 s and the regeneration phase 2 to 10 s. End of loading phase can be detected by a NO_x storage model or by a NO_x sensor after the catalyst and end of regeneration respective discharging phase by a storage model or λ-sensor after the catalyst, indicating $\lambda < 1$. The advantage of NO_x storage catalysts is that they do not need an extra agent. However, they are very sensitive to sulfur compounds, since SO_2 is also stored in the catalyst, leaving less capacity for NO_x with time. Therefore sulfur regeneration may be required by heating the catalyst to temperatures $> 650°C$ by generation of rich exhaust gas $\lambda < 1$ after longer driving distances. The efficiency of the NSC is about 50–80% reduction of NO_x. Hence, the control of this accumulator type NO_x catalyst consists in manipulating the loading and regeneration phase with regard to NO_x and at longer time intervals with regard to SO_2.

The dynamic behavior of NO_x-storage catalyst can be modeled by balance equations for O_2, NO_x and heat and results in a set of nonlinear differential equations, see e.g. Canova et al (2010). Investigations on the regeneration are described e.g. in Pischinger et al (2009) and Ketfi-Cherif et al (2000).

Selective catalytic reduction (SCR) reduces NO_x to nitrogen N_2 and water vapor H_2O by adding ammonia NH_3. The reduction requires temperatures $> 250°C$ and a suitable catalyst. NH_3 has a large sensitivity to be oxidized from NO_x. It is generated by injecting a 32.5% urea/water solution (called AdBlue or Diesel Exhaust Fluid, DEF) into the exhaust pipe, see Fig. 8.9.16b). The solution is then, after composing an intermediate product, converted by thermolysis and hydrolysis into ammonia. The dosing of the solution depends on the molar feed ratio α of the NH_3 equivalent to the present NO_x in the exhaust gas. If $\alpha > 1$, NH_3 will leave the SCR catalyst unconverted after exceeding its NH_3 storage capacity. This so called NH_3 slip or leakage has to be avoided because it is an odor disturbance to the environment. SCR catalysts have a good NH_3 storage capacity, such that small short-term deviations from the theoretical feed ratio do not cause NH_3 slip. Frequently AdBlue SCR system contains a DOC after the SCR catalyst that oxidizes NO to NO_2 and reduces NH_3 in the case of slip, such improving the conversion.

Allowing a NH_3 slip of 20 ppm, a NO_x reduction of up to 90% can be reached with an urea/water solution of about 5% of the diesel fuel consumption.

The instrumentation of the SCR system usually requires temperature sensors upstream and downstream of the SCR catalyst to determine the exhaust gas and the SCR process temperature, an upstream NO_x sensor to determine the injected urea/water solution mass by feedforward control, a downstream NO_x sensor for feedback control and a NH_3 sensor to detect the amount of NH_3 slip. The tank system for AdBlue needs a level sensor and a heater to avoid freezing ($< -11°C$), a pump and a solenoid valve for metering. The urea/water solution may be supported by an air stream in the case of commercial vehicles. The injection of the urea/water solution starts if the SCR catalyst has obtained an operating temperature of 250-450°C. A signal-flow scheme for SCR control according to Guzzella and Onder (2010), Elsener et al (2003) and Schär (2003), is depicted in Fig. 8.9.17.

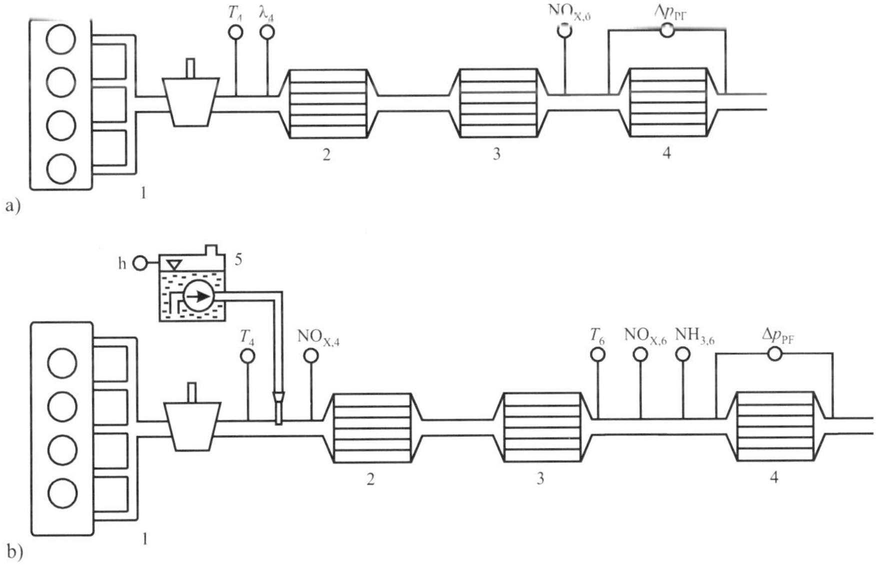

Fig. 8.9.16. Exhaust gas after-treatment systems and possible sensors. **a** 1 exhaust manifold, 2 oxidation catalyst, 3 NO_x storage catalyst, 4 particulate filter. **b** 1 exhaust manifold, 2 SCR catalyst, 3 oxidation catalyst, 4 particulate filter, 5 urea-water tank.

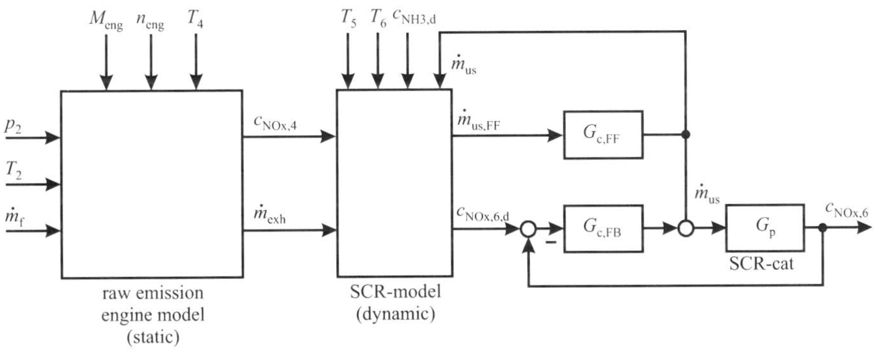

Fig. 8.9.17. Signal flow for a SCR catalyst control system, Guzzella and Onder (2010), Elsener et al (2003).

The concentration of NO_x in the raw emissions varies considerably with the engine torque and speed. On the other side the operating capacity of a SCR catalyst depends on the incoming c_{NOx}, the exhaust gas mass flow rate $\dot{m}_{exh}$ and the catalyst temperature $T_{scr,cat}$. Therefore, the desired reference value $c_{NOx,6,d}$ of the reachable NO_x-concentration after the SCR catalyst has to be adapted to these limitations. This becomes possible by an inverted SCR catalyst model, assuming a certain NH_3 slip $c_{NH3,5}$ as, e.g. 10 or 20 ppm. The set point of $c_{NOx,6}$ control is a function of

$$c_{NOx,6,d} = f(M_{eng}, n_{eng}, \dot{m}_{exh}, T_4, T_6, c_{NOx,4}) \tag{8.9.12}$$

and is stored in a look-up table. The exhaust gas temperatures T_4 and T_6 are used to determine the catalyst temperature $T_{\mathrm{scr,cat}}$ applying a thermal model and a state variable observer. The injected urea/water mass flow $\dot{m}_{\mathrm{us}}$ can be calculated from the SCR model using the NH_3 equivalent $\alpha = \dot{m}_{\mathrm{NH3}}/\dot{m}_{\mathrm{NOx,4}}$ to the determined raw emission $c_{\mathrm{NOx,4}}$, taking the NH_3-slip into account. This value is directly used to manipulate the dosing valve as feedforward control, see Fig. 8.9.17. Measurement of $c_{\mathrm{NOx,6}}$ downstream of the SCR-catalyst allows feedback control with e.g. a PI-controller, see Seher et al (2003). The dynamic behavior of the SCR-catalyst is determined by the storage behavior of NH_3, NOx and heat, which can be modeled by three balance equations with cross couplings. Results with identification methods are described in Tayamon et al (2011), see also van Nieuwstadt and Upadjyay (2000).

The measurement of NOx, e.g. based on sensors with ZrO_2 and oxygen pumps may show a strong cross sensitivity to NH_3 such that special means have to be applied to separate NOx from NH_3 measurements. For more details see e.g. Guzzella and Onder (2010), Hülser et al (2011).

The sequence SCR-cat, DOC, PF depicted in Fig. 8.9.16 is one example. There exist several other arrangements, such as DOC, SCR-cat, PF or DOC, PF, SCR-cat. The selection depends, e.g. on the mutual thermal and emission concentration couplings, see Mollenhauer and Tschöke (2010).

c) Reduction of particulate matter

Particulate filters remove emitted particulates consisting of carbon, ash and unburned hydrocarbons in the size of $< 100\,\mathrm{nm}$. One distinguishes wall-flow filters and open separators. The wall-flow filters are composed of porous ceramics of cordierite or silicon carbide. They consist of a large number of parallel channels which are alternatively closed by ceramic plugs in the front or the back, such that the exhaust gases have to flow through the porous wall with a large surface. The particulates then adhere to the porous wall surface, forming increasingly layers of soot (surface filtration). The retention rate is up to 95% for particles of the size 10 to 1000 nm. Another wall-flow filter type is a sintered metal filter consisting of filter cells with large cross areas at the beginning which then become smaller downstream.

Because the increasing soot layer lets the flow resistance grow the particulate filter must be cleaned by a regeneration process at certain intervals, after about 800 to 2000 km of driving. The collected soot is then burnt off with exhaust gases at temperatures above 600°C. As these temperatures are usually not reached in normal driving the exhaust temperature is increased by retarded main injection, or post injection and reducing the air mass by a throttle valve. Depending on the load these measures have to be combined properly, Mollenhauer and Tschöke (2010). In order not to lose much heat the particulate filters are placed close to the exhaust manifold, frequently combined with the DOC.

8.9.4 Particulate filter control

The *regeneration* of the particulate matter requires the detection of the particulate loading. A first possibility for the detection of the particulate filter loading state

(PFLS) is the increase of the *pressure drop* over the filter, see Fig. 8.9.16,

$$\Delta p_{\mathrm{PF}} = f\left(\dot{V}_{\mathrm{EG}}\right) \tag{8.9.13}$$

which depends on the exhaust gas volume flow rate and the PFLS. The pressure drop in porous structures follows, van Basshuysen and Schäfer (2007)

$$\Delta p = k\, l\, \frac{(1-\epsilon)^2}{\epsilon^3} \left(\frac{S_p}{V_p}\right)^2 \eta\, v \tag{8.9.14}$$

with

l	length of filter	V_p	porous volume
ϵ	porous volume / filter volume	η	dynamic viscosity
S_p	porous surface	v	inlet velocity

If it exceeds the volume flow dependent pressure drop of the empty filter by a certain factor the PF is assumed to be loaded

$$\Delta\Delta p_{\mathrm{PF}} = \Delta p_{\mathrm{PF}}\left(\dot{V}_{\mathrm{EG}}\right) - \Delta p_{\mathrm{PF,empty}}\left(\dot{V}_{\mathrm{EG}}\right). \tag{8.9.15}$$

The pressure drop increases about linearly with the operating time for a constant operating point, e.g. from 50 mbar to 100 mbar in 12 hrs, Hülser et al (2003). However, because the pressure drop also depends on the particulate size population, the location in the porous filter and ashes content (from oil), it is not solely a reliable measure for the PFLC.

Therefore, in addition a *particulate loading model* is required. It is based on a mass-balance equation

$$\dot{m}_{\mathrm{c,st}}(t) = \dot{m}_{\mathrm{c,in}}(t) - \dot{m}_{\mathrm{c,out}}(t) \tag{8.9.16}$$

where $m_{\mathrm{c,st}}$ is the stored particulate mass, and $\dot{m}_{\mathrm{c,in}}$ and $\dot{m}_{\mathrm{c,out}}$ are the incoming and outgoing particulate mass flow rates. $\dot{m}_{\mathrm{c,in}}$ is the engine test bench calibrated particulate generation and $\dot{m}_{\mathrm{c,out}}$ includes the particulate reduction through oxidation with NO_2, thermal burning and fluid mechanic PF storage, Bauder et al (2006). Measurements with truck diesel engines show, that the stored particulate mass is approximately linear to each the fuel consumption, the driving time and the driven distance, Gärtner (2009).

The *regeneration* of the particulate filter is started based on a combined evaluation of the pressure drop, the calculated accumulated particulate mass based on the loading model, the driven distance (e.g. 800 to 2000 km), and the accumulated fuel consumption, Hülser et al (2003), Bauder et al (2006). The regeneration requires an increased exhaust gas temperature of about $600°\mathrm{C}$. As this temperature is usually much lower for diesel engines, it is reached by a combination of late main injection, late additional injections and a reduced air flow by a throttle valve (at low loads). The exhaust temperature at the outlet valve then increases and, in addition, through exothermic reaction of unburned hydrocarbons in the oxidation catalytic converter (DOC). For truck engines extra fuel is injected in front of the DOC to increase the exhaust temperature.

The regeneration process mainly consists of the oxidation of various components such as soot, hydrocarbons, dust etc. The oxidation is a heterogeneous reaction and may be described by reaction kinetics. Therefore the Arrhenius law can be applied to describe the reaction rate

$$\frac{dm_{\mathrm{c,st}}}{dt} = -k_0\, m_{\mathrm{c,st}}^{m}\, p_{\mathrm{O2}}^{n}\, e^{-\frac{E}{R\,T}} \tag{8.9.17}$$

where $m_{\mathrm{c,st}}$ is the relative particulate mass, p_{O2} the partial pressure of oxygen, E the actuation energy (about 140 kJ/mol), R the gas constant, T the gas temperature and k_0 a constant, Mayer et al (1996).

However, these relatively simple equations give only first hints for influencing variables. A more detailed modeling of the loading and regeneration requires the description of the flow and reaction mechanisms and their distribution over the cross sections and length of the filter, see e.g. Konstandopoulos et al (2002), Hoßfeld and Kaiser (2003), Hinterberger et al (2006).

The regeneration management has to be coordinated such, that the torque generation is not influenced and that the exhaust gas after the DOC contains enough oxygen for burning the particulates. Therefore the oxygen content before the PF is measured by a λ-sensor, Bauder et al (2006).

Fig. 8.9.18 depicts a block diagram for the control functions during the regeneration of the PF. The decision for starting the regeneration process follows a combined logic evaluation of the difference pressure and/or the soot loading model and/or the driven distance exceeding certain thresholds. Then the reference value is given to a cascaded control system, where T_6 is the main and T_4 is the minor control variable. The manipulation variables are a combination of retarded main injection, secondary late injection and throttling of intake-air dependent on the torque and engine speed, see Robert Bosch GmbH (2011), Mattes et al (2004).

The supervision of the particulate filter can be based on the difference pressure Δp_{PF}. If it is too low the PF has a defect, e.g. a leak. If it is too high ($>$200 mbar) and regeneration does not reduce the difference pressure the PF is congested. In addition all sensors undergo plausibility checks.

Open particle separators have cells at the entrance where the flow is accelerated to a porous channel wall. The exhaust gases then penetrate into the walls and build up soot at the beginning of the loading phase. As with increasing resistance the penetration of the cells is bypassed, the efficiency of filtering is on average between 30 and 70% only, Mollenhauer and Tschöke (2010).

References

Akihama K, Takatori Y, Inagaki K, Sasaki S, Dean A (2001) Mechanism of the smokeless rich diesel combustion by reducing temperature. In: SAE Technical paper Series, Warrendale, PA, 2001-01-0655

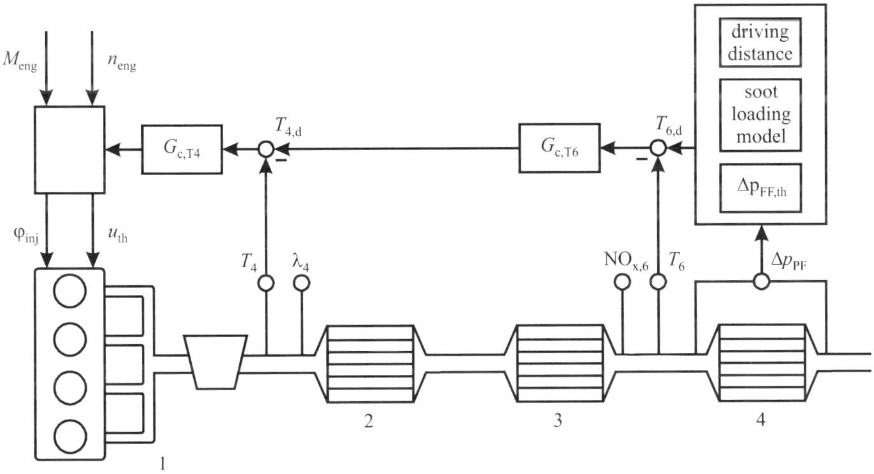

Fig. 8.9.18. Exhaust gas temperature control during particulate filter regeneration.

Alfieri E, Amstutz A, Guzzella L (2009) Gain-scheduled model-based feedback control of the air/fuel ratio in diesel engines. Control Engineering Practice – CEP 17:1417–1425

Barba C (2001) Verbrennungskenntwerte aus Indizierdaten zur Prognose und Simulation des Verbrennungsablaufs. Dissertation. Zürich

van Basshuysen R, Schäfer F (2007) Handbuch Verbrennungsmotor, 4th edn. Vieweg, Wiesbaden

Bauder R, Gruber M, Michels E, Zaccheo-Giovanni P, Schiffgens HJ, Wimmer W (2006) The new audi 4.2 l v8 tdi-engine. part 1: Design and mechanics. part 2: Thermodynamics, application and exhaust after-treatment. MTZ worldwide 66(10 & 11):2–5 & 29–32

Benz M (2010) Model-based optimal emission control of diesel engines. PhD thesis, ETH, Zürich

Benz M, Onder C, Guzzella L (2010) Engine emission modeling using a mixed physics and regression approach. Journal of engineering for gas turbines and power 132(4)

Birk M, Bleile T, Kraus B (2003) Modellbasierte Funktionen im Luftsystem des Dieselmotors. In: 4. Symposium Steuerungsysteme für den Antriebsstrang von Kraftfahrzeugen, Berlin, Germany

Brengtsson J (2004) Closed-loop control of the HCCI engine dynamics. PhD thesis, Dept. of Automatic Control, Lund

Canova M, Midlam-Mohler S, Pisu P, Soliman A (2010) Model-based fault detection and isolation for a diesel lean nox trap aftertreatment system. Control engineering practice 18(11):1307–1317

Catanese A, Landsmann G, Neunteufl K, Stolz M (2009) An optimized approach for closed-loop combustion control of low-nox diesel engines. In: 7th Symposium Powertrain Control Systems for Motor Vehicles, Berlin, Germany, pp 347–358

Caton P, Simon A, Gerdes J, Edwards C (2003) Residual-effected homogeneous charge compression ignition at a low compression ratio using exhaust reinduction. International Journal of Engine Research 4(3):163–177

Chiang CJ, Stefanopoulou A, Jankovic M (2007) Nonlinear observer-based control of load transitions in homogeneous charge compression ignition engines. IEEE Trans on Control Systems Technology 15(3):438–448

Chmela F, Orthaber G (1999) Rate of heat release prediction for direct injection diesel engines based on purely mixing controlled combustion. SAE transactions 108(3):152–160

Chmela F, Dimitrov D, Pirker G, Wimmer A (2006) A consistent simulation methodology for combustion in piston engines. MTZ worldwide 67(6):24–28

Christen U, Vantine K, Collings N (2001) Event-based mean-value modeling of DI diesel engines for controller design. SAE Transactions 110:1472–1482

Eijk v, Hakstege B, Ruhkamp L (2011) The paccar mx engine for heavy duty trucks in north america. ATZautotechnology 11(3):50–57

Elsener M, Geering H, Jaussi F, Koebel M, Kröcher O, Mangold M, Onder C, Röthlisberger R, Schär C (2003) Development and evaluation of a denox system based on urea scr. MTZ worldwide 64(11):28–31

Gärtner U (2001) Die Simulation der Stickoxid Bildung in Nutzfahrzeug-Dieselmotoren. PhD thesis, Technische Universität, Darmstadt

Gärtner U (2009) Regelung der Abgasnachbehandlung von Nutzfahrzeug-Dieselmotoren. In: Elektronisches Management von Fahrzeugantrieben, Haus der Technik, Darmstadt, Germany

Gerhardt J, Höninger H, Bischof H (1998) A new approach to functional and software structure for engine management systems - Bosch ME97. In: SAE 980801, Aachen, Germany

Guzzella L, Amstutz A (1998) Control of diesel engines. IEEE Control Systems 18(5):53–71

Guzzella L, Onder C (2010) Introduction to modeling and control of internal combustion engine systems, 2nd edn. Springer, Berlin

Guzzella L, Boulouchos K, Tschanz F, Barro C (2012) Control-oriented modeling of soot emissions and model-based emission control. In: Heft R 558, FVV Informationstagung Motoren, Frühjahr 2012, pp 251–277

Hadler J, Rudolph F, Dorenkamp R, Stehr H, Hilzendeger J, Kranzusch S (2008) Volkswagen's new 2.0 l TDI engine for the most stringent emission standards. MTZ worldwide edition 69(5, 6):12–12, 54–59

Hafner M (2002) Modellbasierte stationäre und dynamische Optimierung von Verbrennungsmotoren am Motorenprüfstand unter Verwendung neuronaler Netze. Dissertation Technische Universität Darmstadt. Fortschr.-Ber. VDI Reihe 12, 482. VDI Verlag, Düsseldorf

Hafner M, Schüler M, Nelles O, Isermann R (2000) Fast neural networks for diesel engine control design. Control Engineering Practice – CEP 8:1211–1221

Hinterberger C, Kaiser R, Olesen M (2006) 3d-simulation of soot loading and regeneration of diesel particulate filters. MTZ worldwide 67(4):2–5

Hippe P (2006) Windup in Control: Its Effects and Their Prevention. Springer, Berlin

Hoßfeld C, Kaiser R (2003) Soot loading in diesel particulate filtersystems. MTZ worldwide 64(9):6–9

Hülser H, Cartus T, Roduner C, Pfahl U (2003) Electronics enables innovations in engine engineering. MTZ worldwide 64(1):13–16

Hülser H, Schüßler M, Hollauf B, Breitschädel B (2011) Model-based control for highest nox conversion in scr catalysts. ATZelektronik worldwide 6(3):18–21

Isermann R (1987) Digitale Regelsysteme, Band II: Stochastische Regelungen, Mehrgrößenregelungen, Adaptive Regelungen, Anwendungen. Springer, Berlin

Isermann R (1989) Digital control systems, 2nd edn. Springer, Berlin

Isermann R (1991) Digital control systems, vol 2, 2nd edn. Springer, Berlin

Isermann R, Münchhof M (2011) Identification of dynamic systems. Springer, Berlin

Jeschke J (2002) Konzeption und Erprobung eines zylinderdruckbasierten Motormanagements für PKW-Dieselmotoren. PhD thesis, Otto-von-Guericke-Universität Magdeburg, Universitätsbibliothek

Jung M (2003) Mean-value modeling and robust control of the airpath of a turbocharged diesel engine. University of Cambridge

Justi E (1938) Spezifische Wärme, Enthalpie, Entropie und Dissoziation technischer Gase. Springer, Berlin

Kato N, Hameda Y, Kurachi H (1997) Performance of thick film no_x sensor on diesel and gaosline engines. In: 1997 SAE International Congress and Exposition, Cobo Center, Detroit, Michigan, USA, 970858

Ketfi-Cherif A, von Wissel D, Beurthey S, Sorinem M (2000) Modeling and control of a nox trap catalyst. SAE transactions 109(4):798–807

Kimmich F (2004) Modellbasierte Fehlererkennung und Diagnose der Einspritzung und Verbrennung von Dieselmotoren. Dissertation Technische Universität Darmstadt. Fortschr.-Ber. VDI Reihe 12, 549, VDI Verlag, Düsseldorf

Kirchen P, Boulouchos K (2008) Phänomenologisches Mittelwertmodell für Ruß in transientem Motorbetrieb. MTZ-Motortechnische Zeitschrift 69(7-8):624–631

Knippschild C (2011) Zylinderindividuelle Regelung des Gaszustands bei Pkw-Dieselmotoren

Kohlhase M (2011) Brennraumdruckbasiertes Motormanagement für Otto- und Dieselmotoren zur Verbrauchs- und Emissionsreduktion. Dissertation Technische Universität Darmstadt. Fortschr.-Ber. VDI Reihe 12, 743, VDI Verlag, Düsseldorf

Kohlhase M, Isermann R (2009) Self-tuning extremum value control based on cylincer pressure sensors for direct injection spark ignition engines (in German). at - Automatisierungstechnik 57(1):23–31

Kohlhase M, Pfeil K, Zimmerschied R, Isermann R (2008) Cylinder-pressure-based closed-loop-control of the (partial-) homogeneous diesel combustion for passenger cars using data-based nonlinear models. In: 9th International Symposium on Advanced Vehicle Control, Kobe, Japan

Konstandopoulos A, Skaperdas E, Masoudi M (2002) Microstructural properties of soot deposits in diesel particulate traps. SAE TRANSACTIONS 111(4):434–442

Kroll A, Bernd T, Trott S (2000) Fuzzy network model-based fuzzy state controller design. IEEE Transactions on Fuzzy Systems 8(5):632–644

Landsmann G (2010) Control of diesel engines – status and requirements (in German). In: Isermann R (ed) Elektronisches Management motorischer Fahrzeugantriebe, Vieweg-Teubner, Wiesbaden

Larink J (2005) Zylinderdruckbasierte auflade-und abgasrückführregelung für pkw-dieselmotoren. PhD thesis, Otto-von-Guericke-Universität Magdeburg, Universitätsbibliothek

Law D, Kemp D, Allen J, Kirkpatrick G, Copland T (2001) Controlled combustion in an ic-engine with a fully variable valve train. In: SAE Technical paper Series, Warrendale, 2001-01-0251

Lee J, Edgar TF (2004) ISE tuning rule revisted. Automatica 40:1455–1458

Lohmann B (1991a) Vollständige und teilweise Führungsentkopplung dynamischer Systeme durch konstante Zustandsrückführung. Automatisierungstechnik 39(10):376–378

Lohmann B (1991b) Vollständige und teilweise Führungsentkopplung im Zustandsraum. Fortschr.-Ber. VDI Reihe 8, 244. VDI Verlag, Düsseldorf

Magnussen B, Hjertager B (1977) Onmathematical modeling of turbulent combustion with special emphasis on soot formation and combustion. Int Symp on Comubstion 16(1):719–729

Mattes W, Mayr K, Neuhauser W, Steinparzer F (2004) The BMW six-cylinder diesel engine Euro 4 technology. MTZ worldwide edition 65(7–8):7–11

Mayer A, Emig G, Gmehling B, Popovska N, Holemann K, Buck A (1996) Passive regeneration of catalyst-coated knitted-fiber diesel particulate traps. SAE transactions 105(4):36–44

Merker G, Schwarz C, Stiesch G, Otto F (2006) Simulating combustion. Springer, Berlin

MJ van Nieuwstadt, Kolmanovsky I, Moraal P, Stefanopoulou A, Jankovic M (2000) Egr-vgt control schemes: experimental comparison for a high-speed diesel engine. Control Systems, IEEE 20(3):63–79

Mollenhauer K, Tschöke H (2010) Handbook of diesel engines. Springer, Berlin

Motz S, Naber D, Krüger M, Gerhardt J (2011) Approaches for optimizing the transient behavior of diesel engines in terms of emissions and fuel consumption. In: 11. Internationales Stuttgarter Symposium

Mrosek M (2014) Modeling of diesel engines with high and low pressure EGR. Internal Report. Institute of Automatic Control, Technische Universität Darmstadt

Mrosek M, Sequenz H, Isermann R (2010) Control oriented NO_x and soot models for diesel engines. In: IFAC Symposium Advances in Automotive Control, Munich

Mrosek M, Sequenz H, Isermann R (2011) Identification of emission measurement dynamics for diesel engines. In: IFAC Worldcongress, Milano, Italy

Nelles O (2001) Nonlinear system identification. Springer, Heidelberg

Nieuwstadt M, Upadjyay D (2000) Control of urea SCR systems for US diesel applications. Oil and gas science and technology 66(4):655–665

Nitsche R, Bleile T, Birk M, Dieterle W, Rothfuß R (2004) Modellbasierte Ladedruckregelung eines PKW-Dieselmotors. In: AUTOREG 2004, VDI, Wiesloch, Germany, vol VDI-Berichte 1828

Nitzke H, Rebohl T, Jelden H (1999) Emissionsoptimierte abgasrückführkonzepte und dynamische luftmassenerfassung für direkteinspritzende dieselmotoren. In: Symposium Steuerungsysteme für den Antriebsstrang, Berlin, Germany

Pfeil K (2011) Ladedruck- und Luftmassenregelung von aufgeladenen Dieselmotoren mit lokal linearen Modellen und Optimierung des dynamischen Emissionsverhaltens im Rauchbetrieb. Dissertation Technische Universität Darmstadt. Fortschr.-Ber. VDI Reihe 12, 744, VDI Verlag, Düsseldorf

Pfeil K, Zimmerschied R, Isermann R (2007a) Automated parameterization of the egr-/vgt-controller of turbo-charged diesel engines by means of nonlinear identification of the air- and exhaust manifold. In: 7. Internationales Stuttgarter Symposium. Automobil- und Motorentechnik, Stuttgart

Pfeil K, Zimmerschied R, Isermann R (2007b) Nichtlineare Identifikation des Luftpfads von aufgeladenen Dieselmotoren und automatisierter AGR-/VTG-Reglerentwurf. at - Automatisierungstechnik 55

Pischinger R, Kell M, Sams T (2009) Thermodynamik der Verbrennungskraftmaschinen, 3rd edn. Springer, Wien and New York

Pöttker S, Eckert P, Merker G (2005) Homogeneous charge compression ignition with synthetic fuels. MTZ worldwide 66(12):27–29

Rausen D, Stefanopoulou A, Kang JM, Eng J, Kuo TW (2005) A mean-value model for control of homogeneous charge compression ignition (HCCI) engines. J of Dyn Meas and Control 127:355–362

Ravi N, Liao HH, Jungkunz A, Widd A, Gerdes J (2012) Model predictive control of HCCI using variable valve actuation and fuel injection. Control Engineering Practice 20(4):421–430

Renninger P, Weber M, Pfeil K, Hofmann D, Isermann R (2005) Dynamic models and their application to heavy duty engine optimization. In: 14. Aachener Kolloquium Fahrzeug- und Motorentechnik, Aachen, Germany, pp 1205–1222

Renninger P, Weirich M, Pfeil K, Isermann R (2006) Optimierungsstrategien für den transienten Betrieb eines Dieselmotors innerhalb der Rauchbegrenzung. In: Motorenentwicklung auf dynamischen Prüfständen. ATZ/MTZ Konferenz, Wiesbaden, Germany

Rether D, Bargende M, Lämmle C, Boulouchos K (2012) Verbrennung – Physikalisches Brennratenmodell für Dieselmotoren mit teilhomogener Ladung. MTZ - Motortechnische Zeitschrift 73(10):310

Robert Bosch GmbH (ed) (2011) Automotive Handbook, 8th edn. Bentley publishers, Cambridge

Rückert J, Richter F, Schloßer A, Abel D, Herrmann O, Pischinger S, Pfeifer A (2004) A model-based predictive attempt to control boost pressure and egr-rate in a heavy duty diesel engine. In: IFAC Symposium on Advances in Automotive Control, Salerno, Italy, pp 127–133

Schär CM (2003) Control of a selective catalytic reduction process. Diss., Technische Wissenschaften ETH Zürich, Nr. 15221, 2003

Schüler M (2001) Stationäre Optimierung der Motorsteuerung von PKW-Dieselmotoren mit Abgasturbolader durch Einsatz schneller neuronaler Netze. Fortschr.-Ber. VDI Reihe 12, 461. VDI Verlag, Düsseldorf

Schüler M, Hafner M, Isermann R (2000) Model-based optimization of ic engines by means of fast neural networks. part 1: Modeling the engine and emission behavior. part 2: Static and dynamic optimisation of fuel consumption versus emissions. MTZ worldwide 45(10 & 11):29–32 & 28–31

Seher D, Reichelt M, Wickert S (2003) Control strategy for no_x-emission reduction with SCR. In: SAE Technical paper Series, Warrendale, 2003-01-3362

Sequenz H (2013) Emission Modelling and Model-Based Optimisation of the Engine Control. Dissertation Technische Universität Darmstadt. Fortschr.-Ber. VDI Reihe 12, 1222. VDI Verlag, Düsseldorf

Sequenz H, Isermann R (2011) Emission model structures for an implementation on engine control units. In: Preprints of the 18th IFAC World Congress, Milano, Italy

Stanglmaier R, Roberts C (1999) Homogeneous charge compression ignition (hcci): benefits, compromises, and future engine applications. SAE transactions 108(3):2138–2145

Stein J, Dürnholz M, Wirbeleit F, Kopp C, Benz C (2004) Homogene dieselmotorische verbrennung zur darstellung niedrigster emissionen. In: 13. Aachener Kolloquium Fahrzeug-und Motorentechnik

Stiesch G (2003) Modeling Engine Spray and Combustion Processes. Springer, Berlin

Suzuki H, Koike N, Ishi H, Odaka M (1997) Exhaust purification of diesel engines by homogeneous charge with compression ignition. In: SAE Technical paper Series, Warrendale, PA, 970313

Tayamon S, adn T Wigren DZ, Carlsson B (2011) Nonlinear black box identification of a selective catalytic reduction system. In: Preprints of the 18th IFAC World Congress, Milano, Italy

Traver M, Atkinson R, Atkinson C (1999) Neural network-based diesel engine emissions prediction using in-cylinder combustion pressure. SAE transactions 108(4):1166–1180

Wang Y, Haskara I, Yaniv O (2011) Quantitive feedback design of air and boost pressure control system for turbocharged engines. Control Engineering Practice – CEP 19(6):626–637

Weißbäck W, Csató J, Glensvig M, Sams T, Herzog P (2003) Alternative combustion. an approach for future hsdi diesel engines. MTZ worldwide 64(12):17–20

Weirich M, Wassenhoven K, Kemmner M (2006) Die Thermodynamikentwicklung des neuen Mercedes-Benz Medium Duty Nutzfahrzeugmotors für das Abgasgesetz EPA'07. 15. In: Aachener Kolloquium Fahrzeug- und Motorentechnik

Wellers M, Elicker M (2004) Regelung der AGR-Rate und des Ladedrucks mit Hilfe eines nichtlinearen modellbasierten prädiktiven Reglers. In: AUTOREG 2004, VDI, Wiesloch, Germany, vol VDI-Berichte 1828

Weßlau M, Bargende M, Haas S, Boulouchos K, Barroso G, Escher A (2006) Homogeneous diesel combustion process for reducing emissions. MTZ worldwide 67(10):21–24

Wright R, Kravaris C (2005) Two-degree-of-freedom output feedback controllers for nonlinear systems. Chemical Engineering Science 60:4323–4336

Zahn S (2012) Arbeitsspielaufgelöste Modellbildung und Hardware-in-the Loop Simulation von Pkw-Dieselmotoren mit Abgasturbolader. Dissertation Technische Universität Darmstadt. Fortschr.-Ber. VDI Reihe 12, 760. VDI Verlag, Düsseldorf

Zimmerschied R (2008) Identifikation nichtlinearer Prozesse mit dynamischen lokal-affinen Modellen. Dissertation Technische Universität Darmstadt. Fortschr.-Ber. VDI Reihe 8, 1150, VDI Verlag, Düsseldorf

Concluding remarks

This book treats methods for modeling and control of gasoline and diesel engines in order to cope with the increasing complexity of advanced engine management systems. The models of the components and processes of the engines are as far as possible based on physical/chemical principles and are combined with experimental methods to describe the stationary as well as the dynamic behavior. These models are dedicated to be used for the control design, real-time simulation and control calibration and adaptation. Therefore, the models are simplified as far as possible, making compromises between precision and computational effort.

A focus was set on the experimental way of modeling with special identification methods to obtain local and global, mostly nonlinear models and to represent them in grid-based maps, polynomial equations and differential or difference equations. Whereas in the past engine control was mainly oriented on the stationary behavior, the dynamic behavior has become important too, to meet the increasing requirements with regard to real driving. Physical and semi-physical models have been derived for the components of combustion engines, including the air system, fuel system, combustion process, torque generation, exhaust system, with emissions, coolant and lubrication system and VVT, related actuators, and sensors. A signal-flow-based structure of the engine control system allows to develop and to calibrate various digital feedforward and feedback controllers systematically, making use of the models and optimization methods for gasoline and diesel engines. The various steps of the workflow for the control design and calibration is outlined as well as the related software development, including, for example, hardware-in-the-loop simulation and computer-based control prototyping. Results are shown by simulations and measurements with engines on test benches.

A further demanding application of signal and process models is the *monitoring* and *diagnosis* of the engine components and processes. This will be treated in another book with the title

Powertrain Diagnosis

– Model-based condition monitoring of internal combustion engines and electrical drives –

which will also be published with Springer Verlag around 2015/16.

R. Isermann, *Engine Modeling and Control*,
DOI: 10.1007/978-3-642-39934-3

Index

R. Isermann, *Engine Modeling and Control*,
DOI: 10.1007/978-3-642-39934-3

Printed by Printforce, the Netherlands